# Nonlinear Waves in Solid State Physics

# NATO ASI Series

**Advanced Science Institutes Series**

*A series presenting the results of activities sponsored by the NATO Science Committee, which aims at the dissemination of advanced scientific and technological knowledge, with a view to strengthening links between scientific communities.*

The series is published by an international board of publishers in conjunction with the NATO Scientific Affairs Division

| | | |
|---|---|---|
| **A** | **Life Sciences** | Plenum Publishing Corporation |
| **B** | **Physics** | New York and London |
| **C** | **Mathematical and Physical Sciences** | Kluwer Academic Publishers Dordrecht, Boston, and London |
| **D** | **Behavioral and Social Sciences** | |
| **E** | **Applied Sciences** | |
| **F** | **Computer and Systems Sciences** | Springer-Verlag |
| **G** | **Ecological Sciences** | Berlin, Heidelberg, New York, London, |
| **H** | **Cell Biology** | Paris, and Tokyo |


***Recent Volumes in this Series***

*Volume 240*—Global Climate and Ecosystem Change
edited by Gordon J. MacDonald and Luigi Sertorio

*Volume 241*—Applied Laser Spectroscopy
edited by Wolfgang Demtröder and Massimo Inguscio

*Volume 242*—Light, Lasers, and Synchrotron Radiation: A Health Risk Assessment
edited by M. Grandolfo, A. Rindi, and D. H. Sliney

*Volume 243*—Davydov's Soliton Revisited: Self-Trapping of Vibrational Energy in Protein
edited by Peter L. Christiansen and Alwyn C. Scott

*Volume 244*—Nonlinear Wave Processes in Excitable Media
edited by Arun V. Holden, Mario Markus, and Hans G. Othmer

*Volume 245*—Differential Geometric Methods in Theoretical Physics: Physics and Geometry
edited by Ling-Lie Chau and Werner Nahm

*Volume 246*—Dynamics of Magnetic Fluctuations in High-Temperature Superconductors
edited by George Reiter, Peter Horsch, and Gregory C. Psaltakis

*Volume 247*—Nonlinear Waves in Solid State Physics
edited by A. D. Boardman, M. Bertolotti, and T. Twardowski

*Series B: Physics*

# Nonlinear Waves in Solid State Physics

Edited by

A. D. Boardman

University of Salford
Salford, United Kingdom

M. Bertolotti

University of Rome
Rome, Italy

and

T. Twardowski

University of Salford
Salford, United Kingdom

Plenum Press
New York and London
Published in cooperation with NATO Scientific Affairs Division

Proceedings of a NATO Advanced Study Institute
on Nonlinear Waves in Solid State Physics,
held July 1-15, 1989,
in Erice, Sicily, Italy

Library of Congress Cataloging-in-Publication Data

NATO Advanced Study Institute on Nonlinear Waves in Solid State Physics (1989 : Erice, Italy)
Nonlinear waves in solid state physics / edited by A.D. Boardman, M. Bertolotti, and T. Twardowski.
p. cm. -- (NATO ASI series. Series B, Physics ; v. 247)
"Proceedings of a NATO Advanced Study Institute on Nonlinear Waves in Solid State Physics, held July 1-15, 1989, in Erice, Sicily, Italy"--T.p. verso.
"Published in cooperation with NATO Scientific Affairs Division."
Includes bibliographical references and index.
(hard)
1. Nonlinear waves--Congresses. 2. solid state physics--Congresses. I. Boardman, A. D. II. Bertolotti, Mario. III. Twardowski, T. IV. North Atlantic Treaty Organization. Scientific Affairs Division. V. Title. VI. Series.
QC176.8.W3N37 1989
530.4'1--dc20 90-25112
CIP

ISBN 978-1-4684-5900-5 ISBN 978-1-4684-5898-5 (eBook)
DOI 10.1007/978-1-4684-5898-5


Softcover reprint of the hardcover 1st edition 1990

A Division of Plenum Publishing Corporation
233 Spring Street, New York, N.Y. 10013

## PREFACE

This book is based on the contributions to the 17th International School of Materials Science and Technology, entitled *Nonlinear Waves in Solid State Physics*. This was held as a NATO Advanced Study Institute at the Ettore Majorana Centre in Erice, Sicily between the 1st and 15th July 1989, and attracted almost 100 participants from over 20 different countries.

The book covers the fundamental properties of nonlinear waves in solid state materials, dealing with both theory and experiment. The aim is to emphasise the methods underpinning the important new developments in this area. The material is organised into subject areas that can broadly be classified into the following groups: the theory of nonlinear surface and guided waves in self-focusing magnetic and non-magnetic materials; nonlinear effects at interfaces; nonlinear acoustoelectronic and surface acoustic waves; Lagrangian and Hamiltonian formulations of nonlinear problems; nonlinear effects in optical fibres; resonance phenomena; and nonlinear integrated optics. The chapters have been grouped together according to these classifications as closely as possible, but it should be borne in mind that although there is much overlap of ideas, each chapter is essentially independent of the others.

We would like to acknowledge the sponsorship of the NATO Scientific Affairs Division, the European Physical Society, the National Science Foundation of the USA, the European Research Office, the Italian Ministry of Education, the Italian Ministry of Scientific and Technological Research, the Sicilian Regional Government and the Ugo Bordoni Foundation. Also, we would like to thank Prof. M Balkanski, the Director of the International School of Materials Science and Technology, for his help with the staging of the school. Finally, we would like to thank the director of the Ettore Majorana Centre, Prof. A Zichichi, and all the staff, for their great hospitality and efficiency, and for making all aspects of the school so enjoyable.

July 1990 A. D. Boardman, M. Bertolotti, and T. Twardowski

# CONTENTS

CONTENTS

# NONLINEAR SURFACE-GUIDED WAVES IN SELF-FOCUSING OPTICAL MEDIA

A D Boardman, P Egan, T Twardowski and M Wilkins

Applied Optics Group, Department of Physics
University of Salford
Salford, M5 4WT, UK

## 1. INTRODUCTION

All-optical signal processors will possess a high bandwidth and perform parallel processing functions. The anticipated demand for this powerful capability is directing the attention of many laboratories to the creation of new nonlinear materials with self-focusing properties.[1]

The behaviour of guided and surface TE or TM waves at the boundaries between linear and nonlinear materials is therefore of great topical interest.[2-20] Although the area is a wide one, as can be seen from the contents of this volume, this chapter will concentrate entirely on the basic solid, non-magnetic, planar interface structures. These, in principle, consist of one or more planar waveguide elements bounded by an infinite substrate and cladding. It is immediately interesting, however, that the waveguides may be of the single surface, open, kind or the more conventional bounded planar guiding type. In all cases the material will be assumed to be isotropic and the nonlinearity will be third-order permitting all-optical signal processing with equal input and output frequencies. Indeed, of the optical nonlinearities that are now accessible it is the isotropic third-order one that has attracted the greatest recent attention, because it is characterised by the minimum number of possible parameters. Also the phenomena and associated processing functions that can be achieved with third-order nonlinearities include four-wave mixing, phase gratings, the Kerr effect, optical bistability and a number of other effects.[21]

Since all-optical signal processing will probably be in integrated-optics form,[1] direct production of new nonlinear materials as waveguides is desirable. Although the desired strength of nonlinearity has not been achieved to satisfy all the projected applications, outstanding waveguide structures can now be achieved with the newly established and emerging technologies. These enable waveguide structures with one or more optically nonlinear elements to be created in a variety of ways. For example, only the cladding of a planar waveguide needs to be fabricated from nonlinear material for interesting effects to appear. Also, quite apart from the advantages of an integrated-optics format, nonlinear guiding structures possess powerful advantages over bulk waves. Those of primary importance are the maintenance of very high optical power densities and the ability of the waveguide dispersion to offset mode dispersion.

Clearly then, an important design parameter for any new nonlinear material is that it can sustain high optical power densities without optical damage. Hence the major interest in isotropic optical waveguides with self-focusing properties. These studies were stimulated by early pioneering work on matched waveguide structures. In such guides the achievable nonlinearities are the same order of magnitude as the change in the linear refractive index at the guide interface(s) or, at least, have values for which the power dependence of the effective guide index distorts the eigenfunctions from their linear forms. The performance of this type of nonlinear guide cannot be accounted for within a coupled-mode formalism.

The principal motivation in this chapter is to obtain a number of exact solutions of the equations governing nonlinear waveguides. Even if suitable materials do not exist and, as will be seen later on, there is a long way to go yet, it is still desirable to solve such equations exactly in order to avoid the inconvenience of being trapped into coupled mode formalism or

*Nonlinear Waves in Solid State Physics*
Edited by A.D. Boardman *et al.*, Plenum Press, New York, 1990

some other form of perturbation theory. The exact results show clearly the regimes of applicability of the simple and often too simplistic, perturbation schemes. They will also show what is possible if the experimental situation changes to yield waveguides that go even minor distances beyond the reach of conventional perturbation theory.

The development of nonlinear TE solutions demonstrated the strong theoretical advantage of reducing Maxwell's equations to a single fundamental nonlinear equation that, under the assumption of an optical Kerr effect and real electric fields, has an exact solution. This enables an interpretation of any possible experiment to be made in readily available field amplitude shapes. For example, it is easy to show, without even solving the nonlinear equation, that, for a positive nonlinear coefficient, a guided wave field maximum will eventually establish itself in a nonlinear medium bounding a linear dielectric thin film waveguide. This maximum will remain as the film is made thicker and the system approaches the single interface, semi-infinite configuration. In this limit the electric field assumes the character of a nonlinear TE surface wave. This type of wave is quite fascinating because it has no linear counterpart.

The TE nonlinear differential equation for the electric field component carried by this kind of guide has, therefore, an elegant and exact solution.[19,22] This has enabled many benchmarks to be developed of both an analytical and numerical kind. Indeed the development has entrained many detailed solutions.[4,15]

For TM waves the situation is quite different. Theoretical formulations of nonlinear TM surface-guided wave propagation that are free of oversimplifying approximations have only recently been produced[5,6] with the exact theory of guided p-polarised optical waves being given in terms of a numerical algorithm. In a TM wave propagating in an isotropic medium in the xz-plane and along the x-axis, for example, electric field components ($E_x$, $E_z$) exist that are $\pi/2$ out of phase. The earliest calculations[24,25] either neglect $E_x$ or $E_z$ or give them equal weight in the nonlinear polarisation. For a surface plasmon-polariton, however, $E_z \gg E_x$ in the nonlinear cladding at infra-red frequencies, whereas $E_z \simeq E_x$ in the visible, but it is never the case that $E_z \ll E_x$. Hence, in an isotropic medium, a TM wave will experience a rather complicated third-order nonlinear polarisation that assumes various forms according to the relative strengths of the electric field components and the nonlinear mechanism involved. If $E_x$ and $E_z$ are given equal weight at all frequencies, this assumption is consistent with an electrostrictive or thermal nonlinearity. This has been given some attention in the literature. For the type of nonlinear materials envisaged for integrated optics, electronic distortion within isotropic materials is a more probable nonlinear mechanism and this will be assumed throughout most of this chapter. Hence neither $E_x$ nor $E_z$ can be assumed to be dominant over all frequency ranges.[5]

More generally it has long been realised that the polarisation of an electromagnetic wave plays a non-trivial role in the self-focusing of light beams.[26] Stationary nonlinear TM waves in bulk nonlinear media have been analysed in some detail, and it has been shown that solitary TM waves form solitons and that the solitary TM waveguide channel is wider than the corresponding channel for TE waves. Many systems exist that, in the linear low power limit, only support TM surface-guided waves. They are waveguide structures in which one or more components have negative dielectric functions and include guides with metal films or substrates. Such structures, because of the strong frequency dependence of the dielectric function of a metal, or a semiconductor, behave rather differently in the infra-red and visible regions of the spectrum.

In all of the theoretical work referred to above it is assumed that either purely TE or TM waves propagate. An inspection of the nonlinear dielectric tensor for isotropic materials will reveal, however, the possibility that both polarizations can propagate simultaneously and that one polarization can sometimes act as a channel for the other.[27]

The question of nonlinear channeling raises a number of interesting and challenging problems, some of which have been addressed before. For example, strong waves can propagate in an otherwise opaque medium through a kind of nonlinear bleaching action. Of more immediate interest is the ability of one strong polarization to create a channel for a weak wave with a different polarization. In particular, the propagation of a weak TM wave in the channel created by a strong TE wave has been considered recently.[27] In this case 100% modulation of the weak wave by the strong wave should be possible. New developments of a general theory show that weak-wave propagation in the channel of a strong wave is simply one limit of a whole spectrum of mixed TE-TM stationary states.[6] In this case the nonlinear TE-TM interaction is discussed in terms of the nonlinear dielectric tensor together with a discussion of a first integral and a detailed set of results for thin-film guiding structures.

Finally, the question of nonlinear guided wave stability must be addressed. This is done by considering the amplitude of a nonlinear wave, as it progresses down the guide, to be slowly varying. In so doing, non-stationary states are generated. The stability, or otherwise, of such states will depend upon their initial value. The degree of success in generating general yet simple rules to assess the stability of a guided nonlinear wave, for a given power level and wave number, will be discussed at the end of this chapter. Simple ideas on the stability of plane waves will, however, be discussed as a precursor to the work on guided waves.

## 2. THE WAVE EQUATION: TE AND TM WAVES

The study of the propagation characteristics of electromagnetic waves in a bulk nonlinear medium is made relatively straightforward by the introduction of a phenomenological nonlinear polarisation term into the familiar set of Maxwell's equations. If the linear polarisation vector is written as $\mathbf{P}^{(L)}$ and the nonlinear part of the total polarisation is written as $\mathbf{P}^{(NL)}$ then the local displacement vector $\mathbf{D}(\mathbf{r},t)$ becomes

$$\mathbf{D}(\mathbf{r},t) = \epsilon_0\mathbf{E}(\mathbf{r},t) + \mathbf{P}^{(L)}(\mathbf{r},t) + \mathbf{P}^{(NL)}(\mathbf{r},t) \tag{2.1}$$

where $\epsilon_0$ is the permittivity of free space, nonlocal spatial effects are ignored, $\mathbf{r}$ is a position vector and t is time. The Maxwell equations required at this stage are

$$\nabla\times\mathbf{E} = -\frac{\partial\mathbf{B}}{\partial t} \qquad \nabla\times\mathbf{B} = \mu_0\frac{\partial\mathbf{D}}{\partial t} = \frac{1}{c^2\epsilon_0}\frac{\partial\mathbf{D}}{\partial t} \tag{2.2}$$

where $\mathbf{B}$ is the magnetic flux vector, $\mu_0$ is the permeability of free space and c is the velocity of light *in vacuo*. These lead to

$$\nabla\times(\nabla\times\mathbf{E}) = -\frac{\partial(\nabla\times\mathbf{B})}{\partial t} = -\frac{1}{c^2\epsilon_0}\frac{\partial^2\mathbf{D}}{\partial t^2} \tag{2.3}$$

and hence

$$\nabla(\nabla\cdot\mathbf{E}) - \nabla^2\mathbf{E} + \frac{1}{c^2}\frac{\partial^2\mathbf{E}}{\partial t^2} = -\frac{1}{c^2\epsilon_0}\left(\frac{\partial^2\mathbf{P}^{(L)}}{\partial t^2} + \frac{\partial^2\mathbf{P}^{(NL)}}{\partial t^2}\right) \tag{2.4}$$

From this equation many things will flow and it will be seen that the nonlinear part $\mathbf{P}^{(NL)}(t)$ will introduce many interesting features. The precise form of $\mathbf{P}^{(NL)}(t)$ will be discussed later. It should be noted that Eq. (2.4) is entirely equivalent to an infinite set of equations which are independent of time. This set of equations describes the electric field as represented in the frequency domain. This description is implemented by using the following expressions for the electric field and the linear and nonlinear contributions to the polarisation:

$$E_i(\mathbf{r},t) = \frac{1}{2\pi}\int_0^\infty \xi_i(\mathbf{r},\omega)e^{-i\omega t}d\omega + \text{c.c.} \tag{2.5}$$

$$P_i^{(L)}(\mathbf{r},t) = \frac{1}{2\pi}\int_0^\infty p_i^{(L)}(\mathbf{r},\omega)e^{-i\omega t}d\omega + \text{c.c.} \tag{2.6}$$

$$P_i^{(NL)}(\mathbf{r},t) = \frac{1}{2\pi}\int_0^\infty p_i^{(NL)}(\mathbf{r},\omega)e^{-i\omega t}d\omega + \text{c.c.} \tag{2.7}$$

where c.c. denotes the complex conjugate of the immediately preceding quantity and the possibility that $\xi_i(\mathbf{r},\omega)$, $p_i^{(L)}(\mathbf{r},\omega)$ or $p_i^{(NL)}(\mathbf{r},\omega)$ are purely symbolic, i.e. that any of these consists of a set of Dirac delta functions, is not excluded. Noting that $e^{i\omega t}$ and $e^{-i\omega' t}$ $(\omega \neq \omega')$ are formally orthogonal and that any frequency component of the linear polarisation is related to the frequency component of the field at the same frequency via

$$p_i^{(L)}(\mathbf{r},\omega) = \epsilon_0[\epsilon_{ij}(\mathbf{r},\omega) - \delta_{ij}]\xi_j(\mathbf{r},\omega) \tag{2.8}$$

where $\epsilon_{ij}(\mathbf{r},\omega)$ is the linear dielectric tensor (at $\omega$) and $\delta_{ij}$ is the Kronecker delta function, one may obtain

$$\nabla[\nabla\cdot\xi(\mathbf{r},\omega)]_i - \nabla^2\xi_i(\mathbf{r},\omega) - \frac{\omega^2}{c^2}\epsilon_{ij}(\mathbf{r},\omega)\xi_j(\mathbf{r},\omega) = \mu_0\omega^2 p_i^{(NL)}(\mathbf{r},\omega) \qquad 0 \le \omega < \infty \tag{2.9}$$

where $\nabla[\nabla\cdot\xi(\mathbf{r},\omega)]_i$ denotes the $i^{th}$ component of $\nabla[\nabla\cdot\xi(\mathbf{r},\omega)]$.

Before any significant progress can be made in solving these equations, however, a thorough discussion of $\mathbf{p}^{(NL)}(\mathbf{r},\omega)$ must be given. This detailed study of $\mathbf{p}^{(NL)}(\mathbf{r},\omega)$ will be presented shortly but it is appropriate, at this stage, to give an introductory argument based upon a field component of the form

$$E_i(\mathbf{r},t) = \frac{1}{2}[\xi_i(\mathbf{r},\omega)e^{-i\omega t} + \xi_i^*(\mathbf{r},\omega)e^{i\omega t}] \tag{2.10}$$

For simple plane waves with constant amplitudes, $\xi_i(\mathbf{r},\omega)$ = constant × $e^{i\mathbf{k}\cdot\mathbf{r}}$ with **k** as a wave vector and **r** as a position vector.

Intuitively, it is expected that the second-order contribution to $\mathbf{P}^{(NL)}(t)$ will somehow involve the products $E_j(\mathbf{r},t)E_k(\mathbf{r},t)$ and the third-order term the products $E_j(\mathbf{r},t)E_k(\mathbf{r},t)E_l(\mathbf{r},t)$ where, if $\mathbf{P}^{(NL)}(t)$ is regarded as a perturbation, these products involve only the fields as calculated from the *linear* wave equation. They will, in turn, introduce connecting tensors, in the frequency domain, of the form $\chi^{(2)}_{ijk}(\omega_1+\omega_2;\omega_1,\omega_2)$, $\chi^{(3)}_{ijkl}(\omega_1+\omega_2+\omega_3;\omega_1,\omega_2,\omega_3)$ where $\omega_i$ may take on either sign. The notation means, for example, that $\omega_1$ and $\omega_2$ could combine to create $\omega_1 + \omega_2$. At this point, it should be emphasised that only materials with a centre of symmetry will be studied in this chapter so that $\chi^{(2)}_{ijk}$ is identically zero. This can be seen by considering second harmonic generation with $\omega_1 = \omega_2 = \omega$ so that, in the frequency domain, we define

$$p_i(\mathbf{r},2\omega) = \chi^{(2)}_{ijk}(2\omega;\omega,\omega)\xi_j(\mathbf{r},\omega)\xi_k(\mathbf{r},\omega) + \text{c.c.} \tag{2.11}$$

Physically the dielectric medium, the applied electric fields and the polarisation always remain fixed in space but they can be described in a new coordinate system. This new coordinate system can be generated from the old one by a 3x3 transformation matrix that represents rotation, inversion or any other point group symmetry operation $a_{ij}$. Hence, if a dash denotes representation in the new frame of reference,[28]

$$p_i'(2\omega) = a_{is}p_s(2\omega) \tag{2.12}$$

$$E_j'(\omega) = a_{jm}E_m(\omega) \tag{2.13}$$

$$\chi^{(2)'}_{ijk}(2\omega;\omega,\omega) = a_{is}a_{jm}a_{k\gamma}\chi^{(2)}_{sm\gamma}(2\omega;\omega,\omega) \tag{2.14}$$

Even though $a_{ij}$ is restricted by symmetry, $\chi_{ijk}$, being a property tensor, is the *same* in both frames so that

$$\chi^{(2)}_{ijk}(2\omega;\omega,\omega) = a_{is}a_{jm}a_{k\gamma}\chi^{(2)}_{sm\gamma}(2\omega;\omega,\omega) \tag{2.15}$$

The thirty-two crystal classes are defined by the transformation $a_{ij}$ and they will restrict the elements of $\chi^{(2)}_{ijk}$ to certain values. Inversion symmetry is the strongest restriction and is mathematically expressed as $a_{ij} = -\delta_{ij}$. For this symmetry operation Eq. (2.15) becomes

$$\chi^{(2)}_{ijk}(2\omega;\omega,\omega) = (-\delta_{is})(-\delta_{jm})(-\delta_{k\gamma})\chi^{(2)}_{sm\gamma}(2\omega;\omega,\omega) = -\chi^{(2)}_{ijk}(2\omega;\omega,\omega) = 0 \tag{2.16}$$

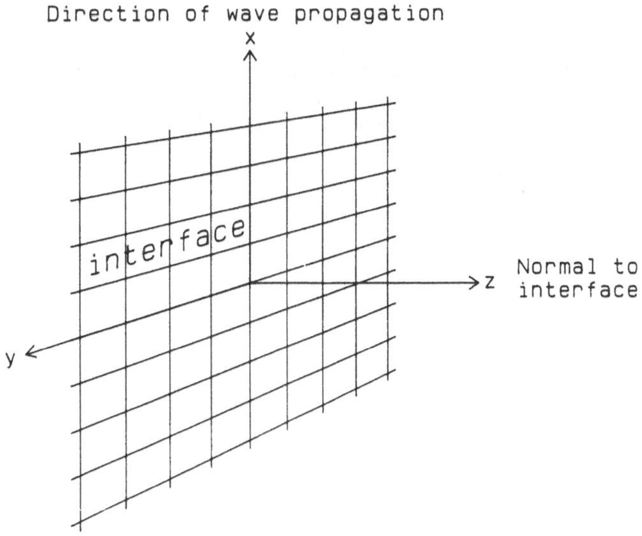


Fig. 1 Geometry of the guiding structures.

Thus, for any material having inversion symmetry, $\chi^{(2)}_{ijk}(2\omega;\omega,\omega)$ is zero. Such a restriction does not apply to any third-order nonlinearity represented by $\chi^{(3)}_{ijkl}$.

It is to $\chi^{(3)}_{ijkl}$ that attention is confined here and much more detail will be presented in section 3. Also, since a lot of time will be devoted to discussions of both TE and TM waves propagating along a single interface, or through a guide that has more than one interface, it is useful now to make some preliminary remarks about these waves.

Firstly, Fig. 1 defines the coordinate system that will be used throughout this chapter. The interfaces are in the xy-plane and propagation is along the x-axis with no dependence upon y. Therefore the TE waves considered here have components

$$\mathbf{E} = (0, E_y, 0) \qquad \mathbf{H} = (H_x, 0, H_z) \tag{2.17}$$

and, since for these waves there is no variation in the y direction, it is always the case that

$$\nabla(\nabla\cdot\mathbf{E}) = \nabla\left(\frac{\partial E_y}{\partial y}\right) = 0 \tag{2.18}$$

The nonlinear wave equation then becomes

$$\frac{1}{c^2}\frac{\partial^2 E_y}{\partial t^2} - \nabla^2 E_y = -\frac{1}{c^2\epsilon_0}\left(\frac{\partial^2 \mathbf{P}^{(L)}}{\partial t^2} + \frac{\partial^2 \mathbf{P}^{(NL)}}{\partial t^2}\right) \tag{2.19}$$

In contrast, TM waves are defined as

$$\mathbf{E} = (E_x, 0, E_z) \qquad \mathbf{H} = (0, H_y, 0) \tag{2.20}$$

for which

$$\nabla(\nabla\cdot\mathbf{E}) = \nabla\left(\frac{\partial E_x}{\partial x} + \frac{\partial E_z}{\partial z}\right) \not\equiv 0 \tag{2.21}$$

i.e. $\nabla(\nabla\cdot\mathbf{E})$ is not necessarily zero. In this case, therefore, the nonlinear wave equation is much more of a problem to solve and is made even more so by the more complicated form of $\mathbf{P}^{(NL)}$ that will be revealed below.

A substantial part of this chapter concerns surface-guided waves in which a single surface, or a group of surfaces, guide a nonlinear TM or TE wave. This guiding, it is emphasised once again, is along the x-axis with any transverse variation of the field components being along the z-axis. No variation along y will be permitted. As has been defined already, TE and TM waves have characteristic field components that fall naturally into groups. For TE waves $H_x$ is $\pi/2$ out of phase with $E_y$ and $H_z$ and for TM waves $E_x$ is $\pi/2$ out of phase with $H_y$ and $E_z$. This can be proved as follows.

Any guiding action is to be associated with the existence of stationary eigenstates of the interface structure and this must mean that power flow is entirely in the x direction. The average energy flow (over one cycle) parallel to the z axis must be identically zero. For TM waves the power flow in the z direction is proportional to $\int E_x H_y dt$ where t is time and the integral is over one cycle of oscillation ($= 2\pi/\omega$).

Now, for example, if

$$E_x = a\cos(k_x x - \omega t) \qquad H_y = b\cos(k_x x - \omega t + \phi)$$

then $\int E_x H_y dt = 0$ only if $\phi = \pi/2$. (A similar argument holds for the TE case.) It is therefore easy to see that $\phi = \pi/2$ is a necessary condition for waveguiding to occur.

## 3. THIRD-ORDER NONLINEAR POLARISATION

It is necessary now to give a more detailed presentation of the polarisation of a non-magnetic dielectric material and its relationship to the components of the electric field that it is supporting. Initially, the linear polarisation will be used as a means of introducing some important concepts. The form of the linear polarisation $P_i^{(1)}(\mathbf{r},t)$ can be written as

$$P_i^{(1)}(\mathbf{r},t) = \epsilon_0 \int d\mathbf{r}' \int dt' \kappa_{ij}^{(1)}(\mathbf{r},t;\mathbf{r}',t') E_j(\mathbf{r}',t') \tag{3.1}$$

where $\kappa_{ij}^{(1)}$ is a linear susceptibility and is a response function of the dielectric material. In other words, if the electric field $\mathbf{E}$ is regarded as an input signal then the linear polarisation $\mathbf{P}^{(1)}(\mathbf{r},t)$ is the corresponding output signal. At this point a spatial dependence $\mathbf{r}$ (conventional shorthand for a functional dependence on x,y,z, with $d\mathbf{r} = dxdydz$) is included. The integrations are over the dummy variables $\mathbf{r}'$ and $t'$ with the response function being dependent upon both. Both the dependences on $\mathbf{r}'$ and $t'$ are forms of nonlocality in which the output signal $P_i^{(1)}(\mathbf{r},t)$ does not depend only on the *local* field component at $\mathbf{r}$ and t but upon fields at all points $\mathbf{r}'$ and upon what happened at previous times $t' \leq t$. The latter dependence is obviously a type of memory. In most of what will be said in this chapter, however, spatial nonlocality is neglected since many processes involve only characteristic radii of nonlocality $|\mathbf{r} - \mathbf{r}'| << \lambda$ where $\lambda$ is the wavelength of the electromagnetic wave. An exception to this will be diffusion-like processes in which excitations are transferred over significant distances during a relaxation time. The properties of the dielectric medium do not depend upon the absolute times t and $t'$ but only on their difference. This invariance under temporal translations is useful since, after dropping the $\mathbf{r}$ dependence because it is not of interest at the moment, it allows the more economical notation defined by

$$\kappa_{ij}^{(1)}(\mathbf{r},t;\ \mathbf{r}',t') = \kappa_{ij}^{(1)}(t - t') \tag{3.2}$$

Hence

$$P_i^{(1)}(\mathbf{r},t) = \epsilon_0 \int_{-\infty}^{\infty} dt' \kappa_{ij}^{(1)}(t - t') E_j(t') \tag{3.3}$$

The polarisation can only depend on the electric field values at times $t' < t$ so that $\kappa_{ij}^{(1)}(t - t') = 0$ if $t' > t$. In other words

$$P_i^{(1)}(\mathbf{r},t) = \epsilon_0 \int_{-\infty}^{t} dt' \kappa_{ij}^{(1)}(t - t') E_j(t') \tag{3.4}$$

The expression for $P_i^{(1)}(\mathbf{r},t)$ can, using the variable $\tau_1 = t - t'$, be written as

$$P_i^{(1)}(\mathbf{r},t) = \epsilon_0 \int_0^{\infty} d\tau_1 \kappa_{ij}^{(1)}(\tau_1) E_j(t-\tau_1) \tag{3.5}$$

This integral only converges independently of the behaviour of the electric field for materials with a finite memory time.

In general the polarisation vectors for the $n^{th}$ order, after neglecting spatial dispersion (nonlocality), are[29]

$$\mathbf{P}^{(n)}(t) = \epsilon_0 \int_0^{\infty} d\tau_1 .... \int_0^{\infty} d\tau_n \kappa_{ijkl....}^{(n)}(\tau_1,\tau_2,\tau_3,....\tau_N) E_j(t-\tau_1) E_k(t-\tau_2) E_l(t-\tau_3).... \tag{3.6}$$

where $n = 1,2,3....$ includes both the linear and the nonlinear terms.

The third-order polarisation vector is

$$P_i^{(3)}(t) = \epsilon_0 \int_0^{\infty} d\tau_1 \int_0^{\infty} d\tau_2 \int_0^{\infty} d\tau_3 \kappa_{ijkl}^{(3)}(\tau_1,\tau_2,\tau_3) E_j(t-\tau_1) E_k(t-\tau_2) E_l(t-\tau_3) \tag{3.7}$$

In various publications it will be seen that this expression for $P_i^{(3)}(t)$ is often approximated at this stage by assuming that $E_j(t)$ varies very little during any characteristic time interval of $\kappa_{ijkl}^{(3)}(\tau_1,\tau_2,\tau_3)$. This means that $E_j$, $E_k$, $E_l$ can be removed from under the integration signs and the following simple expression is then immediately obtained

$$P_i^{(3)}(t) = \epsilon_0 \chi_{ijkl}^{(3)} E_j(t) E_k(t) E_l(t) \tag{3.8}$$

and the implication is that a third-order nonlinear parameter exists with the form

$$\chi_{ijkl}^{(3)} = \int_0^{\infty}\int_0^{\infty}\int_0^{\infty} d\tau_1 d\tau_2 d\tau_3 \kappa_{ijkl}^{(3)}(\tau_1,\tau_2,\tau_3) \tag{3.9}$$

For Eq. (3.8) to be exactly true, $\chi_{ijkl}^{(3)}$ must be independent of frequency with $\kappa_{ijkl}^{(3)}(\tau_1,\tau_2,\tau_3)$ having a Dirac delta function type of behaviour. This frequency independence prevents the introduction of information concerning the mechanism of the nonlinearity. Symmetry operations can still be used, however, to reduce $\chi_{ijkl}^{(3)}$ to a manageable number of elements. For example, in an isotropic crystal all tensor elements can be expressed in terms of $\chi_{xxxx}^{(3)}$ since $\chi_{ijkl}^{(3)}$ has no frequency dependence. Hence only one constant can be fitted to experiment.

Since the fields in the above equation are, by definition, real quantities, they can be written, for monochromatic waves with an angular frequency $\omega$, in the form

$$E_i(\mathbf{r},t) = \frac{1}{2}[\xi_i(\mathbf{r},\omega)e^{-i\omega t} + \xi_i^*(\mathbf{r},\omega)e^{i\omega t}] \tag{3.10}$$

where $\xi_i(\mathbf{r},\omega)$ contains, through $\mathbf{r}$, all spatial variations. For example, for propagation along the x-axis, $\xi_x(\mathbf{r},\omega)$ may be simply $\xi_x e^{ik_x x}$, where $\xi_x$ is a constant and $k_x$ is a wavenumber. Also, the polarisation in Eq. (3.8), for an isotropic material, can be written as

$$\mathbf{P}(t) = \epsilon_0 \chi^{(3)}_{xxxx}(\mathbf{E}\cdot\mathbf{E})\mathbf{E} \tag{3.11}$$

showing that **P** is *parallel* to **E**. From Eq. (3.10)

$$(\mathbf{E}\cdot\mathbf{E})\mathbf{E} = \frac{1}{8}[\boldsymbol{\xi}(\mathbf{r},\omega)\cdot\boldsymbol{\xi}(\mathbf{r},\omega)e^{-2i\omega t} + 2\boldsymbol{\xi}(\mathbf{r},\omega)\cdot\boldsymbol{\xi}^*(\mathbf{r},\omega) + \boldsymbol{\xi}^*(\mathbf{r},\omega)\cdot\boldsymbol{\xi}^*(\mathbf{r},\omega)e^{2i\omega t}] \times [\boldsymbol{\xi}(\mathbf{r},\omega)e^{-i\omega t} + \boldsymbol{\xi}^*(\mathbf{r},\omega)e^{i\omega t}]$$

$$= \frac{1}{8}\{[\boldsymbol{\xi}(\mathbf{r},\omega)\cdot\boldsymbol{\xi}(\mathbf{r},\omega)]\boldsymbol{\xi}(\mathbf{r},\omega)e^{-3i\omega t} + [\boldsymbol{\xi}(\mathbf{r},\omega)\cdot\boldsymbol{\xi}(\mathbf{r},\omega)]\boldsymbol{\xi}^*(\mathbf{r},\omega)e^{-i\omega t} + 2[\boldsymbol{\xi}(\mathbf{r},\omega)\cdot\boldsymbol{\xi}^*(\mathbf{r},\omega)]\boldsymbol{\xi}(\mathbf{r},\omega)e^{-i\omega t}\} + \text{c.c.} \tag{3.12}$$

Equation (3.12) shows that there is a harmonic wave at $3\omega$ and a wave with the fundamental frequency $\omega$. The contributing terms to $\mathbf{P}^{(3)}(t)$ are

$$\frac{1}{8}[\boldsymbol{\xi}(\mathbf{r},\omega)\cdot\boldsymbol{\xi}(\mathbf{r},\omega)]\boldsymbol{\xi}(\mathbf{r},\omega) + \text{c.c.} \qquad \text{at } 3\omega$$

$$\frac{1}{4}\left\{[\boldsymbol{\xi}(\mathbf{r},\omega)\cdot\boldsymbol{\xi}^*(\mathbf{r},\omega)]\boldsymbol{\xi}(\mathbf{r},\omega) + \frac{1}{2}[\boldsymbol{\xi}(\mathbf{r},\omega)\cdot\boldsymbol{\xi}(\mathbf{r},\omega)]\boldsymbol{\xi}^*(\mathbf{r},\omega)\right\} + \text{c.c.} \qquad \text{at } \omega \tag{3.13}$$

The nonlinear polarisation can be written as

$$\mathbf{P}^{(3)}(t) = \frac{1}{2}[\mathbf{p}^{(3)}(\omega)e^{i(k_\omega x-\omega t)} + \mathbf{p}^{(3)}(3\omega)e^{i(k_{3\omega}x-3\omega t)}] + \text{c.c.}$$

$$= \frac{1}{2}[\mathbf{p}^{(3)}(\mathbf{r},\omega)e^{-i\omega t} + \mathbf{p}^{(3)}(\mathbf{r},3\omega)e^{-3i\omega t}] + \text{c.c.} \tag{3.14}$$

where $k_\omega$ and $k_{3\omega}$ are wavenumbers associated with frequencies $\omega$ and $3\omega$ respectively. The generation of the $3\omega$ harmonic polarisation can generally be ignored when considering the behaviour of the fundamental wave. For example, because the product of $\boldsymbol{\xi}(\mathbf{r},\omega)$ in the first part of Eq. (3.13) produces $3k_\omega$, the excitation at $3\omega$ will have a phase factor $e^{i\Delta kx}$ [$\Delta k = 3k_\omega - k_{3\omega}$]. In general $3k_\omega$ is not close to $k_{3\omega}$ and the efficiency of generation of the harmonic is proportional to something like $\sin(\Delta kx)/\Delta k$ which is approximately equal to zero. This is why the change in the polarisation at $\omega$ due to the generated third harmonic field can be neglected.

For the moment consider the case where the electric field is linearly polarised, say, along the y-axis, i.e. $\boldsymbol{\xi}(\mathbf{r},\omega) = \xi_y(\mathbf{r},\omega)\mathbf{y}$. The polarisation at $\omega$ will be of the form

$$\mathbf{p}^{(3)}(\mathbf{r},\omega) = \frac{3\epsilon_0}{4}\chi^{(3)}_{xxxx}|\xi_y(\mathbf{r},\omega)|^2\xi_y(\mathbf{r},\omega) \tag{3.15}$$

Writing the displacement vector $\mathbf{D}(\mathbf{r},t)$ in the form $\frac{1}{2}[\mathbf{d}(\mathbf{r},\omega)e^{-i\omega t} + \text{c.c.}]$ it is seen that

$$\mathbf{d}(\mathbf{r},\omega) = \epsilon_0\boldsymbol{\xi}(\mathbf{r},\omega) + \mathbf{p}(\mathbf{r},\omega) \tag{3.16}$$

where, for an isotropic material,

$$\mathbf{p}(\mathbf{r},\omega) = \epsilon_0[\chi^{(L)}(\omega)\xi_y(\mathbf{r},\omega) + \mathbf{p}^{(3)}(\mathbf{r},\omega)] \tag{3.17}$$

$\chi^{(L)}(\omega)$ being the linear susceptibility function. Hence the nonlinear dielectric function for the isotropic material in question is the scalar function[30]

$$\epsilon^{(NL)}(\omega) = [1 + \chi^{(L)}(\omega)] + \frac{3}{4}\chi^{(3)}_{xxxx}|\xi_y(\mathbf{r},\omega)|^2$$

$$= \epsilon_L(\omega) + \frac{3}{4}\chi^{(3)}_{xxxx}|\xi_y(\mathbf{r},\omega)|^2 \tag{3.18}$$

If a wave has two electric field components, as it will do for a TM guided wave, $\mathbf{P}^{(3)}(t)$ also has the general form given by Eqs. (3.11) and (3.12). This time, however, since

$$\mathbf{E} = \frac{1}{2}[(\xi_x \mathbf{x} + \xi_z \mathbf{z})e^{-i\omega t} + \text{c.c.}] \tag{3.19}$$

where the explicit dependence of $\xi$ upon $\mathbf{r}$ and $\omega$ has been omitted for convenience, Eq. (3.18) is not generally applicable. For example, the scalar product of $\mathbf{P}^{(3)}(t)$ with $\mathbf{x}$ has terms of the form

$$(\xi_x^2\xi_x^* + \xi_z^2\xi_x^* + 2\xi_x^2\xi_x^* + 2\xi_x\xi_z\xi_z^*)\mathbf{x}e^{-i\omega t}$$

so that

$$p_x^{(3)}(\mathbf{r},\omega) = \frac{3}{4}\epsilon_0\chi_{xxxx}^{(3)}\left[\left(|\xi_x|^2 + \frac{2}{3}|\xi_z|^2\right)\xi_x + \frac{1}{3}\xi_z^2\xi_x^*\right] \tag{3.20}$$

In order to establish this result it has been assumed that $\chi_{xxxx}^{(3)}$ is independent of frequency and the observation that[31]

$$\chi_{xxxx}^{(3)} = \chi_{xxzz}^{(3)} + \chi_{xzxz}^{(3)} + \chi_{xzzx}^{(3)} \tag{3.21}$$

has been used. For an isotropic material, intrinsic permutation symmetry permits suffices to be moved about freely giving the additional relationships

$$\chi_{xxzz}^{(3)} = \chi_{xzxz}^{(3)} = \chi_{xzzx}^{(3)} = \frac{1}{3}\chi_{xxxx}^{(3)} \tag{3.22}$$

Since it is assumed at the moment that $\chi_{ijkl}^{(3)}$ is *not* a function of frequency, intrinsic permutation of the subscripts of $\chi_{ijkl}^{(3)}$ can be carried out without any reference to the frequency associated with them. Since it is not true, in general, that $\chi_{ijkl}^{(3)}$ is independent of frequency, a great deal can be lost by failing to recognise this. Indeed, it is not much more trouble to work directly in the frequency domain. In so doing information will be included regarding the dynamics of the nonlinearity. The introduction of a frequency dependence requires a new definition of $\chi_{ijkl}^{(3)}$ that will lead to the identification of components of $\chi_{ijkl}^{(3)}$ that are now not necessarily equal to each other, even in an isotropic material. The net result will be to end up, in the isotropic case, with two independent components of $\chi_{ijkl}^{(3)}$ ($\chi_{xyyx}^{(3)}$, $\chi_{xxyy}^{(3)}$ for example) instead of one, as was used above. It is a somewhat curious fact that these components are degenerate for the type of nonlinearity arising from electronic distortion, effectively sending us back to the treatment described above. It is also just such a nonlinearity that is both rather fast and dominant in the majority of optical fibre materials such as silica glass. In other materials, however, the dominant nonlinearity may be of a different origin and the two components are not necessarily degenerate so it is appropriate now to develop more general formulae.

The form of the third-order polarisation vector is

$$P_i^{(3)}(t) = \epsilon_0\int_0^\infty d\tau_1\int_0^\infty d\tau_2\int_0^\infty d\tau_3\kappa_{ijkl}^{(3)}(\tau_1,\tau_2,\tau_3)E_j(t-\tau_1)E_k(t-\tau_2)E_l(t-\tau_3) \tag{3.23}$$

and the electric field components can be written, more generally, as

$$E_i(\mathbf{r},t) = \frac{1}{2}[\xi_i(\mathbf{r},\omega,t)e^{-i\omega t} + \xi_i^*(\mathbf{r},\omega,t)e^{i\omega t}] \tag{3.24}$$

where t is introduced into $\xi$ to allow for a possible *slow* variation of the amplitude with time. Only frequencies close to the carrier frequency $\omega$ will be present. Thus $\mathbf{E}(t)$ is expressed in terms of a *fast* time-dependence through $e^{-i\omega t}$ and a *slow* time dependence through the slowly varying amplitude factor $\xi(\mathbf{r},\omega,t)$. $\xi(\mathbf{r},\omega,t-\tau_i)$ has a slow enough time dependence for it to be Taylor expanded in the following way[29]

$$\xi(\mathbf{r},\omega,t-\tau_i) = \xi(\mathbf{r},\omega,t) - \tau_i\left[\frac{d}{dt'}\xi(\mathbf{r},\omega,t')\right]_{t'=t} + \ldots\ldots \qquad i = 1,2,3 \tag{3.25}$$

Terms like the second one lead to frequency derivatives of $\chi^{(3)}_{ijkl}$ due to the factors $\tau_i$, $\tau_i^2$, ... that occur during the Taylor expansion. These derivatives come from the process called dispersion and, while it is generally very important in the linear part, it is usually assumed to be negligible when dealing with the third-order nonlinearity. The slowly varying nature of $\xi(\mathbf{r},\omega,t)$, once taken outside the integration signs, will be neglected for the moment, so that $\xi(\mathbf{r},\omega,t) \equiv \xi(\mathbf{r},\omega)$. The introduction of $\mathbf{E}(t)$ in the form (3.24) after using Eq. (3.25) involves the triple multiplication $E_j(t-\tau_1)E_k(t-\tau_2)E_l(t-\tau_3)$. This leads after some patient labour to

$$P_i^{(3)}(t) = \frac{\epsilon_0}{8}\left[ e^{-i3\omega t}\int_0^\infty d\tau_1\int_0^\infty d\tau_2\int_0^\infty d\tau_3\kappa^{(3)}_{ijkl}(\tau_1,\tau_2,\tau_3)\xi_j(\mathbf{r},\omega,t-\tau_1)\xi_k(\mathbf{r},\omega,t-\tau_2)\xi_l(\mathbf{r},\omega,t-\tau_3)e^{i\omega[\tau_1+\tau_2+\tau_3]}\right.$$

$$\left. + 3e^{-i\omega t}\int_0^\infty d\tau_1\int_0^\infty d\tau_2\int_0^\infty d\tau_3\kappa^{(3)}_{ijkl}(\tau_1,\tau_2,\tau_3)\xi_j(\mathbf{r},w,t-\tau_1)\xi_k^*(\mathbf{r},\omega,t-\tau_2)\xi_l(\mathbf{r},\omega,t-\tau_3)e^{i\omega[\tau_1-\tau_2+\tau_3]}\right] + \text{c.c.}$$

$$= \frac{\epsilon_0}{8}[e^{-i3\omega t}\chi^{(3)}_{ijkl}(3\omega;\omega,\omega,\omega)\xi_j(\mathbf{r},\omega,t)\xi_k(\mathbf{r},\omega,t)\xi_l(\mathbf{r},\omega,t)$$

$$+ 3e^{-i\omega t}\chi^{(3)}_{ijkl}(\omega;\omega,-\omega,\omega)\xi_j(\mathbf{r},\omega,t)\xi_k^*(\mathbf{r},\omega,t)\xi_l(\mathbf{r},\omega,t)] + \text{c.c.} \tag{3.26}$$

where the slowly varying $\xi(\mathbf{r},\omega,t)$ has been brought outside the integration signs, all time derivatives of $\xi(\mathbf{r},\omega,t)$ have been neglected and

$$\chi^{(3)}_{ijkl}(3\omega;\omega,\omega,\omega) = \int_0^\infty d\tau_1\int_0^\infty d\tau_2\int_0^\infty d\tau_3\kappa^{(3)}_{ijkl}(\tau_1,\tau_2,\tau_3)e^{i\omega[\tau_1+\tau_2+\tau_3]} \tag{3.27}$$

$$\chi^{(3)}_{ijkl}(\omega;\omega,-\omega,\omega) = \int_0^\infty d\tau_1\int_0^\infty d\tau_2\int_0^\infty d\tau_3\kappa^{(3)}_{ijkl}(\tau_1,\tau_2,\tau_3)e^{i\omega[\tau_1-\tau_2+\tau_3]} \tag{3.28}$$

The notation used here admits the frequency dependence of $\chi^{(3)}_{ijkl}$ and has the general form $\chi^{(3)}_{ijkl}(\omega_1+\omega_2+\omega_3;\omega_1,\omega_2,\omega_3)$. Up to now attention has concentrated upon the time dependence of the nonlinearity because it is via this that the nonlocality or memory effects, if any, are going to arise. Naturally, since it is wave propagation that concerns us most here the wave number dependence must not be forgotten. It is always there through the $\mathbf{r}$ dependence of $\xi(\mathbf{r},\omega)$, i.e. $\xi(\mathbf{r},\omega) = \xi'(\omega)e^{i\mathbf{k}\cdot\mathbf{r}}$ where the slow time-dependence in $\xi(\mathbf{r},\omega,t)$ has been dropped. The $3\omega$ generation will now be neglected. It should be recalled that for the well-known second-harmonic ($2\omega$) generation in lower symmetry materials a great deal of ingenuity is needed to produce the $2\omega$ energy. This is even more the case for the $3\omega$ generation implied in Eq. (3.26) so it is safe to assume that the important third-order processes occur only at the fundamental frequency. Obviously steps could be taken to devise phase-matching schemes for $3\omega$ generation but they do not currently form an important activity in integrated optics. The component of the third-order polarisation at the fundamental frequency is therefore, approximately

$$\mathbf{P}^{(3)}(t) = \frac{3}{8}\epsilon_0[e^{-i\omega t}\chi^{(3)}_{ijkl}(\omega;\omega,-\omega,\omega)\xi_j(\mathbf{r},\omega,t)\xi_k^*(\mathbf{r},\omega,t)\xi_l(\mathbf{r},\omega,t)] + \text{c.c.} \tag{3.29}$$

The isotropic materials to be considered here include gases, liquids or amorphous solids for which $\chi^{(3)}_{ijkl}$ is invariant under the action of *any* rotation or reflection.

This properly restricts $\chi^{(3)}_{ijkl}$ to the values[31]

$$
\begin{aligned}
\chi^{(3)}_{xxxx} = \chi^{(3)}_{yyyy} = \chi^{(3)}_{zzzz} &= \chi^{(3)}_{\{1\}} + \chi^{(3)}_{\{2\}} + \chi^{(3)}_{\{3\}} \\
\chi^{(3)}_{yyzz} = \chi^{(3)}_{zzyy} = \chi^{(3)}_{zzxx} = \chi^{(3)}_{xxzz} = \chi^{(3)}_{xxyy} = \chi^{(3)}_{yyxx} &= \chi^{(3)}_{\{1\}} \\
\chi^{(3)}_{yzyz} = \chi^{(3)}_{zyzy} = \chi^{(3)}_{zxzx} = \chi^{(3)}_{xzxz} = \chi^{(3)}_{xyxy} = \chi^{(3)}_{yxyx} &= \chi^{(3)}_{\{2\}} \\
\chi^{(3)}_{yzzy} = \chi^{(3)}_{zyyz} = \chi^{(3)}_{zxxz} = \chi^{(3)}_{xzzx} = \chi^{(3)}_{xyyx} = \chi^{(3)}_{yxxy} &= \chi^{(3)}_{\{3\}}
\end{aligned}
\qquad (3.30)
$$

which implies that spatial symmetry leads to three independent components of $\chi^{(3)}_{ijkl}$.

In addition to this there is also intrinsic permutation symmetry. This means that the tensor $\chi^{(3)}_{ijkl}$ is invariant under *simultaneous* permutation of a suffix *and* its associated frequency, i.e.

$$\chi^{(3)}_{ijkl}(\omega_1+\omega_2+\omega_3;\omega_1,\omega_2,\omega_3) = \chi^{(3)}_{ijlk}(\omega_1+\omega_2+\omega_3;\omega_1,\omega_3,\omega_2) \qquad (3.31)$$

or, noting that it is the subscript $jk\ell$ that is moved,

$$\begin{Bmatrix} jk\ell \\ \omega_1\omega_2\omega_3 \end{Bmatrix} \rightarrow \begin{Bmatrix} kj\ell \\ \omega_2\omega_1\omega_3 \end{Bmatrix} \rightarrow \begin{Bmatrix} j\ell k \\ \omega_1\omega_3\omega_2 \end{Bmatrix} \qquad (3.32)$$

Even for an isotropic material it is most important that the suffices are only moved *together* with the frequencies. Hence the frequency dependence of the $\chi^{(3)}_{ijkl}$ is not a trivial matter and it leads, in the end, to two independent coefficients rather than one as used in the earlier presentation, where the $\chi^{(3)}_{ijkl}$ were assumed to be independent of frequency. In order to see how this operates consider $P^{(3)}_x(t)$ which can be immediately written as

$$P^{(3)}_x(t) = \frac{1}{2}[p^{(3)}_x(\mathbf{r},\omega)e^{-i\omega t} + p^{(3)*}_x(\mathbf{r},\omega)e^{i\omega t}] \qquad (3.33)$$

where, again dropping the explicit $(\mathbf{r},\omega)$ dependence of $\xi$ for convenience, we obtain from Eq. (3.29)

$$
\begin{aligned}
p^{(3)}_x(\mathbf{r},\omega) = \frac{3\epsilon_0}{4}\Big[ & \chi^{(3)}_{xxxx}(\omega;\omega,-\omega,\omega)\xi_x\xi^*_x\xi_x + \chi^{(3)}_{xyyx}(\omega;\omega,-\omega,\omega)\xi_y\xi^*_y\xi_x \\
& + \chi^{(3)}_{xzzx}(\omega;\omega,-\omega,\omega)\xi_z\xi^*_z\xi_x + \chi^{(3)}_{xxyy}(\omega;\omega,-\omega,\omega)\xi_x\xi^*_y\xi_y + \chi^{(3)}_{xxzz}(\omega;\omega,-\omega,\omega)\xi_x\xi^*_z\xi_z \\
& + \chi^{(3)}_{xyxy}(\omega;\omega,-\omega,\omega)\xi_y\xi^*_x\xi_y + \chi^{(3)}_{xzxz}(\omega;\omega,-\omega,\omega)\xi_z\xi^*_x\xi_z\Big] \\
= \frac{3\epsilon_0}{4}\Big[ & \chi^{(3)}_{xxxx}(\omega;\omega,-\omega,\omega)\xi_x\xi^*_x\xi_x + \{\chi^{(3)}_{xyyx}(\omega;\omega,-\omega,\omega) \\
& + \chi^{(3)}_{xxyy}(\omega;\omega,-\omega,\omega)\}|\xi_y|^2\xi_x + \{\chi^{(3)}_{xzzx}(\omega;\omega,-\omega,\omega) + \chi^{(3)}_{xxzz}(\omega;\omega,-\omega,\omega)\}|\xi_z|^2\xi_x \\
& + \{\chi^{(3)}_{xyxy}(\omega;\omega,-\omega,\omega)\xi^2_y + \chi^{(3)}_{xzxz}(\omega;\omega,-\omega,\omega)\xi^2_z\}\xi^*_x\Big]
\end{aligned}
\qquad (3.34)
$$

From the symmetry of the material

$$\chi^{(3)}_{xzzx}(\omega;\omega,-\omega,\omega) = \chi^{(3)}_{xyyx}(\omega;\omega,-\omega,\omega) \qquad (3.35)$$

$$\chi^{(3)}_{xxzz}(\omega;\omega,-\omega,\omega) = \chi^{(3)}_{xxyy}(\omega;\omega,-\omega,\omega) \qquad (3.36)$$

$$\chi^{(3)}_{xzxz}(\omega;\omega,-\omega,\omega) = \chi^{(3)}_{xyxy}(\omega;\omega,-\omega,\omega) \tag{3.37}$$

$$\chi^{(3)}_{xxxx}(\omega;\omega,-\omega,\omega) = \chi^{(3)}_{xyyx}(\omega;\omega,-\omega,\omega) + \chi^{(3)}_{xxyy}(\omega;\omega,-\omega,\omega) + \chi^{(3)}_{xyxy}(\omega;\omega,-\omega,\omega) \tag{3.38}$$

The relationships (3.35) to (3.38) permit $p_x^{(3)}(\mathbf{r},\omega)$ to be written as

$$p_x^{(3)}(\mathbf{r},\omega) = \frac{3\epsilon_0}{4}\Big[ \{\chi^{(3)}_{xyyx}(\omega;\omega,-\omega,\omega) + \chi^{(3)}_{xxyy}(\omega;\omega,-\omega,\omega)\} (|\xi_x|^2 + |\xi_y|^2 + |\xi_z|^2)\xi_x + \chi^{(3)}_{xyxy}(\omega;\omega,-\omega,\omega)(\xi_x^2 + \xi_y^2 + \xi_z^2)\xi_x^*\Big] \tag{3.39}$$

Now, from permutation symmetry

$$\chi^{(3)}_{xyyx}(\omega;\omega,-\omega,\omega) = \chi^{(3)}_{xxyy}(\omega;\omega,-\omega,\omega) = \chi^{(3)}_{xxyy}(\omega;\omega,\omega,-\omega) \tag{3.40}$$

where the subscripts *and* their associated frequencies have been interchanged. Similarly

$$\chi^{(3)}_{xyxy}(\omega;\omega,-\omega,\omega) = \chi^{(3)}_{xyyx}(\omega;\omega,\omega,-\omega) \tag{3.41}$$

Note that, in general,

$$\chi^{(3)}_{xyyx}(\omega;\omega,-\omega,\omega) \neq \chi^{(3)}_{xyyx}(\omega;\omega,\omega,-\omega) \tag{3.42}$$

Therefore

$$\begin{aligned} p_x^{(3)}(\mathbf{r},\omega) &= \frac{3\epsilon_0}{4}[2\chi^{(3)}_{xxyy}(\omega;\omega,\omega,-\omega)(|\xi_x|^2 + |\xi_y|^2 + |\xi_z|^2)\xi_x + \chi^{(3)}_{xyyx}(\omega;\omega,\omega,-\omega)(\xi_x^2 + \xi_y^2 + \xi_z^2)\xi_x^*] \\ &= \frac{3\epsilon_0}{4}[2\chi^{(3)}_{xxyy}(\omega;\omega,\omega,-\omega)|\xi|^2\xi_x + \chi^{(3)}_{xyyx}(\omega;\omega,\omega,-\omega)\xi^2\xi_x^*] \end{aligned} \tag{3.43}$$

and there are two nonlinear coefficients $\chi^{(3)}_{xxyy}(\omega;\omega,\omega,-\omega)$ and $\chi^{(3)}_{xyyx}(\omega;\omega,\omega,-\omega)$.

For TM guided waves $\mathbf{E} = (E_x, 0, E_z)$, and $E_x$ is $\pi/2$ out of phase with $E_z$. If the transformations $\xi_x \rightarrow \xi_x e^{ik_x x}$, $\xi_z \rightarrow i\xi_z e^{ik_x x}$, $p_x^{(3)} \rightarrow p_x^{(3)} e^{ik_x x}$ and $p_z^{(3)} \rightarrow ip_z^{(3)} e^{ik_x x}$ are adopted, where $k_x$ is the wavenumber of the wave and $\xi_x$ and $\xi_z$ are now real quantities, then $\mathbf{p}^{(3)}(\omega)$ is given by

$$p_x^{(3)}(\omega) = \frac{3\epsilon_0}{4}[(2\chi^{(3)}_{xxyy} + \chi^{(3)}_{xyyx})\xi_x^2 + (2\chi^{(3)}_{xxyy} - \chi^{(3)}_{xyyx})\xi_z^2]\xi_x \tag{3.44}$$

$$p_z^{(3)}(\omega) = \frac{3\epsilon_0}{4}[(2\chi^{(3)}_{xxyy} + \chi^{(3)}_{xyyx})\xi_z^2 + (2\chi^{(3)}_{xxyy} - \chi^{(3)}_{xyyx})\xi_x^2]\xi_z \tag{3.45}$$

where the wavenumber dependence has been factored out.

For TE guided waves $\mathbf{E} = (0, E_y, 0)$ and, adopting the transformation $\xi_y \rightarrow \xi_y e^{ik_x x}$ and $p_y^{(3)} \rightarrow p_y^{(3)} e^{ik_x x}$ with $\xi_y$ now a real quantity, the $p_y^{(3)}(\omega)$ component is

$$\begin{aligned} p_y^{(3)}(\omega) &= \frac{3\epsilon_0}{4}(2\chi^{(3)}_{xxyy} + \chi^{(3)}_{xyyx})\xi_y^2\xi_y \\ &= \frac{3\epsilon_0}{4}\chi^{(3)}_{yyyy}\xi_y^2\xi_y \end{aligned} \tag{3.46}$$

where again the wavenumber dependence has been eliminated.

The dielectric functions or sub-tensors that are needed to characterise nonlinear TE and TM waves propagating in a dielectric medium are, therefore,[5,6]

$$\epsilon = \epsilon_L + a\xi_y^2 \qquad \text{for TE waves} \tag{3.47}$$

$$\underline{\underline{\epsilon}} = \begin{bmatrix} \epsilon_L + a\xi_x^2 + \gamma a\xi_z^2 & 0 \\ 0 & \epsilon_L + a\xi_z^2 + \gamma a\xi_x^2 \end{bmatrix} \qquad \text{for TM waves} \tag{3.48}$$

where Eqs. (3.16) and (3.17) have been used and

$$a = \frac{3}{4}(2\chi^{(3)}_{xxyy} + \chi^{(3)}_{xyyx})$$

$$= \frac{3}{2}\chi^{(3)}_{xxyy}\left[1 + \frac{\chi^{(3)}_{xyyx}}{2\chi^{(3)}_{xxyy}}\right] \tag{3.49}$$

$$\gamma = \frac{\left[1 - \dfrac{\chi^{(3)}_{xyyx}}{2\chi^{(3)}_{xxyy}}\right]}{\left[1 + \dfrac{\chi^{(3)}_{xyyx}}{2\chi^{(3)}_{xxyy}}\right]} \tag{3.50}$$

The ratio $\chi^{(3)}_{xyyx}/2\chi^{(3)}_{xxyy}$ can be modelled for a number of nonlinear mechanisms, but this will not be done here. It is enough to point out, for example, that for a nonlinear polarisation that arises in solids from electronic distortion $\gamma = 1/3$, for a nonlinearity arising from the molecular orientational Kerr effect $\gamma = -1/2$, and for a thermal or electrostrictive nonlinearity $\gamma = 1$.

## 4. STATIONARY AND NON-STATIONARY STATES

The simplest, and perhaps the most illustrative, case to consider here is that of a TE (s-polarised) wave $\mathbf{E} = \frac{1}{2}(0, E, 0)e^{-i\omega t}$ + c.c. propagating along the x-axis with angular frequency $\omega$. Suppose that there is no variation in the y-direction and that the linear dielectric function $\epsilon(z)$ and the nonlinear coefficient $a(z)$ vary only in the z-direction. Then, Eq. (2.4) becomes

$$\frac{\partial^2 E}{\partial x^2} + \frac{\partial^2 E}{\partial z^2} + k_0^2\left[\epsilon(z) + a(z)|E|^2\right]E = 0 \tag{4.1}$$

where $k_0 = \omega/c$ and Eq. (4.1) is an elliptic equation. The solution of a partial differential equation like this requires boundary values for E. If only values of E are given they are called Dirichlet conditions (D). If values of the normal derivatives of E are given they are called Neumann conditions (N) and if values of E and the normal derivative are *both* given at the same points they are called Cauchy conditions (C). A boundary is considered to be closed if it completely surrounds the solution in the region of investigation. It is open if part of the boundary goes to infinity, but no boundary conditions are imposed along the part at infinity. An elliptic equation requires a closed boundary with Dirichlet or Neumann conditions whilst a parabolic equation only needs an open boundary with Dirichlet conditions.[32] It actually is much easier to solve parabolic equations numerically so the elliptic equation (4.1) is usually transformed, via a Fresnel approximation, to parabolic form before a numerical analysis of non-stationary states begins. Before this transformation is defined and effected let us consider an exact stationary state of Eq. (4.1) for a bulk nonlinear medium for which $\epsilon(z) = \epsilon_B$, $a(z) = a_B$, $-\infty < z < \infty$. If only stationary waves are considered the field has the 'beam' form

$$E(x, z) = U(z, k_x)e^{ik_x x} \tag{4.2}$$

so that Eq. (4.1) reduces to the following equation for $U \equiv U(z, k_x)$

$$\frac{\partial^2 U}{\partial z^2} - [k_x^2 - k_0^2\epsilon_B]U + k_0^2 a_B|U|^2 U = 0 \tag{4.3}$$

For plane waves U is independent of z so that

$$k_x^2 = k_0^2\left[\epsilon_B + a_B|U|^2\right] \tag{4.4}$$

If $U(z, k_x)$ is z-dependent then the function

$$U(z) = \frac{\gamma}{k_0}\left(\frac{2}{a_B}\right)^{1/2} \operatorname{sech}(\gamma z) \tag{4.5}$$

where $a_B > 0$ and $\gamma^2 = k_x^2 - k_0^2\epsilon_B > 0$, is a solution of Eq. (4.3). This is perhaps best shown by substitution. The terms of Eq. (4.3) are, using Eq. (4.5),

$$\frac{\partial^2 U}{\partial z^2} = -\frac{\gamma^3}{k_0}\left(\frac{2}{a_B}\right)^{1/2}\left[\operatorname{sech}(\gamma z) - 2\sinh^2(\gamma z)\operatorname{sech}^3(\gamma z)\right] \tag{4.6}$$

$$-\gamma^2 U = -\frac{\gamma^3}{k_0}\left(\frac{2}{a_B}\right)^{1/2} \operatorname{sech}(\gamma z) \tag{4.7}$$

$$k_0^2 a_B|U|^2 U = \frac{2\gamma^3}{k_0}\left(\frac{2}{a_B}\right)^{1/2} \operatorname{sech}^3(\gamma z) \tag{4.8}$$

These collected together give

$$-\operatorname{sech}(\gamma z) + 2\sinh^2(\gamma z)\operatorname{sech}^3(\gamma z) - \operatorname{sech}(\gamma z) + 2\operatorname{sech}^3(\gamma z) = 0 \tag{4.9}$$

The function (4.5) is, therefore, indeed an exact stationary state solution of Eq. (4.3). Furthermore, the limit of U as z goes to $\pm\infty$ is zero. It is called a *stationary state* because $U(z, k_x)$ does not vary as the wave progresses down the propagation x-axis.

This solution for a bulk wave in a nonlinear medium is called a *spatial soliton*. If non-stationary states are required they are solutions of (4.1) of the form

$$E(x, z) = \phi(x, z)e^{ik_1 x} \tag{4.10}$$

where, in general, the amplitude $\phi(x, z)$ evolves as the wave travels along the x-axis. Also since the state is non-stationary then, whereas $k_x$ in Eq. (4.2) is an eigen-number of the propagation, $k_1$ is, at this stage, arbitrary. Hence, using Eq. (4.10) in Eq. (4.1)

$$\frac{\partial^2\phi}{\partial z^2} + \frac{\partial^2\phi}{\partial x^2} + 2ik_1\frac{\partial\phi}{\partial x} - [k_1^2 - k_0^2\epsilon(z)]\phi + k_0^2 a(z)|\phi|^2\phi = 0 \tag{4.11}$$

where once again the possibility of $\epsilon(z)$ and $a(z)$ varying with z is included but their derivatives are neglected. In practice, this feature permits the investigation of discontinuous structures such as multilayers. For a set of homogeneous layers $\epsilon(z)$ and $a(z)$ would be constant within each layer and this is the case for all the examples introduced later on. Equation (4.11) is still elliptic but it reduces to a parabolic one through the Fresnel approximation

$$\left|\frac{\partial^2\phi}{\partial x^2}\right| << \left|2ik_1\frac{\partial\phi}{\partial x}\right| \quad \text{when } |\partial\phi/\partial x| \neq 0 \tag{4.12}$$

$$\left|\frac{\partial^2\phi}{\partial x^2}\right| = 0 \Big) \qquad \text{when } |\partial\phi/\partial x| = 0$$

so that, at the very least, $k_1$ must be consistent with this assumption. The parabolic form of Eq. (4.1) is therefore

$$\frac{\partial^2\phi}{\partial z^2} + 2ik_1\frac{\partial\phi}{\partial x} - [k_1^2 - k_0^2\epsilon(z)]\phi + k_0^2 a(z)|\phi|^2\phi = 0 \tag{4.13}$$

Since $\phi(x,z)$ is slowly varying compared to $e^{ik_1x}$, it is probably the case that

$$\left|\frac{\partial\phi(x,z)}{\partial x}\right| << k_1|\phi(x,z)| \tag{4.14}$$

Hence, almost everywhere in the region of interest, in order to be consistent with the neglect of terms in Eq. (4.11), $k_1$ should be *chosen* to satisfy the inequality

$$|k_1^2 - k_0^2\epsilon_B| << 2k_1 \tag{4.15}$$

In this context it is useful to remember that physically, $k_0^2 a_B|\phi(x,z)|^2 << k_0^2\epsilon_B$.

A natural choice for $k_1$, therefore, is

$$k_1 = k_0\sqrt{\epsilon_B} \tag{4.16}$$

For this choice, Eq. (4.11) becomes

$$2ik_0\sqrt{\epsilon_B}\,\frac{\partial\phi(x,z)}{\partial x} + \frac{\partial^2\phi(x,z)}{\partial z^2} + k_0^2 a_B|\phi(x,z)|^2\phi(x,z) = 0 \tag{4.17}$$

Although Eq. (4.17) would only be used if non-stationary solutions were being sought, it is obvious that the stationary states of (4.17) will be approximately a solution of (4.1) when $k_x$ is *sufficiently close* to $k_0\sqrt{\epsilon_B}$. In order to appreciate this a solution of the form

$$\phi(x,z) = U'(z,k_x)\exp\left[i\left(\frac{k_x^2 - k_0^2\epsilon_B}{2k_0\sqrt{\epsilon_B}}\right)x\right] \tag{4.18}$$

can be tried. This form gives directly the equation

$$\frac{d^2U'(z,k_x)}{dz^2} - (k_x^2 - k_0^2\epsilon_B)U'(z,k_x) + k_0^2 a_B|U'(z,k_x)|^2U'(z,k_x) = 0 \tag{4.19}$$

This equation is *exactly* the same as the previous equation for $U(z,k_x)$.

Hence, if $U'(z,k_x)$ and $U(z,k_x)$ are subject to the same boundary conditions then

$$U'(z,k_x) = U(z,k_x) \tag{4.20}$$

Self-consistency demands that any solution of Eq. (4.17) is only a good approximation to the solution of Eq. (4.1) if it satisfies the conditions (4.12). Hence, using the form (4.18), it is necessary that

$$\frac{\left(k_x^2 - k_0^2\epsilon_B\right)^2}{4k_0^2\epsilon_B} << |k_x^2 - k_0^2\epsilon_B| \tag{4.21}$$

for any nontrivial solution to exist. Since $k_x^2 > k_0^2\epsilon_B$ in a self-focusing medium, Eq. (4.21) becomes

$$\frac{k_x^2 - k_0^2\epsilon_B}{4k_0^2\epsilon_B} << 1 \tag{4.22}$$

A Taylor expansion then gives

$$k_x = k_0\sqrt{\epsilon_B}\left[1 + \frac{k_x^2 - k_0^2\epsilon_B}{2k_0^2\epsilon_B} - \frac{1}{8}\left(\frac{k_x^2 - k_0^2\epsilon_B}{k_0^2\epsilon_B}\right)^2 \cdots\right]$$

$$\approx k_0\sqrt{\epsilon_B} + \frac{k_x^2 - k_0^2\epsilon_B}{2k_0\sqrt{\epsilon_B}} + \ldots \tag{4.23}$$

Therefore, to a good approximation

$$E(x,z) = U(z,k_x)e^{ik_0\sqrt{\epsilon_B}\,x}\exp\left[i\left(\frac{k_x^2 - k_0^2\epsilon_B}{2k_0\sqrt{\epsilon_B}}\right)x\right]$$

$$= U(z,k_x)e^{ik_x x} \tag{4.24}$$

This once again is the special solution[33-36] called the *spatial soliton*. The characteristic length over which U changes as the wave progresses down the x-axis is obviously much greater than the distance over which a change in the transverse direction occurs. This spatial soliton can be called a beam soliton since it is a self-focused beam (to be defined later) propagating down the x-axis. It is also the end product of a transverse modulational instability in which a modulation transverse to the propagation direction initially grows with propagation distance and is quite distinct from the pulse envelope soliton that is extensively discussed elsewhere in this book. The pulse envelope soliton is the end product of a longitudinal modulational instability.

If $\phi$ is a constant in Eq. (4.11) then

$$k_1^2 = k_0^2\left[\epsilon(z) + a|\phi|^2\right] \tag{4.25}$$

and the plane wave stationary state is recovered. Some comments on the stability of plane waves in a nonlinear medium will now be made.

## 5. STABILITY OF NONLINEAR PLANE WAVES

For a constant nonlinear coefficient $a(z) = a$ characterising an infinite bulk medium and for a scalar non-stationary electric field $E(x,z) = U(x,z)e^{ik_1x}$, where $k_1$ is arbitrary, the equation for $U(x,z)$ is, writing $\epsilon \equiv \epsilon_B$, $a \equiv a_B$,

$$\frac{\partial^2 U}{\partial z^2} + 2ik_1\frac{\partial U}{\partial x} - [k_1^2 - k_0^2\epsilon]U + k_0^2 a|U|^2U = 0 \tag{5.1}$$

within a parabolic assumption. Here $U(x,z) \equiv U$ and $\epsilon$ is the constant, linear, bulk dielectric constant.

If the field $E(x,z)$ is *assumed* to be a simple plane wave with a real constant amplitude $U_0$ then

$$U(x,z) = U_0 \qquad U_0 \text{ real} \tag{5.2}$$

$$k_1^2 = \frac{\omega^2}{c^2}(\epsilon + aU_0^2) \tag{5.3}$$

where $U_0$ could have been taken as complex if desired. $k_1$ in this case is *exactly* the stationary state eigenvalue. The question now arises as to how stable this solution is. A simple, yet revealing, way to find out is to take $U_0$ as an *initial condition* but with an added perturbation. The exact plane wave solution can be defined as having

$$U(0,z) = U_0 \tag{5.4}$$

as the initial field distribution on the plane x = 0.

This initial condition can then be perturbed to

$$U(0,z) = U_0 + \Gamma g(z) \tag{5.5}$$

where Γ is a real arbitrarily small parameter. Equation (5.1) can now be used to study the propagation of this initial condition along the x-axis. At some distance x from the plane x = 0, U(x, z) can be written as

$$U(x,z) = U_0 + \Gamma d(x,z) \qquad d(0,z) = g(z) \tag{5.6}$$

This on substitution into the parabolic wave equation, and after the neglect of powers of Γ higher than the first, leads to the following equations associated with $\Gamma^0$ (zeroth order) and Γ (first order)

$$(k_1^2 - k_0^2\epsilon - k_0^2 a U_0^2)U_0 = 0 \qquad \Gamma^0 \tag{5.7}$$

$$\frac{\partial^2 d}{\partial z^2} + 2ik_1\frac{\partial d}{\partial x} - (k_1^2 - k_0^2\epsilon - k_0^2 a U_0^2)d + k_0^2 a U_0^2(d + d^*) = 0 \qquad \Gamma \tag{5.8}$$

where $d \equiv d(x,z)$ and $d^*$ is its complex conjugate. These equations together reduce to

$$\frac{\partial^2 d}{\partial z^2} + 2ik_1\frac{\partial d}{\partial x} + k_0^2 a U_0^2(d + d^*) = 0 \tag{5.9}$$

The solution of this equation can be best understood in terms of the spatial spectral properties of d(x, z). Hence the substitution of d(x, z) into Eq. (5.9) in the form of the Fourier integral

$$d(x,z) = \frac{1}{2\pi}\int_{-\infty}^{\infty} b(k,x)e^{ikz}dk \tag{5.10}$$

leads immediately to

$$\frac{1}{2\pi}\int_{-\infty}^{\infty}\left(2ik_1\frac{\partial}{\partial x} - k^2 + \frac{\omega^2}{c^2}aU_0^2\right)b(k,x)e^{ikz}dk + \frac{1}{2\pi}\int_{-\infty}^{\infty}\frac{\omega^2}{c^2}aU_0^2 b^*(k,x)e^{-ikz}dk = 0 \tag{5.11}$$

The second integration, however, can be written as

$$\frac{1}{2\pi}\int_{-\infty}^{\infty}\frac{\omega^2}{c^2}aU_0^2 b^*(k,x)e^{-ikz}dk = \frac{1}{2\pi}\int_{-\infty}^{\infty}\frac{\omega^2}{c^2}aU_0^2 b^*(-k,x)e^{ikz}dk \tag{5.12}$$

so that

$$\frac{\omega^2}{c^2}aU_0^2 b^*(-k,x) = -Lb(k,x) \tag{5.13}$$

where

$$L = 2ik_1\frac{\partial}{\partial x} - \left(k^2 - \frac{\omega^2}{c^2}aU_0^2\right) \tag{5.14}$$

$$L^* = -\left(2ik_1\frac{\partial}{\partial x} + k^2 - \frac{\omega^2}{c^2}aU_0^2\right) \tag{5.15}$$

Now,

$$L^*Lb(k,x) = -\frac{\omega^2}{c^2}aU_0^2L^*b^*(-k,x) \tag{5.16}$$

where

$$L^*b^*(-k,x) = -\frac{\omega^2}{c^2}aU_0^2b(k,x) \tag{5.17}$$

and

$$LL^* = \left(2k_1\right)^2\frac{\partial^2}{\partial x^2} + \left(k^2 - \frac{\omega^2}{c^2}aU_0^2\right)^2 \tag{5.18}$$

Therefore

$$\left\{\left(2k_1\right)^2\frac{\partial^2}{\partial x^2} + \left(k^2 - \frac{\omega^2}{c^2}aU_0^2\right)^2\right\}b(k,x) = \left(\frac{\omega^2}{c^2}aU_0^2\right)^2 b(k,x) \tag{5.19}$$

This kind of equation can readily be solved by trying a solution of the form $e^{\pm qx}$. If this is done then q must satisfy the equation

$$\left(2k_1\right)^2q^2 + \left(k^2 - \frac{\omega^2}{c^2}aU_0^2\right)^2 - \left(\frac{\omega^2}{c^2}aU_0^2\right)^2 = 0 \tag{5.20}$$

In simpler form this gives

$$q^2 = \left(\frac{k}{2k_1}\right)^2\left(2\frac{\omega^2}{c^2}aU_0^2 - k^2\right) \tag{5.21}$$

This equation shows that q is real and non-zero if, and only if,

$$0 < k^2 < 2\frac{\omega^2}{c^2}aU_0^2 \tag{5.22}$$

Hence if the nonlinear coefficient is greater than zero a definite range of k exists for which q can be real and growth of the perturbation is at least a possibility. For negative a there is no range of k for which q can be real and the plane wave always remains stable. The regimes $a > 0$ and $a < 0$ are known as self-focusing and self-defocusing[37] respectively.

The concept of self-focusing needs now to be addressed briefly since such solutions will appear throughout the rest of this account. Firstly, however, it should be emphasised that up to now we have viewed the self-focusing situation as the outcome of an unstable plane wave. This type of modulational instability, however, can evolve to the stationary spatial soliton sech type solution. This is a steady state outcome in a self-focusing medium and need not be unstable itself.

Figure 2 shows a simple picture of self-focusing in two dimensions covering the (x,z) plane. The initial uniform $|E(0,z)|^2 = |U_0|^2$ intensity distribution is considered to be slightly disturbed. If we concentrate on the position $z = z_0$ then it is clear that because in a self-focusing medium $a|U_0|^2$ is *added* to the existing linear dielectric function, the effective refractive index increases at that point. This region then acts as its own waveguide. It channels power into the region $z = z_0$. This causes the local refractive index to increase even more. A runaway effect (an instability) is therefore possible. The only thing that will oppose this runa-

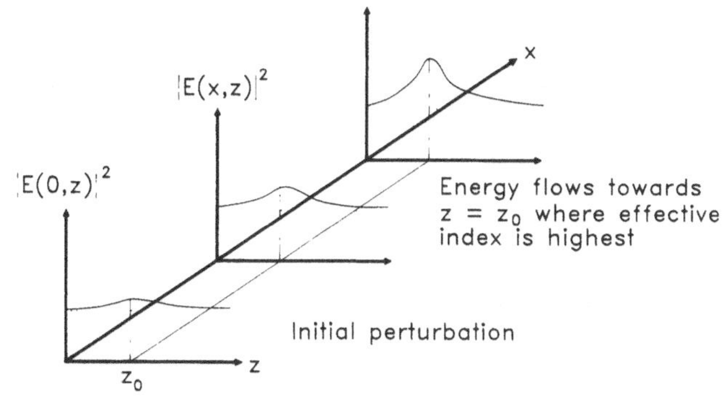


Fig. 2 Illustration of self-focusing.

way effect is diffraction. As the self-focusing (or channelling) in the region $z = z_0$ progresses then the channel itself will act as an aperture to diffract the electromagnetic waves. It is the existence of self-focusing in the *presence* of diffraction that leads to the stationary spatial soliton solution. This is similar to self-phase modulation acting in the *presence* of group velocity dispersion in a way that leads to stable pulse envelope solitons.

## 6. NONLINEAR TE WAVES

The nonlinearity considered here is so strong that the shape of the stationary fields of nonlinear guided TE waves depends upon the power flow. Since this is the case the usual coupled-mode approach to nonlinear waveguide analysis cannot be used.

Although the nonlinearity now to be used may saturate at some fairly modest value it can easily be accounted for. As discussed earlier in this chapter, the nonlinear polarisation may arise from electronic distortion, molecular alignment, or electrostriction (thermal) effects and it lies in what is referred to as the Kerr regime. Most of the general arguments apply both to self-focusing and self-defocusing nonlinearities.

The nonlinear cw TE waves are polarised along the y-direction, with a wave number $k_x$ and an angular frequency $\omega$. The electromagnetic field of these waves is defined as

$$\mathbf{E} = [0, E_y(k_x, \omega, z), 0]e^{i(k_x x - \omega t)} \qquad \mathbf{H} = [H_x(k_x, \omega, z), 0, H_z(k_x, \omega, z)]e^{i(k_x x - \omega t)} \tag{6.1}$$

and the nonlinear dielectric function is

$$\epsilon^{NL}(\omega, |E|^2) = \epsilon(\omega) + a(\omega)|E_y(k_x, \omega, z)|^2 \tag{6.2}$$

where $\epsilon(\omega)$ is a purely linear part.

Even though $\epsilon^{NL}(\omega, |E|^2)$ is generally a function of $\omega$, in practice, it makes no difference to the generality of the formulation if the $\omega$ dependence is dropped. Similarly the nonlinear coefficient $a(\omega)$ can also be assumed to be a simple constant 'a'. The final caveat is that the nonlinearity is made local in space and time. If we do introduce nonlocality at any time it will be through processes such as diffusion. This can be quickly referred to later on.

The components of Maxwell's equations are

$$\frac{d}{dz}E_y(k_x, \omega, z) = -i\omega\mu_0 H_x(k_x, \omega, z) \tag{6.3}$$

$$k_x E_y(k_x, \omega, z) = \omega\mu_0 H_z(k_x, \omega, z) \tag{6.4}$$

$$\frac{d}{dz}H_x(k_x, \omega, z) = ik_x H_z(k_x, \omega, z) - i\omega\epsilon_0\epsilon^{NL}(\omega, |E_y|^2)E_y(k_x, \omega, z) \tag{6.5}$$

At this stage $E_y(k_x, \omega, z)$ must be considered to be complex. Indeed it can be expressed as

$$E_y(k_x,\omega,z) = E(k_x,\omega,z)e^{i\phi(k_x,\omega,z)} \tag{6.6}$$

where $E(k_x,\omega,z)$ and $\phi(k_x,\omega,z)$ are now real functions.

Equations (6.3) - (6.5) can be combined to give[2]

$$\frac{d^2E_y}{dz^2} + (k_0^2\epsilon - k_x^2)E_y + k_0^2a|E_y|^2E_y = 0 \tag{6.7}$$

where $k_0 = \omega/c$. If $E_y$, in the form given in Eq. (6.6), is substituted into Eq. (6.7) we obtain

$$\frac{d^2E}{dz^2} - E\left(\frac{d\phi}{dz}\right)^2 + 2i\frac{d\phi}{dz}\frac{dE}{dz} + i\frac{d^2\phi}{dz^2}E + (k_0^2\epsilon - k_x^2)E + k_0^2aE^3 = 0 \tag{6.8}$$

The real and imaginary parts of this equation can be equated to zero and this gives[38]

$$\frac{d^2E}{dz^2} - E\left(\frac{d\phi}{dz}\right)^2 + (k_0^2\epsilon - k_x^2)E + k_0^2aE^3 = 0 \tag{6.9}$$

$$2\frac{d\phi}{dz}\frac{dE}{dz} + \frac{d^2\phi}{dz^2}E = 0 \tag{6.10}$$

The latter equation is equivalent to

$$\frac{d}{dz}\left[E^2\frac{d\phi}{dz}\right] = 0 \tag{6.11}$$

and can be integrated at once to give

$$\frac{d\phi}{dz} = \frac{K}{E^2} \tag{6.12}$$

where K is a constant of integration.

The equation involving K is a statement of the conservation of energy flux in the nonlinear medium and the complete equation for the real quantity E is

$$\frac{d^2E}{dz^2} + (k_0^2\epsilon - k_x^2)E + k_0^2aE^3 - \frac{K^2}{E^3} = 0 \tag{6.13}$$

As indicated qualitatively earlier on, for guided or surface waves there can be no flux of energy either to or from the guiding surfaces so that $E \to 0$ as $z \to \pm\infty$. This means that for surface or guided waves $K = 0$.

This can also be seen from a consideration of the time averaged Poynting vector whose z-component is[38]

$$\langle S\rangle_z = -\frac{1}{2}Re(E_y^*H_x) = \frac{c^2}{2\omega^2}\epsilon_0E^2\frac{d\phi}{dz} = \frac{c^2}{2\omega^2}\epsilon_0K \tag{6.14}$$

where *Re* denotes the real part and $^*$ denotes the complex conjugate. It is quite transparent from Eq. (6.14) that if $\langle S\rangle_z = 0$ then the constant $K = 0$. This implies in turn that $d\phi/dz = 0$.

A scalar nonlinear wave equation remains if $K = 0$ and it can be integrated once to give

$$\left(\frac{dE}{dz}\right)^2 + (k_0^2\epsilon - k_x^2)E^2 + \frac{a}{2}k_0^2E^4 = C \tag{6.15}$$

This is often called the first integral and C is another integration constant. In fact C can be expressed in terms of the fields evaluated at the boundaries of any given TE guiding situation. It can be shown that the constant C is non-zero, or zero, according to the application that is

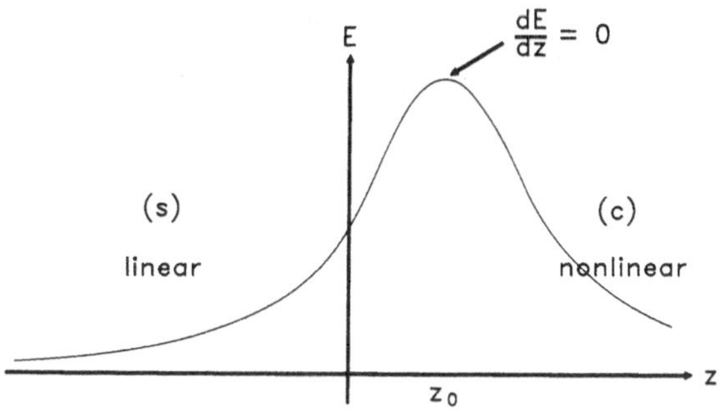


Fig. 3 Self-focused TE wave guided by a single interface between a linear substrate (s) and nonlinear cladding (c).

required.[4] In a nonlinear thin film C is non-zero. For nonlinear waves travelling along an interface between two semi-infinite media, however, C = 0. In general, in a medium in which a boundary approaches infinity, C = 0 whenever both E and dE/dz → 0 as z → ±∞.

In guides for which C = 0, using $\kappa^2 = k_x^2 - k_0^2\epsilon$ and $\Lambda = \omega^2 a/(2c^2)$, Eq. (6.15) factorises to

$$\left(\frac{dE}{dz} - E\sqrt{\kappa^2 - \Lambda E^2}\right)\left(\frac{dE}{dz} + E\sqrt{\kappa^2 - \Lambda E^2}\right) = 0 \tag{6.16}$$

and, for a > 0, the solution required is the integral

$$\int \frac{dE}{E\sqrt{\kappa^2 - \Lambda E^2}} = z \pm z_0 \tag{6.17}$$

where again $z_0$ (> 0) is an integration constant. This integration is quite elementary and yields a form of E that we encountered earlier, namely

$$E = \frac{\kappa}{\sqrt{\Lambda}}\text{sech}[\kappa(z \pm z_0)] \tag{6.18}$$

where the sign depends upon the waveguide structure.

A single interface between two semi-infinite media, one of which is nonlinear, is illustrated in Fig. 3. The linear substrate is labelled 's' and the nonlinear cladding is labelled 'c'. For the cladding, using the subscript c

$$\left(\frac{dE_c}{dz}\right)^2 - \kappa_c^2 E_c^2 + \Lambda_c E_c^4 = 0 \tag{6.19}$$

At the point $dE_c/dz = 0$ (for E > 0)

$$E_c = \frac{\kappa_c}{\sqrt{\Lambda_c}} \tag{6.20}$$

which shows that $a_c > 0$ is a necessary condition for real E to exist. This confirms the earlier conclusions based upon a plane wave analysis. Furthermore when $dE_c/dz = 0$

$$\frac{d^2E_c}{dz^2} = (\kappa_c^2 - 2\kappa_c^2)\frac{\kappa_c}{\Lambda_c} = -\frac{\kappa_c^3}{\Lambda_c} < 0 \tag{6.21}$$

proving that there is a maximum in the field amplitude. This is, of course, what would be expected with a self-focused peak in the electric field so, as shown in Fig. 3, there is self-focused field peaked at $z_0$ in the nonlinear medium. Nevertheless it is now propagating along the interface between the two media.

The fields in the respective semi-infinite media are, therefore,

$$E_s = E_0 e^{\kappa_s z} \qquad z < 0 \text{ (substrate)} \tag{6.22}$$

$$E_c = \frac{\kappa_c}{\sqrt{\Lambda_c}} \operatorname{sech}[\kappa_c (z - z_0)] \qquad z > 0 \text{ (cladding)} \tag{6.23}$$

At z = 0 the boundary conditions are

$$E_s = E_c \qquad \frac{dE_s}{dz} = \frac{dE_c}{dz} \tag{6.24}$$

Hence,

$$E_0 = \frac{\kappa_c}{\sqrt{\Lambda_c}} \operatorname{sech}(\kappa_c z_0) \tag{6.25}$$

and

$$\kappa_s E_0 = \frac{\kappa_c^2}{\sqrt{\Lambda_c}} \tanh(\kappa_c z_0) \operatorname{sech}(\kappa_c z_0) \tag{6.26}$$

These boundary conditions give

$$\tanh(\kappa_c z_0) = \frac{\kappa_s}{\kappa_c} \tag{6.27}$$

which implies that

$$\operatorname{sech}(\kappa_c z_0) = \frac{\left(\kappa_c^2 - \kappa_s^2\right)^{1/2}}{\kappa_c} \tag{6.28}$$

The boundary field value is, therefore,[19,22,39]

$$E_0^2 = \frac{2}{a_c} \kappa_c^2 \left\{ \frac{\kappa_c^2 - \kappa_s^2}{\kappa_c^2} \right\} \tag{6.29}$$

$$= \frac{2}{a_c} (\epsilon_s - \epsilon_c) \tag{6.30}$$

Hence $E_0^2$ is independent of wave number. Once the constants $a_c$, $\epsilon_s$ and $\epsilon_c$ of the medium are fixed, $E_0^2$ remains fixed no matter what power is flowing along the interface.

A formula for the power flow is obtained from $S_x$, the x-component of the Poynting vector. For this coordinate system, and for TE waves,

$$S_x = \frac{1}{2} Re(E_y^* H_z) \tag{6.31}$$

Since $H_z = k_x E_y / (\omega \mu_0)$, the time-averaged power flow along the x-axis is

$$P = \frac{1}{2} \frac{c^2 \epsilon_0 k_x}{\omega} \int_{z_1}^{z_2} E_y^2 dz \tag{6.32}$$

where z is considered, initially, to lie in the range $z_1 < z < z_2$. A little later on the limit $z_1 \rightarrow -\infty$, $z_2 \rightarrow \infty$ is taken. The reason for this step will now be demonstrated.

Each half-space can, as indicated above, be nonlinear as far as deriving a general formula is concerned so this will be assumed for now. Denoting i = c or s and using the first integral allows us to write[4]

$$\left(\frac{1}{E_i}\frac{dE_i}{dz}\right)^2 = \kappa_i^2 - \Lambda_i E_i^2 \tag{6.33}$$

which on differentiating with respect to z gives

$$\left(\frac{1}{E_i}\frac{dE_i}{dz}\right)\frac{d}{dz}\left(\frac{1}{E_i}\frac{dE_i}{dz}\right) = -\Lambda_i E_i \frac{dE_i}{dz} \tag{6.34}$$

i.e.

$$\frac{d}{dz}\left(\frac{1}{E_i}\frac{dE_i}{dz}\right) = -\Lambda_i E_i^2 \tag{6.35}$$

so that the time-averaged power flow along x involves the integral[4]

$$I = \int_{z_1}^{z_2} E_i^2 dz = -\left[\frac{1}{\Lambda_i E_i}\frac{dE_i}{dz}\right]\Bigg|_{z_1}^{z_2} \tag{6.36}$$

Calculation of the power flow therefore does not need a full scale integration of the actual field components over the whole of the space but only the evaluation of the limits of the ratio of $dE_i/dz$ to $E_i$ as $z \rightarrow \mp\infty$. These can be found easily from (6.33) and give, paying due respect to the sign of the square root

$$\left.\frac{1}{E_c}\frac{dE_c}{dz}\right|_{z\rightarrow\infty} = -\kappa_c \tag{6.37}$$

$$\left.\frac{1}{E_s}\frac{dE_s}{dz}\right|_{z\rightarrow-\infty} = \kappa_s \tag{6.38}$$

At the interface $E_s \rightarrow E_c \rightarrow E_0$ so that

$$\left.\frac{1}{E_i}\frac{dE_i}{dz}\right|_{z=0} = \pm\left(\kappa_s^2 - \Lambda_s E_0^2\right)^{1/2} = \pm\left(\kappa_c^2 - \Lambda_c E_0^2\right)^{1/2} \tag{6.39}$$

The power flow in each half-space is, respectively

$$P_s = \frac{c^2\epsilon_0 k_x}{2\omega}\int_{-\infty}^{0} E_s^2 dz = -\frac{c^2\epsilon_0 k_x}{2\omega\Lambda_s}\left[\frac{1}{E_s}\frac{dE_s}{dz}\right]\Bigg|_{-\infty}^{0}$$

$$= \frac{c^2\epsilon_0 k_x}{k_0^2\omega a_s}\left[\kappa_s \pm \left(\kappa_s^2 - \Lambda_s E_0^2\right)^{1/2}\right] \tag{6.40}$$

and

$$P_c = \frac{c^2\epsilon_0 k_x}{k_0^2 \omega a_c}\left[\kappa_c \pm \left(\kappa_c^2 - A_c E_0^2\right)^{1/2}\right] \tag{6.41}$$

For a linear substrate $a_s \to 0$ and

$$\lim_{a_s \to 0} \frac{1}{a_s}\left[\kappa_s - \left(\kappa_s^2 - A_s E_0^2\right)^{1/2}\right] = \frac{k_0^2 E_0^2}{4\kappa_s} \tag{6.42}$$

so that taking the appropriate sign in (6.40) the power flow becomes

$$P_s = \frac{c^2\epsilon_0 k_x}{\omega}\frac{E_0^2}{4\kappa_s} = \frac{c^2\epsilon_0 k_x}{2\omega\kappa_s a_c}(\epsilon_s - \epsilon_c). \tag{6.43}$$

Another way to view the power flow in the cladding is to substitute for $E_0^2$ in (6.41) and obtain, for example,[4]

$$P_c = \frac{c^2\epsilon_0 k_x}{k_0^2 a_c \omega}\left[\kappa_c + \left(k_x^2 - k_0^2\epsilon_c - k_0^2\epsilon_s + k_0^2\epsilon_c\right)^{1/2}\right] = \frac{c^2\epsilon_0 k_x}{k_0^2 a_c \omega}(\kappa_s + \kappa_c) \tag{6.44}$$

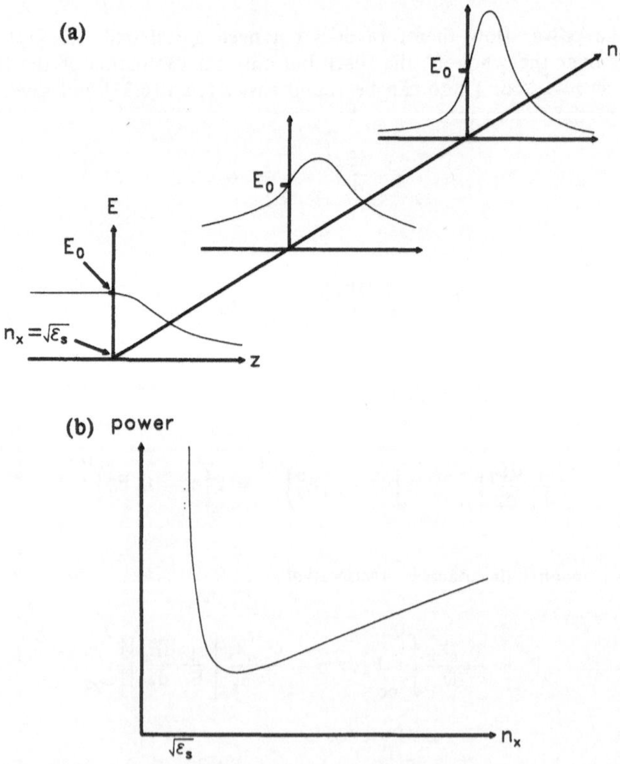


Fig. 4 Variation in (a) the field profile and (b) the power flow of the nonlinear TE wave of Fig. 3 with increasing wave index. $\epsilon_s$ is the dielectric constant of the linear substrate.

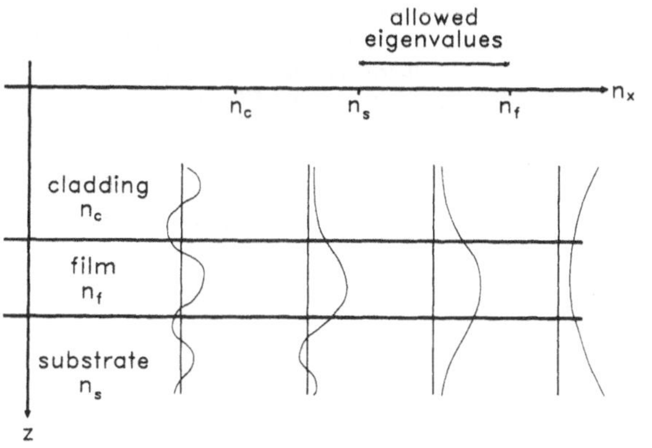


Fig. 5 Typical field profiles as a function of wave index for a linear thin film waveguide.

Using $n_x = ck_x/\omega$ this gives for the total power flow

$$P_{total} = \frac{c^2\epsilon_0 n_x}{k_0 a_c \omega}\left[\kappa_s + \kappa_c + k_0^2 \frac{\epsilon_s - \epsilon_c}{2\kappa_s}\right]. \tag{6.45}$$

Figure 4 summarises a single surface case between a linear substrate and a nonlinear cladding. In Fig. 4a the field component $E \equiv E_y$ is sketched for various values of $n_x$ as $n_x$ approaches $\sqrt{\epsilon_s}$. The figure emphasises the fact that the field pivots about $E_0$ and yet the boundary conditions are still maintained at $z = 0$, i.e. both $E_y$ and $dE_y/dz$ are always continuous. As $n_x$ approaches $\sqrt{\epsilon_s}$ the field in the linear substrate approaches cut-off. As it does so the field falls off less and less rapidly as $z \to -\infty$. In the end, an infinite amount of power is needed to drive this case which is exactly what is meant by cut-off. In Fig. 4b this feature is demonstrated on the power-$n_x$ plot. The plot shows the power tending to infinity as $n_x \to \sqrt{\epsilon_s}$. It also shows that a minimum exists and hence a minimum level of power (a threshold value) must be provided before any TE nonlinear wave can be created at a single interface. This feature also explains the existence of a finite, yet fixed, value of $E_0$. Finally, then, this case has no linear limit. It is a special example of how new nonlinear features arise in the theory of nonlinear self-focused waves.

A thin film guiding structure is shown in Fig. 5 together with a diagram showing the range of permissible eigenvalues and eigenfunctions. Naturally it does not matter whether the discussion is developed for TE waves or TM waves. The essential features remain the same. In Fig. 5 it is postulated that the refractive indices of the waveguide are $n_f > n_s > n_c$. The transverse variation is along the z-axis and propagation with angular frequency $\omega$ and wave number $k_x$ takes place down the x-axis. Given that $n_f > n_s$, $n_c$ guiding can occur in the film. The guided wave equation for a linear material is Eq. (6.13), with $a = 0$, $K = 0$, i.e.

$$\frac{d^2E}{dz^2} + (k_0^2 n^2 - k_x^2)E = 0 \tag{6.46}$$

where $n^2 = \epsilon$. The general solution for the field profile in medium i ($i = c,f,s$) is

$$E = Ae^{\kappa_i z} + Be^{-\kappa_i z} \tag{6.47}$$

where A and B are constants. For a guided wave solution the fields in the substrate and cladding must decay to zero as $z \to \mp\infty$. For this to occur it is necessary that $n_x > n_s$. The only solution in the film that can be matched to decaying solutions in the substrate and cladding is a sinusoidal one. Hence $n_x < n_f$. Radiative modes will occur in the substrate when $n_c < n_x < n_s$ and in both the cladding and the substrate when $n_x < n_c$.

The allowed eigenvalues for this linear guide are therefore located between $n_s$ and $n_f$. A question about the power dependence of these eigenvalues can now be posed. For a linear guide does it matter what power is flowing down the guide? It is evident from the earlier sections of this chapter that a linear guide has a refractive index profile that does not depend on the power flow. As the power is increased the eigenvalues remain locked into their linear values. For a nonlinear guide the situation is quite different, as is shown in Fig. 6. Some solutions will have a linear limit and some will not. It has already been shown that a single interface between a linear and nonlinear medium where the linear values of the dielectric constant are positive can support nonlinear waves but not linear waves. This is an example of a guide without a linear limit. Another example of such a guide is a nonlinear film that is, in the low power limit, unable to guide but as the power increases it is able to reach a guiding state. The other class of nonlinear guides does have a linear limit and includes the usual layered structure in which one or more elements become nonlinear. In this class is also included a metal/dielectric interface which will support surface plasmons in the linear limit. Since linear surface plasmons carry TM fields discussion of these will be deferred until later in the chapter.

In order to interpret the kind of power dependence shown in Fig. 6 it is necessary to be reminded of exactly what we mean by nonlinearity. If a material is nonlinear then any power level above zero power drives the material into nonlinearity. Mathematically, then, zero power corresponds to the linear limit and the linear eigenvalue. This is how the figures are drawn but naturally a material that is really linear or quasi-linear will have a power flow-$n_x$ curve with an infinite, or very steep slope.

The sketches in Fig. 6 are merely intriguing at this stage since no proof has been given up to now. A lot of analytical progress can be made for TE waves, however, and much of this depends upon a boundary field relationship in which field values at the respective boundaries can be linked elegantly together.[4] This development gives a conceptual path to the understanding of nonlinear waves without any demand for a knowledge of the detailed form of the field solution.

The most general planar waveguiding structure to be considered here consists of a nonlinear planar layer (film) of thickness d bounded by dissimilar nonlinear media. It is the relationship between the square of the field amplitude at the lower boundary, $E_0^2$, and the square of the field amplitude at the upper boundary, $E_d^2$, that is of fundamental interest. This $(E_0^2, E_d^2)$ relationship can be used to great effect in the nonlinear generalisation of the dispersion relationships of asymmetric and symmetric waveguides that are familiar from linear solid state optics.

The first integrals associated with the substrate (s), the film (f) and the cladding (c) of the waveguide structure are, for constant refractive indices

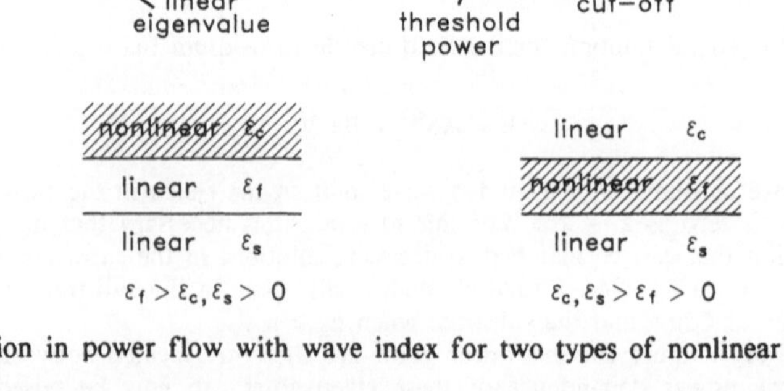

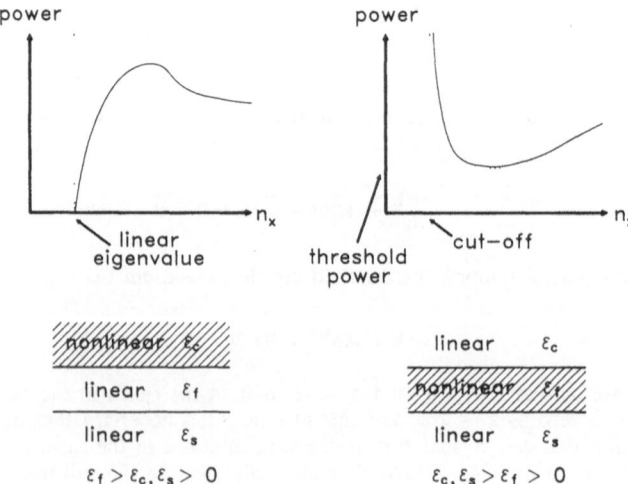

Fig. 6 Variation in power flow with wave index for two types of nonlinear thin film waveguide.

$$\left(\frac{dE_s}{dz}\right)^2 - [\kappa_s^2 - \Lambda_s E_s^2]E_s^2 = 0 \qquad -\infty \le z \le 0 \tag{6.48}$$

$$\left(\frac{dE_f}{dz}\right)^2 - [\kappa_f^2 - \Lambda_f E_f^2]E_f^2 = C_f \qquad 0 \le z \le d \tag{6.49}$$

$$\left(\frac{dE_c}{dz}\right)^2 - [\kappa_c^2 - \Lambda_c E_c^2]E_c^2 = 0 \qquad d \le z \le \infty \tag{6.50}$$

The continuity conditions for the tangential electric fields E and their derivatives with respect to z give

$$E_c = E_f \equiv E_d \tag{6.51}$$

$$\frac{dE_c}{dz} = \frac{dE_f}{dz} \tag{6.52}$$

at $z = d$

$$E_s = E_f \equiv E_0 \tag{6.53}$$

$$\frac{dE_s}{dz} = \frac{dE_f}{dz} \tag{6.54}$$

at $z = 0$

The expression for the integration constant $C_f$, obtained from these boundary conditions, is

$$C_f = k_0^2E_0^2\left(\epsilon_f - \epsilon_s + \frac{a_f - a_s}{2}E_0^2\right) = k_0^2E_d^2\left(\epsilon_f - \epsilon_c + \frac{a_f - a_c}{2}E_d^2\right) \tag{6.55}$$

and the boundary field relationship is[42]

$$E_0^2\left(\epsilon_f - \epsilon_s + \frac{a_f - a_s}{2}E_0^2\right) = E_d^2\left(\epsilon_f - \epsilon_c + \frac{a_f - a_c}{2}E_d^2\right) \tag{6.56}$$

This equation is very important because it is independent of wave number and thickness of the central guiding film. It depends only on the intrinsic parameters of the three nonlinear dielectric materials. Hence Eq. (6.56), together with the fact that the field amplitudes must also be real, will considerably lessen the burden of the theoretical analysis.

Equation (6.56) is actually a conic section with coordinates $(|\delta_s|E_0^2, |\delta_c|E_d^2)$, one branch of which passes through the point $E_0^2 = 0$; $E_d^2 = 0$. After setting $\eta_i = \epsilon_f - \epsilon_i$ and $\delta_i = a_f - a_i$ where $i = c$ or $s$ appropriately, this conic section is more clearly written as[4,41]

$$\frac{\left(\delta_s E_0^2 + \eta_s\right)^2}{\left(\eta_s^2 - \frac{\delta_s}{\delta_c}\eta_c^2\right)} - \frac{\left(\delta_c E_d^2 + \eta_c\right)^2}{\left(\frac{\delta_c}{\delta_s}\eta_s^2 - \eta_c^2\right)} = 1 \tag{6.57}$$

the centre of which is at the point

$$\mathbf{R} = \left(-\frac{\eta_s}{|\delta_s|}\delta_s, -\frac{\eta_c}{|\delta_c|}\delta_c\right) \tag{6.58}$$

Since $E_0$, $E_d$ are real and one branch passes through the origin, only the part lying in the first quadrant is physically significant. Hence, for a given $E_0^2$, there may be two, one, or zero corresponding values of $E_d^2$ depending upon the shape of the conic section and the position of its centre.

Table 1 Dispersion relations for TE waves in a layered structure consisting of a linear film bounded by two identical nonlinear media

| | $\cos(p_f D)$ | |
|---|---|---|
| | Even Parity ($E_0 \times E_d > 0$) | Odd Parity ($E_0 \times E_d < 0$) |
| field maximum in only one of the nonlinear media | $\dfrac{p_f^2 + q_s q_c}{M}$ | $\dfrac{-p_f^2 - q_s q_c}{M}$ |
| field maxima or exponential decay in both nonlinear media | $\dfrac{p_f^2 - q_s q_c}{M}$ | $\dfrac{-p_f^2 + q_s q_c}{M}$ |

One of the most striking points to emerge from the theoretical investigation[2,4,41] of strongly nonlinear waveguides is the prediction that *asymmetric* TE waves exist in a geometrically *symmetric* layered structure consisting of a linear film bounded by two identical nonlinear Kerr media. Their existence is due entirely to the nonlinearity and a simple conceptual path to this result can now be established. Setting $\epsilon_c = \epsilon_s$, $a_c = a_s$, $a_f = 0$ in Eq. (6.56) yields the relationship

$$(E_0^2 - E_d^2)\left[\epsilon_f - \epsilon_s + (E_0^2 + E_d^2)\left(\frac{a_f - a_s}{2}\right)\right] = 0 \tag{6.59}$$

The solutions $E_0^2 = E_d^2$ correspond to symmetric and antisymmetric waves that are expected for a symmetric system. There is another physical solution, however, for which[42]

$$E_0^2 + E_d^2 = \frac{2(\epsilon_f - \epsilon_s)}{a_s - a_f} \qquad E_0^2 \neq E_d^2 \tag{6.60}$$

These are *asymmetric* modes. The $E_0^2 = E_d^2$ waves have field distributions that are symmetric with respect to the centre of the film while asymmetric waves do not. Also, since asymmetric waves are entirely nonlinearly induced they can be initiated only above a certain power threshold.

Since for this waveguide $a_f = 0$, the guided waves inside the film have fields that are not hyperbolic. If the definition $p_f^2 = \epsilon_f - n_x^2$ is used then the nonlinear TE dispersion equation for the guide can readily be found. After applying the boundary conditions that the field and its derivative are continuous at each boundary and after using the boundary field relationship the following dispersion equation[4] is obtained:

$$\cos(p_f D) = \pm \frac{p_f^2 \pm q_s q_c}{\left(q_s^2 + p_f^2\right)^{1/2}\left(q_c^2 + p_f^2\right)^{1/2}} \tag{6.61}$$

where $D = \omega d/c$ and $q_{s,c}^2 = n_x^2 - \epsilon_{s,c} - \frac{1}{2}a_{s,c}E_{0,d}^2$.

The four nonlinear eigenvalue equations implied by Eq. (6.61), have even and odd parity solutions where *odd* and *even* means solutions in which $E_0$ and $E_d$ have opposite or equal signs, respectively. Using $M = (q_s^2 + p_f^2)^{1/2}(q_c^2 + p_f^2)^{1/2}$, these four equations are given in Table 1. All the possibilities concerning the existence of a self-focusing bulge, or otherwise, can be deduced without actually using the quantitative form of the fields in the nonlinear media.

Accepting, then, that increasing the power results in the field maximum (the bulge) being drawn closer to and eventually being pushed through the film/substrate boundary, a further property of the nonlinear system emerges. Once the bulge has passed into the substrate, increasing the power to higher levels has little effect on the amplitude profile of the guided wave

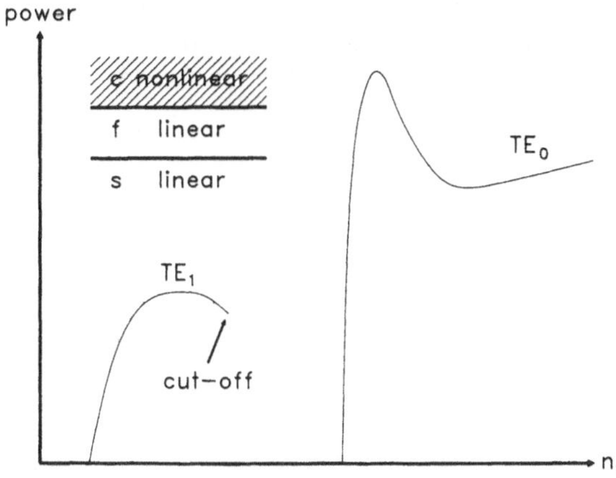


Fig. 7 Power flow versus wave index for nonlinear TE$_0$ and TE$_1$ waves in a linearly symmetric thin film guiding structure with $\epsilon_f > \epsilon_c = \epsilon_s$. The dotted line shows a power flow that is associated with three different field profiles.

inside the film. The intensity at the boundary approaches a limiting value, equal to $2(\epsilon_f - \epsilon_s)/a_s$, that leads to a decrease in the power transported by the linear film as the curvature of the amplitude profile decreases. Eventually the curvature becomes zero and changes sign. The field profile then exhibits a surface wave pattern of behaviour with the bulge in the substrate increasing in size with $n_x$. Thus, at high power levels, most of the energy for TE$_0$ (and TE$_1$) waves is carried outside the linear film. This will be very significant if the nonlinear substrate happens to be lossy, since it gives a possible mechanism for depositing power from a linear guide into a nonlinear environment.

The linear and nonlinear materials are assumed here to be non-absorptive with the net power flow being parallel to the boundary. This power should, in principle, be calculated by integrating the time-averaged Poynting vector $\langle S \rangle$ over $-\infty \leq z \leq \infty$ but, as has been pointed out earlier in this chapter, the explicit integration can be avoided by working in terms of the boundary fields and their derivatives. Naturally, actually carrying out the integrations yields exactly the same results.

Typical behaviour for TE$_0$ and TE$_1$ waves in a less complicated thin film guiding structure, where only the cladding is nonlinear, is shown in Fig. 7. The results given are for typi-

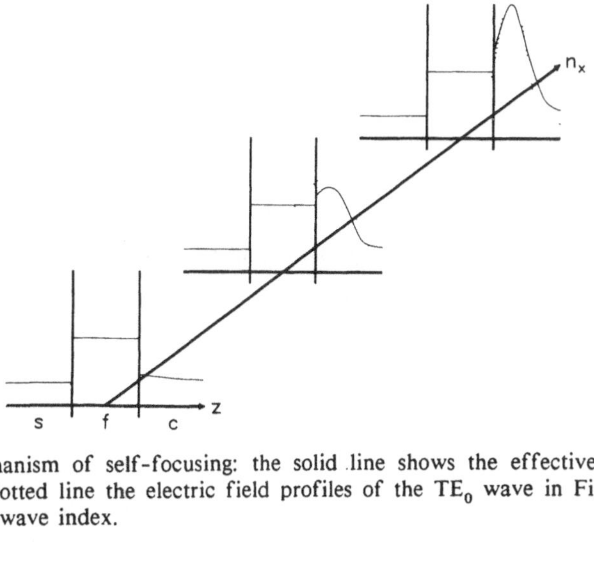


Fig. 8 The mechanism of self-focusing: the solid line shows the effective refractive index and the dotted line the electric field profiles of the TE$_0$ wave in Fig. 7 for different values of wave index.

cal data that permit distortions of the modal fields due to increasing the power level. The nonlinearity may allow a variety of field profile shapes for one particular power level, thus opening up the possibility of multistable states that could be used in switching devices. In such a case the power curves exhibit local maxima and/or multiple branches.

It is interesting now to consider the effect that increasing the value of the boundary field intensity has upon the nonlinear refractive index distribution of the guide. Figure 8 shows the nonlinear refractive index distribution for several different values of the power flow for the $TE_0$ wave of Fig. 7. It can be seen that the appearance of a self-focused peak in the field amplitude within the nonlinear medium is due to the nonlinearly induced formation of a guiding channel whose refractive index is higher than that of the adjoining medium.

## 7. NONLINEAR TM WAVES

Theoretical formulations of nonlinear TM surface/guided wave propagation need to be free of oversimplifying approximations. The nonlinear medium assumed here is linearly isotropic and the waves propagate along the x-axis with a wave number $k_x$ and angular frequency $\omega$ with the field vectors

$$\mathbf{E} = [E_x(z), 0, E_z(z)]e^{i(k_x x - \omega t)}$$
$$\mathbf{H} = [0, H_y(z), 0]e^{i(k_x x - \omega t)} \tag{7.1}$$

As pointed out in section 3, the components $E_x(z)$ and $E_z(z)$ are $\pi/2$ out of phase so the transformations $H_y \rightarrow H_y$, $E_x \rightarrow iE_x$, $E_z \rightarrow E_z$ can be made which results in the reduction of Maxwell's equations to

$$\frac{\partial E_x}{\partial z} - k_x E_z = \omega\mu_0 H_y \tag{7.2}$$

$$\frac{\partial H_y}{\partial z} = -\omega\epsilon_0\epsilon_{xx} E_x \tag{7.3}$$

$$k_x H_y = -\omega\epsilon_0\epsilon_{zz} E_z \tag{7.4}$$

where, for notational convenience, the dependence of $E_x$, $E_z$ and $H_y$ upon z is suppressed, as it will be in all subsequent expressions.

The nonlinear dielectric tensor is

$$\epsilon^{NL} = \begin{pmatrix} \epsilon_{xx} & 0 \\ 0 & \epsilon_{zz} \end{pmatrix} \tag{7.5}$$

After taking into account the $\pi/2$ phase difference between $E_x$ and $E_z$ for a surface/guided wave, the components of this nonlinear dielectric tensor are, from Eq. (3.48)

$$\epsilon_{xx} = \epsilon + a|E_x|^2 + \gamma a|E_z|^2 \tag{7.6}$$

$$\epsilon_{zz} = \epsilon + \gamma a|E_x|^2 + a|E_z|^2 \tag{7.7}$$

where the coefficients a and $\gamma$ have been defined in Eqs. (3.49) and (3.50) respectively. The first integral of Eqs. (7.2) to (7.4) is

$$\frac{c^2}{\omega^2}\left(\frac{\partial E_x}{\partial z}\right)^2 + \frac{a}{2}(E_x^4 + E_z^4) + \gamma a E_x^2 E_z^2 + \beta^2 E_z^2 - (\beta^2 - \epsilon)(E_x^2 + E_z^2) = C \tag{7.8}$$

where $\beta = ck_x/\omega$ and C is the constant of integration.

For a single interface between, for example, a nonlinear substrate and a linear cladding, information about the TM waves can be obtained elegantly from a 'phase-plot' of $E_x$ versus

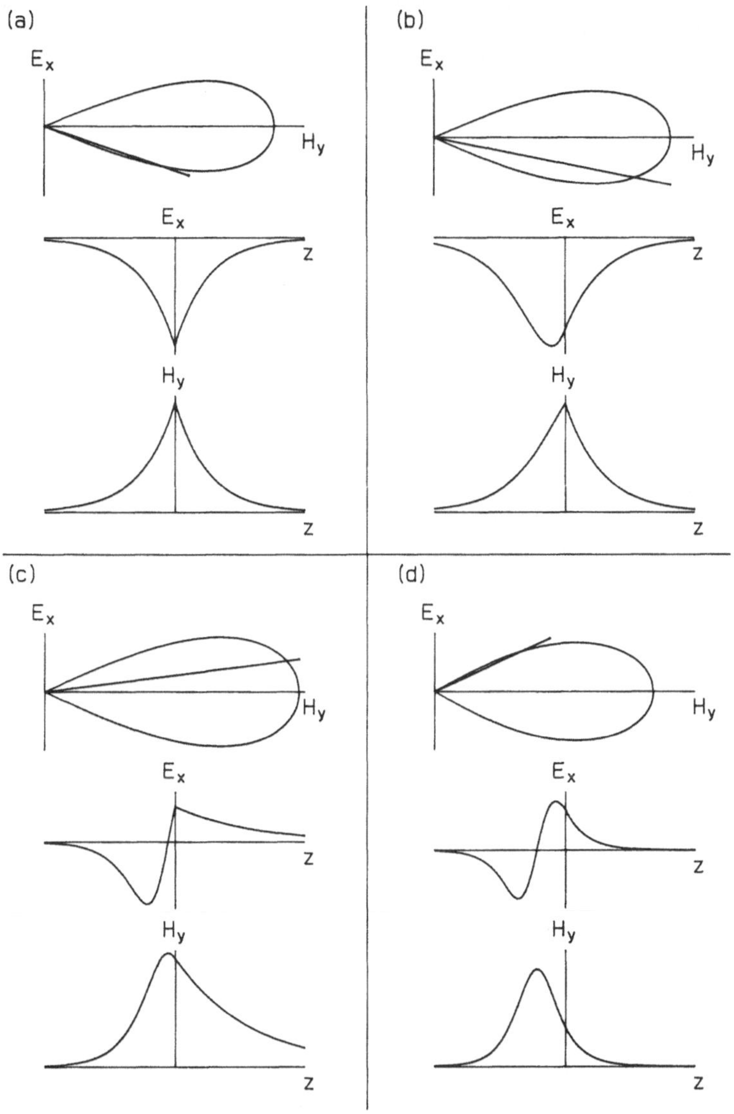


Fig. 9 Phase plots for TM waves guided by a single interface between a nonlinear ($z < 0$) and linear ($z > 0$) medium. By following the phase trajectories the qualitative field profiles shown below each phase plot can be obtained. The nonlinear medium is positive and self-focusing, and the linear medium is negative in (a) and (b) and positive in (c) and (d).

$H_y$ using values obtained directly from the first integral. The point of intersection of the phase curves for the nonlinear and linear media gives the field amplitudes of the surface or guided wave at the interface. By proceeding along one of these 'phase-paths' a qualitative idea of the shape of the field profiles can also be obtained. This is demonstrated in Fig. 9 for two cases of a single interface between a positive or negative semi-infinite linear medium and a self-focusing nonlinear medium.

The origin of the $E_x$ versus $H_y$ plot corresponds to the point where both these fields are zero, i.e. it corresponds to $z = \pm\infty$. Equation (7.3) shows that if $\epsilon_{xx}$ is positive, a positive gradient of $H_y$ implies that $E_x$ will begin to decrease as the phase plot evolves from the origin ($z = -\infty$). The phase trajectory follows the negative $E_x$ loop of the phase plot for the nonlinear medium until the intersection with the straight line phase plot of the linear solution. This intersection corresponds to the interface position at z = 0 and the trajectory joins the line at this point and follows it back to the origin again (z = +∞). Figure 9 gives several examples of how the general shape of the field profiles can be readily deduced from these phase plots. The solutions divide into two main groups:

(i) if the linear medium has $\epsilon < 0$, i.e. it is a metal, the $H_y$ component peaks at the interface and the solution can be classed as a surface wave. This type of wave has a linear limit that is the familiar surface plasmon-polariton;

(ii) if the linear medium has $\epsilon > 0$, a bulge occurs in the $H_y$ field component similar to that described in the section on TE waves, and a nonlinear wave is guided by the interface. In common with single interface guided TE waves of earlier sections, this type of wave has no linear limit and is a purely nonlinear phenomenon.

Additionally, as shown in the Fig. 9, each main group can be further subdivided into two groups according to the number of maxima that occur in $E_x$ in the nonlinear medium. Nonlinear TM waves on a single interface, therefore, will have quite different power flow-$n_x$ dependences[5,6,43,44] according to whether a metal or a positive dielectric is involved.

The method outlined above is useful for obtaining boundary field amplitudes and a rough idea of the field profiles in single interface guiding structures. However, in general one must use a numerical approach to solve Maxwell's equations directly if exact field distributions and power flows are to be obtained. To facilitate the numerical approach the Maxwell equations should first be cast into a more suitable form.

Equation (7.4), after differentiation with respect to z, becomes

$$k_x \frac{\partial H_y}{\partial z} = -\omega\epsilon_0 \left( \frac{\partial \epsilon_{zz}}{\partial z} E_z + \epsilon_{zz} \frac{\partial E_z}{\partial z} \right)$$

$$= -\omega\epsilon_0 \left( 2a\gamma E_x \frac{\partial E_x}{\partial z} + 2aE_z \frac{\partial E_z}{\partial z} \right) E_z - \omega\epsilon_0\epsilon_{zz} \frac{\partial E_z}{\partial z} \tag{7.9}$$

Therefore,

$$(\epsilon_{zz} + 2aE_z^2) \frac{\partial E_z}{\partial z} = k_x \epsilon_{xx} E_x - 2a\gamma E_x E_z \frac{\partial E_x}{\partial z} \tag{7.10}$$

In its final form Eq. (7.10) is

$$\frac{\partial E_z}{\partial z} = \frac{k_x \epsilon_{xx} E_x - (k_x E_z + \omega\mu_0 H_y) E_z \dfrac{\partial \epsilon_{zz}}{\partial E_x}}{\epsilon_{zz} + E_z \dfrac{\partial \epsilon_{zz}}{\partial E_z}} \tag{7.11}$$

Consider now a nonlinear semi-infinite medium filling the half space z > 0. If z is transformed through the relationship $Z = 1 - e^{-z/L}$ where L is a scaling factor then the basic equations can be re-expressed as[5,6,43]

$$\frac{\partial E_x}{\partial Z} = \frac{L}{1 - Z}(k_x E_z + \omega\mu_0 H_y) \tag{7.12}$$

$$\frac{\partial H_y}{\partial Z} = \frac{L}{1 - Z}(-\omega\epsilon_0\epsilon_{xx} E_x) \tag{7.13}$$

$$\frac{\partial E_z}{\partial Z} = \frac{L}{1 - Z} \left( \frac{k_x \epsilon_{xx} E_x - (k_x E_z + \omega\mu_0 H_y) \dfrac{\partial \epsilon_{zz}}{\partial E_x}}{\epsilon_{zz} + E_z \dfrac{\partial \epsilon_{zz}}{\partial E_z}} \right) \tag{7.14}$$

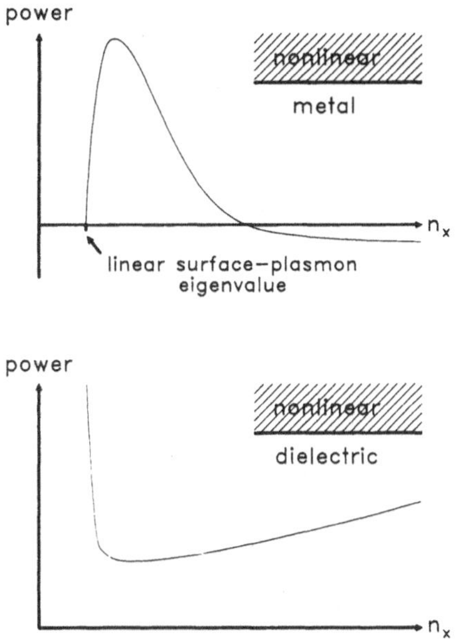


Fig. 10 Power flow versus wave index for TM surface/guided waves at a single interface between a positive self-focusing nonlinear medium and a metal or positive dielectric.

$$\frac{\partial \mathcal{P}}{\partial Z} = - \frac{L}{1 - Z} \frac{E_z H_y}{2} \tag{7.15}$$

where $\mathcal{P}$ is the power flow down the guide. Note that the transformation converts the range $0 < z < \infty$ to $0 < Z < 1$. Equations (7.12) - (7.15) are a set of 'standard' first order coupled ordinary differential equations having the general form $dy_i/dZ = f_i(Z, y_1, y_2, y_3)$. Since the equations are required to satisfy boundary conditions at more than one point, e.g.

$$E_x(0) = E_0 \qquad E_x(1) = H_y(1) = 0 \tag{7.16}$$

this is termed a boundary value problem. The equations can now be solved numerically using a finite element technique or some other suitable method.

On a practical note, the equations become indeterminate at $Z = 1$ but the solution of the nonlinear equations becomes effectively linear in the neighbourhood of $Z = 1$. Hence, by choosing L so that the linear solution is of the form $A = A_0(1 - Z)$ (where A has been used to represent $E_x$, $H_y$ or $E_z$) we can evaluate the equations at $Z = 1$ by simple linear interpolation. Since the linear solution has the functional form $e^{-\kappa z}$, where $\kappa^2 = k_x^2 - \omega^2\epsilon/c^2$, the L should be set equal to $1/\kappa$.

Typical results for a single interface structure are illustrated in Fig. 10. Nonlinear TM waves in layered structures containing positive dielectric material have power-$n_x$ characteristics that look like those for TE waves (Fig. 7). If metal (linear) films are involved, however, much more interesting phenomena occur and these will now be discussed.

From studies of the propagation of TM waves on thin linear metal films emerged the conclusions that there are two types of surface plasmon-polariton. These are called the long-range plasmon (LRP) and the short-range plasmon (SRP). The field of the long-range plasmon lies mainly outside the film while for the short-range plasmon the energy is mainly transported inside the film. The short-range plasmon has a higher eigenvalue $\beta$ than the long-range plasmon, and it is $E_z$ that changes sign in the metal film for the short-range plasmon while it is $E_x$ that changes sign in the film for the long range plasmon. Finally short-range plasmons exist for all film thicknesses whereas long-range plasmons are cut-off if the film gets thin enough.

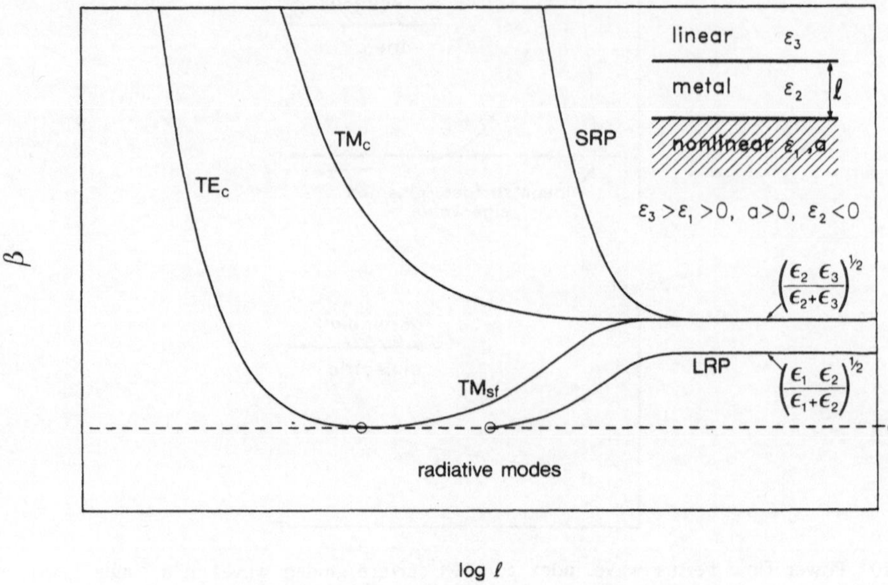


Fig. 11 The effects of film thickness on the waves supported by a thin metal film structure. The curves labelled SRP and LRP are the linear eigenvalues of short- and long-range plasmons, $TM_c$ and $TE_c$ are the loci of the cut-off points of pseudoplasmons and TE plasmons, and $TM_{sf}$ is the locus of values of wave index $\beta$ at which self-focusing of long-range plasmons and pseudoplasmons sets in.

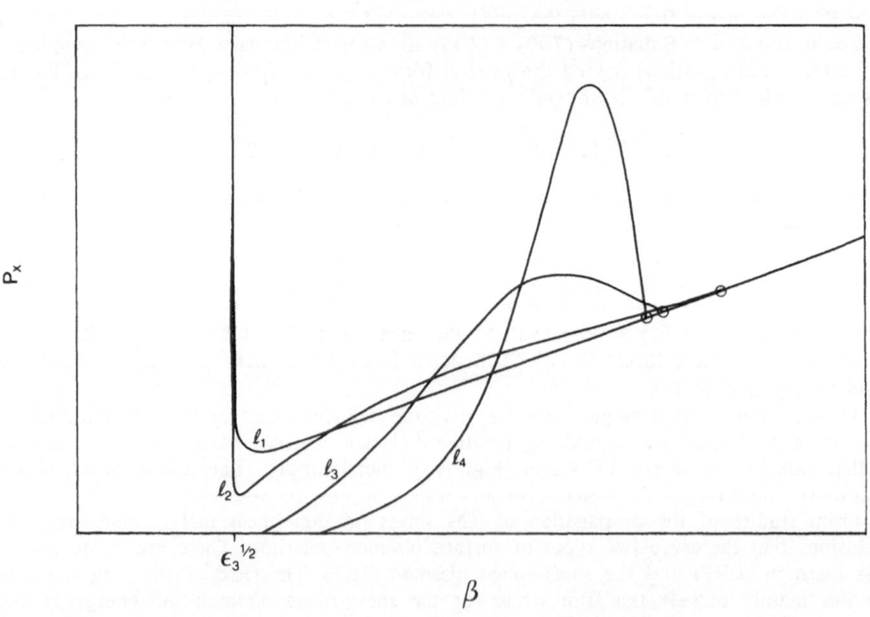


Fig. 12 Power flow versus wave index for TM waves in the thin metal film structure of Fig. 11 with $\ell_4 > \ell_3 > \ell_2 > \ell_1 = 0$. The circles show cut-off points.

Table 2 Analytical results for TM polarised waves in a metal film structure with one nonlinear bounding medium

| Field solutions $-\infty < z < 0$ | $0 < z < \ell$ | $\ell < z < \infty$ |
|---|---|---|
| $E_{x0}$, $H_{y0}$ | $E_x = \frac{\kappa_2}{\omega\epsilon_0\epsilon_2}(be^{-\kappa_2 z} - ae^{\kappa_2 z})$ | $E_x = \frac{\kappa_3}{\omega\epsilon_0\epsilon_3}H_{y\ell}e^{-\kappa_3(z-\ell)}$ |
| (obtained numerically) | $H_y = ae^{\kappa_2 z} + be^{-\kappa_2 z}$ | $H_y = H_{y\ell}e^{-\kappa_3(z-\ell)}$ |

| Boundary conditions | at $z = 0$ | at $z = \ell$ |
|---|---|---|
| $\Delta H_y = 0$ | $H_{y0} = a + b$ | $ae^{\kappa_2\ell} + be^{-\kappa_2\ell} = H_{y\ell}$ |
| $\Delta E_x = 0$ | $E_{x0} = \frac{\kappa_2}{\omega\epsilon_0\epsilon_2}(b - a)$ | $\frac{\kappa_2}{\epsilon_2}(be^{-\kappa_2\ell} - ae^{\kappa_2\ell}) = \frac{\kappa_3}{\epsilon_3}H_{y\ell}$ |

$$\frac{E_{x0}}{H_{y0}} = \frac{\kappa_2[\kappa_2\epsilon_3\tanh(\kappa_2\ell) + \kappa_3\epsilon_2]}{\omega\epsilon_0\epsilon_2[\kappa_2\epsilon_3 + \kappa_3\epsilon_2\tanh(\kappa_2\ell)]} \qquad \text{(T1)}$$

We will now examine the effects of nonlinearity by considering a nonlinear/metal/linear-dielectric thin film structure.[6] The dielectric constant of the linear medium, $\epsilon_3$, is assumed to be greater than $\epsilon_1$, the dielectric constant of the nonlinear medium, and the nonlinear medium is assumed to be self-focusing (a > 0). Figure 11 shows the linear eigenvalues of the short-range plasmon and the long-range plasmon as a function of film thickness $\ell$. The figure shows that $\beta_{SRP} > \beta_{LRP}$ for all $\ell$ but it also illustrates the fact that, as $\ell$ decreases, $\beta_{LRP}$ falls until it cuts-off into a radiative mode at a certain minimum film thickness. On the other hand linear $\beta_{SRP}$ values exist for all film thicknesses. As the film gets thicker $\beta_{LRP}$ and $\beta_{SRP}$ asymptotically approach the single-interface values associated with the nonlinear/metal and metal/linear surfaces respectively. Figure 11 also shows three other (nonlinear) loci two of which refer to TM waves and one rather surprisingly to TE waves again. We will now discuss these 'nonlinear' loci in more detail.

An important change to the nonlinear long-range plasmon characteristics occurs as the metal film thickness is decreased. Figure 12 shows power dispersion curves for a series of decreasing thicknesses. First an upper $\beta$ cut-off is introduced. Reducing the film thickness below the range of existence of linear long-range plasmons introduces a new kind of plasmon with a minimum power threshold, which degenerates into a single interface self-focused wave as the film thickness is reduced to zero.

In terms of the field distributions this behaviour can be explained as follows. For values of $\beta$ near the long-range plasmon linear eigenvalue the field profiles have the normal type of long-range plasmon distribution. However, as the wavenumber $\beta$ is increased, a self-focused peak forms in the nonlinear medium and moves away from the interface with the metal film, causing a surge in the power flow. The wave eventually cuts off when the self-focused peak has moved out an infinite distance from the metal film.

Decreasing the film thickness still further to values which no longer support linear long-range plasmons has no effect on the form of the field profiles other than introducing a minimum into the field amplitudes at the interfaces; this is responsible for the formation of the minimum power threshold. From this behaviour it is clear that, as the metal film thickness is decreased, there is a continuous transition from single interface nonlinear long-range plasmons to self-focused TM waves, via nonlinear long-range plasmons with an upper $\beta$ cut-off. Because of the close similarity of self-focused TM waves in a metal film structure to nonlinear long-range plasmons, the self-focused waves have been named 'pseudoplasmons'.[6]

Although the power dispersion curves must be found numerically, the onset of self-focusing and the cut-off point for nonlinear long-range plasmons and pseudoplasmons can be found analytically. The onset of self-focusing corresponds to $dH_y/dz = 0$ (i.e. $E_x = 0$) at the nonlinear medium/metal interface. From Table 2 this is given by

$$\tanh(\kappa_2 \ell) = -\frac{\kappa_3 \epsilon_2}{\kappa_2 \epsilon_3} \tag{7.17}$$

Hence Eq. (7.17) is the locus on Fig. 11 that delineates the onset of self-focusing for TM waves. This threshold falls with decreasing film thickness until it merges with the radiative cut-off line, which, since $\epsilon_3 > \epsilon_1$ in this example, is defined by $\kappa_3 = 0$ (i.e. $\beta^2 = \epsilon_3$).

The upper cut-off can be obtained by matching the field ratio given in Table 2, Eq. (T1), to that obtained from the nonlinear TM first integral, Eq. (7.8). Since the cut-off point corresponds to the self-focused peak of the wave being an infinite distance from the metal film, the field amplitudes at the nonlinear medium/metal film interface must be zero and hence only the lowest order terms need be retained.

Hence neglecting quartic terms in Eq. (7.8) and setting C = 0, since the nonlinear region is semi-infinite, gives, to begin with,

$$\frac{c^2}{\omega^2}\left(\frac{\partial E_x}{\partial z}\right)^2\Bigg|_0 + \beta^2 E_{x0}^2 - (\beta^2 - \epsilon_1)(E_{x0}^2 + E_{z0}^2) = 0 \tag{7.18}$$

where the subscript 0 denotes evaluation at z=0.

From Eqs. (7.2) and (7.4), $(\partial E_x/\partial z)$ is given by

$$\frac{\partial E_x}{\partial z}\Bigg|_0 = \left(k_x - \frac{\omega^2}{c^2}\frac{\epsilon_1}{k_x}\right)E_{z0} \tag{7.19}$$

Therefore,

$$\frac{c^2}{\omega^2}\left(\frac{\partial E_x}{\partial z}\right)^2\Bigg|_0 = \beta^2\left(1 - \frac{\epsilon_1}{\beta^2}\right)^2 E_{z0}^2 \tag{7.20}$$

Equation (7.18), after using (7.20), is

$$\frac{\left(\beta^2 - \epsilon_1\right)^2}{\beta^2}E_{z0}^2 - (\beta^2 - \epsilon_1)E_{z0}^2 + \epsilon_1 E_{x0}^2 = 0 \tag{7.21}$$

This immediately simplifies to

$$(\beta^2 - \epsilon_1)E_{z0}^2 = \beta^2 E_{x0}^2 \tag{7.22}$$

or, substituting for $E_z$ in terms of $H_y$ from Eq. (7.4)

$$\frac{E_{x0}}{H_{y0}} = \pm\frac{\kappa_1}{\omega\epsilon_0\epsilon_1} \tag{7.23}$$

Equating the positive root of Eq. (7.23) with Eq. (T1) gives the following expression for cut-off

$$\tanh(\kappa_2 \ell) = \frac{\kappa_2\epsilon_2(\kappa_1\epsilon_3 - \kappa_3\epsilon_1)}{\kappa_2^2\epsilon_1\epsilon_3 - \kappa_1\kappa_3\epsilon_2^2} \tag{7.24}$$

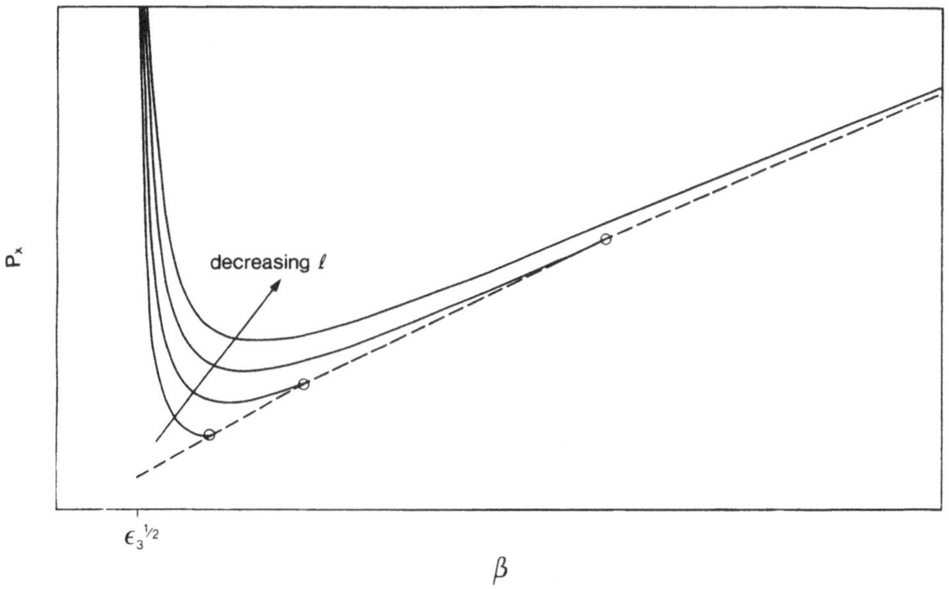


Fig. 13 Power flow versus wave index for TE plasmons in the thin metal film structure of Fig. 11, with circles showing the cut-off points. The top curve corresponds to zero metal film thickness and the dashed line gives the locus of power flow at cut-off.

Since the above equation does not contain any nonlinear parameters, the cut-off is independent of the magnitude and mechanism of the nonlinearity. This is an interesting point in view of the fact that at the cut-off point the waves are very highly nonlinear.

The loci given by Eqs. (7.17) and (7.24) are shown in Fig. 11 as a function of film thickness. Notice that as the film thickness is reduced to zero, the cut-off point increases to infinity and the onset of self-focusing meets the lower $\beta$ threshold for the existence of surface/guided waves. This is consistent with the properties of a single interface self-focused nonlinear TM wave. For large film thicknesses, both lines converge onto the linear eigenvalue of the short-range plasmon.

Interestingly, Eq. (T1) of Table 2 and the negative root of Eq. (7.23) lead to the linear dispersion relations of long- and short-range plasmons since these have identical boundary conditions to those used above for the self-focused waves at cut-off (i.e. field amplitudes which are zero at infinity and tending to zero at the metal film interfaces).

There is a TE equivalent of the pseudoplasmon which exists over a limited range of film thicknesses and has been termed a TE-plasmon in the literature.[45-47] In this case the position of the self-focused peak is given by Eq. (T2) in Table 3. Hence the locus of TE-plasmon cut-off, obtained by setting $z_0 = -\infty$, is given by

$$\tanh(\kappa_2 \ell) = \frac{\kappa_2(\kappa_1 - \kappa_3)}{\kappa_2^2 - \kappa_1 \kappa_3} \tag{7.25}$$

This is shown in Fig. 11 as a function of film thickness. The critical film thickness above which TE-plasmons do not exist can be obtained simply by setting $\kappa_3 = 0$ in Eq. (7.25).

Figure 13 shows several power dispersion curves for TE-plasmons in very thin metal film structures of different thicknesses. For zero film thickness (the single interface case) there is no upper $\beta$ cut-off point. As the metal film thickness is increased the power threshold decreases and an upper $\beta$ cut-off appears. Since at the upper cut-off point the self-focused peak is at an infinite distance from the metal film and the field amplitudes in the metal and linear dielectric are zero, the power at the upper $\beta$ cut-off is carried entirely within the nonlinear medium, and is given analytically from Eq. (6.32) by the integral

Table 3 Analytical results for TE polarised waves in a metal film structure with one nonlinear bounding medium

| Field solutions | | |
|---|---|---|
| $-\infty < z < 0$ | $0 < z < \ell$ | $\ell < z < \infty$ |
| $H_x = -\dfrac{\kappa_1^2\sinh[\kappa_1(z-z_0)]}{\omega\mu_0\sqrt{A_1}\cosh^2[\kappa_1(z-z_0)]}$ | $H_x = \dfrac{\kappa_2}{\omega\mu_0}(ae^{\kappa_2 z} - be^{-\kappa_2 z})$ | $H_x = -\dfrac{\kappa_3}{\omega\mu_0}E_{y\ell}e^{-\kappa_3(z-\ell)}$ |
| $E_y = \dfrac{\kappa_1}{\sqrt{A_1}\cosh[\kappa_1(z-z_0)]}$ | $E_y = ae^{\kappa_2 z} + be^{-\kappa_2 z}$ | $E_y = E_{y\ell}e^{-\kappa_3(z-\ell)}$ |

| Boundary conditions | | |
|---|---|---|
| | at $z = 0$ | at $z = \ell$ |
| $\Delta H_x = 0$ | $\dfrac{\kappa_1^2\sinh(\kappa_1 z_0)}{\sqrt{A_1}\cosh^2(\kappa_1 z_0)} = \kappa_2(a - b)$ | $\kappa_2(be^{-\kappa_2\ell} - ae^{\kappa_2\ell}) = \kappa_3 E_{y\ell}$ |
| $\Delta E_y = 0$ | $\dfrac{\kappa_1}{\sqrt{A_1}\cosh(\kappa_1 z_0)} = a + b$ | $ae^{\kappa_2\ell} + be^{-\kappa_2\ell} = E_{y\ell}$ |

$$\tanh(\kappa_1 z_0) = -\frac{\kappa_2[\kappa_2\tanh(\kappa_2\ell) + \kappa_3]}{\kappa_1[\kappa_2 + \kappa_3\tanh(\kappa_2\ell)]} \tag{T2}$$

$$\mathscr{P}_{\text{cut-off}} = \frac{k}{2\omega\mu_0}\int_{-\infty}^{\infty} E_y^2 dz$$

$$= \frac{k\kappa_1^2 c^2}{\omega^3\mu_0 a}\int_{-\infty}^{\infty}\frac{dz}{\cosh^2(\kappa_1 z)}$$

$$= \frac{2k\kappa_1 c^2}{\omega^3\mu_0 a} \tag{7.26}$$

which is independent of the data describing the metal film and linear dielectric. The locus of the power at cut-off given by Eq. (7.26) is also shown in Fig. 13.

## 8. INTERACTING TE AND TM WAVES

In the interpretations of experiments, it is usually assumed that either purely TE or TM waves propagate. An inspection of the nonlinear dielectric tensor for isotropic materials reveals, however, the possibility that both polarisations can propagate simultaneously and that one polarisation can sometimes act as a channel for the other. Indeed, the propagation of a weak TM wave in the channel created by a strong TE wave has been considered recently.[27] It is possible, however, to give a full account of the way in which nonlinear TE-TM interactions for a more general class of nonlinear dielectric tensor occur.

In order to do this the interaction of the two waves in question, namely TE and TM, with *different* wavenumbers p and q but the *same* frequency $\omega$ is considered.[6] The relevant nonlinear dielectric tensor, for isotropic media, will now be derived from first principles.

As before we consider a planar guiding structure whose interfaces lie in the x-y Cartesian plane supporting a surface/guided wave whose wavevector is parallel to the x-axis.

In general for a nonlinear medium the third order electric polarisation at position $\mathbf{r}$ and time t can be written, as shown by Eq. (3.26), as

$$P_i^{(3)}(\mathbf{r},t) = \frac{3}{8}\epsilon_0 e^{-i\omega t}\chi_{ijkl}^{(3)}(\omega;\omega,-\omega,\omega)\xi_j(\mathbf{r},\omega)\xi_k^*(\mathbf{r},\omega)\xi_l(\mathbf{r},\omega) + \text{c.c.} \tag{8.1}$$

where the $\xi_i$ are defined in Eq. (3.24) and are not assumed to be slowly varying in time. The case

$$\boldsymbol{\xi} = \mathbf{A}e^{ipx} + \mathbf{B}e^{iqx} \tag{8.2}$$

can now be considered, where

$$\mathbf{A} = (\mathcal{E}_x, 0, \mathcal{E}_z) \tag{8.3}$$

and

$$\mathbf{B} = (0, \mathcal{E}_y, 0) \tag{8.4}$$

i.e. the electric field has a single frequency $\omega$ but the TM and TE components have different wavenumbers, p and q respectively. The polarisation $P_i^{(3)}(\mathbf{r},t)$ is then given by

$$P_i^{(3)}(\mathbf{r},t) = \frac{3}{8}\epsilon_0\chi_{ijkl}^{(3)}(\omega;\omega,-\omega,\omega)(A_j e^{ipx} + B_j e^{iqx})(A_k^* e^{-ipx} + B_k^* e^{-iqx})(A_l e^{ipx} + B_l e^{iqx})e^{-i\omega t} \tag{8.5}$$

After multiplying out, neglecting non-phase-matched terms and moving around dummy suffices using intrinsic permutation symmetry [Eq. (3.31)], Eq. (8.5) becomes

$$P_i^{(3)}(\mathbf{r},t) = \frac{3}{8}\epsilon_0\chi_{ijkl}^{(3)}(\omega;\omega,\omega,-\omega)(A_j A_k A_l^* + 2A_j B_k B_l^*)e^{i(px-\omega t)}$$
$$+ \frac{3}{8}\epsilon_0\chi_{ijkl}^{(3)}(\omega;\omega,\omega,-\omega)(B_j B_k B_l^* + 2B_j A_k A_l^*)e^{i(qx-\omega t)} + \text{c.c.} \tag{8.6}$$

The summations over repeated indices in Eq. (8.6) for an isotropic medium gives, using Eq. (3.33),

$$p_x^{(3)}(\mathbf{r},\omega) = \frac{3}{4}\epsilon_0\left\{2\chi_{xxyy}\left[|\mathcal{E}_x|^2 + |\mathcal{E}_y|^2 + |\mathcal{E}_z|^2\right]\mathcal{E}_x + \chi_{xyyx}(\mathcal{E}_x^2 + \mathcal{E}_z^2)\mathcal{E}_x^*\right\}e^{ipx} \tag{8.7a}$$

$$p_y^{(3)}(\mathbf{r},\omega) = \frac{3}{4}\epsilon_0\left\{2\chi_{xxyy}\left[|\mathcal{E}_x|^2 + |\mathcal{E}_y|^2 + |\mathcal{E}_z|^2\right]\mathcal{E}_y + \chi_{xyyx}\mathcal{E}_y^2\mathcal{E}_y^*\right\}e^{iqx} \tag{8.7b}$$

$$p_z^{(3)}(\mathbf{r},\omega) = \frac{3}{4}\epsilon_0\left\{2\chi_{xxyy}\left[|\mathcal{E}_x|^2 + |\mathcal{E}_y|^2 + |\mathcal{E}_z|^2\right]\mathcal{E}_z + \chi_{xyyx}(\mathcal{E}_x^2 + \mathcal{E}_z^2)\mathcal{E}_z^*\right\}e^{ipx} \tag{8.7c}$$

The $\chi_{xxyy}$ describes an 'isotropic' part of the nonlinear polarisation and this term alone gives a nonlinear refractive index change that is the same in all directions. The coefficient $\chi_{xyyx}$ describes an 'anisotropic' part of the nonlinear polarisation even though Eqs. (8.7) refer to the nonlinear polarisation in an *isotropic* medium. Hence the second term gives rise to a nonlinear birefringence. Notice that if the TE and TM components had the same wave number Eq. (8.7) would take the form given by Eq. (3.43), i.e. $p_i^{(3)}(\mathbf{r},\omega) = (3\epsilon_0/4)\times(2\chi_{xxyy}\xi_j\xi_j^*\xi_i + \chi_{xyyx}\xi_j\xi_j\xi_i^*)$.

For a surface or guided wave $\mathcal{E}_x$ and $\mathcal{E}_z$ are $\pi/2$ out of phase with each other. Introducing the substitutions,

$$\mathcal{E}_x = iE_x \qquad \mathcal{E}_y = E_y \qquad \mathcal{E}_z = E_z \tag{8.8}$$

where $E_x$, $E_y$ and $E_z$ are all real, and the definitions

$$\alpha = 2\chi_{xxyy} + \chi_{xyyx} \tag{8.9}$$

$$\gamma = \left(\frac{2\chi_{xxyy} - \chi_{xyyx}}{2\chi_{xxyy} + \chi_{xyyx}}\right) \tag{8.10}$$

$$\eta = \frac{2\chi_{xxyy}}{2\chi_{xxyy} + \chi_{xyyx}} \tag{8.11}$$

leads to

$$\begin{aligned} p_x^{(3)}(r,\omega) &= \epsilon_0\alpha(E_x^2 + \eta E_y^2 + \gamma E_z^2)E_x e^{ipx} \\ p_y^{(3)}(r,\omega) &= \epsilon_0\alpha(\eta E_x^2 + E_y^2 + \eta E_z^2)E_y e^{iqx} \\ p_z^{(3)}(r,\omega) &= \epsilon_0\alpha(\gamma E_x^2 + \eta E_y^2 + E_z^2)E_z e^{ipx} \end{aligned} \tag{8.12}$$

where $\gamma$ = 1/3, -1/2 and 1, and $\eta$ = 2/3, 1/4 and 1 for electronic distortion, molecular orientational or thermal nonlinear mechanisms respectively. The dielectric tensor for a nonlinear isotropic material can then be written as[6]

$$\underline{\underline{\epsilon}} = \begin{pmatrix} \epsilon_{xx} & & 0 \\ & \epsilon_{yy} & \\ 0 & & \epsilon_{zz} \end{pmatrix}$$

$$= \begin{pmatrix} \epsilon + \alpha(E_x^2 + \eta E_y^2 + \gamma E_z^2) & 0 & 0 \\ 0 & \epsilon + \alpha(\eta E_x^2 + E_y^2 + \eta E_z^2) & 0 \\ 0 & 0 & \epsilon + \alpha(\gamma E_x^2 + \eta E_y^2 + E_z^2) \end{pmatrix}. \tag{8.13}$$

where $\epsilon$ is the linear part of the dielectric function and $\alpha$ is the nonlinear coefficient. The presence of all three electric field components in each diagonal term causes the TE and TM polarisations to interact with each other. In materials of lower symmetry TE-TM coupling may occur via off-diagonal elements leading to hybrid modes, but this is not appropriate in the present case and hybrid modes do not occur. In general the stationary mixed TE-TM waves will have associated with them two wavenumbers, one for the TE and the other for the TM component, and they must be considered as interacting but *discrete*.

For TM waves Maxwell's equations are

$$\begin{aligned} \frac{\partial E_x}{\partial z} - pE_z &= \omega\mu_0 H_y \\ \frac{\partial H_y}{\partial z} &= -\omega\epsilon_0\epsilon_{xx}E_x \\ pH_y &= -\omega\epsilon_0\epsilon_{zz}E_z \end{aligned} \tag{8.14}$$

These can be manipulated to give

$$\frac{\partial E_x}{\partial z}\frac{\partial^2 E_x}{\partial z^2} = p^2\left(1 - \frac{\epsilon_{zz}}{\beta_{TM}^2}\right)E_z\frac{\partial E_z}{\partial z} - \frac{\omega^2}{c^2}\epsilon_{xx}E_x\frac{\partial E_x}{\partial z} \tag{8.15}$$

where $\beta_{TM} = pc/\omega$. After integration with respect to z the following integral is obtained

$$\frac{c^2}{\omega^2}\left(\frac{\partial E_x}{\partial z}\right)^2 + \frac{\alpha}{2}(E_x^4 + E_z^4) + \gamma\alpha E_x^2E_z^2 + \beta_{TM}^2E_x^2 - (\beta_{TM}^2 - \epsilon)(E_x^2 + E_z^2)$$

$$= -2\alpha\eta\int(E_y^2E_z dE_z + E_y^2E_x dE_x) + C_{TM} \qquad (8.16)$$

where $C_{TM}$ is the integration constant.

The TE set of Maxwell's equations is

$$\frac{\partial E_y}{\partial z} = \omega\mu_0 H_x \qquad (8.17a)$$

$$qE_y = \omega\mu_0 H_z \qquad (8.17b)$$

$$\frac{\partial H_x}{\partial z} - qH_z = -\omega\epsilon_0\epsilon_{yy}E_y \qquad (8.17c)$$

Hence

$$\frac{\partial^2 E_y}{\partial z^2} = \left(q^2 - \frac{\omega^2}{c^2}\epsilon_{yy}\right)E_y \qquad (8.18)$$

which integrates to

$$\frac{c^2}{\omega^2}\left(\frac{\partial E_y}{\partial z}\right)^2 - (\beta_{TE}^2 - \epsilon)E_y^2 + \frac{\alpha}{2}E_y^4 = -2\alpha\eta\int(E_x^2E_y dE_y + E_z^2E_y dE_y) + C_{TE} \qquad (8.19)$$

where $C_{TE}$ is another integration constant and $\beta_{TE}$=qc/$\omega$.

The unresolved integrals in Eqs. (8.16) and (8.19) can be evaluated by adding them together, i.e.

$$2\int(E_y^2E_z dE_z + E_y^2E_x dE_x + E_x^2E_y dE_y + E_z^2E_y dE_y) = E_x^2E_y^2 + E_y^2E_z^2 \qquad (8.20)$$

The first integral for stationary TE-TM waves is therefore

$$\alpha(\eta E_y^2E_z^2 + \eta E_x^2E_y^2 + \gamma E_x^2E_z^2) + \frac{\alpha}{2}(E_x^4 + E_y^4 + E_z^4) + \epsilon(E_x^2 + E_y^2 + E_z^2)$$

$$= \beta_{TM}^2E_z^2 + \beta_{TE}^2E_y^2 - \frac{c^2}{\omega^2}\left[\left(\frac{\partial E_x}{\partial z}\right)^2 + \left(\frac{\partial E_y}{\partial z}\right)^2\right] + C_{TM} + C_{TE} \qquad (8.21)$$

The procedure adopted earlier was to set the integration constants to zero for semi-infinite media. However, it must be noted that for surface/guided waves in linear semi-infinite media (i.e. $\alpha = 0$) the constants $C_{TM}$ and $C_{TE}$ must both *separately* be zero in order that all the field components are zero at infinity. This condition *must also apply in nonlinear media* and hence the condition that the sum of $C_{TE}$ and $C_{TM}$ is zero in Eq. (8.21) is not sufficient because it does not disallow the possibility that $C_{TE} = -C_{TM}$ and that both are finite. For example, the application of single interface boundary conditions to the first integral for TE-TM waves will yield a continuum of eigenvalues corresponding to the continuum of values that $C_{TE} = -C_{TM}$ can take, with only the eigenvalue corresponding to $C_{TE} = C_{TM} = 0$ being correct. It is not possible to treat the two constants separately so that the first integral in this case is of very limited value. It is useful however for checking any numerical results.[6]

There are no TE-TM waves at a single interface between a metal and a nonlinear dielectric since this supports only TM polarised waves. We consider instead a single interface between a linear substrate and a self-focusing nonlinear cladding (Fig. 14), whose linear dielectric functions, $\epsilon_s$ and $\epsilon_c$ respectively, are both positive. As discussed in sections 6 and 7, this guide will support both TE and TM polarisations, and hence can permit nonlinear interaction between the two. Guiding takes place due to the nonlinear formation of a channel of relatively high refractive index in the nonlinear medium. The channel formed by, for example, a nonlinear TE wave will, if somehow 'frozen' into the nonlinear medium, support a *linear* TM wave with an eigenvalue related (via the shape and size of the channel) to the nonlinear eigenvalue of the TE wave. This corresponds in practice to a nonlinear TE wave and a zero amplitude nonlinear (i.e. effectively linear) TM wave propagating simultaneously along a single interface. The TM wave is guided by the channel created by the TE wave, but the TE wave is unaffected by the TM wave because of the zero amplitude of the TM field components. It is therefore possible to consider a $\beta_{TE}$, $\beta_{TM}$ plane and draw on it a locus that represents a pure nonlinear TE wave and a zero amplitude TM wave. There is also a second locus that corresponds to a pure nonlinear TM wave and a zero amplitude TE wave. Solutions to Maxwell's equations should be found within the area between the two loci and these are the desired stationary TE-TM nonlinear surface/guided waves. Points outside this region correspond to non-stationary states and will not be considered here.

The two loci on the $\beta_{TE}$, $\beta_{TM}$ plane can be found by incorporating an expression for the nonlinear guiding channel created by one polarisation into the linear dielectric function used in calculating the linear eigenvalue of the other polarisation. For example, the dielectric tensor for the nonlinear cladding used in calculating the linear eigenvalue of TM waves in a TE induced channel has the form

$$\underline{\underline{\epsilon}} = \begin{pmatrix} \epsilon_c + \alpha\eta E_y^2(z) & 0 \\ 0 & \epsilon_c + \alpha\eta E_y^2(z) \end{pmatrix} \tag{8.22}$$

where, from Eqs. (6.18) and (6.27),

$$E_y^2(z) = \frac{\kappa_{TE,c}^2}{\Lambda \cosh^2[\kappa_{TE,c}(z-z_0)]} \tag{8.23}$$

$$\Lambda = \frac{\omega^2\alpha}{2c^2} \tag{8.24}$$

$$\tanh(\kappa_{TE,c} z_0) = \frac{\kappa_{TE,s}}{\kappa_{TE,c}} \tag{8.25}$$

and

$$\kappa_{TE,i}^2 = \frac{\omega^2}{c^2}(\beta_{TE}^2 - \epsilon_i) \qquad i = s,c \tag{8.26}$$

The corresponding dielectric tensor for a TM induced channel cannot be expressed analytically, but, by solving the TM wave equations, can be generated numerically in the same form as the tensor (8.22). The linear eigenvalues for this modified structure can then be calculated.

Figure 14 shows schematically the region on the $\beta_{TE}/\beta_{TM}$ plane in which nonlinear TE-TM stationary states occur for a single interface between a linear substrate and a nonlinear cladding. The two loci converge to a point at $\beta_{TE}^2 = \beta_{TM}^2 = \epsilon_s$, below which the nonlinear waves become oscillatory in the linear dielectric and the power flow becomes infinite. No stationary TE-TM wave solutions exist for $\beta_{TE} = \beta_{TM}$. This means that in the stationary TE-TM wave the two polarisations maintain their separate identities by having different guided wavelengths. This type of interaction is unique to nonlinear waves since it is due purely to the nonlinear terms in the dielectric function. The upper locus shows the linear eigenvalues of TM waves in the presence of a nonlinear TE wave, and the lower locus describes zero amplitude TE waves in a structure modified by a nonlinear TM wave.

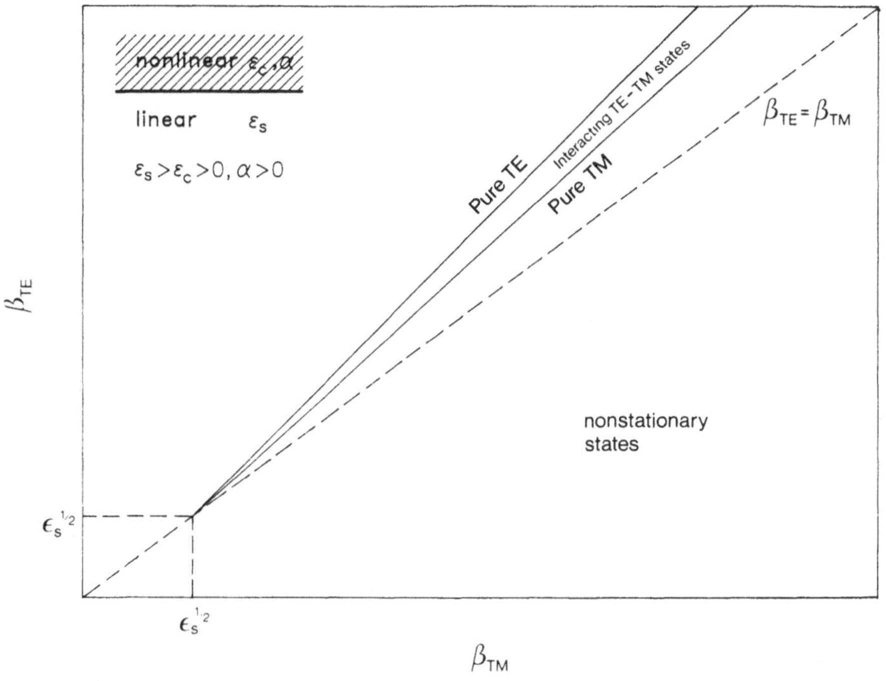


Fig. 14 The $\beta_{TE}$-$\beta_{TM}$ plane for nonlinear surface/guided waves at a single interface between a positive self-focusing nonlinear and positive linear medium.

The possibility for the existence of stationary nonlinear TE-TM waves increases greatly if a thin metal film or dielectric layer is inserted between the nonlinear medium and linear dielectric. Here we will concentrate on the dielectric layer case as it illustrates clearly the basic features.

Figure 15a shows the loci on the $\beta_{TE}$-$\beta_{TM}$ plane delineating the regions of stationary TE-TM waves in a linearly symmetric dielectric layer structure, i.e. $\epsilon_f > \epsilon_s = \epsilon_c$, whose parameters are such that only the two lowest order guided modes exist. This leads to four regions of interaction, namely $TE_0$-$TM_0$, $TE_0$-$TM_1$, $TE_1$-$TM_0$ and $TE_1$-$TM_1$. The loci corresponding to pure $TM_0$ and $TE_0$ modes increase without limit while those corresponding to pure $TM_1$ and $TE_1$ have a cut-off. In principle this configuration can support the greatest number of different regions of nonlinear TE-TM stationary states. Notice that the locus of linear eigenvalues of $TM_0$ modes in a $TE_1$ channel begins and ends with the same value of $\beta_{TM}$: at the lower value of $\beta_{TE}$ the $TE_1$ mode has zero amplitude and at the upper value the self-focused peak of the $TE_1$ mode has moved out to infinity. Hence in both limits there is no nonlinear TE-TM interaction and therefore $\beta_{TM}$ is the same. For interacting $TE_0$-$TM_1$ stationary waves it is $\beta_{TE}$ that has the same value at the two extremes of the pure TM locus.

Figure 15b shows the power flow for interacting $TE_1$-$TM_0$ stationary states for values of $\beta_{TE}$, $\beta_{TM}$ that lie on the dashed line in Fig. 15a. Interacting states exist for values of $\beta_{TE}$ well above the normal cut-off for pure nonlinear $TE_1$ waves and hence in principle the $TM_0$ wave could be used for switching the $TE_1$ wave 'on' and 'off'. (Similarly, $TE_0$ waves could be used to switch $TM_1$ waves.) The evolution of the TE and TM field profiles with increasing total power is shown in Fig. 15c. Notice that as the self-focused peak of the $TE_1$ wave moves away from the dielectric layer, it creates a separate nonlinear guiding channel that causes the peak in the $TM_0$ field amplitude to split into two separate peaks. At the same time, the $TM_0$ components distort the effective index seen by the $TE_1$ wave, allowing the amplitude of $E_y$ to change sign in the nonlinear medium. It is this completely new phenomenon of symmetry breaking inside the nonlinear medium that allows the $TE_1$ wave to exist above its normal cut-off point and the same effect comes into play with all nonlinear interacting TE-TM stationary states.[6]

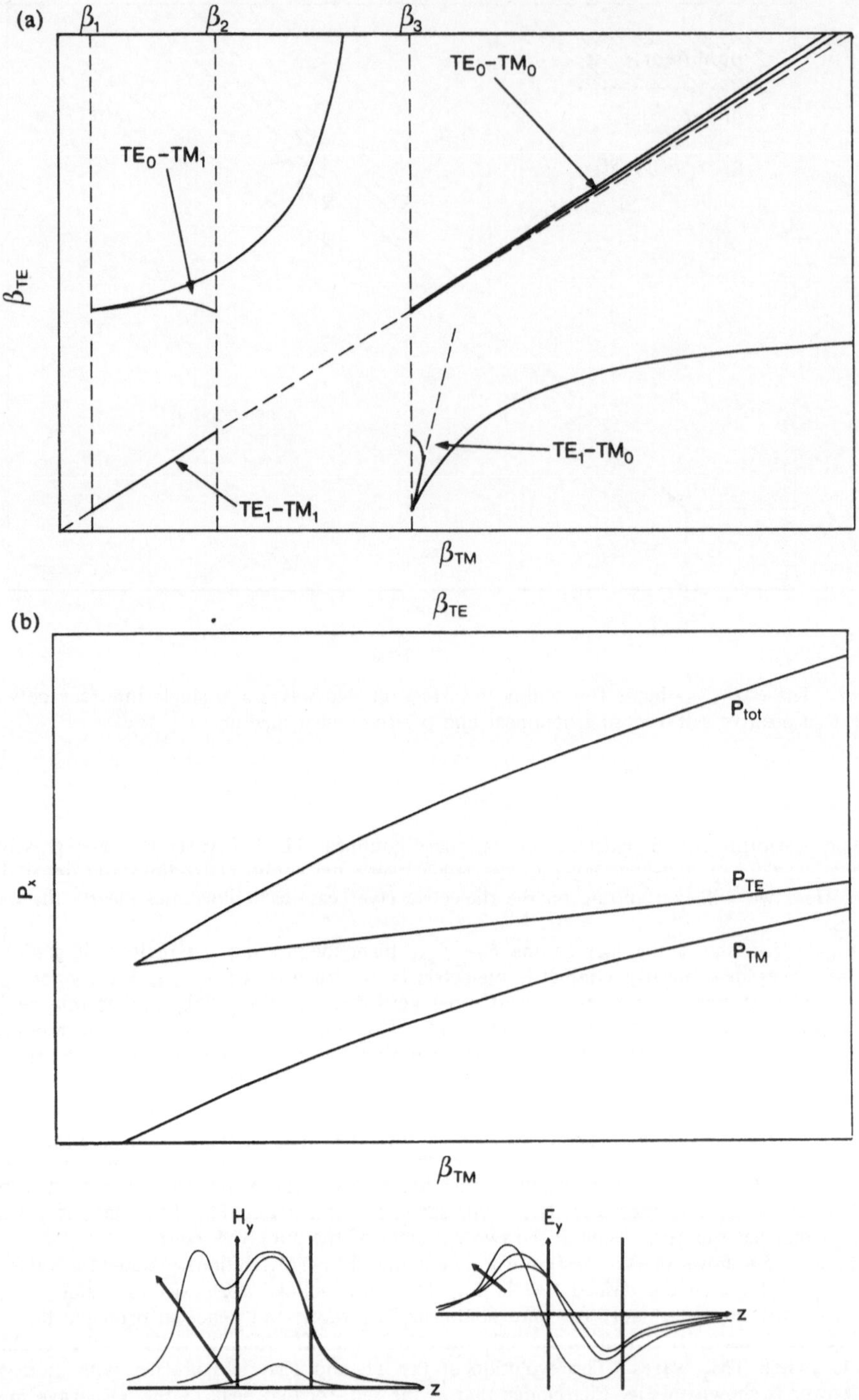


Fig. 15 (a) The $\beta_{TE}$-$\beta_{TM}$ plane for interacting nonlinear surface/guided waves in the thin film structure of Fig. 7. $\beta_1$ is the linear eigenvalue of the TM$_1$ mode, $\beta_2$ its nonlinear cut-off and $\beta_3$ the linear eigenvalue of the TM$_0$ mode. (b) Power flow for interacting TE$_1$-TM$_0$ waves for values of $\beta_{TE}$ and $\beta_{TM}$ lying along the dashed line in the TE$_1$-TM$_0$ region of the $\beta_{TE}$-$\beta_{TM}$ plane. P$_{tot}$, P$_{TE}$ and P$_{TM}$ are the total power flow and contributions from the TE$_1$ and TM$_0$ modes, respectively. (c) Field profiles for interacting TE$_1$-TM$_0$ modes. The arrows show the direction of increasing $\beta_{TE}$ and $\beta_{TM}$.

## 9. STABILITY OF NONLINEAR SURFACE/GUIDED WAVES

The preceding sections concentrate upon the propagation of 'stationary' electromagnetic waves along nonlinear guiding structures. A stationary wave is characterised by a transverse profile that is independent of both time and distance propagated and also by a definite frequency of oscillation. For example the electric field of a stationary electromagnetic wave has the form

$$E_i(\mathbf{r}_T, x, t) = \phi(\mathbf{r}_T)\cos(k_x x - \omega t) \tag{9.1}$$

where $E_i$ is the $i^{th}$ component of the field, x is the coordinate along the axis of the propagation, $\mathbf{r}_T$ is the transverse position vector, t is the time and $k_x$ and $\omega$ are the wavenumber and angular frequency of the wave respectively. Obviously these waves exist under certain circumstances but nonstationary waves also exist that do not exhibit a stationary transverse profile but do exhibit a definite frequency of oscillation, e.g. electromagnetic waves of the form

$$E_i(\mathbf{r}_T, x, t) = \phi(\mathbf{r}_T, x)\cos[k_x x + \theta(\mathbf{r}_T, x) - \omega t] \tag{9.2}$$

Other types may exhibit a stationary profile (at least to a good approximation) but not a definite frequency, e.g. waves of the form

$$E_i(\mathbf{r}_T, x, t) = A(x, t)\phi(\mathbf{r}_T)\cos[k_x x + \theta(x, t) - \omega t] \tag{9.3}$$

or there may be types that exhibit neither a stationary profile nor a definite frequency.

A fundamental concern about stationary nonlinear electromagnetic waves that was not addressed in the preceding sections is whether it is possible to generate and observe them. A full discussion cannot be given here but one particular facet will be examined, namely, the question of whether these waves can ever be excited in a realistic situation. This is equivalent to asking how critical are the values of the experimental parameters with respect to the existence of a particular stationary wave.

In order to answer this question, it is necessary to consider the evolution of a nonstationary wave as it propagates along the waveguide. For brevity, attention will be restricted to a class of nonstationary waves that possesses a definite frequency of oscillation. Consideration will also be limited to the (relatively) simple case of TE-polarised waves propagating in planar dielectric structures.

### 9.1 Evolution equation

For (effectively) two-dimensional excitation of TE waves, the wave equation is, as explained in section 4, the *elliptic* wave equation,

$$\frac{\partial^2 A(x,z)}{\partial x^2} + \frac{\partial^2 A(x,z)}{\partial z^2} + \frac{\omega^2}{c^2}\epsilon(x, z, |A(x,z)|)A(x,z) = 0 \tag{9.4}$$

where $\epsilon$ is now the nonlinear dielectric function and the y-component of the electric field, $E_y$, is given by

$$E_y(x, z, t) = \frac{1}{2}[A(x, z)e^{-i\omega t} + \text{c.c.}] \tag{9.5}$$

The variable

$$u(x, z) = A(x, z)e^{-ik_x x} \tag{9.6}$$

can now be introduced together with the Fresnel approximation

$$\left|\frac{\partial^2 u}{\partial x^2}\right| << 2k_x\left|\frac{\partial u}{\partial x}\right| \tag{9.7}$$

Under this approximation Eq. (9.4) reduces to the *parabolic* wave equation[48-51]

$$2ik_x \frac{\partial u(x,z)}{\partial x} + \frac{\partial^2 u(x,z)}{\partial z^2} + \left[\frac{\omega^2}{c^2}\epsilon(x,z,|u(x,z)|) - k_x^2\right]u(x,z) = 0 \qquad (9.8)$$

In order to be consistent with the Fresnel approximation the effective index $n_x$ must be close to the (average) linear refractive index but it is otherwise arbitrary.

Eq. (9.8) is the appropriate evolution equation for nonstationary waves and is the basic equation used for the investigation of the stability of nonlinear surface/guided waves. Stationary solutions of this equation are approximately the solutions of Eq. (9.4) but this approximation is only very good if the solutions are consistent with Eq. (9.7).

## 9.2 Types of perturbations

The question of the stability of stationary waves is basically one of how sensitive is the solution of Eq. (9.4), along with the associated boundary and initial conditions, to arbitrarily small changes (perturbations) in this system. There are three types of perturbation of interest:

*Type A: perturbations in the profile of the incident wave.* The boundary conditions imposed for surface/guided waves are

$$\lim_{z \to \pm\infty} u(x,z) = 0 \qquad (9.9)$$

and these, being perfectly natural ones, are retained when stability is being investigated. For a perturbation in the initial wave profile the initial condition for the wave $u(0,z)$ (which corresponds to the incident wave) imposed on the line $x = 0$ (for all z), is changed from that of the stationary wave solution $u_0(0,z)$ by a small quantity [compared to $u_0(0,z)$], i.e. the initial condition used is

$$u(0,z) = u_0(0,z) + \delta u(0,z) \qquad \text{for all } z \qquad (9.10)$$

where

$$|\delta u(0,z)| << |u_0(0,z)| \qquad (9.11)$$

*Type B: perturbations in the material parameters of the waveguide.* In this case the boundary and the initial conditions are those corresponding to the stationary wave, i.e. the condition

$$u(0,z) = u_0(0,z) \qquad (9.12)$$

and Eq. (9.9) are imposed but the perturbation amounts to small changes made in the linear dielectric constants and/or the nonlinear coefficients. These changes may include the addition of both linear and nonlinear absorption.

*Type C: perturbations in the geometry of the waveguide.* The perturbation here is a change imposed upon the environment of the wave. It could be a variation in the refractive index profile with propagation distance, with the boundary and initial conditions remaining as for type B perturbations. The length scale of this variation will be made much smaller than the spatial period of the corresponding stationary wave which is equal to $2\pi/k_x$, where $k_x$ is the propagation constant of the stationary wave.

By studying the behaviour of the wave as it evolves away from the initial condition and perceiving whether or not the wave profile and wavefront remain close to the corresponding stationary wave, the stability of a surface/guided wave may be determined. The literature deals almost exclusively with stability under type A perturbations and this is reflected in the treatment presented here but stability under type C perturbations will also be discussed.

## 9.3 Approaches to the problem of stability

There are two main methods for investigating stability that are commonly used in the literature:

(i) *Determination of eigenvalues.* As was done earlier for the plane wave this method involves the linearisation of the evolution equation about the stationary wave. Eigenfunctions are then sought for this linearised equation, i.e. solutions of the form

$$d(x,z) = f(z)e^{\lambda x} \tag{9.13}$$

where d(x,z) is the perturbation, are sought. If $\lambda$ has a real positive part, this solution will grow with increasing propagation distance and, hence, the stationary wave is unstable with respect to perturbations in the initial condition. It might be thought that if only solutions of the form (9.13) with $\lambda$ pure imaginary can be found then the wave is stable. However, the eigenfunctions of the linearised equation do not, in general, form a complete set, i.e. every form of the perturbation cannot be represented as an eigenfunction expansion (one exception to this is the plane wave case discussed earlier). Due to this limitation stationary waves cannot be definitely identified as being stable.

This method of analysis has only been utilised for investigating stability under type A perturbations but there is no reason why it cannot be extended to deal with type B perturbations. For type C perturbations, however, this method would be inappropriate in general.

(ii) *Numerical simulation.* The exact behaviour of the wave as it propagates can be simulated using a suitable numerical technique, e.g. explicit finite difference schemes, the propagating beam method, pseudospectral techniques, etc. The initial field distribution is perturbed away from that of the stationary wave and the evolution of the wave from this is computed. The stability of the stationary wave, under this particular perturbation may be decided by examining its numerical behaviour. Questions more general than that of stability of the wave may also be answered. For instance, the final state (if any) of the perturbed stationary wave might be of interest and this could not be predicted by method (i). Unfortunately every form of perturbation cannot be tested and, hence, the wave may not be classified as definitely stable.

## 9.4 Stability under type A perturbations

(i) *A single interface system.* This structure, consisting of a semi-infinite linear medium adjacent to a semi-infinite nonlinear medium (Fig. 3), supports only nonlinear surface waves. Eigenvalue analysis [approach (i) in section 9.3] shows that under two-dimensional perturbations nonlinear surface waves are unstable if[51]

$$\frac{dI(\beta)}{d\beta} < 0 \tag{9.14}$$

where $\beta$ is the effective index of the nonlinear surface waves and $I(\beta)$ is given by

$$I(\beta) = \int_{-\infty}^{\infty} |u(z,\beta)|^2 dz \tag{9.15}$$

Various numerical simulations have confirmed that the above is the condition for instability. These have also revealed that unstable nonlinear surface waves decay either by evolving to low-flux nonlinear surface waves, radiating off the excess power per unit length, or by a complete emission of the nonlinear surface waves into the linear medium with the emitted wave having a nearly Gaussian intensity distribution.

Stability of nonlinear surface waves in this structure under y-dependent perturbations has also been addressed (under the scalar wave approximations). It has been shown that all nonlinear surface waves are unstable (under this type of perturbation) and decay into separate beams for propagation lengths greater than $L_y$, where $L_y$ is given by

$$L_y \simeq \lambda\pi\epsilon_c(\epsilon_s - \epsilon_c) \tag{9.16}$$

Here $\lambda$ is the free-space wavelength and $\epsilon_s$ and $\epsilon_c$ are the linear dielectric constants of the linear and nonlinear medium respectively. $L_y$ is typically the order of a millimetre.

(ii) *A linear film with a nonlinear cover but with a linear substrate.* This structure (shown in Fig. 7) supports both nonlinear guided waves and nonlinear surface waves. From eigenvalue analysis the instability criterion for nonlinear surface waves and $TE_0$ nonlinear guided waves under two-dimensional perturbations is again given by Eq. (9.14). Numerical simulation studies indicate that low power $TE_0$ nonlinear guided waves show nearly linear features, as is obvious, and that they are unconditionally stable. Perturbed unstable $TE_0$ nonlinear guided waves and nonlinear surface waves may exhibit a weak oscillatory behaviour in which the intensity maximum of the wave bounces off each interface periodically. It appears that this sort of behaviour is always associated with the section of an $I(\beta)$ vs. $\beta$ curve having a linear limit which has a negative slope independently of the geometry of the waveguide. Unstable $TE_0$ nonlinear guided waves or nonlinear surface waves also decay by the emission of a spatial soliton into the nonlinear cover.

For $TE_1$ nonlinear guided waves there have been no successful eigenvalue analyses of the stability. However, numerical simulation studies indicate that $TE_1$ nonlinear guided waves are nearly stable with a typical evolution length of a few thousand wavelengths. On the negatively sloped section of the $TE_1$ $I(\beta)$ vs. $\beta$ curve, the instability is manifested by soliton emission into the nonlinear cover.

(iii) *A linear film with both cover and substrate being nonlinear*. If the structure is symmetric then both symmetric and asymmetric $TE_0$ nonlinear guided waves exist. From an eigenvalue analysis, it is found that if the effective index is such that asymmetric $TE_0$ nonlinear guided waves cannot be supported then the corresponding symmetric $TE_0$ nonlinear guided waves are stable. If the effective index is such that asymmetric $TE_0$ nonlinear guided waves can be supported the corresponding symmetric $TE_0$ nonlinear guided waves will be unstable. From numerical stability studies it seems that unstable symmetric nonlinear guided waves decay into stable asymmetric ones. The instability criterion for asymmetric $TE_0$ nonlinear guided waves is found analytically once again to be Eq. (9.14) and it seems that unstable asymmetric nonlinear guided waves exhibit the same sort of oscillatory behaviour as was described in the last sub-subsection.

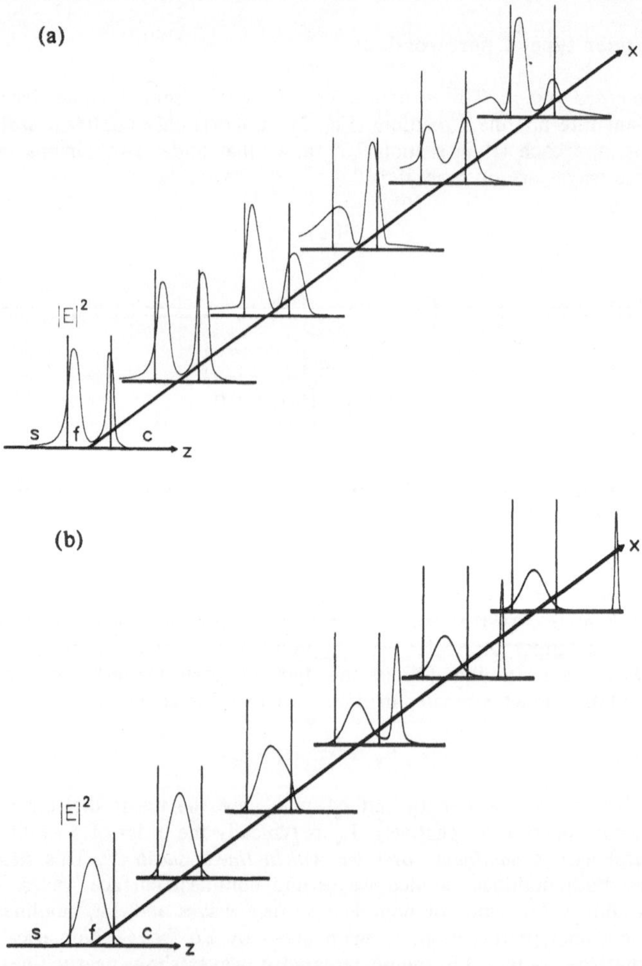


Fig. 16 Effect of microbending on the propagation of (a) a $TE_1$ wave and (b) a $TE_0$ wave in the thin film guide of Fig. 7.

## 9.5 Stability under type C perturbations

In this subsection the propagation of TE waves in a 3-layer step-index structure, in which the position of the interfaces (in the transverse plane) between the media varies with the propagation distance is considered. Guided waves propagating along such a guide will experience discontinuities in the dielectric constant. These discontinuities act as partial mirrors reflecting some of the incident power flux backwards along the guide. If a large fraction of the power is so reflected, the Fresnel approximation will not be valid. Also the incident wave would not be well defined because of the invalidity of linear superposition (due to the nonlinearity). Hence, in the model analysed here, the width of these "mirrors" will be quite small compared to the effective width of the incident wave. The Fresnel approximation will also be inapplicable if the characteristic length scale of the variation of position of the interfaces is comparable to the characteristic length scale of the incident wave. Only situations for which the parabolic wave equation is appropriate will be considered, however.

The only way to investigate stability in these systems is by direct numerical simulation. This obviously involves discretising the position of the interfaces but the details of this procedure would be out of place here. Firstly the effect of microbending upon the waveguide is investigated. A numerical simulation shows that microbending has a cumulative effect on the behaviour of the wave, i.e. the wave propagates tens of wavelengths before its behaviour changes appreciably from that of the stationary wave. The behaviour of the wave depends upon the effective index, $\beta \equiv n_x = ck_x/\omega$, of the corresponding stationary wave. The $TE_1$ wave, as it propagates, radiates power predominantly into the linear substrate, as shown in Fig. 16a. An intensity maximum may in fact form in the nonlinear cover but this maximum eventually re-enters the linear film.

For $TE_0$ waves the primary decay mechanism is via the emission of a soliton into the nonlinear cover, as shown in Fig. 16b. This behaviour occurs even when the corresponding stationary wave does not have an intensity maximum near the nonlinear cover. Finally, the behaviour in the same guide but with *rough* interfaces displays the same basic features; the only significant difference is a slight change in the characteristic evolution length-scales.

## REFERENCES


1. Stegeman G I, Wright E M, Finlayson N, Zanoni R, Seaton C T, J. Lightwave Tech. **6**, 953-970 (1988)
2. Akhmediev N N, Sov. Phys. JETP **56**, 299-303 (1982)
3. Akhmediev N N, Zvezdin A K, Korneev V I, Mitskevich N V, Sov. Phys. Sol. St. **30**, 1337-1339 (1988)
4. Boardman A D, Egan P, IEEE J. Quantum Electron. **QE-21**, 1701-1713 (1985)
5. Boardman A D, Maradudin A A, Stegeman G I, Twardowski T, Wright E M, Phys. Rev. A **35**, 1159-1164 (1987)
6. Boardman A D, Twardowski T, Phys. Rev. A **39**, 2481-2492 (1989)
7. Chen W, Mills D L, Phys. Rev. B **35**, 524-532 (1987)
8. Shivarova A, Dimitrov N, Plasma Phys. and Controlled Fusion **27**, 219-224 (1985)
9. Langbein U, Lederer F, Ponath H E, Trutschel U, Appl. Phys. B **38**, 263-268 (1985)
10. Langbein U, Lederer F, Ponath H E, Trutschel U, Appl. Phys. B **36**, 187-193 (1985)
11. Mihalache D, Mazilu D, Totia H, Phys. Scr. **30**, 335-340 (1984)
12. Mihalache D, Mazilu D, Appl. Phys. B **41**, 119-123 (1986)
13. Mihalache D, Wang R-P, Boardman A D, Solid State Commun. **71**, 613-617 (1989)
14. Boardman A D, Twardowski T, Wright E M, Opt. Commun. **74**, 347-352 (1990)
15. Seaton C T, Valera J D, Shoemaker R L, Stegeman G I, Chilwell J T, Smith S D, IEEE J. Quantum Electron. **QE-21**, 774-783 (1985)
16. Stegeman G I, Seaton C T, in *Phenomena at surfaces, interfaces and superlattices*, Nizzoli, Rieder and Willis eds. (Springer-Verlag, 1985)
17. Stegeman G I, Wright E M, Seaton C T, Moloney J V, Shen T P, Maradudin A A, Wallis R F, IEEE J. Quantum Electron. **QE-22**, 977-983 (1986)
18. Wright E M, Stegeman G I, Seaton C T, Moloney J V, Boardman A D, Phys. Rev. A **34**, 4442-4444 (1986)
19. Maradudin A A, Z. Phys. B **41**, 341-344 (1981)
20. Maradudin A A, in *Optical and acoustic waves in solids - Modern Topics*, Borissov M ed. (World Scientific, 1983)

21. Zyss J, J. Mol. Elec. **1**, 25-45 (1985)
22. Litvak A G, Mironov V A, Izvestiya VUZ Radiofizika **11**, 1911-1912 (1968)
23. Miyagi M, Nishida S, Rep. Res. Inst. Elect. Comm. Tohoku Univ. **24**, 53-67 (1972)
24. Agranovich V M, Babichenko V S, Chernyak V Ya, JETP Lett. **32**, 512-515 (1980)
25. Stegeman G I, Seaton C T, Opt. Lett. **9**, 235-237 (1984)
26. Pohl D, Phys. Rev. A **5**, 1906-1909 (1970)
27. Shen T P, Stegeman G I, Maradudin A A, Appl. Phys. Lett. **52**, 1-3 (1988)
28. Franken P A, Ward J F, Rev. Mod. Phys. **35**, 23-39 (1963)
29. Schubert M and Wilhelmi B, *Nonlinear Optics and Quantum Electronics* (Wiley, New York, 1986)
30. Agrawal G P, *Nonlinear Fiber Optics* (Academic, London, 1989)
31. Butcher P N, *Nonlinear optical phenomena* (Bulletin 200, Eng. Exp. Stn., Ohio State Univ., 1965)
32. Morse P M, Feshbach H, *Methods of Theoretical Physics* (McGraw-Hill, New York, 1953)
33. Miyagi M, Nishida S, Rep. Res. Inst. Elect. Comm. Tohoku Univ. **25**, 53-66 (1973)
34. Miyagi M, Nishida S, Proc. IEEE **62**, 1284-1285 (1974)
35. Miyagi M, Nishida S, Electron. and Commun. JPN (USA) **58**, 104-110 (1975)
36. Miyagi M, Nishida S, Radio Science **10** (1975)
37. Svelto O, in *Progress in Optics XII*, Wolf E ed. (North Holland, 1974)
38. Chen W, Maradudin A A, J. Opt. Soc. Am. B **5** (1988)
39. Tomlinson W J, Opt. Lett. **5**, 323-325 (1980)
40. Yariv A, *Optical Electronics* (Holt, Rinehart and Winston, New York, 1985)
41. Boardman A D, Egan P, IEEE J. Quantum Electron. **QE-22**, 319-324 (1986)
42. Boardman A D, Egan P, J. de Phys. **45**, C5-291-303 (1984)
43. Boardman A D, Twardowski T, Shivarova A, Stegeman G I, IEE Proc. J **134**, 152-160 (1987)
44. Akhmediev N N, Sov. Phys. JETP **57**, 1111-1116 (1983)
45. Stegeman G I, Valera J D, Seaton C T, Sipe J, Maradudin A A, Solid State Commun. **52**, 293-297 (1984)
46. Lederer F, Mihalache D, Solid State Commun. **59**, 151-153 (1986)
47. Mihalache D, Mazilu D, Lederer F, Opt. Commun. **59**, 391-394 (1986)
48. Akhmediev N N, Korneev V I, Kuzmenko Yu V, Sov. Phys. JETP **61**, 62-67 (1985)
49. Wright E M, Stegeman G I, Seaton C T, Moloney J V, Appl. Phys. Lett. **49**, 435-436 (1986)
50. Leine L, Wachter C, Langbein U, Lederer F, J. Opt. Soc. Am. B **5** (1988)
51. Akhmediev N N, Korneev V I, Kuzmenko Yu V, Sov. Tech. Phys. Lett. **10**, 327-329 (1984)

NONLINEAR ELECTROMAGNETIC SURFACE WAVES ON GYROTROPIC MEDIA

R. F. Wallis

Department of Physics and Institute for Surface and Interface Science
University of California
Irvine, CA 92717, USA

A. D. Boardman and M. Shabat

Applied Optics Group, Department of Physics
University of Salford
Salford M5 4WT, UK

## 1. INTRODUCTION

In this chapter we analyze the behavior of electromagnetic waves propagating along the interface between a linear substrate and a nonlinear cladding when the materials involved are either magnetic materials or electrically conducting materials with an external magnetic field present. Optically nonlinear systems of this sort are of great interest in the development of solid state optical devices for applications to optical computing or signal processing.[1-15]

The outline of this chapter is as follows. First, we will present an introductory development of the linear response of both a magnetic system and a magnetoplasma system to an electromagnetic wave, as characterized by the magnetic permeability tensor and the magneto-dielectric tensor, respectively. The dispersion relations for electromagnetic surface waves will be derived for these two cases. We will then discuss nonlinearities in magnetic systems and will analyze the properties of electromagnetic waves guided by the interface between a linear ferromagnetic (or ferrimagnetic) material and an isotropic nonlinear magnetic material. Finally, we will present a similar development for electromagnetic waves guided by the interface between a linear magnetoplasma and an isotropic nonlinear dielectric material.

## 2. MAGNETIC PERMEABILITY TENSOR

Consider a system of atoms with total angular momentum $\hbar\mathbf{J}$ and magnetic moment $\mathbf{M}$. The relationship between the magnetic moment and the total angular momentum is given by

$$\mathbf{M} = \gamma\hbar\mathbf{J} \tag{2.1}$$

where $\gamma$ is the magnetogyric ratio. The latter is related to the Bohr magneton $\mu_b$ and the Landé splitting factor g by

$$\gamma\hbar = -g\mu_b \tag{2.2}$$

If the system of atoms is placed in an external magnetic field characterized by magnetic induction $\mathbf{B} = \mu_0\mathbf{H}$, where $\mu_0$ is the permeability constant, the time rate of change of its angular momentum is specified by the equation

$$\hbar \frac{d\mathbf{J}}{dt} = \mathbf{M}\times\mathbf{B} = \mu_0\mathbf{M}\times\mathbf{H} \tag{2.3}$$

in the absence of damping. Multiplying Eq. (2.3) by $\gamma$ and using Eq. (2.1) gives

$$\frac{d\mathbf{M}}{dt} = \gamma\mu_0\mathbf{M}\times\mathbf{H} \tag{2.4}$$

which is the equation of motion for the magnetic moment.

We are now in a position to calculate the response of the magnetic moment to an electromagnetic wave of frequency $\omega$. Let $\mathbf{H}_0 = (0,-H_0,0)$ be a constant, uniform external magnetic field oriented in the negative y-direction. The total magnetic field $\mathbf{H}$ is the sum of $\mathbf{H}_0$ and the contribution from the electromagnetic field $\mathbf{h} = (h_x, h_y, h_z)$:

$$\mathbf{H} = \mathbf{H}_0 + \mathbf{h} \tag{2.5}$$

Similarly, we can write the total magnetic moment as

$$\mathbf{M} = \mathbf{M}_0 + \mathbf{m} \tag{2.6}$$

where $\mathbf{M}_0 = (0,-M_0,0)$ is the static part of $\mathbf{M}$ and $\mathbf{m} = (m_x, m_y, m_z)$ is the dynamic part of $\mathbf{M}$.

We assume that the electromagnetic field is weak in the sense that $\mathbf{h}$ and $\mathbf{m}$ are small compared to $\mathbf{H}_0$ and $\mathbf{M}_0$ respectively. We can therefore substitute Eqs. (2.5) and (2.6) into Eq. (2.4) and linearize the latter with respect to $\mathbf{h}$ and $\mathbf{m}$ to yield the equations of motion in component form

$$\frac{dm_x}{dt} = -\gamma\mu_0(M_0h_z - m_zH_0) \tag{2.7a}$$

$$\frac{dm_y}{dt} = 0 \tag{2.7b}$$

$$\frac{dm_z}{dt} = -\gamma\mu_0(m_xH_0 - M_0h_x) \tag{2.7c}$$

where we have made use of the fact that

$$\frac{d\mathbf{M}_0}{dt} = 0 \tag{2.8}$$

Since the electromagnetic field is characterized by frequency $\omega$, we can write

$$\mathbf{h} = \mathbf{h}_0e^{-i\omega t} \tag{2.9}$$

We anticipate that $\mathbf{m}$, which represents the response of the atomic moment to the electromagnetic field, should also vary in time with frequency $\omega$, so that we can write

$$\mathbf{m} = \mathbf{m}_0e^{-i\omega t} \tag{2.10}$$

Substituting Eq. (2.10) into Eqs. (2.7) we obtain

$$i\omega m_x = \gamma\mu_0(M_0h_z - m_zH_0) \tag{2.11a}$$

$$i\omega m_y = 0 \tag{2.11b}$$

$$i\omega m_z = \gamma\mu_0(m_xH_0 - M_0h_x) \tag{2.11c}$$

Solving Eqs. (2.11) for $m_x$ and $m_z$, we get

$$m_x = \frac{\omega_0\omega_m}{\omega_0^2-\omega^2}h_x + i\frac{\omega\omega_m}{\omega_0^2-\omega^2}h_z \tag{2.12a}$$

$$m_z = -i\frac{\omega\omega_m}{\omega_0^2-\omega^2}h_x + \frac{\omega_0\omega_m}{\omega_0^2-\omega^2}h_z \tag{2.12b}$$

where $\omega_0 = \gamma\mu_0 H_0$ and $\omega_m = \gamma\mu_0 M_0$.

The magnetic induction **b** is defined by

$$\mathbf{b} = \mu_0(\mathbf{h} + \mathbf{m}) \tag{2.13}$$

$$\equiv \mu_0 \underline{\underline{\mu}}\cdot\mathbf{h} \tag{2.14}$$

where $\underline{\underline{\mu}}$ is the permeability tensor. Substituting Eqs. (2.12) into Eq. (2.13) and comparing with Eq. (2.14), we find that the elements of the permeability tensor are given by

$$\mu_{xx} = \mu_{zz} = 1 + \frac{\omega_0\omega_m}{\omega_0^2-\omega^2} \tag{2.15a}$$

$$\mu_{xz} = -\mu_{zx} = i\frac{\omega\omega_m}{\omega_0^2-\omega^2} \tag{2.15b}$$

$$\mu_{yy} = 1 \tag{2.15c}$$

$$\mu_{xy} = \mu_{yx} = \mu_{yz} = \mu_{zy} = 0 \tag{2.15d}$$

## 3. SURFACE MAGNON POLARITONS

We now consider electromagnetic surface waves localized at the interface between a linear ferromagnetic medium and vacuum. We refer to these excitations as surface magnon polaritons. The dielectric function of the medium is taken to be unity; however, the generalization of the treatment to a frequency-dependent dielectric tensor is straightforward.[16] The geometry of the system is shown in Fig. 1. Our task is to solve Maxwell's equations for the fields in the magnetic medium and in the vacuum and then satisfy the electromagnetic boundary conditions at the interface between the two media.

We first direct our attention to the magnetic medium. If we take Maxwell's curl equations and eliminate the electric field, we obtain the equation

$$\nabla^2\mathbf{H} - \nabla(\nabla\cdot\mathbf{H}) - \epsilon_0\mu_0\underline{\underline{\mu}}\cdot\ddot{\mathbf{H}} = 0 \tag{3.1}$$

where the double dots over the **H** denote a second time derivative. For isotropic magnetic media in which the permeability tensor $\underline{\underline{\mu}}$ is diagonal with equal diagonal elements, $\nabla\cdot\mathbf{H} = 0$, and the second term in Eq. (3.1) vanishes. For the ferromagnetic media that we are considering, however, $\underline{\underline{\mu}}$ is not diagonal and the divergence equation

$$\nabla\cdot\mathbf{B} = \mu_0\nabla\cdot\left(\underline{\underline{\mu}}\cdot\mathbf{H}\right) = 0 \tag{3.2}$$

does not imply that $\nabla\cdot\mathbf{H} = 0$. Hence, we must retain the second term in Eq. (3.1).

We seek a solution of Eq. (3.1) that is localized at the surface $z = 0$ and corresponds to a TE wave propagating in the x-direction with wavenumber k and frequency $\omega$. Since the ferromagnetic medium occupies the region $z < 0$, such a solution can be written as

$$\mathbf{H} = (H_{2x}, 0, H_{2z})e^{ikx}e^{\alpha_2 z}e^{-i\omega t} \tag{3.3}$$

where $\alpha_2$ is a decay constant to be determined. Substitution of Eq. (3.3) into Eq. (3.1) leads to the algebraic equations

$$(\alpha_2^2 + k_0^2\mu_{xx})H_x - (ik\alpha_2 - k_0^2\mu_{xz})H_z = 0 \tag{3.4a}$$

$$(ik\alpha_2 - k_0^2\mu_{zx})H_x + (k^2 - k_0^2\mu_{zz})H_z = 0 \tag{3.4b}$$

where $k_0^2 = \omega^2\epsilon_0\mu_0$. Setting the determinant of the coefficients of $H_x$ and $H_z$ to zero gives the result

$$\alpha_2^2 = k^2 - k_0^2\mu_V \tag{3.5}$$

where

$$\mu_V = \mu_{xx} + \frac{\mu_{xz}^2}{\mu_{xx}} \tag{3.6}$$

is the permeability for propagation in the Voigt configuration ($\mathbf{H_0} \parallel$ surface, $\mathbf{H_0} \perp \mathbf{k}$).

We now turn our attention to the fields in the vacuum region $z > 0$. Here, the permeability tensor $\underline{\underline{\mu}}$ is given by $\underline{\underline{\mu}} = \underline{\underline{I}}$ where $\underline{\underline{I}}$ is the 3x3 unit tensor. The divergence equation

$$\nabla\cdot\mathbf{B} = 0 \tag{3.7}$$

reduces to

$$\nabla\cdot\mathbf{H} = 0 \tag{3.8}$$

and Eq. (3.1) simplifies to the wave equation

$$\nabla^2\mathbf{H} - \epsilon_0\mu_0\ddot{\mathbf{H}} = 0 \tag{3.9}$$

We seek a solution to Eq. (3.9) of the form

$$\mathbf{H} = (H_{1x}, 0, H_{1z})e^{ikx}e^{-\alpha_1 z}e^{-i\omega t} \tag{3.10}$$

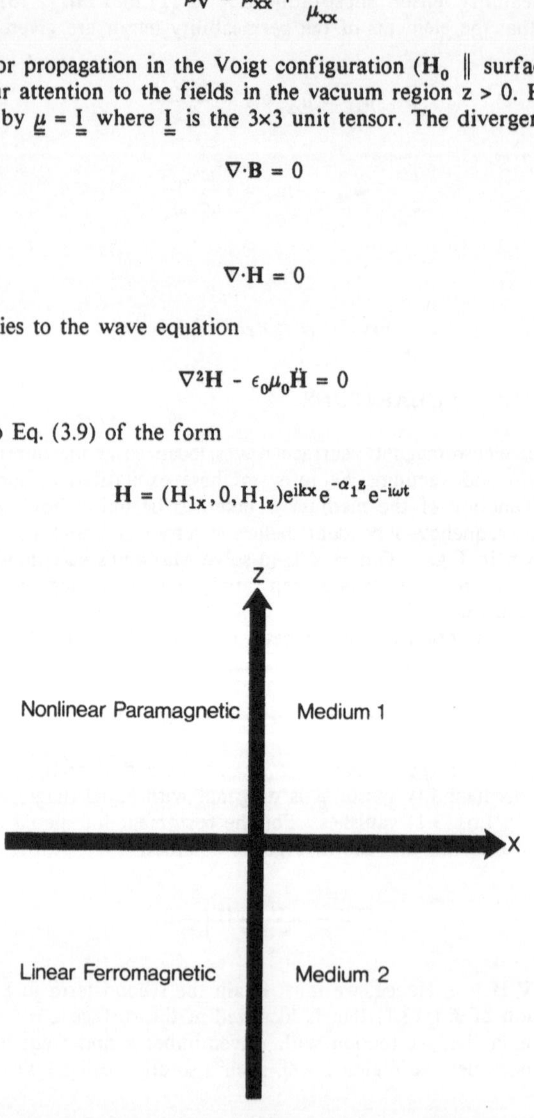


Fig. 1 Coordinate system for the single interface between a nonlinear paramagnetic cladding and a linear ferromagnetic substrate.

corresponding to a solution in which the field components are localized at the surface z = 0. Substitution of Eq. (3.10) into Eq. (3.9) yields the equation

$$\alpha_1^2 = k^2 - k_0^2 \tag{3.11}$$

which must be satisfied for a nontrivial solution of Eq. (3.9) to exist.

In order to complete the problem we must satisfy the electromagnetic boundary conditions at the surface z = 0. We express these boundary conditions as the continuity of the tangential component of **H** and the normal component of **B**:

$$H_{1x} = H_{2x} \qquad \text{at } z = 0 \tag{3.12a}$$

$$B_{1z} = B_{2z} \qquad \text{at } z = 0 \tag{3.12b}$$

In order to exploit these equations, we need to establish relationships between $H_{1x}$ and $H_{1z}$ and between $H_{2x}$ and $H_{2z}$. They can be obtained by substituting Eqs. (3.3) and (3.10) into Eqs. (3.2) and (3.8), respectively. Taking account of the fact that in vacuum, $\mathbf{B} = \mu_0\mathbf{H}$, we obtain

$$H_{2z} = \frac{\alpha_2\mu_{xz} - ik\mu_{xx}}{\alpha_2\mu_{xx} + ik\mu_{xz}} H_{2x} \tag{3.13a}$$

$$H_{1z} = \frac{ik}{\alpha_1} H_{1x} \tag{3.13b}$$

If we re-express Eq. (3.12b) in terms of the z-component of **H** using $\mathbf{B} = \mu_0\underline{\underline{\mu}}\cdot\mathbf{H}$, we find that

$$\mu_{zx}H_{2x} + \mu_{zz}H_{2z} = H_{1z} \tag{3.14}$$

Substitution of Eqs. (3.13) into Eq. (3.14) and use of Eq. (3.12a) yields the equation

$$\alpha_2 + ik\frac{\mu_{xz}}{\mu_{xx}} + \alpha_1\mu_V = 0 \tag{3.15}$$

which is the dispersion relation for surface magnon polaritons. Eliminating $\alpha_1$ and $\alpha_2$ from Eq. (3.15) with the aid of Eqs. (3.5) and (3.11), we obtain an explicit expression for k as a function of $\omega$:

$$k^2 = k_0^2 \frac{(\mu_V-1)(\mu_{xx}\mu_V-1)\mu_{xx} + 2\mu_{xz}^2 \pm 2i\mu_{xz}\left(\mu_{xx}(2\mu_{xx}-\mu_{xx}\mu_V-1)\right)^{1/2}}{(\mu_{xx}\mu_V-1)^2 + 4\mu_{xz}^2} \tag{3.16}$$

In general, Eq. (3.16) has two physical and two nonphysical solutions for $\omega(k)$, the dispersion relation. The two physical solutions correspond to nonreciprocal propagation in the **+k** and in the **-k** directions. This feature of nonreciprocal propagation will be discussed in more detail later.

Equation (3.13a) allows us to determine the polarization of the magnetic field **H**(t) at a fixed point in the medium. A simple application of boundary conditions enables us to determine the polarization of the field in the vacuum. The polarizations of the magnetic fields of the surface polariton in both media are elliptically polarized in the sagittal plane. When the limits $k \rightarrow \pm\infty$ are allowed by Eq. (3.16) we find $\alpha_1 = \alpha_2 = |k|$ and Eq. (3.13a) reduces to $H_{2z} = -iH_{2x}$ for the +**k** direction and $H_{1z} = iH_{1x}$ for the -**k** direction. This corresponds to left (-) and right (+) circular polarization of **H** about the $\hat{z}$ axis for +k and -k respectively. These polarizations each correspond to an $\hat{n}\times\mathbf{k}$ circular polarization for +k, and $-\hat{n}\times\mathbf{k}$ circular polarization for -k as illustrated in Fig. 2a. The axial vector $\hat{n}\times\mathbf{k}$ refers to the sense of rotation of **H**, and we represent these rotations as $\mathbf{H}_{\rho\pm}$ for the $\pm\mathbf{k}$ directions. An $\hat{n}\times\mathbf{k}$ rotation corresponds to a rolling motion of $H_2$ along the surface in space. This $\hat{n}\times\mathbf{k}$ sense of rotation in the medium and $-\hat{n}\times\mathbf{k}$ sense of rotation in the vacuum have previously been termed prograde, while the opposite senses of rotation have been termed retrograde. This motion is clockwise about $\hat{z}$ in the medium and counterclockwise about $\hat{y}$ in the vacuum for +k (see Fig. 2c). The condition for these $\mathbf{k} \to \pm\infty$ solutions is given by

$$\mu_{xx} \mp i\mu_{xz} = \mu_{\rho\pm}(\omega) = -1 \tag{3.17}$$

where $\mu_{\rho\pm}(\omega)$ are the permeability functions associated with $\pm\hat{n}\times\mathbf{k}$ rotations $\mathbf{H}_{\rho\pm}$ in the $\pm\mathbf{k}$ directions respectively. Note that the magnetic permeability functions $\mu_\pm(\omega) = \mu_{xx} \mp i\mu_{xz}$ are associated with the left and right circularly polarized modes about $\mathbf{k}$ propagating in the Faraday configuration ($\mathbf{H}_0 \parallel \mathbf{k}$), and that $\mu_+\mu_- = \mu_{xx}\mu_V$. When $k \neq \pm\infty$, Eq. (3.13a) describes elliptically polarized fields.

To further illustrate the nature of the surface polaritons we consider the permeability associated with the magnons in a single domain ferromagnetic insulator. The permeability tensor Eq. (2.15) applies in the configuration $\mathbf{H}_0$ perpendicular to the sagittal plane in the absence of high frequency magnetic dipole excitations. The important components of this tensor when $\mathbf{H}_0$ is along -y are given below neglecting both spatial dispersion and damping, but including the contribution of other magnetic dipole excitations,

$$\mu_{xx}(\omega) = \mu_B\left[1 + \frac{\omega_s\omega_0}{\omega_0^2 - \omega^2}\right] \tag{3.18a}$$

$$\mu_{xz}(\omega) = \mu_B\frac{\omega_s\omega}{\omega_0^2 - \omega^2}i \tag{3.18b}$$

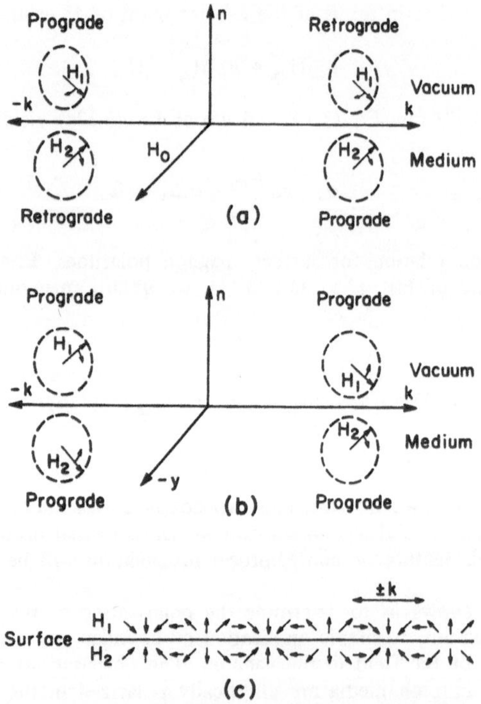


Fig. 2 The time variation of the magnetic fields at fixed points in space above and below the surface for surface polaritons propagating in the +k and -k directions on (a) the gyromagnetic medium and (b) an isotropic medium. Note that on a gyromagnetic medium the magnetic fields are all rotating in the same sense. (c) The spatial variation of the magnetic field at a fixed instant of time for an isotropic medium.

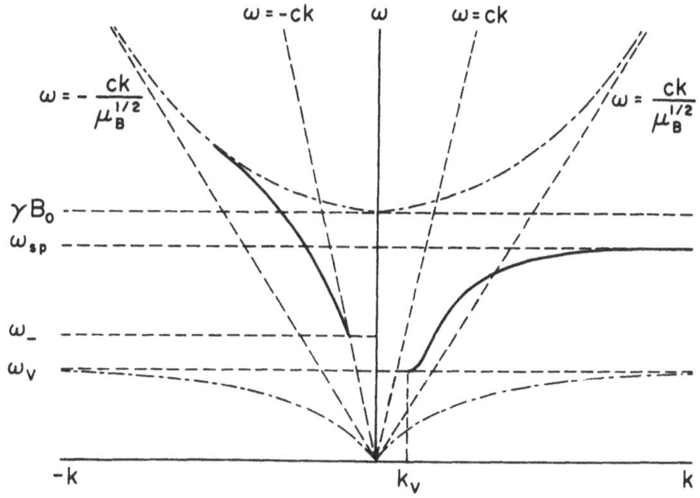


Fig. 3 Dispersion curves for surface polaritons for the two non-equivalent directions of propagation on a gyromagnetic medium (full curves) and for bulk polaritons (dot-dashed curves).

where $\omega_0 = \gamma\mu_0 H_0$ and $\omega_s = \omega_m/\mu_B$ in terms of the static magnetic field and screened magnetization $M_0/\mu_B$ of the sample. Note that the sign of $\mu_{xz}$ depends on the direction of $\mathbf{H}_0$, through $\omega_s$, and the sign of $\gamma$, the gyromagnetic ratio. $\mu_B$ is the permeability caused by other magnetic dipole excitations in the solid (optical magnons, etc.). If these excitations occur at a much higher frequency than $\omega_0$, we can take $\mu_B$ to be frequency independent to a very good approximation. We estimate $\mu_B \simeq 1.25$ in YIG by taking the theoretical optical magnon spectra[17] and $M_0/\mu_B$ = 500 G. As we shall see, the magnitude of $\mu_B$ plays an important role in the surface polariton dispersion relation.

This dispersion relation is shown in Fig. 3 for both +**k** and -**k** directions. The nonreciprocal nature of the propagation is clearly evident. As is illustrated in Fig. 2a, the +**k** surface polariton mode has the elliptically polarized magnetic field in the medium rotating in the $\hat{n}\times\mathbf{k}$ sense, while the -**k** mode has the field rotating in the $-\hat{n}\times\mathbf{k}$ sense. The fields in the vacuum rotate in the same sense in space as the fields in the medium, although the ellipticity of the rotation is different in the two media. In contrast, the normal features of surface polaritons in nongyromagnetic media are that the polarizations of the magnetic fields exhibit $\hat{n}\times\mathbf{k}$ rotation in the medium and $-\hat{n}\times\mathbf{k}$ rotation in the vacuum (Fig. 2b) and propagation is independent of the sign of **k**.

The discussion of the $k \to \pm\infty$ limit and Eq. (3.17) can now be treated in detail. Using Eqs. (3.18a) and (3.18b), we find

$$\mu_{\rho\pm}(\omega) = \mu_B\left(1 + \frac{\omega_s}{\omega_0 \mp \omega}\right) \tag{3.19}$$

where the upper signs all go together. The component $\mu_{\rho+}(\omega)$ goes with the resonant left circular rotation of **M** (and **H**) about $\mathbf{H}_0$; that is, given by $d\mathbf{M}/dt = \gamma\mu_0\mathbf{M}\times\mathbf{H}_0$. For the $k \to +\infty$ limit, Eqs. (3.19) and (3.17) give us the asymptotic surface polariton frequency

$$\omega_{sp} = \gamma\left(\frac{\mu_0 H_0 + \mu_B B_s}{1 + \mu_B}\right) \qquad B_s = \mu_0\left(H_0 + \frac{M_0}{\mu_B}\right) \tag{3.20}$$

When $\mu_B$ = 1, this reduces to $\omega_{sp} = \frac{1}{2}\gamma(B_0 + \mu_0 H_0)$ where $B_0 = \mu_0(H_0 + M_0)$. This is the frequency of the unretarded surface magnon mode discussed by Damon and Eshbach.[18] The frequency is the same in both semi-infinite and slab geometries since the penetration depth goes to zero as $k \to \infty$. This mode is $\hat{n}\times k$ circularly polarized and is a normal surface polariton solution in that it is asymptotic to the actual surface magnon excitation frequency $\omega_{sp}$. In the vacuum the magnetic fields are $\hat{n}\times k$ elliptically polarized (Fig. 2a).

If we consider the other limit $k \to -\infty$, we immediately see from Eq. (3.18c) that $\mu_{\rho-}(\omega)$ is always positive. This means that the condition $\mu_{\rho-}(\omega) = -1$ is never satisfied and no $k \to -\infty$ surface polariton mode can exist. It is just this difference between $\mu_+(\omega)$ and $\mu_-(\omega)$ which gives rise to nonreciprocal propagation.

Let us for the moment consider only the +k solution. Its existence does not depend on the value of $\mu_B$ and as we have already seen, $\omega_{sp}$ is simplest when $\mu_B = 1$. For this direction the lower frequency in the dispersion relation is given by

$$\omega_V = \gamma\left(\mu_0 B_0 H_0\right)^{1/2} \tag{3.21a}$$

which corresponds to the frequency in this configuration at which $\mu_{xx} \to 0$ and $\mu_V(\omega) \to -\infty$. It occurs at a wavevector given by

$$k_V = \frac{\omega_V}{c}\left(\frac{B_0}{B_0 - \mu_0 H_0}\right)^{1/2} > \frac{\omega_V}{c} \tag{3.21b}$$

$\omega_V$ is just the point as we decrease frequency where bulk polariton propagation begins. The magnetic field in the medium of the surface polariton is $\hat{n}\times k$ elliptically polarized in the intermediate frequency range ($\omega_V < \omega < \omega_{sp}$) and becomes linearly polarized ($\mathbf{H} \parallel \hat{n}$) at $\omega_V$. For the case $H_0 = 0$, the Voigt frequency $\omega_V$ is equal to zero, and the dispersion relations start smoothly from the origin.

The photon content of this mode, which can be determined most readily from the magnitudes of the electric field of the TE modes, goes to zero at both limiting frequencies $\omega_V$ and $\omega_{sp}$. At these frequencies the surface polariton consists of pure surface magnetostatic fields with no photon content ($\mathbf{E} = 0$). At these frequencies the Poynting vector $\mathbf{S} = (c/4\pi)(\mathbf{E}\times\mathbf{H})$ shows that there can be no energy flow. This is further confirmed by the fact that the group velocity ($v_g = \partial\omega/\partial k$) is equal to zero. The penetration depth ($\lambda_2 = 2\pi/\alpha_2$) can be related to the wavelength ($\lambda = 2\pi/k$) of the surface polariton through Eq. (3.5). We find that $\lambda_2 \leq \lambda$ with the equal sign occurring when $k \to \infty$. In the same way, $\lambda_1 = 2\pi/\alpha_1 \geq \lambda$ as we expect. This result holds for the -k mode also. We note that in the large wavevector, small wavelength limit, the penetration depth is of the same order of magnitude as the wavelength.

The -k surface polariton mode only exists when $\mu_B > 1$. As can be seen in Fig. 3, there is no $k \to -\infty$ solution. The -k mode has an appreciable photon content at all frequencies. The dispersion relation of the surface polariton is sandwiched between the curves $\omega = ck$ and $\omega = ck\mu_B^{-1/2}$. It begins on the dispersion curve for electromagnetic radiation in the vacuum ($\omega = ck$) at the frequency $\omega_-$ where $\mu_{xx}(\omega_-) = 1$ and terminates on the upper branch of the bulk polariton dispersion curve. The frequency $\omega_-$ is given by

$$\omega_- = \gamma\left(\frac{\mu_0\mu_B H_0 B_0 - \mu_0^2 H_0^2}{\mu_B - 1}\right)^{1/2} \tag{3.22}$$

We note that it is strongly dependent on $\mu_B$. We also note that when $\mu_B \to 1$, the mode never begins ($\omega_- \to \infty$) and the two limiting curves ($\omega = ck$ and $\omega = ck\mu_B^{-1/2}$) coalesce into one. This surface mode has a further distinction that it extends into the frequency region ($\omega \geq \gamma B_0$) where bulk modes also exist.

The polarization of the -k mode in the medium is $-\hat{n}\times k$ elliptical throughout. This means that the magnetic field $\mathbf{H}$ of both the +k and -k modes rotates in the same sense about the static magnetic field (Fig. 2a), corresponding to the sense for which magnon is resonant in the magnetic field. The $\hat{n}\times k$ rotation of the +k mode corresponds to both the precession of the magnetization in the static magnetic field and the natural rotation of the magnetic field of a surface polariton. The $-\hat{n}\times k$ rotation of the -k mode, while resonant with the static field, opposes the normal $\hat{n}\times\mathbf{k}$ rotation of a surface polariton and must be driven by the associated photon fields. This photon content is appreciable and is responsible for the -k solution occurring at wave vectors closer to the vacuum photon than the more easily driven +k mode.

## 4. GYRODIELECTRIC TENSOR

In this section we consider a gyrodielectric medium consisting of a solid containing free charge carriers (e.g., a metal or semiconductor) in the presence of a constant, uniform external magnetic field $\mathbf{H}_0$. In essence, we are dealing with a magnetoplasma. If an electromagnetic wave is impressed on the system, the electric field $\mathbf{E}$ of the wave exerts a force on each free charge carrier which then undergoes displacement. In calculating the response of the system to the electric field, we shall assume that only one type of charge carrier is present and that the charge carriers have an isotropic effective mass $m^*$.

We shall follow the Drude procedure in which all the charge carriers are taken to have the mean velocity $\mathbf{v}$. The Lorentz equation of motion for a charge carrier has the form

$$m^* \frac{d\mathbf{v}}{dt} + \frac{m^*}{\tau}\mathbf{v} = -e\left(\mathbf{E} + \frac{1}{c}\mathbf{v}\times\mathbf{H}\right) \tag{4.1}$$

where $\mathbf{H}$ is the total magnetic field strength, $\tau$ is the carrier scattering time and we have assumed that the current carriers are electrons. For not too high intensity of the electromagnetic radiation, the magnetic field of the radiation is negligible compared to the uniform external magnetic field, and we can take $\mathbf{H} = \mathbf{H}_0$ in Eq. (4.1).

If we let $\mathbf{H}_0 = (0, H_0, 0)$ as in section 2 and set

$$\mathbf{v} = \mathbf{v}_0 e^{-i\omega t} \tag{4.2a}$$

$$\mathbf{E} = \mathbf{E}_0 e^{-i\omega t} \tag{4.2b}$$

the equations of motion, Eq. (4.1), can be written in component form as

$$\left(-i\omega + \frac{1}{\tau}\right)v_x = -\frac{e}{m^*}\left(E_x + \frac{1}{c}H_0 v_z\right) \tag{4.3a}$$

$$\left(-i\omega + \frac{1}{\tau}\right)v_y = -\frac{e}{m^*}E_y \tag{4.3b}$$

$$\left(-i\omega + \frac{1}{\tau}\right)v_z = -\frac{e}{m^*}\left(E_z - \frac{1}{c}H_0 v_x\right) \tag{4.3c}$$

Solving Eqs. (4.3), we obtain

$$v_x = \frac{\omega + i\nu}{[(\omega + i\nu)^2 - \omega_c^2]}\frac{e}{m^*}\left(-iE_x - \frac{\omega_c}{\omega + i\nu}E_z\right) \tag{4.4a}$$

$$v_y = -\frac{e}{m^*(\omega + i\nu)}E_y \tag{4.4b}$$

$$v_z = \frac{\omega + i\nu}{[(\omega + i\nu)^2 - \omega_c^2]}\frac{e}{m^*}\left(-\frac{\omega_c}{\omega + i\nu}E_x - iE_z\right) \tag{4.4c}$$

where $\nu = 1/\tau$ and $\omega_c = eH_0/(m^* c)$ is the cyclotron frequency. The current density $\mathbf{j}$ is related to $\mathbf{v}$ by

$$\mathbf{j} = -ne\mathbf{v} \tag{4.5}$$

where the current carriers are electrons. From Ohm's law, on the other hand,

$$\mathbf{j} = \underline{\underline{\sigma}}\mathbf{E} \tag{4.6}$$

when $\underline{\underline{\sigma}}$ is the conductivity tensor. The dielectric tensor $\underline{\underline{\epsilon}}$ is given by

$$\underline{\underline{\epsilon}} = \epsilon_{\infty}\underline{\underline{I}} + \frac{i}{\omega}\underline{\underline{\sigma}} \tag{4.7}$$

where $\epsilon_{\infty}$ is the background dielectric constant due to interband electronic transitions. Substitution of Eqs. (4.4) into Eq. (4.5) and comparison of the results with Eqs. (4.6) and (4.7) yields the elements of the dielectric tensor:

$$\epsilon_{xx} = \epsilon_{zz} = \epsilon_{\infty}\left\{1 - \frac{(\omega + i\nu)\omega_p^2}{\omega[(\omega + i\nu)^2 - \omega_c^2]}\right\} \tag{4.8a}$$

$$\epsilon_{xz} = -\epsilon_{zx} = i\epsilon_{\infty}\frac{\omega_c\omega_p^2}{\omega[(\omega + i\nu)^2 - \omega_c^2]} \tag{4.8b}$$

$$\epsilon_{xy} = \epsilon_{yx} = \epsilon_{yz} = \epsilon_{zy} = 0 \tag{4.8c}$$

$$\epsilon_{yy} = \epsilon_{\infty}\left\{1 - \frac{\omega_p^2}{\omega(\omega + i\nu)}\right\} \tag{4.8d}$$

where

$$\omega_p^2 = \frac{ne^2}{\epsilon_{\infty}m^*} \tag{4.9}$$

is the square of the plasma frequency. We shall refer to the dielectric tensor with elements specified by Eq. (4.8) as the gyrodielectric tensor.

## 5. SURFACE MAGNETOPLASMON POLARITONS

The gyrodielectric tensor derived in the preceding section can be used to investigate the dispersion relation for surface magnetoplasmon polaritons in metals or doped semiconductors.[19-21] The geometry is the same as that shown in Fig. 1, except that in the region $z < 0$, the linear ferromagnetic medium is replaced by a linear gyrodielectric medium. The external magnetic field is again taken to be in the -y direction.

We first consider TM modes and focus our attention on the electric vector **E** which, after eliminating the magnetic field from Maxwell's curl equations, satisfies the equation

$$\nabla^2\mathbf{E} - \nabla(\nabla\cdot\mathbf{E}) = \epsilon_0\mu_0\underline{\underline{\epsilon}}\cdot\frac{\partial^2\mathbf{E}}{\partial t^2} \tag{5.1}$$

For gyrodielectric media with elements of the dielectric tensor $\underline{\underline{\epsilon}}$ given by Eqs. (4.8) and the displacement **D** given by $\mathbf{D} = \epsilon_0\underline{\underline{\epsilon}}\cdot\mathbf{E}$, the divergence equation

$$\nabla\cdot\mathbf{D} = 0 \tag{5.2}$$

does not lead to the conclusion $\nabla\cdot\mathbf{E} = 0$. Hence, the second term on the left-hand side of Eq. (5.1) must be retained.

For a TM wave propagating in the x-direction with wavenumber k and frequency $\omega$, we can write the solution to Eq. (5.1) in the form

$$\mathbf{E} = (E_{2x}, 0, E_{2z})e^{ikx}e^{\alpha_2 z}e^{-i\omega t} \tag{5.3}$$

where $\alpha_2$ is the decay constant. Substituting Eq. (5.3) into Eq. (5.1), we obtain

$$(\alpha_2^2 + k_0^2\epsilon_{xx})E_x - (ik\alpha_2 - k_0^2\epsilon_{xz})E_z = 0 \tag{5.4a}$$

$$(ik\alpha_2 - k_0^2\epsilon_{zx})E_x + (k^2 - k_0^2\epsilon_{zz})E_z = 0 \tag{5.4b}$$

The condition for a nontrivial solution to Eqs. (5.4) is obtained by setting the determinant of the coefficients of $E_x$ and $E_z$ to zero to give the result

$$\alpha_2^2 = k^2 - k_0^2\epsilon_V \tag{5.5}$$

where

$$\epsilon_V = \epsilon_{xx} + \frac{\epsilon_{xz}^2}{\epsilon_{xx}} \tag{5.6}$$

is the Voigt dielectric constant.

In the vacuum region, $z \geq 0$, the dielectric tensor is simply $\underline{\underline{\epsilon}} = \underline{\underline{I}}$, and the divergence equation

$$\nabla \cdot \mathbf{D} = 0 \tag{5.7}$$

reduces to

$$\nabla \cdot \mathbf{E} = 0 \tag{5.8}$$

Equation (5.1) then becomes the wave equation

$$\nabla^2 \mathbf{E} - \epsilon_0 \mu_0 \ddot{\mathbf{E}} = 0 \tag{5.9}$$

A solution to Eq. (5.9) localized at the surface $z = 0$ can be written in the form

$$\mathbf{E} = (E_{1x}, 0, E_{1z})e^{ikx}e^{-\alpha_1 z}e^{-i\omega t} \tag{5.10}$$

where the decay constant $\alpha_1$ is given by Eq. (3.11).

The next task is to satisfy the electromagnetic boundary conditions at the surface, which we take to be

$$E_{1x} = E_{2x} \tag{5.11a}$$

$$D_{1z} = D_{2z} \tag{5.11b}$$

We use the divergence equation, Eq. (5.2), to establish the relationship between $E_{1x}$ and $E_{1z}$ and between $E_{2x}$ and $E_{2z}$. The result is

$$E_{2z} = \frac{\alpha_2\epsilon_{xz} - ik\epsilon_{xx}}{\alpha_2\epsilon_{xx} + ik\epsilon_{xz}} E_{2x} \tag{5.12a}$$

$$E_{1z} = \frac{ik}{\alpha_1} E_{1x} \tag{5.12b}$$

Eliminating $D_{1z}$ and $D_{2z}$ from Eq. (5.11b) in favor of $E_{1z}$ and $E_{2z}$ using the appropriate forms of the relation $\mathbf{D} = \epsilon_0\underline{\underline{\epsilon}} \cdot \mathbf{E}$, we find that

$$\epsilon_{zx}E_{2x} + \epsilon_{zz}E_{2z} = E_{1z} \tag{5.13}$$

We now substitute Eqs. (5.12) into Eq. (5.13) and use Eq. (5.11a) to yield the result

$$\alpha_2 + ik\frac{\epsilon_{xz}}{\epsilon_{xx}} + \alpha_1\epsilon_V = 0 \tag{5.14}$$

which is the dispersion relation for surface magnetoplasmon polaritons.

One notes that this dispersion relation is non-reciprocal, i.e. positive and negative values of the wavenumber k are not equivalent. In the unretarded limit, $\alpha_1 \rightarrow |k|$, $\alpha_2 \rightarrow |k|$, and the dispersion relation reduces to

$$1 + \epsilon_{xx} - i\epsilon_{xz}\,\mathrm{sgn}(k) = 0 \tag{5.15}$$

This is a cubic equation in $\omega$. For small $\omega_c/\omega_p$, one obtains the approximate value for the unretarded surface magnetoplasmon frequency

$$\omega_{sm} \simeq \omega_{sp} + \frac{1}{2}\omega_c\,\mathrm{sgn}(k) \tag{5.16}$$

where $\omega_{sp}$ is the zero field non-retarded surface plasmon frequency given by

$$\omega_{sp} = \omega_p\left(1 + \frac{1}{\epsilon_\infty}\right)^{-1/2} \tag{5.17}$$

Calculations of the dispersion relation for this configuration have been carried out for n-InSb. The case $k > 0$ with $\omega_c/\omega_p = 0.5$ is plotted in Fig. 4. A very interesting feature is immediately apparent, namely the dispersion curve consists of two parts with a gap between them. The low portion starts from the origin, rises just to the right of the light line, $\omega = kc$, bends over, and terminates when the curve intersects the dispersion curve for bulk magnetoplasmons (bulk polaritons) defined by $\alpha_2^2 = k^2 - \omega^2\epsilon_V/c^2 = 0$. The upper branch starts on the line defined by $\epsilon_{xx} = 0$, rises, and then approaches the asymptotic frequency for unretarded surface magnetoplasmons defined by the equation $1 + \epsilon_{xx} - i\epsilon_{xz} = 0$. In the large wave vector, unretarded limit, the electric vector of the upper branch executes a circular motion in the sagittal plane. At the small wavevector extremity of this branch, the electric vector is plane polarized perpendicular to the surface. Note that this branch stops before it reaches the light line.

The reduced wave vector at which the upper branch starts is specified by the equation[19,20]

$$\zeta_s^2 = \frac{1 + \eta^2}{1 - (1 + \eta^2)/(\epsilon_\infty^2\eta^2)} \tag{5.18}$$

where $\zeta = |k|c/\omega_p$ and $\eta = \omega_c/\omega_p$. If $\zeta_s^2$ is to be finite and positive, then $\epsilon_\infty$ and $\eta$ must satisfy the inequality

$$\epsilon_\infty > \frac{(1 + \eta^2)^{1/2}}{\eta} = \frac{\left(\omega_c^2 + \omega_p^2\right)^{1/2}}{\omega_c} \tag{5.19}$$

That a gap can exist in the dispersion curve is evident from a consideration of Fig. 5, where $\epsilon_V$ is plotted against $\omega$. There is a region to the left of the line $\epsilon_{xx} = 0$ where $\epsilon_V$ is large and positive, so that $\alpha^2$ is negative for finite wave vectors and no surface wave exists. If the unretarded surface magnetoplasmon frequency lies above the line $\epsilon_{xx} = 0$, as it does for InSb with $\eta = 0.5$, then a surface wave exists both above and well below this line, and a gap must exist just below this line. To have a surface-wave branch above the line $\epsilon_{xx} = 0$, the ratio $\omega_c/\omega_p$ must exceed a critical value specified by Eq. (5.19). For InSb, this critical value is 0.064.

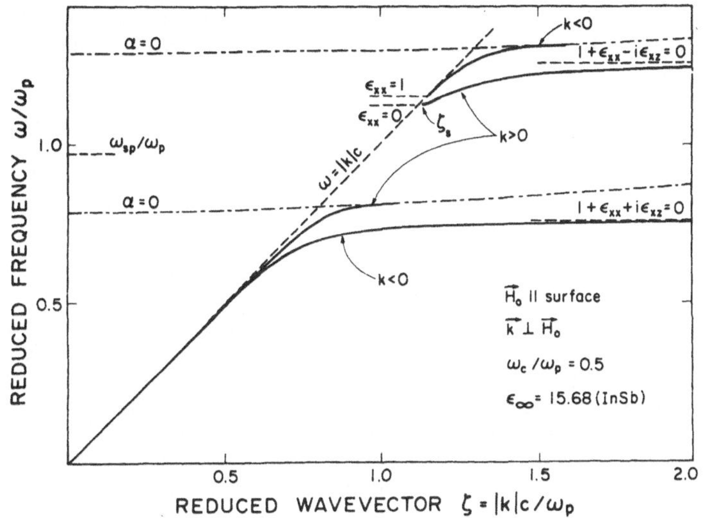


Fig. 4 Dispersion curves for surface magnetoplasmon polaritons in n-type InSb. The dashed lines labelled $\alpha = 0$ are bulk polariton dispersion curves.

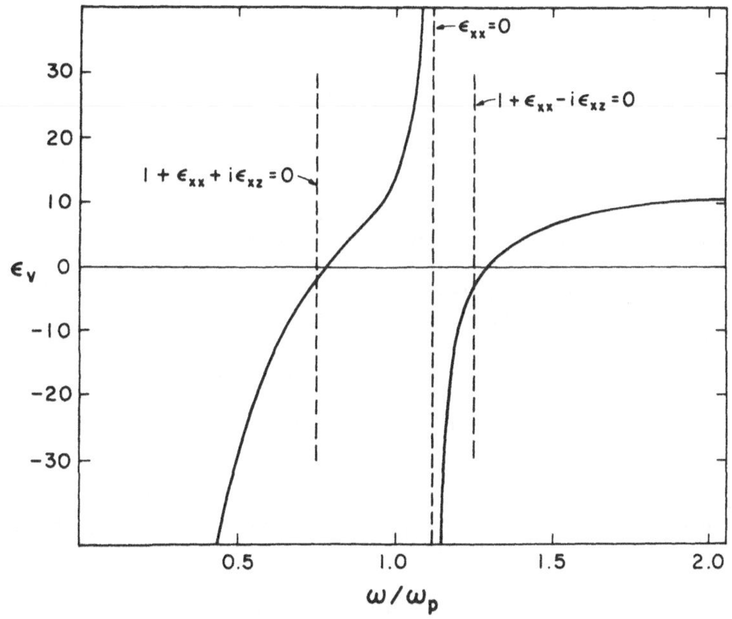


Fig. 5 Voigt dielectric constant $\epsilon_V$ versus frequency.

For a value of $\epsilon_\infty$ just below the right-hand side of Eq. (5.19), the upper branch lies below the line $\epsilon_{xx} = 0$ and to the right of the lower bulk dispersion curve. As $\epsilon_\infty$ decreases, the gap rapidly closes, and only a single branch remains.

Turning now to the case of $k < 0$, we plot in Fig. 4 the surface-polariton dispersion curve for n-type InSb with $\eta = 0.5$ and $k < 0$. It is clear that the situation is qualitatively different from that where $k > 0$. We now have a complete lower branch running from the origin out to the asymptotic value specified by the equation $1 + \epsilon_{xx} + i\epsilon_{xz} = 0$. In addition, however, we have an upper branch which starts at the light line, where $\epsilon_{xx} = 1$, rises to the right of the light line, and ends when it meets the upper bulk-polariton dispersion curve, $\alpha = 0$. The upper branch for $k < 0$ exists whenever $\epsilon_\infty > 1$ and $\eta > 0$. It may be remarked that, for both $k > 0$ and $k < 0$, there are frequencies at which both a bulk wave and a surface wave can propagate, a situation which does not exist in the absence of a magnetic field.

## 6. NONLINEAR MAGNETODYNAMIC WAVES ON MAGNETIC MATERIALS

The purpose of this section is to examine the behaviour of surface nonlinear magnetodynamic (polariton) waves that are propagating along the interface between a weakly, semi-infinite, nonlinear substrate and the kind of dominant strongly nonlinear magnetic, semi-infinite, cladding that was defined above. The assumption of such a guiding system permits the neglect of the weak, intrinsic, nonlinearity, compared to the nonlinearity in the cladding, and envisages self-focusing in the frequency ranges for which no linear eigenvalue exists. Since the expected propagation frequency band is quite narrow only third-order nonlinearity needs to be considered with all harmonic generation being ignored because of lack of phase-matching and the fact that they are weak, driven, harmonic waves lying outside the linear eigenfrequency band. Although the mathematical infrastructure of the application reported here is unique, there will be a similarity of attack between the methods employed and some previous work on non-magnetic materials. This is also true of many linear problems, however.

### 6.1 TM Waves

The guiding structure to be considered consists of a semi-infinite ferri(ferro)magnetic substrate, such as YIG (yttrium iron garnet), and a semi-infinite nonlinear magnetic cladding in contact everywhere on the $z = 0$ plane as shown in Fig. 1. The substrate, even if it is YIG, will be assumed to be operating in a frequency window that makes it appear to be ferromagnetic. The appropriate form of the permeability tensor of the magnetic substrate medium, for this geometry, is given by Eq. (2.15). We introduce a factor $\mu_B$ into Eq. (2.15) which can be interpreted as the background permeability caused by other magnetic dipole excitations, such as optical magnons. It is interesting that because $\mu_B$ is close to unity it is usually taken as unity in practical device designs without apparent penalty. $\mu_B \simeq 1.25$ in practical frequency ranges for YIG, however, and the background dielectric constant is also not equal to unity although it is sometimes assumed to be so. Both $\mu_B$ and the background dielectric constant can lead to branches in the dispersion relationship that depend upon their deviation from unity.

We seek a solution to Maxwell's equations corresponding to a TM electromagnetic wave propagating along the x-axis in the xz-plane with wavenumber k and angular frequency $\omega$. The electric and magnetic vectors of the electromagnetic field then take the form

$$\mathbf{E} = [E_x(z), 0, E_z(z)]e^{i(kx-\omega t)} \tag{6.1}$$

$$\mathbf{H} = [0, H_y(z), 0]e^{i(kx-\omega t)} \tag{6.2}$$

The nonlinear magnetic cladding is assumed to be isotropic with a permeability given by

$$\mu^{NL} = \mu_L + \alpha\left|\mathbf{H}\right|^2 = \mu_L + \alpha H_y^2 \tag{6.3}$$

where $\mu_L$ is the linear part of the permeability and $\alpha$ is the nonlinear coefficient. This expression arises from an expansion of the permeability about an applied static field $H_0$, and terms that could lead to harmonic generation are neglected. Hence, $\mathbf{H}$ is the ac magnetic field carried by the TM wave. $H_y$ is real because only stationary, non-radiating waves will be considered.[3]

Let us first investigate the nonlinear medium for which Maxwell's curl equations have the form

$$\nabla\times\mathbf{E} = i\omega\mu_0\mu^{NL}\mathbf{H} \tag{6.4}$$

$$\nabla\times\mathbf{H} = -i\omega\epsilon_0\epsilon_1\mathbf{E} \tag{6.5}$$

where $\epsilon_1$ is the relative dielectric constant of the nonlinear medium. Substituting Eqs. (6.1) and (6.2) into Eqs. (6.4) and (6.5) yields

$$ikE_z - \frac{\partial E_x}{\partial z} = -i\omega\mu_0\mu^{NL}H_y \tag{6.6}$$

$$\frac{\partial H_y}{\partial z} = i\omega\epsilon_0\epsilon_1 E_x \tag{6.7a}$$

$$kH_y = -\omega\epsilon_0\epsilon_1 E_z \tag{6.7b}$$

Eliminating $E_x$ and $E_z$ from Eq. (6.6) using Eqs. (6.7) gives the result

$$\frac{\partial^2 H_y}{\partial z^2} - k^2 H_y + k_0^2\epsilon_1\mu^{NL}H_y = 0 \tag{6.8}$$

where $k_0^2 = \omega^2/c^2 = \omega^2\epsilon_0\mu_0$, and $\epsilon_0$ and $\mu_0$ are the dielectric permittivity and magnetic permeability of free space, respectively.

If we eliminate $\mu^{NL}$ from Eq. (6.8) using Eq. (6.3), we obtain

$$\frac{\partial^2 H_y}{\partial z^2} - k^2 H_y + k_0^2\epsilon_1(\mu_L + \alpha H_y^2)H_y = 0 \tag{6.9}$$

We can rewrite Eq. (6.9) as

$$\frac{\partial^2 H_y}{\partial z^2} - (k_1^2 - k_0^2\epsilon_1\alpha H_y^2)H_y = 0 \tag{6.10}$$

where

$$k_1^2 = k^2 - k_0^2\epsilon_1\mu_L \tag{6.11}$$

Equation (6.10) is a nonlinear differential equation that can be solved exactly.[1-8] We obtain a first integral by multiplying through by $\partial H_y/\partial z$ and integrating over z,

$$\left(\frac{\partial H_y}{\partial z}\right)^2 - \left(k_1^2 - \frac{1}{2}k_0^2\epsilon_1\alpha H_y^2\right)H_y^2 = C \tag{6.12}$$

where C is a constant of integration. For the surface waves under investigation here, $H_y \to 0$ and $\partial H_y/\partial z \to 0$ as $z \to \infty$ so that $C = 0$. This conclusion also implies that there is no radiation leaving the surface or returning to it.

The solution to Eq. (6.12) has the form well known for TE waves propagating along the boundary between two nonmagnetic dielectric media.[4,5] It is

$$H_y(z) = \frac{1}{k_0}\left(\frac{2}{\alpha\epsilon_1}\right)^{1/2}\frac{k_1}{\cosh[k_1(z - z_0)]} \tag{6.13}$$

where $z_0$ is a constant of integration that defines the position of a self-focused peak in $H_y$.

We now consider the linear ferri(ferro)magnetic medium for which the permeability tensor is given by Eq. (2.15) and obtain from the curl equations, together with Eqs. (6.1) and (6.2), the following equation for $H_y$

$$\frac{\partial^2 H_y}{\partial z^2} - k_2^2 H_y = 0 \tag{6.14}$$

where

$$k_2^2 = k^2 - k_0^2 \epsilon_2 \mu_B \tag{6.15}$$

and $\epsilon_2$ is now the relative dielectric constant of the ferrimagnetic medium. The solution to Eq. (6.14), corresponding to a surface wave localized at the plane z = 0, can be written as

$$H_y(z) = H_L e^{k_2 z} \tag{6.16}$$

where $H_L$ is the amplitude of $H_y$ at z = 0.

The electromagnetic boundary conditions that express the continuity of the tangential components of **E** and **H** and the normal component of the electric displacement **D** at the boundary z = 0, are

$$E_{x1}(0) = E_{x2}(0) \tag{6.17}$$

$$H_{y1}(0) = H_{y2}(0) \tag{6.18}$$

$$D_{z1}(0) = D_{z2}(0) \tag{6.19}$$

where, according to Fig. 1, the subscripts 1 and 2 designate the nonlinear cladding and the linear substrate medium respectively. Substituting Eqs. (6.13) and (6.16) into Eq. (6.18) yields

$$H_L = \frac{k_1}{k_0}\left(\frac{2}{\alpha\epsilon_1}\right)^{1/2} \frac{1}{\cosh(k_1 z_0)} \tag{6.20}$$

With the aid of Eq. (6.7a), and using Eqs. (6.13) and (6.16), we can transform Eq. (6.17) to the form

$$\frac{k_1^2}{k_0\epsilon_1}\left(\frac{2}{\alpha\epsilon_1}\right)^{1/2} \frac{\sinh(k_1 z_0)}{\cosh^2(k_1 z_0)} = \frac{k_2}{\epsilon_2} H_L \tag{6.21}$$

Eliminating $H_L$ from Eqs. (6.10) and (6.21) gives a dispersion relation for the nonlinear TM surface waves in the form

$$\tanh(k_1 z_0) = \frac{k_2\epsilon_1}{k_1\epsilon_2} \tag{6.22}$$

The dispersion relation can now be used to obtain bounds on the range of allowed values of the wavenumber k. Since we assume that the dielectric constants $\epsilon_1$ and $\epsilon_2$ are positive and since the decay constants $k_1$ and $k_2$ must also be positive for a surface wave, the right-hand side of Eq. (6.22) is positive. The positive values of the hyperbolic tangent on the left-hand side, therefore lie in the range from zero to unity, so we must have

$$0 < \frac{k_2\epsilon_1}{k_1\epsilon_2} \leq 1 \tag{6.23}$$

Squaring each side of this inequality, using Eqs. (6.13) and (6.15), and rearranging, leads to the result

$$k^2 < \frac{k_0^2\epsilon_1\epsilon_2(\epsilon_1\mu_B - \epsilon_2\mu_L)}{\epsilon_1^2 - \epsilon_2^2} \tag{6.24}$$

which defines an upper cut-off on the value of the wavenumber k that approaches infinity as $\epsilon_1 \rightarrow \epsilon_2$. Actually as $z_0 \rightarrow \infty$, $\tanh(k_1z_0) \rightarrow 1$ so that the cut-off corresponds to a self-focused peak in the field moving out to infinity. It will be shown later that this cut-off leads to a corresponding cut-off on the power flow. The requirements that $k_1^2 > 0$ and $k_2^2 > 0$ for a surface wave together with Eqs. (6.11) and (6.15) give the relations

$$\begin{aligned} k^2 &> k_0^2\epsilon_1\mu_L \\ k^2 &> k_0^2\epsilon_2\mu_B \end{aligned} \tag{6.25}$$

so we have a lower cutoff on k as well. We must also note from Eq. (6.25) that, if $\epsilon_1 > \epsilon_2$, we must have $\epsilon_1\mu_B > \epsilon_2\mu_L$.

Yet another inequality can be obtained by eliminating $z_0$ from the boundary conditions, Eqs. (6.20) and (6.21), to yield the result

$$\epsilon_2^2k_1^2 - \epsilon_1^2k_2^2 = \frac{1}{2}\epsilon_2^2k_0^2\alpha\epsilon_1H_L^2 \tag{6.26}$$

Eliminating $k_1^2$ and $k_2^2$, with the aid of Eqs. (6.11) and (6.15), and solving for $k^2$ gives

$$k^2 = \frac{k_0^2\epsilon_1\epsilon_2}{\epsilon_1^2 - \epsilon_2^2}\left(\epsilon_1\mu_B - \epsilon_2\mu_L - \frac{1}{2}\epsilon_2\alpha H_L^2\right) \tag{6.27}$$

Again taking $\epsilon_1 > \epsilon_2$ and noting that $k^2$ must be positive, we obtain the inequality

$$\frac{1}{2}\epsilon_2\alpha H_L^2 < \epsilon_1\mu_B - \epsilon_2\mu_L \tag{6.28}$$

Combining the inequality $k^2 > \epsilon_2\mu_B$, Eq. (6.25), with Eq. (6.27) leads to the result

$$\frac{1}{2}\epsilon_1\alpha H_L^2 < \epsilon_2\mu_B - \epsilon_1\mu_L \tag{6.29}$$

Thus, depending on the material parameters, there is a maximum value of the magnetic field $H_L$.

We now turn our attention to the power flow in the direction of propagation (x-direction) given by

$$P = \frac{1}{2}\int(E\times H^*)_x dz = -\frac{1}{2}\int E_z H_y^* dz \tag{6.30}$$

Eliminating $E_z$ by means of Eq. (6.7b) we obtain the power flow in the nonlinear medium in the form

$$P^{NL} = \frac{k}{2\omega\epsilon_0\epsilon_1}\int_0^\infty H_y^2 dz$$

which, with the aid of Eq. (6.13), becomes

$$P^{NL} = \frac{kk_1^2}{\alpha\omega k_0^2\epsilon_0\epsilon_1^2}\int_0^\infty \frac{dz}{\cosh^2[k_1(z-z_0)]} = \frac{kk_1^2}{\alpha\omega k_0^2\epsilon_0\epsilon_1^2}\int_{-z_0}^\infty \frac{dz'}{\cosh^2(k_1z')} \tag{6.31}$$

where $z' = z - z_0$. The quantity $z_0$ is the position of maximum power density and when it moves to infinity, the power flow reaches

$$P^{NL} = \frac{kk_1^2}{\alpha\omega k_0^2\epsilon_0\epsilon_1^2}\int_{-\infty}^{\infty}\frac{dz'}{\cosh^2(k_1z')} = \frac{kk_1}{\alpha\omega k_0^2\epsilon_0\epsilon_1^2} \tag{6.32}$$

Substituting the maximum value of k, as specified by Eq. (6.24), into Eq. (6.32), gives the following cut-off power flow

$$P_c^{NL} = \frac{2k_1}{\alpha\omega k_0\epsilon_0\epsilon_1^2}\left(\frac{\epsilon_1\epsilon_2(\epsilon_1\mu_2 - \epsilon_2\mu_1)}{\epsilon_1^2 - \epsilon_2^2}\right)^{1/2} \tag{6.33}$$

The power flow in the linear medium is given by

$$P^L = \frac{k}{2\omega\epsilon_0\epsilon_2}\int_{-\infty}^{0}H_y^2dz \tag{6.34}$$

If we substitute Eq. (6.16) into (6.34) and integrate, we obtain

$$P^L = \frac{k}{4\omega\epsilon_0\epsilon_2k_2}H_L^2$$

and eliminating $H_L$, with the aid of (6.20), gives the result

$$P^L = \frac{kk_1^2}{2\alpha\omega k_0^2\epsilon_0\epsilon_1\epsilon_2k_2\cosh^2(k_1z_0)} \tag{6.35}$$

from which it can be seen that when $z_0 \to \infty$, $P^L \to 0$ and all of the power flow is in the nonlinear medium.

## 6.2 TE Waves

In this section we consider the same structure as that considered in section 6.1 for TM waves. The electric and magnetic field vectors of the electromagnetic field for the TE waves take the form

$$\mathbf{E} = [0, E_y(z), 0]e^{i(kx-\omega t)} \tag{6.36}$$

$$\mathbf{H} = [H_x(z), 0, H_z(z)]e^{i(kx-\omega t)} \tag{6.37}$$

Substitution of Eqs. (6.36) and (6.37) into (6.4) and (6.5) for the nonlinear medium yields

$$\frac{\partial E_y}{\partial y} = -i\omega\mu_0\mu^{NL}H_x \tag{6.38a}$$

$$kE_y = \omega\mu_0\mu^{NL}H_z \tag{6.38b}$$

$$\frac{\partial H_x}{\partial z} - ikH_z = -i\omega\epsilon_0\epsilon_1E_y \tag{6.39}$$

In order to make contact with previous work on linear TE waves, we eliminate $E_y$ rather than $H_x$ and $H_z$ from Eqs. (6.38) and (6.39) and obtain the equations

$$\frac{\partial^2 H_x}{\partial z^2} - ik\frac{\partial H_z}{\partial z} = -k_0^2\epsilon_1\mu^{NL}H_x \tag{6.40}$$

$$\frac{\partial H_x}{\partial z} = \frac{i}{k}(k^2 - k_0^2\epsilon_1\mu^{NL})H_z \tag{6.41}$$

Since there is a $\pi/2$ phase difference between $H_x$ and $H_z$ for TE waves, we can write

$$H_x = h_x \qquad H_z = ih_z$$

and express Eqs. (6.40) and (6.41) in the alternative forms

$$\frac{\partial^2 h_x}{\partial z^2} + k\frac{\partial h_z}{\partial z} = -k_0^2\epsilon_1\mu^{NL}h_x \tag{6.42}$$

$$\frac{\partial h_x}{\partial z} = -\frac{1}{k}(k^2 - k_0^2\epsilon_1\mu^{NL})h_z \tag{6.43}$$

The first integral is obtained from Eq. (6.42) by multiplying it by $\partial h_x/\partial z$ and utilizing Eq. (6.43) to give

$$\frac{\partial}{\partial z}\left(\frac{\partial h_x}{\partial z}\right)^2 - k^2\frac{\partial h_z^2}{\partial z} + k_0^2\epsilon_1\mu_L\frac{\partial h^2}{\partial z} + \frac{1}{2}k_0^2\epsilon_1\alpha\frac{\partial h^4}{\partial z} = 0 \tag{6.44}$$

where

$$h^2 = h_x^2 + h_z^2 \tag{6.45}$$

Substituting Eq. (6.43) into Eq. (6.44) and integrating with respect to z gives

$$\frac{1}{k^2}\left(k^2 - k_0^2\epsilon_1\mu^{NL}\right)^2 h_z^2 - k^2h_z^2 + k_0^2\epsilon_1\mu_L h^2 + \frac{1}{2}k_0^2\epsilon_1\alpha h^4 = C \tag{6.46}$$

where C is a constant of integration. The requirement that the magnetic field components ($h_x$, $h_z$) describe a surface wave localized at the interface $z = 0$ yields the result $C = 0$. Letting $n^2 = k^2/k_0^2$ we can rewrite Eq. (6.46) as

$$\frac{1}{n^2}\left(n^2 - \epsilon_1\mu^{NL}\right)^2 h_z^2 - n^2h_z^2 + \epsilon_1\mu_L h^2 + \frac{1}{2}\epsilon_1\alpha h^4 = 0 \tag{6.47}$$

An alternative form of the first integral can be obtained by eliminating $\mu_L$ from Eq. (6.47) with the aid of Eq. (6.1) to give

$$[2n^2 - \epsilon_1(\mu_L + \alpha h^2)]h_z^2 - n^2h^2 + \frac{\alpha n^2h^4}{2(\mu_L + \alpha h^2)} = 0 \tag{6.48}$$

If we regard the nonlinear parameter $\alpha$ as small, we can obtain an approximate expression for the first integral correct to first order in $\alpha$ that has the form

$$(2n^2 - \epsilon_1\mu_L)h_z^2 - n^2h^2 + \frac{\alpha h^2}{2\mu_L}(n^2h^2 - 2\epsilon_1\mu_L h_z^2) = 0 \tag{6.49}$$

We now turn our attention to the linear medium. Using the permeability tensor elements given by Eqs. (3.18) and the expressions for the field components given by Eqs. (6.36) and (6.37), we can write the curl equations as

$$-\frac{\partial E_y}{\partial z} = i\omega\mu_0(\mu_{xx}H_x + \mu_{xz}H_z) \tag{6.50a}$$

$$ikE_y = i\omega\mu_0(-\mu_{xz}H_x + \mu_{xx}H_z) \tag{6.50b}$$

$$\frac{\partial H_x}{\partial z} - ikH_z = -i\omega\epsilon_0 E_y \tag{6.50c}$$

Eliminating first $H_z$ and then $H_x$ from Eqs. (6.50) yields

$$-\mu_{xx}\frac{\partial E_y}{\partial z} - ik\mu_{xz}E_y = i\omega\mu_0(\mu_{xx}^2 + \mu_{xz}^2)H_x \tag{6.51}$$

$$-\mu_{xz}\frac{\partial E_y}{\partial z} + ik\mu_{xx}E_y = i\omega\mu_0(\mu_{xx}^2 + \mu_{xz}^2)H_z \tag{6.52}$$

Differentiating Eq. (6.51) with respect to z and then eliminating $H_z$ and $\partial H_x/\partial z$ from Eq. (6.50c) gives the differential equation satisfied by $E_y$

$$\frac{\partial^2 E_y}{\partial z^2} - \gamma_2^2 E_y = 0 \tag{6.53}$$

where

$$\gamma_2^2 = k^2 - k_0^2\mu_V \tag{6.54}$$

$$\mu_V = \mu_{xx} + \frac{\mu_{xz}^2}{\mu_{xx}} \tag{6.55}$$

The solution of Eq. (6.53) corresponding to a surface wave localized at z = 0 is given by

$$E_y(z) = E_L e^{\gamma_2 z} \tag{6.56}$$

Using the $\pi/2$ phase relation between the field components, we set

$$H_x = h_x \qquad H_z = ih_z \qquad E_y = ie_y$$

and obtain from Eqs. (6.51) and (6.52)

$$h_x = -\frac{\gamma_2\mu_{xx} + ik\mu_{xz}}{\omega\mu_0\mu_{xx}\mu_V}e_y \tag{6.57a}$$

$$h_z = \frac{k\mu_{xx} + i\gamma_2\mu_{xz}}{\omega\mu_0\mu_{xx}\mu_V}e_y \tag{6.57b}$$

We are now in a position to obtain the nonlinear dispersion relation by applying the electromagnetic boundary equations. From the continuity of tangential H at z = 0 we get, from Eq. (6.57a),

$$-\frac{\gamma_2\mu_{xx} + ik\mu_{xz}}{\omega\mu_0\mu_{xx}\mu_V}e_L = h_{x1}(0) \tag{6.58}$$

where $e_L = -iE_L$. From the continuity of normal B at z = 0 we get

$$-\mu_{xz}h_{x2}(0) + i\mu_{xx}h_{z2}(0) = i[\mu_L + \alpha h_1^2(0)]h_{z1}(0) \tag{6.59}$$

The subscripts 1 and 2 in Eqs. (6.58) and (6.59) denote the nonlinear and linear media, respectively. Eliminating $h_{x2}(0)$ and $h_{z2}(0)$ from Eq. (6.59) with the aid of Eqs. (6.57) yields

$$\left(\frac{\mu_{xz}(\gamma_2\mu_{xx} + ik\mu_{xz})}{\omega\mu_0\mu_{xx}\mu_V} + \frac{i\mu_{xx}(k\mu_{xx} + i\gamma_2\mu_{xz})}{\omega\mu_0\mu_{xx}\mu_V}\right)e_L = i[\mu_L + \alpha h_1^2(0)]h_{x1}(0) \tag{6.60}$$

which can be simplified to

$$\frac{k}{\omega\mu_0}e_L = [\mu_L + \alpha h_1^2(0)]h_{z1}(0) \tag{6.61}$$

We can eliminate $e_L$ from Eqs. (6.58) and (6.61) and obtain the relation

$$-(\gamma_2\mu_{xx} + ik\mu_{xz})\{\mu_L + \alpha[A_0^2 + h_{z1}^2(0)]\}h_{z1}(0) = k\mu_{xx}\mu_V A_0 \tag{6.62}$$

where $A_0 = h_{x1}(0)$ can be regarded as a nonlinear parameter.

Equation (6.62) together with the first integral given by Eq. (6.49) determines the nonlinear dispersion relation. To obtain this relation explicitly, we must eliminate $h_{z1}(0)$ from these two equations. Although this is difficult to do for the general nonlinearity, it can be done rather simply for the case of weak nonlinearity.

Let us rewrite Eq. (6.62) as

$$\alpha h_{z1}^3(0) + \alpha A_0^2 h_{z1}(0) + \mu_L h_{z1}(0) + \frac{k\mu_{xx}\mu_V}{\gamma_2\mu_{xx} + ik\mu_{xz}}A_0 = 0 \tag{6.63}$$

and assume that

$$\alpha h_{z1}^2(0) << \mu_L \qquad \alpha A_0^2 << \mu_L \tag{6.64}$$

We set

$$h_{z1}(0) = h_{z1}^{(1)} + \Delta \tag{6.65}$$

where

$$h_{z1}^{(1)} = -\frac{k\mu_{xx}\mu_V A_0}{\mu_L(\gamma_2\mu_{xx} + ik\mu_{xz})} \tag{6.66}$$

and $\Delta$ is $O(\alpha)$, and substitute into Eq. (6.63). Neglecting terms of higher order than the first in $\alpha$, we find that

$$\Delta = -\frac{\alpha h_{z1}^{(1)}}{\mu_L}\left(\left[h_{z1}^{(1)}\right]^2 + A_0^2\right) \tag{6.67}$$

and hence

$$h_{z1}(0) \simeq h_{z1}^{(1)}\left[1 - \frac{\alpha}{\mu_L}\left(\left[h_{z1}^{(1)}\right]^2 + A_0^2\right)\right] \tag{6.68}$$

If we define

$$G = -\frac{k\mu_{xx}\mu_V}{\mu_L(\gamma_2\mu_{xx} + ik\mu_{xz})} \tag{6.69}$$

then $h_{z1}^{(1)} = GA_0$ and Eq. (6.68) becomes

$$h_{z1}(0) \simeq GA_0\left[1 - \frac{\alpha A_0^2}{\mu_L}(1 + G^2)\right] \tag{6.70}$$

Eliminating $h_{x1}$ and $h_{z1}$ from Eq. (6.49) in favour of $A_0$ and G yields the nonlinear dispersion relation in the form

$$n^2(G^2 - 1) - \epsilon_1\mu_L G^2 - \frac{\alpha}{2\mu_L}(G^2 + 1)[(3G^2 - 1)n^2 - 2\epsilon_1\mu_L G^2]A_0^2 = 0 \tag{6.71}$$

where we have discarded terms of second order and higher in $\alpha$. If we let $\alpha \to 0$ we obtain

$$n^2 = \frac{\epsilon_1\mu_L G^2}{G^2 - 1} \tag{6.72}$$

which corresponds to the linear dispersion relation of Hartstein *et al.*[16] In Eq. (6.71) the dominant k terms can be determined by letting $k \to \pm\infty$ to give $G^2 \to [G^{(\pm)}]^2$, where now

$$(G^2 - 1) - \frac{\alpha}{2\mu_L}(G^2 + 1)(3G^2 - 1)A_0^2 = 0 \tag{6.73}$$

Solving Eq. (6.73) gives

$$G^2 \simeq \frac{\mu_L}{3\alpha A_0^2} - \frac{1}{3} \pm \left[\frac{\mu_L}{3\alpha A_0^2} - \frac{4}{3} - \frac{2\alpha A_0^2}{\mu_L} \dots\right] = 1 + \frac{2\alpha A_0^2}{\mu_L} = 1 + \Gamma \tag{6.74}$$

### 6.3 Asymptotic limits for large numbers

As $k \to \pm\infty$, $\gamma_2 \to |k|$ and

$$G^{(+)} = -\frac{\mu_{xx}\mu_V}{\mu_L(\mu_{xx} + i\mu_{xz})} \qquad G^{(-)} = \frac{\mu_{xx}\mu_V}{\mu_L(\mu_{xx} - i\mu_{xz})} \tag{6.75}$$

so that the large k limits of the nonlinear dispersion equation are given by

k > 0:
$$\left(\mu_{xx} - i\mu_{xz}\right)^2 - \mu_L^2 = \mu_L^2\Gamma \tag{6.76}$$

k < 0:
$$\left(\mu_{xx} + i\mu_{xz}\right)^2 - \mu_L^2 = \mu_L^2\Gamma \tag{6.77}$$

If $\Gamma = 0$ and the nonlinearity disappears, these limits yield

k > 0:
$$\mu_{xx} - i\mu_{xz} = \pm\mu_L \tag{6.78}$$

which has possible solutions

$$\omega_+ = \omega_0 + \frac{\mu_B}{\mu_B - \mu_L}\omega_s \tag{6.79a}$$

$$\omega_- = \omega_0 + \frac{\mu_B}{\mu_B + \mu_L}\omega_s \tag{6.79b}$$

The original unsquared dispersion relation is

$$\mu_{xx} - i\mu_{xz} = -\mu_L \tag{6.79c}$$

however, so that only $\omega_-$ is an acceptable solution.

k < 0:
$$\mu_{xx} + i\mu_{xz} = \pm\mu_L \tag{6.80}$$

which has possible solutions

$$\omega_+ = -\omega_0 - \frac{\mu_B}{\mu_B - \mu_L}\omega_s \tag{6.81a}$$

$$\omega_- = -\omega_0 - \frac{\mu_B}{\mu_B + \mu_L}\omega_s \tag{6.81b}$$

The original unsquared dispersion equation is

$$\mu_{xx} + i\mu_{xz} = -\mu_L \tag{6.81c}$$

so that $\omega_+$ must be disregarded. In this case, however, since $\omega_-$ is negative, no large k limit exits at all.

The inclusion of nonlinearity changes Eqs. (6.78) and (6.80) to

k > 0:

$$[\mu_B^2 - \mu_L^2(1 + \Gamma)]\omega^2 - 2[\mu_B^2(\omega_s + \omega_0) - \omega_0\mu_L^2(1 + \Gamma)]\omega + \mu_B^2\left(\omega_0 + \omega_s\right)^2 - \omega_0^2\mu_L^2(1 + \Gamma) = 0 \tag{6.82}$$

k < 0:

$$[\mu_B^2 - \mu_L^2(1 + \Gamma)]\omega^2 + 2[\mu_B^2(\omega_s + \omega_0) - \omega_0\mu_L^2(1 + \Gamma)]\omega + \mu_B^2\left(\omega_0 + \omega_s\right)^2 - \omega_0^2\mu_L^2(1 + \Gamma) = 0 \tag{6.83}$$

The possible solutions of these quadratic equations, filtering out the extra solutions that appear through squaring, are

k > 0:
$$\omega = \omega_0 + \frac{\mu_B^2\omega_s}{\mu_B^2 - \mu_L^2(1 + \Gamma)} - \frac{\mu_B\omega_s\sqrt{\mu_L^2(1 + \Gamma)}}{\mu_B^2 - \mu_L^2(1 + \Gamma)} \tag{6.84}$$

k < 0:
$$\omega = -\left[\omega_0 + \frac{\mu_B^2\omega_s}{\mu_B^2 - \mu_L^2(1 + \Gamma)}\right] + \frac{\mu_B\omega_s\sqrt{\mu_L^2(1 + \Gamma)}}{\mu_B^2 - \mu_L^2(1 + \Gamma)} \tag{6.85}$$

Several cases can now be developed, taking as an example a self-focusing situation in which $\alpha > 0$ and $\Gamma > 0$. The k > 0 cases are

Case A:
$$\mu_B \simeq \mu_L \simeq 1 \qquad |\mu_B^2 - \mu_L^2| >> \mu_L^2\Gamma \qquad \Gamma << 1$$

$$\omega \simeq \omega_0 + \frac{\omega_s}{2} + O(\Gamma) \tag{6.86}$$

Case B:
$$\mu_B > \mu_L \qquad 0 < \mu_L^2\Gamma \leq |\mu_B^2 - \mu_L^2| \qquad \Gamma << 1$$

$$\omega \simeq \omega_0 + \frac{(\mu_B - \mu_L)\mu_B\omega_s}{(\mu_B^2 - \mu_L^2) - \mu_L^2\Gamma} \tag{6.87}$$

It appears that when $\mu_L^2\Gamma \simeq (\mu_B^2 - \mu_L^2)$, the nonlinearity will significantly alter the large k > 0 limit.

For k < 0 we have

Case A:

$$\mu_B \simeq \mu_L \simeq 1 \qquad |\mu_B^2 - \mu_L^2| >> \mu_L^2\Gamma \qquad \Gamma << 1$$

$$\omega \simeq -\omega_0 - \frac{\omega_s}{2} + O(\Gamma) \tag{6.88}$$

Case B:

$$\mu_B > \mu_L \qquad |\mu_B^2 - \mu_L^2| \leq \mu_L^2\Gamma \qquad \Gamma << 1$$

$$\omega \simeq -\omega_0 - \frac{(\mu_B - \mu_L)\mu_B\omega_s}{(\mu_B^2 - \mu_L^2) - \mu_L^2\Gamma} \tag{6.89}$$

Typically $\omega_0/\omega_s \simeq 2$ so that here we see the possibility of propagation induced in the large $|k|$ limit for the k < 0 direction simply by balancing the value of $\Gamma$ with $\mu_B - \mu_L$.

Our next objective in this section is to calculate the power flow P, in the direction of propagation. From the general expression in (6.30) we see with the aid of Eqs. (6.36) and (6.37) that

$$P = \frac{1}{2}\int E_y H_z^* dz \tag{6.90}$$

For the nonlinear medium, $P = P^{NL}$. We can eliminate $E_y$ from Eq. (6.90) using Eq. (6.38b) to obtain

$$P^{NL} = \frac{\omega\mu_0}{2k}\int_0^\infty \mu^{NL}|H_z|^2 dz$$

$$= \frac{\omega\mu_0}{2k}\int_0^\infty \mu^{NL} h_z^2 dz \tag{6.91}$$

We shall return to the evaluation of the integral in Eq. (6.91) in a moment, but first we consider $P^L$, the power flow in the linear medium. Using Eq. (6.57b) relating $e_y$ and $h_z$, we find that

$$P^L = \frac{\omega\mu_0\mu_{xx}\mu_V}{2(k\mu_{zz} + i\gamma_2\mu_{xz})}\int_{-\infty}^0 h_z^2 dz \tag{6.92}$$

It is convenient to eliminate $h_z$ in favour of $h_x$ by means of Eqs. (6.57) and evaluate the integral in Eq. (6.92) using the z-dependence of $h_x$ in the linear medium given by

$$h_x(z) = A_0 e^{\gamma_2 z} \tag{6.93}$$

The result for the power flow in the linear medium is

$$P^L = \frac{\omega\mu_0\mu_{xx}\mu_V(k\mu_{xx} + i\gamma_2\mu_{xz})A_0^2}{4\gamma_2(\gamma_2\mu_{xx} + ik\mu_{xz})^2} \tag{6.94}$$

Let us now go back to the calculation of the power flow for the nonlinear medium given by Eq. (6.91). The exact evaluation of $P^{NL}$ can only be done numerically for $\mu_V < 0$ but, for $\mu_V > 0$, an approximate analytic treatment can be carried out that gives rather good results.[4] We proceed by using Eq. (6.38b) to eliminate $h_z$ in favour of $e_y$ in Eq. (6.91) to give

$$P^{NL} = \frac{k}{2\omega\mu_0}\int_0^{\infty}\frac{e_y^2}{\mu^{NL}}dz \tag{6.95}$$

Restricting our attention to the self-focusing case ($\alpha > 0$), the electric field component $e_y$ reaches its maximum value $e_{ym}$ at the value of z where $P^{NL}$ is a maximum.[15] We can therefore write to a good approximation that

$$P^{NL} \simeq \frac{k\Delta z e_{ym}^2}{2\omega\mu_0\left(\mu_L + \dfrac{\alpha k^2}{\omega^2\mu_0^2\mu_L^2}e_{ym}^2\right)} \tag{6.96}$$

where $\Delta z$ is a suitable interval about the maximum.[5]

It remains to evaluate $e_{ym}$. At the maximum of $e_y$, $\partial e_y/\partial z = 0$ and hence from Eq. (6.38a) we have $h_x = 0$. If we take the first integral in the form given by Eq. (6.48) and carry out some elementary manipulations, we obtain

$$(2k^2 - k_0^2\epsilon_1\mu^{NL})e_y^2 - \omega^2\mu_0^2\mu^{NL}\left(\mu_L h^2 + \frac{\alpha}{2}h^4\right) = 0 \tag{6.97}$$

Setting $h_x = 0$ and making the substitution

$$h_z = \frac{k}{\omega\mu_0\mu^{NL}}e_y \tag{6.98}$$

from Eq. (6.38b), we can transform Eq. (6.97) to

$$(2k^2 - k_0^2\epsilon_1\mu^{NL})e_{ym}^2 - \frac{k^2}{\mu^{NL}}\left(\mu_L e_{ym}^2 + \frac{\alpha k^2}{2\omega^2\mu_0^2(\mu^{NL})^2}e_{ym}^4\right) = 0 \tag{6.99}$$

with

$$\mu^{NL} = \mu_L + \frac{\alpha k^2}{\omega^2\mu_0^2(\mu^{NL})^2}e_{ym}^2 \tag{6.100}$$

The solution of Eqs. (6.99) and (6.100) gives $e_{ym}$.

## 6.4 Numerical results for TM waves

Nonlinear TM waves propagating along the interface of the system shown in Fig. 1 carry the magnetic field (0, $H_y$, 0). This means that, even though the permeability tensor of a ferri(ferro)magnet has frequency-dependent terms, it is the frequency-independent background permeability term $\mu_2 = \mu_B$ that enters into the nonlinear dispersion equation.

This dispersion equation, in this case, coincides with the dependence of the total power flow in the x-direction upon the surface-guided wave index $n = ck/\omega$. The interface, in its linear state, will not support TM surface waves so, from the experience gained with nonlinear TE waves on non-magnetic materials, it is not unexpected[1-7] that a power threshold has to be exceeded before nonlinear waves can propagate. This is shown in Fig. 6, together with two other features. Firstly, the power approaches an infinite value as the wave index becomes smaller on the left of the minimum. Again this is a feature that is found for TE nonlinear

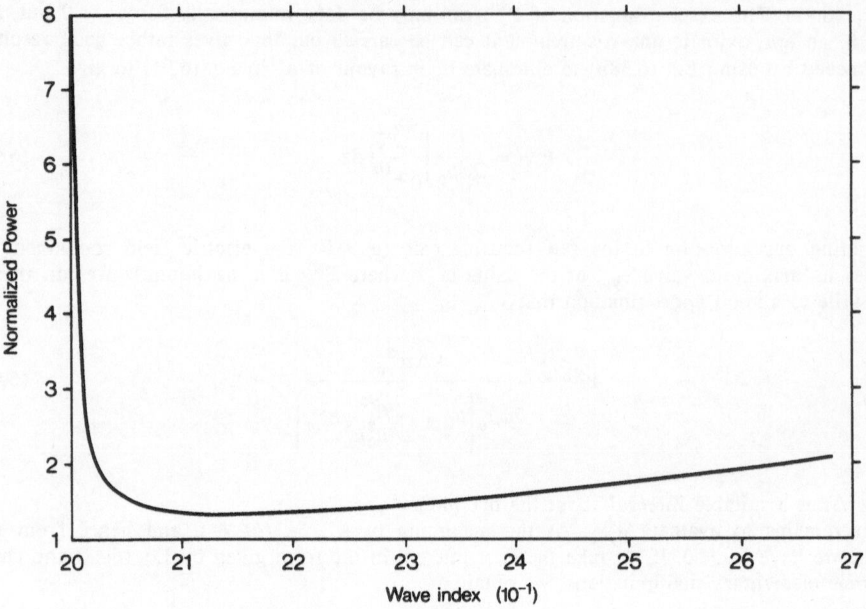


Fig. 6 Normalised power flow along the x-direction as a function of wave index ($n = ck/\omega$) for TM waves. $\mu_L = 1.29$, $\epsilon_L = 2.3$, $\mu_B = 2$, $\epsilon_2 = 2$, frequency f = 90 GHz, $\alpha = 8.869\times10^{-8}$ $m^2A^{-2}$. The normalized power is $P/P_0$ where $P_0 = 1/(2\omega\alpha\epsilon_0)$ and $\epsilon_0$ is the permittivity of free space.

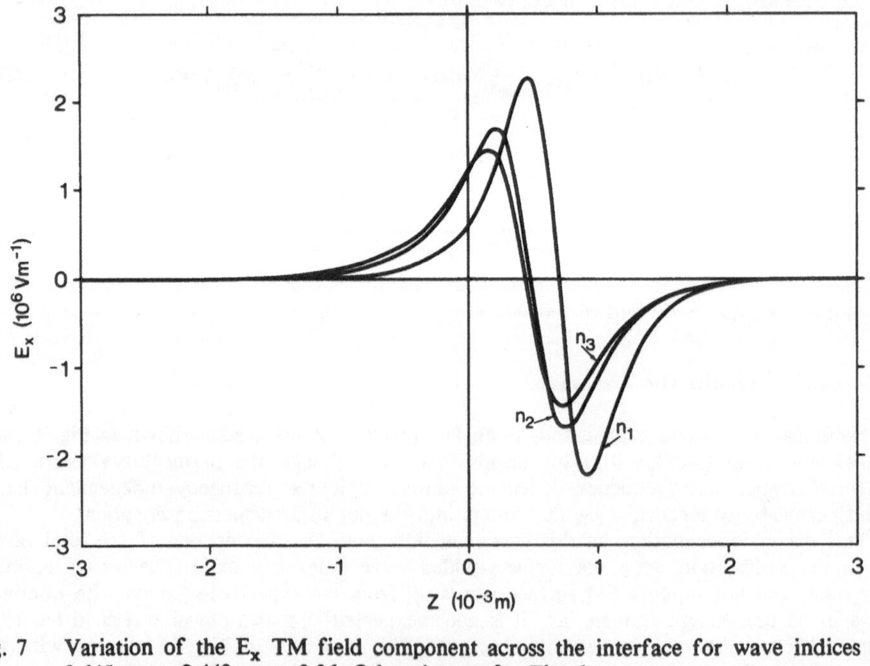


Fig. 7 Variation of the $E_x$ TM field component across the interface for wave indices $n_1 = 2.645$, $n_2 = 2.449$, $n_3 = 2.36$. Other data as for Fig. 6.

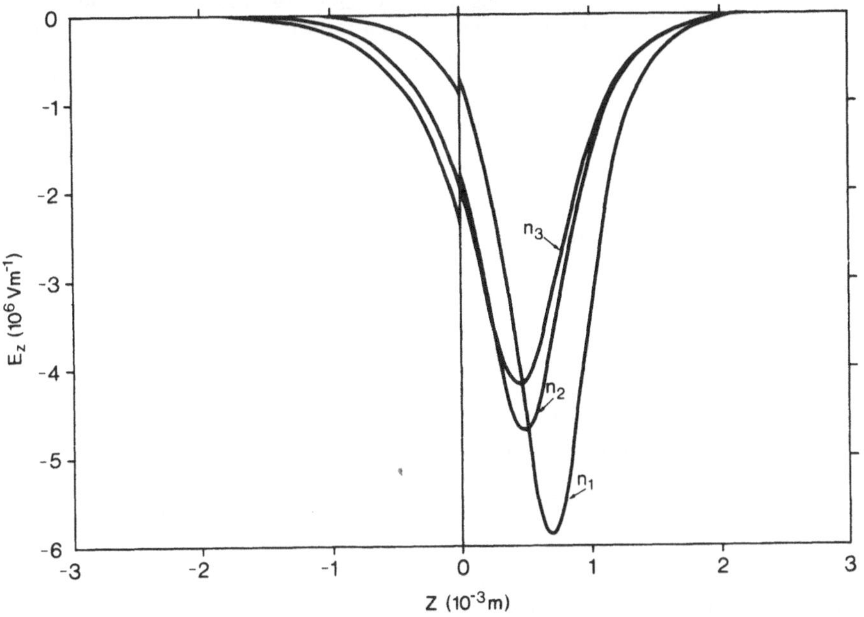


Fig. 8 Variation of the $E_z$ TM field component across the interface for the data of Fig. 7.

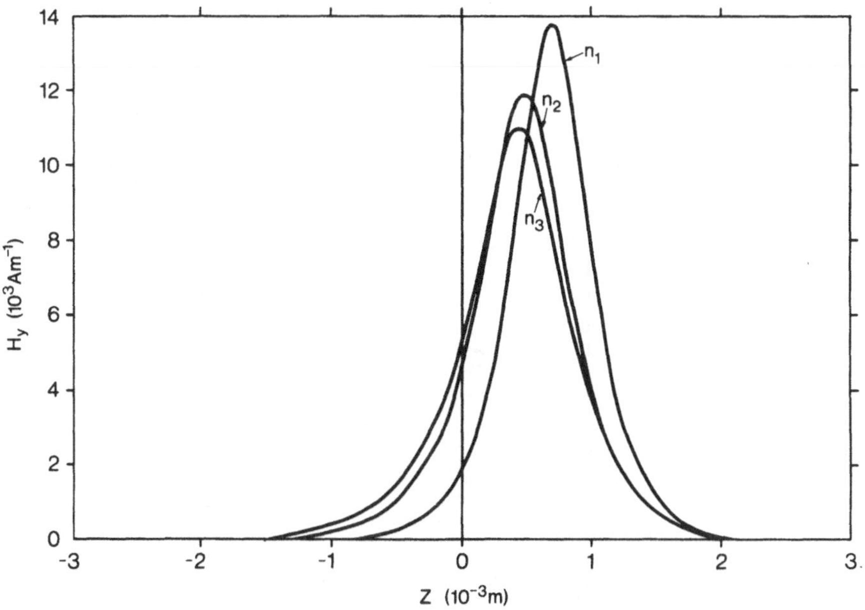


Fig. 9 Variation of the $H_y$ TM field component across the interface for the data of Fig. 7.

waves at a single interface between non-magnetic materials. Equation (6.25) shows that this lower limit on n is $\sqrt{\mu_B \epsilon_2} = 2$ for the data chosen. The data required to illustrate the theory given in this chapter is not easy to locate, since in many cases new materials are required, so we have selected parameter values that have been referred to in the literature, at least as possibilities.[16,22-29] These values are shown in the table. As $\sqrt{\mu_B \epsilon_2}$ is approached, most of the power resides in the linear medium necessitating very large powers to drive the nonlinear wave. As n increases, the power passes through a minimum and finally reaches the upper wave number limit given by Eq. (6.24) and finite power limit given by Eq. (6.33). At this extinction point any self-focused peak has moved out to infinity and all the power flow is in the nonlinear medium.

Figures 7, 8 and 9 show, respectively, the shape of nonlinear $E_x$, $E_z$ and $H_y$ electromagnetic field components in the vicinity of the interface for the n values $\sqrt{5}$, $\sqrt{6}$ and $\sqrt{7}$, conveniently chosen to represent typical values in the permitted range $2 < n < 2.68$, where $n = 2.68$ is the upper wave number limit specified by Eq. (6.24).

### 6.5 Numerical results for TE Waves with $\mu_V > 0$

In the Voigt propagation geometry of Fig. 1, non-reciprocal TE waves can exist in the zero power linear limit and are controlled by the effective permeability $\mu_V$ that is defined in Eq. (3.6). Such linear surface polaritons, however, can propagate provided only that $\mu_V < 0$. A dominant feature of nonlinear wave propagation, as has already been shown for the TM case above, is the possibility of wave propagation in regimes for which there is no clear limit. Accordingly, the frequency region to the left of the resonance in Fig. 10, for which $\mu_V > 0$, is also of interest here and will permit nonlinear waves to be sustained, once again, above a threshold of power. Since TE waves are associated with both $H_x$ and $H_z$ magnetic field components, they will depend upon the frequency-dependent $\mu_{xx}$ and $\mu_{xz}$ components of the permeability tensor as expressed through $\mu_V$.

The power curves shown in Fig. 11 show the threshold referred to but they now approach infinity both at a lower cut-off of $n = \sqrt{\mu_V}$ and as $n \to \infty$. Hence, the peak of the self-focused field moves out to infinity for both TE and TM modes, but the distinction between the two modes is that n has no bound for TE modes; i.e., as the self-focused field

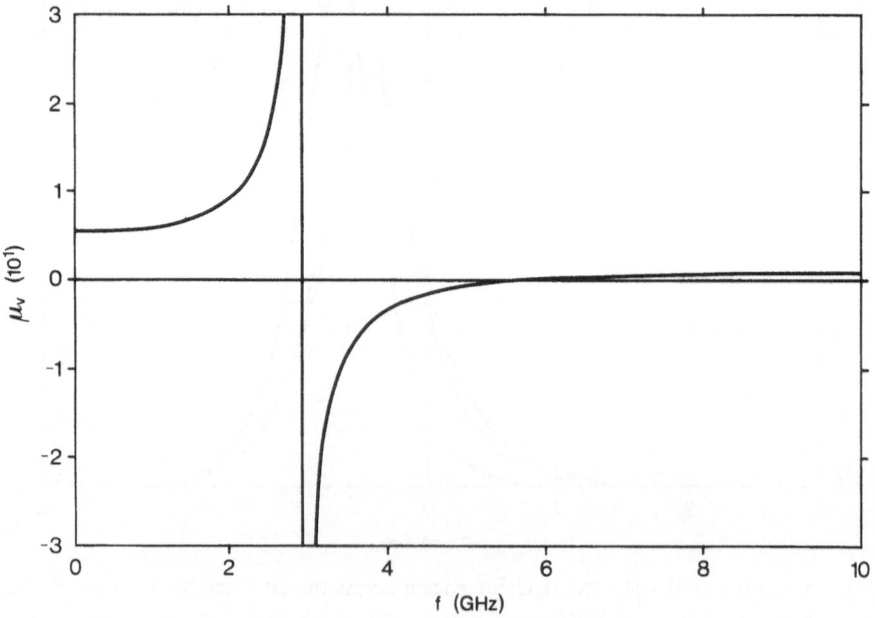


Fig. 10 The effective permeability $\mu_V$ as a function of frequency f. $\mu_0 H_0 = 500$ G, $\mu_0 M_0 = 1750$ G, $\gamma = 2\pi(2.8)\times 10^6$ rad $s^{-1}G^{-1}$.

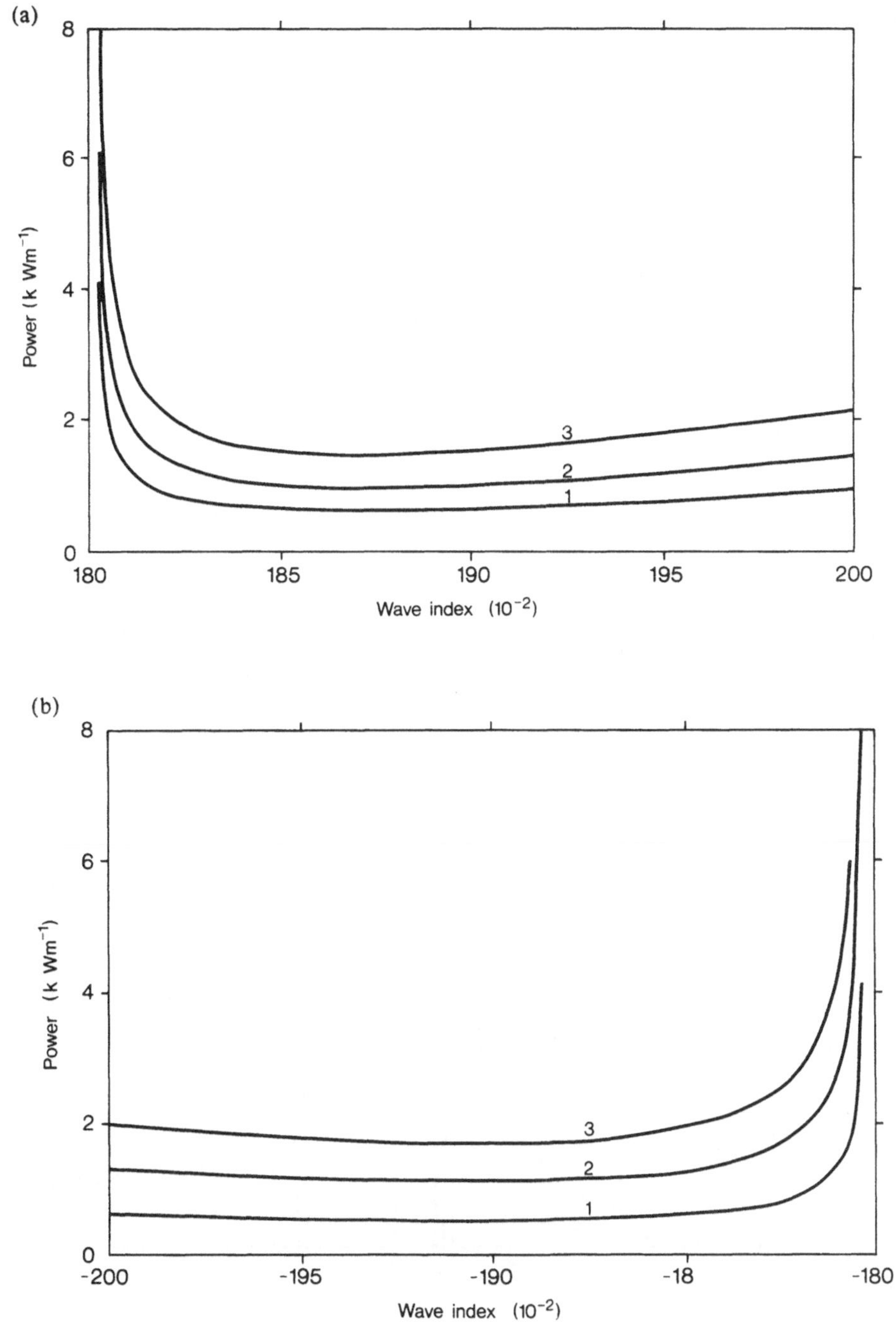


Fig. 11 Power flow as a function of wave index for TE waves when $\mu_V > 0$. (a) $k > 0$ (b) $k < 0$. $\mu_0 H_0 = 500$ G, $\mu_0 M_0 = 1000$ G, $\mu_V = 3.24$, $\mu_L = 1.29$, $\epsilon_L = 2.3$, $\epsilon_2 = 1$, $\alpha = 8.869 \times 10^{-8}$ m$^2$A$^{-2}$. The curves are labelled (1) $\alpha$, (2) $\alpha/2$, (3) $\alpha/3$.

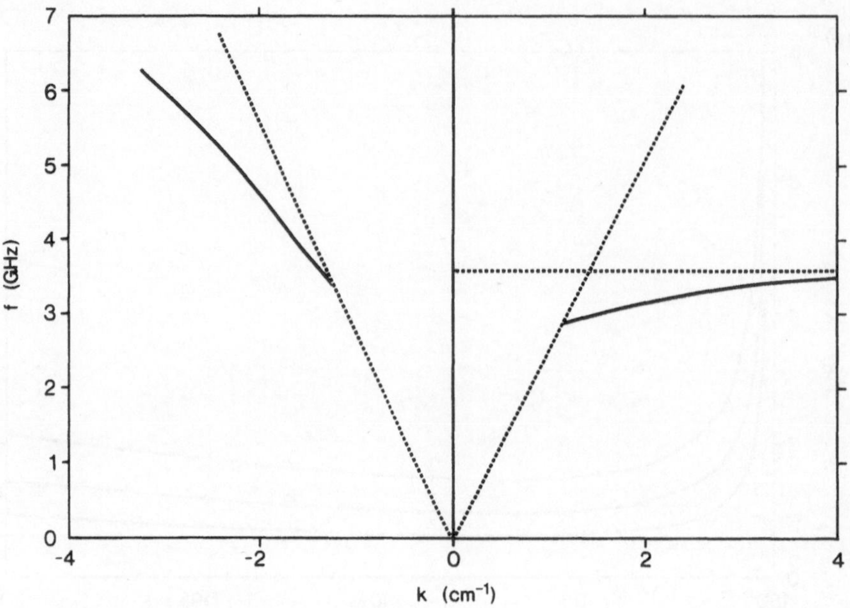


Fig. 12 Typical linear dispersion curves for TE waves in a vacuum-bounded semi-infinite ferromagnetic substrate when $\mu_V < 0$. The curves exhibit three endpoints and the right-hand branch is asymptotic to the horizontal dotted line.[16] $\mu_0 M_0$ = 1750 G, $\mu_0 H_0$ = 500 G, $\epsilon_2$ = 16, $\mu_B$ = 1, $\epsilon_L$ = 2.3, $\mu_L$ = 1.29.

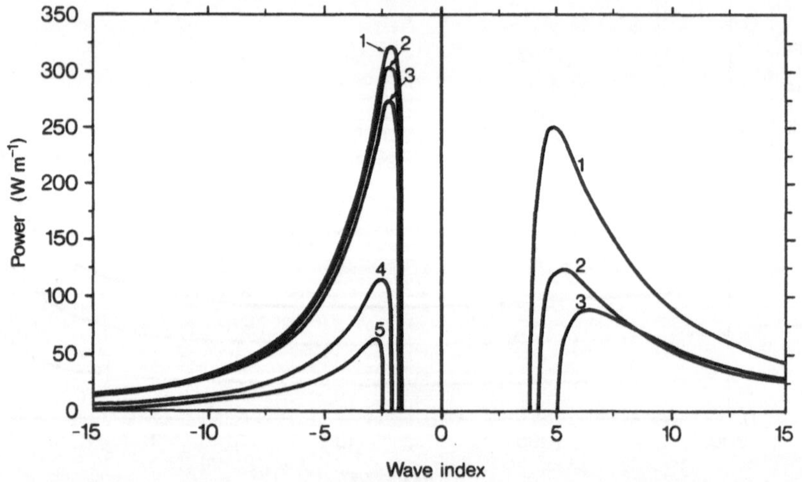


Fig. 13 Power flow P versus wave index for TE waves when $\mu_V < 0$. The curves are labelled (1) P($10^{-1}$): f = 3.25 GHz, $\mu_V$ = -16.72; (2) P($10^{-1}$): f = 3.35 GHz, $\mu_V$ = -11.85; (3) P($10^{-1}$)(k < 0), P($10^{-2}$)(k > 0): f = 3.45 GHz, $\mu_V$ = -9.01; (4) P($10^{-2}$): f = 3.50 GHz, $\mu_V$ = -8.00; (5) P($10^{-3}$): f = 3.55 GHz, $\mu_V$ = -7.16. Other data as for Fig. 12.

moves out to infinity, $n \to \infty$. The non-reciprocal behaviour, often associated with magnetic materials, is present but, as can be seen in Fig. 11, it is not very strong in this case with only a slight asymmetry being discernible.

### 6.6 Numerical results for TE Waves with $\mu_V < 0$

The quantity $\mu_V$ is less than zero in a frequency range to the right of the singularity in Fig. 10. For these frequencies the interface can support linear surface waves. For $\mu_0 H_0 = 500$ G and $\mu_0 M_0 = 1750$ G, the resonance frequency is $f_r = 3$ GHz. The linear dispersion curves shown in Fig. 12 are well-known[16,23,24,30] when the cladding is vacuum (see Fig. 3), but have been recalculated here for the case that the cladding is not a vacuum. They possess several main points of interest. These are

(1) definite end or cut-off points
(2) propagation characteristics to the left ($k < 0$) and to the right ($k > 0$) that are not symmetrical
(3) only the right-hand propagation dispersion curve has a large wave number limit and this tends to the magnetostatic limit.

For a sequence of frequencies, labelled (1) - (5), that cause $\mu_V$ to lie in the range $-7.16 \leq \mu_V \leq -16.72$ the variation of the wave index with total power flow is given in Fig. 13. The curves labelled (1), (2) and (3) lie on both the $k > 0$ and $k < 0$ branches of Fig. 12, while (4) and (5) lie only on the $k < 0$ branch. The expected nonreciprocal behaviour shows a strong nonlinear dependence upon power. For a given frequency, the power flow can be significantly greater in one direction than in the other. This could lead to some interesting experimental possibilities involving non-reciprocal nonlinear power transfer. Figures 14 and 15 show, on an expanded scale, the $k > 0$ power curves as the nonlinear coefficient $\alpha$ or $\omega(\mu_V)$ is varied. Since the nonlinearity is weakened by diminishing $\alpha$ or $\omega$, higher peak values must be reached in both cases to enter the strongly nonlinear regimes.

The variation of $H_x$, $H_z$, $E_y$ field components is shown in the sequence Fig. 16(a), (b) and (c) with the nonlinear medium on the right-hand side. Several values of frequency f are selected corresponding to different positions on the $(\mu_V, f)$ curve.

As a final numerical example, Fig. 17 shows the nonlinearity, defined as $\alpha|H|^2$, as a function of z measured from the interface for $\mu_V = -16.716$ and several values of n. The nonlinearity increases in strength as n increases. The absolute values of the nonlinearity appear to be within the reach of experimental observation.

## 7. NONLINEAR ELECTROMAGNETIC SURFACE WAVES GUIDED BY A GYRODIELECTRIC MEDIUM

In this section we analyze the propagation of electromagnetic surface waves along the interface between a uniaxial nonlinear dielectric medium and a linear gyrodielectric medium. The former occupies the half-space $z > 0$ and the latter the half-space $z < 0$. We shall restrict our attention to the TM case for which surface waves exist for the gyrodielectric medium as discussed in section 5.

The nonlinear medium is assumed to be uniaxial and characterized by an intensity-dependent dielectric tensor with elements

$$\epsilon_{xx} = \epsilon_L + \alpha_1 |E_x|^2 + \alpha_2 |E_z|^2 \tag{7.1a}$$

$$\epsilon_{zz} = \epsilon_L + \alpha_2 |E_x|^2 + \alpha_1 |E_z|^2 \tag{7.1b}$$

where $\alpha_1$ and $\alpha_2$ are constants characterizing the nonlinearity. The gyrodielectric medium is taken to be a magnetoplasma with dielectric tensor given by Eq. (4.8).

For TM waves the electric and magnetic vectors can be written in the form

$$\mathbf{E} = [E_x(z), 0, E_z(z)] e^{i(kx - \omega t)} \tag{7.2a}$$

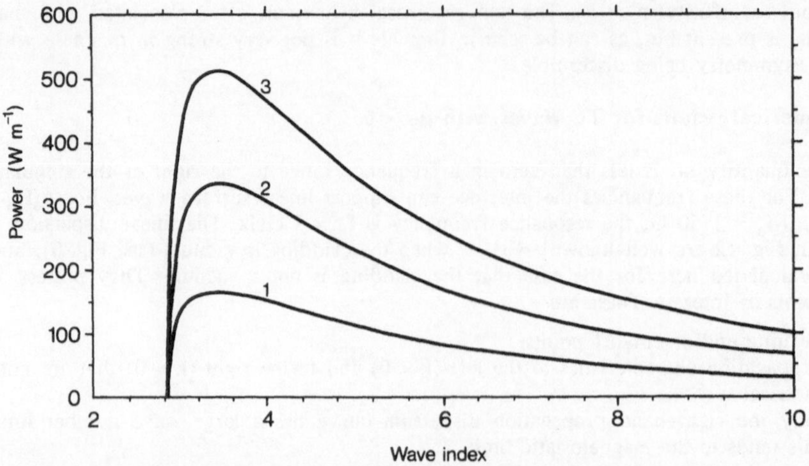


Fig. 14 Power flow versus wave index for TE waves when $\mu_V < 0$. The curves, in ascending order, correspond to $\alpha$, $\alpha/2$ and $\alpha/3$. $\mu_V = -13.91$, f = 3.3 GHz, other data as for Fig. 12.

$$H = [0, H_y(z), 0]e^{i(kx-\omega t)} \tag{7.2b}$$

Maxwell's curl equations then simplify to

$$\frac{\partial H_y}{\partial z} = i\omega\epsilon_0\epsilon_{xx}E_x \tag{7.3a}$$

$$H_y = -\frac{\omega}{k}\epsilon_0\epsilon_{zz}E_z \tag{7.3b}$$

$$\frac{\partial E_x}{\partial z} + ikE_z = -i\omega\mu_0 H_y \tag{7.3c}$$

Eliminating the magnetic field component $H_y$ from Eqs. (7.3b) and (7.3c) leads to the result

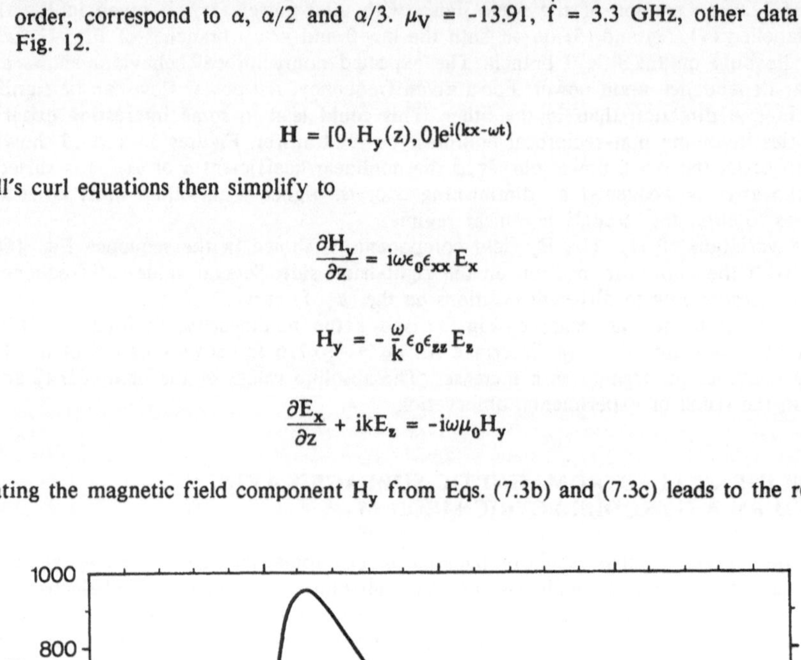

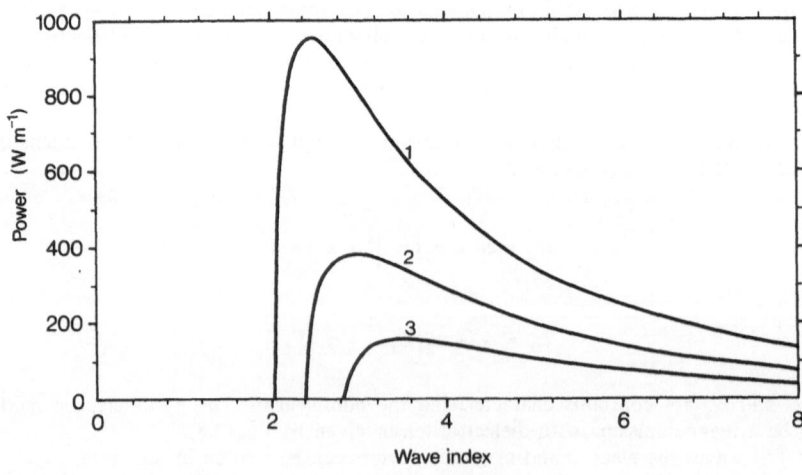

Fig. 15 Power flow versus wave index for TE waves when $\mu_V < 0$. The curves, in ascending order, correspond to the frequency, ($\mu_V$) pairs: 2.97 GHz, (-4511); 3.0 GHz, (-170); 3.05 GHz, (-62.98).

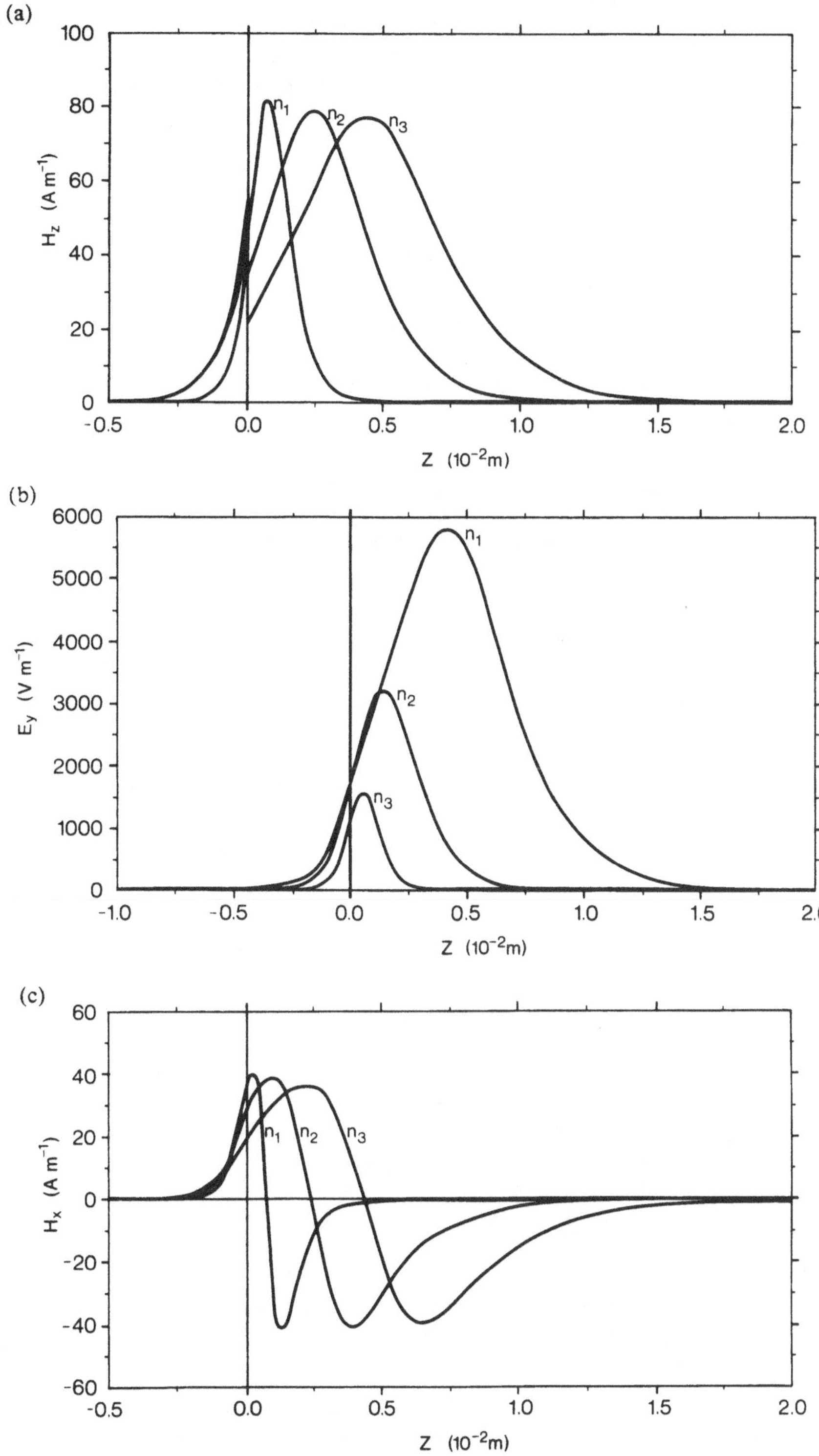


Fig. 16 Field profiles for TE waves when $\mu_V < 0$. (a) $H_z$ (b) $E_y$ (c) $H_x$. $n_1 = 2.62$, $n_2 = 9$, $n_3 = 12$, $f = 3.25$ GHz, $\mu_V = -6.72$, $\alpha = 8.869\times10^{-8}\ m^2A^{-2}$.

$$\frac{\partial E_x}{\partial z} = \frac{i}{k}(k^2 - k_0^2\epsilon_{zz})E_z \tag{7.4}$$

If we differentiate Eq. (7.3c) with respect to z and eliminate $\partial H_y/\partial z$ using Eq. (7.3a), we obtain

$$\frac{\partial^2 E_x}{\partial z^2} - ik\frac{\partial E_z}{\partial z} = -k_0^2\epsilon_{xx}E_x \tag{7.5}$$

Multiplying Eq. (7.5) by $\partial E_x/\partial z$ and rearranging gives

$$\frac{\partial E_x}{\partial z}\frac{\partial^2 E_x}{\partial z^2} = -k_0^2\epsilon_{xx}E_x\frac{\partial E_x}{\partial z} + ik\frac{\partial E_x}{\partial z}\frac{\partial E_z}{\partial z} \tag{7.6}$$

We now eliminate $\partial E_x/\partial z$ from the last term on the right hand side of Eq. (7.6) using Eq. (7.4) and obtain

$$\frac{\partial E_x}{\partial z}\frac{\partial^2 E_x}{\partial z^2} = -k_0^2\epsilon_{xx}E_x\frac{\partial E_x}{\partial z} - (k^2 - k_0^2\epsilon_{zz})E_z\frac{\partial E_z}{\partial z} \tag{7.7}$$

Converting to real electric field components by taking $E_x$ real and setting $\overline{E}_z = iE_z$ where $\overline{E}_z$ is real, yields the result

$$\frac{\partial E_x}{\partial z}\frac{\partial^2 E_x}{\partial z^2} = -k_0^2\epsilon_{xx}E_x\frac{\partial E_x}{\partial z} + (k^2 - k_0^2\epsilon_{zz})\overline{E}_z\frac{\partial \overline{E}_z}{\partial z} \tag{7.8}$$

If we now substitute the expressions for $\epsilon_{xx}$ and $\epsilon_{zz}$ given by Eqs. (7.2) into Eq. (7.8), we find that

$$\begin{aligned}\frac{\partial E_x}{\partial z}\frac{\partial^2 E_x}{\partial z^2} &= -k_0^2(\epsilon_L + \alpha_1 E_x^2 + \alpha_2\overline{E}_z^{\,2})E_x\frac{\partial E_x}{\partial z} + [k^2 - k_0^2(\epsilon_L + \alpha_2 E_x^2 + \alpha_1\overline{E}_z^{\,2})]\overline{E}_z\frac{\partial \overline{E}_z}{\partial z}\\ &= -k_0^2\epsilon_L E_x\frac{\partial E_x}{\partial z} + (k^2 - k_0^2\epsilon_L)\overline{E}_z\frac{\partial \overline{E}_z}{\partial z} - k_0^2\alpha_1\left[E_x^3\frac{\partial E_x}{\partial z} + \overline{E}_z^{\,3}\frac{\partial \overline{E}_z}{\partial z}\right]\\ &\qquad - k_0^2\alpha_2\left[\overline{E}_z^{\,2}E_x\frac{\partial E_x}{\partial z} + E_x^2\overline{E}_z\frac{\partial \overline{E}_z}{\partial z}\right]\end{aligned} \tag{7.9}$$

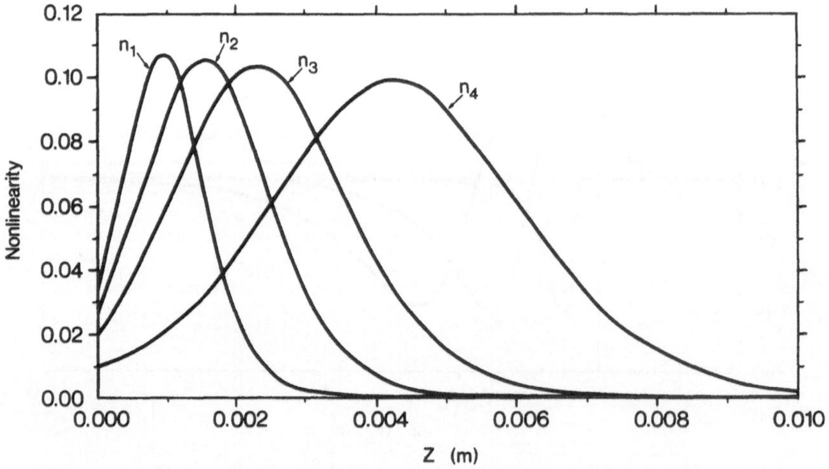


Fig. 17 Nonlinearity as a function of distance from the interface for TE waves when $\mu_V < 0$. $n_4 = 18$, other data as for Fig. 16.

Rewriting this equation as

$$\frac{1}{2}\frac{\partial}{\partial z}\left(\frac{\partial E_x}{\partial z}\right)^2 = -\frac{1}{2}\left\{k_0^2\epsilon_L \frac{\partial}{\partial z}(E_x^2) - (k^2 - k_0^2\epsilon_L)\frac{\partial}{\partial z}(\overline{E}_z^{\,2}) + \frac{1}{2}k_0^2\alpha_1\frac{\partial}{\partial z}(E_x^4 + \overline{E}_z^4) + k_0^2\alpha_2\frac{\partial}{\partial z}(E_x^2\overline{E}_z^{\,2})\right\}$$

and integrating leads to the first integral

$$\left(\frac{\partial E_x}{\partial z}\right)^2 = -k_0^2\epsilon_L E_x^2 + (k^2 - k_0^2\epsilon_L)\overline{E}_z^{\,2} - \frac{1}{2}k_0^2\alpha_1(E_x^4 + \overline{E}_z^4) - k_0^2\alpha_2 E_x^2\overline{E}_z^{\,2} + C \tag{7.10}$$

where C is a constant of integration. For surface waves, the electric field components and their derivatives must approach zero as $z \to \infty$, so $C = 0$. If we express Eq. (7.4) in the form

$$\frac{\partial E_x}{\partial z} = \frac{1}{k}(k^2 - k_0^2\epsilon_{zz})\overline{E}_z$$

and eliminate $\partial E_x/\partial z$ from Eq. (7.10), the first integral can be rewritten as

$$\frac{1}{n_x^2}\left[n_x^2 - (\epsilon_L + \alpha_2 E_x^2 + \alpha_1\overline{E}_z^{\,2})\right]^2\overline{E}_z^{\,2} = -\epsilon_L E_x^2 + (n_x^2 - \epsilon_L)\overline{E}_z^{\,2} - \frac{1}{2}\alpha_1(E_x^4 + \overline{E}_z^4) - \alpha_2 E_x^2\overline{E}_z^{\,2} \tag{7.11}$$

where $n_x = k/k_0$. Equation (7.11) may be solved for $E_x^2$ in terms of $\overline{E}_z^{\,2}$:

$$E_x^2 = \frac{1}{2A}\{-B - \sqrt{B^2 - 4AC}\} \tag{7.12}$$

$$A = \frac{1}{n_x^4}\alpha_2^2\overline{E}_z^{\,2} + \frac{1}{2}\alpha_1 \tag{7.13a}$$

$$B = \frac{2\alpha_2}{n_x^2}\overline{E}_z^{\,2}[n_x^2 - (\epsilon_L + \alpha_1\overline{E}_z^{\,2})] - (\epsilon_L + \alpha_2\overline{E}_z^{\,2}) \tag{7.13b}$$

$$C = \frac{1}{n_x^2}\left[n_x^2 - (\epsilon_L + \alpha_1\overline{E}_z^{\,2})\right]^2\overline{E}_z^{\,2} - \left[n_x^2 - \left(\epsilon_L + \frac{\alpha_1}{2}\overline{E}_z^{\,2}\right)\right]\overline{E}_z^{\,2} \tag{7.13c}$$

We now turn to the determination of the dispersion relation for nonlinear surface magnetoplasmon polaritons by applying the electromagnetic boundary conditions

$$E_{1x} = E_{2x} \qquad \text{at } z = 0 \tag{7.14a}$$

$$D_{1z} = D_{2z} \qquad \text{at } z = 0 \tag{7.14b}$$

where, as before the subscript 1 refers to the nonlinear medium and the subscript 2 to the linear medium. The linear medium is gyrodielectric with dielectric tensor given by Eq. (4.8). From the second boundary condition we obtain the relation

$$\epsilon_{zz}^{NL}(0)E_{1z}(0) = \epsilon_{zx}^{L}E_{2x}(0) + \epsilon_{zz}^{L}E_{2z}(0)$$

$$= E_{2x}(0)\left[\epsilon_{zx}^{L} + \epsilon_{zz}^{L}\frac{E_{2z}(0)}{E_{2x}(0)}\right]$$

$$= E_{1x}(0)\left[\epsilon^L_{zx} + \epsilon^L_{zz}\frac{E_{2z}(0)}{E_{2x}(0)}\right] \tag{7.15}$$

For the linear magnetoplasma,

$$\frac{E_{2z}(0)}{E_{2x}(0)} = \frac{k_2\epsilon^L_{xz} - ik\epsilon^L_{xx}}{k_2\epsilon^L_{xx} + ik\epsilon^L_{xz}}$$

so Eq. (7.15) can be rewritten as

$$\epsilon^{NL}_{zz}(0)\frac{E_{1z}(0)}{E_{1x}(0)} = \epsilon^L_{zx} + \epsilon^L_{zz}\frac{k_2\epsilon^L_{xz} - ik\epsilon^L_{xx}}{k_2\epsilon^L_{xx} + ik\epsilon^L_{xz}} \tag{7.16}$$

Rearranging,

$$\epsilon^{NL}_{zz}(0)\frac{E_{1z}(0)}{E_{1x}(0)} = -ik\frac{\epsilon^L_{xx}\epsilon^L_v}{k_2\epsilon^L_{xx} + ik\epsilon^L_{xz}}$$

or

$$E_{1x}(0) = i\,\frac{k_2\epsilon^L_{xx} + ik\epsilon^L_{xz}}{k\epsilon^L_{xx}\epsilon^L_v}\epsilon^{NL}_{zz}(0)E_{1z}(0) \tag{7.17}$$

Now from Eq. (7.1),

$$\epsilon^{NL}_{zz}(0) = \epsilon_L + \alpha_2\left|E_{1x}(0)\right|^2 + \alpha_1\left|E_{1z}(0)\right|^2 \tag{7.18}$$

We can eliminate $E_{1x}(0)$ from Eq. (7.18) using Eq. (7.17) and solve for $\epsilon^{NL}_{zz}(0)$. Setting $\overline{E}_{1z}(0) = iE_{1z}(0)$ and $\epsilon^L_{12} = i\epsilon^L_{xz}$, we obtain

$$\epsilon^{NL}_{zz}(0) = \frac{k^2\left(\epsilon^L_{xx}\epsilon^L_v\right)^2}{2\alpha_2(k_2\epsilon^L_{xx} + k\epsilon^L_{12})^2}\left\{1 - \left[1 - 4[\epsilon_L + \alpha_1\overline{E}^2_{1z}(0)]\alpha_2\left(\frac{k_2\epsilon^L_{xx} + k\epsilon^L_{12}}{k\epsilon^L_{xx}\epsilon^L_v}\right)^2\overline{E}^2_{1z}(0)\right]^{1/2}\right\} \tag{7.19}$$

Using the general expression for $E_{1x}$ in the nonlinear medium given by Eq. (7.12), we have

$$E_{1x}(0) = \left[\frac{1}{2A_1(0)}\left(-B_1(0) - \left[B^2_1(0) - 4A_1(0)C_1(0)\right]^{1/2}\right)\right]^{1/2} \tag{7.20}$$

The second boundary condition expressed by Eq. (7.17) now becomes

$$\left[\frac{1}{2A_1(0)}\left(-B_1(0) - \sqrt{B^2_1(0) - 4A_1(0)C_1(0)}\right)\right]^{1/2} = \frac{k_2\epsilon^L_{xx} + k\epsilon^L_{12}}{k\epsilon^L_{xx}\epsilon^L_v}\epsilon^{NL}_{zz}(0)\overline{E}_{1z}(0) \tag{7.21}$$

where the quantities $A_1(0)$, $B_1(0)$, and $C_1(0)$ are expressed in terms of $\overline{E}_{1z}(0)$ through Eqs. (7.13), and $\epsilon^{NL}_{zz}(0)$ is expressed in terms of $\overline{E}_{1z}(0)$ through Eq. (7.19). Equation (7.21) is the dispersion relation for nonlinear surface magnetoplasmon polaritons. It is a functional relationship of the frequency $\omega$, the wave vector k, and the amplitude $\overline{E}_{1z}(0)$.

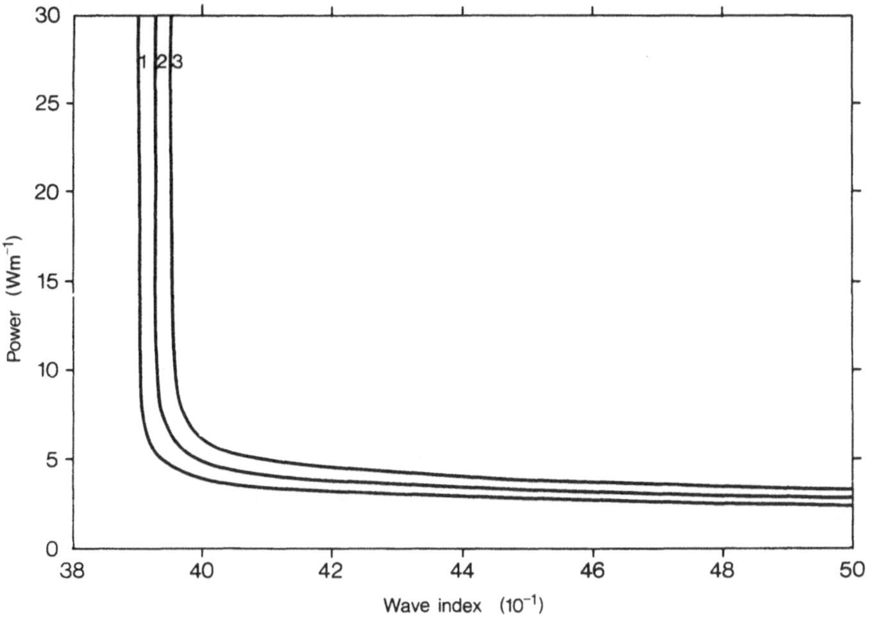


Fig. 18 Total power flow variation with wave index for k > 0 and $\epsilon_V > 0$ when no linear limit exists with $\omega_p = 1.8\times10^{15}$ rad $s^{-1}$, $\epsilon_2 = 5.3361$, $\alpha_1 = \alpha_2 = 1.21\times10^{-8}$ $m^2W^{-1}$, $\epsilon_\infty = 15.68$, $\lambda = 1.06$ $\mu$m, (1) $\omega_c/\omega_p = 0.9$, $\epsilon_V = 15.2$, (2) $\omega_c/\omega_p = 1.2$, $\epsilon_V = 15.4$, (3) $\omega_c/\omega_p = 2.0$, $\epsilon_V = 15.6$.

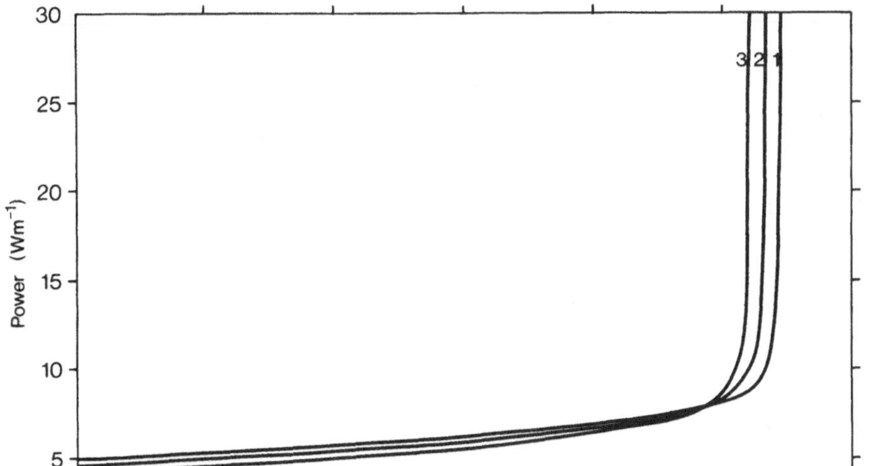


Fig. 19 Total power flow variation with wave index for k < 0 and $\epsilon_V > 0$ when no linear limit exists. Data as for Fig. 18.

The power flow in the nonlinear medium can be written as

$$P^{NL} = -\frac{1}{2}\int_0^{\infty} E_{1z}H_{1y}^{*}dz \tag{7.22}$$

Using Eq. (7.3b) to eliminate $H_{1y}^{*}$, we obtain

$$P^{NL} = \frac{\omega\epsilon_0}{2k}\int_0^{\infty} \epsilon_{zz}^{NL}\left|E_{1z}\right|^2 dz$$

$$= \frac{\omega\epsilon_0}{2k}\int_0^{\infty}\left(\epsilon_L + \alpha_2\left|E_{1x}\right|^2 + \alpha_1\left|E_{1z}\right|^2\right)\left|E_{1z}\right|^2 dz \tag{7.23}$$

In the linear gyrodielectric medium, the non-vanishing magnetic field component $H_y$ can be expressed as

$$H_{2y} = H_L e^{k_2 z} e^{i(kx-\omega t)} \tag{7.24}$$

where $k_2$ must be given by

$$k_2^2 = k^2 - k_0^2\epsilon_V \tag{7.25}$$

in order for Maxwell's curl equations to be satisfied. The power flow $P^L$ takes the form

$$P^L = -\frac{1}{2}\int_{-\infty}^{0} E_{2z}H_{2y}^{*}dz \tag{7.26}$$

where $E_{2z}$ can be expressed in terms of $H_{2y}$ with the aid of Maxwell's curl equations as

$$E_{2z} = -\frac{k\epsilon_{xx}^L + ik_2\epsilon_{xz}^L}{\omega\epsilon_0\epsilon_{xx}^L\epsilon_V^L}H_{2y} \tag{7.27}$$

The power flow can now be written as

$$P^L = \int_{-\infty}^{0}\frac{k\epsilon_{xx}^L + ik_2\epsilon_{xz}^L}{2\omega\epsilon_0\epsilon_{xx}^L\epsilon_V^L}\left|H_{2y}\right|^2 dz \tag{7.28}$$

Using Eq. (7.24), we can evaluate the integral to yield

$$P^L = \frac{k\epsilon_{xx}^L + k_2\epsilon_{12}^L}{4\omega\epsilon_0\epsilon_{xx}^L\epsilon_V^L k_2}H_L^2 \tag{7.29}$$

The power flow has been computed numerically as a function of wave index $n_x$ for a case appropriate to InSb with $\epsilon_V > 0$ when no linear limit exists and $\lambda = 1.06$ $\mu$m. The results are plotted in Fig. 18 for $k > 0$ and in Fig. 19 for $k < 0$ for various values of the magnetic field. In both cases a threshold power is required to establish the nonlinear wave. The nonreciprocal nature of the propagation is evident from a comparison of the two figures. The situ-

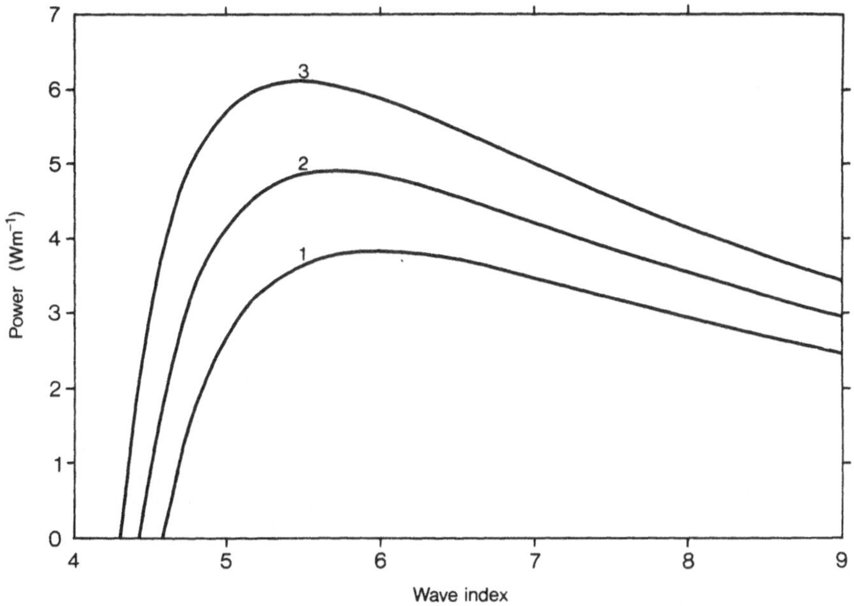


Fig. 20 Total power flow variation with wave index for $\epsilon_V > 0$ when a linear limit exists with $\omega_p = 1.8\times10^{14}$ rad s$^{-1}$, $\epsilon_2 = 15.21$, $\alpha_1 = \alpha_2 = 5\times10^{-7}$ m$^2$W$^{-1}$, $\epsilon_\infty = 15.68$, $\lambda = 10.6$ $\mu$m, (1) $\omega_c/\omega_p = 0.94$, $\epsilon_V = 15.23$, (2) $\omega_c/\omega_p = 1.05$, $\epsilon_V = 15.32$, (3) $\omega_c/\omega_p = 1.2$, $\epsilon_V = 15.4$.

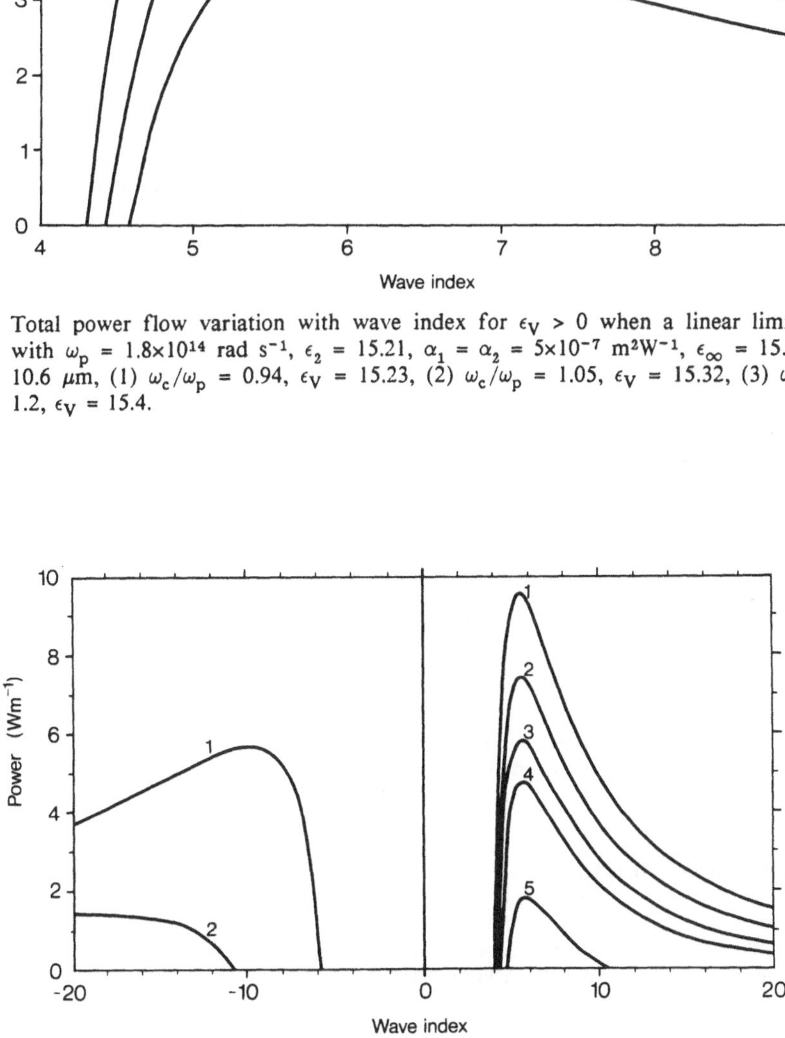


Fig. 21 Total power flow variation with wave index for the lower branch of Fig. 4 with $\omega_p = 2.962\times10^{14}$ rad s$^{-1}$, $\epsilon_\infty = 15.68$, $\omega_c/\omega_p = 0.5$, (1) $\omega/\omega_p = 0.4$, $\epsilon_V = -60.0$, (2) $\omega/\omega_p = 0.48$, $\epsilon_V = -35.7$, (3) $\omega/\omega_p = 0.55$, $\epsilon_V = -22.4$, (4) $\omega/\omega_p = 0.60$, $\epsilon_V = -15.4$, (5) $\omega/\omega_p = 0.75$, $\epsilon_V = -2.0$.

Table of Parameters

| Nonlinear Cladding | | | | |
|---|---|---|---|---|
| $\alpha(m^2A^{-2})$ | $\mu_L$ | $\epsilon_1$ | Fig. No. | Ref. |
| $8.869 \times 10^{-8}$ | 1.29 | 2.3 | 6,7,8,9 | 22,28 |

| Linear Substrate | | | | | |
|---|---|---|---|---|---|
| $\mu_0H_0$(G) | $\mu_0M_0$(G) | $\mu_B$ | $\epsilon_2$ | Fig. No. | Ref. |
| - | - | 2 | 2 | 6 | 27,29 |
| 500 | 1000 | 1.25 | 1 | 11 | 16 |
| 500 | 1750 | 1 | 16 | 12-17 | 24-26 |

ation for $\epsilon_V > 0$ when a linear limit exists is illustrated in Fig. 20 for $k > 0$, $\lambda = 10.6$ $\mu$m, and various values of the magnetic field. One sees that the power flow for given $n_x$ increases as $\omega_c/\omega_p$ increases and that the power flow reaches zero for values of $n_x$ corresponding to the linear limit. The power flow corresponding to the lower branch of Fig. 4 with $\omega_c/\omega_p = 0.5$ and $\epsilon_V < 0$ is plotted against wave index in Fig. 21 for various values of the frequency. The power flow for given $n_x$ increases as the frequency decreases. The limit for which the power flow vanishes is evident.

## 8. CONCLUSIONS

The theory of nonlinear TM and TE waves propagating along the interface between two semi-infinite magnetic media has been presented. Exact solutions are given for a particular type of nonlinear magnetic medium that may be created artificially.[22] It is shown that the excitation of TM waves requires a threshold power and that nonreciprocal behavior is exhibited by both TM and TE waves. A corresponding development has been carried out for TM waves propagating along the interface between a nonlinear dielectric medium and a linear gyrodielectric medium such as a magnetoplasma in a semiconductor. Nonreciprocal behavior and the requirement of a threshold power under certain conditions are demonstrated.

The single interface configuration considered in this chapter is very useful for establishing the general nature of the phenomena to be expected. The extension of our treatment to thin film configurations is of some interest, because such configurations are important for experimental investigations of practical applications such as optical switching. The potential availability of artificial magnetic[22] and dielectric materials[11,12] with large nonlinear coefficients enhances the prospect for practical devices based on the phenomena discussed in this chapter.

## ACKNOWLEDGEMENTS

R. F. Wallis and A. D. Boardman wish to thank NATO for support from collaborative grant RG.86/0474 on "Nonlinear Electromagnetic Guided Waves on Magnetooptic and Magnetic Structures". R. F. Wallis also acknowledges support from NSF Grant No. DMR-8815866.

## REFERENCES

1. A. D. Boardman, and P. Egan, IEEE J. Quantum Electron. **QE-21**, 1701 (1985).
2. D. Mihalache, D. Mazilu and F. Lederer, Opt. Commun. **B59**, 107 (1986).
3. A. D. Boardman, A. A. Maradudin, G. I. Stegeman, T. Twardowski, and E. M. Wright, Phys. Rev. A **35**, 1159 (1987).
4. A. D. Boardman, and T. Twardowski, Phys. Rev. A **39**, 2481 (1988).

5. A. D. Boardman, T. Twardowski, A. Shivarova, and G. I. Stegeman, IEEE Proc. J. **134**, 152 (1987).
6. G. I. Stegeman and C. T. Seaton, Opt. Lett. **9**, 235 (1984).
7. C. T. Seaton, J. D. Valera, R. L. Shoemaker, G. I. Stegeman, J. T. Chilwell, and S. D. Smith, IEEE J. Quantum Electron. **QE-21**, 774 (1985).
8. T. P. Shen, G. I. Stegeman, and A. A. Maradudin, J. Opt. Soc. Am. B **5**, 1391 (1988).
9. J. Zyss, J. Mol. Electron **1**, 25 (1985).
10. P. W. Smith, A. Ashkin, and M. J. Tomlinson, Opt. Lett. **6**, 284 (1981).
11. P. W. Smith, P. J. Maloney, and A. Ashkin, Opt. Lett. **7**, 347 (1982).
12. B. Bobbs, R. Shih, and H. R. Fetterman, App. Phys. Lett. **52**, 4 (1988).
13. C. B. Galanti, and C. L. Giles, SPIE **517**, 219 (1984).
14. A. D. Boardman, and S. A. Nikitov, Phys. Rev. B, (to be published).
15. A. K. Zvezdin and A. F. Popkov, Sov. Phys. JETP **57**, 350 (1985).
16. A. Hartstein, E. Burstein, A. A. Maradudin, R. Brower, and R. F. Wallis, J. Phys. C **6**, 1266 (1973).
17. R. L. Douglass, Phys. Rev. **120**, 1612 (1960).
18. R. W. Damon and J. R. Eshbach, J. Phys. Chem. Solids **19**, 308 (1960).
19. J. J. Brion, R. F. Wallis, A. Hartstein, and E. Burstein, Phys. Rev. Lett. **28**, 1455 (1972).
20. R. F. Wallis, J. J. Brion, E. Burstein, and A. Hartstein, Phys. Rev. B **9**, 3424 (1974).
21. R. F. Wallis, in *Electromagnetic Surface Modes*, A. D. Boardman ed. (Wiley, New York, 1982), p. 575.
22. M. Bertolotti, C. Sibilia, and I. Fuli, Int. Journ. Infrared and Millimeter Waves, **8**, 723 (1987).
23. A. D. Karsano, and D. R. Tilley, J. Phys. **11**, 3487 (1978).
24. J. P. Parekh, and S. R. Ponamgi, J. Appl. Phys. **44**, 1384 (1973).
25. T. J. Gerson, and J. S. Nadan, IEEE Trans. Microwave Theory and Tech. **MTT**-22, 757 (1974).
26. J. P. Parekh, J. App. Phys. **46**, 5040 (1975).
27. M. Fukui, H. Dohi, J. Matsuura and O. Tada, J. Phys. C: Solid State **17**, 1783 (1984).
28. N. Marcuvitz, *Waveguide Handbook* (McGraw Hill, New York, 1951).
29. C. Thibaudeau and A. Caillé, Phys. Rev. B **32**, 5911 (1985).
30. A. J. Palmer, Opt. Lett. **5**, 54 (1980).

NONLINEAR WAVES AND SWITCHING EFFECTS AT NONLINEAR INTERFACES

A. E. Kaplan

Department of Electrical and Computer Engineering
The Johns Hopkins University
Baltimore, MD 21218, USA

P. W. Smith and W. J. Tomlinson

Bell Communications Research
Red Bank, NJ 07701-7020, USA

## 1. INTRODUCTION

Over the past several years, a new direction has arisen in nonlinear optics, which can be described as self-action at nonlinear interfaces (i.e. an interface between linear and nonlinear transparent media). In the most studied configuration, light is incident from a linear medium (with susceptibility $\epsilon_0$) onto the plane surface of a nonlinear medium with a nonlinear refractive index (Fig. 1). The studies of such an interface have followed an almost classical progression of alternating theoretical and experimental advances, which have resulted in an increasingly detailed understanding of the behavior, features, and limitations of nonlinear interfaces, and which have brought into focus the remaining unsolved problems.

Studies of the behavior of nonlinear interfaces originated in theoretical analyses[1-6] which assumed that the incident and reflected fields (i.e. the fields in the linear medium) can be represented as infinite plane waves and that the field in the nonlinear medium is either an infinite plane wave or a nonlinear evanescent wave, or a mixture of both. These studies resulted in closed-form analytical expressions. We refer to them as the "plane wave theory". A major result of the plane-wave theory was that under appropriate conditions the reflectivity of a nonlinear interface would exhibit a threshold behavior as a function of the intensity of the incident field: for intensities below the threshold value the interface would be totally reflecting, and for intensities above the threshold the interface would have a significant transmission. The plane-wave theory also predicted that for a range of input intensities just below the threshold value the reflectivity of the interface can be bistable.

The behavior predicted by the plane-wave theory suggested several applications (e.g. ultra-fast optical switches, scanners, and logic elements), and experiments were initiated to study these predicted phenomena. The initial experimental results were obtained using $CS_2$ as the nonlinear medium (with the so-called positive or self-focusing nonlinearity, see below), and a high-power mode-locked ruby laser as the source.[7-9] The experimental results clearly displayed the predicted threshold behavior. They also seemed to show certain evidence of the predicted bistability.[1,2] However, because of the limited time resolution of the detection system, and the fact that the experiments were done with only a single input pulse width, this evidence was inconclusive.

The interpretation of the experimental results was also complicated by the fact that the experiments were done with a moderately tightly focused Gaussian beam (not an infinite plane wave). Various numerical simulations were reported for a one-dimensional Gaussian input beam (i.e. a beam with a Gaussian distribution in the plane perpendicular to the interface, but which extends uniformly to infinity in the direction perpendicular to the plane of incidence),[8-13] with contradictory results. Some of them (Ref. 10 and to some extent Ref. 11) endorsed the existence of bistability for the Gaussian beam case, while others[12] disputed it. Although some numerical simulations[8,11,14] predicted the appearance of nonlinear surface waves, later results[13,15] showed that these surface waves are prohibited under conditions implied in Refs. 8 and 10-12. The most recent modeling study confirmed the threshold behavior, but showed that the detailed behavior of the interface is quite complex.[13] The reflectivity exhibits not a single threshold, but a series of thresholds at increasingly higher input intensities. The transmitted light was shown to form self-trapped channels, which can easily be identified with the two-dimensional self-trapping[16-19] well known in the theory of light propagation in a semi-infinite medium with a positive nonlinear refractive index. In the case of grazing incidence, though, these channels propagate along complicated curved paths, with each step in the reflectivity associated with the formation of an additional channel. The simulations used a steady-state model, which is not capable of proving, or disproving, the existence of bistability, but it was postulated that the nonlinear interface could not exhibit bistability, because of the lack of any memory mechanism to store the information on the previous field configuration.[13]

The next set of experiments made use of an "artificial" nonlinear medium, with a very large, but also very slowly responding, effective nonlinearity.[20] This made it possible to do experiments with a cw laser, and relatively detailed data were obtained on the reflectivity of a nonlinear interface as a function of the input beam intensity. The results clearly displayed the multiple thresholds, and were in good agreement with the (one dimensional) Gaussian beam simulations. Bistability was *not* observed. However, it was found that when the input intensity was lowered through a threshold value, the high transmission state would persist for of the order of 300 times the response time of the nonlinearity. This quasistability has not yet been accounted for theoretically. Hysteresis and, in fact, quasistability have also been reported recently[21] in an experiment on the nonlinear reflection of picosecond pulses from a glass-liquid interface; the authors of Ref. 21 attribute this effect to probable thermal changes in the near-surface layers in both the glass and the liquid.

The case of a negative (self-defocusing) nonlinearity has been much less thoroughly explored in experiments, but for this case, the plane-incident-wave theory predicts[3-5] an excitation of nonlinear waves of a new kind (described in Refs. 3-5 as longitudinally inhomogeneous traveling waves), which possess interesting physical features.

All of the nonlinear interface effects are caused by the competition between the mismatch in linear susceptibilities $\Delta\epsilon_L$ and the nonlinear component $\Delta\epsilon_{NL}$. If the linear difference is sufficiently small, this competition can result in a strong change of reflection even if the nonlinear component is small as well (as it usually is in an optical range). As the intensity of the incident field increases, the penetrating (or evanescent) field in the nonlinear medium changes the index difference at the interface. At a critical intensity, an abrupt switch from total internal reflection to partial transmission occurs if the nonlinearity is positive, or *vice versa* if the nonlinearity is negative. In order for these effects to occur, it is necessary to have $|\Delta\epsilon_L/\epsilon_0| \ll 1$, and the angle of incidence, $\psi \ll 1$ (see below, Section 2).[1-4] The lack of these conditions leads to absence of the phenomenon. This may be the reason why such effects were not discovered earlier, either in theoretical or experimental research, e.g. in the work[22] devoted to harmonic generation due to reflection of light from nonlinear dielectrics.

The optical hysteresis and bistable behavior suggested by the plane-wave theory[1,2] can be viewed as phenomena quite typical for many nonlinear systems. The concept of bistability (multistability, in general) implies that the system can assume one of two (or more) stable stationary states under the same external conditions. Usually it correlates with hysteresis which occurs as some of the external parameters of the system (e.g. the intensity of incident radiation) vary. This phenomenon is well known in physics, e.g. in the field of phase transitions, in the theory of nonlinear mechanical or electronic oscillators,[23] etc.

One of the reasons that the predictions of bistable and hysteretic reflectivity at a nonlinear interface attracted attention is that in recent years optical bistability has become a very promising field in nonlinear optics.[24] Elements based on optical bistability can be utilized for optically-controlled memory and logic operations in high-speed optical signal processing systems.[25] Most known bistable optical devices comprise a Fabry-Pérot resonator filled with a nonlinear medium, as first proposed in Refs. 26 and 27; bistability in these systems was first observed in Ref. 28. In these systems, bistability is due to the presence of a resonator which provides a feedback. The media used might have a resonant saturated absorption[26-28] or a non-resonant Kerr nonlinearity.[29] The use of Fabry-Pérot resonators causes these devices to be strongly selective to the frequency of the incident light. In contrast to these devices, a nonlinear interface does not employ any resonators. Therefore, no resonant tuning of the laser frequency is required; moreover, a broad spectrum of input light can be used.

A number of new effects, which do not exist in resonators, have also been predicted for nonlinear interfaces, and have stimulated various proposals for applications. New optical switching devices related to the nonlinear interfaces have been proposed in Refs. 6 and 9. It was suggested in Refs. 1 and 2, and studied in detail in Ref. 30, that nonlinear interfaces should be used as bistable mirrors for self-pulsing lasers. Most recently, reversible optical computing circuits based on the plane wave switching effect at nonlinear interfaces were proposed.[31] Interesting modifications of the nonlinear interface idea have appeared, including optical bistability in a frustrated-total-reflection optical cavity,[32] and in a dielectric cladded thin film near the total internal reflection state.[33] An enhancement of nonlinear interface effects by exploiting surface plasmons at metallic surfaces was proposed in Ref. 34 and experimentally explored in Ref. 35. Optical bistability using long-range surface plasmons was proposed in Ref. 36, and surface resonance modes (attributed to dielectric spheres placed on a metal substrate) in Ref. 37. An interesting analogy of bistable total internal reflection at nonlinear interfaces was found[38] in the bistability of the reflection of light from metallic diffraction gratings (positioned at a nonlinear interface) in the vicinity of the Wood anomaly. One trend in the theory and applications of nonlinear interfaces is the study of nonlinear waveguides, e.g. dielectric waveguides formed by a linear dielectric layer sandwich between two nonlinear layers or *vice versa* (see e.g. Refs. 9, 39-41 and references therein), which have promising characteristics.

Despite its deficiencies, the plane-wave theory of nonlinear interfaces has proven to be a useful tool, and a powerful stimulant to research in this field. The theory has suggested a number of interesting nonlinear effects:

(i) multistability and hysteresis jumps in the reflection coefficient;[1-6,42,43]
(ii) change and scanning of the refraction angle and reflection coefficient by varying the intensity of the incident light;[1,2,4]
(iii) optically-induced transparency of the interface by incident light with specific intensity;[1,2,4]
(iv) change of penetration depth of the field into a reflecting medium in the regime of total internal reflection;[2]
(v) excitation of nonlinear waves of a new kind[2-6] [longitudinally inhomogeneous traveling waves (LITW)] and effects related to longitudinally inhomogeneous traveling waves, namely:
(vi) strong nonlinear self-parallax of refracted rays along the interface,[4] and
(vii) self-limitation of the energy flux of the refracted light.[4,43]

The theory has also been used to describe other situations, for example bistability at an electro-optically driven interface,[44] which has recently been observed experimentally,[45] and bistable reflection of light at a nonlinear-optical waveguide junction.[6]

The main results of the plane-wave theory are described in Section 2. The effects predicted by this theory in the particular cases of positive and negative nonlinearities are described in Sections 3 and 4, respectively. The experimental and computer simulation results are discussed in Section 5.

## 2. PLANE-WAVE THEORETICAL FORMALISM

Let a plane wave with amplitude $E_{in}$ be incident from a linear medium with susceptibility $\epsilon_0$ at the glancing angle $\psi$ (Fig. 1) onto the boundary of a nonlinear medium whose susceptibility, $\epsilon_{NL}$, depends on the field amplitude E in the medium according to the equation:

$$\epsilon_{NL}(E) = \epsilon_0 + \Delta\epsilon_L + \Delta\epsilon_{NL}|E|^2 \tag{1}$$

where

$$\Delta\epsilon_{NL}|E|^2 = \epsilon_2|E|^2 \tag{2}$$

$\epsilon_2$ is the nonlinear coefficient (this is usually referred to as a Kerr nonlinearity), and $\Delta\epsilon_L$ is the field-independent mismatch between susceptibilities of the two media. The most interest-

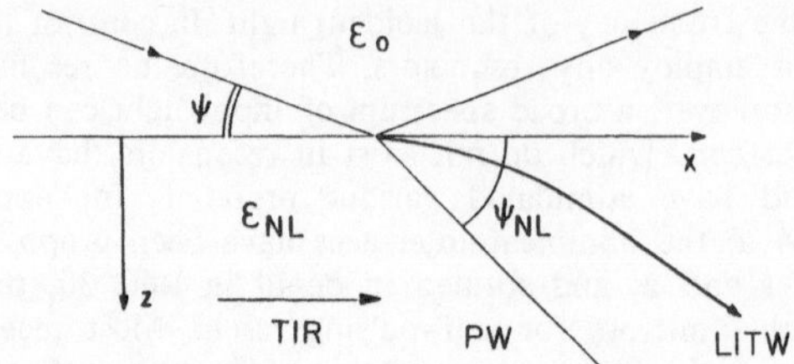


Fig. 1 Wave diagram of the nonlinear interface. Ray trajectories in the nonlinear medium: (1) traveling plane wave (PW), (2) longitudinally inhomogeneous traveling wave (LITW), (3) total internal reflection (TIR).

ing self-action effects are expected when the nonlinear component $\Delta\epsilon_{NL}$ (even if it is small) can compete with the linear mismatch $\Delta\epsilon_L$. Only then can the nonlinear component (caused by the field penetrating into the nonlinear medium) dramatically change the total difference $\Delta\epsilon = \Delta\epsilon_L + \Delta\epsilon_{NL}$ between the susceptibilities of the two media. Such a change, in turn, can result in an abrupt jump of the reflectivity of the interface at some critical incident intensity, e.g. the switch from the total internal reflection state into the transmission state. For the same reason, the incidence of the light has to be almost grazing, because the required magnitude of the glancing angle $\psi$ should be of the order of the critical angle for total internal reflection, which results in the following conditions:[1,2]

$$\left|\frac{\Delta\epsilon_L}{\epsilon_0}\right| \sim \left|\frac{\Delta\epsilon_{NL}}{\epsilon_0}\right| << 1 \qquad \psi \sim \left(\frac{\Delta\epsilon}{\epsilon_0}\right)^{1/2} << 1 \tag{3}$$

These conditions give $\psi \sim 1°$ for $|\Delta\epsilon_{NL}/\epsilon_0| \sim 10^{-4}$. For most of the rest of this chapter we will assume that conditions (3) are valid. This will simplify the theory and make the phenomena independent of the polarization of the incident light.[1,2,4]

The field in the linear medium is represented in the form of two homogeneous plane waves, one of which is the incident wave and the other is the reflected wave, determined by an unknown complex reflection coefficient r. The wave equation for the complex amplitude of the field, E, in a nonlinear medium in the two-dimensional case is written as

$$\frac{d^2E}{dz^2} + k_0^2E\left[\frac{\epsilon_{NL}|E|^2}{\epsilon_0} - \cos^2\psi\right] = 0 \tag{4}$$

where $k_0 = \omega\sqrt{\epsilon_0}/c$, and the z-axis is taken to be perpendicular to the boundary. The total field is $\frac{1}{2}Ee^{-i\omega t}$ + c.c. By equalizing tangential components of the fields on both sides of the boundary, the generalized boundary condition for the refracted wave E can be obtained:[2-4]

$$i\frac{dE(0)}{dz} + k_0\sin\psi[2E_{in} - E(0)] = 0 \tag{5}$$

The expression for the reflection coefficient r is

$$r = \frac{E(0)}{E_{in}} - 1 \tag{6}$$

which also follows from the continuity of tangential components of the fields at the interface. The boundary conditions are a very important part of any problem related to nonlinear interfaces. The failure to define these conditions adequately could result in missing important effects. For example, inadequate conditions for the refracted plane wave in an earlier work[46] made it impossible to determine the wave amplitude and refraction angle separately in the case of $\epsilon_2 > 0$. This led the authors of Ref. 46 to controversial conclusions regarding the existence of optical bistability even within the plane-wave approach.

The field E in the nonlinear medium is written in the form[2-5]

$$E = u(z)\exp\left[ik_0\int_0^z \xi(z)dz + i\phi + ik_0 x\cos\psi\right] \tag{7}$$

where u(z) and ξ(z) are real quantities, u being the real amplitude of the field, and ξ being the angle between the rays and the x-axis at a given point z; φ is a constant phase term. For plane waves, u and ξ are constant, and for surface (or evanescent) waves, when total internal reflection occurs, $\xi = 0$. In general, u and ξ are not constant.

In the theory of reflection from a semi-infinite medium, the conditions at infinity must be defined. Since there are no sources inside the nonlinear medium, traveling waves should only propagate *away* from the interface in the nonlinear medium (surface waves do not carry any energy along the z-axis at all). Therefore, in Eq. (7) for $z \to \infty$, the following conditions have to be satisfied:[1-3]

$$u \to \text{const} \equiv u_\infty \geq 0 \qquad \xi \to \text{const} \equiv \xi_\infty \geq 0 \tag{8}$$

which is, essentially, the Sommerfeld radiation condition (the absence of the backward traveling wave). Substituting the field in the form of Eq. (7) into the wave equation [Eq. (4)], we obtain its first integral

$$I = \xi u^2 = \text{const} \equiv \xi_\infty u_\infty^2 \geq 0 \tag{9}$$

(which expresses the conservation of the energy flux), as well as the equation for the real amplitude u,

$$\frac{d^2u}{dz^2} + k_0^2 u\left(\frac{\epsilon_{NL}(u^2)}{\epsilon_0} - \cos^2\psi - \frac{I^2}{u^4}\right) = 0 \tag{10}$$

The first integral of this equation, which satisfies the radiation condition [Eq. (8)], can be written in the form:[3]

$$\left(\frac{du}{dz}\right)^2 = k_0^2\int_{u_\infty}^{u} [F(u) - F(u_\infty)]d\left(\frac{1}{u^2}\right) \tag{11}$$

where the nonlinear "characteristic" function F(u) is introduced:

$$F(u) = u^4\left[\frac{\epsilon_{NL}(u^2)}{\epsilon_0} - \cos^2\psi\right] \tag{12}$$

Integrating Eq. (11) and taking the boundary condition Eq. (5) into account, one can obtain all possible wave solutions for the problem.

## 3. POSITIVE NONLINEARITIES

In the simplest case of cubic nonlinearity [Eq. (2)] with $\epsilon_2 > 0$ (i.e. Kerr nonlinearity), only two kinds of wave modes, with their intensities depending on z only, can exist:

(i) the homogeneous plane wave, $u(z) = \text{const}$, which corresponds to the transmission regime ($|r| < 1$), and

(ii) the surface wave, or evanescent wave, $\xi = 0$, which corresponds to total internal reflection ($|r| = 1$).

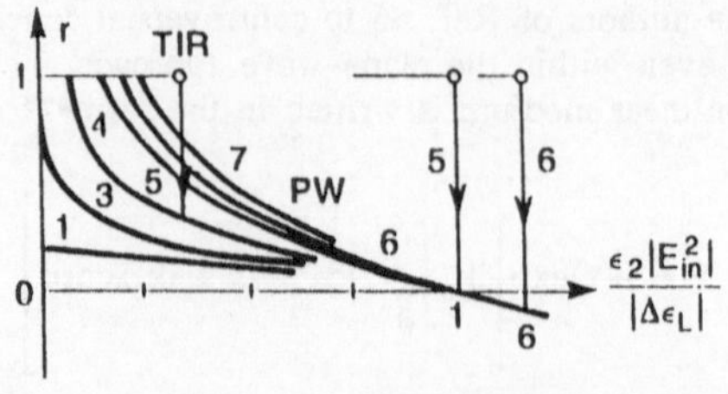


Fig. 2 Reflection coefficient r versus the incident light intensity $|E_{in}|^2$ at different glancing angles $\psi$ in the case of negative linear mismatch of susceptibilities ($\Delta\epsilon_L < 0$) and positive nonlinearity ($\epsilon_2 > 0$). Curves (1) $\psi > \psi_{cr}$, (2) $\psi = \psi_{cr}$, (3), (4) $\psi/\psi_{cr} > 1/(2\sqrt{2})$, (5) $\psi = \psi_{cr}/(2\sqrt{2})$ (nonlinear bleaching), (6), (7) $\psi < \psi_{cr}/(2\sqrt{2})$. After Ref. 1.

Both of these waves are nonlinear analogs of corresponding linear waves, but differ now from those in their dependence on the incident light intensity. (We should note that in two, or more, dimensions, a homogeneous plane wave is not a stable solution in a medium with a positive Kerr nonlinearity,[17] and thus we would only expect the plane-wave theory to apply to the transmitted field in a region sufficiently close to the interface). Using Eqs. (4)-(7) with u and $\xi$ constant, one readily obtains "Snell's nonlinear formula" for the transmission angle $\xi$:[1,2]

$$\left(1 + \frac{\xi}{\psi}\right)^2 \left(\xi^2 - \psi^2 - \frac{\Delta\epsilon_L}{\epsilon_0}\right) = 4\frac{\epsilon_2}{\epsilon_0}|E_{in}|^2 \tag{13}$$

or "Fresnel's nonlinear formula" for the reflection coefficient r,

$$4r\psi^2 + \frac{\Delta\epsilon_L}{\epsilon_0}(1 + r)^2 + \frac{\epsilon_2}{\epsilon_0}|E_{in}|^2(1 + r)^4 = 0 \tag{14}$$

(see Fig. 2). It was shown in Ref. 5 that in the general case of arbitrary magnitude of the nonlinearity $\epsilon_2|E_{in}|^2$ and the incident angle $\psi$ [i.e. in the case when Eq. (3) is not satisfied any more], "Snell's formula" for the angle $\psi_{NL}$ of a refracted plane wave is

$$\left(\frac{\tan^2\psi_{NL}}{\tan^2\psi} - 1 - \frac{\Delta\epsilon_L}{\epsilon_0\sin^2\psi}\right) = \frac{4\epsilon_2|E_{in}|^2}{\epsilon_0\sin^2\psi} \tag{15}$$

(where now $\tan\psi_{NL} = \xi/\cos\xi$), while "Fresnel's formula" for the amplitude u of a reflected plane wave is

$$4\sin^2\psi\left(1 - \frac{u}{|E_{in}|}\right) = \frac{u^2}{\epsilon_0|E_{in}|^2}(\Delta\epsilon_L + \epsilon_2 u^2) \tag{16}$$

The formula for the reflection coefficient r follows from Eq. (16) via Eq. (6) which again yields Eq. (14) where one has to replace $\psi^2$ by $\sin^2\psi$.

The nonlinear total internal reflection state can be excited if the linear mismatch is negative ($\Delta\epsilon_L < 0$) and $\psi < \psi_{cr}$, where $\psi_{cr} = (|\Delta\epsilon_L|/\epsilon_0)^{1/2}$ is the critical angle of linear total internal reflection. The profile of wave intensity now is not exponential: it is obtained from Eq. (11) with $u_\infty = 0$. In the case of a Kerr nonlinearity [Eq. (2)], the general equation (11) under the conditions given by Eq. (3) can then be written in the form

$$\left(\frac{du}{dz}\right)^2 = k_0^2u^2\left(\psi_{cr}^2 - \psi^2 - \frac{u^2}{2}\right) \tag{17}$$

Integration of this equation yields[1,2] the amplitude profile u(z) of a total internal reflection evanescent field (surface wave) which has a "self-channel" shape well known in the soliton theory of two-dimensional self-trapping:

$$u = \left(\frac{2\epsilon_0}{\epsilon_2}\right)^{1/2} \frac{\gamma}{\cosh(k_0\gamma z + C)} \tag{18}$$

where $\gamma = (\psi_{cr}^2 - \psi^2)^{1/2}$. The constant C is determined from the boundary conditions [Eq. (5)], and can have from one to four values (for different $\psi$ and $E_{in}$) which causes multistability and hysteretic jumps (see Fig. 2). Indeed, from the complex boundary conditions [Eq. (5)], one obtains a condition for the real amplitude u(0) of an evanescent field at the boundary:

$$\left[\frac{du(0)}{dz}\right]^2 + k_0^2\psi^2\left[u^2(0) - 4|E_{in}|^2\right] = 0 \tag{19}$$

as well as a formula for the phases of the evanescent field $\phi$ and reflected wave $\phi_r$ [where $\phi_r$ is determined by the relationship $r = \exp(i\phi_r)$]:

$$\phi = \frac{\phi_r}{2} = \mp\arccos\left(\frac{2u(0)}{E_{in}}\right) \tag{20}$$

where the upper sign corresponds to $u'(0) < 0$ and the lower to $u'(0) > 0$. Now, eliminating du(0)/dx from Eq. (19), and from Eq. (17) (taken at z = 0), one obtains an amplitude u(0):[2]

$$|u(0)|^2 = \frac{\epsilon_0}{\epsilon_2}\left[\psi_{cr}^2 \pm \left(\psi_{cr}^4 - 8\psi^2\frac{\epsilon_2|E_{in}|^2}{\epsilon_0}\right)^{1/2}\right] \tag{21}$$

Stipulating that the amplitude u(0), Eq. (21), and the phase $\phi$, Eq. (20), be real quantities, one obtains the conditions for nonlinear total internal reflection to occur:

$$0 \le u^2(0) \le 4|E_{in}|^2 \qquad 8\psi^2\epsilon_2|E_{in}|^2 \ge \psi_{cr}^4\epsilon_0 \tag{22}$$

Based on these conditions, one can show that the threshold grazing angle $\psi_{th}$, at which the switch from nonlinear total internal reflection to partial transmission occurs, is given by[1,2]

$$\left(\frac{\psi_{th}}{\psi_{cr}}\right)^2 = \begin{cases} 1 - \dfrac{2\epsilon_2|E_{in}|^2}{|\Delta\epsilon_L|} & 4\epsilon_2|E_{in}|^2 < |\Delta\epsilon_L| \\[2ex] \dfrac{1}{8}\dfrac{|\Delta\epsilon_L|}{\epsilon_2|E_{in}|^2} & 4\epsilon_2|E_{in}|^2 > |\Delta\epsilon_L| \end{cases} \tag{23}$$

Therefore for a fixed grazing angle $\psi$, the threshold input intensity $|E_{th1}|^2$, at which such switching occurs, is given by

$$\frac{\epsilon_2|E_{th1}|^2}{\Delta\epsilon_L} = \begin{cases} \dfrac{1}{2}\left(1 - \dfrac{\psi}{\psi_{cr}}\right)^2 & \dfrac{1}{\sqrt{2}} \le \dfrac{\psi}{\psi_{cr}} < 1 \\[2ex] \dfrac{1}{8}\left(\dfrac{\psi_{cr}}{\psi}\right)^2 & 0 < \dfrac{\psi}{\psi_{cr}} \le \dfrac{1}{\sqrt{2}} \end{cases} \tag{24}$$

Thus, if starting from total internal reflection at low input intensity and gradually increasing the intensity, the interface switches from the total internal reflection state ($|r| = 1$) to a partial transmission state ($|r| \leq 1$, see Fig. 2) at some intensity given by Eq. (24). The reflection coefficient r at this point is determined by Eq. (14), where the value of $\epsilon_2|E_{th1}|^2$ should be substituted from Eq. (24). In the reverse process (i.e. where the input intensity is decreased), switching from the partial transmission state to the total internal reflection state occurs at the lower intensity, $|E_{th2}|^2$, which is determined from Eq. (13) [or Eq. (14)] by assuming $\xi = 0$ (or $r = 1$):

$$\frac{\epsilon_2|E_{th2}|^2}{\Delta\epsilon_L} = \frac{1}{4}\left[1 - \left(\frac{\psi}{\psi_{cr}}\right)^2\right]^2 \tag{25}$$

i.e. $|E_{th2}|^2 < |E_{th1}|^2$, (see Fig. 2). A jump in the reflection coefficient occurs only during the switch from the total internal reflection to the transmission state; when moving in the opposite direction, there is only a jump in the phase of reflection.

One of the interesting features of nonlinear reflection is what can be called nonlinear bleaching (or optically-induced transparency).[1,2] It is seen from Eq. (14) that at $\Delta\epsilon_L < 0$, a "bleaching" input intensity, $|E_{bl}|^2$, exists:

$$|E_{bl}|^2 \equiv \left|\frac{\Delta\epsilon_L}{\epsilon_2}\right| \qquad \text{i.e.} \quad \Delta\epsilon_L = -\Delta\epsilon_{NL} \tag{26}$$

at which there is no reflection at all ($r = 0$) for any glancing angle $\psi$ (see Fig. 2). This phenomenon cannot be regarded as a nonlinear analog of Brewster bleaching (since it does not depend on either the angle or polarization of incident light). The nonlinear bleaching is attributed to the fact that the field equalizes the susceptibilities of the two media and, therefore, makes the boundary completely transparent. In this case, the jump from total reflection to total transmission occurs at $\psi = \psi_{cr}/(2\sqrt{2})$ (see curve 5 in Fig. 2). For $CS_2$, where $n_2 \sim 10^{-11}$ cgs esu, we have $E_{bl} \sim 1.8\times10^6$ V/cm at $\Delta\epsilon_L \sim -10^{-3}$, in which case $\psi = 0.54°$.

## 4. NEGATIVE NONLINEARITIES; LONGITUDINALLY INHOMOGENEOUS TRAVELING WAVES (LITW)

In a number of physical situations, the nonlinearity can be negative (i.e. $\epsilon_2 < 0$), for instance due to nonlinear resonant interactions (see e.g. Ref. 47), or due to giant nonlinearities, recently observed in some semiconductors (e.g. in InSb, Ref. 48). In such cases, the nonlinear analogs of linear waves (i.e. plane waves and total internal reflection) can exist as well. For plane waves, Snell's, [Eqs. (13), (15)], as well as Fresnel's [Eqs. (14), (16)], formulae remain valid (one has to take into consideration that now the sign of $\epsilon_2$ has changed). In this regime it is also possible to vary the value of the transmission angle $\xi$ and the reflection coefficient r by changing the incident light intensity (Fig. 3). Similarly to the "positive" nonlinearity, a nonlinear total bleaching of the interface also exists; the "bleaching" incident light intensity is given by the same expression Eq. (26) but now the value of the linear mismatch should be positive (Fig. 3, curves 3 and 4).

In contrast to the "positive" nonlinear case, the surface (evanescent) waves at "negative" nonlinear interfaces in the total internal reflection regime exist in exactly the same range of linear mismatches and glancing angles (i.e. $\Delta\epsilon_L < 0$ and $\psi < \psi_{cr} = |\Delta\epsilon_L/\epsilon_0|^{1/2}$) as in the linear case. However, the amplitude profile of the surface waves, given by integration of Eq. (11), is now:[2]

$$u = \left(2\frac{\epsilon_0}{|\epsilon_2|}\right)^{1/2} \frac{\gamma}{\sinh(k_0\gamma z + C)} \tag{27}$$

where $\gamma = (\psi_{cr}^2 - \psi^2)^{1/2}$, and C is a constant determined by the boundary conditions.

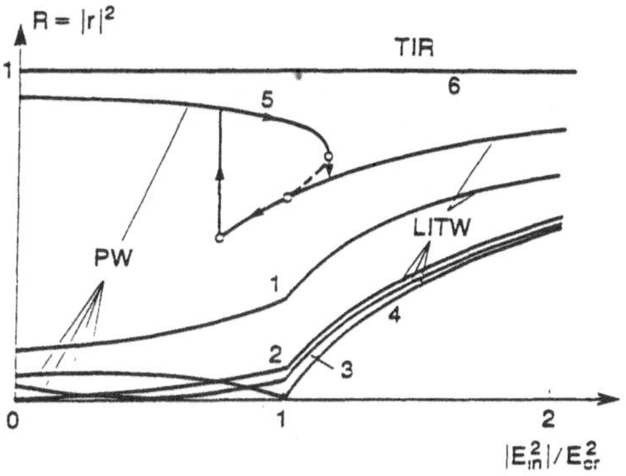


Fig. 3 Reflectivity $R = |r|^2$ as a function of the incident light intensity $|E_{in}|^2$, for a negative nonlinearity ($\epsilon_2 < 0$), and $D \equiv \Delta\epsilon_L/\epsilon_0$. Curves: (1) $D < 0$, $\psi^2 > |D|$; (2) $D = 0$; (3) $\psi^2 > D/2 > 0$; (4) $\psi^2 = D/2$; (5) $\psi^2 < D/2$; (6) $D < 0$, $\psi^2 < |D|$. After Ref. 4.

The amplitude of the field u(0) at the boundary is given by Eq. (21), where one has to take the negative sign in front of the radical. As $z \to \infty$, the evanescent field, Eq. (27), decreases as $\sim \exp(-K_0\gamma z)$, just as in the linear case. But the depth of field penetration into the medium (at the half-intensity level), $L_{NL}$, which is determined by the amplitude profile, Eq. (27), and the value of $u(0) \equiv u_0$, decreases:[2]

$$L_{NL} = L_L \left[ 1 + \frac{2}{\ell n 2} \ell n \left( \frac{1 + \left(1 + u_0^2/4\gamma^2\right)^{1/2}}{1 + \left(1 + u_0^2/2\gamma^2\right)^{1/2}} \right) \right] \tag{28}$$

where $L_L = (\ell n\ 2)/(2k_0\gamma)$ is the depth of the linear penetration. In a strong field, i.e. for $u_0^2 >> \gamma^2$, one has $L_{NL} \simeq 0.6\gamma L_L/u_0 << L_L$.

The most interesting feature in the case of negative nonlinearity is the feasibility of excitation (under some special conditions) of waves of a new kind, so called longitudinally inhomogeneous traveling waves, obtained analytically in Refs. 3-5. Being inhomogeneous and nonplanar near the boundary, these waves reduce to plane waves sufficiently far from the boundary. The concept of longitudinal inhomogeneity was introduced in Ref. 3 to distinguish this phenomenon from the self-action effects due to transversely inhomogeneous waves, such as self-focusing[16-19] and self-bending[49] of beams with limited cross-section.

Examination of Eq. (11) shows that under the conditions

$$\frac{2}{3}q < \frac{|\epsilon_2|}{\epsilon_0}u_\infty^2 < q \qquad q \equiv \frac{\Delta\epsilon_L}{\epsilon_0} + \sin^2\psi \tag{29}$$

excitation of longitudinally inhomogeneous traveling waves is possible. Integration of Eq. (11) yields two possible intensity profiles for longitudinally inhomogeneous traveling waves:[3,4]

$$u^2 = u_\infty^2 \pm 2B^2 \begin{Bmatrix} \sinh^{-2} \\ \cosh^{-2} \end{Bmatrix} (Bk_0 z + C) \qquad B \equiv \left[ \frac{3}{2}\frac{|\epsilon_2|}{\epsilon_0}u_\infty^2 - q \right]^{1/2} \tag{30}$$

For the minimum possible energy density at infinity,

$$u_\infty^2 = u_m^2 = \frac{2}{3}\frac{\epsilon_0}{|\epsilon_2|}q \tag{31}$$

only one (limiting) type of longitudinally inhomogeneous traveling wave remains:

$$u_{lim}^2 = 2\frac{\epsilon_0}{|\epsilon_2|}\left[\frac{q}{3} + \frac{1}{(k_0z + C)^2}\right] \tag{32}$$

A unique "continuum" problem arises in relation to the new solutions: we have only two conditions [which are equivalent to a single complex boundary condition, Eq. (5)] to determine three unknown constants [$u_\infty$ and C in Eq. (32), and $\phi$ in Eq. (7)]. The radiation conditions, Eq. (8), have already been used for the construction of the longitudinally inhomogeneous traveling waves, Eqs. (11) and (30). Therefore, there is a continuum of solutions even when the boundary and radiation conditions are completely specified.

This situation differs drastically from the situation in a linear medium and a "positive" nonlinear medium. An energy criterion for selecting a unique kind of physically realized longitudinally inhomogeneous traveling wave was proposed[4] which is based on the minimization of wave energy density (simultaneously, it results in a maximum energy flux). The only "surviving" kind of longitudinally inhomogeneous traveling wave turns out to be a limiting type, Eq. (32), where $u_\infty^2$ reduces to $u_m^2$, the minimum possible value for longitudinally inhomogeneous traveling waves [Eq. (31)]. This choice was verified analytically[4] by an asymptotic method whereby a weak absorption is introduced and the asymptotic results are obtained when the absorption tends to zero. A small parameter of this limiting transition turned out to be $(k_0L_{ab})^{-2/5}$, where $L_{ab}$ is a characteristic length of the wave absorption.

Now, when the unique "surviving" *type* of longitudinally inhomogeneous traveling wave (but not yet unique *solution*) is found, one can obtain all the characteristics of these waves and, in particular, their amplitude at the boundary, $u_0$.[4,5] Based on this result and using the relationship for $u_0$ for plane waves, Eq. (16), it is easy to find the domains of existence of all of the possible waves (i.e. plane wave, longitudinally inhomogeneous traveling wave and total internal reflection) in the system.[5] These domains are shown in Fig. 4 in the space of the generalized parameters P and Q:

$$P = \frac{\epsilon_2|E_{in}|^2}{\epsilon_0\sin^2\psi} \qquad Q = \frac{\Delta\epsilon_L}{\epsilon_0\sin^2\psi} + 1 \tag{33}$$

The longitudinally inhomogeneous traveling wave excitation gives rise to new effects. First of all, as in the "positive" nonlinear case, it leads to the hysteretic behavior of a system under definite conditions, but now the hysteresis jumps occur between states of two different traveling waves, plane waves and longitudinally inhomogeneous traveling waves (in contrast to "positive" nonlinearity, where they occur between plane waves and total internal reflection). Hysteresis occurs only under the conditions:[4,5]

$$|\epsilon_2|\cdot|E_{in}|^2 > \Delta\epsilon_L > 2\epsilon_0\sin^2\psi$$

The behavior of the reflectivity for a system with several different parameters is shown in Fig. 3 (in particular, one can see hysteresis, curve 5, and self-bleaching, curves 3 and 4). In this figure, $E_{cr}^2$ is the critical intensity of the incident light required for excitation of longitudinally inhomogeneous traveling waves for each particular case, and $D \equiv \Delta\epsilon_L/\epsilon_0$.

Another effect, which is of interest for applications, consists of the self-limitation of the energy flux of longitudinally inhomogeneous traveling waves penetrating into a nonlinear medium.[4,43] This is a direct consequence of the selection of a unique value $u_\infty = u_m$, Eq. (31). The behavior of the energy flux I is shown in Fig. 5 for hysteretic and non-hysteretic situations.

One of the most interesting effects attributed to the excitation of longitudinally inhomogeneous traveling waves is self-parallax,[4] i.e. a displacement of the refracted rays along the interface. This effect is not a nonlinear analog of Goos-Hänchen displacement of Gaussian

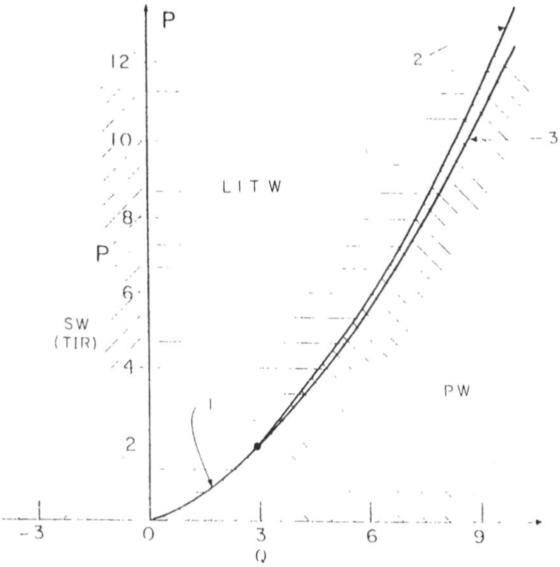


Fig. 4 Diagram of wave states in the space of parameters P and Q. Curve (1) is the boundary between plane wave and longitudinally inhomogeneous traveling wave states; curve (2) is the boundary of the transition jump plane wave → longitudinally inhomogeneous traveling wave; curve (3) is the boundary of the reverse transition jump longitudinally inhomogeneous traveling wave → plane wave. The space between curves 2 and 3 is the hysteresis region. After Ref. 5.

beams under total internal reflection. This is because, in contrast to the Goos-Hänchen effect, self-parallax is now valid for plane waves and for the transmission regime, rather than for total internal reflection. It corresponds to the initial "sticking" of the rays to the interface (Fig. 1, curve 2) which occurs because of conservation of the energy flux [Eq. (9): the angle $\xi$ should increase as the intensity $u^2$ decreases], and it is not related to the existence of hysteresis. The ray trajectory in space is determined by the equation[4]

$$x = z\left(\frac{3}{Q}\right)^{1/2} + \ell\left(1 + \frac{z_\ell}{z}\right)^{-1} \tag{34}$$

where $\ell$ is the parallax (ray displacement) for $z \to \infty$ and $z_\ell$ is the characteristic depth for half of the parallax; $\ell$ and $z_\ell$ are determined by $E_{in}$, $\psi$ and $\Delta\epsilon_L$. Let $k_0 = 10^5$ cm$^{-1}$, $\Delta\epsilon_L/\epsilon_0 = -10^{-4}$, $\psi \sim 0.7°$ and $|\epsilon_2/\epsilon_0| = 10^{-10}$ cgs esu; the critical value of $E_{in}$ for excitation of longitudinally inhomogeneous traveling waves is then $E_{cr} = 1.15\times10^5$ V/cm, and if $E_{in} = 2E_{cr}$, one obtains $\ell = 1.2$ cm and $z_\ell = 0.01$ mm; the parallax is very strong and occurs at a very small depth.

The existence of longitudinally inhomogeneous traveling waves in a system with "negative" nonlinearity and its absence in a "positive" nonlinear system poses a number of questions concerning the kind of nonlinearity that may permit longitudinally inhomogeneous traveling waves. For example, is it possible to predict what kinds of nonlinearities and system characteristics give rise to longitudinally inhomogeneous traveling waves and what kinds of nonlinearities do not allow them? What are the parameters of longitudinally inhomogeneous traveling waves if these are allowed?

These questions have been addressed by a theorem formulated and proved in Refs. 3 and 4. This theorem relates the existence of different kinds of traveling waves to the behavior of a "characteristic" nonlinear function F(u), Eq. (12) above. The main conclusion of this theorem

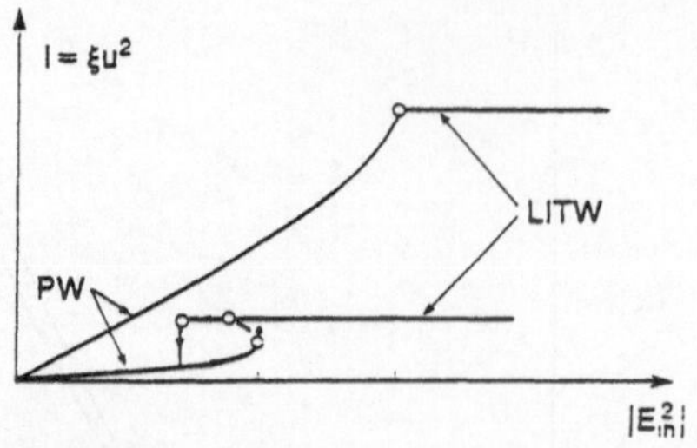


Fig. 5 Energy flux I of transmitted waves as a function of the incident light intensity $|E_{in}|^2$ for hysteresis and nonhysteresis situations. After Ref. 4.

is that the existence of longitudinally inhomogeneous traveling waves in a transparent medium is possible if, and only if:

(i) There is a range V, for which $F(u) > 0$, $u \in V$,

(ii) F(u) decreases at least somewhere in V (i.e. there is at least one interval $W \subset V$ where F(u) falls monotonically),

(iii) If these conditions are satisfied, the longitudinally inhomogeneous traveling wave, if excited, can have a value of $u_\infty$ which can belong only to the interval where this fall occurs, i.e. $u_\infty \in W$ always.

The proof of this theorem is based on an investigation of the behavior of the integral on the right-hand side of Eq. (11).

One of the main implications of this theorem is that longitudinally inhomogeneous traveling waves can be exçited only in a nonlinear medium whose nonlinearity $\epsilon_{NL}(u^2)$ has at least one interval of fall. One the other hand, longitudinally inhomogeneous traveling waves are prohibited in a medium with an increasing function $\epsilon_{NL}(u^2)$. This is the reason why excitation of longitudinally inhomogeneous traveling waves is prohibited for "positive" nonlinearities.

The "continuum problem" arises for all kinds of longitudinally inhomogeneous traveling waves. Therefore, the special issue in the theory of two-dimensional nonlinear waves is related to the "survival" of nonlinear transmitted waves in the case of arbitrary nonlinearity. Several theorems related to this problem were formulated in Ref. 4. In particular, a complete class of functions F(u) were found which allow for the "surviving" longitudinally inhomogeneous traveling wave, and it was proved that the principle of minimization of wave energy remains valid throughout this class.

## 5. EXPERIMENTS

For the purpose of describing the experiments, it is convenient to use a refractive index n, rather than the susceptibility $\epsilon$. We write the index of the nonlinear medium as

$$n(I) = n_0 + \Delta + n_2 I \tag{35}$$

where $n_0$ is the index of the linear medium, $\Delta$ is the mismatch between the refractive indices of the two media, which does not depend on the light intensity I, and $n_2$ is the nonlinear index. In terms of the parameters used earlier in this chapter, we can write (for $\Delta \ll 1$; $n_2 I \ll 1$)

$$\Delta = \frac{\Delta\epsilon_L}{2n_0} \qquad n_2 I = \frac{\epsilon_2 |E|^2}{2n_0} \tag{36}$$

The first experiment to observe the effects suggested by theory[1,2] was reported in Refs. 7 and 9. In this experiment, the liquid $CS_2$ was utilized as a nonlinear medium with positive nonlinearity ($\epsilon_2 > 0$), and glass was used as the linear medium. The temperature regulation of

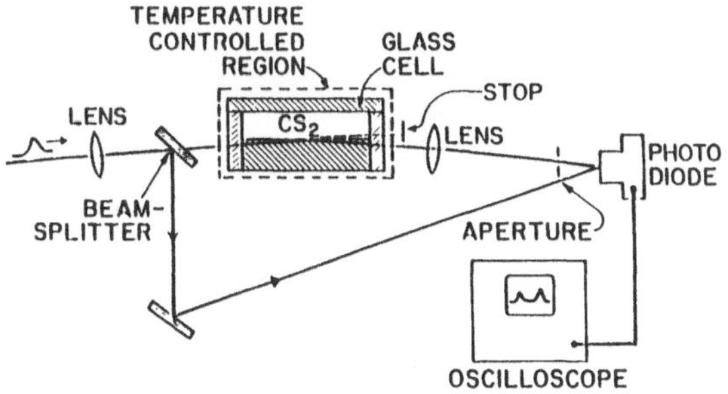


Fig. 6 Experimental setup. After Ref. 7.

the cell served as a fine control of the difference $\Delta\epsilon_L$ between both susceptibilities required by the relationships in Eq. (3). The experiment is briefly described below following Refs. 7-9.

Experiments were performed with the apparatus shown in Fig. 6. The input pulse was generated by a mode-locked traveling-wave ruby laser. Internal Fabry-Pérot mode selection was employed so that the laser output consisted of a train of $\simeq$ 1 ns pulses. The absence of internal structure in these pulses was verified with a streak camera. A single pulse was selected from this train and directed onto the setup shown in Fig. 6. A portion of the beam was sampled and directed with an optical delay of $\simeq$ 6 ns onto a fast photodiode. The remainder of the beam was focused into the glass cell containing $CS_2$. The glass was chosen to have an index of refraction close to that of $CS_2$ at room temperature in order to satisfy the conditions in Eq. (3). The entire cell was placed in a temperature controlled holder and the temperature was adjusted so that for the 694.3 nm wavelength of the ruby laser, the index difference $\Delta \simeq 10^{-3}$. This corresponded to $\psi_{cr} \simeq 2.0°$ in the glass cell.

The reflected beam was monitored by the same photodiode used to monitor the incident pulse, and the diode output was displayed on a fast oscilloscope. The detector-oscilloscope combination had a measured response time of 320 ps. Fig. 7a shows the measured incident and reflected pulse shapes for an incident intensity slightly above the threshold for hysteresis. The dashed curve gives the results of calculations based on the plane-wave theory[1,2] [see Eq. (14)] and was fitted to the data by adjusting the ratio of the peak intensity to the threshold intensity $I_0 = |E_{th1}|^2$, which corresponds to switching from a total internal reflection to partial

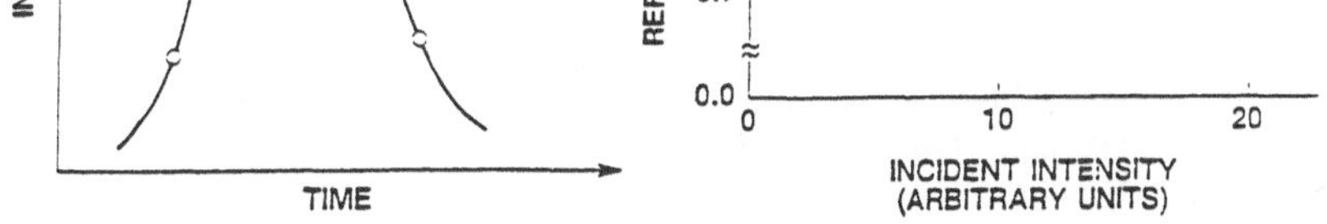

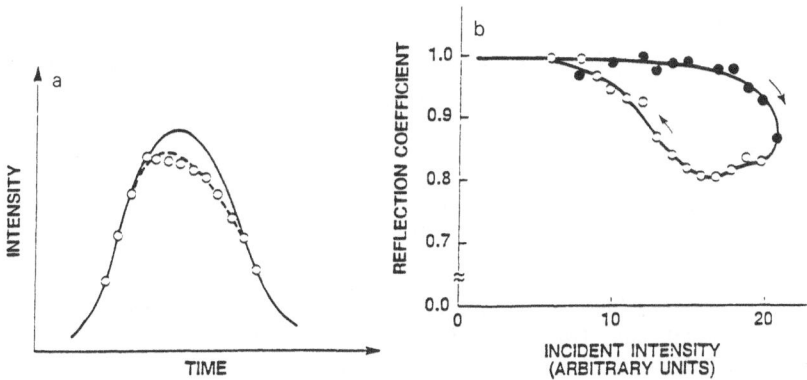

Fig. 7 Experimental measurements of incident and reflected pulse shapes for $\psi/\psi_{cr}$ = 0.707. (a) The solid curve is the incident pulse, and the dots are the experimental measurements of the reflected pulse. The dashed line is the reflected pulse calculated from an incoherent plane-wave analysis for $I/I_0$ = 1.125 (see text). (b) The experimental data in (a) plotted in a way that demonstrates the optical hysteresis observed. After Ref. 7.

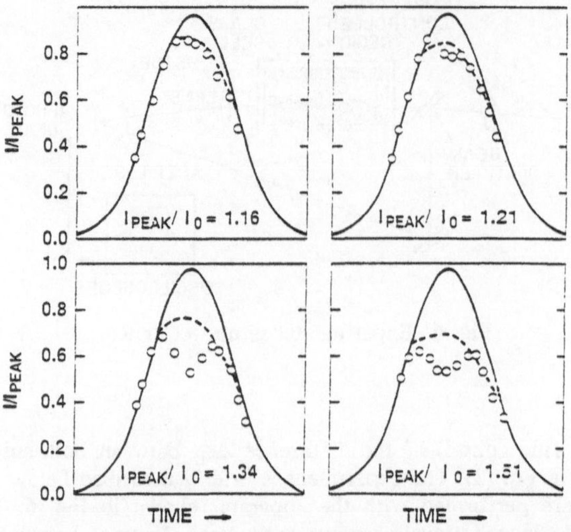


Fig. 8 Experimental measurements of incident and reflected pulse shapes for $\psi/\psi_{cr} = 0.707$. The solid curves are the incident pulses, and the dots are the experimental measurements of the reflected pulses. The dashed lines are from the incoherent plane-wave analysis fitted to all four curves with a single value for the adjustable parameter, $I_0$. After Ref. 8.

transmission state. The fit becomes less good as the peak input intensity is increased (see Fig. 8), however, the plane-wave theory[2] still accurately predicts the threshold intensity $I_0 = |E_{th1}|^2$ [Eq. (24) above] for the onset of switching from total internal reflection to transmission. Moreover, one can see from Figs. 7 and 8, that always[50]

$$I_2 \simeq \frac{I_0}{2} \tag{37}$$

which is also consistent with the plane-wave theory prediction [see Eq. (25) as compared with Eq. (24)].

From the published value[51] $n_2 = 3\times10^{-8}$ $(MW/cm^2)^{-1}$ for $CS_2$, the theoretical value of $I_0 = 8.1\times10^9$ $W/cm^2$ was computed based on Eq. (24) above. The experimentally measured value was $I_0 = 7.5\times10^9$ $W/cm^2$, in very good agreement with the theoretical value. In Fig. 7b, the same experimental data as in Fig. 7a are shown but replotted in the form of reflectivity versus input intensity. This form emphasizes the optical hysteresis observed. However, because of the limited time resolution of the detection system, and the fact that the experiments were only done for a single input pulse width, the observed hysteresis does not prove that the reflectivity is bistable.

Finally, the measured values of the threshold intensity $I_0$ as a function of angle of incidence $\psi$ are shown in Fig. 9. We see that the results are again in good agreement with the predictions of the plane-wave theory [Eq. (24) above].[2]

Some of the experimental results show substantial disagreement with plane-wave calculations. For instance, for sufficiently large peak input intensities, the measured reflected pulse shapes deviate significantly from those obtained from calculations. It was believed and seemingly backed by computer simulations[8,9,11] that at least some of the disagreements could be caused by the excitation of a nonlinear surface wave propagating along the interface with no decay, beginning from the point of reflection of the Gaussian beam. Indeed, it was shown[14] that a nonlinear surface wave in the form of a stationary channel can propagate along the interface, traveling from $x = -\infty$ to $+\infty$, in the absence of any wave incident from either linear or nonlinear media. (Similar waves were discussed previously[52,53] in application to plasmas.) However, using a general theory of the stability of solitons of a cubic-nonlinear Schrödinger equation,[54] it was pointed out[15] that if the surface nonlinear wave (in fact, a soliton solution)

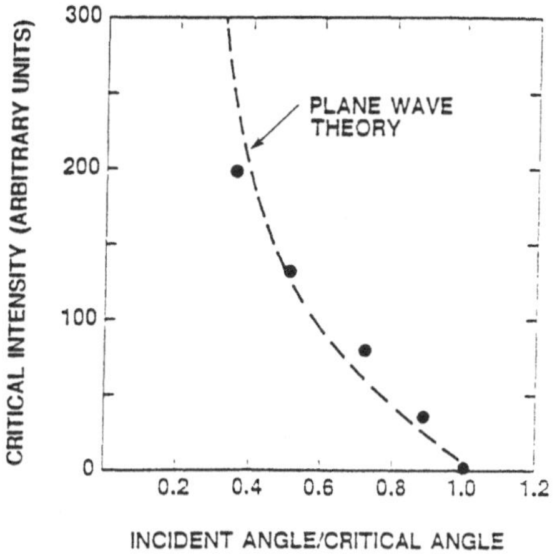


Fig. 9 Experimentally measured critical intensity, $I_0$, as a function of $\psi/\psi_{cr}$. The dashed curve shows the critical intensity predicted by the plane-wave theory. After Refs. 8, 9.

propagates in the direction $x \to \infty$ beyond the point where the incident Gaussian beam hits the interface, then the same soliton-like surface wave (albeit with a spatially shifted maximum) must exist even before this point, i.e. this wave can be observed even when $x \to -\infty$. This would be inconsistent with the formulation of the problem (indeed, it is presumed that at $x \to -\infty$, the entire energy of the field is concentrated in the linear semi-space); therefore the "semi-infinite" nonlinear surface wave[9,11] is prohibited.

A computer simulation[13] confirmed this result. Furthermore, it showed that when the intensity of a Gaussian laser beam exceeds a critical level, the total internal reflection state is drastically disturbed. The laser beam, instead of being totally reflected, breaks through the interface and propagates in the nonlinear material away from the interface. The intensity of the beam at this moment is sufficiently high for the beam to form a self-trapped channel, as shown in Fig. 10. If the incident intensity increases further, a second channel is formed.[13,55] It is natural to expect the formation of more and more channels as the intensity increases, which would correspond to a multisoliton solution.[53] The formation of each self-trapped channel is accompanied by a downward jump in the reflection coefficient.[13] Since the self-trapped channels form right at the interface, resulting in a strong spatial modulation of the intensity at the interface, it is clear that this phenomenon cannot be described by the plane-wave theory, in which it is assumed that the transmitted field is a homogeneous plane wave. The simulations used a steady-state model, which is not capable of proving, or disproving, the existence of bistability, but it was postulated that a nonlinear interface cannot exhibit bistability, because of the lack of any memory mechanism to store the information on the previous field configuration. (This argument does not exclude bistable behavior for the case of other nonlinearities, such as either thermal or diffusion nonlinearities, that are either non-instantaneous and/or nonlocal).

In later experimental research,[20] an "artificial" nonlinear medium (dielectric spheres suspended in a liquid) was used in order to observe switching at a nonlinear interface in the cw regime. The results of these observations were very close to the computer results.[13] Fig. 11 shows the results of the computer calculations for a one-dimensional Gaussian beam. The experimental reflectivity data shown in Fig. 12 are in good qualitative agreement with this model. Perhaps more striking is the excellent agreement between the experimental data for the critical intensity for reflectivity "jumps" and the predictions of the Gaussian-beam model (Fig. 13).

For short observation times, hysteresis was observed. Only the upper branch of the reflection hysteresis characteristic was stable, however. The lower branch would decay to the upper branch with a surprisingly long time constant of ~ $10^3$ x the response time of the nonlinear medium. The nature of this effect remains unclear, although it suggests, at least, that for any application with a pulsed mode of operation, the nonlinear interface will demonstrate

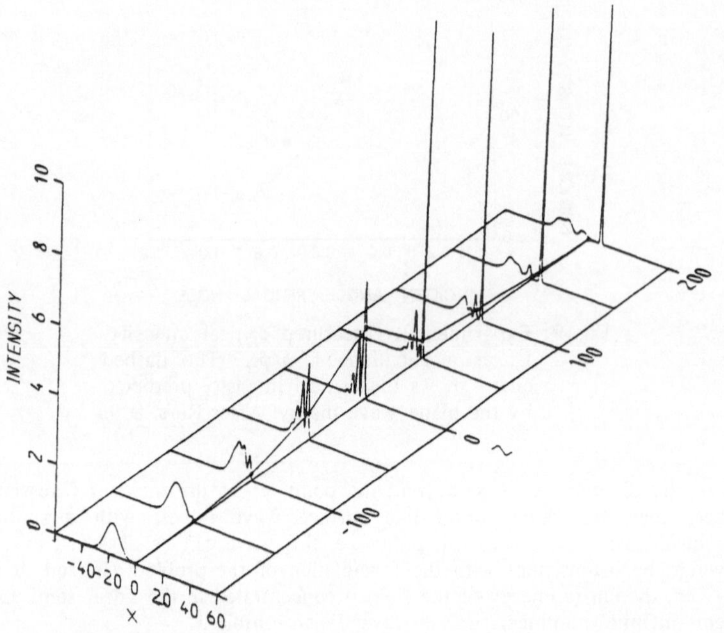


Fig. 10 Perspective plot showing the results of numerical simulations of the behavior of a nonlinear interface for an incident one-dimensional Gaussian beam. This plot is for an input intensity such that a self-focused channel forms in the nonlinear medium. For this figure, from Ref. 13, the definitions of the x and z coordinates are interchanged from those used in the rest of this chapter.

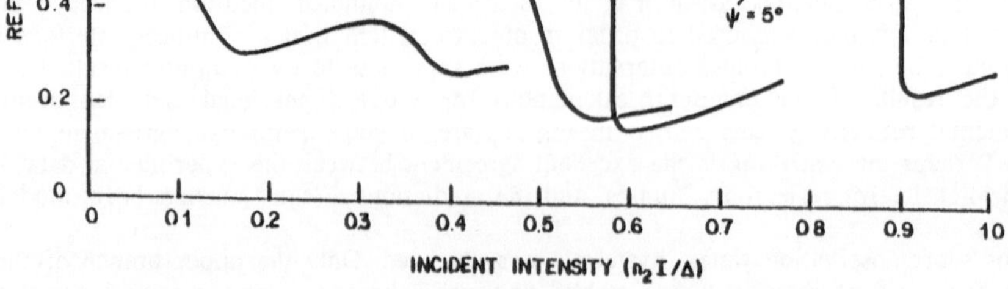

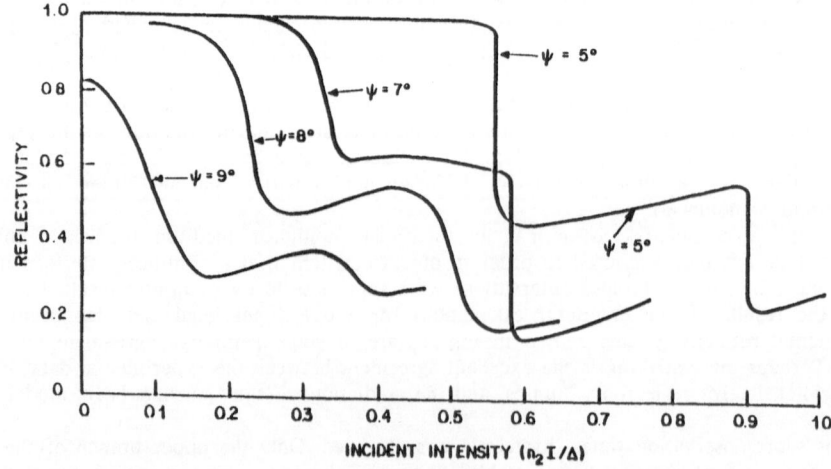

Fig. 11 Reflectivity as a function of intensity for a nonlinear interface; one-dimensional Gaussian beam theory. After Ref. 20.

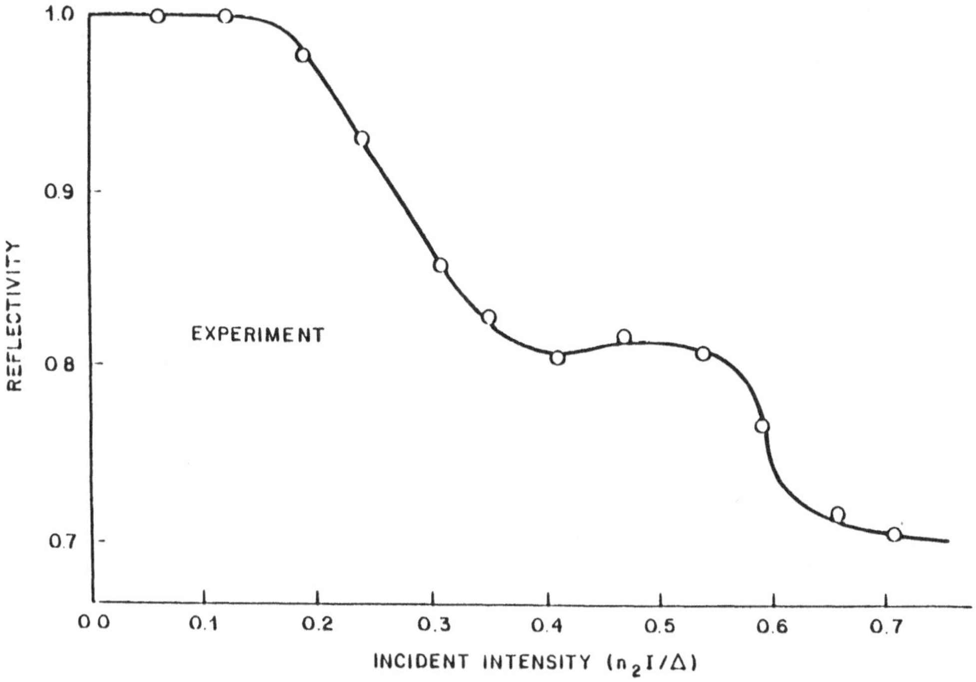


Fig. 12 Experimental plot of reflectivity versus incident intensity for $\psi = 6°$. After Ref. 20.

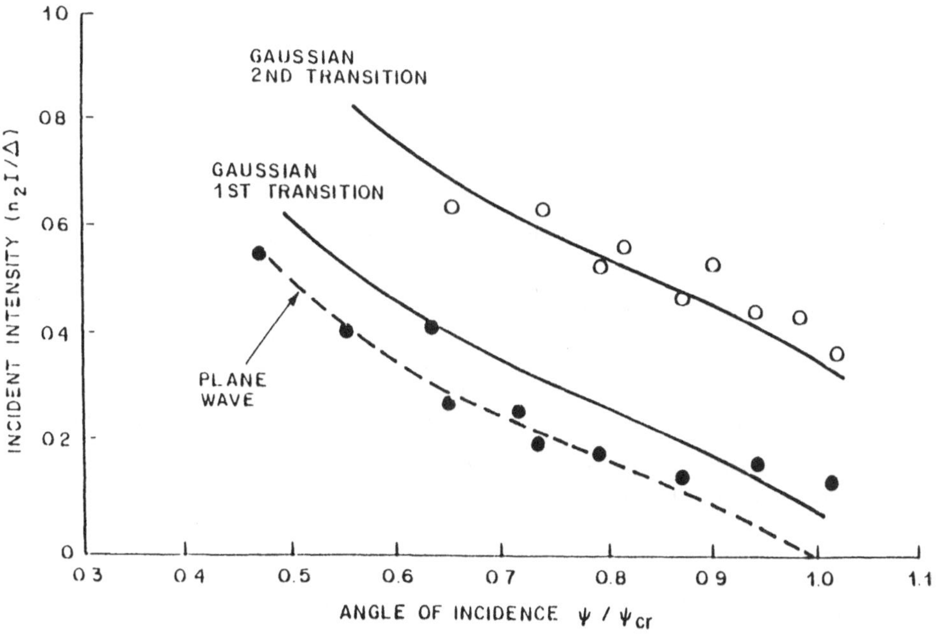


Fig. 13 Critical intensity for reflectivity "jumps" as a function of incidence angle. After Ref. 20.

hysteresis. Indeed, the most recent experimental work[56] done with essentially the same kind of media as in Ref. 20 and using a long pulse mode of operation, revealed very distinct hysteretic jumps. The authors of Ref. 56 have also experimentally observed the large nonlinear Goos-Hänchen effect suggested in Ref. 2 and obtained in a computer simulation.[13] The Goos-Hänchen effect consists of the displacement of a reflected beam along the interface with regard to the point of incidence (see e.g. Ref. 57). It was shown in Ref. 13 that for intensities of the incident beam slightly lower than the critical intensity of formation of the first self-trapped channel, the point of reflection at the nonlinear interface moves very far away from the point of incidence as the incident intensity increases.

## 6. CONCLUSIONS

In this chapter, we have attempted to review the present status of the theoretical analysis of nonlinear interface phenomena, and to present some of the available experimental data. In general, the experimental results are in good agreement with theoretical predictions. The major area where uncertainty remains is the question of optical bistability. No clear evidence from either theory or experiment exists for the presence of two completely stable states when finite sized optical beams are involved.

There are many phenomena predicted by the theoretical analyses that have not yet been observed experimentally. This is primarily due to the difficulty of finding suitable nonlinear media. Recent studies of nonlinear coefficients in highly transparent optical glasses[58] may lead to all-glass interface structures that will allow detailed experimental verification of many of the novel phenomena predicted by the analysis presented in this chapter.

## 7. ACKNOWLEDGMENTS

The work by A. E. Kaplan is supported by AFOSR.

## REFERENCES

1. A E Kaplan, JETP Lett. **24**, 114 (1976)
2. A E Kaplan, Sov. Phys. JETP **45**, 896 (1977)
3. A E Kaplan, Sov. J. Quantum. Electron. **8**, 95 (1978)
4. A E Kaplan, Radiophys. Quantum Electron. **22**, 229 (1979)
5. A E Kaplan, IEEE J. Quantum Electron. **QE-17**, 336 (1981)
6. A E Kaplan in *Optical Bistability*, C W Bowden, M Ciftan and M R Robl eds. (Plenum, New York, 1981), p. 447.
7. P W Smith, J-P Hermann, W J Tomlinson and P L Maloney, Appl. Phys. Lett. **35**, 846 (1979)
8. P W Smith, W J Tomlinson, P J Maloney and J-P Hermann, IEEE J. Quantum Electron. **QE-17**, 340 (1981)
9. P W Smith and W J Tomlinson, in *Optical Bistability*, C W Bowden, M Ciftan and M R Robl eds. (Plenum, New York, 1981), p 463
10. N N Rosanov, Opt. Spectr. **47**, 335 (1979)
11. D Marcuse, Appl. Opt. **19**, 3130 (1980)
12. A A Kolokolov and A I Sukov, Radiophys. Quantum Electron. **21**, 1013 (1978)
13. W J Tomlinson, J P Gordon, P W Smith and A E Kaplan, Appl. Opt. **21**, 2041 (1982)
14. W J Tomlinson, Opt. Lett. **5**, 323 (1980)
15. A E Kaplan, J. Opt. Soc. Am. **71**, 1640 (1981)
16. G A Askar'yan, Sov. Phys. JETP **15**, 1088 (1962)
17. V I Talanov, Izv. Vuzov. Radiofizika 7, 564 (1964); JETP Lett. 2, 138 (1965)
18. R Chiao, E Garmire and C H Townes, Phys. Rev. Lett. **13**, 479 (1964)
19. P L Kelley, Phys. Rev. Lett. **15**, 1005 (1965)
20. P W Smith and W J Tomlinson, IEEE J. Quantum Electron. **QE-20**, 30 (1984)
21. G B Altshuller, V S Ermolaev, K I Krylov, M A Makarov and L I Pavlov, Optics Commun. **56**, 131 (1985)
22. N Bloembergen and D S Pershan, Phys. Rev. **128**, 606 (1962);
N Bloembergen and J Ducuing, Phys. Lett. **6**, 5 (1963);

N Bloembergen and C H Lee, Phys. Rev. Lett. **19**, 835 (1967)
23. J J Stoker, *Nonlinear Vibrations in Mechanical and Electrical Systems* (Interscience, 1950);
A E Kaplan, Yu A Kravtsov and V A Rylov, *Parametric Oscillators and Frequency Dividers* (Sov. Radio, Moscow, 1966) (in Russian)
24. H M Gibbs, *Optical Bistability*, (Academic Press, New York, 1985)
25. P W Smith and W J Tomlinson, IEEE Spectrum **18**, No. 6, 26 (June, 1981)
26. H Seidel, U S Patent No. 3,610,731 (1969)
27. A Szöke, V Daneu, T Goldhar and N A Kurnit, Appl. Phys. Lett. **15**, 376 (1969)
28. H M Gibbs, S L McCall and T N C Venkatesan, Phys. Rev. Lett. **36**, 1135 (1976)
29. F S Felber and J H Marburger, Appl. Phys. Lett. **28**, 731 (1976)
30. T S Dlodlo, Phys. Lett. **84A**, 107 (1981); Phys. Rev. A **27**, 2 (1983)
31. R Cuykendall and D R Andersen, Opt. Lett. **12**, 542 (1987)
32. B Bosacchi and L M Narducci, Opt Lett. **8**, 324 (1983)
33. I C Khoo and J Y Hou, J. Opt. Soc. Am. B 2, 761 (1985)
34. G M Vysin, H J Simon and R T Deck, Opt. Lett. **6**, 30 (1981)
35. P Martinot, A Koster and S Laval, IEEE J. Quantum Electron. **QE-21**, 1140 (1985)
36. R K Hickernell and D Sarid, J. Opt. Soc. Am. B **3**, 1059 (1986)
37. M Inoue, Phys. Rev. Lett. **58**, 871 (1987)
38. G Ya. Slepyan, Radio Eng. & Electron. Phys. **8** (1985)
39. G I Stegeman, IEEE Trans. Micr. Theory and Applications **MTT-30**, 1598 (1982)
40. C T Seaton, J D Valera, R L Shoemaker, G I Stegeman, J T Chilwell and S D Smith, IEEE J. Quantum Electron. **QE-21**, 774 (1985)
41. H Vach, C T Seaton, G I Stegeman and I C Khoo, Opt. Lett. **9**, 238 (1984)
42. N N Rosanov, Sov. Tech. Phys. Lett. **3**, 583 (1977); ibid **4**, 30 (1978)
43. V A Permyakov and O V Bagdasaryan, Radiophys. Quantum Electron. **21**, 92 (1978)
44. A E Kaplan, Appl. Phys. Lett. **38**, 67 (1981)
45. P W Smith, W J Tomlinson, P J Maloney and A E Kaplan, Opt. Lett. **7**, 57 (1982)
46. B B Boiko, I Z Dzhilavdari and N S Petrov, Appl. Spectr. **23**, 1511 (1975)
47. V S Butylkin, A E Kaplan, Yu. G Khronopulo and E I Yakubovick, *Resonant Nonlinear Interactions of Light With Matter*, to be published by Springer-Verlag
48. D A B Miller, S D Smith and B S Wherett, Opt. Commun. **35**, 221 (1980)
49. A E Kaplan, JETP Lett. **9**, 33 (1969)
50. A E Kaplan, P W Smith and W J Tomlinson, **SPIE 317**, 305 (1982)
51. M J Moran, C-Y She and R L Carmen, IEEE J. Quantum Electron. **QE-11**, 259 (1975)
52. A G Litvak and V A Mironov, Izv. Vyssh. Uchehn. Zaved. Radiofiz. **11**, 1911 (1968)
53. Yu. R Alanakyan, Sov. Phys. Tech. Phys. **12**, 587 (1967)
54. V E Zakharov and A B Shabat, Sov. Phys. JETP **38**, 62 (1971)
55. Similar multisoliton emission from a nonlinear waveguide has recently been demonstrated in a computer simulation by E M Wright, G I Stegeman, C T Seaton, J V Moloney and A D Boardman, to be published
56. G Delfino and P Mormile, Opt. Lett. **10**, 618 (1985)
57. H Lotsch, Optik **32**, 776, 189, 299, 553 (1970/72)
58. S Friberg and P W Smith, IEEE J. Quantum Electron. (to be published)

## SURFACE ACOUSTIC WAVES ON NONLINEAR SUBSTRATES

A. A. Maradudin and A. P. Mayer

Department of Physics and Institute for Surface and Interface Science
University of California
Irvine, CA 92717, USA

### 1. INTRODUCTION

All elastic media are nonlinear to a greater or lesser extent. In this chapter we examine some of the consequences of this nonlinearity on the propagation of surface acoustic waves of sagittal and predominantly shear horizontal polarizations across planar and periodically corrugated solid surfaces. These consequences can involve changes in the frequency of the surface acoustic wave, such as occur in the acoustical rectification of a surface acoustic wave, in harmonic generation, and in nonlinear mixing of surface acoustic waves. The nonlinearity can also affect a surface acoustic wave in ways that involve no change in its frequency. It can give rise to nonlinear surface acoustic waves, and to associated surface acoustic solitary waves. Although both kinds of consequences of elastic nonlinearity for surface acoustic waves will be considered in this chapter, the emphasis will be on effects that involve no change in the frequency of the wave. In addition, in the calculations described in this chapter we will confine ourselves primarily to non-piezoelectric materials. However, their extension to piezoelectric media will be indicated briefly.

We begin by outlining the derivation of the equations of motion and the associated boundary conditions for nonlinear elastic media in the forms in which we will use them. We then apply these results to the determination of the Grüneisen constants of surface acoustic waves on anisotropic, homogeneous, semi-infinite substrates and on layered structures. These yield the fractional change in the frequencies of the surface waves due to an externally imposed strain. On the one hand the confrontation of the results of this discussion with experimental results provides information about the nonlinearity of the elastic medium on which the surface acoustic wave propagates. On the other hand, the application of external stresses can be used to tune the frequencies of surface acoustic waves used in technical devices. Our principal motivation in presenting this discussion, however, is that it serves to introduce a projection method for the study of the effects of weak nonlinearity on surface acoustic waves that will be used throughout this chapter. A determination of the nonlinearity parameter for the second harmonic generation of Rayleigh waves by this approach constitutes a second application of this method.

As an introduction to the study of nonlinear surface acoustic waves we next examine the somewhat simpler case of the propagation of nonlinear elastic bulk waves in the presence of spatial dispersion. The dispersion relation for these waves is obtained exactly, in the presence of third order anharmonicity, and by perturbation theory when fourth order anharmonicity is included as well. The latter calculation serves to introduce the method that is used subsequently to obtain the dispersion relation for nonlinear surface acoustic waves. On the basis of the dispersion relation for the nonlinear elastic bulk waves, we derive a nonlinear Schrödinger equation for the slow variation of the complex amplitude of these waves by an eikonal method due to Karpman and Krushkal',[1] and compare the derivation with the one provided by a multiple scales approach. Although the latter approach is more cumbersome than the former, it provides information not obtainable by the eikonal method.

*Nonlinear Waves in Solid State Physics*
Edited by A.D. Boardman *et al.*, Plenum Press, New York, 1990

At this point we are ready to study the propagation of weakly nonlinear surface acoustic waves of both sagittal and predominantly shear horizontal polarization over a nonlinear, spatially dispersive, semi-infinite elastic medium. The spatial dispersion is introduced into the system either by coating the semi-infinite substrate by a thin film whose material properties differ from those of the substrate, or by taking into account the discreteness of the underlying crystal structure of the substrate. The principal effect of the spatial dispersion is to modify the boundary conditions satisfied by the nonlinear elastic displacement field. The character of the weakly nonlinear waves propagating under these conditions is very different from that of the nonlinear surface waves in the absence of dispersion as studied by Parker.[2] Amplitude-dependent dispersion relations are obtained for the surface acoustic waves in the systems considered. By the use of a method developed by Karpman and Krushkal' we obtain an equation governing the slow variations of the amplitudes of these waves which is of the type of the nonlinear Schrödinger equation. The derivation is compared with a multiple scales approach to this problem. Explicit conditions for the formation of envelope solitons, as a consequence of the competition between dispersion and nonlinearity, are presented, as well as the condition for the appearance of self-focusing of the surface acoustic waves.

A periodic corrugation of the surface of a nonlinear elastic substrate is used finally to introduce spatial dispersion into the system. In addition to effects similar to those obtained when spatial dispersion is introduced either by coating the surface by a thin film or by taking into account the underlying crystal structure of the substrate, phenomena resulting from the combination of nonlinearity and periodicity are discussed. An amplitude dependence of the stop bands for the propagation of weakly nonlinear surface acoustic waves across the resulting periodic structure and an amplitude-dependent attenuation in the case of sagittal polarization is found.

## 2. EQUATIONS OF MOTION AND BOUNDARY CONDITIONS

In this section we derive the equations of motion and boundary conditions for the displacement field $\mathbf{u}(\mathbf{x})$ in a system consisting of a nonlinear elastic medium filling the half-space $z < 0$, and a nonlinear elastic medium occupying the region $0 < z < d$ (Fig. 1). The choice of system is prompted by the following considerations. We will be interested in surface acoustic waves of both sagittal and shear horizontal polarizations. However, in general a semi-infinite, homogeneous, linear elastic medium, bounded by a planar, stress-free surface does not support a surface acoustic wave of pure shear horizontal polarization. The deposition of a layer with different material properties on top of the semi-infinite substrate gives rise to surface acoustic waves of this polarization called Love waves, provided that the speed of the appropriate transverse acoustic waves in the layer is smaller than the corresponding speed in the substrate. Because there is now a characteristic length in this system, namely the thickness d of the adlayer, the Love waves are dispersive. This means that their phase velocity is a function of the wave vector in the plane of the surface that describes their propagation across it. For the same reason, the surface acoustic waves of sagittal polarization, Rayleigh waves, that exist even in the absence of the adlayer, and are nondispersive in that case, also become dispersive in the presence of the adlayer. The fact that the system being considered supports dispersive surface acoustic waves in the linear approximation is useful because it is well known that a balance between the spreading tendency of dispersion and the contracting tendency of nonlinearity in the propagation of a wave through a nonlinear dispersive medium can result in the formation of a stable solitary wave, which travels through the medium without change of shape and has some localized form. Consequently, one may expect that under suitable conditions a surface acoustic solitary wave (or surface acoustic soliton†) can exist. We will investigate the conditions under which such surface acoustic excitations can exist.

In the formulation of the theory that follows, we always refer to a fixed coordinate system associated with the undeformed elastic media, i.e. to the material frame.[3] By doing so we avoid many of the complications arising in connection with the nonlinear stress-strain relation in formulations which use different coordinate systems to define the stress and the strain components.

†Strictly speaking, a soliton is a solitary wave that has the property that when it collides with another solitary wave each passes through the other with no change of form and only a small change in the phase of each. However, since the term "surface acoustic soliton" has entered the literature, and since it is less cumbersome than the term "surface acoustic solitary wave", we will use the former here.

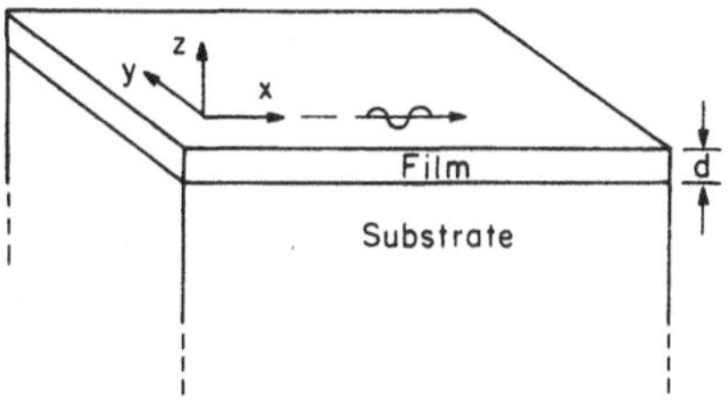


Fig. 1. Geometry considered for the study of surface acoustic envelope solitons.

The starting point for the following derivations is the Lagrange function for the elastic media[3]

$$L = T - V . \tag{2.1}$$

The kinetic energy T is given by

$$T = \frac{1}{2}\int d^3x \; \rho(\mathbf{x})\dot{u}_\alpha(\mathbf{x})\dot{u}_\alpha(\mathbf{x}) , \tag{2.2}$$

where $\rho(\mathbf{x})$ is the mass density of the undeformed elastic media, and the integration is performed over the volume of the undeformed media. To simplify the notation, we employ the summation convention and sum over repeated Cartesian indices in the same term. We write the potential energy V as

$$V = \int d^3x \; \Phi(\mathbf{x}) . \tag{2.3}$$

The potential energy density $\Phi(\mathbf{x})$ is usually written as an expansion in the Lagrangian, finite strain parameters $\eta_{\alpha\beta}$:

$$\Phi = C_{\alpha\beta}\eta_{\alpha\beta} + \frac{1}{2!}C_{\alpha\beta\mu\nu}\eta_{\alpha\beta}\eta_{\mu\nu} + \frac{1}{3!}C_{\alpha\beta\mu\nu\gamma\delta}\eta_{\alpha\beta}\eta_{\mu\nu}\eta_{\gamma\delta} + O(\eta^4) , \tag{2.4}$$

where

$$\eta_{\alpha\beta} = \frac{1}{2}(\epsilon_{\alpha\beta} + \epsilon_{\beta\alpha} + \epsilon_{\mu\alpha}\epsilon_{\mu\beta}) = \eta_{\beta\alpha} \tag{2.5}$$

and $\epsilon_{\alpha\beta}$ is the displacement gradient

$$\epsilon_{\alpha\beta} = \frac{\partial}{\partial x_\beta}u_\alpha \equiv u_{\alpha|\beta} . \tag{2.6}$$

The expansion coefficients are the linear ($C_{\alpha\beta\mu\nu}$) and nonlinear ($C_{\alpha\beta\mu\nu\gamma\delta,...}$) elastic moduli. The $\{C_{\alpha\beta}\}$ are initial stresses, if they are present in the system. For many purposes it is more convenient to expand the potential energy density in terms of the displacement gradients[4]

$$\Phi = S_{\alpha\beta}\epsilon_{\alpha\beta} + \frac{1}{2!}S_{\alpha\beta\mu\nu}\epsilon_{\alpha\beta}\epsilon_{\mu\nu} + \frac{1}{3!}S_{\alpha\beta\mu\nu\gamma\delta}\epsilon_{\alpha\beta}\epsilon_{\mu\nu}\epsilon_{\gamma\delta} + O(\epsilon^4) , \tag{2.7}$$

where the S-tensors can be related to the C-tensors by substituting Eq. (2.5) into Eq. (2.4) and manipulating the resulting expansion into the form (2.7). We will need these conversion relations for the first, second, third, and fourth order tensors:

$$S_{\alpha\beta} = C_{\alpha\beta} \tag{2.8a}$$

$$S_{\alpha\beta\mu\nu} = C_{\alpha\beta\mu\nu} + \delta_{\alpha\mu} C_{\beta\nu} \tag{2.8b}$$

$$S_{\alpha\beta\mu\nu\gamma\delta} = C_{\alpha\beta\mu\nu\gamma\delta} + \delta_{\alpha\mu} C_{\beta\nu\gamma\delta} + \delta_{\alpha\gamma} C_{\mu\nu\beta\delta} + \delta_{\mu\gamma} C_{\alpha\beta\nu\delta} \tag{2.8c}$$

$$\begin{aligned} S_{\alpha\beta\mu\nu\gamma\delta\sigma\lambda} = {} & C_{\alpha\beta\mu\nu\gamma\delta\sigma\lambda} + \delta_{\alpha\mu} C_{\sigma\lambda\gamma\delta\nu\beta} + \delta_{\alpha\gamma} C_{\sigma\lambda\mu\nu\delta\beta} + \delta_{\gamma\mu} C_{\sigma\lambda\alpha\beta\delta\nu} + \delta_{\sigma\alpha} C_{\gamma\delta\mu\nu\lambda\beta} \\ & + \delta_{\sigma\mu} C_{\gamma\delta\alpha\beta\lambda\nu} + \delta_{\sigma\gamma} C_{\mu\nu\alpha\beta\lambda\delta} + \delta_{\sigma\mu}\delta_{\gamma\alpha} C_{\lambda\nu\beta\delta} + \delta_{\gamma\mu}\delta_{\sigma\alpha} C_{\lambda\beta\nu\delta} + \delta_{\sigma\gamma}\delta_{\alpha\mu} C_{\lambda\delta\beta\nu} \, . \end{aligned} \tag{2.8d}$$

It should be noticed, that while both types of tensors are symmetric with respect to permutations of pairs of indices, the S-tensors are in general not symmetric with respect to interchanges of the two indices in a pair, in contrast to the elastic moduli, for which consequently the contracted Voigt notation can be used.

In performing the variation of the Lagrange function with respect to the displacement field, 'one has to take account of the discontinuities of the S-tensors at the interface $z = 0$ and at the surface $z = d$,

$$S_{\alpha\beta\ldots}(z) = \begin{cases} S^{(S)}_{\alpha\beta\ldots} & \text{for } z < 0 \\ S^{(F)}_{\alpha\beta\ldots} & \text{for } 0 < z < d \\ 0 & \text{for } d < z \, , \end{cases} \tag{2.9}$$

where the $S^{(j)}_{\alpha\beta\ldots}$ are constants. These discontinuities lead to $\delta$-functions in the Lagrange equations, whose coefficients must be required to vanish. This yields the boundary conditions

$$T^{(S)}_{\alpha z}(\mathbf{R},0_-;t) = T^{(F)}_{\alpha z}(\mathbf{R},0_+;t) \tag{2.10}$$

$$T^{(F)}_{\alpha z}(\mathbf{R},d;t) = 0 \, , \tag{2.11}$$

where we have decomposed $\mathbf{x} = (\mathbf{R}, z)$ and introduced the stress tensor

$$T^{(j)}_{\alpha\beta} = \frac{\partial\Phi}{\partial\epsilon_{\alpha\beta}} = S^{(j)}_{\alpha\beta} + S^{(j)}_{\alpha\beta\mu\nu} u_{\mu|\nu} + \frac{1}{2} S^{(j)}_{\alpha\beta\mu\nu\gamma\delta} u_{\mu|\nu} u_{\gamma|\delta} + O(u^3) \, . \tag{2.12}$$

In addition, we have to require the displacement field to be continuous across the interface:

$$u_\alpha(\mathbf{R},0_-;t) = u_\alpha(\mathbf{R},0_+;t) \, . \tag{2.13}$$

This condition does not follow from the variational calculation, but is a consequence of the assumption that the film is bonded to the substrate. Finally, the Lagrange equations of motion for $\mathbf{x}$ in the medium j (= F,S) take the form

$$\rho_j \ddot{u}_\alpha(\mathbf{x};t) = T^{(j)}_{\alpha\beta|\beta}(\mathbf{x};t) \, , \tag{2.14}$$

where the $\rho_j$ are constants. Equations (2.10)-(2.14) constitute the complete set of equations of motion and boundary conditions for the displacement field in the system under consideration.

## 2.1 Thin film approximation

At this point, we recall that the reason for introducing a thin film on the surface of a semi-infinite elastic medium is to generate dispersion of the frequencies of the surface acoustic waves of the medium and to ensure the existence of a surface wave of shear horizontal polarization. These two purposes will still be met under the following restrictions, which will greatly simplify the subsequent calculations:

1. Only those solutions of the equations of motion and boundary conditions are considered, for which the displacement field varies little on a length scale given by the film thickness. This means in the weakly nonlinear case, that the wavelength $\lambda$ of the nonlinear wave is much larger than the film thickness d.

2. The nonlinearity in the film is neglected. This condition is not essential. Indeed, by extending the elimination scheme outlined below, it can be shown that if the linear and nonlinear elastic moduli and the mass density of the film are of the same order of magnitude as those of the substrate, the nonlinearity of the film enters the boundary conditions with a prefactor $d/\lambda$ which, in point 1 above, we have assumed to be small.

Under these conditions, the displacement field in the film can be eliminated in favour of one effective boundary condition for the displacement field in the substrate at z=0.[5†] In doing this, we assume in the rest of this section, that external stresses are absent.

To this end, we turn to Eq. (2.11), which has the explicit form

$$C^{(F)}_{\alpha z \mu\nu} u_{\mu|\nu}(z{=}d) = 0 \tag{2.15}$$

and expand it in powers of d:

$$0 = C^{(F)}_{\alpha z \mu\nu} u_{\mu|\nu}(z{=}0_+) + dC^{(F)}_{\alpha z \mu\nu} u_{\mu|\nu z}(z{=}0_+) + O(d^2) . \tag{2.16}$$

Equation (2.10) then becomes

$$T^{(S)}_{\alpha z}(z{=}0_-) = -dC^{(F)}_{\alpha z \mu\nu} u_{\mu|\nu z}(z{=}0_+) + O(d^2) . \tag{2.17}$$

The derivatives of the displacement field with respect to z may have discontinuities at z = 0. In order to eliminate these derivatives, we use the equations of motion in the film:

$$C^{(F)}_{\alpha z \mu\nu} u_{\mu|\nu z} = \rho_F \ddot{u}_\alpha - C^{(F)}_{\alpha\Phi\mu\Theta} u_{\mu|\Phi\Theta} - C^{(F)}_{\alpha\Phi\mu z} u_{\mu|\Phi z} . \tag{2.18}$$

The indices $\Theta$ and $\Phi$ run over x and y only. A second use of Eq. (2.10), in the form

$$C^{(F)}_{\alpha z \mu z} u_{\mu|z}(z{=}0_+) = T^{(S)}_{\alpha z}(z{=}0_-) - C^{(F)}_{\alpha z \mu\Theta} u_{\mu|\Theta}(z{=}0_+) , \tag{2.19}$$

enables us to replace the last term on the right hand side of Eq. (2.18) at $z = 0_+$ by

$$-C^{(F)}_{\alpha\Phi\beta z}\Gamma_{\beta\gamma}\{T^{(S)}_{\gamma z|\Phi}(z{=}0_-) - C^{(F)}_{\gamma z\mu\Theta} u_{\mu|\Theta\Phi}(z{=}0_+)\} . \tag{2.20}$$

The 3x3 matrix $(\Gamma_{\alpha\beta})$ is the inverse of the matrix $(C^{(F)}_{\alpha z\beta z})$. If the film is of tetragonal symmetry with the tetragonal axis along the z axis and the two equivalent axes along the x and y axes, this matrix reduces to

$$(\Gamma_{\alpha\beta}) = \begin{pmatrix} C^{(F)^{-1}}_{44} & 0 & 0 \\ 0 & C^{(F)^{-1}}_{44} & 0 \\ 0 & 0 & C^{(F)^{-1}}_{33} \end{pmatrix} . \tag{2.21}$$

When we substitute Eq. (2.18) together with Eqs. (2.19)-(2.20) into Eq. (2.17), and use the fact that because the displacement field is continuous across the interface z = 0 (Eq. (2.13)), so are its time derivatives and its tangential derivatives, i.e. its derivatives with respect to x and y, we obtain the following effective boundary conditions for the displacement field in the substrate,

$$T^{(S)}_{\alpha z}(z{=}0_-) = -d\{\rho_F \ddot{u}_\alpha(z{=}0_-) - D_{\alpha\Phi\mu\Theta} u_{\mu|\Phi\Theta}(z{=}0_-) - C^{(F)}_{\alpha\Phi\beta z}\Gamma_{\beta\gamma} T^{(S)}_{\gamma z|\Phi}(z{=}0_-)\} + O(d^2) , \tag{2.22}$$

where

†References 6-12 represent earlier work concerned with the derivation of effective boundary conditions.

$$D_{\alpha\Phi\mu\Theta} = C^{(F)}_{\alpha\Phi\mu\Theta} - C^{(F)}_{\alpha\Phi\beta z}\Gamma_{\beta\gamma}C^{(F)}_{\gamma z\mu\Theta} . \tag{2.23}$$

The last term in (2.22) is effectively of second order in d and may therefore be disregarded. The form in which we will use this effective boundary condition in what follows is therefore

$$T^{(S)}_{\alpha z}(z=0_-) = -d\{\rho_F \ddot{u}_\alpha(z=0_-) - D_{\alpha\Phi\mu\Theta}u_{\mu|\Phi\Theta}(z=0_-)\} + O(d^2) . \tag{2.24}$$

When it is used, we will suppress the superscript and index S for quantities which refer to the substrate.

## 2.2 Spatial dispersion

We conclude this section with the following observation. For sufficiently small wavelengths $\lambda$, the discreteness of the underlying lattice renders surface acoustic waves dispersive, i.e. their phase velocity becomes a function of their wavelength $\lambda$. This comes about through the inclusion of higher order spatial derivatives of the displacement field in the equations of motion and boundary conditions satisfied by the latter. The presence of these higher order derivatives is referred to as the incorporation of spatial dispersion into the theory of elasticity, by analogy with the use of this term in discussions of the optical properties of solids. The additional length scale in this case is the lattice constant $a_0$.

To first order in $a_0/\lambda$ it is again through the boundary conditions that the dispersion enters into the determination of the frequency of a surface acoustic wave; it enters the equations of motion to second order in $a_0/\lambda$. In obtaining the effects of spatial dispersion on the frequencies of sagittally polarized surface acoustic waves the contribution of spatial dispersion to the equations of motion can be neglected. However, in a study which showed that spatial dispersion can bind an acoustic wave of shear horizontal polarization to the planar, stress-free surface of a homogeneous, semi-infinite medium, it was found necessary to take into account the contributions of spatial dispersion to both the boundary conditions and the equations of motion.[5,13]

On the basis of a lattice dynamical model by Gazis *et al*[14] for the (001) surface of a simple cubic crystal, in which the atoms interact through nearest and next-nearest neighbor forces, effective boundary conditions for the displacement field have been derived.[5,14] They are of the form

$$T_{\alpha z}(z=0) = a_0 D^{(L)}_{\alpha\beta\mu\nu}u_{\mu|\beta\nu}(z=0) + O(a_0^2) , \tag{2.25}$$

where, within this simple model, the tensor elements $D^{(L)}_{\alpha\beta\mu\nu}$ can be entirely expressed in terms of the elastic moduli. We note parenthetically, that Eq. (2.24) can be brought into the general form (2.25), if the equation of motion in the substrate is used and nonlinear terms in the displacement field proportional to d are neglected.

# 3. THE GRÜNEISEN CONSTANTS OF SURFACE ACOUSTIC WAVES

The dependence of the frequencies of surface acoustic waves in elastic media on externally imposed strains is of interest for both fundamental and applicative reasons. On the one hand, it provides information about the nonlinearity of the elastic medium, in which the surface wave propagates. On the other hand, the application of external stresses might serve as a method to tune the frequencies of surface acoustic waves used in technical devices.

Several experiments have been carried out so far on the change of the velocity of Rayleigh waves due to external stresses.[15,16] The problem of calculating the relative frequency shifts of Rayleigh waves to first order in the externally imposed strains, i.e. the Grüneisen constants, or equivalently to first order in the external stresses, has also been addressed by several workers.[17-22] In an early investigation, Hayes and Rivlin[17] have calculated the dependence of the Rayleigh wave velocity in isotropic media on an external stress the principle direction of which is equal to the direction of propagation. Their results have been generalized to account for arbitrary directions of uniaxial stress by Iwashimizu and Kobori[18] and Tverdokhlebov.[19] In their calculations these authors use Rayleigh's method and expand the eigenvectors, amplitudes, and decay constants of the Rayleigh wave displacement field to first order in the static strains or stresses. Recently, this approach has been extended by Delsanto and Clark[20] to include small orthotropy perturbatively. A different approach has been chosen by Mase and Jonson[21] to calculate the stress-dependence of the Rayleigh wave velocity, which is applicable to arbitrarily anisotropic elastic media. They employ the method originally used by Lothe and Barnett[23] to prove the existence of a surface acoustic wave for any direction of propagation in anisotropic elastic media, to obtain an implicit equation for the Rayleigh wave velocity in a pre-stressed crystal.

In the present discussion, an explicit expression for the derivatives of the frequencies of surface acoustic waves on a semi-infinite, nonlinear medium on which a thin film of a second nonlinear medium has been deposited, with respect to static strains is derived. This expression is obtained by projecting the equation of motion for the displacement field in the strained system on its solution in the absence of static strains. The method used is equivalent to the perturbation technique developed by Auld[24] which has been applied to the problem of frequency shifts of surface acoustic waves in piezoelectric media due to external stresses and fields by Tiersten.[25,26] A similar approach has been used by Husson[27] in a study of the acousto-elastic effect for surface acoustic waves. The resulting expression depends linearly on the elastic constants of third order of the different layers and contains, apart from the mass density and second order elastic constants, only the displacement field of the surface acoustic wave in the undeformed media. It therefore seems to be well suited for the analysis of experiments aimed at determining nonlinear elastic constants. The expression can be applied to Rayleigh waves in a homogeneous elastic medium as a special case†, for which numerical results are presented.

As an example for surface acoustic waves in layered media, we consider Love waves in a film stratified on a homogeneous semi-infinite elastic material. Only those strain configurations are considered which correspond to homogeneous deformations in each layer and are compatible with the presence of the interfaces. Explicit formulas for the frequency shifts to first order in the strain components are given for this system.

The starting point for the discussion that follows is the equations of motion of the film and of the substrate, Eqs. (2.14) which, through cubic anharmonic terms, have the explicit forms

$$\rho_F \ddot{u}_\alpha^{(F)} = S^{(F)}_{\alpha\beta\mu\nu} u_{\mu|\beta\nu} + S^{(F)}_{\alpha\beta\mu\nu\gamma\delta} u_{\gamma|\delta} u_{\mu|\beta\nu} + \dots \tag{3.1}$$

for $0 < z < d$, and

$$\rho_S \ddot{u}_\alpha^{(S)} = S^{(S)}_{\alpha\beta\mu\nu} u_{\mu|\beta\nu} + S^{(S)}_{\alpha\beta\mu\nu\gamma\delta} u_{\gamma|\delta} u_{\mu|\beta\nu} + \dots \tag{3.2}$$

for $z < 0$, and the boundary conditions (2.10) and (2.11) which in the same approximation, become

$$\left[ S^{(F)}_{\alpha z} + S^{(F)}_{\alpha z\mu\nu} u_{\mu|\nu} + \frac{1}{2} S^{(F)}_{\alpha z\mu\nu\gamma\delta} u_{\mu|\nu} u_{\gamma|\delta} + \dots \right]_{z=0_+} = \left[ S^{(S)}_{\alpha z} + S^{(S)}_{\alpha z\mu\nu} u_{\mu|\nu} + \frac{1}{2} S^{(S)}_{\alpha z\mu\nu\gamma\delta} u_{\mu|\nu} u_{\gamma|\delta} + \dots \right]_{z=0_-} \tag{3.3}$$

$$\left[ S^{(F)}_{\alpha z} + S^{(F)}_{\alpha z\mu\nu} u_{\mu|\nu} + \frac{1}{2} S^{(F)}_{\alpha z\mu\nu\gamma\delta} u_{\mu|\nu} u_{\gamma|\delta} + \dots \right]_{z=d} = 0 \,. \tag{3.4}$$

In addition, on the assumption that the film is chemically bonded to the substrate, we require that

$$u_\alpha \big|_{z=0_+} = u_\alpha \big|_{z=0_-} \,. \tag{3.5}$$

†It should be possible to derive a similar expression for the case of Rayleigh waves by the method used in Ref. 21 by expanding the matrix elements S with respect to the external stress to first order and solving the implicit equation for the Rayleigh wave velocity to first order in the stress.

We now decompose the displacement field into a part that describes static strains $v^{(j)}_{\alpha\beta}$ (j = F,S), which is homogeneous in the film and in the substrate, respectively, and a dynamical part,

$$u_\alpha(\mathbf{x};t) = v^{(F)}_{\alpha\beta} x_\beta + \tilde{u}_\alpha(\mathbf{x};t) \qquad (3.6)$$

for 0 < z < d, and

$$u_\alpha(\mathbf{x};t) = v^{(S)}_{\alpha\beta} x_\beta + \tilde{u}_\alpha(\mathbf{x};t) \qquad (3.7)$$

for z < 0. When these decompositions are substituted into the potential energy density, Eq. (2.7), and the result is differentiated with respect to $v^{(j)}_{\alpha\beta}$, the condition for static equilibrium in each medium takes the form

$$S^{(j)}_{\alpha\beta} + S^{(j)}_{\alpha\beta\mu\nu} v^{(j)}_{\mu\nu} + \frac{1}{2} S^{(j)}_{\alpha\beta\mu\nu\gamma\delta} v^{(j)}_{\mu\nu} v^{(j)}_{\gamma\delta} + \dots = 0 . \qquad (3.8)$$

The equations of motion (3.1) and (3.2) become

$$\rho_F \ddot{\tilde{u}}_\alpha = [S^{(F)}_{\alpha\beta\mu\nu} + S^{(F)}_{\alpha\beta\mu\nu\gamma\delta} v^{(F)}_{\gamma\delta} + \dots]\tilde{u}_{\mu|\beta\nu} \qquad (3.9)$$

$$\rho_S \ddot{\tilde{u}}_\alpha = [S^{(S)}_{\alpha\beta\mu\nu} + S^{(S)}_{\alpha\beta\mu\nu\gamma\delta} v^{(S)}_{\gamma\delta} + \dots]\tilde{u}_{\mu|\beta\nu} . \qquad (3.10)$$

The boundary conditions at the interface z = 0 and at the surface z = d become

$$\left[S^{(F)}_{\alpha z\mu\nu}\tilde{u}_{\mu|\nu} + S^{(F)}_{\alpha z\mu\nu\gamma\delta} v^{(F)}_{\gamma\delta}\tilde{u}_{\mu|\nu} + \frac{1}{2} S^{(F)}_{\alpha z\mu\nu\gamma\delta}\tilde{u}_{\mu|\nu}\tilde{u}_{\gamma|\delta} + \dots\right]_{z=0_+}$$

$$= \left[S^{(S)}_{\alpha z\mu\nu}\tilde{u}_{\mu|\nu} + S^{(S)}_{\alpha z\mu\nu\gamma\delta} v^{(S)}_{\gamma\delta}\tilde{u}_{\mu|\nu} + \frac{1}{2} S^{(S)}_{\alpha z\mu\nu\gamma\delta}\tilde{u}_{\mu|\nu}\tilde{u}_{\gamma|\delta} + \dots\right]_{z=0_-} \qquad (3.11)$$

$$\left[S^{(F)}_{\alpha z\mu\nu}\tilde{u}_{\mu|\nu} + S^{(F)}_{\alpha z\mu\nu\gamma\delta} v^{(F)}_{\gamma\delta}\tilde{u}_{\mu|\nu} + \frac{1}{2} S^{(F)}_{\alpha z\mu\nu\gamma\delta}\tilde{u}_{\mu|\nu}\tilde{u}_{\gamma|\delta} + \dots\right]_{z=d} = 0 , \qquad (3.12)$$

when we use the equilibrium conditions (3.8).

The continuity of the displacement field across the interface z = 0, Eq. (3.5), now requires that

$$v^{(F)}_{\alpha x} x + v^{(F)}_{\alpha y} y + \tilde{u}_\alpha|_{z=0_+} = v^{(S)}_{\alpha x} x + v^{(S)}_{\alpha y} y + \tilde{u}_\alpha|_{z=0_-} \qquad (3.13)$$

for all x and y. This requirement imposes strong restrictions on $v^{(j)}_{\alpha x}$ and $v^{(j)}_{\alpha y}$, viz.

$$v^{(F)}_{\alpha x} = v^{(S)}_{\alpha x} \qquad (3.14)$$

$$v^{(F)}_{\alpha y} = v^{(S)}_{\alpha y} . \qquad (3.15)$$

An experimental situation in which these conditions are satisfied will be described below. The satisfaction of Eqs. (3.14)-(3.15) requires further that

$$\tilde{u}_\alpha|_{z=0_+} = \tilde{u}_\alpha|_{z=0_-} . \qquad (3.16)$$

### 3.1 Frequency shifts due to homogeneous strains

We are now in a position to calculate the shift in the frequency of linear surface acoustic or interface acoustic waves due to a homogeneous deformation of the system supporting them, to first order in $v^{(j)}_{\alpha\beta}$. Let $u^{(1)}_\alpha(\mathbf{x})e^{-i\omega_1 t}$ be a solution of Eqs. (3.9)-(3.10), of Eqs. (3.11)-(3.12) with only the linear terms in $\tilde{u}$ retained, and Eq. (3.16). Similarly, let $u^{(0)}_\alpha(\mathbf{x})e^{-i\omega_0 t}$ be a solution to the equations of motion and boundary conditions with $v^{(j)}_{\zeta\xi} = 0$. We further assume that

$$u^{(1)}_\alpha(\mathbf{x};t) - u^{(0)}_\alpha(\mathbf{x};t) = O\{v^{(j)}_{\alpha\beta}\} \tag{3.17a}$$

$$\omega_1 - \omega_0 = O\{v^{(j)}_{\alpha\beta}\} . \tag{3.17b}$$

We now substitute $u^{(1)}_\alpha(\mathbf{x})e^{-i\omega_1 t}$ into the equations of motion (3.9)-(3.10), and project on $u^{(0)}_\alpha(\mathbf{x})$, i.e. we multiply by $u^{(0)*}_\alpha(\mathbf{x})$ and integrate over $\mathbf{x}$ to obtain

$$0 = \sum_j \int_{\zeta_j} dz \int d^2R \left\{ \omega_1^2 \rho_j \sum_\alpha u^{(0)*}_\alpha(\mathbf{x}) u^{(1)}_\alpha(\mathbf{x}) + \left[ C^{(j)}_{\alpha\beta\mu\nu} u^{(0)}_\alpha(\mathbf{x})^* u^{(1)}_{\mu|\nu\beta}(\mathbf{x}) + S^{(j)}_{\alpha\beta\mu\nu\zeta\xi} v^{(j)}_{\zeta\xi} u^{(0)*}_\alpha(\mathbf{x}) u^{(1)}_{\mu|\nu\beta}(\mathbf{x}) \right] \right\} \tag{3.18}$$

where $\zeta_S$ is the segment $(-\infty,0)$ of the z-axis, while $\zeta_F$ is the segment $(0,d)$. Since we consider waves localized at a surface or interface, the sum over j and the integrals over z are convergent. In the xy-plane, we impose periodic boundary conditions. We now integrate the second and third terms in Eq. (3.18) by parts. Since $u^{(1)}_\alpha(\mathbf{x})$ satisfies the boundary conditions (3.11)-(3.12) in the presence of the static strains $\{v^{(j)}_{\alpha\beta}\}$ at the surface and at the interface, and $u^{(0)*}_\alpha(\mathbf{x})$ is continuous across the interface, no surface or interface integrals remain, and we obtain

$$0 = \sum_j \int_{\zeta_j} dz \int d^2R \{ \omega_1^2 \rho_j u^{(0)*}_\alpha(\mathbf{x}) u^{(1)}_\alpha(\mathbf{x}) - [C^{(j)}_{\alpha\beta\mu\nu} u^{(0)*}_{\alpha|\beta}(\mathbf{x}) u^{(1)}_{\mu|\nu}(\mathbf{x}) + S^{(j)}_{\alpha\beta\mu\nu\zeta\xi} v^{(j)}_{\zeta\xi} u^{(0)*}_{\alpha|\beta}(\mathbf{x}) u^{(1)}_{\mu|\nu}(\mathbf{x})]\} . \tag{3.19}$$

The second term on the right hand side of Eq. (3.19) may again be integrated by parts without giving rise to surface or interface integrals, because $u^{(0)*}_\alpha(\mathbf{x})$ satisfies the boundary conditions (3.11)-(3.12) in the absence of the static strains $\{v^{(j)}_{\alpha\beta}\}$ and the nonlinear terms, and $u^{(1)}_\alpha(\mathbf{x})$ is continuous across the interface. Therefore, we find

$$\sum_j \int_{\zeta_j} dz \int d^2R \; C^{(j)}_{\alpha\beta\mu\nu} u^{(0)*}_{\alpha|\beta}(\mathbf{x}) u^{(1)}_{\mu|\nu}(\mathbf{x}) = - \sum_j \int_{\zeta_j} dz \int d^2R \; C^{(j)}_{\alpha\beta\mu\nu} u^{(0)*}_{\alpha|\beta\nu}(\mathbf{x}) u^{(1)}_\mu(\mathbf{x})$$

$$= \omega_0^2 \sum_j \rho_j \int_{\zeta_j} dz \int d^2R \; u^{(0)*}_\alpha(\mathbf{x}) u^{(1)}_\alpha(\mathbf{x}) . \tag{3.20}$$

Substituting Eq. (3.20) into Eq. (3.19) and retaining only terms of first order in the $\{v^{(j)}_{\alpha\beta}\}$, we arrive at our basic result

$$\omega_1^2 - \omega_0^2 = N^{-1} \sum_j S^{(j)}_{\alpha\beta\mu\nu\zeta\xi} v^{(j)}_{\zeta\xi} \int_{\zeta_j} dz \int d^2R \; u^{(0)*}_{\alpha|\beta}(\mathbf{x}) u^{(0)}_{\mu|\nu}(\mathbf{x}) , \tag{3.21}$$

where the normalization constant N is given by

$$N = \sum_j \rho_j \int_{\zeta_j} dz \int d^2R\, u_\alpha^{(0)*}(\mathbf{x}) u_\alpha^{(0)}(\mathbf{x}) \,. \qquad (3.22)$$

This general expression for the frequency shift to first order in the static strains contains, apart from the densities $\rho_j$ and the tensors $(S^{(j)}_{\alpha\beta\mu\nu\zeta\xi})$, only the displacement field of the linear wave in the undeformed system. For the semi-infinite homogeneous or layered structures we have in view here, the field $u_\alpha^{(0)}(\mathbf{x})$ is of the form

$$u_\alpha^{(0)}(\mathbf{R},z) = \frac{\sqrt{K}}{L} e^{i\mathbf{K}.\mathbf{R}} \sum_r w_\alpha^{(j)}(r,\mathbf{K}) e^{\alpha_j(r,\mathbf{K})z} \,, \qquad (3.23)$$

where L is the periodicity length in the xy-plane, and the result for the strain induced frequency shifts can therefore be further specified:

$$\delta\omega^2 = N^{-1} \sum_j S^{(j)}_{\alpha\beta\mu\nu\zeta\xi} v^{(j)}_{\zeta\xi} \sum_{rr'} [-iK_\beta(1-\delta_{\beta z}) + \alpha_j^*(r,\mathbf{K})\delta_{\beta z}]$$
$$\times [iK_\nu(1-\delta_{\nu z}) + \alpha_j(r',\mathbf{K})\delta_{\nu z}] w_\alpha^{(j)*}(r,\mathbf{K}) w_\mu^{(j)}(r',\mathbf{K}) F^{(j)}(r,r';\mathbf{K}) \,, \qquad (3.24)$$

where

$$F^{(S)}(r,r';\mathbf{K}) = \frac{K}{\alpha_S^*(r,\mathbf{K}) + \alpha_S(r',\mathbf{K})}$$

$$F^{(F)}(r,r';\mathbf{K}) = \frac{K}{\alpha_F^*(r,\mathbf{K}) + \alpha_F(r',\mathbf{K})}\left[e^{(\alpha_F^*(r,\mathbf{K})+\alpha_F(r',\mathbf{K}))d} - 1\right] \qquad \text{if } \alpha_F^*(r,\mathbf{K}) \neq -\alpha_F(r',\mathbf{K})$$

$$(3.25)$$

$$= Kd \qquad \text{if } \alpha_F^*(r,\mathbf{K}) = -\alpha_F(r',\mathbf{K}) \,,$$

and

$$N = \sum_j \rho_j \sum_{rr'} w_\alpha^{(j)*}(r,\mathbf{K}) w_\alpha^{(j)}(r',\mathbf{K}) F^{(j)}(r,r';\mathbf{K}) \,. \qquad (3.26)$$

In the following, we discuss two special cases of the general expressions (3.24)-(3.26).

By using the relation (2.8c) and the fact that $u_\alpha^{(0)}(\mathbf{x})$ is an eigenvector of the operator

$$\rho_j^{-1} \sum_{\beta\nu} C^{(j)}_{\alpha\beta\mu\nu} \frac{\partial^2}{\partial x_\beta \partial x_\nu}$$

with eigenvalue $\omega_0^2$, it can be shown that only the symmetric parts of the tensors $v^{(j)}_{\alpha\beta}$ enter the equations (3.21) and (3.24). In the above considerations, we have confined ourselves to non-piezoelectric media. They can however easily be extended to account for the presence of the electrostatic potentials in piezoelectric media (see Ref. 2). Using the same projection technique and extending the integrals also over the vacuum, an expression for the frequency shift linear in an external stress as well as an external electric field can be obtained analogous to (3.24) with (3.25), (3.26). It depends on the parameters of the surface wave in the absence of external stresses and fields and linearly on the second and third order elastic, piezoelectric, electrostriction, dielectric, and electro-optic constants.

### 3.2 Grüneisen constants of Rayleigh waves

The Grüneisen constants $\gamma_{\alpha\beta}(\mathbf{K})$ of surface acoustic waves of wave vector **K** parallel to the surface in a semi-infinite elastic medium may be defined in analogy to the case of bulk waves as

$$\gamma_{\alpha\beta}(\mathbf{K}) = -\omega_{\mathbf{K}}^{-1}\frac{\partial}{\partial\eta_{\alpha\beta}}\omega_{\mathbf{K}}|_{(\eta_{\mu\nu})=0} \,. \tag{3.27}$$

The finite or Lagrangian strain parameter $\eta_{\alpha\beta}$ is, to first order in the infinitesimal strain parameters, equal to the symmetric part of $v_{\alpha\beta}$:

$$\eta_{\alpha\beta} = \frac{1}{2}(v_{\alpha\beta} + v_{\beta\alpha}) + O(v^2) \,. \tag{3.28}$$

Specializing Eq. (3.23) to Rayleigh waves, i.e.

$$u_\alpha^{(0)}(\mathbf{x}) = \frac{\sqrt{K}}{L}e^{i\mathbf{K}.\mathbf{R}}\sum_r w_\alpha(r,\hat{\mathbf{K}})e^{K\hat{\alpha}(r,\hat{\mathbf{K}})z} \,, \tag{3.29}$$

where $\hat{\mathbf{K}} = \mathbf{K}/K$, and $\hat{\alpha} = \alpha/K$, we obtain from Eqs. (3.24)-(3.26) (with d = 0)

$$\gamma_{\zeta\xi}(\mathbf{K}) = -\frac{1}{2Nv_{\hat{\mathbf{K}}}^2}S_{\alpha\beta\mu\nu\zeta\xi}\sum_{rr'}[-i\hat{K}_\beta(1-\delta_{\beta z}) + \hat{\alpha}^*(r,\hat{\mathbf{K}})\delta_{\beta z}]$$

$$\times[i\hat{K}_\nu(1-\delta_{\nu z}) + \hat{\alpha}(r',\hat{\mathbf{K}})\delta_{\nu z}]w_\alpha^*(r,\hat{\mathbf{K}})w_\mu(r',\hat{\mathbf{K}})\left[\hat{\alpha}^*(r,\hat{\mathbf{K}}) + \hat{\alpha}(r',\hat{\mathbf{K}})\right]^{-1} \,, \tag{3.30}$$

where

$$N = \rho\sum_{rr'}w_\mu^*(r,\hat{\mathbf{K}})w_\mu(r',\hat{\mathbf{K}})\left[\hat{\alpha}^*(r,\hat{\mathbf{K}}) + \hat{\alpha}(r',\hat{\mathbf{K}})\right]^{-1} \,, \tag{3.31}$$

and $v_{\hat{\mathbf{K}}}$ is the phase velocity of the Rayleigh wave. Equations (3.30) and (3.31) are valid for Rayleigh waves in arbitrarily anisotropic elastic media. Since in a semi-infinite elastic medium, no length scale is present, all quantities on the right hand side of Eqs. (3.30) and (3.31), and therefore also the Grüneisen constants, depend only on $\hat{\mathbf{K}}$. The left hand side of Eq. (3.30) can therefore be replaced by $\gamma_{\zeta\xi}(\hat{\mathbf{K}})$. It is therefore appropriate to consider the relative change of the phase velocity of a Rayleigh wave due to homogeneous deformations. To express this quantity in terms of Grüneisen constants, we have to recall that the difference between the Rayleigh wave frequencies in the deformed and undeformed medium is given by

$$\delta\omega = v'_{\hat{\mathbf{K}}'}K' - v_{\hat{\mathbf{K}}}K \,, \tag{3.32}$$

where

$$K'_\alpha = (\delta_{\alpha\beta} - v_{\beta\alpha})K_\beta \tag{3.33}$$

to first order in $v_{\alpha\beta}$. $\mathbf{K}'$ is the wave vector in the deformed medium, that corresponds to the wave vector **K** in the undeformed medium. $v'_{\hat{\mathbf{K}}'}$ is the Rayleigh wave velocity for the propagation direction $\hat{\mathbf{K}}'$ in the deformed medium. Thus, the following relation holds

$$v_{\hat{K}}^{-1} \frac{\partial}{\partial \eta_{\alpha\beta}} v_{\hat{K}} = -\gamma_{\alpha\beta}(\hat{K}) + \hat{K}_\alpha \hat{K}_\beta (1 - \delta_{\alpha z})(1 - \delta_{\beta z}) . \tag{3.34}$$

For the case of an isotropic, semi-infinite, nonlinear elastic medium, we have compared our results of Eqs. (3.30), (3.31), and (3.34) numerically with the results given by Eqs. (44) and (45) in Ref. 18 by using

$$\hat{K} = (0,1,0) \tag{3.35}$$

$$\hat{\alpha}(r,\hat{K}) =: \alpha(r) = \left[1 - \frac{v_R^2}{v_r^2}\right]^{1/2} \qquad r = L,T , \tag{3.36}$$

where $v_R$ is the Rayleigh wave velocity and $v_L$, $v_T$ are the velocities of longitudinal and transverse bulk waves, respectively,

$$w(L,\hat{K}) = [0,1,-i\alpha(L)] \tag{3.37a}$$

$$w(T,\hat{K}) = \left(0,-\left[\alpha(L)\alpha(T)\right]^{1/2},i\left[\alpha(L)/\alpha(T)\right]^{1/2}\right) , \tag{3.37b}$$

and

$$N = \rho \frac{[\alpha(L) - \alpha(T)][\alpha(L) - \alpha(T) + 2\alpha(L)\alpha^2(T)]}{2\alpha(L)\alpha^2(T)} . \tag{3.38}$$

Furthermore, Eq. (2.8c) and the relations between the Lamé constants and the second and third order elastic moduli, given in Ref. 28, have been used, and perfect agreement has been found.

In Fig. 2, the Grüneisen constants of Rayleigh waves are shown as a function of propagation direction for a (001) surface of a $BaF_2$ crystal. This material has the peculiar property of being virtually isotropic with respect to the second order elastic moduli (its anisotropy factor $2C_{44}/(C_{11}-C_{12}) = 1.02$), while the nonlinear coupling coefficients $S_{\alpha\beta\mu\nu\zeta\xi}$ show a considerable anisotropy. In the numerical evaluation of Eqs. (3.30) and (3.31), we have therefore used the isotropic approximation for the quantities $\alpha(r,\hat{K})$ and $w_\alpha(r,\hat{K})$, with $\rho v_L^2 = C_{11}$ and $\rho v_T^2 = C_{44}$, while the anisotropy has been fully taken into account in the tensor $(S_{\alpha\beta\mu\nu\zeta\xi})$ by using the experimental elastic moduli of Ref. 29 in Eq. (2.8c). It should be emphasized that the isotropic approximation for the linear quantities is not necessary and has only been invoked to simplify the numerical calculation. To illustrate the anisotropy of the Grüneisen constants due to the anisotropy of the tensor $(S_{\alpha\beta\mu\nu\zeta\xi})$, we have calculated $\gamma_{\alpha\alpha}^{(r)}(\phi)$, the Grüneisen constants in a coordinate system rotated with respect to the cubic axes by an angle $\phi$ around the z-axis, so that $\hat{K}$ always points into the x-direction. In a purely isotropic nonlinear elastic medium, these quantities would not depend on $\phi$. In the case of a (001) surface of $BaF_2$, however, a strong angular dependence, in particular of $\gamma_{xx}^{(r)}(\phi)$, can be seen in Fig. 3.

In Table 1, Grüneisen constants are listed for Rayleigh waves propagating in the (100) direction on the (001) surface of various cubic materials. They have been obtained by evaluating (3.30) and (3.31) without making use of the isotropic approximation for the linear Rayleigh wave displacement field.

### 3.3 Love waves

We now consider guided waves propagating in the structure consisting of a semi-infinite substrate of one nonlinear medium in the region $z < 0$, on which a film of thickness d of a second nonlinear medium has been deposited. As we remarked above, the strain parameters $\{v_{\alpha\beta}^{(j)}\}$ in this case must satisfy the conditions (3.14)-(3.15). An experimental situation in which these conditions are satisfied is the following. Let us assume that our layered structure is composed of isotropic or cubic materials, whose cubic axes are parallel to the coordinate axes. If this structure is compressed along the x- and y-axes by walls hard enough compared to the

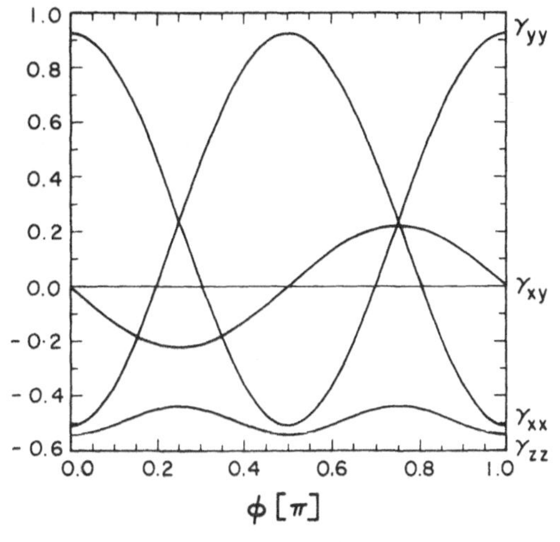


Fig. 2. Grüneisen constants of Rayleigh waves on a (001) surface of $BaF_2$ as a function of the propagation direction $\hat{K} = (\cos\phi, \sin\phi, 0)$. The following values for the elastic constants have been used:[29] (unit: $10^{10}$ $Nm^{-2}$)

| | | |
|---|---|---|
| $C_{11} = 9.07$ | $C_{12} = 4.10$ | $C_{44} = 2.53$ |
| $C_{111} = -58.4$ | $C_{144} = -12.1$ | $C_{112} = -29.9$ |
| $C_{166} = -8.9$ | $C_{123} = -20.6$ | $C_{456} = -2.7$ . |

For the Rayleigh wave displacement field, the isotropic approximation has been used, with the longitudinal sound velocity proportional to $\sqrt{C_{11}}$ and the transverse proportional to $\sqrt{C_{44}}$. For the (001) surface of a cubic crystal, $\gamma_{xz}$ and $\gamma_{yz}$ vanish.

layer materials, that they may approximately be considered as rigid, homogeneous strains $v_{xx}$ and $v_{yy}$ are generated as well as relative changes of the layer thicknesses:

$$v_{zz}^{(j)} = -\frac{c_{12}^{(j)}}{c_{11}^{(j)}}(v_{xx} + v_{yy}) \tag{3.39}$$

to first order in $v_{xx}$ and $v_{yy}$.

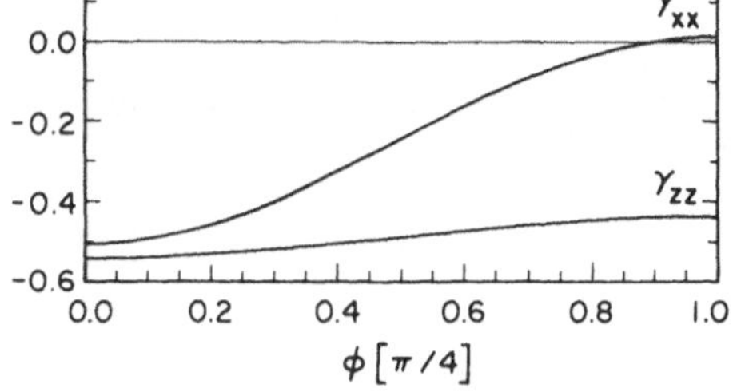

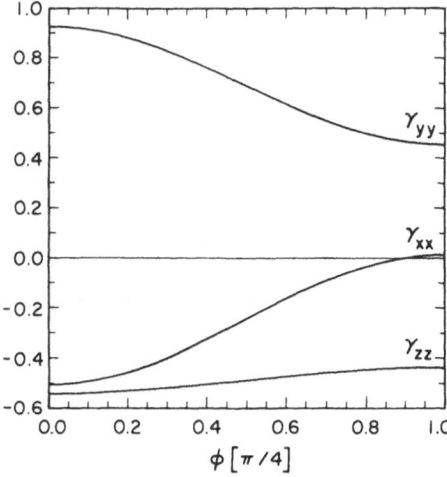

Fig. 3. Grüneisen constants of Rayleigh waves on a (001) surface of $BaF_2$ in a coordinate system rotated with the propagation direction such that $\hat{K} = (1,0,0)$.

Table 1. Grüneisen constants of Rayleigh waves propagating in the (100) direction of the (001) surface

| Substance | $\gamma_{xx}$ | $\gamma_{yy}$ | $\gamma_{zz}$ | $\alpha(r)$ |
|---|---|---|---|---|
| Cu | -0.063 | 2.528 | -0.688 | complex |
| Si | -0.193 | 0.080 | -0.458 | complex |
| GaAs | 0.038 | 0.909 | -0.308 | complex |
| NaF | 0.267 | -0.747 | 0.609 | real |
| KCl | -1.087 | -2.286 | 0.546 | real |
| $BaF_2$ | -0.519 | 0.911 | -0.555 | real |
| $BaF_2$† | -0.507 | 0.926 | -0.541 | real |

The simplest example of localized waves in a layered structure are Love waves in a semi-infinite medium of material (1) covered by a film of thickness d consisting of material (2).[13] The two materials are taken to be isotropic or cubic with the Love waves propagating along the cubic axes. By applying the general formulas of Section 3.1 to this special case, the following expression for the frequency shift can be derived:

$$\frac{\delta\omega^2}{K^2} = \frac{S^{(1)}(+)2d - S^{(1)}(-)|\alpha_1|^{-1}\sin(2|\alpha_1|d) + S^{(2)}(+)2\alpha_2^{-1}\cos^2(|\alpha_1|d)}{\rho_1\left[2d - |\alpha_1|^{-1}\sin(2|\alpha_1|d)\right] + \rho_2 2\alpha_2^{-1}\cos^2(|\alpha_1|d)} . \tag{3.40}$$

Here, K is the modulus of the two-dimensional wave vector of the Love wave, and

$$\alpha_\kappa = \left[K^2 - \frac{\omega^2\rho_\kappa}{C_{44}^{(\kappa)}}\right]^{1/2} \tag{3.41}$$

$$S^{(\kappa)}(\pm) = v_{xx}\left\{C_{166}^{(\kappa)} + C_{11}^{(\kappa)} \pm |\hat{\alpha}_\kappa|^2(C_{144}^{(\kappa)} + C_{12}^{(\kappa)})\right\} + v_{yy}\left\{C_{166}^{(\kappa)} + C_{12}^{(\kappa)} + 2C_{44}^{(\kappa)}\right.$$

$$\left. \pm |\hat{\alpha}_\kappa|^2(C_{166}^{(\kappa)} + C_{12}^{(\kappa)} + 2C_{44}^{(\kappa)})\right\} + v_{zz}^{(\kappa)}\left\{C_{144}^{(\kappa)} + C_{12}^{(\kappa)} \pm |\hat{\alpha}_\kappa|^2(C_{166}^{(\kappa)} + C_{11}^{(\kappa)})\right\} , \tag{3.42}$$

$$\hat{\alpha}_\kappa = \frac{\alpha_\kappa}{K} .$$

In contrast to the case of Rayleigh waves in a semi-infinite medium the strain induced frequency shifts of surface waves of layered structures also depend on K.

## 4. NONLINEARITY PARAMETER FOR RAYLEIGH WAVES

The theory of second harmonic generation of Rayleigh waves has been addressed by several authors[30,31] with partly different approaches. The problem consists in finding a solution to the equations of motion (2.14) and boundary conditions (2.10 - 2.11), specialized to a semi-infinite elastic medium, of the form

$$u_\alpha(\mathbf{x};t) = u_\alpha^{(0)}(\mathbf{x})e^{-i\omega_0 t} + u_\alpha^{(2)}(\mathbf{x};t) , \tag{4.1}$$

†The isotropic approximation with $C_{12} = C_{11} - 2C_{44}$ has been used to calculate the Rayleigh wave velocity and displacement field in the undeformed medium.

where $u_\alpha^{(0)}(\mathbf{x})$ is the displacement field of the input Rayleigh wave, given by (3.29), and $u_\alpha^{(2)}(\mathbf{x};t)$ is of second order in the amplitude of the input wave and is further decomposed as

$$u_\alpha^{(2)}(\mathbf{x};t) = \{\tilde{u}_\alpha^{(2)}(\mathbf{x}) + h(\mathbf{K})\hat{\mathbf{K}}.\mathbf{x}u_\alpha^{(0)}(2\mathbf{x})\}e^{-2i\omega_0 t} \ . \tag{4.2}$$

The existence of a solution of this form has been shown in Refs. 31, 32. The first term in (4.2) is bounded while the second term describes the linear growth of the second harmonic with distance of propagation. This secular term arises because of the absence of spatial dispersion for Rayleigh waves. $u_\alpha^{(2)}(\mathbf{x})$ is a solution of the inhomogeneous equations

$$-\rho\ddot{u}_\alpha^{(2)}(\mathbf{x};t) + C_{\alpha\beta\mu\nu}u_{\mu|\beta\nu}^{(2)}(\mathbf{x};t) = -S_{\alpha\beta\mu\nu\zeta\xi}u_{\zeta|\xi}^{(0)}(\mathbf{x})u_{\mu|\beta\nu}^{(0)}(\mathbf{x})e^{-2i\omega_0 t} \tag{4.3}$$

and boundary conditions

$$C_{\alpha z\mu\nu}u_{\mu|\nu}^{(2)}(\mathbf{R},0;t) = -\frac{1}{2}S_{\alpha z\mu\nu\zeta\xi}u_{\mu|\nu}^{(0)}(\mathbf{R},0)u_{\zeta|\xi}^{(0)}(\mathbf{R},0)e^{-2i\omega_0 t} \ . \tag{4.4}$$

By adding the homogeneous solution

$$u_\mu^{(h)}(\mathbf{x};t) = h(\mathbf{K})\frac{\sqrt{K}}{L}e^{2i(\mathbf{K}.\mathbf{R}-\omega_0 t)}\left\{\left[v_{\hat{\mathbf{K}}}t - \hat{\mathbf{K}}.\mathbf{x}\right]\sum_r w_\mu(r,\hat{\mathbf{K}})e^{2\alpha(r,\mathbf{K})z} - iz\sum_r \hat{\alpha}(r,\hat{\mathbf{K}})w_\mu(r,\hat{\mathbf{K}})e^{2\alpha(r,\mathbf{K})z}\right\} \tag{4.5}$$

of the system of equations (4.3) and (4.4) to $u_\alpha^{(2)}(\mathbf{x},t)$, it is possible to transform from a solution with the second harmonic growing with propagation distance to a solution with a temporal growth to match resonator-like boundary conditions.

The quantity of main interest for second harmonic generation experiments is the growth parameter $h(\mathbf{K})$, or the nonlinearity parameter

$$\hat{h}(\hat{\mathbf{K}}) = h(\mathbf{K})\left[K^5N(\hat{\mathbf{K}})\right]^{-1/2} , \tag{4.6}$$

where $N(\hat{\mathbf{K}})$ is defined in Eq. (3.31), which is independent of the intensity of the input wave and its wavelength. For the calculation of this quantity, the knowledge of $\tilde{u}_\alpha^{(2)}(\mathbf{x})$ is not required, as will be seen in the subsequent derivation of a general expression for $h(\mathbf{K})$, which proceeds in a similar way as in the derivation of Eq. (3.24). Inserting (4.2) into the equations of motion and projecting on $b_\alpha(\mathbf{x}) := u_\alpha^{(0)}(2\mathbf{x})$ yields

$$-4\rho\omega_0^2\int d^3x\, b_\alpha^*(\mathbf{x})\tilde{u}_\alpha^{(2)}(\mathbf{x}) = C_{\alpha\beta\mu\nu}\int d^3x\, b_\alpha^*(\mathbf{x})\{\tilde{u}_{\mu|\beta\nu}^{(2)}(\mathbf{x}) + h(\mathbf{K})[\hat{K}_\beta b_{\mu|\nu}(\mathbf{x}) + \hat{K}_\nu b_{\mu|\beta}(\mathbf{x})]\}$$

$$+ S_{\alpha\beta\mu\nu\zeta\xi}\int d^3x\, b_\alpha^*(\mathbf{x})u_{\zeta|\xi}^{(0)}(\mathbf{x})u_{\mu|\nu\beta}^{(0)}(\mathbf{x}) \ . \tag{4.7}$$

The integrals have to be extended over the half space with $z < 0$, and periodic boundary conditions apply in the xy-plane. We now integrate the first, third and fourth terms on the right hand side of (4.7) by parts and make use of the fact that $u_\alpha^{(2)}(\mathbf{x};t)$ satisfies the boundary conditions (4.4), i.e.

$$C_{\alpha z\mu\nu}\{\tilde{u}_{\mu|\nu}^{(2)}(\mathbf{R},0) + h(\mathbf{K})\hat{K}_\nu b_\mu(\mathbf{R},0)\} + \frac{1}{2}S_{\alpha z\mu\nu\zeta\xi}u_{\mu|\nu}^{(0)}(\mathbf{R},0)u_{\zeta|\xi}^{(0)}(\mathbf{R},0) = 0 \ , \tag{4.8}$$

to obtain

$$4\rho\omega_0^2 \int d^3x\, b^*_\alpha(\mathbf{x})\tilde{u}^{(2)}_\alpha(\mathbf{x}) = C_{\alpha\beta\mu\nu} \int d^3x \big[b^*_{\alpha|\beta}(\mathbf{x})\{\tilde{u}^{(2)}_{\mu|\nu}(\mathbf{x}) + h(\mathbf{K})\hat{K}_\nu b_\mu(\mathbf{x})\} - h(\mathbf{K})\hat{K}_\beta b^*_\alpha(\mathbf{x}) b_{\mu|\nu}(\mathbf{x})\big]$$

$$+ \frac{1}{2} S_{\alpha\beta\mu\nu\zeta\xi} \int d^3x\, b^*_{\alpha|\beta}(\mathbf{x}) u^{(0)}_{\mu|\nu}(\mathbf{x}) u^{(0)}_{\zeta|\xi}(\mathbf{x}) . \qquad (4.9)$$

A further integration by parts, of the first term of the right hand side of (4.9), using the fact that $b^*_\alpha(\mathbf{x})e^{-2i\omega_0 t}$ is a solution of the linear equation of motion satisfying the linear boundary conditions at the surface, leads to the result

$$h(\mathbf{K}) C_{\alpha\beta\mu\nu} \int d^3x \{\hat{K}_\beta b^*_\alpha(\mathbf{x}) b_{\mu|\nu}(\mathbf{x}) - \hat{K}_\nu b^*_{\alpha|\beta}(\mathbf{x}) b_\mu(\mathbf{x})\} = \frac{1}{2} S_{\alpha\beta\mu\nu\zeta\xi} \int d^3x\, b^*_{\alpha|\beta}(\mathbf{x}) u^{(0)}_{\mu|\nu}(\mathbf{x}) u^{(0)}_{\zeta|\xi}(\mathbf{x}) .$$

$$(4.10)$$

This equation no longer contains the function $\tilde{u}^{(2)}_\alpha(\mathbf{x})$ and shows that for the calculation of $h(\mathbf{K})$, only the elastic moduli of second and third order and the displacement field of the linear Rayleigh wave is required. The coefficient of $h(\mathbf{K})$ on the left hand side of (4.10) is related to the velocity of energy transport $\mathbf{V}(\hat{\mathbf{K}})$ of a Rayleigh wave via the relation

$$C_{\alpha\beta\mu\nu} \hat{K}_\beta \int d^3x \{b^*_\alpha(\mathbf{x}) b_{\mu|\nu}(\mathbf{x}) - b_\alpha(\mathbf{x}) b^*_{\mu|\nu}(\mathbf{x})\} = 2iN(\hat{\mathbf{K}})\omega_{\mathbf{K}} \hat{\mathbf{K}}.\mathbf{V}(\hat{\mathbf{K}}) , \qquad (4.11)$$

where $N(\hat{\mathbf{K}})$ is given by (3.31). The velocity of energy transport of Rayleigh waves is equal to their group velocity $\nabla_{\mathbf{K}}\omega_{\mathbf{K}}$ on account of the argument, that the energy associated with a wave packet travelling with the group velocity and being localized in space has to travel with the same velocity vector. Since dispersion of the Rayleigh wave frequencies is disregarded,

$$\hat{\mathbf{K}}.\mathbf{V}(\hat{\mathbf{K}}) = v_{\hat{\mathbf{K}}} , \qquad (4.12)$$

the phase velocity. Invoking (3.29), we arrive at our final result for the growth parameter $h(\mathbf{K})$:

$$h(\mathbf{K}) = -\frac{iK^{5/2}}{2N(\hat{\mathbf{K}})v^2_{\hat{\mathbf{K}}}} S_{\alpha\beta\mu\nu\zeta\xi} \sum_{rr'r''} [-i\hat{K}_\beta(1-\delta_{\beta z}) + \hat{\alpha}^*(r,\hat{\mathbf{K}})\delta_{\beta z}][i\hat{K}_\nu(1-\delta_{\nu z}) + \hat{\alpha}(r',\hat{\mathbf{K}})\delta_{\nu z}]$$

$$\times [i\hat{K}_\xi(1-\delta_{\xi z}) + \hat{\alpha}(r'',\hat{\mathbf{K}})\delta_{\xi z}] w^*_\alpha(r,\hat{\mathbf{K}}) w_\mu(r',\hat{\mathbf{K}}) w_\zeta(r'',\hat{\mathbf{K}}) \Big[2\hat{\alpha}^*(r,\hat{\mathbf{K}}) + \hat{\alpha}(r',\hat{\mathbf{K}}) + \hat{\alpha}(r'',\hat{\mathbf{K}})\Big]^{-1} . \qquad (4.13)$$

Once the quantities $w_\alpha(r,\hat{\mathbf{K}})$ and $\hat{\alpha}(r,\hat{\mathbf{K}})$ and the elastic moduli of second and third order are known, the growth parameter $h(\mathbf{K})$ can be calculated directly from (4.13) without solving any linear equations. The expression (4.13) therefore seems to be well suited for the analysis of experimental data such as is provided in Refs. 33-35, for example, also with respect to extracting combinations of second and third order elastic constants from them. The result (4.13) is exact in the sense that it follows from the exact solution to second order in the amplitude of the input wave without further approximations. The derivation of the growth parameter can be extended to include the electrostatic potential in the case of piezoelectric media. By extending the integrals in (4.7), (4.9)-(4.11) also over the vacuum, the quantity $\tilde{u}^{(2)}_\alpha$ and the corresponding contribution of the electrostatic potential can be eliminated in the same way as in the non-piezoelectric case.

## 5. NONLINEAR CONTINUOUS BULK WAVES

Before we search for solutions of the coupled system of equations of motion (2.14) and boundary conditions (2.10)-(2.11), and (2.13) which correspond to surface waves, we first examine the comparatively easier case of propagation of nonlinear elastic bulk waves. We assume that the elastic medium has cubic symmetry with the cubic axes along the coordinate axes. Dispersion due to the discreteness of the crystal lattice can be included by adding the term

$$\Lambda_{\alpha\beta\mu\nu\gamma\delta}\, u_{\beta|\mu\nu\gamma\delta}(\mathbf{x};t) \tag{5.1}$$

to the right hand side of Eq. (2.14). For a Bravais lattice, the tensor $\underline{\underline{\Lambda}}$ is given in terms of the harmonic force constants $\phi_{\alpha\beta}(\ell\ell')$ and the distance vector $\mathbf{X}(\ell 0)$ between atom $\ell$ and atom 0 by

$$\Lambda_{\alpha\beta\mu\nu\gamma\delta} = -\frac{1}{24v_a}\sum_{\ell}\phi_{\alpha\beta}(\ell 0)X_\mu(\ell 0)X_\nu(\ell 0)X_\gamma(\ell 0)X_\delta(\ell 0)\ , \tag{5.2}$$

where $v_a$ is the volume of the crystal unit cell. The propagation direction of the nonlinear waves under consideration is chosen to be along the x-axis, and we assume that the displacement field depends only on x.

From an inspection of the symmetry properties of the tensors $\underline{\underline{S}}$ and $\underline{\underline{\Lambda}}$ for cubic crystals, it is easily seen that a solution of purely longitudinal polarization should exist which, if only cubic anharmonicity is taken into account, obeys the equation

$$\ddot{u}_x - c u_{x|xx} - \lambda u_{x|xxxx} - S u_{x|x} u_{x|xx} = 0\ , \tag{5.3}$$

where the coefficients c, $\lambda$ and S are defined by

$$c = \frac{C_{11}}{\rho} \qquad \lambda = \frac{\Lambda_{xxxxxx}}{\rho} \qquad S = \frac{S_{xxxxxx}}{\rho}\ . \tag{5.4}$$

Equation (5.3) is closely related to the Korteweg - de Vries equation[36] and possesses stable cnoidal wave solutions for the strain $u_{x|x}$ of the form

$$u_{x|x}(x-vt) = \epsilon_\infty - B\,\mathrm{cn}^2(\alpha[x - vt]|m)\ . \tag{5.5}$$

Here, cn is an elliptic function. Insertion of the Ansatz (5.5) into (5.3) yields, for given v and m, i.e. for given velocity and wave form, two relations between the three parameters $\epsilon_\infty$, B and $\alpha$: $4(1 - 2m)\lambda\alpha^2 = c - v^2 + S\epsilon_\infty$ and $SB + 12m\lambda\alpha^2 = 0$. A third relation is provided by the condition that the average strain has to vanish: $\epsilon_\infty = B/2$. By eliminating $\epsilon_\infty$ and B, the following "wave form dependent" dispersion relation can be derived:

$$\omega^2 = cq^2 - f(m)\lambda q^4\ . \tag{5.6}$$

In writing (5.6), we define $\omega$ as qv and q as $2\pi$ divided by the periodicity length of the wave. The function f(m) is given by

$$f(m) = 4\,\frac{K^2(m)}{\pi^2}\left[3\,\frac{E(m)}{K(m)} + m - 2\right], \tag{5.7}$$

where K(m) and E(m) are the complete elliptic integrals of the first and second kind, respectively. For $m \to 0$, the solution (5.5) develops into a sinusoidal wave whose amplitude is proportional in lowest order to m. By using the relation between B and v,

$$B = -\frac{12mK^2(m)}{\pi^2 f(m)}\left(\frac{c - v^2}{S}\right), \tag{5.8}$$

we may eliminate the parameter m to obtain an amplitude dependent dispersion relation of the form

$$\omega^2 = cq^2 - \lambda q^4 + \frac{1}{96}\frac{S^2}{\lambda}B^2 \tag{5.9}$$

to lowest order in B.

The dispersion relation, which we have deduced from the exact solution of Eq. (5.3), can also be obtained by a perturbative approach in the case of weak nonlinearity, i.e. in the case of small amplitudes. Since we will use this approach in our study of nonlinear surface acoustic waves, we sketch it out here for the simpler case of nonlinear continuous bulk waves. In doing so we generalize the preceding discussion slightly by including fourth order anharmonicity into the equation of motion by adding the term

$$-Tu_{x|x}^2 u_{x|xx} \tag{5.10}$$

with

$$T = \frac{S_{xxxxxxxx}}{2\rho} \tag{5.11}$$

to the left hand side of Eq. (5.3), which now reads

$$\ddot{u}_x - cu_{x|xx} - \lambda u_{x|xxxx} - Su_{x|x}u_{x|xx} - Tu_{x|x}^2 u_{x|xx} = 0\ . \tag{5.12}$$

We seek a solution of Eq. (5.12) of the form

$$u_x(x;t) = f(\theta)\ , \tag{5.13}$$

where

$$\theta = qx - \omega t\ , \tag{5.14}$$

which describes a wave propagating without change of form. The equation satisfied by $f(\theta)$ is

$$(\omega^2 - cq^2)f_{|\theta\theta} - \lambda q^4 f_{|\theta\theta\theta\theta} - Sq^3 f_{|\theta} f_{|\theta\theta} - Tq^4 f_{|\theta}^2 f_{|\theta\theta} = 0\ . \tag{5.15}$$

To solve Eq. (5.15) we expand $f(\theta)$ and $\omega^2$ according to

$$f = f^{(1)} + f^{(2)} + f^{(3)} + \dots \tag{5.16}$$

$$\omega^2 = \omega_0^2 + \omega_2^2 + \dots\ , \tag{5.17}$$

substitute these expansions into Eq. (5.15), and equate to zero the terms of first, second, third, ... order. In this way we obtain the sequence of equations

$$(\omega_0^2 - cq^2)f_{|\theta\theta}^{(1)} - \lambda q^4 f_{|\theta\theta\theta\theta}^{(1)} = 0 \tag{5.18a}$$

$$(\omega_0^2 - cq^2)f_{|\theta\theta}^{(2)} - \lambda q^4 f_{|\theta\theta\theta\theta}^{(2)} = Sq^3 f_{|\theta}^{(1)} f_{|\theta\theta}^{(1)} \tag{5.18b}$$

$$(\omega_0^2 - cq^2)f_{|\theta\theta}^{(3)} - \lambda q^4 f_{|\theta\theta\theta\theta}^{(3)} = -\omega_2^2 f_{|\theta\theta}^{(1)} + Sq^3(f_{|\theta}^{(1)} f_{|\theta\theta}^{(2)} + f_{|\theta}^{(2)} f_{|\theta\theta}^{(1)}) + Tq^4 f_{|\theta}^{(1)2} f_{|\theta\theta}^{(1)}\ . \tag{5.18c}$$

The solution of Eq. (5.18a) has the form

$$f^{(1)}(\theta) = Ae^{i\theta} + \text{c.c.} \tag{5.19}$$

and yields for $\omega_0^2$ the result

$$\omega_0^2 = cq^2 - \lambda q^4 . \tag{5.20}$$

When Eq. (5.19) is substituted into Eq. (5.18b), the latter becomes

$$(\omega_0^2 - cq^2)f^{(2)}_{|\theta\theta} - \lambda q^4 f^{(2)}_{|\theta\theta\theta\theta} = -iSq^3A^2e^{2i\theta} + \text{c.c.} \tag{5.21}$$

The solution of this equation is readily found to be

$$f^{(2)}(\theta) = \frac{iSq^3A^2e^{2i\theta}}{4(\omega_0^2 - cq^2 + 4\lambda q^4)} + \text{c.c.} \tag{5.22}$$

Equation (5.18c) now takes the form

$$(\omega_0^2 - cq^2)f^{(3)}_{|\theta\theta} - \lambda q^4 f^{(3)}_{|\theta\theta\theta\theta} = Ae^{i\theta}\left[\omega_2^2 - \left(\frac{1}{2}\frac{S^2q^6}{\omega_0^2 - cq^2 + 4\lambda q^4} + Tq^4\right)|A|^2\right]$$
$$+ A^3e^{3i\theta}\left[\frac{3}{2}\frac{S^2q^6}{\omega_0^2 - cq^2 + 4\lambda q^4} + Tq^4\right] + \text{c.c.} \tag{5.23}$$

Because $e^{i\theta}$, which is part of the inhomogeneous term in this equation, is a solution of the corresponding homogeneous equation, the solution of Eq. (5.23) will contain a secular term. Since a secular term cannot be present in a continuous wave solution, we must require the coefficient of $e^{i\theta}$ on the right hand side of Eq. (5.23) to vanish. This condition determines $\omega_2^2$:

$$\omega_2^2 = \left(\frac{1}{2}\frac{S^2q^6}{\omega_0^2 - cq^2 + 4\lambda q^4} + Tq^4\right)|A|^2 . \tag{5.24}$$

The substitution of Eqs. (5.20) and (5.24) into Eq. (5.17) yields the nonlinear cw dispersion relation

$$\omega^2 = cq^2 - \lambda q^4 + \left(\frac{S^2}{6\lambda} + Tq^2\right)q^2|A|^2 . \tag{5.25}$$

From the preceding results we see that the expansion parameter in the expansions given by Eqs. (5.16) and (5.17) is in fact the amplitude A of the solution (5.19) of the linear version of Eq. (5.15).

Unlike the quartic nonlinearity, the cubic anharmonicity influences the frequency of a weakly nonlinear wave to lowest order via the second harmonic and therefore produces a contribution quadratic in S in the dispersion relation. A comparison of Eq. (5.5) for small m and Eq. (5.19) suggests the identification

$$|A| = \frac{|B|}{4q} , \tag{5.26}$$

and the nonlinear cw dispersion relation (5.9) is recovered, except for the term resulting from quartic anharmonicity, which had been neglected in the derivation of Eq. (5.9), but appears to be of the same order in the perturbation expansion leading to Eq. (5.25). In fact, this contribution should be negligible in comparison to the cubic terms on the basis of the following argument. In most cases, the coefficients c, S, and T should be roughly of the same order of magnitude, while for the dispersion, we have to assume

$$|\lambda q^2| \ll c \tag{5.27}$$

and therefore

$$\frac{S^2}{6\lambda q^2} >> T . \tag{5.28}$$

Having investigated solutions to the equations of motion corresponding to nonlinear continuous bulk waves of longitudinal polarization, we now briefly comment on solutions involving the transverse components of the displacement field. If nonlinearity of higher than third order is neglected, it follows from the symmetry of the S-coefficients of third order, that circularly polarized sinusoidal waves with frequencies obeying the linear dispersion relation

$$\rho\omega^2 = C_{44}q^2 - \Lambda_{yyxxxx}q^4 \tag{5.29}$$

and arbitrary amplitude solve the equation of motion, i.e. they are not affected by the cubic anharmonicity. It can be shown, that in the presence of higher than cubic anharmonicity, solutions of purely transverse polarization no longer exist.

## 6. MODULATED NONLINEAR CONTINUOUS WAVES

Having found nonlinear continuous waves as solutions of the equations of motion in the preceding section, we now consider modulations of the amplitude a and phase

$$\Theta = Q.x - \omega_Q t + \phi \tag{6.1}$$

of a weakly nonlinear carrier wave. We allow the complex amplitude

$$A = ae^{i\phi} \tag{6.2}$$

to depend slowly on the time and the spatial coordinates. To account for finite beam widths, we also consider a dependence on the coordinates transverse to the propagation direction. Various techniques have been developed to deduce a single differential equation for the slow variation of the complex amplitude from the equations governing wave propagation in weakly nonlinear systems. The method of the averaged Lagrangian due to Whitham[37] has been applied successfully to the problem of deep water waves,[38] and the nonlinear Schrödinger equation has been derived for their complex amplitudes. Within an eikonal approach, Karpman and Krushkal'[1] showed that for a large variety of nonlinear systems it is the nonlinear Schrödinger equation, that governs the temporal and spatial variation of the complex amplitude. These systems have to be characterized by a nonlinear cw dispersion relation of the form

$$\omega_{\mathbf{q}} = \omega_{\mathbf{q}}^{(0)} + \omega^{(2)} a^2 , \tag{6.3}$$

where $\omega_{\mathbf{q}}^{(0)}$ denotes the wavevector dependence of the frequency in the linear limit, and a continuity equation for the energy of the approximate form

$$\frac{\partial}{\partial t}a^2 + \frac{\partial}{\partial x_\alpha}V_\alpha(\mathbf{q})a^2 = 0 , \tag{6.4}$$

where $\mathbf{V}(\mathbf{q})$ is the linear group velocity†. We note, that the system studied in Section 5, namely bulk waves in a nonlinear elastic medium, belongs to this category, and we will show in the next section, that the case of surface acoustic waves in the system described in Section 2 is also included.

We briefly outline the Karpman-Krushkal' derivation here. In the eikonal method, $\omega_{\mathbf{q}}^{(0)}$ in Eq. (6.3) is expanded to second order and $\mathbf{V}(\mathbf{q})$ in Eq. (6.4) to first order in $\mathbf{k} = \mathbf{q} - \mathbf{Q}$ around $\mathbf{Q}$, which is conveniently chosen to point in the x-direction. Subsequently, $k_\alpha$ is identified with $\partial\Theta/\partial x_\alpha$ and $\omega$ with $-\partial\Theta/\partial t$. In this way, the following system of equations is obtained

†The derivation of the nonlinear Schrödinger equation by Karpman and Krushkal' is for isotropic media. Here we have slightly extended their results to include anisotropy of the frequency dispersion.

$$\dot{\phi} + V_\alpha(Q)\phi_{|\alpha} + \frac{1}{2}\Omega_{\alpha\beta}(Q)\phi_{|\alpha}\phi_{|\beta} + \omega^{(2)}a^2 = 0 \tag{6.5}$$

$$\frac{\partial}{\partial t}a^2 + V_\alpha(Q)\frac{\partial}{\partial x_\alpha}a^2 + \Omega_{\alpha\beta}(Q)\frac{\partial}{\partial x_\alpha}(\phi_{|\beta}a^2) = 0\ , \tag{6.6}$$

where we have defined

$$\Omega_{\alpha\beta}(\mathbf{q}) = \frac{\partial^2}{\partial q_\alpha \partial q_\beta}\omega^{(0)}_{\mathbf{q}}\ . \tag{6.7}$$

As for the displacement gradients, we use the abbreviated notation $|\alpha$ for the derivative with respect to $x_\alpha$. The two equations (6.5) and (6.6) can be combined to yield one equation for the complex amplitude A

$$i\dot{A} + iV_\alpha(Q)A_{|\alpha} + \frac{1}{2}\Omega_{\alpha\beta}(Q)A_{|\alpha\beta} - \omega^{(2)}|A|^2A = 0 \tag{6.8}$$

if the term

$$-\frac{1}{2}\Omega_{\alpha\beta}(Q)\frac{a_{|\alpha\beta}}{a} \tag{6.9}$$

is added to the left hand side of Eq. (6.5). The absence of this term is a deficiency of the eikonal approximation. It has to be included on the argument, that in the linear case, the temporal and spatial evolution of the slowly varying complex amplitude is governed by the Schrödinger equation (6.8) with $\omega^{(2)} = 0$. This may be seen by the following simple consideration. A modulated plane wave may be represented as a Fourier integral

$$A(\mathbf{x},t)e^{i(\mathbf{Q}.\mathbf{x}-\omega_{\mathbf{Q}}t)} = \int \frac{d^Dk}{(2\pi)^D}\tilde{A}(\mathbf{k})e^{i([\mathbf{Q}+\mathbf{k}].\mathbf{x}-\omega_{\mathbf{Q}+\mathbf{k}}t)}\ . \tag{6.10}$$

By D, we denote the dimension of wavevector space. D=3 for bulk waves and D=2 for surface waves. If $A(\mathbf{x},t)$ is sufficiently slowly varying, we may expand $\omega_{\mathbf{Q}+\mathbf{k}}$ with respect to k, truncate that expansion after the second order, and obtain

$$A(\mathbf{x},t) = \int \frac{d^Dk}{(2\pi)^D}\tilde{A}(\mathbf{k})e^{i(\mathbf{k}.[\mathbf{x}-\mathbf{V_Q}t]-\frac{1}{2}\Omega_{\alpha\beta}(\mathbf{Q})k_\alpha k_\beta t)}\ , \tag{6.11}$$

which satisfies the linear version of Eq. (6.8). The only parameters, which enter Eq. (6.8) for the complex amplitude are the first and second derivatives of the frequency of the carrier wave with respect to the wave vector, and the quantity $\omega^{(2)}$ resulting from the nonlinearity of the system. A major part of the present chapter will be devoted to the derivation of this nonlinearity parameter for surface acoustic waves.

A different way of deriving a governing equation for the slow variations of the complex amplitude is provided by the method of multiple scales. In this approach, the solution of the wave equations is expanded as

$$u_\alpha(\mathbf{x};t) = \sum_j \epsilon^j U^{(j)}_\alpha(\mathbf{x},t,\xi,\tau)\ , \tag{6.12}$$

where $\epsilon$ is a small expansion parameter and $\xi_\alpha$ and $\tau$ are linear combinations of the variables x, y, z and t multiplied by an appropriately chosen positive power of $\epsilon$ to account for the slowness of the variations. Furthermore

$$U_\alpha^{(1)}(\mathbf{x},t,\xi,\tau) = A(\xi,\tau)\hat{W}_\alpha e^{i(\mathbf{Q}.\mathbf{x}-\omega_\mathbf{Q} t)} , \tag{6.13}$$

and $\hat{\mathbf{W}}$ is the unit polarization vector of the carrier wave. A hierarchy of equations of successive orders in $\epsilon$ is then established. Taniuti and Yajima[39] have developed a reductive perturbation scheme with the Ansatz

$$U_\alpha^{(j)}(\mathbf{x},t,\xi,\tau) = \sum_\ell \tilde{U}_\alpha^{(j,\ell)}(\xi,\tau)e^{i\ell(Qx-\omega_Q t)} \tag{6.14}$$

and the scaling

$$\xi = \epsilon(x - V(Q)t) , \qquad \tau = \epsilon^2 t . \tag{6.15}$$

The propagation direction of the carrier wave has been chosen to be the x-axis. Applying this scheme to the case of elastic bulk waves, the nonlinear Schrödinger equation for A is derived, and it is an easy task to show, that the parameters in that equation are identical to those in (6.8), specialised to the dependence on one spatial coordinate only.

In comparison to the method of Karpman and Krushkal', which requires only the nonlinear cw dispersion relation and the energy conservation condition, the derivation of the nonlinear Schrödinger equation within a multiple scales approach is often more cumbersome. It might, however, provide clearer insight into the details of nonlinear wave propagation in the particular system under consideration. In the case of surface acoustic waves, the derivation of the parameter $\omega^{(2)}$ will be performed in a way very similar to the procedure of the multiple scales methods, which has been applied to the propagation of surface acoustic waves in the presence of nonlinearity by several authors.[40-49] The extra work required in the multiple scales approach of Taniuti and Yajima is rewarded by a perturbative expression for the second derivatives of the frequencies with respect to the wave vector components.

In a series of papers by Sakuma and his colleagues,[46-49] this method has been applied to the study of envelope solitons of nonlinear Rayleigh waves. Their treatment of the problem differs considerably from the one outlined in the next section as they start from a quantum mechanical formulation, introduce dispersion into the equation of motion, and treat the boundary conditions in a different way.

## 7. WEAKLY NONLINEAR SURFACE ACOUSTIC WAVES

In this section, we turn to the investigation of nonlinear waves in the system described in Section 2 (see Fig. 1) with a displacement field that decreases exponentially in the direction normal to the surface, i.e. as z tends to minus infinity. Concerning the symmetry of the system, we make the following specifications:

> both the substrate and the film are of cubic symmetry with their cubic axes identical to the coordinate axes. This implies that we are dealing with an (001) surface. Furthermore, we consider wave propagation along the x-direction.

In the absence of nonlinearity, there exist waves of purely sagittal and purely shear horizontal polarization, which satisfy the boundary condition at infinite depth, i.e. are guided at the substrate surface covered with the thin film. The displacement field of guided waves with shear horizontal polarization (Love waves) is of a standing wave character in the film, while in the case of sagittal polarization, it may also contain components which decrease into the film from the interface and from the surface. The dispersion relation of these guided modes may, depending on the ratio of the film thickness d to the wavelength $\lambda$, consist of several branches. Since we have imposed the condition that $\lambda$ be much larger than d, it is only the lowest branch with the frequency tending to zero as $\lambda \to \infty$, that we are concerned with here. In the sagittal case, the modes of this branch become the Rayleigh waves of the substrate for $d \to 0$. In the case of shear horizontal polarization, no surface waves exist in this case. For $d \to 0$, the Love wave frequencies approach those of the surface skimming bulk transverse wave.

An inspection of the symmetry of the S-tensors in the anharmonic terms of the equations of motion and boundary conditions shows that solutions of purely sagittal polarization exist to arbitrary order of the nonlinearity. This does not apply to the case of shear horizontal polarization. In fact, the component $u_y(x,z;t)$ of the displacement field provides driving terms for the sagittal components in the equations of motion and boundary conditions via the cubic anharmonicity. Since, however, these driving terms are quadratic in $u_y$, we may expect, that for small amplitudes, weakly nonlinear analogs of the Love waves, i.e. nonlinear continuous waves of predominantly shear horizontal polarization, do exist.

In the following two subsections, we shall derive nonlinear dispersion relations of the form (6.3) for weakly nonlinear waves of both purely sagittal and predominantly shear horizontal polarization and thus show, that they belong to the class of systems, to which the method of Karpman and Krushkal' can be applied to derive the nonlinear Schrödinger equation for slow modulations. Unfortunately, unlike in the bulk case, the derivation of explicit analytic expressions for nonlinear acoustic surface waves from the true equations of motion and boundary conditions in the geometry of a substrate covered by a thin film has, to our knowledge, not yet been achieved. Cnoidal waves of sagittal polarization have, however, been found experimentally by Nayanov.[50]

The perturbative derivation of the nonlinear cw dispersion relation follows the scheme outlined in Section 5 for nonlinear bulk waves. We expand the solution in the form

$$u_\alpha(x,z;t) = \sum_j \epsilon^j u_\alpha^{(j)}(x,z,t) \ . \tag{7.1}$$

In a multiple scales approach the functions $u_\alpha^{(j)}(x,z,t)$ also depend on the slow variables $\xi$, $\eta$ and $\tau$, where $\eta = \epsilon z$, and for describing self-modulation, $\xi = \epsilon(x - V(Q)t)$ and $\tau = \epsilon^2 t$, while for describing self-focusing, the choice $\xi = \epsilon y$ and $\tau = \epsilon^2 x$ should be made. $\mathbf{Q}$ is now the two-dimensional wave-vector of the carrier wave and $\mathbf{V}(\mathbf{Q})$ the two-dimensional group velocity. Since we are concerned with waves localized at the surface, the explicit z-dependence of the functions $u_\alpha^{(j)}(\xi,\eta,\tau,x,z,t)$ is dominated by an exponential decrease for $z \to -\infty$. We may therefore formally expand them with respect to $\eta$ and rearrange the power series (7.1) to obtain a new series, in which the functions $u_\alpha^{(j)}$ no longer depend on $\eta$. The new function $u_\alpha^{(1)}$ may also be decomposed in the form

$$u_\alpha^{(1)}(x,z,t) = AU_\alpha(x,z)e^{-i\omega t} + \text{c.c.} \ , \tag{7.2}$$

where A is now the complex amplitude at the surface, i.e. at $\eta = 0$. It is for the slow fluctuations of this quantity that we shall derive the nonlinear Schrödinger equation. The formal procedure outlined above generates secular terms in the functions $u_\alpha^{(j)}$ of the form

$$z^n e^{\alpha z} \tag{7.3}$$

with integer n. The presence of these terms in the perturbative solution has been criticised by Lardner.[43,44] They may be avoidable by obeying certain compatibility relations which, however, at least in the sagittal case, complicate the calculation. We will therefore make use of secular terms in z.

### 7.1 Dispersion relation for weakly nonlinear Rayleigh waves

By substituting the Ansatz (7.1) in the nonlinear equations of motion and boundary conditions and equating terms of the same order in $\epsilon$, a hierarchy of linear equations is generated, which we will truncate after the third order in $\epsilon$. The calculation is then performed in three steps.

The equations of first order in $\epsilon$ are the linear versions of Eqs. (2.12), (2.14), and (2.24), from which we have to determine the Rayleigh wave solution for $u_\alpha^{(1)}$. We do this in the usual way, by using Rayleigh's method, see e.g. Royer and Dieulesaint.[51] The equations of motion are solved with the Ansatz

$$U_\alpha(x,z)e^{-i\omega t} = W_\alpha e^{iqx+\alpha z-i\omega t} \ . \tag{7.4}$$

On insertion of Eq. (7.4) into the equation of motion, a homogeneous linear system is obtained for $W_x$, $W_z$, the solvability condition of which,

$$\alpha^4 C_{11}C_{44} + \alpha^2[C_{11}(\rho\omega_0^2 - C_{11}q^2) + C_{44}(\rho\omega_0^2 - C_{44}q^2) + q^2(C_{44} + C_{12})^2] + (\rho\omega_0^2 - C_{11}q^2)(\rho\omega_0^2 - C_{44}q^2) = 0\,, \tag{7.5}$$

relates $\alpha$ to the zero order frequency $\omega_0$. The two relevant roots of (7.5) can be either both real or complex conjugates of each other. The eigenvectors $(W_\alpha)$, associated with the two roots $\alpha(1)$ and $\alpha(2)$, may be written as

$$\begin{bmatrix} W_x \\ W_z \end{bmatrix} = \begin{bmatrix} 1 \\ p(r) \end{bmatrix}, \tag{7.6}$$

where

$$p(r) = i\,\frac{\rho\omega_0^2 - C_{11}q^2 + C_{44}\alpha^2(r)}{q\alpha(r)(C_{44} + C_{12})}\,. \tag{7.7}$$

We thus obtain

$$U_\mu(x,z) = e^{iqx}\sum_r c(r)e^{\alpha(r)z}\begin{bmatrix} 1 \\ p(r) \end{bmatrix}_\mu\,. \tag{7.8}$$

The ratio of the coefficients c(1) and c(2) is determined by the boundary conditions, which give rise to the following homogeneous system of equations,

$$\sum_{r=1}^{2} c(r)\{C_{44}[\alpha(r) + iqp(r)] - d(\rho_F\omega_0^2 - Dq^2)\} = 0 \tag{7.9a}$$

$$\sum_{r=1}^{2} c(r)\{C_{11}p(r)\alpha(r) + C_{12}iq - d\rho_F\omega_0^2 p(r)\} = 0\,, \tag{7.9b}$$

the solvability condition of which fixes the frequency and with it also the values of $\alpha(r)$. In Eqs. (7.9) D is defined as

$$D = C_{11}^{(F)} - \frac{\left[C_{12}^{(F)}\right]^2}{C_{11}^{(F)}}\,. \tag{7.10}$$

In the numerical calculations we use the convention c(1) = 1 with $|\alpha(1)/\alpha(2)| \geq 1$.

The second order equations of motion are the following inhomogeneous differential equations

$$\rho\ddot{u}_\alpha^{(2)} - C_{\alpha\beta\mu\nu}u_{\mu|\beta\nu}^{(2)} = S_{\alpha\beta\mu\nu\gamma\delta}u_{\mu|\nu}^{(1)}u_{\gamma|\delta\beta}^{(1)}\,, \tag{7.11}$$

and the boundary conditions of the same order in $\epsilon$ are given by

$$[C_{\alpha z\mu\nu}u_{\mu|\nu}^{(2)} + d\rho_F\ddot{u}_\alpha^{(2)} - dD\delta_{\alpha x}u_{x|xx}^{(2)}]_{z=0} = -\frac{1}{2}S_{\alpha z\mu\nu\gamma\delta}u_{\mu|\nu}^{(1)}u_{\gamma|\delta}^{(1)}|_{z=0}\,. \tag{7.12}$$

The right hand sides of Eqs. (7.11) and (7.12) consist of a part independent of x and t and a term proportional to $e^{\pm 2i(qx-\omega t)}$. We first address the latter, which implies the presence of the second harmonic in the complete solution. Our way of solving Eqs. (7.11) and (7.12) is analogous to the theoretical treatment of the problem of second harmonic generation of surface acoustic waves by Normandin *et al*,[32] and Tiersten and Baumhauer.[31] We first consider the non-synchronous part of Eq. (7.11). The corresponding solution is

$$u_\alpha^{(2,n)}(\mathbf{x},t) = \begin{pmatrix} n_x \\ n_z \end{pmatrix} e^{(\alpha(1)+\alpha(2))z} e^{2i(qx-\omega t)} A^2 + \text{c.c.}\,, \tag{7.13}$$

where the coefficients $n_x$, $n_z$ are determined by the linear system

$$\begin{pmatrix} 4\rho\omega_0^2 - C_{11}4q^2 + C_{44}\left[\alpha(1)+\alpha(2)\right]^2 & (C_{44}+C_{12})i2q\left[\alpha(1)+\alpha(2)\right] \\ (C_{44}+C_{12})i2q\left[\alpha(1)+\alpha(2)\right] & 4\rho\omega_0^2 + C_{11}\left[\alpha(1)+\alpha(2)\right]^2 - C_{44}4q^2 \end{pmatrix} \begin{pmatrix} n_x \\ n_z \end{pmatrix} = \begin{pmatrix} F_x^{(n)} \\ F_z^{(n)} \end{pmatrix} \tag{7.14}$$

with

$$F_\alpha^{(n)} = -S_{\alpha\beta\mu\nu\gamma\delta} \begin{pmatrix} 1 \\ p(1) \end{pmatrix}_\mu \begin{pmatrix} 1 \\ p(2) \end{pmatrix}_\gamma \begin{pmatrix} iq \\ \alpha(1) \end{pmatrix}_\nu \begin{pmatrix} iq \\ \alpha(2) \end{pmatrix}_\delta \begin{pmatrix} 2iq \\ \alpha(1)+\alpha(2) \end{pmatrix}_\beta c(1)c(2)\,. \tag{7.15}$$

For the synchronous part, we make the Ansatz

$$u_\alpha^{(2,s)}(\mathbf{x},t) = \sum_{r=1}^{2} e^{2\alpha(r)z} \left\{ b(r)z \begin{pmatrix} 1 \\ p(r) \end{pmatrix}_\alpha + d(r) \begin{pmatrix} p(r) \\ -1 \end{pmatrix}_\alpha \right\} e^{2i(qx-\omega t)} A^2 + \text{c.c.} \tag{7.16}$$

The secular term in z is needed, because

$$e^{2\alpha(r)z} \begin{pmatrix} 1 \\ p(r) \end{pmatrix} e^{2i(qx-\omega t)} \tag{7.17}$$

is a solution of the homogeneous part of Eq. (7.11). This problem arises, because the dispersion comes into play via the boundary conditions and does not directly enter the equations of motion. The coefficients b(r) and d(r) have to be calculated from the two linear systems

$$\begin{aligned}&[4C_{44}\alpha(r) + (C_{44}+C_{12})2iqp(r)]b(r) + [(4\rho\omega_0^2 - 4C_{11}q^2 + 4C_{44}\alpha^2(r))p(r) \\ &\qquad - (C_{44}+C_{12})4iq\alpha(r)]d(r) = F_x^{(r)}\end{aligned} \tag{7.18a}$$

$$\begin{aligned}&[(C_{44}+C_{12})2iq + C_{11}4\alpha(r)p(r)]b(r) + [(C_{44}+C_{12})4iq\alpha(r)p(r) \\ &\qquad - (4\rho\omega_0^2 + 4C_{11}\alpha^2(r) - 4C_{44}q^2)]d(r) = F_z^{(r)}\end{aligned} \tag{7.18b}$$

with

$$F_\alpha^{(r)} = -S_{\alpha\beta\mu\nu\gamma\delta} \begin{pmatrix} 1 \\ p(r) \end{pmatrix}_\mu \begin{pmatrix} 1 \\ p(r) \end{pmatrix}_\gamma \begin{pmatrix} iq \\ \alpha(r) \end{pmatrix}_\nu \begin{pmatrix} iq \\ \alpha(r) \end{pmatrix}_\delta \begin{pmatrix} iq \\ \alpha(r) \end{pmatrix}_\beta c^2(r)\,. \tag{7.19}$$

We now substitute $u_\alpha^{(2,n)} + u_\alpha^{(2,s)}$ into the boundary conditions (7.12) and satisfy them with a homogeneous solution of (7.11) of the form

$$u_\alpha^{(2,b)}(\mathbf{x},t) = \sum_{r=1}^{2} e^{2\alpha(r)z} e(r) \begin{pmatrix} 1 \\ p(r) \end{pmatrix}_\alpha e^{2i(qx-\omega t)} A^2 + \text{c.c.} \tag{7.20}$$

The coefficients e(r) are the solution of the linear system

$$\sum_{r=1}^{2} e(r)\{2C_{44}[\alpha(r) + iqp(r)] - 4d(\rho_F\omega_0^2 - Dq^2)\} = B_x$$

$$\sum_{r=1}^{2} e(r)\{2C_{11}p(r)\alpha(r) + 2C_{12}iq - 4d\rho_F\omega_0^2 p(r)\} = B_z \tag{7.21}$$

with

$$B_\alpha = -\frac{1}{2}S_{\alpha z\mu\nu\gamma\delta}\sum_{r,r'=1}^{2}\begin{pmatrix}1\\p(r)\end{pmatrix}_\mu\begin{pmatrix}1\\p(r')\end{pmatrix}_\gamma\begin{pmatrix}iq\\\alpha(r)\end{pmatrix}_\nu\begin{pmatrix}iq\\\alpha(r')\end{pmatrix}_\delta c(r)c(r')$$

$$+ 4d\rho_F\omega_0^2\left\{n_\alpha + \sum_{r=1}^{2} d(r)\begin{pmatrix}p(r)\\-1\end{pmatrix}_\alpha\right\} - \delta_{\alpha x}4dDq^2\left\{n_x + \sum_{r=1}^{2} d(r)p(r)\right\} + \tilde{B}_\alpha \tag{7.22a}$$

$$\tilde{B}_x = C_{44}\left\{\sum_{r=1}^{2}[(iq - \alpha(r)p(r))2d(r) - b(r)] - (\alpha(1) + \alpha(2))n_x - 2iqn_z\right\}$$

$$\tilde{B}_z = C_{11}\left\{\sum_{r=1}^{2}[2\alpha(r)d(r) - p(r)b(r)] - (\alpha(1) + \alpha(2))n_z\right\} - 2iqC_{12}\left\{\sum_{r=1}^{2} p(r)d(r) + n_x\right\}. \tag{7.22b}$$

To obtain the second harmonic part of the second order solution, we thus have to solve successively four linear inhomogeneous 2×2 systems of equations. Unlike in the bulk case, the second order solution also contains a static part, which is of the form

$$u_\alpha^{(2,0)}(z) = \sum_{rr'}\begin{pmatrix}C_{44}^{-1}F_x^{(0)}(rr')\\C_{11}^{-1}F_z^{(0)}(rr')\end{pmatrix}_\alpha e^{[\alpha(r)+\alpha^*(r')]z}\,|A|^2 \tag{7.23}$$

with

$$F_\alpha^{(0)}(rr') = -S_{\alpha z\mu\nu\gamma\delta}\begin{pmatrix}1\\p(r)\end{pmatrix}_\mu\begin{pmatrix}1\\p^*(r')\end{pmatrix}_\gamma\begin{pmatrix}iq\\\alpha(r)\end{pmatrix}_\nu\begin{pmatrix}-iq\\\alpha^*(r')\end{pmatrix}_\delta\frac{c(r)c^*(r')}{\alpha(r) + \alpha^*(r')}. \tag{7.24}$$

It is easily seen, that (7.23) satisfies both the corresponding parts of the equations of motion and boundary conditions.

The contributions to the equations of motion and boundary conditions of third order in $\epsilon$ are

$$\rho\ddot{u}_\alpha^{(3)} - C_{\alpha\beta\mu\nu}u_{\mu|\beta\nu}^{(3)} = T_{\alpha\beta|\beta}^{(N)} + \rho\omega_2^2u_\alpha^{(1)} \tag{7.25}$$

$$[C_{\alpha z\mu\nu}u_{\mu|\nu}^{(3)} + d\{\rho_F\ddot{u}_\alpha^{(3)} - D_{\alpha\Phi\beta\Theta}u_{\beta|\Phi\Theta}^{(3)}\}]_{z=0} = [d\rho_F\omega_2^2u_\alpha^{(1)} - T_{\alpha z}^{(N)}]_{z=0}\,, \tag{7.26}$$

where

$$T^{(N)}_{\alpha\beta} = S_{\alpha\beta\mu\nu\gamma\delta}u^{(1)}_{\mu|\nu}u^{(2)}_{\gamma|\delta} + \frac{1}{6}S_{\alpha\beta\mu\nu\gamma\delta\sigma\lambda}u^{(1)}_{\mu|\nu}u^{(1)}_{\gamma|\delta}u^{(1)}_{\sigma|\lambda} . \tag{7.27}$$

By the introduction of an amplitude-dependent contribution to the frequency, secular terms in x or t can be avoided for $u^{(3)}_\alpha$, and a continuous nonlinear wave solution is made possible. The functions $u^{(3)}_\alpha$ can then be obtained in a way analogous to the calculation of $u^{(2)}_\alpha$. To calculate the correction $\omega_2^2$ to the frequency square, the explicit form of $u^{(3)}_\alpha$ is, however, not required. Instead, we project the part of (7.25) proportional to $e^{-i\omega t}$ on the first order solution $U_\alpha$,

$$\rho\omega_2^2\int d^3x\, U^*_\alpha\tilde{u}^{(1)}_\alpha + \rho\omega_0^2\int d^3x\, U^*_\alpha\tilde{u}^{(3)}_\alpha + C_{\alpha\beta\mu\nu}\int d^3x\, U^*_\alpha\tilde{u}^{(3)}_{\mu|\beta\nu} + \int d^3x\, U^*_\alpha\tilde{T}^{(N)}_{\alpha\beta|\beta} = 0 , \tag{7.28}$$

and make use of the fact that the part of the complete continuous wave solution proportional to $e^{-i\omega t}$ satisfies the equations of motion and boundary conditions separately. The tilde on a symbol indicates the prefactor of $e^{-i\omega t}$ in the respective quantity. We now integrate the last term once and the third term twice by parts, employing periodic boundary conditions in the xy-plane. Since $U_\alpha$ is a solution of the first order equation of motion, this yields

$$\rho\omega_2^2\int d^3x\, U^*_\alpha\tilde{u}^{(1)}_\alpha - \int d^3x\, U^*_{\alpha|\beta}\tilde{T}^{(N)}_{\alpha\beta} = \int d^2R\{C_{\alpha\beta\mu z}U^*_{\alpha|\beta}\tilde{u}^{(3)}_\mu - C_{\alpha z\mu\nu}U^*_\alpha\tilde{u}^{(3)}_{\mu|\nu} - U^*_\alpha\tilde{T}^{(N)}_{\alpha z}\}_{z=0} . \tag{7.29}$$

If we now make use of the boundary conditions

$$C_{\alpha z\mu\nu}U_{\mu|\nu} - d\{\rho_F\omega_0^2U_\alpha + D_{\alpha\Phi\beta\Theta}U_{\beta|\Phi\Theta}\} = 0 \tag{7.30}$$

$$C_{\alpha z\mu\nu}\tilde{u}^{(3)}_{\mu|\nu} - d\{\rho_F\omega_0^2\tilde{u}^{(3)}_\alpha + D_{\alpha\Phi\beta\Theta}\tilde{u}^{(3)}_{\beta|\Phi\Theta}\} = -\tilde{T}^{(N)}_{\alpha z} + d\rho_F\omega_2^2U_\alpha A \tag{7.31}$$

the boundary terms on the right hand side of (7.29) may be reduced to

$$\left[-\omega_2^2 d\rho_F\int d^2R U^*_\alpha U_\alpha A\right]_{z=0} . \tag{7.32}$$

Thus, the third order solution is eliminated from Eq. (7.29), and by inserting (7.8), (7.13), (7.16), (7.20) and (7.23) into (7.27), we arrive at the final result for the nonlinear dispersion relation for surface waves of sagittal polarization:

$$\omega = \omega_0(\mathbf{q}) + \omega^{(2)}_{\mathbf{q}}|A|^2 , \tag{7.33}$$

where

$$\omega^{(2)}_{\mathbf{q}}|A|^2 = \frac{\omega_2^2}{2\omega_0} = N^{-1}2\int d^3x U^*_{\alpha|\beta}\tilde{T}^{(N)}_{\alpha\beta} = N^{-1}(K + L + M)|A|^2 \tag{7.34}$$

$$K = S_{\alpha\beta\mu\nu\gamma\delta}\sum_{rr'}\left[\left(\frac{1}{p^*(r)}\right)_\alpha\left(\frac{1}{p^*(r')}\right)_\mu\left(\frac{-iq}{\alpha^*(r)}\right)_\beta\left(\frac{-iq}{\alpha^*(r')}\right)_\nu\left\{n_\gamma\left(\frac{2iq}{\alpha(1)+\alpha(2)}\right)_\delta\left[\alpha^*(r)+\alpha^*(r')+\alpha(1)+\alpha(2)\right]^{-1}\right.\right.$$

$$+\sum_{r''}\left[d(r'')\begin{pmatrix}p(r'')\\-1\end{pmatrix}_\gamma + e(r'')\begin{pmatrix}1\\p(r'')\end{pmatrix}_\gamma\right]\begin{pmatrix}2iq\\2\alpha(r'')\end{pmatrix}_\delta\left[\alpha^*(r)+\alpha^*(r')+2\alpha(r'')\right]^{-1}$$

$$+\sum_{r''}b(r'')\begin{pmatrix}1\\p(r'')\end{pmatrix}_\gamma\left[\begin{pmatrix}0\\1\end{pmatrix}_\delta\left[\alpha^*(r)+\alpha^*(r')+2\alpha(r'')\right]^{-1}\right.$$

$$\left.\left.\left.-\begin{pmatrix}2iq\\2\alpha(r'')\end{pmatrix}_\delta\left[\alpha^*(r)+\alpha^*(r')+2\alpha(r'')\right]^{-2}\right]\right\}c^*(r)c^*(r')\right] + \text{c.c.} \qquad (7.35)$$

$$L = S_{\alpha\beta\mu\nu\gamma z}\sum_{r_1r_2r_3r_4}\left\{\begin{pmatrix}1\\p^*(r_1)\end{pmatrix}_\alpha\begin{pmatrix}1\\p(r_2)\end{pmatrix}_\mu\begin{pmatrix}-iq\\\alpha^*(r_1)\end{pmatrix}_\beta\begin{pmatrix}iq\\\alpha(r_2)\end{pmatrix}_\nu\right.$$

$$\left.\times\begin{pmatrix}C_{44}^{-1}F_x(r_3r_4)\\C_{11}^{-1}F_z(r_3r_4)\end{pmatrix}_\gamma\frac{\alpha(r_3)+\alpha^*(r_4)}{\alpha^*(r_1)+\alpha(r_2)+\alpha(r_3)+\alpha^*(r_4)}c^*(r_1)c(r_2)\right\} + \text{c.c.} \qquad (7.36)$$

$$M = S_{\alpha\beta\mu\nu\gamma\delta\sigma\lambda}\sum_{r_1r_2r_3r_4}\left\{\begin{pmatrix}1\\p^*(r_1)\end{pmatrix}_\alpha\begin{pmatrix}1\\p^*(r_2)\end{pmatrix}_\mu\begin{pmatrix}1\\p(r_3)\end{pmatrix}_\gamma\begin{pmatrix}1\\p(r_4)\end{pmatrix}_\sigma\right.$$

$$\times\begin{pmatrix}-iq\\\alpha^*(r_1)\end{pmatrix}_\beta\begin{pmatrix}-iq\\\alpha^*(r_2)\end{pmatrix}_\nu\begin{pmatrix}iq\\\alpha(r_3)\end{pmatrix}_\delta\begin{pmatrix}iq\\\alpha(r_4)\end{pmatrix}_\lambda\left[\alpha^*(r_1)+\alpha^*(r_2)+\alpha(r_3)+\alpha(r_4)\right]^{-1}$$

$$\left.\times c^*(r_1)c^*(r_2)c(r_3)c(r_4)\right\} \qquad (7.37)$$

$$N = 4\omega_0(q)\sum_{rr'}\left\{\rho\left[\alpha^*(r)+\alpha(r')\right]^{-1}+\rho_F d\right\}[1+p^*(r)p(r')]c^*(r)c(r')\,. \qquad (7.38)$$

Taking account of Eq. (7.9), it is easily seen, that the determinant of the linear system (7.21) is proportional to d and therefore $e(r) \sim d^{-1}$. While L and M in Eq. (7.34) are of order $d^0$, the term proportional to $e(r'')$ in the expression for K, Eq. (7.35) is of order $d^{-1}$ and therefore represents the leading contribution to the quantity $\omega^{(2)}$, being proportional to $q^2$. As in the case of nonlinear bulk waves (Section 5), we find that, for small dispersion, the quartic anharmonicity is negligible in comparison to the cubic terms.

We also state the second derivative of the frequency with respect to q to first order in d:

$$\Omega_{xx}(q\hat{x}) = -dv_0^{-1}N_0^{-1}\sum_{r,r'=1}^{2}c_\alpha^{(0)*}(r)c_\mu^{(0)}(r')\{\delta_{\alpha\mu}\rho_F v_0^2 + D_{\alpha\beta\mu\nu}[\delta_{\beta x}i+\delta_{\beta z}\hat{\alpha}(r')][\delta_{\nu x}i+\delta_{\nu z}\hat{\alpha}(r')]\}$$

$$(7.39)$$

with

$$N_0 = \rho\sum_{r,r'=1}^{2}c_\alpha^{(0)*}(r)c_\alpha^{(0)}(r')\left[\hat{\alpha}^*(r)+\hat{\alpha}(r')\right]^{-1} \qquad (7.40)$$

(see Appendix A.) The quantities $v_0$, $\alpha(r{=}1,2) = q\hat{\alpha}(r{=}1,2)$, $c_\mu(r{=}1,2)$ are the Rayleigh wave phase velocity, decay constants and relative amplitudes of the substrate in the absence of the film. We note that, unlike in the bulk case, $\Omega_{xx}(q\hat{x})$ is of zero order in q.

### 7.2 Dispersion relation for weakly nonlinear Love waves

To derive an amplitude-dependent dispersion relation for weakly nonlinear surface waves of predominantly shear horizontal polarization, we proceed in essentially the same way as in the preceding subsection. The first order solution is of the form

$$u_\alpha^{(1)}(\mathbf{x},t) = e^{i(qx-\omega t)+\beta z}\delta_{\alpha y}A + \text{c.c.} \tag{7.41}$$

For given q, the zero order frequency $\omega_0$ and decay parameter $\beta$ are fixed by the equations of motion and boundary conditions, which yield the relations

$$\rho\omega_0^2 + C_{44}(\beta^2 - q^2) = 0 \tag{7.42}$$

$$C_{44}\beta - d\{\rho_F\omega_0^2 - C_{44}^{(F)}q^2\} = 0 . \tag{7.43}$$

To first order in the film thickness d, we obtain the following expressions for $\beta$ and $\omega_0$:

$$\beta = q^2 dM_h + O(d^2) \tag{7.44}$$

$$\omega_0^2 = q^2\frac{C_{44}}{\rho}\{1 - d^2q^2M_h^2 + O(d^4)\} , \tag{7.45}$$

where

$$M_h = \frac{\rho_F}{\rho} - \frac{C_{44}^{(F)}}{C_{44}} . \tag{7.46}$$

To have exponential decay into the substrate, the condition $M_h > 0$ has to be fulfilled.

The second order solution involves only the sagittal components. The second order equations of motion and boundary conditions are therefore identical to Eqs. (7.11) and (7.12) with $u_\alpha^{(1)}$ now given by Eq. (7.41). Again, the second order solution consists of a static part and the second harmonic. We first determine the latter. Since there are no synchronous terms in this case, we may use the Ansatz

$$u_\alpha^{(2,n)}(\mathbf{x},t) = \begin{pmatrix} m_x \\ 0 \\ m_z \end{pmatrix} e^{2i(qx-\omega t)+2\beta z}A^2 + \text{c.c.} \tag{7.47}$$

for an inhomogeneous solution of the equation of motion. The coefficients $m_x$, $m_z$ are then the solutions of the linear system

$$\begin{pmatrix} \rho\omega_0^2 - C_{11}q^2 + C_{44}\beta^2 & (C_{44} + C_{12})iq\beta \\ (C_{44} + C_{12})iq\beta & \rho\omega_0^2 + C_{11}\beta^2 - C_{44}q^2 \end{pmatrix}\begin{pmatrix} m_x \\ m_z \end{pmatrix} = -\frac{1}{4}\begin{pmatrix} F_x \\ F_z \end{pmatrix} , \tag{7.48}$$

where

$$F_x = S_{xxyxyx}(iq)^3 + (S_{xxyzyz} + 2S_{xzyxyz})iq\beta^2 \tag{7.49a}$$

$$F_z = (S_{zzyxyx} + 2S_{zxyxyz})(iq)^2\beta + S_{zzyzyz}\beta^3 . \tag{7.49b}$$

To satisfy also the boundary conditions, we use a homogeneous solution of Eq. (7.11),

$$u_\alpha^{(2,b)}(\mathbf{x},t) = \sum_{r=1}^{2} e^{2\alpha(r)z}s(r)\begin{pmatrix} 1 \\ p(r) \end{pmatrix} e^{2i(qx-\omega t)}A^2 , \tag{7.50}$$

where $\alpha(r)$ and $p(r)$ are determined by $\omega_0$ and q via the relations (7.5) and (7.7). The coefficients s(r) are the solutions of the linear system

$$\sum_{r=1}^{2} s(r)\{2C_{44}[\alpha(r) + iqp(r)] - 4d(\rho_F\omega_0^2 - Dq^2)\} = -B_x$$

$$\sum_{r=1}^{2} s(r)\{2C_{11}p(r)\alpha(r) + 2C_{12}iq - 4d\rho_F\omega_0^2 p(r)\} = -B_z \tag{7.51}$$

with

$$B_x = S_{xzyxyz}\, iq\beta + m_x\{2C_{44}\beta - 4d(\rho_F\omega_0^2 - Dq^2)\} + m_z 2C_{44}iq$$

$$B_z = \frac{1}{2}\{S_{zzyzyz}\beta^2 - S_{zzyxyx}q^2\} + m_x 2C_{12}iq + m_z\{2C_{11}\beta - 4d\rho_F\omega_0^2\} . \tag{7.52}$$

The static part of the second order solution of the equation of motion can easily be shown to be

$$u_\alpha^{(2,0)}(z) = -e^{2\beta z}\delta_{\alpha z}\frac{1}{2\beta C_{11}}\{S_{zzyxyx}q^2 + S_{zzyzyz}\beta^2\}\left|A\right|^2 . \tag{7.53}$$

It also satisfies the static part of the boundary conditions to second order in $\epsilon$.

In order to obtain the frequency correction due to the nonlinearity to second order in the amplitude, we do not need to determine the explicit form of the third order solution, but may apply the projection method used in the sagittal case. Equations (7.28) to (7.34) can be directly taken over from Section 7.1. The quantities K, L, M, and N in Eq. (7.34) are now given by

$$K = S_{y\mu y\nu\gamma\delta}\begin{pmatrix}-iq\\0\\\beta\end{pmatrix}_\mu\begin{pmatrix}-iq\\0\\\beta\end{pmatrix}_\nu\left\{\begin{pmatrix}m_x\\0\\m_z\end{pmatrix}_\gamma\begin{pmatrix}2iq\\0\\2\beta\end{pmatrix}_\delta\frac{1}{4\beta} + \sum_{r=1}^{2}s(r)\begin{pmatrix}1\\0\\p(r)\end{pmatrix}_\gamma\begin{pmatrix}2iq\\0\\2\alpha(r)\end{pmatrix}_\delta\left[2\beta + 2\alpha(r)\right]^{-1}\right\} + \text{c.c.} \tag{7.54}$$

$$L = -\left(S_{zzyxyx}q^2 + S_{zzyzyz}\beta^2\right)^2\frac{1}{2\beta C_{11}} \tag{7.55}$$

$$M = S_{yxyxyxyx}(q^4 + \beta^4)\frac{1}{4\beta} \tag{7.56}$$

$$N = 4\omega_0(q)\left(\frac{\rho}{2\beta} + \rho_F d\right) . \tag{7.57}$$

To lowest order in qd, we obtain

$$\alpha_0^2(1) = -\frac{C_{11}(C_{44} - C_{11}) + \left(C_{44} + C_{12}\right)^2}{C_{11}C_{44}}q^2 = O(d^0) \tag{7.58}$$

$$\alpha_0^2(2) = \frac{C_{44}(C_{44} - C_{11})}{C_{11}(C_{44} - C_{11}) + (C_{44} + C_{12})^2}\beta^2 =: \bar{\alpha}^2\beta^2 = O(d^2) \tag{7.59}$$

$$p_0(1) = i\,\frac{(C_{44} - C_{11})q^2 + C_{44}\alpha_0^2(1)}{q\alpha_0(1)(C_{44} + C_{12})} = O(d^0) \tag{7.60}$$

$$p_0(2) = i\,\frac{q}{\beta}\,\frac{C_{44} - C_{11}}{C_{44} + C_{12}}\left[\frac{C_{11}(C_{44} - C_{11}) + (C_{44} + C_{12})^2}{C_{44}(C_{44} - C_{11})}\right]^{1/2} =: i\bar{p}\,\frac{1}{\beta} = O(d^{-1}) \tag{7.61}$$

$$m_x^{(0)} = \left[C_{12}^2 - C_{11}^2 + 2C_{44}(C_{12}+C_{11})\right]^{-1}\frac{-iq}{4}\{(C_{44}-C_{11})S_{xxzxzx} + (C_{12}+C_{44})(2S_{zxyxyz} + S_{zzyxyx})\}$$
$$=: i\bar{m}_x q = O(d^0) \tag{7.62}$$

$$m_z^{(0)} = \left[C_{12}^2 - C_{11}^2 + 2C_{44}(C_{12}+C_{11})\right]^{-1}\frac{q^2}{4\beta}\{(C_{44}-C_{11})[2S_{zxyxyz} + S_{zzyxyx}] + (C_{12}+C_{44})S_{xxzxzx}\}$$
$$=: \bar{m}_z\,\frac{q^2}{\beta} = O(d^{-1}) \tag{7.63}$$

$$s_0(2) = -\frac{m_z}{p_0(2)} =: i\bar{s}q = O(d^0)\ , \tag{7.64}$$

and for the parameter $\omega^{(2)}$

$$\omega^{(2)} = \frac{q^3}{\sqrt{\rho C_{44}}}\left\{S_{yxyxxx}\left(\frac{1}{2}\bar{m}_x + \frac{\bar{s}}{1+\bar{\alpha}}\right) - S_{yxyxzz}\left(\frac{1}{2}\bar{m}_z - \frac{\bar{s}\bar{\alpha}\bar{p}}{1+\bar{\alpha}}\right)\right.$$
$$\left. + S_{yxyzzx}\left(\bar{m}_z - 2\,\frac{\bar{s}\bar{p}}{1+\bar{\alpha}}\right) - \frac{1}{4C_{11}}S_{zzyxyx}^2 + \frac{1}{8}S_{yxyxyxyx}\right\} + O(d)\ . \tag{7.65}$$

It depends only on quantities referring to the substrate, and in contrast to the case of purely sagittal polarization, the limit $d \to 0$ can be performed. In order to ensure the surface wave character of the nonlinear wave, the squares of the decay constants $\alpha(1,2)$ have to be positive. This leads to the condition

$$C_{11}(C_{11} - C_{44}) > (C_{44} + C_{12})^2\ . \tag{7.66}$$

If this condition is not fulfilled, the surface wave becomes leaky due to the nonlinear coupling to the second harmonic of sagittal polarization, and an amplitude dependent damping constant can be attributed to it. The elements of the S-tensors of third and fourth order occuring in (7.65) are related to the elastic moduli via

$$S_{yxyxxx} = C_{166} + C_{11} \tag{7.67}$$

$$S_{yxyxzz} = C_{144} + C_{12} \tag{7.68}$$

$$S_{yxyzzx} = C_{456} + C_{44} \tag{7.69}$$

$$S_{yxyxyxyx} = C_{4444} + 6C_{166} + 3C_{11}\ . \tag{7.70}$$

In the case of isotropic elastic media, the expressions (7.58) - (7.65) simplify considerably, and the following short formula is obtained for $\omega^{(2)}$:

$$\omega^{(2)} = \frac{q^3}{\sqrt{\rho C_{44}}}\left\{\frac{1}{8}(C_{4444}+6C_{166}+3C_{11}) - \frac{1}{4C_{11}}\left(C_{144}+C_{12}\right)^2 - \frac{1}{4(C_{11}+C_{12})}\left(C_{166}+C_{11}\right)^2\right\} + O(d). \tag{7.71}$$

Finally, we note that

$$\Omega_{xx}(q\hat{x}) = -3\left(\frac{C_{44}}{\rho}\right)^{1/2} d^2 M_h^2 q + O(d^4) . \tag{7.72}$$

## 8. THE NONLINEAR SCHRÖDINGER EQUATION

Having derived the amplitude dependent dispersion relations for surface acoustic waves of the form (7.33), the argumentation of Section 6 can be carried through to obtain the nonlinear Schrödinger equation for the complex amplitude A at the surface. The Cartesian indices in (6.4) to (6.11) now run only over x and y. A continuity equation for the energy of the approximate form (6.4) clearly holds also in the case of surface acoustic waves. We therefore obtain equation (6.8) as the governing equation for slow variations of the complex amplitude of surface acoustic waves, as well. For the geometry under consideration, in which the wavevector **Q** of the carrier wave is along a cubic axis on a (001) surface, the group velocity points in the same direction as **Q**, and the matrix $\Omega_{\alpha\beta}(\mathbf{Q})$ is diagonal.

In the following, we consider two special cases of Eq. (6.8):

1. *Modulation of the carrier wave along its propagation direction.* In this case, we obtain, after transforming to the variable $X = x - V_x(\mathbf{Q})t$,

$$i\frac{\partial}{\partial t}A + \frac{1}{2}\Omega_{xx}(\mathbf{Q})\frac{\partial^2}{\partial X^2}A - \omega^{(2)}\left|A\right|^2 A = 0 . \tag{8.1}$$

2. *A stationary beam with a slowly varying profile in the y-direction* (slowly on the length scale of the carrier wave length). Here, we are interested in the slow variation of this profile along the propagation direction.[52] To describe this situation approximately, we disregard the time derivative, since we want to describe a stationary situation, and neglect higher than first derivatives with respect to x, in Eq. (6.8) to obtain

$$iV_x(\mathbf{Q})\frac{\partial}{\partial x}A + \frac{1}{2}\Omega_{yy}(\mathbf{Q})\frac{\partial^2}{\partial y^2}A - \omega^{(2)}\left|A\right|^2 A = 0 . \tag{8.2}$$

A direct derivation of this equation in the framework of the multiple scales approach is given in Appendix B.

In both of these cases, we are led to the nonlinear Schrödinger equation, which can be brought into the dimensionless form

$$i\dot{\psi} + \psi_{|xx} + \kappa\left|\psi\right|^2\psi = 0 . \tag{8.3}$$

(The variables x and t in Eq. (8.3) need not necessarily be interpreted as space and time.) The solution

$$\psi(x,t) = \psi_0 e^{i\kappa|\psi_0|^2 t} \tag{8.4}$$

corresponds to the spatially unmodulated continuous wave. By linearising around the solution (8.4), it can be easily shown, that for $\kappa > 0$ this solution is unstable against fluctuations with a wavelength larger than $4\pi/[(2\kappa)^{1/2}|\psi_0|]$.[1]

The properties of the nonlinear Schrödinger equation in the case $\kappa > 0$ have been investigated in a fundamental work by Zakharov and Shabat[53] using the inverse scattering method. The specification to the two cases (8.1) and (8.2) will be made later. From the results of these authors, it may be concluded, that an initial pulse of the form $\psi(x,t{=}0)$ disintegrates into a number of solitons of the form

$$\psi(x,t) = 2\left(\frac{2}{\kappa}\right)^{1/2} \eta e^{-4i(\xi^2-\eta^2)t-2i\xi x+i\phi} \cosh^{-1}(2\eta(x-x_0) + 8\eta\xi t) \tag{8.5}$$

and a continuous background. The soliton velocities $-4\xi$ and amplitudes $\eta$ can be calculated from the initial pulse as the real and imaginary parts of the eigenvalues $\zeta$ of the system of linear differential equations

$$\begin{aligned} v_1' + i\zeta v_1 &= i\sqrt{\frac{\kappa}{2}}\,\psi(x,t)v_2 \\ v_2' - i\zeta v_2 &= i\sqrt{\frac{\kappa}{2}}\,\psi^*(x,t)v_1 \end{aligned} \tag{8.6}$$

with normalizable eigenfunctions. If two or more real parts coincide, bound states of solitons are formed. For the case of a rectangular initial pulse† of spatial width $\Delta$ and height H, it is shown in Appendix C, that the quantities $\zeta$ are the solutions of the equation

$$\tan(z) = -\frac{z}{[\frac{1}{2}\kappa(|H|\Delta)^2 - z^2]^{1/2}}, \tag{8.7}$$

where

$$z = \Delta\left[\zeta^2 + \frac{1}{2}\kappa|H|^2\right]^{1/2}. \tag{8.8}$$

It then follows, that a necessary condition for a soliton with $\xi = 0$ to develop is

$$|H|\Delta > \frac{\pi}{\sqrt{2\kappa}}. \tag{8.9}$$

For a bound state of N solitons to be formed, the condition

$$(2N-1)\frac{\pi}{\sqrt{2\kappa}} < |H|\Delta < (2N+1)\frac{\pi}{\sqrt{2\kappa}} \tag{8.10}$$

has to be satisfied. It cannot be excluded *a priori*, that Eq. (8.7) also has solutions with $\zeta$ having a finite real part. The condition (8.9) can be understood on an intuitive basis.[54] If a soliton of the form (8.5) is contained in the initial pulse, the area covered by the latter should exceed the one of the soliton, which is of the order of magnitude of $\kappa^{-1/2}$. For the special case, that only one soliton is contained in the initial pulse, and the "non-soliton" background gives rise to only small reflection coefficients in the continuous spectrum of the associated

†Strictly speaking, a rectangular pulse is not slowly varying. We may, however, expect that in the present context it approximates pulses of nearly rectangular shape with smoothed boundaries, such that they are slowly varying on the length scale of the wavelength of the carrier wave.

linear eigenvalue problem (8.6), Zakharov and Shabat have shown, that with increasing time t, this background decreases as $t^{-1/2}$. The time, which is required for a pulse to develop into one soliton may be defined as the time, after which the integrated "non-soliton" background has decreased by a certain factor. We have not carried out such a calculation, but use as an estimate the characteristic time $t_c$ for an efficient rearrangement of the Fourier-components of a pulse of width $\Delta^{-1}$ in reciprocal space in the linear case,

$$t_c \approx \frac{\Delta^2}{\partial^2\omega/\partial q^2} . \tag{8.11}$$

The general results on the behaviour of the solutions of the nonlinear Schrödinger equation (8.3) are now interpreted in terms of the two physical situations which we have specified at the beginning of this section.

1. If $\Omega_{xx}/\omega^{(2)} < 0$, an initial modulation of the complex amplitude of the carrier wave of sufficient size develops within the time t into one or more solitons. If the eigenvalues $\zeta_j$ in Eqs. (8.6) (for a rectangular pulse the solutions of Eq. (8.7) with Eq. (8.8)) have no real part and the imaginary parts are distinct, only one stable pulse is formed, which may, however, have a complex shape. It travels with the group velocity of the carrier wave. The real parts of $\zeta_j$ have the meaning of relative velocities with respect to this group velocity.
2. If $\Omega_{yy}/\omega^{(2)} < 0$, an initial beam profile develops within a distance of the order of $Vt_c$ into stable channels. If the $\zeta_j$, calculated from Eq. (8.6), have no real parts, only one channel in the x-direction is formed, which may display a complex shape, if several eigenvalues $\zeta_j = i\eta_j$ are involved. If there are eigenvalues with non-vanishing real parts, channels are formed, which show a finite angle of inclination towards the x-axis. In the geometry chosen here, the group velocity points in the same direction as the wavevector. If, however, the wave vector of the carrier wave is chosen to lie off the symmetry directions, this will no longer be the case, and the phenomenon of beam steering occurs due to the anisotropy of the elastic medium. In contradistinction to this effect, the self-focusing phenomenon, which has just been described, is caused by the interplay of nonlinearity and dispersion. The two effects may occur simultaneously and interfere with each other.

In the case of isotropic elastic media, the relation

$$\Omega_{yy}(q\hat{x}) = q^{-1}V_x(q\hat{x}) \tag{8.12}$$

holds, and thus $\Omega_{yy} > 0$, while $\Omega_{xx}$ is supposed to be negative. The self-modulation instability and the self-channeling instability then occur alternatively. This need not be the case, if the elastic media are anisotropic.

## 9. NUMERICAL ESTIMATES

In order to find out whether the self-modulation instability or the self-channeling instability occurs, and to estimate the evolution length of a soliton or a channel, the sign and magnitude of the parameters entering the nonlinear Schrödinger equations (8.1) and (8.2) have to be known. In this section, numerical estimates are provided for a number of substances.

In the case of purely sagittal polarization, we may decompose

$$\omega^{(2)} = \frac{v_R q^2}{dM_s} Q , \tag{9.1}$$

where $M_s$ is the sagittal mismatch factor

$$M_s = \frac{\rho_F}{\rho_S} - \frac{C_F}{C_S} , \qquad C_j = C_{11}^{(j)} - \frac{C_{12}^{(j)2}}{C_{11}^{(j)}} \tag{9.2}$$

and Q is a dimensionless quantity, which depends only on the elastic moduli of second and

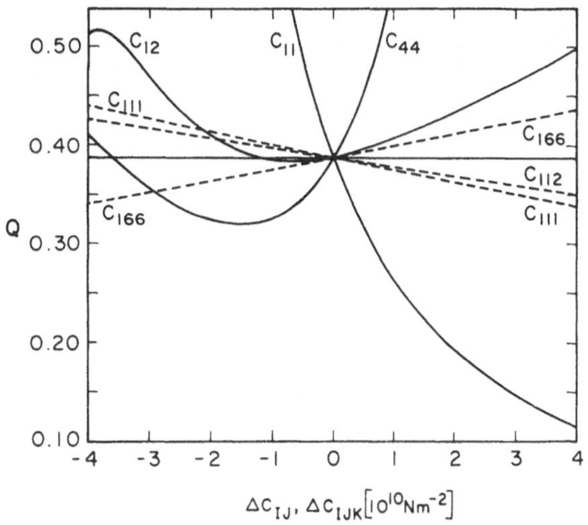


Fig. 4. Dependence of the dimensionless quantity Q in the coefficient $\omega^{(2)}$ on the linear and nonlinear elastic moduli of the substrate (copper) [see Eq. (9.1)].

third order of the substrate to lowest order in qd. In the same way, we write

$$\Omega_{xx}(q\hat{x}) = PM_s d v_R \ , \tag{9.3}$$

where the dimensionless quantity P depends only on the elastic moduli of the substrate, to lowest order in qd. In Figs. 2 and 3 we show Q and P, respectively, as functions of the elastic moduli they depend on, in the case of copper as the substrate material. We note that the sign of the ratio $\omega^{(2)}/\Omega_{xx}$ does not depend on the mismatch parameter. From the data in Table 2 we conclude that for the substances we have chosen as substrate materials, it is negative, i.e. the self-modulation instability takes place. The Rayleigh wave velocity and the quantity $\Omega_{yy}q/v_R$ are also given in the table. (If not stated otherwise, the data for the elastic moduli are taken from Landolt-Börnstein.[29]) The latter turned out to be positive for the substances considered. The self-channeling instability can be generated for these substances by choosing a negative mismatch parameter.

For a rough estimate of the soliton evolution length $x_c$ on the basis of Eq. (8.11), we recall, that for our approximation to be valid, the following inequalities have to be fulfilled for the four lengths involved:

$$\Delta >> \lambda >> \bar{d} >> |A| \ , \tag{9.4}$$

where $\bar{d} = M_s d$ is the effective film thickness. The first has to be required, because the nonlinear Schrödinger equation has been derived for slow variations of the carrier wave amplitude, the second because of the approximate elimination of the displacement field in the film, and the third, because the smallness of the higher harmonics has to be ensured. If we fix the length scale by the choice $\lambda = 10^{-6}$m and allow the lengths in (9.4) to differ successively by one order of magnitude (which implies $|A|/\lambda \approx 0.01$), we arrive at an evolution length $x_c \approx 10^{-2} - 10^{-1}$m, if $|P| \approx 0.1$.

In the case of shear horizontal polarization, we define the dimensionless quantities

$$R: = \frac{vq^3}{\omega^{(2)}} \tag{9.5}$$

and

$$S: = \Omega_{yy} \frac{q}{v} \ , \tag{9.6}$$

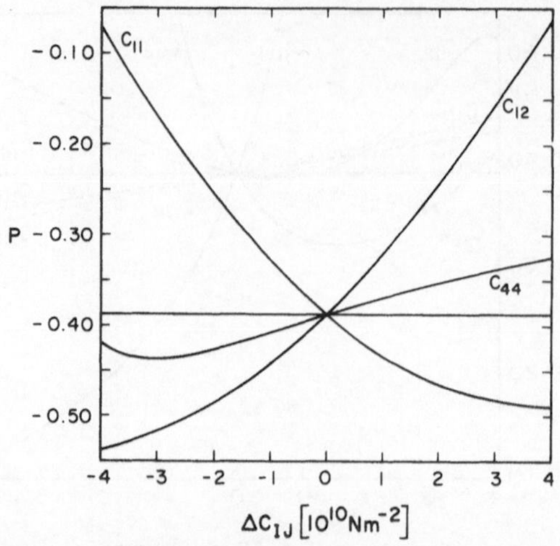


Fig. 5. Dependence of the dimensionless quantity P governing the dispersion of Rayleigh waves, propagating in the geometry of Fig. 1, on the elastic moduli of the substrate (copper). [See Eq. (9.3)].

where

$$v = \sqrt{\frac{C_{44}}{\rho}} \,. \tag{9.7}$$

Furthermore,

$$\Omega_{xx} = -3qvM_h^2 d^2 < 0 \,, \tag{9.8}$$

where $M_h$ is the mismatch parameter in the shear horizontal case defined in (7.46). To lowest order in d, R and S only depend on the elastic moduli of the substrate. For the latter, the following simple expression is obtained:

$$S = \frac{C_{11}}{C_{44}} + 2\frac{\left(C_{12} + C_{44}\right)^2}{(C_{11} + C_{12})(C_{44} - C_{11})}\frac{(C_{44} + C_{12})\bar{\alpha} - (C_{44} - C_{11})}{C_{44}(1 + \bar{\alpha})} \,, \tag{9.9}$$

where $\bar{\alpha}$ is defined in (7.59). In Table 2, numerical values for R and S are listed for a number of substances, for which the condition (7.66) is satisfied. Since strong compensations can be expected to occur between elastic moduli of different orders in the relations (2.8), we have chosen a consistent set of theoretical data for the elastic moduli of second to fourth order for the alkali halides in Table 3 (Ref. 55). Unfortunately, we are not aware of a cubic material, for which all elastic moduli needed in our calculation have been measured.

For all substances listed in Table 3, S is positive and R negative. Since the mismatch parameter $M_h$ has to be positive for a Love wave to exist, we may conclude that the self-channeling instability occurs for nonlinear surface waves of predominantly shear horizontal polarization, while these waves are stable against modulation. From the decomposition of $R^{-1}$ into the contributions resulting from the fourth order anharmonicity, the static part of the solution, and the second harmonic, it can be seen that the fourth order contribution, which is found to be positive in all cases, is overcompensated by the terms resulting from the cubic anharmonicity, the dominant part of which is provided by the second harmonic.

Table 2. Numerical values for the quantities governing the parameters in the nonlinear Schrödinger equations (8.1) and (8.2) in the case of sagittal polarization

| Substance | Q | -P | $\Omega_{yy}q/v_R$ | $v_R(10^3 ms^{-1})$ | $\alpha(r)$ |
|---|---|---|---|---|---|
| NaF | 0.079 | 0.337 | 0.889 | 3.045 | real |
| Cu | 0.387 | 0.388 | 1.761 | 2.015 | complex |
| KCl | 0.014 | 0.148 | 0.787 | 1.752 | real |
| Si | 0.023 | 0.579 | 1.189 | 4.917 | complex |

Table 3. Numerical values for the quantities governing the parameters in the nonlinear Schrödinger equation (8.2) in the case of predominantly shear horizontal polarization

| Substance | R | $\delta R^{-1}$ | | | S |
|---|---|---|---|---|---|
| | | 4th order | static | 2nd harmonic | |
| NaCl | -0.56 | 1.23 | -0.65 | -2.36 | 1.538 |
| KCl | -0.55 | 1.21 | -0.79 | -2.23 | 2.223 |
| RbCl | -1.83 | 1.78 | -0.70 | -1.63 | 4.067 |
| CsI | -0.10 | 3.16 | -3.86 | -9.10 | 5.212 |
| Al | -0.16 | 3.33 | -0.12 | -9.61 | 0.115 |

For an order of magnitude estimate of the length $x_c$, over which an initial beam profile of width $\Delta$ develops a channel on the basis of (8.11), we approximate $x_c \approx \Delta^2/\lambda$, which yields $x_c \approx 10^{-3}$m for $\Delta = 10^{-4}$m and $\lambda = 10^{-5}$m.

## 10. SURFACE ACOUSTIC WAVES ON PERIODIC GRATINGS

In addition to the deposition of a film on the surface and spatial dispersion due to the discreteness of the underlying lattice, surface corrugation is a third way of introducing dispersion of the surface acoustic wave frequencies, since it provides in general new length scales. In the case of a periodic corrugation profile of the form

$$\zeta(x) = \zeta_0 \cos(Gx) \tag{10.1}$$

these are the corrugation depth $\zeta_0$ and the periodicity length $a = 2\pi/G$. In the following, we consider only profiles which are periodic in the x-direction and independent of y.

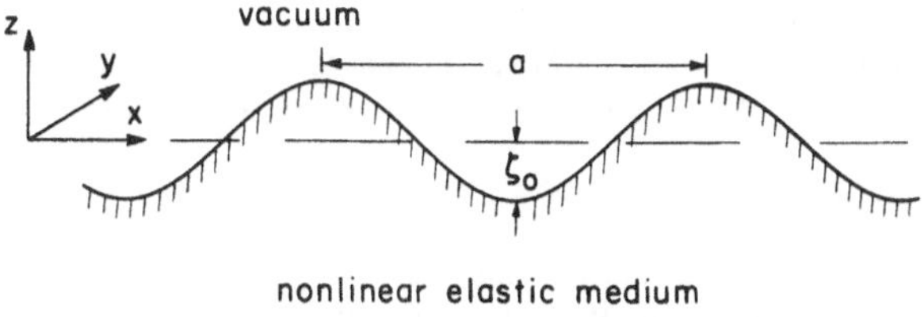


Fig. 6. Geometry considered for the study of nonlinear effects on the propagation of surface acoustic waves on periodically corrugated surfaces.

Before discussing some effects of nonlinearity on surface wave propagation in this geometry, a few remarks on the properties of linear surface wave solutions are made, which will be of use in the subsequent derivations.

The propagation of linear surface waves on periodic gratings has been investigated by several authors.[56-66] It is again via the boundary conditions, that the dispersion enters the calculation of the frequencies of surface acoustic waves. To first order of the corrugation profile and its derivative they take the form

$$0 = T_{\alpha z}(x, z{=}0) + \zeta(x)T_{\alpha z|z}(x, z{=}0) - \zeta'(x)T_{\alpha x}(x, z{=}0) + O(\zeta^2, \zeta\zeta') . \tag{10.2}$$

To determine the displacement field associated with linear surface acoustic waves of sagittal polarization, one may choose solutions of the equations of motion of the form

$$\mathbf{w}^{(r)}(q,\omega|\mathbf{x})\ e^{-i\omega t} = \begin{pmatrix} 1 \\ p_r(q,\omega) \end{pmatrix} e^{iqx+\alpha_r(q,\omega)z-i\omega t} . \tag{10.3}$$

Here, we have slightly changed the notation from that of the previous sections to express explicitly the dependence of $\alpha$ and p on $\omega$. In the case of propagation along the (100) direction on the (001) surface of a cubic crystal, to which we confine our discussion here, the quantities $\alpha_r(q,\omega)$ are, for a given frequency $\omega = \omega_0$, the two solutions of Eq. (7.5) with positive real part, and the coefficients $p_r(q,\omega_0)$ are then given by (7.7). To satisfy the boundary conditions with a periodic corrugation profile, the displacement field is written as a linear combination of the functions $\mathbf{w}^{(r)}(q,\omega|\mathbf{x})$ in the form of a Bloch function

$$\mathbf{u}(\mathbf{x};t) = e^{-i\omega t}\,\mathbf{U}(\mathbf{x}) \tag{10.4a}$$

$$\mathbf{U}(\mathbf{x}) = \sum_n \sum_r \mathbf{w}^{(r)}(q + nG,\omega|\mathbf{x})\ A_r(n) , \tag{10.4b}$$

where q can now be restricted to the first Brillouin zone, i.e. $|q| \le G/2$. In the framework of Rayleigh's method, the Ansatz (10.4) is inserted into the boundary conditions, which lead to a homogeneous linear equation for the amplitudes $A_r(n)$. The solvability condition for this equation then yields the frequencies of the surface acoustic waves. Numerical calculations of the frequency dispersion curves[62,63] show, among others, the following three important phenomena:

1. At the Brillouin zone boundary, a frequency gap appears corresponding to a stop band for the propagation of the surface waves.
2. Since the branches of the disperion curve have to reach the Brillouin zone boundary with zero slope, the group and phase velocity of the surface waves of sagittal polarization corresponding to the lowest frequency branch become smaller than those of the Rayleigh waves on a flat surface with the same wavelength. This is the phenomenon of wave slowing.
3. The surface waves associated with the higher branches in the reduced zone scheme become leaky, since some of the Fourier components in the solution (10.4) enter the radiative regime (see Fig. 7). This effect has been studied both theoretically[56,58,63] and experimentally.[57]

We also note that the dispersion introduced by the periodic corrugation enables a surface wave of shear horizontal polarization to exist in the geometry under consideration.[5,13] In the following, we confine ourselves to a corrugation profile of the form (10.1).

## 10.1 Stress dependence of the stop band

In view of technical applications, it appears desirable to be able to tune the stop band width and position. This can be achieved by external stresses or, in the case of piezoelectric media, by an external electric field. In the discussion of this section we confine ourselves to non-piezoelectric media for simplicity.

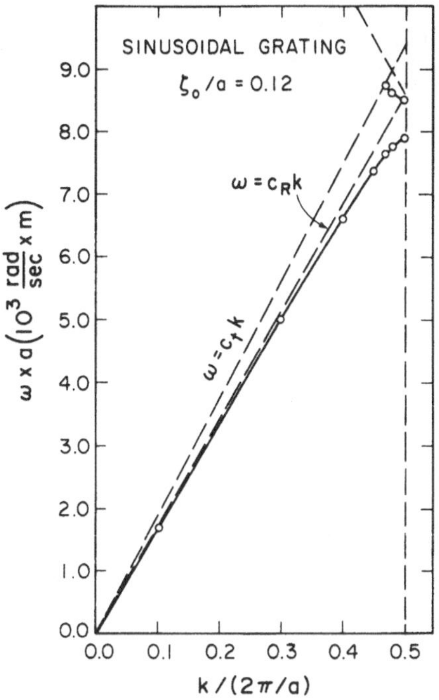


Fig. 7. Dispersion curves for Rayleigh waves on a sinusoidal grating with corrugation strength $\zeta_0/a = 0.12$. The elastic medium is isotropic and characterized by the speeds of transverse sound $c_t(\hat{=}\ v_t) = 3\times10^3\ ms^{-1}$ and $c_\ell(\hat{=}\ v_\ell) = 5\times10^3\ ms^{-1}$ ($c_R \hat{=} v_R$). From Ref. 62.

If a uniaxial stress $\sigma$ is applied at right angles to the propagation direction in the plane of the surface, i.e. in the y-direction, the induced strains can be expected to be homogeneous. In this case, only diagonal components of the deformation tensor are nonzero and obey the relations

$$v_{xx} = v_{zz} = -\frac{C_{12}}{C_{11} + C_{12}} v_{yy} \tag{10.5}$$

$$v_{yy} = S_{||}\sigma , \tag{10.6}$$

where we define

$$S_{||} = \frac{C_{11} + C_{12}}{C_{11}^2 + C_{11}C_{12} - 2C_{12}^2} . \tag{10.7}$$

Using the projection method introduced in Section 3 and applying Green's theorem, we arrive at Eq. (3.21) with (3.22) for the frequency shift to first order in the static strains $v_{\alpha\beta}$, where the integrals now have to be extended to the corrugated surface of the undeformed medium. To evaluate these formulas for the frequencies limiting the gap at the Brillouin zone boundary, we need the solution (10.4) for the first and second branch at $q = G/2$. For a result correct to first order in the corrugation depth $\zeta_0$ of the profile (10.1), we may confine the sum over n to $n = 0, -1$. The amplitudes $A_r(0)$ and $A_r(-1)$ are then obtained from the system of equations

$$\sum_r \left\{ M_{\alpha r}^{(0)}\left(\frac{1}{2}G,\omega\right)A_r(0) + \zeta_0 M_{\alpha r}^{(-)}\left(\frac{1}{2}G,\omega\right)A_r(-1)\right\} = 0 \tag{10.8a}$$

$$\sum_r \left\{ M_{\alpha r}^{(0)}\left(-\frac{1}{2}G,\omega\right)A_r(-1) + \zeta_0 M_{\alpha r}^{(+)}\left(-\frac{1}{2}G,\omega\right)A_r(0)\right\} = 0 \,. \tag{10.8b}$$

Here, we have introduced the matrices

$$M_{xr}^{(0)}(k,\omega) = C_{44}\{\alpha_r(k,\omega) + ikp_r(k,\omega)\} \tag{10.9a}$$

$$M_{zr}^{(0)}(k,\omega) = C_{11}p_r(k,\omega)\alpha_r(k,\omega) + C_{12}ik \tag{10.9b}$$

and

$$M_{xr}^{(\pm)}(k,\omega) = \frac{1}{2}C_{44}\alpha_r(k\pm G,\omega)\{\alpha_r(k\pm G,\omega) + i(k\pm G)p_r(k\pm G,\omega)\}$$
$$\pm \frac{i}{2}G\{C_{11}i(k\pm G) + C_{12}p_r(k\pm G,\omega)\alpha_r(k\pm G,\omega)\} \tag{10.10a}$$

$$M_{zr}^{(\pm)}(k,\omega) = \frac{1}{2}\alpha_r(k\pm G,\omega)\{C_{11}p_r(k\pm G,\omega)\alpha_r(k\pm G,\omega)+C_{12}i(k\pm G)\}$$
$$\pm \frac{i}{2}GC_{44}\{\alpha_r(k\pm G,\omega) + p_r(k\pm G,\omega)i(k\pm G)\} \,. \tag{10.10b}$$

In the following, we treat only substances for which the $\alpha_r$ in a Rayleigh wave are purely real. The 4x4 system (10.8) can be decoupled into two 2x2 systems by the transformation to the variables

$$B_{s,a}(r) = A_r(0) \pm A_r(-1) \,. \tag{10.11}$$

The solvability conditions of these two homogeneous systems yield the frequencies in the undeformed state. The amplitude ratios $A_2(n)/A_1(n)$ correct to first order in $\zeta_0$ are given by

$$\frac{A_2(n)}{A_1(n)} = - \frac{M_{x1}^{(0)}\left(\frac{1}{2}G,\omega_\pm\right) \pm \zeta_0 M_{x1}^{(-)}\left(\frac{1}{2}G,\omega_\pm\right)}{M_{x2}^{(0)}\left(\frac{1}{2}G,\omega_\pm\right) \pm \zeta_0 M_{x2}^{(-)}\left(\frac{1}{2}G,\omega_\pm\right)} \,. \tag{10.12}$$

The upper sign refers to the symmetric solution ($A_r(0) = A_r(-1)$), the lower sign to the antisymmetric solution ($A_r(0) = -A_r(-1)$). Equation (10.12) holds for $n = 0$ and $n = -1$ separately. By adding to the frequencies in the absence of external stress obtained from (10.8) the correction (3.21), we obtain for the frequencies to first order in $\zeta_0$ and $\sigma$:

$$\omega_\pm^2 = \left(\frac{1}{2}Gv_R\right)^2 \left\{1 \pm G\zeta_0 Z + \left(M_0 \pm \frac{1}{2}G\zeta_0 M_1\right)S_{\|}\sigma\right\}. \tag{10.13}$$

Numerical values for the dimensionless quantities Z, $M_0$ and $M_1$, the Rayleigh wave velocity $v_R$, and the effective compliance $S_{\|}$ are given for several substances in Table 4.

### 10.2 Intensity dependence of the gap width

With increasing amplitude, the nonlinearity of the elastic medium causes higher harmonics to develop. Because of the dispersion provided by the periodic grating, continuous, peri-

Table 4. Parameters determining the dependence of the gap of the dispersion curve of Rayleigh waves propagating on a sinusoidal grating on stress and intensity. See Eqs. (10.13) and (10.26)

| Substance | $Z\times 10^2$ | $M_0$ | $M_1$ | $v_R(10^3 ms^{-1})$ | $S_{\|\|}(10^{-12}m^2N^{-1})$ | $T\times 10^2$ |
|---|---|---|---|---|---|---|
| $BaF_2$ | -6.1 | -4.7 | +2.0 | 2.108 | 15.34 | -6.5 |
| NaF | -4.3 | +0.84 | -2.4 | 3.045 | 9.95 | -31.0 |
| KCl | -1.1 | +3.2 | -3.9 | 1.752 | 25.98 | -8.8 |
| $CaF_2$ | -2.5 | -2.8 | +1.7 | 3.171 | 6.90 | -3.7 |

odic, weakly nonlinear waves can form which, with decreasing intensity, develop continuously into the linear surface waves. They may be viewed as a nonlinear generalization of Bloch waves, since the corresponding displacement field is of the form

$$u_\alpha(\mathbf{x};t) = f_\alpha(x - vt, x, z) \tag{10.14}$$

with

$$f_\alpha(\Theta, x+a, z) = f_\alpha(\Theta, x, z) \tag{10.15a}$$

$$f_\alpha(\Theta+\lambda, x, z) = f_\alpha(\Theta, x, z)\ . \tag{10.15b}$$

The length $\lambda = 2\pi/q$ is the wavelength of the fundamental component in the nonlinear wave. In the same fashion as has been outlined in Section 7 for nonlinear sagittally polarized surface waves on a coated surface, an amplitude dependent correction to the frequency can be derived, where the amplitude $A_0 = A_{r=1}(0)$ serves as the effective expansion parameter. After having determined the frequency and amplitude $A_r(n)$ relative to $A_0$ in the linear solution (10.4), the second harmonic

$$u_\alpha^{(2)}(\mathbf{x};t) = \hat{u}_\alpha^{(2)}(\mathbf{x})e^{-2i\omega t} \tag{10.16}$$

has to be calculated. The equation of motion for the second harmonic takes the form

$$4\rho\omega^2\hat{u}_\alpha^{(2)}(\mathbf{x}) + C_{\alpha\beta\mu\nu}\hat{u}_{\mu|\beta\nu}^{(2)}(\mathbf{x}) = \sum_{nn'}\sum_{rr'} F_\alpha(nn',rr')e^{i(2q+[n+n']G)x}e^{[\alpha_r(q+nG,\omega)+\alpha_{r'}(q+n'G,\omega)]z}A_0^2\ . \tag{10.17}$$

The coefficients $F_\alpha(nn',rr')$ are obtained by inserting (10.4) into the nonlinear part of (2.14). A particular solution $u_\alpha^{(2,p)}$ of (10.17) can be constructed as the sum of particular solutions for each Fourier component separately:

$$\hat{u}_\alpha^{(2)}(\mathbf{x}) = u_\alpha^{(2,h)}(\mathbf{x}) + \sum_{nn'}\sum_{rr'} u_\alpha^{(2,p)}(\mathbf{x}|nn',rr')\ , \tag{10.18}$$

where

$$u_\alpha^{(2,p)}(\mathbf{x}|nn',rr') = e^{i(2q+[n+n']G)x}e^{\alpha_r(q+nG,\omega)z+\alpha_{r'}(q+n'G,\omega)z}N_\alpha(nn',rr')A_0^2 \tag{10.19}$$

if $n \neq n'$ or $r \neq r'$. When $n = n'$ and $r = r'$, i.e. for the synchronous terms, a particular solution can be found by making use of secular terms in z:

$u_\alpha^{(2,p)}(\mathbf{x}|nn,rr)$

$$= e^{i2(q+nG)x}e^{2\alpha_r(q+nG,\omega)z}\left\{b(n,r)z\begin{bmatrix}1\\ p_r(q+nG,\omega)\end{bmatrix}_\alpha + d(n,r)\begin{bmatrix}p_r(q+nG,\omega)\\ -1\end{bmatrix}_\alpha\right\}A_0^2 \,. \quad (10.20)$$

The coefficients $N_\alpha(nn',rr')$, $b(n,r)$ and $d(n,r)$ are obtained as the solutions of 2×2-systems of linear inhomogeneous equations.

Now the contributions of the boundary conditions of second order in $A_0$ proportional to $e^{-2i\omega t}$ have to be considered. Substituting (10.4), (10.18), (10.19), and (10.20) into (2.24) we are led to equations of the form

$$C_{\alpha z\mu\nu}u_{\mu|\nu}^{(2,h)}(x,z{=}0) + \zeta(x)C_{\alpha z\mu\nu}u_{\mu|\nu z}^{(2,h)}(x,z{=}0) - \zeta'(x)C_{\alpha x\mu\nu}u_{\mu|\nu}^{(2,h)}(x,z{=}0) + O(\zeta^2,\zeta\zeta')$$

$$= \sum_n B_\alpha(n)e^{i(2q+nG)x}A_0^2 \,. \quad (10.21)$$

The homogeneous solution $u_\alpha^{(2,h)}$ has to be of the form

$$u_\alpha^{(2,h)}(\mathbf{x}) = \sum_n\sum_r E_r(n)w_\alpha^{(r)}(2q+nG,2\omega|\mathbf{x})\ A_0^2 \,. \quad (10.22)$$

Insertion of (10.22) into (10.21) leads to a system of coupled linear inhomogeneous equations for the coefficients $E_r(n)$,

$$\sum_{n'}\sum_{r'} M(n\alpha,n'r')E_{r'}(n') = B_\alpha(n) \,. \quad (10.23)$$

It is important to note, that the determinant of the matrix $M(n\alpha,n'r')$ is of second order in $\zeta_0$ for $q \ll G/2$, and is of first order in $\zeta_0$ for $q = G/2$. It therefore vanishes in the limit of a flat surface, manifesting thereby the essential role of dispersion for the existence of weakly nonlinear waves. For weak corrugations, the part $u_\alpha^{(2,h)}$ is therefore the dominant contribution to the second harmonic, and we therefore neglect the static part of $u_\alpha^{(2)}$. In the last step of the calculation, the correction $\omega_2^2$ of second order in $A_0$ is calculated by projecting the part of the equation of motion proportional to $e^{-i\omega t}A_0^3$ on the linear solution, using Green's theorem and making use of the boundary conditions in the same way as in Section 7.1 to obtain for $\omega_2^2$ an expression of the form (7.34), which contains only the linear solution and its second harmonic.

Here, we confine ourselves to the calculation of the corrections to the frequencies limiting the gap to leading order in $A_0$ and $\zeta_0$. To this end, we set $q{=}G/2$ and retain only the Fourier components $n{=}0$ and $n{=}{-1}$ in the linear solution $u_\alpha^{(0)}$. The relevant coefficients $E_r(0) = E_r(-2)$ are obtained to leading order $O(\zeta_0^{-1})$ from the equation

$$\sum_r M_{\alpha r}^{(0)}(G,2\omega)E_r(0) = B_\alpha(0) \,, \quad (10.24)$$

where $B_\alpha(0)$ need only be calculated to zeroth order in $\zeta_0$, i.e. with displacement fields corresponding to Rayleigh waves propagating on a flat surface. The determinant of the matrix on the left hand side of (10.24) is given by

$$\det(M_{\alpha r}^{(0)}(G,2\omega)) = \mp 4\zeta_0 \left\{ M_{x1}^{(0)}\left(\frac{1}{2}G,\omega_0\right) M_{z2}^{(-)}\left(\frac{1}{2}G,\omega_0\right) + M_{z2}^{(0)}\left(\frac{1}{2}G,\omega_0\right) M_{x1}^{(-)}\left(\frac{1}{2}G,\omega_0\right) \right.$$

$$\left. - M_{x2}^{(0)}\left(\frac{1}{2}G,\omega_0\right) M_{z1}^{(-)}\left(\frac{1}{2}G,\omega_0\right) - M_{z1}^{(0)}\left(\frac{1}{2}G,\omega_0\right) M_{x2}^{(-)}\left(\frac{1}{2}G,\omega_0\right) \right\} + O(\zeta_0^2) \,, \qquad (10.25)$$

where $\omega_0 = v_R G/2$, and the upper and lower signs refer to the frequency corresponding to the symmetric and antisymmetric solutions, respectively. Using the coefficients $E_r(0)$, $E_r(-2)$ in (10.22), (10.18), (10.16), (7.34), and (7.27), and retaining only terms to lowest order in $|A_0|^2$ and $\zeta_0$, we obtain

$$\frac{\delta\omega_\pm}{\omega_0} = \pm \frac{1}{2} T \frac{\left|GA_0\right|^2}{G\zeta_0} + O(\zeta_0^0) \,. \qquad (10.26)$$

The data for the dimensionless quantity T given in Table 4 show, that for the substances analyzed, the gap becomes larger with increasing intensity.

In the case of shear horizontal polarization of the fundamental component in the nonlinear wave, the calculation will be largely analogous to the one outlined above, since the second harmonic will also be of sagittal polarization. The determinant of the matrix M in (10.23) however, will not vanish in this case, and the contributions to the frequency correction resulting from the second harmonic, the static part of the second order solution, and the fourth order anharmonicity, are expected to be of equal importance.

We conclude this section with the following supplementary remarks. The second harmonic contains the Fourier component (2q - G) which, for $q > q_t$ with

$$q_t \approx \frac{1}{2} G \frac{v_t}{v_t + v_R} \,, \qquad (10.27)$$

becomes leaky even for the lowest frequency branch. This means that for $q > q_t$, an imaginary part has to be added to the frequency correction $\omega_2^2/2\omega_0$, which plays the role of an amplitude dependent damping constant.

One may finally ask the question, whether the weakly nonlinear Rayleigh waves on a periodic grating are stable against modulation of the complex amplitude. As has been pointed out in Section 8, a criterion for the stability is the sign of the ratio of $\omega_2^2$ and the second derivative of the linear frequency with respect to q. While the latter quantity is always negative for the lowest frequency branch, calculations to leading order in $\zeta_0$ show that the sign of $\omega_2^2$ depends on q/G. The modulation instability may therefore occur in certain parts of the Brillouin zone.

## NOTE ADDED IN PROOF

A number of useful techniques for the theory of nonlinear effects in connection with surface acoustic waves, including those used in sections 3 and 5, are demonstrated in Ref. 67.


## ACKNOWLEDGEMENT

One of us (APM) gratefully acknowledges discussions with Profs. A. D. Boardman, L. Palmieri, G. Socino and R. K. Wehner, the hospitality of the Physics Department of the University of California, Irvine and financial support of the Deutsche Forschungsgemeinschaft. The work of AAM was supported in part by NSF Grant No. DMR 88-15866.

## APPENDIX A: DISPERSION OF RAYLEIGH WAVES

The following equations of motion and boundary conditions for the displacement field are considered:

$$\rho \ddot{u}_\alpha = C_{\alpha\beta\mu\nu} u_{\mu|\beta\nu} \qquad (A.1)$$

$$0 = [C_{\alpha z\mu\nu} u_{\mu|\nu} + d\{\rho_F \ddot{u}_\alpha - D_{\alpha\beta\mu\nu} u_{\mu|\beta\nu}\}]_{z=0} \; . \qquad (A.2)$$

This Ansatz for the boundary conditions includes the two cases of dispersion generated by the presence of a thin film of thickness d or by the underlying crystal lattice (see Section 2). We denote the Rayleigh wave solution of (A.1) and (A.2) proportional to $e^{iqx}$ for $d \neq 0$ by $u_\alpha^{(1)}$ and its frequency by $\omega$, and the corresponding solution for $d = 0$ by $u_\alpha^{(0)}$ and its frequency by $\omega_0$. On insertion of $u_\alpha^{(1)}$ into (A.1) and projecting on $u_\alpha^{(0)}$, we obtain

$$-\rho\omega^2 \int d^3x \; u_\alpha^{(0)*} u_\alpha^{(1)} = C_{\alpha\beta\mu\nu} \int d^3x \; u_\alpha^{(0)*} u_{\mu|\beta\nu}^{(1)} \; . \qquad (A.3)$$

Integrating the right hand side of (A.3) twice by parts with periodic boundary conditions in the x-y plane and making use of the boundary conditions for $u_\alpha^{(1)}$ and the equations of motion and boundary conditions for $u_\alpha^{(0)}$, we arrive at

$$\rho(\omega_0^2 - \omega^2) \int d^3x \; u_\alpha^{(0)*} u_\alpha^{(0)} = d \int d^2R \; u_\alpha^{(0)*} \{\rho_F \omega_0^2 u_\alpha^{(0)} + D_{\alpha\beta\mu\nu} u_{\mu|\beta\nu}^{(0)}\}|_{z=0} \qquad (A.4)$$

to first order in d. Since $u_\alpha^{(0)}$ is of the form

$$u_\alpha^{(0)} = \sum_{r=1}^{2} c_\alpha^{(0)}(r) e^{q\hat{\alpha}(r)z} e^{i(qx-\omega_0 t)} \qquad (A.5)$$

we finally obtain

$$\omega = \omega_0 + \frac{1}{2}\Omega_{xx} q^2 \; , \qquad \omega_0 = v_0 q \; , \qquad (A.6)$$

with $\Omega_{xx}$ given in Eq. (7.39).

## APPENDIX B: DERIVATION OF THE NONLINEAR SCHRÖDINGER EQUATION BY A MULTIPLE SCALE APPROACH

To demonstrate how the nonlinear Schrödinger equation can be obtained within the method of multiple scales with a scaling of the slow variables analogous to that introduced by Taniuti and Yajima, we choose the case of the evolution of a beam profile for nonlinear waves of predominantly shear horizontal polarization. As a useful byproduct of the derivation of the nonlinear Schrödinger equation, we will implicitly calculate the quantity $\Omega_{yy}(q\hat{x})$. As the slow variables, we choose

$$\xi := \epsilon y \; , \qquad \tau := \epsilon^2 x \; . \qquad (B.1)$$

The equations of first order in $\epsilon$ and their solution are identical to those given in Section 7 (with $\omega = \omega_0$). The second order displacement field has to be supplemented by the solution $u_\alpha^{(2,m)}$ of

$$\rho\omega_0^2 u_\alpha^{(2,m)} + C_{\alpha\beta\mu\nu} u_{\mu|\beta\nu}^{(2,m)} = -[C_{\alpha\beta yy} + C_{\alpha yy\beta}] A_{|\xi} U_{y|\beta} \tag{B.2}$$

$$[C_{\alpha z\mu\nu} u_{\mu|\nu}^{(2,m)} + d\{-\rho_F \omega_0^2 u_\alpha^{(2,m)} - D_{\alpha\Phi\mu\Theta} u_{\mu|\Phi\Theta}^{(2,m)}\}]_{z=0} = [-C_{\alpha zyy} A_{|\xi} U_y + d(D_{\alpha yy\Theta} + D_{\alpha\Theta yy}) A_{|\xi} U_{y|\Theta}]_{z=0} \tag{B.3}$$

The solution of (B.2) and (B.3) involves only sagittal components. An inhomogeneous solution of (B.2) is given by

$$u_\alpha^{(2,mi)} = A_{|\xi} \begin{pmatrix} n_x \\ 0 \\ n_z \end{pmatrix} e^{i(qx-\omega_0 t)+\beta z} . \tag{B.4}$$

The coefficients $n_x$, $n_z$ solve the linear system

$$\begin{pmatrix} \rho\omega_0^2 - q^2 C_{11} + \beta^2 C_{44} & iq\beta\, [C_{12} + C_{44}] \\ iq\beta\, [C_{12} + C_{44}] & \rho\omega_0^2 + \beta^2 C_{11} - q^2 C_{44} \end{pmatrix} \begin{pmatrix} n_x \\ n_z \end{pmatrix} = -[C_{12} + C_{44}] \begin{pmatrix} iq \\ \beta \end{pmatrix} . \tag{B.5}$$

To satisfy the boundary conditions (B.3), we choose a homogeneous solution of (B.2) of the form

$$u_\alpha^{(2,mh)} = A_{|\xi} e^{i(qx-\omega_0 t)} \sum_{r=1}^{2} f(r) \begin{pmatrix} 1 \\ 0 \\ p(r) \end{pmatrix} e^{\alpha(r) z} . \tag{B.6}$$

The quantities p(r) and $\alpha$(r) are the same as those in Section 7. To lowest order in d, they are given explicitly in Eqs. (7.58)-(7.61). Because of (7.58) and (7.59), we are again led to the condition (7.66). The displacement field $u_\alpha^{(2,m)}$ can alternatively be interpreted as the first order fields in a calculation of the quantity $\Omega_{yy}$ in perturbation theory in the absence of non-linearity. The condition (7.66) then has the following meaning. If it is fulfilled, the lowest Love wave branch can be continued to directions off the x-direction with the displacement field remaining localized at the surface. Otherwise, the surface waves become leaky as their wave vector is tilted with respect to the x-direction. The coefficients f(r) can be determined from the linear system

$$\begin{aligned} &\sum_{r=1}^{2} f(r)\{C_{44}(\alpha(r) + iqp(r)) - d(\rho_F \omega_0^2 - Dq^2)\} = G_x \\ &\sum_{r=1}^{2} f(r)\{C_{11} p(r)\alpha(r) + C_{12} iq - d\rho_F \omega_0^2 p(r)\} = G_z , \end{aligned} \tag{B.7}$$

where

$$\begin{aligned} G_x &= -C_{44}(\beta n_x + iqn_z) + d(\rho_F \omega_0^2 - Dq^2) n_x + d \left\{ C_{12}^{(F)} + C_{44}^{(F)} - \frac{C_{12}^{(F)2}}{C_{11}^{(F)}} \right\} iq \\ G_z &= -C_{11}\beta n_z - C_{12} iqn_x + d\rho_F \omega_0^2 n_z - C_{12} . \end{aligned} \tag{B.8}$$

The equation of motion of third order in $\epsilon$, projected on $U_\alpha$, takes the form

$$\rho\omega_0^2\int d^3x\; U^*_\alpha\tilde{u}^{(3)}_\alpha + C_{\alpha\beta\mu\nu}\int d^3x\; U^*_\alpha\tilde{u}^{(3)}_{\mu|\beta\nu} + \int d^3x\; U^*_\alpha\tilde{T}^{(N)}_{\alpha\beta|\beta} + A_{|\tau}(C_{\alpha x\mu\nu} + C_{\alpha\nu\mu x})\int d^3x\; U^*_\alpha U_{\mu|\nu}$$

$$+ A_{|\xi\xi}C_{\alpha y\mu y}\int d^3x\; U^*_\alpha U_\mu + A_{|\xi\xi}(C_{\alpha y\mu\nu} + C_{\alpha\nu\mu y})\int d^3x\; U^*_\alpha U^{(2)}_\mu = 0\;, \qquad \text{(B.9)}$$

where we have defined

$$u^{(2,m)}_\alpha = A_{|\xi}U^{(2)}_\alpha e^{-i\omega_0 t}\;. \qquad \text{(B.10)}$$

In order to eliminate $\tilde{u}^{(3)}_\alpha$, we proceed as in Section 7, and integrate by parts, making use of the boundary conditions, the third order part of which reads

$$[C_{\alpha z\mu\nu}\tilde{u}^{(3)}_{\mu|\nu} - d\{\rho_F\omega_0^2\tilde{u}^{(3)} + D_{\alpha\Phi\beta\Theta}\tilde{u}^{(3)}_{\beta|\Phi\Theta}\} + \tilde{T}^{(N)}_{\alpha z} + A_{|\tau}C_{\alpha z\mu x}U_\mu + A_{|\xi\xi}C_{\alpha z\mu y}U^{(2)}_\mu$$

$$- A_{|\tau}2dD_{\alpha x\beta x}U_{\beta|x} - A_{|\xi\xi}d\{D_{\alpha y\beta x} + D_{\alpha x\beta y}\}U^{(2)}_{\beta|x} - A_{|\xi\xi}dD_{\alpha y\beta y}U_\beta]_{z=0} = 0\;. \qquad \text{(B.11)}$$

In this way, (B.9) is transformed into the nonlinear Schrödinger equation

$$\bar{\alpha}A_{|\tau} + \bar{\beta}A_{|\xi\xi} + \bar{\gamma}|A|^2A = 0 \qquad \text{(B.12)}$$

with coefficients

$$\bar{\alpha} = (C_{\alpha x\mu\nu} + C_{\alpha\nu\mu x})\int d^3x\; U^*_\alpha U_{\mu|\nu} - C_{\alpha z\mu x}\int d^2R\;[U^*_\alpha U_\mu]_{z=0} + 2dD_{\alpha x\beta x}\int d^2R\;[U^*_\alpha U_{\beta|x}]_{z=0}$$

$$\text{(B.13)}$$

$$\bar{\beta} = C_{\alpha y\mu y}\int d^3x\; U_\alpha U^*_\mu + (C_{\alpha y\mu\nu} + C_{\alpha\nu\mu y})\int d^3x\; U^*_\alpha U^{(2)}_\mu - C_{\alpha z\mu y}\int d^2R\;[U^*_\alpha U^{(2)}_\mu]_{z=0}$$

$$+ (D_{\alpha y\beta x} + D_{\alpha x\beta y})\int d^2R\;[U^*_\alpha U^{(2)}_{\beta|x}]_{z=0} + D_{\alpha y\beta y}\int d^2R\;[U^*_\alpha U_\beta]_{z=0} \qquad \text{(B.14)}$$

$$\bar{\gamma}|A|^2A = -\int d^3x\; U^*_{\alpha|\beta}\tilde{T}^{(N)}_{\alpha\beta}\;. \qquad \text{(B.15)}$$

Equation (B.12) has to be identified with (8.2).

## APPENDIX C: EQUATIONS FOR THE HEIGHTS AND VELOCITIES OF THE SOLITONS RESULTING FROM A RECTANGULAR PULSE WITH COMPLEX AMPLITUDE

The initial pulse, that we consider here, is of the form

$$\psi_0(x,t=0) = \begin{cases} H & \text{for } |x| < \Delta/2 \\ 0 & \text{otherwise}\;, \end{cases} \qquad \text{(C.1)}$$

where H may be complex. The solution of the system of linear differential equations (8.6) with $\zeta = \xi + i\eta$, $\eta > 0$, for $|x| > \Delta/2$, which vanishes for $|x| \to \infty$, is

$$\begin{pmatrix} v_1 \\ v_2 \end{pmatrix} = \begin{pmatrix} A \\ 0 \end{pmatrix} e^{-i\zeta x} \qquad \text{for } x < -\Delta/2$$

$$\begin{pmatrix} v_1 \\ v_2 \end{pmatrix} = \begin{pmatrix} 0 \\ B \end{pmatrix} e^{i\zeta x} \qquad \text{for } x > \Delta/2 \, . \tag{C.2}$$

For $|x| < \Delta/2$, the system (8.6) takes the form

$$\begin{aligned} v_1' + i\zeta v_1 - ihv_2 &= 0 \\ v_2' - i\zeta v_2 - ih^* v_1 &= 0 \, , \end{aligned} \tag{C.3}$$

where

$$h = H\sqrt{\frac{\kappa}{2}} \, . \tag{C.4}$$

The Ansatz

$$\begin{pmatrix} C \\ D \end{pmatrix} e^{\lambda x} \tag{C.5}$$

leads to the linear homogeneous system

$$\begin{aligned} (\lambda + i\zeta)C - ihD &= 0 \\ -ih^* C + (\lambda - i\zeta)D &= 0 \, , \end{aligned} \tag{C.6}$$

the solvability condition of which yields

$$\lambda_{1,2} = \pm i\left[\zeta^2 + |h|^2\right]^{1/2} \, . \tag{C.7}$$

The solution for $|x| < \Delta/2$ may then be written as

$$\begin{pmatrix} v_1 \\ v_2 \end{pmatrix} = C_1 \begin{pmatrix} 1 \\ (\lambda_1 + i\zeta)/ih \end{pmatrix} e^{\lambda_1 x} + C_2 \begin{pmatrix} 1 \\ (\lambda_2 + i\zeta)/ih \end{pmatrix} e^{\lambda_2 x} \, . \tag{C.8}$$

The requirement of continuity of the solution at $x = \pm\Delta/2$ leads to a homogeneous system of four linear equations for $A$, $B$, $C_1$, $C_2$, only two of which contain $C_{1,2}$. The solvability condition of this subsystem for $h \neq 0$ is:

$$(\lambda_2 + i\zeta)e^{(\lambda_1 - \lambda_2)\Delta/2} - (\lambda_1 + i\zeta)e^{(\lambda_2 - \lambda_1)\Delta/2} = 0 \, . \tag{C.9}$$

By making the change of variable

$$z = \Delta\left[\zeta^2 + |h|^2\right]^{1/2} , \tag{C.10}$$

(C.9) can be cast into the form

$$\tan(z) = -\frac{z}{[\frac{1}{2}\kappa(|H|\Delta)^2 - z^2]^{1/2}} \qquad \text{(C.11)}$$

with the additional condition that the square root with positive real part has to be chosen. The solution $z = 0$ of (C.11) has to be excluded, because in this case $A = B = 0$, and the corresponding solution of (8.6) cannot be obtained by analytic continuation of the Jost function with the asymptotics $\begin{pmatrix}1\\0\end{pmatrix}e^{-i\xi x}$ for $x \rightarrow -\infty$. This condition, however, has to be fulfilled for a solution to correspond to a soliton.

To find solutions with $\xi_j = 0$, $1 \leq j \leq N$, which for $N = 1$ correspond to a single stationary soliton and for $N > 1$ for a soliton bound state, i.e. a stationary soliton of a more complicated shape, we first note, that (C.11) can only be fulfilled for $\eta < |h|$. We therefore have to search for real solutions of (C.11) in the interval $0 < z < \sqrt{\kappa/2}|H|\Delta$. They can easily be found graphically. There is a solution for $|H|\Delta$ larger than the threshold $\pi/\sqrt{2\kappa}$. More precisely, there are N pairwise distinct solutions for

$$(2N - 1)\frac{\pi}{\sqrt{2\kappa}} < |H|\Delta < (2N + 1)\frac{\pi}{\sqrt{2\kappa}}. \qquad \text{(C.12)}$$

There may exist further solutions of (C.11) corresponding to values of $\zeta$ with nonzero imaginary and real part, corresponding to solitons with nonzero drift velocity, which we, however, have not searched for.

## REFERENCES


1. V. I. Karpman and E. M. Krushkal', Zh. Eksp. Teor. Fiz. **55**, 530 (1968) [Soviet Physics - JETP **28**, 277 (1969)].
2. D. F. Parker, Physica **16D**, 385 (1985).
3. D. F. Nelson, *Electric, Optic & Acoustic Interactions in Dielectrics* (Wiley, New York, 1979).
4. G. Leibfried and W. Ludwig, Z. Phys. **160**, 80 (1960).
5. A. A. Maradudin, in *Physics of Phonons*, T. Paszkiewicz ed. (Springer, New York, 1987).
6. H. F. Tiersten, J. Appl. Phys. **40**, 770 (1969).
7. S. V. Bogdanov, M. D. Levin, and I. B. Yakovkin, Akust. Zhur. **15**, 12 (1969).
8. A. I. Murdoch, J. Mech. Phys. Solids **24**, 137 (1976).
9. G. W. Farnell, in *Acoustic Surface Waves*, A. A. Oliner ed. (Springer, New York, 1978), p. 13.
10. V. R. Velasco and F. Garcia-Moliner, Physica Scripta **20**, 111, (1979).
11. B. Djafari-Rouhani, L. Dobrzynski, V. R. Velasco and F. Garcia-Moliner, Surf. Sci. **110**, 129 (1981).
12. L. Dobrzynski and B. Djafari-Rouhani, in *Vibrations at Surfaces*, R. Caudano, J. M. Gilles, and A. A. Lucas eds. (Plenum, New York, 1982), p.1.
13. A. A. Maradudin, in *Nonequilibrium Phonon Dynamics*, W. E. Bron ed. (Plenum, New York, 1985), p. 395.
14. D. C. Gazis, R. Herman, and R. F. Wallis, Phys. Rev. **119**, 533, (1960).
15. G. R. Gerhart, J. Acoust. Soc. Am. **60**, 1085 (1976).
16. K. Jassby and D. Saltoun, Mater. Eval. **40**, 198 (1982).
17. M. Hayes and R. S. Rivlin, Arch. Rat. Mech. Anal. **8**, 15 (1961).
18. Y. Iwashimizu and O. Kobori, J. Acoust. Soc. Am. **64**, 910 (1978).
19. A. Tverdokhlebov, J. Acoust. Soc. Am. **73**, 2006 (1983).
20. P. P. Delsanto and A. V. Clark, Jr., J. Acoust. Soc. Am. **81**, 952 (1987).
21. G. T. Mase and G. C. Johnson, J. Appl. Mech. **54**, 127 (1987).
22. A. J. C. Ladd and W. G. Hoover, Preprint.
23. J. Lothe and D. M. Barnett, J. Appl. Phys. **47**, 428 (1976).
24. B. A. Auld, *Acoustic Fields and Waves in Solids*, vol. II (Wiley, New York, 1973).
25. H. F. Tiersten, J. Acoust. Soc. Am. **64**, 832 (1978).
26. H. F. Tiersten and B. K. Sinha, J. Appl. Phys. **49**, 87 (1978).
27. D. Husson, J. Appl. Phys. **57**, 1562 (1985).

28. R. N. Thurston and K. Brugger, Phys. Rev. **133**, A1604 (1964).
29. Landolt-Börnstein New Series, Group III, vol. 11, K.-H. Hellwege ed. (Springer, Berlin, 1979); vol. 18, K.-H. Hellwege and O. Madelung eds. (Springer, Berlin, 1984).
30. For a review see: F. Nizzoli and G. I. Stegeman, in *Surface Excitations*, V. M. Agranovich and R. Loudon eds. (North-Holland, Amsterdam, 1985). For a detailed discussion see M. Planat, *Propagation nonlinéaire des ondes acoustiques dans les solides*, thesis, Université de Franche-Comté (1984).
31. H. F. Tiersten and J. C. Baumhauer, J. Appl. Phys. **58**, 1867 (1985).
32. R. Normandin, M. Fukui, and G. I. Stegeman, J. Appl. Phys. **50**, 81 (1979).
33. P. O. Løpen, J. Appl. Phys. **39**, 5400 (1968).
34. J. W. Gibson and P. Meijer, J. Appl. Phys. **45**, 3288 (1974).
35. A. Alippi, A. Palma, L. Palmieri, and G. Socino, J. Appl. Phys. **48**, 2182 (1977).
36. M. Toda, *Theory of Nonlinear Lattices* (Springer, Berlin, 1981).
37. G. B. Whitham, *Linear and Nonlinear Waves* (Wiley, New York, 1974).
38. H. C. Yuen and B. M. Lake, Phys. Fluids **18**, 956 (1975) and references therein.
39. T. Taniuti and N. Yajima, J. Math. Phys. **10**, 1369 (1969).
40. N. Kalyanasundaram, Int. J. Engng. Sci. **19**, 279 (1981).
41. N. Kalyanasundaram, Int. J. Engng. Sci. **19**, 435 (1981).
42. N. Kalyanasundaram, R. Ravindran, and P. Prasad, J. Acoust. Soc. Am. **72**, 488 (1982).
43. R. W. Lardner, Int. J. Engng. Sci. **21**, 1331 (1983).
44. R. W. Lardner, J. Appl. Phys. **55**, 3251 (1984).
45. M. Planat, J. Appl. Phys. **57**, 4911 (1985).
46. T. Sakuma and Y. Kawanami, Phys. Rev. **B29**, 869 (1984).
47. T. Sakuma and Y. Kawanami, Phys. Rev. **B29**, 880 (1984).
48. T. Sakuma and T. Miyazaki, Phys. Rev. **B33**, 1036 (1986).
49. T. Sakuma and O. Saito, Phys. Rev. **B35**, 1294 (1987).
50. V. I. Nayanov, Pis'ma Zh. Eksp. Teor. Fiz. **44**, 245 (1986) [JETP Letters **44**, 314 (1986)].
51. D. Royer and E. Dieulesaint, J. Acoust. Soc. Am. **76**, 1438 (1984).
52. V. I. Karpman, *Non-Linear Waves in Dispersive Media*, (Pergamon, Oxford, 1974) [Russ. edition: Nauka, Moscow, 1973].
53. V. E. Zakharov and A. B. Shabat, Zh. Eksp. Teor. Fiz. **61**, 118 (1971) [Soviet Physics - JETP **34**, 62 (1972)].
54. A. D. Boardman, G. S. Cooper, A. A. Maradudin, and T. P. Shen, Phys. Rev. **B34**, 8273 (1986).
55. R. C. Hollinger and G. R. Barsch, J. Phys. Chem. Solids **37**, 845 (1976).
56. L. M. Brekhovskikh, Akust. Zhur. **5**, 288 (1959) [Soviet Physics - Acoustics **5**, 288 (1960)].
57. F. Rischbieter, Acustica **16**, 75 (1965).
58. P. V. H. Sabine, Electronics Lett. **6**, 149 (1970).
59. B. A. Auld, J. J. Gagnepain, and M. Tan, Electronics Lett. **12**, 650 (1976).
60. Yu. V. Gulyaev and V. P. Plessky, Zh. Tekh. Fiz. **48**, 447 (1978) [Soviet Physics - Technical Physics **23**, 266 (1978)].
61. N. E. Glass and A. A. Maradudin, Electronics Lett. **17**, 773, (1981).
62. N. E. Glass, R. Loudon, and A. A. Maradudin, Phys. Rev. **B24**, 6843 (1981).
63. N. E. Glass and A. A. Maradudin, J. Appl. Phys. **54**, 796 (1983).
64. H. F. Tiersten, J. T. Song, and D. V. Shick, J. Appl. Phys. **62**, 1154 (1987).
65. B. Hillebrands and G. I. Stegeman, unpublished work.
66. For a review and introduction, see also Ref. 5.
67. G. A. Maugin, *Nonlinear Electromechanical Effects and Applications* (World Scientific, Singapore, 1985).

# NONLINEAR SURFACE ACOUSTIC WAVES ON HOMOGENEOUS MEDIA

D. F. Parker†

Department of Theoretical Mechanics
University of Nottingham
Nottingham, NG7 2RD, U.K.

## 1. DERIVATION OF THE EVOLUTION EQUATION

### 1.0 Introduction

This chapter on the theoretical treatment of nonlinear elastic and piezoelectric surface waves will concentrate on those features which distinguish waves on a homogeneous half-space from other guided nonlinear waves. These differences arise principally because the linear theory of such waves predicts no dispersion, so that all frequencies and wavelengths travel at the same speed. This allows ample time for nonlinear interactions to reinforce each other constructively. Except in special cases, attention should not be confined to interactions between just a small number of modes. Although derivation of approximate nonlinear evolution equations is based on multiple-scale asymptotics (as in many other contributions in these proceedings) the derivation procedure is not a formal expansion procedure. It is one which has been found useful not just in the present context of surface acoustic waves, but also in treatments of nonlinear fibre optics.

The first section reviews the linear theory of elastic surface acoustic waves, including the classical Rayleigh wave as a special case. Then, by incorporating quadratic nonlinearity, an evolution equation governing the Fourier transform of the surface elevation is derived. The special case of periodic waveforms is discussed, showing how sinusoidal waveforms first generate second and third harmonics, but later excite all harmonics. Computations showing typical break-up of a profile are illustrated.

The second section deals with nonlinear waves of permanent form. Again these have some unusual features, since they are never close to sinusoidal. They have an unfamiliar relationship between phase speed, amplitude and wavelength. This allows a simple prediction of the propagation of amplitude modulations. After a brief review of this theory, some numerical predictions of the stability of these waveforms is presented. Although the mathematical theory of the stability of these waveforms is still being developed, some preliminary theory and predictions are mentioned.

The final section deals with a variety of other phenomena. Firstly, the extension to include piezoelectricity and other electro-mechanical interactions is given. This causes little additional theoretical difficulty, but does allow the extra possibility of purely transverse, non-dispersive modes - the Bleustein-Gulyaev waves. Then, the pertinent differences for waves on magnetic materials are described. The underlying theory for the application of surface acoustic waves to filters and convolvers is summarized. These involve, respectively, two co-propagating and two counter-propagating signals. Finally, the equations describing spreading of a beam of surface acoustic waves are given.

†New address: Department of Mathematics, University of Edinburgh, The King's Buildings, Mayfield Road, Edinburgh, EH9 3JZ, U.K.

*Nonlinear Waves in Solid State Physics*
Edited by A.D. Boardman *et al.*, Plenum Press, New York, 1990

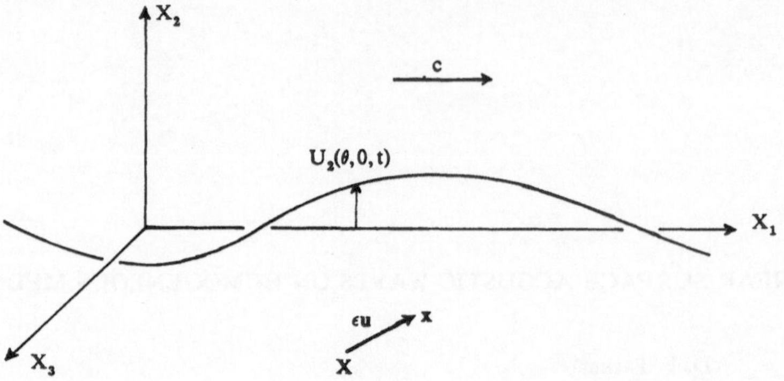


Fig. 1 Reference coordinate system $X_1$, $X_2$, $X_3$, displacement $\epsilon\mathbf{u}$ and disturbed free surface $X_2 = 0$.

### 1.1 Linear elastic surface waves

We consider waves travelling along the surface $X_2 = 0$ of a homogeneous elastic half-space $X_2 < 0$, where $X_L$ $(L = 1,2,3)$ denote Lagrangian coordinates and $x_j$ $(j = 1,2,3)$ are the current (Eulerian) coordinates of a material point. When thermal effects are neglected, the components $\tau_{Lj}$ of the Piola-Kirchhoff stress tensor are related to the components $F_{jL}$ of the deformation gradient tensor $\mathbf{F}$ by

$$\tau_{Lj} = \frac{\partial W}{\partial F_{jL}}, \qquad F_{jL} = x_{j,L} \equiv \frac{\partial x_j}{\partial X_L},$$

where $W = W(\mathbf{F})$ is the isentropic strain energy density. In the absence of body forces, the momentum equations and (traction-free) boundary conditions are

$$\rho \ddot{x}_j = \tau_{Lj,L} \qquad \text{in } X_2 < 0, \qquad (1.1)$$

$$\tau_{2j} = 0 \qquad \text{on } X_2 = 0, \qquad (1.2)$$

where $\rho$ is the (uniform) reference density, dots denote time derivatives and the comma denotes partial differentiation.

Typically the displacements $x_j - \delta_{jJ}X_J$ and deformation gradient $F_{jJ} - \delta_{jJ}$ are small, so we write

$$x_j = \delta_{jJ}X_J + \epsilon u_j(X_L, t), \qquad F_{jL} - \delta_{jL} = \epsilon u_{j,L}, \qquad (1.3)$$

where $\epsilon$ is a formal small parameter. In linear theory, the Piola Kirchhoff stress $\tau_{Lj}$ is indistinguishable from the (symmetric) Cauchy stress and so has the form $\tau_{Lj} = \epsilon c_{jLmM} u_{m,M}$. More generally, we could write

$$\tau_{Lj} = \epsilon c_{jLmM} u_{m,M} + N_{Lj}(\epsilon u_{k,K}),$$

though for simplicity we follow standard practice and replace the nonlinear term $N_{Lj}$ by a quadratic approximation, so writing

$$\tau_{Lj} = \epsilon c_{jLmM} u_{m,M} + \epsilon^2 c_{jLmMnN} u_{m,M} u_{n,N}. \qquad (1.4)$$

Here, the coefficients $c_{jLmMnN} = c_{mMjLnN} = c_{jLnNmM}$ are symmetric under pairwise interchanges of indices and depend on both the first-order elastic constants $c_{jLmM}$ and the standard second-order constants.[1]

We seek waves travelling parallel to the $X_1$-axis and so consider generalised plane-strain disturbances $u_j = u_j(X_1, X_2, t)$. Then, linearization of Eqs. (1.1), (1.2) and (1.4) gives the boundary value problem

$$\rho \ddot{u}_j = c_{j\alpha m\beta} u_{m,\alpha\beta} \qquad \text{in } X_2 < 0 , \tag{1.5}$$

$$c_{j2m\beta} u_{m,\beta} = 0 \qquad \text{on } X_2 = 0 , \tag{1.6}$$

together with the decay condition

$$u_j(X_\alpha, t) \to 0 \qquad \text{as } X_2 \to -\infty \tag{1.7}$$

which arises because surface acoustic waves are disturbances travelling near the surface. We adopt the convention that Greek subscripts range only over the values 1 and 2. These equations possess (complex) solutions periodic in the phase variable $\theta \equiv X_1 - ct$. Such solutions are a superposition of 'partial waves'[2]

$$\mathbf{u} = \mathbf{a}e^{ik(\theta + sX_2)} , \qquad k \text{ real} ,$$

where the components $a_j$ of $\mathbf{a}$ and the depth factors s satisfy the algebraic system

$$L_{jm}(s)a_m \equiv \{c_{j1m1} - \rho c^2\delta_{jm} + s(c_{j1m2} + c_{j2m1}) + s^2 c_{j2m2}\}a_m = 0 . \tag{1.8}$$

For each choice of c, Eq. (1.8) has a compatibility condition which is a sextic polynomial having real coefficients. Any real roots s describe plane waves, which violate the decay requirement (1.7). We are concerned with the situation in which the six roots occur in complex conjugate pairs

$$s^{(p)} = \alpha^{(p)} - i\beta^{(p)} , \qquad s^{(p)*} = \alpha^{(p)} + i\beta^{(p)} , \qquad \beta^{(p)} > 0$$

with corresponding eigenvectors $\mathbf{a} = \mathbf{a}^{(p)}, \mathbf{a}^{(p)*}$ for p = 1,2,3, where * denotes a complex conjugate. Each partial wave

$$\mathbf{u} = \mathbf{a}^{(p)} e^{ik(\theta + s^{(p)} X_2)} = \mathbf{a}^{(p)} e^{k\beta^{(p)} X_2} e^{ik(X_1 + \alpha^{(p)} X_2 - ct)} \tag{1.9}$$

describes a sinusoidal disturbance travelling with speed c and associated with surface traction $\tau_{2j} = ik(c_{j2m1} + s^{(p)} c_{j2m2})a_m^{(p)} e^{ik\theta}$. The traction-free boundary condition (1.6) and decay condition (1.7) are satisfied for k > 0 only when some linear combination

$$\mathbf{u} = \sum_{p=1}^{3} B^{(p)} \mathbf{a}^{(p)} e^{ik(\theta + s^{(p)} X_2)}$$

of partial waves (1.9) gives vanishing traction on $X_2 = 0$. This yields the boundary matrix equation

$$M_{jp} B^{(p)} \equiv \sum_{p=1}^{3} (c_{j2m1} + s^{(p)} c_{j2m2}) a_m^{(p)} B^{(p)} = 0 \tag{1.10}$$

for the multipliers $B^{(p)}$. The condition $\det(M_{jp}) = 0$ is an algebraic equation for $c^2$, whose structure has been elucidated by a number of authors,[3] using a formalism due originally to Stroh.[4] Typically it has just one root, the square of the Rayleigh speed. Without loss of generality, the corresponding solutions $B^{(p)}$ may be chosen to give $\Sigma B^{(p)} a_2^{(p)} = 1$. Then the displacements within a surface wave having wavelength $2\pi/k$ are $\mathbf{u} = \mathbf{A}(kX_2)e^{ik\theta}$, where

$$\mathbf{A}(kX_2) = \sum_{p=1}^{3} B^{(p)} \mathbf{a}^{(p)} e^{iks^{(p)} X_2} , \qquad A_2(0) = 1, \quad k > 0 .$$

For k < 0, the corresponding depth-dependence is

$$\mathbf{A}(kX_2) = \mathbf{A}(-|k|X_2) = \mathbf{A}^*(|k|X_2) = \sum_{p=1}^{3} B^{(p)*}\mathbf{a}^{(p)*}e^{iks^{(p)*}X_2} .$$

Since these waves are non-dispersive, the phase speed c is independent of wavelength and the depth-dependence scales directly as the wavelength. Different wavelengths and phases may be superposed as

$$\mathbf{u}(X_1, X_2, t) = \int_{-\infty}^{\infty} C(k)\mathbf{A}(kX_2)e^{ik\theta}dk = \tilde{\mathbf{u}}(X_1 - ct, X_2) . \tag{1.11}$$

In this general disturbance travelling at speed c without distortion the surface elevation profile $\tilde{u}_2(X_1 - ct, 0) \equiv u_0(X_1 - ct)$ is *arbitrary*. It is related directly to C(k) by the Fourier transform pair

$$u_2(X_1, 0, t) = \int_{-\infty}^{\infty} C(k)e^{ik\theta}dk = u_0(\theta) ,$$

$$C(k) = \frac{1}{2\pi}\int_{-\infty}^{\infty} u_0(\theta)e^{-ik\theta}d\theta , \tag{1.12}$$

so showing that the condition ensuring that displacements are real is

$$C(-k) = C^*(k) . \tag{1.13}$$

The fact that linearized waves travel without frequency dispersion is the distinguishing feature of waves on a homogeneous half-space, when boundary conditions do not introduce a natural scale of length or time. It is also the reason why a general nonlinear theory should use displacements of the form (1.11) as a basic approximation. Before showing this, the special case of isotropic elasticity will be illustrated, so as to fix ideas.

In isotropic elasticity, we have $c_{j\alpha m\beta} = \lambda\delta_{j\alpha}\delta_{m\beta} + \mu(\delta_{jm}\delta_{\alpha\beta} + \delta_{j\beta}\delta_{m\alpha})$ so that $L_{3m} = L_{m3} = (\mu - \rho c^2 + s^2\mu)\delta_{3m}$ which uncouples $a_3$ from $a_1$ and $a_2$ in (1.8). The solutions $s^{(p)}$ are

$$s^{(1)} = -i\beta_1 , \qquad s^{(2)} = -i\beta_2 , \qquad s^{(3)} = -i\beta_1 ,$$

where

$$\beta_1 = \left\{1 - \frac{\rho c^2}{\mu}\right\}^{1/2} , \qquad \beta_2 = \left\{1 - \frac{\rho c^2}{\lambda + 2\mu}\right\}^{1/2} ,$$

with corresponding eigenvectors

$$\mathbf{a}^{(1)} = \begin{bmatrix} i\beta_1 \\ 1 \\ 0 \end{bmatrix} , \qquad \mathbf{a}^{(2)} = \begin{bmatrix} i \\ \beta_2 \\ 0 \end{bmatrix} , \qquad \mathbf{a}^{(3)} = \begin{bmatrix} 0 \\ 0 \\ 1 \end{bmatrix} .$$

Since only the $\mathbf{a}^{(3)}$ partial wave gives non-zero $\tau_{23}$, the boundary matrix equation (1.10) shows that $B^{(3)} = 0$ and so surface waves involve no transverse displacements. Either material stratification (Love waves) or anisotropy (e.g. piezoelectricity - Bleustein-Gulyaev waves) is necessary for transverse motions in surface acoustic waves. The depth-dependence is found to be given by

$$\mathbf{A}(kX_2) = B^{(1)}\mathbf{a}^{(1)}e^{k\beta_1 X_2} + B^{(2)}\mathbf{a}^{(2)}e^{k\beta_2 X_2} \qquad (k > 0)\ ,$$

where

$$B^{(1)} = \frac{2}{1-\beta_1^2}\ , \qquad B^{(2)} = -\frac{4\beta_1}{1-\beta_1^4} = -\frac{2\beta_1}{1+\beta_1^2}B^{(1)} = -\sqrt{\frac{\beta_1}{\beta_2}}\,B^{(1)}\ .$$

and where $\beta_1$ and $\beta_2$ satisfy the condition

$$\left(1+\beta_1^2\right)^2 = 4\beta_1\beta_2\ .$$

This is equivalent to the standard *secular equation*

$$\left(\frac{\rho c^2}{\mu}\right)^3 - 8\left(\frac{\rho c^2}{\mu}\right)^2 + 8\,\frac{3\lambda+4\mu}{\lambda+\mu}\,\frac{\rho c^2}{\mu} - 16\,\frac{\lambda+\mu}{\lambda+2\mu} = 0 \qquad (1.14)$$

which defines only one [subsonic, i.e. $c < (\mu/\rho)^{1/2}$] speed of sinusoidal Rayleigh waves. The corresponding particle motions

$$u_1 = \frac{2\beta_1}{1-\beta_1^4}\left\{2e^{k\beta_2 X_2} - (1+\beta_1^2)e^{k\beta_1 X_2}\right\}\sin[k(X_1 - ct)]$$

$$u_2 = \frac{1}{1-\beta_1^2}\left\{2e^{k\beta_1 X_2} - (1+\beta_1^2)e^{k\beta_2 X_2}\right\}\cos[k(X_1 - ct)]$$

in a sinusoidal wave of unit amplitude are elliptical, with eccentricity varying with depth. The sense of the motion is *retrograde* near the surface but switches to *prograde* at a depth of approximately one sixth of a wavelength (see e.g. Achenbach[5]). Notice that significant displacements arise only within one or two wavelengths of the surface [since $\beta_1$ and $\beta_2$ are $O(1)$ quantities]. Consequently any boundaries or flaws at substantial depth influence only wavelengths at least as large as that depth.

Clearly sinusoidal waves may be superposed, as in (1.11), to give much more complex particle motions, even in periodic signals.

## 1.2 The evolution equation for nonlinear surface acoustic waves

Inclusion of the quadratic terms from Eq. (1.4) in the governing equations (1.1), (1.2) and (1.7) causes $O(\epsilon)$ perturbations to the distribution of displacements (1.11). Moreover, it can cause gradual distortion over a 'long' scale measured by the variable $X \equiv \epsilon X_1$.

Since linear theory shows that every surface acoustic wave travelling in the $X_1$ direction can be expressed in terms of just two coordinates $\theta$ and $X_2$, we seek to express nonlinear surface acoustic waves in terms of the three coordinates $\theta$, $X_2$ and $X$. Consequently, in (1.2), (1.3) and (1.4) the displacement gradient and velocity components become

$$u_{j,1} = \frac{\partial u_j}{\partial\theta} + \epsilon\frac{\partial u_j}{\partial X}\ , \qquad u_{j,2} = \frac{\partial u_j}{\partial X_2}\ , \qquad u_{j,3} = 0\ , \qquad \dot{u}_j = -c\frac{\partial u_j}{\partial\theta}$$

so giving the boundary value problem

$$\rho c^2 u_{j,11} - c_{j\alpha m\beta}u_{m,\alpha\beta} = \epsilon c_{j\alpha m\beta n\gamma}(u_{m,\beta}u_{n,\gamma})_{,\alpha} + \epsilon(c_{j1m\beta} + c_{j\beta m1})u_{m,\beta X} + O(\epsilon^2)$$

$$\text{in } X_2 < 0\ , \qquad (1.15)$$

$$c_{j2m\beta} u_{m,\beta} = -\epsilon c_{j2m\beta n\gamma} u_{m,\beta} u_{n,\gamma} - \epsilon c_{j2m1} u_{m,X} \qquad \text{on } X_2 = 0 , \qquad (1.16)$$

together with the decay condition (1.7), namely $u_j \to 0$ as $X_2 \to -\infty$. Here, and subsequently, the subscripts 1, 2 and X appearing after a comma denote differentiation with respect to $\theta$, $X_2$ and X, respectively. Since the nonlinear terms and X-derivatives appear only in $O(\epsilon)$ terms in the boundary value problem, it is readily seen that the expression

$$\mathbf{u} = \mathbf{U}(\theta, X_2, X) \equiv \int_{-\infty}^{\infty} C(k, X)A(kX_2)e^{ik\theta}dk \qquad (1.17)$$

is an $O(1)$ solution, for all $C(k,X)$. It is real when $C(-k,X) = C^*(k,X)$. When correction terms are constructed, using either a perturbation or an iterative scheme, compatibility considerations lead to an evolution equation for the X-dependence of $C(k,X)$. However, it is unnecessary to *construct* these terms in order to derive the evolution equation. Neither is it necessary to use a multiple-scales method. The procedure to be described here requires only three coordinates $\theta$, $X_2$ and X, even if carried to *all* orders in $\epsilon$.

The exact solution is first written as

$$\mathbf{u} = \mathbf{U}(\theta, X_2, X) + \epsilon\mathbf{w}(\theta, X_2, X; \epsilon)$$

where $\epsilon\mathbf{w}$ is the total correction to the expression (1.17). Substitution into (1.15) and (1.16) then gives a system of the form:

$$L_j\mathbf{w} \equiv \rho c^2 w_{j,11} - c_{j\alpha m\beta} w_{m,\alpha\beta} = P_j(\theta, X_2; X, \epsilon) \qquad (1.18)$$

$$B_j\mathbf{w} \equiv c_{j2m\beta} w_{m,\beta} = -Q_j(\theta; X, \epsilon) \qquad \text{on } X_2 = 0 \qquad (1.19)$$

together with the decay condition

$$|\mathbf{w}| \to 0 \qquad \text{as } X_2 \to -\infty . \qquad (1.20)$$

Here $P_j$ and $Q_j$ are given by

$$P_j \equiv c_{j\alpha m\beta n\gamma}(u_{m,\beta} u_{n,\gamma})_{,\alpha} + (c_{j1m\beta} + c_{j\beta m1})u_{m,\beta X} + O(\epsilon) ,$$

$$Q_j \equiv c_{j2m\beta n\gamma} u_{m,\beta} u_{n,\gamma} + c_{j2m1} u_{m,X} + O(\epsilon) ,$$

where the $O(\epsilon)$ terms are combinations of derivatives of both $\mathbf{U}$ and $\mathbf{w}$. The semi-colons in $P_j$ and $Q_j$ emphasize that the differential operators $L_j$ and $B_j$ involve differentiations with respect to $\theta$ and $X_2$ only. Thus, when solutions are constructed either iteratively or by perturbation expansions, the boundary value problem has the form of (1.18)-(1.20) with appropriate approximations to $P_j$ and $Q_j$ involving $C(k,X)$ and previous expressions for $\mathbf{w}$. In each boundary value problem, the propagation range X may be treated as a parameter. The resulting two dimensional boundary value problem is solvable only when the source-like terms $P_j$ and $Q_j$ satisfy a *compatibility condition*, and so at each stage of an approximation process will determine one member of a hierarchy of evolution equations for $C(k,X)$.

To determine the compatibility condition, we take Fourier transforms with respect to $\theta$, in the form

$$\mathcal{F}\{f\} = \bar{f}(k, X_2, X) = \frac{1}{2\pi}\int_{-\infty}^{\infty} f(\theta, X_2, X)e^{-ik\theta}d\theta .$$

Equations (1.18) and (1.19) then yield the ordinary differential system

$$\mathcal{L}_{jm}\bar{w}_m = \bar{P}_j \qquad \text{for } -\infty < X_2 < 0 , \qquad (1.21)$$

$$\mathcal{B}_{jm}\bar{w}_m = -\bar{Q}_j \qquad \text{on } X_2 = 0 , \qquad (1.22)$$

subject to the decay condition $\overline{w}_m \to 0$ as $X_2 \to -\infty$. Here, the ordinary differential operators $\mathscr{L}_{jm}$ and $\mathscr{B}_{jm}$ are

$$\mathscr{L}_{jm} \equiv k^2(c_{j1m1} - \rho c^2 \delta_{jm}) - ik(c_{j1m2} + c_{j2m1})\frac{d}{dX_2} - c_{j2m2}\frac{d^2}{dX_2^2}$$

$$\mathscr{B}_{jm} \equiv ikc_{j2m1} + c_{j2m2}\frac{d}{dX_2} \, .$$

Now, if $v_j(k,X_2)$ are any suitably differentiable functions, it is straightforward to verify the identity

$$\int_{-D}^{0} v_j \overline{P}_j dX_2 = \int_{-D}^{0} \overline{w}_m \mathscr{L}^*_{mj} v_j dX_2 - \left[ v_j \mathscr{B}_{jm} \overline{w}_m - \overline{w}_m \mathscr{B}^*_{mj} v_j \right]_{-D}^{0} \, .$$

Then, if $v_j$ is chosen to be any solution of the homogeneous system

$$\mathscr{L}^*_{mj} v_j \equiv k^2(c_{m1j1} - \rho c^2 \delta_{mj})v_j + ik(c_{m1j2} + c_{j2m1})\frac{dv_j}{dX_2} - c_{m2j2}\frac{d^2 v_j}{dX_2^2} = 0$$

$$\mathscr{B}^*_{mj} v_j \equiv -ikc_{m2j1} v_j + c_{m2j2}\frac{dv_j}{dX_2} = 0 \qquad \text{on } X_2 = 0$$

which is adjoint to (1.21) and (1.22), we obtain

$$\int_{-D}^{0} v_j \overline{P}_j dX_2 = v_j(k,0)\overline{Q}_j + \left[ v_j \mathscr{B}_{jm} \overline{w}_m - \overline{w}_m \mathscr{B}^*_{mj} v_j \right]_{X_2=-D} \, .$$

If, additionally, the condition $v_j \to 0$ as $X_2 \to -\infty$ is imposed, this relationship becomes

$$\int_{-\infty}^{0} v_j(k,X_2)\overline{P}_j dX_2 = v_j(k,0)\overline{Q}_j \, .$$

At this stage, we note that the equations defining $v_j$ are virtually identical to those defining the Fourier transform $A_j(kX_2)$ of the linearized periodic disturbance. The only difference is the complex conjugate * (or, equivalently, a change of sign $k \to -k$), so showing that the relevant functions $v_j(k,X_2)$ have the form

$$v_j(k,X_2) = \lambda \sum_{p=1}^{3} B^{(p)*} a_j^{(p)*} e^{-iks^{(p)*}X_2} = \lambda A_j^*(kX_2) \, .$$

Thus the *solvability condition* for the system (1.18)-(1.20) is

$$A^*_{0j}\overline{Q}_j(k;X,\epsilon) = \int_{-\infty}^{0} A^*_j(kX_2)\overline{P}_j(k,X_2;X,\epsilon)dX_2 \, , \qquad (1.23)$$

where $A_{0j}$ are the components of

$$A_0 \equiv \sum_{p=1}^{3} B^{(p)} a^{(p)} = \lim_{kX_2 \to 0_-} A(kX_2)$$

Condition (1.23) has the form obtained by the *projection method* of Maradudin and Mayer.[6]

In any approximation procedure, Eq. (1.23) must apply to the approximations used for **P** and **Q** at each stage. In the fundamental approximation, $P_j$ and $Q_j$ have the forms

$$P_j = (c_{j1m\beta} + c_{j\beta m1})U_{m,\beta X} + c_{j\alpha m\beta n\gamma}(U_{m,\beta}U_{n,\gamma})_{,\alpha} ,$$

$$Q_j = c_{j2m1}U_{m,X} + c_{j2m\beta n\gamma}U_{m,\beta}U_{n,\gamma} .$$

Using (1.17) it is found that the Fourier transforms of typical terms are

$$\mathscr{F}\{(c_{j1m\beta} + c_{j\beta m1})U_{m,\beta X}\} = k\{2ic_{j1m1}A_m(kX_2) + (c_{j1m2} + c_{j2m1})A'_m(kX_2)\}C_{,X}(k,X) ,$$

$$\mathscr{F}\{c_{j\alpha m\beta n\gamma}U_{m,\beta}U_{n,\gamma}\} = \int_0^\infty K_{j\alpha}(kX_2;k^{-1}\ell)(k-\ell)\ell C(k-\ell,X)C(\ell,X)d\ell ,$$

where the kernel $K_{j\alpha}$ in this convolution integral is defined in both $0 < \ell < k$ and in $k < \ell$ as linear combinations of products of pairs of the complex exponential factors occurring in $A(kX_2)$. Details are given in Ref. 1. This then shows that both $\overline{P}_j$ and $\overline{Q}_j$ are composed of just two types of term - those involving $C_{,X}$ and those involving a weighted convolution integral - since

$$\overline{P}_j(k,X_2,X) = k\int_0^\infty (iK_{j1} + K'_{j2})(k-\ell)\ell C(k-\ell,X)C(\ell,X)d\ell$$
$$+ k\{2ic_{j1m1}A_m(kX_2) + (c_{j1m2} + c_{j2m1})A'_m(kX_2)\}C_{,X}(k,X) ,$$

$$\overline{Q}_j(k,X) = \int_0^\infty K_{j2}(0;k^{-1}\ell)(k-\ell)\ell C(k-\ell,X)C(\ell,X)d\ell + c_{j2m1}A_{0m}C_{,X}(k,X) .$$

Insertion of these expressions into (1.23) finally yields the evolution equation for $C(k,X)$ in the form

$$JC_{,X}(k,X) = i\int_0^\infty K\left[\frac{\ell}{k}\right](k-\ell)\ell C(k-\ell,X)C(\ell,X)d\ell \qquad (k > 0) . \qquad (1.24)$$

In Eq. (1.24) the *real* constant J and the kernel $K(k^{-1}\ell)$ are given by

$$J = \int_{-\infty}^0 \{2c_{j1m1}A_j^*(\xi)A_m(\xi) + ic_{j2m1}A_j^{*\prime}(\xi)A_m(\xi) - ic_{j1m2}A_j^*(\xi)A'_m(\xi)\}d\xi ,$$

$$(1.25)$$

$$K(k^{-1}\ell) = \int_{-\infty}^0 \{iA_j^*(\xi)K_{j1}(\xi;k^{-1}\ell) - A_j^{*\prime}(\xi)K_{j2}(\xi;k^{-1}\ell)\}d\xi .$$

This shows that interaction between any two wavelengths involves details of the displacements at all depths (a feature typical of guided modes). Also, Eq. (1.24) shows that the amplitude at one wavelength interacts with that at *all* other wavelengths. An analogous equation was first obtained by Lardner[7] using a multiple-scales procedure leading to a more complicated representation of the 'reduced kernel' K/J.

In evaluating the integral in (1.24), it is appropriate to write $C(k-\ell,X) = C^*(\ell-k,X)$ for $\ell > k$ and to use the identity

$$K\left(\frac{\ell}{k}\right) = -\frac{2k}{\ell}K^*\left(\frac{k}{\ell}\right) \qquad \text{for } \ell > k > 0 . \qquad (1.26)$$

Then the evolution equation becomes

$$JC_{,X}(k,X) = i\int_0^k K(k^{-1}\ell)(k-\ell)\ell C(k-\ell,X)C(\ell,X)d\ell + 2i\int_k^\infty K^*(\ell^{-1}k)k(\ell-k)C^*(\ell-k,X)C(\ell,X)d\ell , \qquad (1.27)$$

in which the first integral is the contribution from pairs of waves for which the *sum* of the wavenumbers equals k, whilst the second involves pairs for which the *difference* equals k. In both cases, the contribution from wavenumber $\ell$ to the evolution of $C(k,X)$ depends on the magnitude of the kernel K at appropriate values of its argument.

## 1.3 Periodic waveforms

Predictions from evolution equations of the type (1.24) [or (1.27)] have been investigated mainly for periodic waves. Waveforms of fundamental wavelength $2\pi$ are described by replacing (1.17) by

$$\mathbf{U}(\theta, X_2, X) = \sum_{n=-\infty}^{\infty} C_n(X)\mathbf{A}(nX_2)e^{in\theta} .$$

When $C_n(X)$ is written in terms of real coefficients $D_n(X)$, $E_n(X)$ $(n > 0)$ as

$$C_n(X) = \frac{1}{2}[D_n(X) + iE_n(X)] , \qquad \text{for } n > 0 ,$$

with $C_n(X) = C^*_{-n}(X)$, the surface elevation profile is

$$U_2(\theta, 0, X) = \sum_{n=1}^{\infty}\{D_n(X)\cos(n\theta) - E_n(X)\sin(n\theta)\} . \qquad (1.28)$$

The evolution equation (1.27) is replaced by the coupled amplitude equations

$$J\frac{dC_n}{dX} = i\sum_{m=1}^{n-1} K\left(\frac{m}{n}\right)(n-m)mC_{n-m}C_m + 2i\sum_{m=n+1}^{\infty} K^*\left(\frac{n}{m}\right)n(m-n)C^*_{m-n}C_m .$$

Clearly, as for (1.27), it is sufficient to determine the kernel K for values of its argument lying in $0 < \ell/k < 1$. In fact, the identity

$$K\left(\frac{\ell-k}{k}\right) = K\left(\frac{\ell}{k}\right)$$

shows that the graph of K is symmetric about $k^{-1}\ell = \frac{1}{2}$, so that the behaviour of $K(k^{-1}\ell)$ in $0 < k^{-1}\ell \le \frac{1}{2}$ determines K everywhere. A further simplification arises if the elastic material has reflectional symmetry in planes $X_1$ = constant. This makes all roots $s^{(p)}$ purely imaginary and makes the kernel K *real*. The coupled amplitude equations then clearly possess solutions in which each Fourier coefficient $C_n(X)$ is purely imaginary, so that the surface profile (1.28) evolves as an anti-symmetric profile described by a Fourier sine series. Its coefficients $E_n = 2\mathit{Im}C_n$ then satisfy

$$2\frac{dE_n}{dX} = -\sum_{m=1}^{n-1} \Gamma_{mn}(n-m)mE_mE_{n-m} + \sum_{m=1}^{\infty} \Delta_{mn}m(n+m)E_mE_{n+m}\,, \tag{1.29}$$

where

$$\Gamma_{mn} = J^{-1}K\left(\frac{m}{n}\right),$$

$$\Delta_{mn} = -J^{-1}K[(n+m)n^{-1}] = 2n(n+m)^{-1}J^{-1}K[n(n+m)^{-1}] = 2n(n+m)^{-1}\Gamma_{n,n+m}\,.$$

For the general isotropic material, $\Gamma_{mn}$ and $\Delta_{mn}$ are given by explicit algebraic functions of m/n, similar to the formulae given in Ref. 1.

For small values of the propagation range X, .solutions to the system (1.29) may be expressed as power series in X. In particular, the generation of higher harmonics by an initially sinusoidal signal of unit amplitude [$E_1(0) = 1$, $E_n(0) = 0$ for $n > 1$] is described by

$$E_1 = 1 - \tfrac{1}{4}\Gamma_{12}^2X^2 + \dots\,,$$

$$E_2 = -\tfrac{1}{2}\Gamma_{12}X + \tfrac{1}{12}\Gamma_{12}(\Gamma_{12}^2 + 4\Gamma_{13}^2)X^3 + \dots\,,$$

$$E_3 = \tfrac{1}{2}\Gamma_{12}\Gamma_{13}X^2 + \dots\,,$$

$$E_4 = -\tfrac{1}{6}\Gamma_{12}(\Gamma_{12}^2 + 3\Gamma_{13}\Gamma_{14})X^3 + \dots\,.$$

This shows how the higher harmonics are successively introduced. Computations have been carried out by a number of authors and a typical example is shown in Fig. 2. The full and dashed lines correspond to elastic coefficients taken from two different sets of experimental data. Although the characteristic length scales for the range X differ substantially, the linear growth of $E_2$ and the quadratic growth of $E_3$ is evident in both cases.

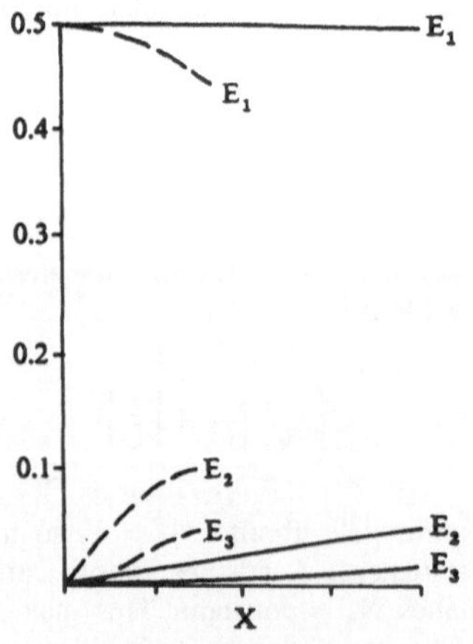


Fig. 2 Growth and decay of the first three harmonics on Y-cut $LiNbO_3$ (after Harvey, Craine and Syngellakis[8]).

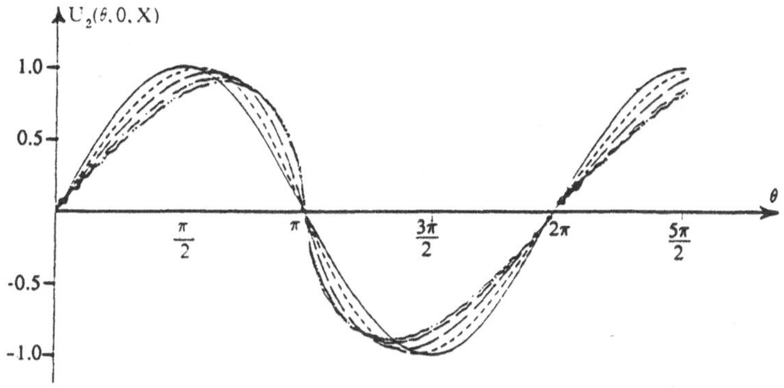


Fig. 3 Profiles of surface elevation $U_2(\theta,0,X)$ for an initially sinusoidal wave $U_2(\theta,0,0) = \sin\theta$ at successive ranges X = 0 (solid line), X = 1 (short dash line), X = 2 (long dash line), X = 3 (double dot dash line), X = 3.5 (triple dot dash line). Calculations for 80 sine coefficients $E_n(X)$ from Eq. (1.29) truncated at N = 80.

At longer ranges, the power series become invalid. Computations of truncated sets of equations analogous to (1.29) have been computed by a number of authors. Figure 3 (based on Ref. 1) concerns a special class of isotropic materials for which the $\Gamma_{mn}$ involve only two material constants, the Lamé constants. This shows that the waveform gradually distorts until at a certain range X, the waveform rapidly becomes disordered. At this stage, the truncation of the system (1.29) to N terms (here, N = 80) is clearly inappropriate. However, for a number of other sets of coupled amplitude equations (for anisotropic and piezoelectric materials) a similar behaviour has been observed. The 'break-up' range is usually well-defined and not too sensitive to the truncation number N. It is usually clearly evident from the behaviour of the profile of the vertical speed $-cU_{2,\theta}$. It is inversely proportional to the amplitude and directly proportional to the wavelength of the initial sinusoidal signal. In practice, additional physical effects important on the length scale comparable with the wavelength of the higher harmonics would normally be important once the signal starts to break up.

## 2. NONDISTORTING WAVETRAINS AND THEIR STABILITY

### 2.1 Waves of permanent form

In many branches of physics and mechanics the only types of nonlinear wave which can be analysed exactly are either 'simple waves' or 'waves of permanent form'. In simple waves the relevant vector fields take only a one-parameter family of values and are constant on each member of a set of propagating planes. Clearly this is much too restrictive to describe surface acoustic waves, in which the displacement field in linear theory must be a superposition of at least two partial waves. However, the displacement field (1.11) does represent a 'wave of permanent form' for *each* choice of C(k), or equivalently for each choice of $u_0(\theta)$. Thus it is relevant to enquire whether nonlinear theories of surface acoustic waves can predict the existence of waveforms which have displacement fields

$$\mathbf{u} = \mathbf{u}(X_1 - \hat{c}t, X_2) \qquad (2.1)$$

travelling without distortion at some speed $\hat{c}$.

During the 1960's and 1970's, when Whitham's treatment of modulated wavetrains[9] and the theory of solitons[10] were being developed, both using special waves of permanent form as basic ingredients, it was the possible existence of nonlinear surface waves of permanent form which first aroused the present author's interest in surface acoustic waves. However, the analysis of nondistorting waves proved to be much more intricate than in most other theories since the governing equations do not reduce to a nonlinear ordinary differential equation, but remain as an elliptic eigenvalue problem in the half-space $X_2 \le 0$. Also, unlike gravity waves on water, non-distorting waves of moderate amplitude are not close to sinusoidal and so cannot be obtained by the procedures giving Stokes' waves. Nevertheless, it is not now necessary to follow the original development,[11,12] since the structure of the nondistorting waves can be quickly deduced from the general analysis of Section 1.3.

In (1.17), any choice of C(k, X) of the form

$$C(k,X) = \tilde{C}(k)e^{-ik\bar{c}X} \tag{2.2}$$

gives a leading order surface elevation profile

$$U_2(\theta,0,X) = \int_{-\infty}^{\infty} \tilde{C}(k)e^{ik(\theta-\bar{c}X)}\,dk \tag{2.3}$$

which is a function of $\theta - \bar{c}X = (1 - \epsilon\bar{c})X_1 - ct$ only. Consequently this choice represents a disturbance travelling without distortion at speed $c(1 - \epsilon\bar{c})^{-1} \simeq c(1 + \epsilon\bar{c})$.

Nonlinear waves having displacements of the form (2.1) must correspond to C(k, X) of the form (2.2), with $\hat{c} = c(1 - \epsilon\bar{c})^{-1}$. Such a choice of C is compatible with (1.24) and merely reduces the evolution equation to the integral equation

$$\bar{c}Jk\tilde{C}(k) + \int_0^{\infty} K\left(\frac{\ell}{k}\right)(k-\ell)\ell\tilde{C}(k-\ell)\tilde{C}(\ell)d\ell = 0 \qquad (k > 0)\ .$$

With use of the subsidiary condition $\tilde{C}(-k) = \tilde{C}^*(k)$, this may be written equivalently as

$$\bar{c}Jk\tilde{C}(k) + \int_0^{k} K\left(\frac{\ell}{k}\right)(k-\ell)\ell\tilde{C}(k-\ell)\tilde{C}(\ell)d\ell + 2\int_k^{\infty} K^*\left(\frac{k}{\ell}\right)k(\ell-k)\tilde{C}^*(\ell-k)\tilde{C}(\ell)d\ell = 0\ . \tag{2.4}$$

Dimensional arguments show that if $\tilde{C}(k)$ is a solution of (2.4), then for all choices of real constants a and b the function $ab\tilde{C}(bk)$ is also a solution, provided only that $\bar{c}$ is replaced by $ab^{-1}\bar{c}$. Thus, once one solution to (2.4) and its related displacement field

$$U = \int_{-\infty}^{\infty} \tilde{C}(k)A(kX_2)e^{ik(\theta-\bar{c}X)}\,dk$$

is found, a related two-parameter family of waves with amplitude scaled by a factor a and all length scales multiplied by the factor b is also known. The corresponding perturbation of wavespeed is directly proportional to amplitude and is inversely proportional to the characteristic wavelength. [Of course, solutions $\tilde{C}(k)$ may also be multiplied by $e^{ik\psi}$, corresponding to a shift of the waveform along the $X_1$-axis].

Currently no method is known for solving (2.4) analytically. Computations of the kernel $K(k^{-1}\ell)$ in a number of elastic (and piezoelectric) cases show that K is typically a smooth (symmetric) function over the range $0 < k^{-1}\ell < 1$, (see Fig. 4). Nevertheless limited attempts to compute solutions numerically have met with convergence difficulties - perhaps because of the freedom represented by the rescaling parameters a and b. For periodic waves of fundamental wavelength $2\pi/\bar{k}$, waveforms have been computed by writing

$$U(\theta,X_2,X) = \sum_{n=-\infty}^{\infty} \tilde{C}_n e^{-in\bar{k}\bar{c}X} A\left(n\bar{k}X_2\right)e^{in\bar{k}\theta}$$

so that the coupled amplitude equations show that the quantities

$$P_n \equiv \bar{c}^{-1}\bar{k}n\tilde{C}_n$$

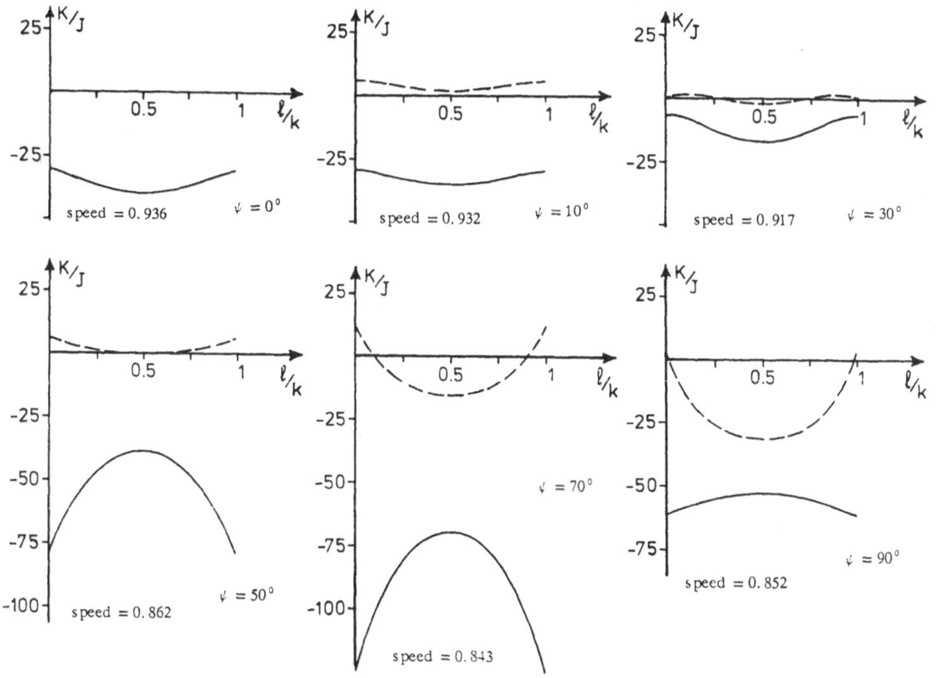


Fig. 4 The reduced kernel $J^{-1}K(k^{-1}\ell)$ and dimensionless wavespeed $c(\rho_0/c_{66})^{1/2}$ for surface acoustic waves propagating at various angles $\psi$ to the 'X' direction on 'Y-cut' $LiNbO_3$ having a traction-free earthed boundary. Real part (solid line), imaginary part (dashed line). Results for $\ell > k > 0$ are obtained using Eq. (1.26). (Figure taken from Ref. 13).

must be solutions to the infinite set of quadratic algebraic equations

$$JP_n + \sum_{m=1}^{n-1} K\left(\frac{m}{n}\right)P_{n-m}P_m + 2\sum_{m=n+1}^{\infty} \frac{n}{m}K^*\left(\frac{n}{m}\right)P^*_{m-n}P_m = 0 . \tag{2.5}$$

The solution strategy is to satisfy the first N equations ($n = 1,2,...,N$) when the summations are truncated by setting $P_n = 0$ for $n > N$. For $N = 3$, the truncated system may be solved analytically, since it may be written as

$$P_1 + \Gamma^*_{12}P^*_1P_2 + \frac{2}{3}\Gamma^*_{13}P^*_2P_3 = 0 ,$$

$$P_2 + \Gamma_{12}P_1^2 + \frac{4}{3}\Gamma^*_{13}P^*_1P_3 = 0 ,$$

$$P_3 + 2\Gamma_{13}P_1P_2 = 0 ,$$

again using the notation $J^{-1}K(m/n) = \Gamma_{mn}$, even when K is complex. This reduces to the real quadratic equation

$$\left(\Gamma_{12}\Gamma^*_{12} + \frac{16}{3}\Gamma_{13}\Gamma^*_{13}\right)\left\{\frac{4}{3}\Gamma_{13}\Gamma^*_{13}\left(P_1P^*_1\right)^2 - P_1P^*_1\right\} + 1 = 0 \tag{2.6}$$

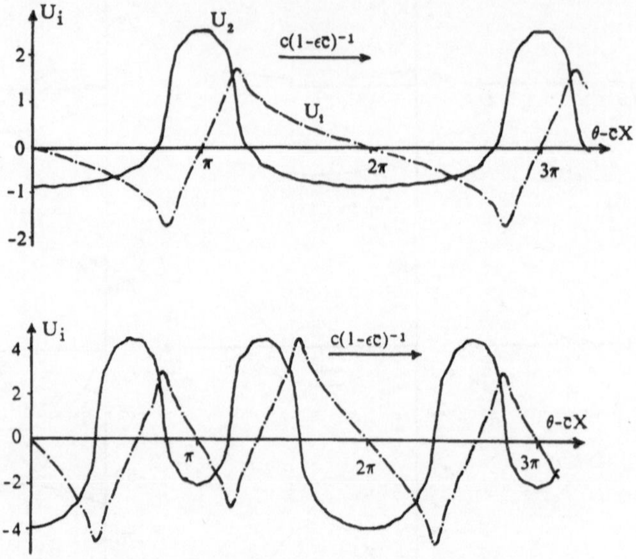


Fig. 5 Computed nondistorting waveforms based on 43 and 48 Fourier coefficients respectively (after Ref. 12).

for $P_1P_1^*$. Without loss of generality, $P_1$ may be chosen to be the positive square root of either of the solutions $|P_1|^2$, then giving

$$P_2 = -\frac{3\Gamma_{12}P_1^2}{3 + 8\Gamma_{13}\Gamma_{13}^*P_1^2}, \qquad P_3 = -2\Gamma_{13}P_1P_2 .$$

Each of these two sets is then used, in turn, as an initial guess for iterative solution of a larger truncated set obtained from Eqs. (2.5). Experience, using a modified Newton-Raphson scheme, has shown that convergence to a solution of the enlarged set of equations is usually obtained when N is increased in steps of 5 with $P_{N+1},\ldots,P_{N+5}$ initially set equal to zero. For kernels $K(m/n)$, both real and complex, the waveforms

$$U_2(\theta,0,X) = \frac{\bar{c}}{\bar{k}}\sum_{n=1}^{N}\frac{1}{n}\left\{P_n e^{in\bar{k}(\theta-\bar{c}X)} + P_n^* e^{-in\bar{k}(\theta-\bar{c}X)}\right\} \tag{2.7}$$

typically have either one or two sharp humps per period. Typical examples are shown in Fig. 5. The small ripples on the waveform are due to the truncation of the Fourier series in (2.7).

## 2.2 Modulations to periodic wavetrains

Each set of coefficients $\{P_n\}$ obtained by the methods of the previous section yields the Fourier coefficients of an approximation to the $O(\epsilon)$ displacements

$$U(\theta, X_2, X) = \hat{U}(\Theta,\xi) = \hat{U}(\Theta+2\pi,\xi), \qquad \xi < 0$$

$$\Theta \equiv \bar{k}(1 - \epsilon\bar{c})X_1 - \bar{k}ct, \qquad \xi = \bar{k}X_2$$

in a periodic wavetrain of 'frequency' $\bar{k}c$ and wavespeed

$$(1 - \epsilon\bar{c})^{-1}c = -\frac{\Theta_{,t}}{\Theta_{,X_1}} .$$

Since wavetrains having differing amplitude and wavelength have displacement fields obtained by simple scaling laws, it is possible to apply Whitham's theory of modulated wavetrains[9] to analyse the manner in which gradual variations in amplitude, wavelength and frequency propagate. The theory is restricted to the situation in which the waveform in any locality is described approximately by a scaling of the profile (2.7) corresponding to a particular set of solutions $\{P_n\}$ to (2.5).

In such a periodic wavetrain having amplitude a, frequency $\omega$ and wavenumber k, the $O(\epsilon)$ displacements have the form

$$U = U(\Theta, \xi; \omega, k, a) \tag{2.8}$$

where

$$\frac{\partial\Theta}{\partial X_1} = k , \qquad -\frac{\partial\Theta}{\partial t} = \omega , \qquad \xi = kX_2 \leq 0 .$$

In modulated wavetrains it is assumed that displacements everywhere are given approximately by (2.8), but that $\omega$, k and a may depend on 'slow' variables $Y \equiv \bar{\epsilon}X_1$, $T \equiv \bar{\epsilon}t$, for some small parameter $\bar{\epsilon}$. Consequently, $\omega$ and k must satisfy the compatibility condition

$$\frac{\partial k}{\partial T} + \frac{\partial\omega}{\partial Y} = 0 \tag{2.9}$$

(known as the equation of conservation of waves).

Whitham's procedure is to form the 'averaged Lagrangian' $\mathcal{L}(\omega, k, a)$ taken over one period $2\pi$ of the phase variable $\Theta$. In this case, integration with respect to depth $\xi$ is also necessary. Vanishing of the first variations of $\mathcal{L}$ with respect to both a and $\Theta$ then yields:

$\delta a$: $$\mathcal{L}_{,a} = 0 , \tag{2.10}$$

$\delta\Theta$: $$\frac{\partial\mathcal{L}_{,\omega}}{\partial T} + \frac{\partial\mathcal{L}_{,k}}{\partial Y} = 0 . \tag{2.11}$$

Now, since the $O(\epsilon)$ displacements (2.8) give distributions of deformation gradient and velocity which have the scaling factors ak and a$\omega$ respectively, the Lagrangian $\mathcal{L}$ has the property

$$\mathcal{L}(\omega, k, a) = \mathcal{L}\left(\frac{\omega}{\ell}, \frac{k}{\ell}, a\ell\right)$$

for all values of the parameter $\ell$. In view of (2.10), the resulting Euler identity $\omega\mathcal{L}_{,\omega} + k\mathcal{L}_{,k} - a\mathcal{L}_{,a} = 0$ reduces to

$$\mathcal{L}_{,k} = -c\mathcal{L}_{,\omega} \qquad c \equiv \omega/k \tag{2.12}$$

Also, setting $\ell = k$ gives

$$\mathcal{L}(\omega, k, a) = \mathcal{L}(c, 1, ak) \equiv \bar{\mathcal{L}}(c, ak) .$$

This, together with the *nonlinear dispersion relation* (2.10) implies that the wavespeed c is related only to ak, the *wave steepness* which is proportional to the ratio (peak amplitude/wavelength).

Assuming that this relationship may be expressed as

$$ak = F(c) ,$$

the values of the Lagrangian may be written as

$$\mathcal{L}(\omega, k, a) = \bar{\mathcal{L}}(c)$$

so that we have

$$\mathcal{L}_{,\omega} = k^{-1}\tilde{\mathcal{L}}_{,c} = k^{-1}\tilde{\mathcal{L}}'(c) , \qquad \mathcal{L}_{,k} = -ck^{-1}\tilde{\mathcal{L}}_{,c} = -ck^{-1}\tilde{\mathcal{L}}'(c) . \tag{2.13}$$

With the use of (2.12) and (2.13), it is found that Eqs. (2.9) and (2.11) may be put into the elegant characteristic form

$$\frac{\partial c}{\partial T} + G(c)\frac{\partial c}{\partial Y} = 0 , \qquad \frac{\partial}{\partial T}\{k^{-2}\tilde{\mathcal{L}}'(c)\} + c\frac{\partial}{\partial Y}\{k^{-2}\tilde{\mathcal{L}}'(c)\} = 0 , \tag{2.14}$$

where $G(c) \equiv c + 2\tilde{\mathcal{L}}'(c)/\tilde{\mathcal{L}}''(c)$ may be identified as the *nonlinear group velocity*.

Equations (2.14) show that each locus c = constant (i.e. ak = constant) is always a straight characteristic line $dY/dT = G(c)$. The second equation then shows that $k^{-2}\tilde{\mathcal{L}}'(c)$ is conserved along characteristics travelling at the local phase speed c. As shown in Ref. 14, implicit expressions for the general solution for $c(Y,T)$ and $k(Y,T)$ may be obtained. The speed c is determined as in a simple wave by

$$Y - G(c)T = \alpha , \qquad c = c(Y,T) = c_0(\alpha) \tag{2.15}$$

where $c_0(Y) = c(Y,0)$. When the phase curves $dY/dT = c$ are labelled by the parameter $\beta$, such that $\beta = Y$ at $T = 0$, then

$$k(Y,T) = k_0(\alpha)\left\{\frac{\tilde{\mathcal{L}}'[c_0(\beta)]}{\tilde{\mathcal{L}}'[c_0(\alpha)]}\right\}^{1/2} , \qquad a = k^{-1}F\{c_0(\alpha)\} . \tag{2.16}$$

The location of these $\beta$-characteristics is given implicitly by

$$T(\alpha,\beta) = \frac{\tilde{\mathcal{L}}''[c_0(\alpha)]}{2\left[\tilde{\mathcal{L}}'[c_0(\alpha)]\right]^{1/2}} \int_\alpha^\beta \left[\tilde{\mathcal{L}}'[c_0(\gamma)]\right]^{1/2} d\gamma .$$

Figure 6 (adapted from Ref. 14) indicates how the transition from a large amplitude to a smaller amplitude wavetrain evolves, when k is initially uniform. It shows how c makes a transition from a larger to a smaller value through a spreading simple wave (2.15). Assuming that G(c) exceeds the linearized Rayleigh speed (here denoted by $c_R$) and that $G(c) - c_R > c - c_R$ [as predicted by the quadratic approximation (1.4)] the phase curves $\beta$ = constant travel more slowly than the $\alpha$-characteristics. Consequently, the transition process splits into four regions, while (2.16) shows that k does not remain uniform. The wavelength is shortened in the final non-distorting region, IV.

### 2.3 Stability of periodic waveforms

The predictions of the modulation theory are not applicable if the 'waves of permanent form' are unstable to perturbations in the initial data. Since waveform distortion is governed by the evolution equation (1.24) or the equivalent coupled amplitude equations for periodic waveforms, the stability of the computed periodic waves is tested by computing solutions to the coupled amplitude equations with initial conditions corresponding to slight perturbations from values $\check{C}_n$ resulting from solution of a truncated set of equations (2.5).

In Ref. 1, solutions $C_n(X)$ to a set of 43 coupled amplitude equations were computed for an isotropic elastic material, for which K is real. Initial conditions $C_n(0)$, n = 1,...,18 were taken as the values $\check{C}_n$ derived from a truncation of the set (2.5) to 43 equations with $\bar{k}\bar{c}$ chosen so that $C_1(0) = \frac{1}{2}$ as in a unit amplitude cosine wave. The remaining initial conditions were taken as $C_n(0) = 0$, n = 19,...,43. Thus, initial conditions describe a waveform given by just the first 18 harmonics of the solution satisfying the set of 43 simultaneous algebraic equations. Not surprisingly, numerical integration of the 43 complex coupled amplitude equations shows that the higher coefficients are excited. However, the resulting waveforms are not significantly degraded even after they have travelled to ranges about eight times the break-up

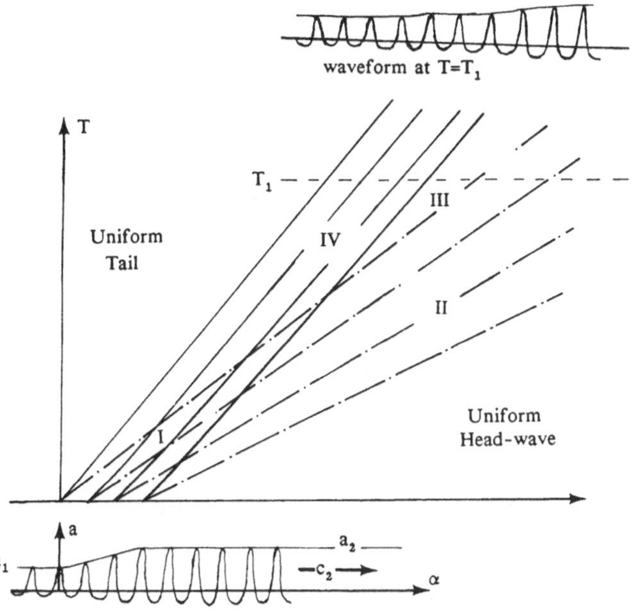


Fig. 6 Dispersion of a wavetrain initially having a uniform wavenumber $k_0$, with a wavetrain of amplitude $a_1$ preceded by a wavetrain of larger amplitude $a_2$. Phase curves $\beta$ = constant (solid line), straight rays $\alpha$ = constant (dot dash line) carrying constant c. Region III is a uniform wavetrain of amplitude $a = a_2(c_2/c_1)^{1/2}$. Region IV is a nondistorting wave.

range of the unit amplitude cosine wave. This is despite the fact that the maximum surface steepness and material strain considerably exceed those in the cosine wave. Figure 7 shows waveforms at ranges approximately 2, 4, 6 and 8 times the cosine wave break-up range. Although high frequency ripples develop fairly rapidly, the waveform retains its dominant feature which is one sharp hump per period. Calculations have been carried out for a number of different truncation numbers N and for coupled amplitude equations with kernels corresponding to different materials. Typically they show that the 'non-distorting profile' having one hump per period is relatively stable to perturbations of the initial conditions, while the profile having two humps per period [and arising from iteration starting from the second solution of the quadratic (2.6)] degrades much more rapidly. Figure 8 shows this difference clearly. It also indicates a tendency for the waveform with two humps to evolve towards one with a single hump. Reference 15, which concerns piezoelectric surface acoustic waves, includes calculations for crystal orientations in which the kernel K is complex. The resulting single-hump profile is asymmetric, yet introduces ripples at a rate comparable with those shown for the single-hump waveform in Fig. 8.

### 2.4 Stability analysis and some model equations

Recently, Hunter[16] has studied the qualitative behaviour of solutions to evolution equations of the type (1.24) from another standpoint. He considers general hyperbolic systems in two space coordinates and time, with boundary conditions which, like (1.2), allow scale invariant waves. In particular, he shows that if **u** satisfies the hyperbolic conservation law and boundary conditions

$$\mathbf{u}_t + \mathbf{f}(\mathbf{u})_x + \mathbf{g}(\mathbf{u})_y = \mathbf{0} \qquad \text{in } y > 0\,,$$

$$C\mathbf{u} = \mathbf{0} \qquad \text{on } y = 0\,,$$

$$\mathbf{u} \to \mathbf{0} \qquad \text{as } y \to +\infty\,,$$

then the canonical equation for weakly nonlinear surface waves may be written as

$$u_t + \Phi[u]_x = 0 , \tag{2.17}$$

where $\Phi$ is a quadratically nonlinear, *nonlocal* flux

$$\Phi[u](x,t) \equiv \int_{-\infty}^{\infty}\int_{-\infty}^{\infty} G(x-y, x-z)u(y,t)u(z,t)dydz . \tag{2.18}$$

(In this section, subscripts denote partial derivatives.) Here, the kernel $G(x,x')$ is a real, symmetric distribution which satisfies the scale invariance requirement

$$G(\alpha x, \alpha x') = \alpha^{-2}G(x,x') , \qquad \alpha > 0 .$$

Hunter observes that the special choice $G(x,x') = \frac{1}{2}\delta(x)\delta(x')$ makes $\Phi$ into the *local* flux $\frac{1}{2}u^2$ and so reduces (2.17) to the simple form $u_t + uu_x = 0$, which is readily solved and which, for most initial data $u(x,0)$, predicts *shock formation.* Of major interest is the extent to which the nonlocal contributions to $\Phi$ alter the shock formation tendency.

Since Fourier transformation of (2.17) and (2.18) gives

$$U_t(k,t) + ik\int_{-\infty}^{\infty} \Lambda(k-\ell,\ell)U(k-\ell,t)U(\ell,t)d\ell = 0 , \tag{2.19}$$

where

$$U(k,t) = \frac{1}{2\pi}\int_{-\infty}^{\infty} u(x,t)e^{-ikx}dx ,$$

$$\Lambda(k,\ell) = \frac{1}{(2\pi)^2}\int_{-\infty}^{\infty}\int_{-\infty}^{\infty} G(x,x')e^{-i(kx+\ell x')}dxdx' ,$$

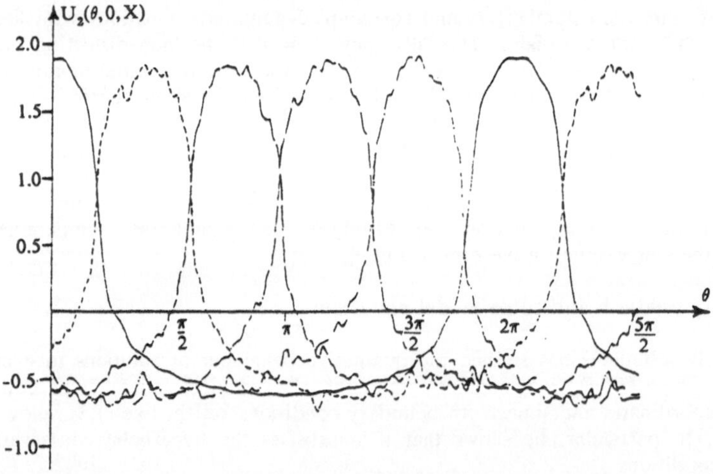


Fig. 7 Evolution of a waveform initially having 18 non-zero cosine coefficients, corresponding to a 'non-distorting' waveform of Ref. 12. Surface elevations corresponding to truncation of Eqs. (2.5) and (2.7) at N = 43 for ranges X=0 (solid line), X=5 (short dash line), X=10 (long dash line), X=15 (dash dot line), X=20 (dash double dot line). (Figure taken from Ref. 1).

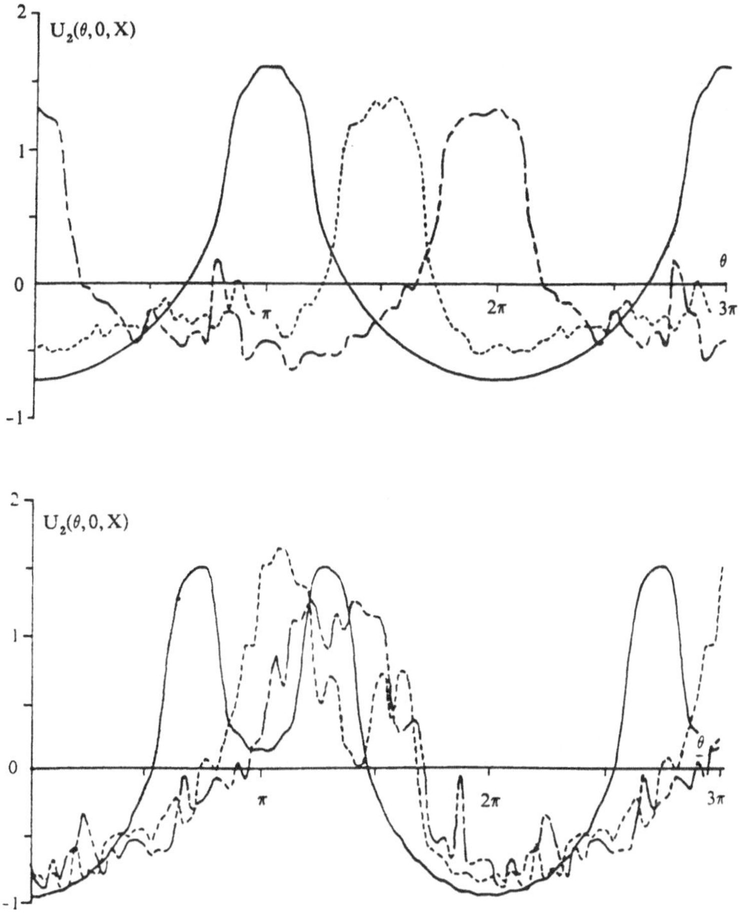


Fig. 8 Degradation of typical 'non-distorting' profiles.

the similarity with the evolution equation (1.24) is immediately evident. Here U(k, t) plays the role of -kC(k, x) and the kernel $\Lambda(k-\ell,\ell)$ replaces $K(k^{-1}\ell)$. The kernel has the properties

| | | |
|---|---|---|
| $\Lambda(k,\ell) = \Lambda(\ell,k)$ | | symmetry |
| $\Lambda(-k,-\ell) = \Lambda^*(k,\ell)$ | | reality of U |
| $\Lambda(\alpha k,\alpha\ell) = \Lambda(k,\ell)$ , | $\alpha > 0$ | scale invariance . |

Hunter's derivation of (2.18) proceeds through many of the steps used in Section 1.2. However, at various stages he interprets results involving the Fourier transform U(k, t) of the signal directly in terms of the signal u(x, t), identified as a component of **u** at the surface y = 0. This is analogous, in the treatment in Section 1.2, to eliminating C(k, X) in favour of the elevation signal $U_2(\theta,0,X)$. For example, the leading order solution (1.17) may be interpreted in the form

$$\mathbf{U}(\theta,X_2,X) = \int_{-\infty}^{\infty} U_2(\varsigma,0,X)\mathbf{r}(\theta-\varsigma,X)d\varsigma \ ,$$

where **r** is a Green's function for the half-space and is readily defined in terms of the quantities involved in $\mathbf{A}(kX_2)$.

Although the kernels $G(x,x')$ and $\Lambda(k,\ell)$ arising from elasticity are too complicated to allow much analytical progress, Hunter has made some useful observations. He tests the well-posedness of Eq. (2.19) by considering small, rapidly oscillating perturbations $v(x,t)$ to a solution $u_0(x,t)$ of (2.17). Expressing the Fourier transforms as

$$U(k,t) = U_0(k,t) + V(k,t)$$

and linearizing (2.19) then gives

$$V_t + 2ik\int_{-\infty}^{\infty} \Lambda(k-\ell,\ell)U_0(\ell,t)V(k-\ell,t)d\ell = 0 .$$

Assuming that $V(k,t)$ is significant *only* at large wavenumbers [greatly exceeding those at which $U_0(k,t)$ is significant] it is justifiable to approximate this equation as

$$V_t + 2ik\left[\int_{-\infty}^{\infty} \Lambda(k,\ell)U_0(\ell,t)d\ell\right] V(k,t) = 0 . \tag{2.20}$$

Further analysis, exploiting the scale invariance of $\Lambda(k,\ell)$, shows that the term in brackets is given asymptotically by

$$\int_{-\infty}^{\infty} \Lambda(k,\ell)U_0(\ell,t)d\ell \sim \Lambda_0(t) \equiv \Lambda(1,0+)\int_0^{\infty} U_0(\ell,t)d\ell + \Lambda(1,0-)\int_0^{\infty} U_0^*(\ell,t)d\ell .$$

Substituting $\Lambda_0(t)$ into (2.20) and (for large k) treating $\Lambda_0$ as constant shows that

$$V(k,t) \sim V_0(k)e^{-ik\Lambda_0 t} .$$

This indicates that (2.17) is ill-posed unless $Im\,\Lambda_0 \le 0$, since otherwise $u_0(x,t)$ is unstable to ultrashort perturbations, corresponding to the rapid, unbounded growth of $V(k,t)$ as $k \to +\infty$. From the above definition of $\Lambda_0(t)$, it can be seen that the only way to ensure that $Im\,\Lambda_0 \le 0$ for arbitrary $U_0$ is to insist that

$$\Lambda(1,0+) = \Lambda^*(1,0-) . \tag{2.21}$$

This requirement makes $\Lambda_0(t)$ real, indicating neutral stability for $u_0(x,t)$.

Hunter applies the test (2.21) to a number of *model equations* as well as to the evolution equation for the isotropic material of Ref. 1. He concludes that the initial value problem for the evolution equation (1.24) is *linearly well-posed* in this elastic case. This corroborates the predictions in Fig. 7.

By considering piecewise constant kernels $\Lambda(k,\ell)$ some families of simple kernels $G(x,x')$ leading to special nonlocal fluxes (2.17) are identified. These include both well-posed and ill-posed examples. One well-posed choice is

$$u_t + \left\{\frac{1}{2}\alpha u^2 + \frac{1}{2}\beta H\left[H[u^2]\right]\right\}_x = 0 \tag{2.22}$$

where H denotes the Hilbert transform. It can be shown that if u is $2\pi$-periodic in x with zero mean, then

$$\int_0^{2\pi} u^2(x,t)dx = \text{constant} .$$

Also, using h(x,t) to denote the Hilbert transform H[u](x,t), it can be shown that (2.22) leads to

$$(\alpha u - \beta h)_t + \{\tfrac{1}{2}(\alpha u - \beta h)^2\}_x = 0 \ .$$

Since this equation is readily solved, we choose to denote a solution by

$$\alpha u - \beta h \equiv f \ .$$

Taking Hilbert transforms of this equation gives

$$\alpha h + \beta u = H[f] \ .$$

Eliminating h from these two equations then allows us to write

$$u(x,t) = \frac{\alpha f + \beta H[f]}{\alpha^2 + \beta^2} \ , \tag{2.23}$$

where f(x,t) is any solution of

$$f_t + (\tfrac{1}{2}f^2)_x = f_t + f f_x = 0 \ . \tag{2.24}$$

The solution of (2.23) and (2.24) with sinusoidal initial data $u(x,0) = \sin x$ gives an interesting comparison with Fig. 3. At $t = 0$, this requires

$$f = \alpha \sin x - \beta \cos x = \rho \sin(x - \delta) \qquad \alpha \equiv \rho\cos\delta,\ \beta \equiv \rho\sin\delta \ .$$

The corresponding solution f(x,t) to (2.24) may be expressed in Fubini's form[17] as

$$f(x,t) = \sum_{n=1}^{\infty} \frac{2(-1)^{n+1}}{nt} J_n(\rho n t)\sin\{n(x-\delta)\} \qquad \text{for } \rho t < 1 \ . \tag{2.25}$$

Finding H[f], inserting it into (2.23) and then rearranging gives

$$u(x,t) = \sum_{n=1}^{\infty} \frac{2(-1)^{n+1}}{\rho n t} J_n(\rho n t)\sin\{n(x-\delta) + \delta\} \qquad \text{for } \rho t < 1 \ . \tag{2.26}$$

In solution (2.25), the value $t = \rho^{-1}$ corresponds to the *shock-formation* time familiar in nonlinear acoustics. Since (2.26) differs from (2.25) only by a multiplicative factor $\rho^{-1}$ and by a phase shift in each Fourier component, it seems plausible that the value $t = \rho^{-1}$ is the greatest at which the representation (2.26) converges. This would then correspond to the break-up range revealed by numerical computations. Also, the phase shifts have significance. As solutions to (2.25) form shocks, failure of convergence usually is revealed as a Gibbs' phenomenon of oscillatory overshoot near the abrupt transition. The computations in Fig. 3 (and many others) reveal that *break-up* occurs over the whole period of the waveform. Indeed it is often most marked midway between the abrupt jumps. This feature shows how the *nonlocal* behaviour in the evolution equation has a dominant influence.

## 3. ELECTROMECHANICAL INTERACTIONS AND THEIR APPLICATIONS

### 3.1 Nonlinear electro-elastic surface waves

A number of electro-mechanical interaction effects (e.g. piezoelectricity, electrostriction, electroelasticity) have been recognized by physicists since the nineteenth century. During the 1960's, extensive mathematical treatments of these couplings were performed. Perhaps the most striking prediction was the discovery, independently by Bleustein and Gulyaev,[18] that in some crystalline materials cut at special orientations the *linear* phenomenon of piezoelectricity allows the existence of surface acoustic waves, in which the only displacements are parallel to the surface but transverse to the propagation direction. These (like Love waves in layered isotropic elastic media) are examples of shear horizontal surface acoustic waves.

More generally, when surface acoustic waves propagate in materials exhibiting piezoelectricity, the linear theory predicts that the depth-dependence of the displacements and the electric potential are coupled. However, both the linear theory and the quadratically nonlinear theory of surface acoustic waves may be developed by suitably adapting the methods used in Sections 1.1 and 1.2. Moreover, a number of different boundary conditions may be imposed on the electric field at the material surface, causing only minor amendments to the form of the equations.

Since the speed of light vastly exceeds acoustic speeds, electric effects are usually treated by using a quasi-static approximation, with electric potential $\epsilon\phi(X_1, X_2, t)$. Then, to quadratic approximation, equation (1.4) is replaced by expressions

$$\tau_{Lj} = \epsilon\{c_{jLmM}u_{m,M} + e_{M.jL}\phi_{,M}\} + \epsilon^2\{d_{jLmMnN}u_{m,M}u_{n,N} + 2d_{jLmMN}u_{m,M}\phi_{,N} + d_{jLMN}\phi_{,M}\phi_{,N}\} \tag{3.1}$$

$$D_L = \epsilon\{-\epsilon_{LM}\phi_{,M} + e_{L.mM}u_{m,M}\} + \epsilon^2\{d_{mMnNL}u_{m,M}u_{n,N} + 2d_{mMLN}u_{m,M}\phi_{,N} + \tfrac{1}{2}\chi_{LMN}\phi_{,M}\phi_{,N}\} \tag{3.2}$$

for the Piola-Kirchhoff stress $\tau$ and the material electric field $\mathscr{D}$ (see Maugin[19]). Here $e_{M.jL}$ are the *piezoelectric* coefficients, $\epsilon_{LM}$ and $\chi_{LMN}$ are the *linear* and *second-order permittivity* coefficients, while $d_{jLmMnN}$, $d_{jLmMN}$ and $d_{jLMN}$ are respectively the *second-order elastic*, *electroelastic* and *electrostriction* coefficients, adjusted by the inclusion of certain linear combinations of the first-order coefficients. Within the material, the governing equations are the momentum and Gauss equations

$$\tau_{Lj,L} = \epsilon\rho\ddot{u}_j , \qquad D_{L,L} = 0 . \tag{3.3}$$

By writing $\phi \equiv u_4$, $D_L \equiv \tau_{L4}$ and allowing all lower case subscripts to range over the values 1, 2, 3 and 4, it is possible to recast (3.1) - (3.3) in a form closely analogous to (1.4) and (1.5), namely

$$\tau_{Lj} = \epsilon c_{jLmM}u_{m,M} + \epsilon^2 d_{jLmMnN}u_{m,M}u_{n,N} , \tag{3.4}$$

$$\tau_{Lj,L} = \epsilon\rho\eta_{jm}\ddot{u}_m , \tag{3.5}$$

where $\eta_{jm} = \delta_{jm} - \delta_{j4}\delta_{m4}$. By comparison with (3.1) and (3.2) it is seen that the first-order coefficients for which j = 4 or m = 4 are given by $c_{jL4M} = e_{M.jL} = c_{4MjL}$, $c_{4L4M} = -\epsilon_{LM}$, while the second-order coefficients are defined analogously. The relevant 'partial waves' satisfying (3.4) and (3.5) are

$$u_j = a_j^{(p)}e^{ik(\theta+s^{(p)}X_2)} \qquad p = 1,2,3,4 \tag{3.6}$$

where the vectors $\mathbf{a}^{(p)}$ and depth factors $s = s^{(p)} = \alpha^{(p)} - i\beta^{(p)}$ ($\beta^{(p)} > 0$) satisfy Eq. (1.8) with $\delta_{jm}$ replaced by $\eta_{jm}$ (since the Gauss equation contains no inertia term). As in Section 1.1, the speed c is determined by applying the boundary conditions at $X_2 = 0$. In Ref. 20, three different electric boundary conditions are considered:

| | |
|---|---|
| Case E (earthed, conducting boundary) | $u_4 \equiv \phi = 0$ , |
| Case F (material adjoining freespace) | $[\![\phi]\!] = 0$, $[\![D_2]\!] = 0$ , |
| Case O (open circuit) | $D_2 = 0$ . |

Cases E and O require only simple modifications to Sections 1.1 and 1.2, while Case F is a little more complicated since it involves matching across $X_2 = 0$ to a potential $\hat{\phi}$ and a derived electric displacement $\mathscr{D}$ in the free-space region $X_2 > 0$. In the linearized treatment these conditions lead to the boundary matrix equations (j=1,2,3,4)

Case E: $$M_{jp}B^{(p)} \equiv \sum_{p=1}^{4} \{\eta_{jm}(c_{m2n1} + s^{(p)}c_{m2n2}) + \delta_{j4}\delta_{n4}\}a_n^{(p)}B^{(p)} = 0 , \tag{3.7}$$

Case F: $$\hat{M}_{jp}B^{(p)} \equiv \sum_{p=1}^{4} \{c_{j2m1} + s^{(p)}c_{j2m2} + i\epsilon_0\delta_{j4}\delta_{m4}\}a_m^{(p)}B^{(p)} = 0 , \tag{3.8}$$

Case O: $$\mathring{M}_{jp}B^{(p)} \equiv \sum_{p=1}^{4} \{c_{j2m1} + s^{(p)}c_{j2m2}\}a_m^{(p)}B^{(p)} = 0 . \tag{3.9}$$

(In Case F, the partial wave for the free-space potential is $\hat{\phi} = a_5 e^{ik\theta-|k|X_2}$ but the matching conditions $\phi = \hat{\phi}$, $D_2 = -\epsilon_0\hat{\phi}_{,2}$, readily allow $a_5$ to be eliminated.)

Bleustein-Gulyaev waves arise in the special case when the first-order coefficients $c_{j\alpha m\beta}$ vanish in all the situations:

$$m = 1 \text{ or } 2 \quad \text{with} \quad j = 3 \text{ or } 4 ,$$

or

$$m = 3 \text{ or } 4 \quad \text{with} \quad j = 1 \text{ or } 2 .$$

Two of the partial waves (3.6) then involve disturbances only in $u_1$ and $u_2$, while the remaining two involve disturbances only in the transverse displacement $u_3$ and the potential $\phi$ (and also $\hat{\phi}$ in Case F). Without loss of generality, we may choose $\mathbf{a}^{(1)}$ and $\mathbf{a}^{(2)}$ of the form $(*,*,0,0)^T$, so that $\mathbf{a}^{(3)}$ and $\mathbf{a}^{(4)}$ have the form $(0,0,*,*)^T$.

One solution to any of (3.7)-(3.9) then has $B^{(3)} = 0 = B^{(4)}$. The corresponding linearized surface acoustic wave then involves displacements only in the $X_1$, $X_2$ plane (the saggittal plane), just as in isotropic elasticity (see Section 1.1). This wave excites no electric fields. An additional solution has $B^{(1)} = 0 = B^{(2)}$ and so describes a coupling between shear horizontal displacements $u_3$ and the field components $D_1$, $D_2$ arising from $\phi(X_1,X_2,t)$ and $\hat{\phi}(X_1,X_2,t)$. This second possibility is the *Bleustein-Gulyaev* wave and occurs in a piezoelectric crystal which possesses no centre of symmetry, yet for which the $X_3$-axis may be chosen as a six-fold symmetry axis. In this crystal orientation, it is clear that two distinct surface acoustic waves can propagate. One is essentially similar to the Rayleigh wave of isotropic elasticity, having displacement amplitudes which decay on a scale comparable with a wavelength. The second wave, the Bleustein-Gulyaev wave, involves a *weak* coupling between transverse (shear horizontal) displacements and electric fields. Consequently the monotonic decay of $A_3(\xi)$ is gradual and the depth of penetration is many wavelengths. The disturbance is only a weak perturbation of plane elastic shear horizontal waves, which satisfy the boundary condition $\tau_{2j} = 0$. The propagation speed c is close to the corresponding speed $c_T$ of transverse shear waves.

Since Bleustein-Gulyaev waves are special examples of piezoelectric surface acoustic waves they will not be treated separately in the nonlinear theory. They do, of course, lead to some manipulative simplifications.

In Cases E and O, the derivation of the evolution equation follows closely the procedure outlined in Section 1.2. The only differences are that lower case indices take the values 1,2,3 *and* 4, while the depth dependences $A_j(\xi)$ must be determined by using (3.7), or (3.9), respectively. With these adjustments, it is found that the transform C(k,X) of the surface elevation profile satisfies the evolution equation (1.24) with the constant J and kernel $K(k^{-1}\ell)$ defined by (1.25), where the identity (1.26) still holds.

In Case F, allowance must be made for the deformation of the boundary $X_2 = 0$ of the free-space region. One procedure (following Tiersten and Baumhauer[21]) is to write $\hat{\phi}$ in terms of Eulerian coordinates and to deduce values on $X_2 = 0$ by Taylor expansion about $x_2 = 0$. In Ref. 20 an alternative procedure was adopted, by introducing *pseudo-displacements* $\hat{v}_j$ (j=1,2,3) in $X_2 \geq 0$ so that $x_j = \delta_{jJ}X_J + \epsilon\hat{v}_j(X_1,X_2,t)$ for $X_2 \geq 0$ with

$$\hat{v}_j(X_1,0,t) = u_j(X_1,0,t) , \qquad \hat{v}_j \to 0 \text{ as } X_2 \to +\infty , \qquad j=1,2,3 . \tag{3.10}$$

To second order in $\epsilon$, this converts Laplace's equation governing $\epsilon\hat{\phi}$ into

$$\epsilon\hat{\phi}_{,JJ} - \epsilon^2\delta_{jL}\hat{\phi}_{,L}\hat{v}_{j,JJ} - 2\epsilon^2\delta_{jM}\hat{\phi}_{,JM}\hat{v}_{j,J} = 0 \qquad \text{in } X_2 > 0 .$$

The compatibility condition (1.23) is modified by the inclusion of terms involving fields in $0 \le X_2 < \infty$. However, these terms may be rearranged to involve $\hat{v}_j$ only over the boundary $X_2 = 0$, where it is given by (3.10). Consequently, the resulting evolution equation

$$\tilde{J}C_{,X}(k,X) = i\int_0^\infty \tilde{K}(k^{-1}\ell)(k-\ell)\ell C(k-\ell,X)C(\ell,X)d\ell \tag{3.11}$$

does not involve the fictitious displacements $\epsilon\hat{v}_j$.

The constant $\tilde{J}$ in (3.11) is given by

$$\tilde{J} = J - 2\epsilon_0 A_{04}A_{04}^* ,$$

where J is defined through (1.25). The additional term describes the free-space Maxwell stress evaluated at $X_2 = 0$ and has been found to be numerically insignificant. Similarly, the kernel $\tilde{K}$ is simply related to K, defined as in (1.25). It is given by

$$\tilde{K}(k^{-1}\ell) = \begin{cases} K(k^{-1}\ell) + \epsilon_0 A_{04}^2(A_{02}^* - iA_{01}^*) & \text{for } 0 < \ell < k \\ K(k^{-1}\ell) & \text{for } k < \ell \end{cases}$$

and so differs from $K(k^{-1}\ell)$ by no more than a complex constant.

In Fig. 9 the reduced kernel $J^{-1}K(k^{-1}\ell)$ and the corresponding speed (non-dimensionalized with respect to density and a representative elastic modulus) are shown for all three boundary conditions, for waves travelling along either the 'X'-axis or the 'Z'-axis of 'Y-cut' $LiNbO_3$ (c.f. Fig. 4, which uses the same material and 'cut' to illustrate the effect of anisotropy). For 'X-propagation', all three boundary conditions preserve the symmetry which makes K real. For propagation in the perpendicular direction, it is seen that the boundary condition has significant influence on both the real and imaginary parts of $J^{-1}K$. Since $J^{-1}K(k^{-1}\ell)$ is (apart from the free-space contributions in Case F) the only material-dependent term arising in the evolution equation, these differences manifest themselves as different forms of distortion of a sinusoidal waveform. In particular, the distortion of $U_2(\theta,0,X)$ in Fig. 10 is virtually symmetric, since for 'Z-propagation' in Case F the kernel has a small real part (see Fig. 9). This distortion should be compared with the asymmetric distortion in Fig. 3, which corresponds to a real kernel.

Other examples, including the evolution of a waveform of the electric potential $\phi$, are given in Ref. 20. Some periodic waves of permanent form corresponding to this material and these boundary conditions are included in Ref. 15 (see also Fig. 8).

### 3.2 Surface waves on magnetic materials

In principle, a nonlinear treatment of magneto-elastic couplings could be developed analogously to that in the previous section, since the governing equations in perfect insulators may be written as

$$\tau_{Lj,L} = \epsilon\rho\ddot{u}_j , \qquad B_{L,L} = 0 , \qquad H_L = -\Phi_{,L} . \tag{3.12}$$

Here, as in (3.3), lower case indices range only over the values 1, 2 and 3. However, when the stress components $\tau_{Lj}$ and magnetic induction components $B_L$ are expanded in powers of the gradients of the displacements $\epsilon u_j$ and of the magnetic potential $\Phi$, many of the terms equivalent to those in (3.1) and (3.2) disappear. For example, the linear phenomenon of piezomagnetism is extremely rare, since magnetization is due essentially to orbital motions within molecules and so lacks invariance under time reversal. In crystals, the combination of crystal symmetry and time-reversal symmetry requires that stress is an even function and $\mathcal{B}$ is an odd function of $\mathcal{H}$. Consequently there is no linearized coupling between stress and magnetic fields.

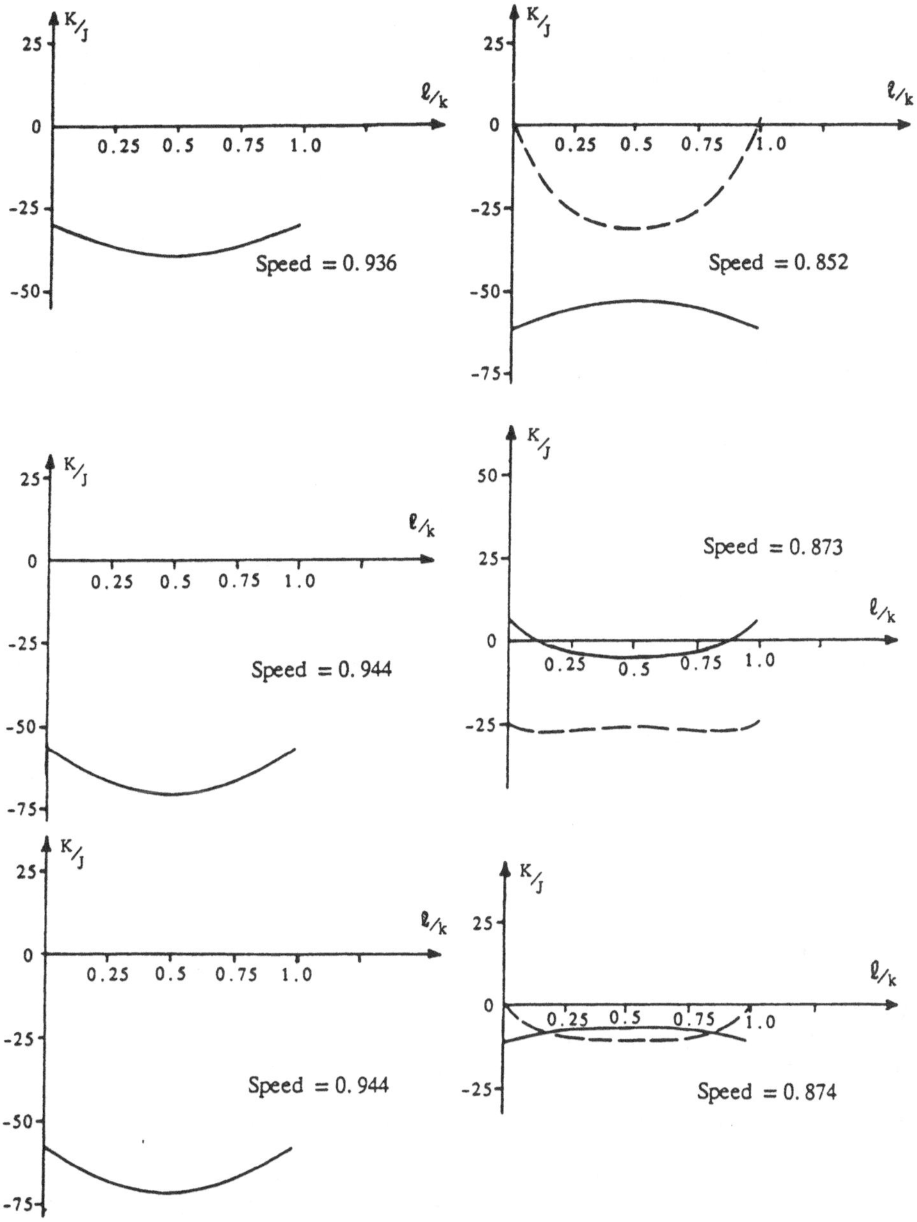


Fig. 9 The reduced kernel $J^{-1}K(k^{-1}\ell)$ and dimesionless wavespeed $c(\rho/c_{66})^{1/2}$ for surface acoustic waves on 'Y-cut' $LiNbO_3$.

A general treatment of constitutive equations for magnetoacoustics has been given by Abd-Alla and Maugin[22] and various treatments of magnetoacoustic surface acoustic waves are due to Maugin and his co-workers.[23] In particular, they analyse surface acoustic waves on a material subjected to a uniform 'bias' magnetic field, which induces a linear coupling similar to piezomagnetism.

If the (uniform) bias field $\mathcal{H}^0$ has components $H_L^0$, the components of the magnetic induction $\mathcal{H}$ are written as

$$H_L = H_L^0 - \epsilon\phi_{,L} ,$$

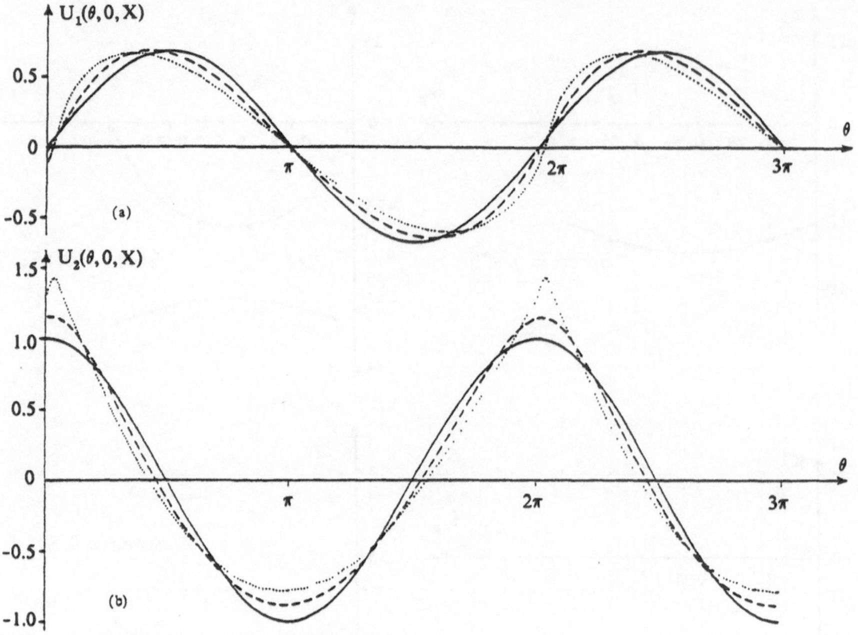


Fig. 10 Distortion of an initially sinusoidal (a) horizontal displacement $U_1(\theta,0,X)$ and (b) vertical displacement $U_2(\theta,0,X)$ for propagation in the 'Z' direction, Case F. Profiles at ranges X = 0 (solid line), X = 0.01 (dashed line), X = 0.02 (dotted line) (after Ref. 20).

where $\phi = \phi(X_1, X_2, t)$. It is then found that the linearized constitutive laws resemble the linear part of (3.1), since to $O(\epsilon)$ they become

$$\tau_{Lj} = \epsilon\{\bar{c}_{jLmM} u_{m,M} - F'_{MjL}\phi_{,M}\} ,$$
$$B_L = B_L^0 - \epsilon\{\bar{\mu}_{LM}\phi_{,M} - \hat{F}_{LjM} u_{j,M}\} . \tag{3.13}$$

Here $B_L^0$, $\bar{c}_{jLmM}$ and $\bar{\mu}_{LM}$ denote the uniform magnetic induction, effective first-order elastic coefficients and effective permeability coefficients corresponding to the uniform field $\mathcal{H}^0$. However, the terms $F'_{MjL}$ and $\hat{F}_{LjM}$ playing the role of piezomagnetic coefficients typically have $F'_{PjQ} \neq \hat{F}_{PjQ}$. Nevertheless, the corresponding linearized theory of surface acoustic waves may be constructed as in Sections 1.1 and 3.1, using the notation $u_4 = \phi$ as in Eqs. (3.4) - (3.9), for convenience.

Similarly, terms of second degree in $u_j$ and $\phi$ may be retained in the constitutive laws, so that Eqs. (3.12) and associated boundary, or matching, conditions at $X_2 = 0$ may be treated just as in preceding sections. To the author's knowledge, this programme has not been carried through, but should encounter little conceptual difficulty. The resulting evolution equation should again be of the form (1.24).

However, identification of all the coefficients required both in the linearized problem arising from (3.12) and (3.13), and in the subsequent quadratic problem, is daunting. They clearly depend on the uniform state of magnetization induced by $\mathcal{H}^0$. In Ref. 23, Abd-Alla and Maugin specify $\tau_{Lj}$ and $B_L$ up to terms of third degree jointly in components $u_{m,M}$ and $H_L$. They then investigate orders of magnitudes of the non-vanishing terms in this expansion around the state of zero strain and zero field. After certain approximations, taking account of the magnitude of the field $\mathcal{H}^0$ yielding levels of linear coupling appropriate for exploitation in devices such as electromagnetic transducers, they relate the coefficients in (3.13) and the corresponding $O(\epsilon^2)$ corrections to linear combinations of the coefficients in the expansion around the zero-state. They also analyse the generation of higher-harmonics from a sinusoidal input. For further details, the reader should consult the references cited.

### 3.3 Filters and convolvers

Among the devices utilizing piezoelectric coupling are surface wave convolvers and filters. Both rely on the fact that an electromagnetic input to a transducer creates surface waves travelling at a minute fraction of the speed of light so that a compact device (dimensions < 1 cm) can accommodate many cycles of a signal in the MHz range.

In surface acoustic wave filters two signals are launched independently in the same direction across the surface of a crystal. Nonlinearity is an undesirable feature, since the phase matching between distinct frequencies $\omega_1$ and $\omega_2$ in the two signals leads to the cumulative interactions discussed in Section 1.2. This is known as intermodulation. Although the effect is well-documented, until recently[24] theoretical analysis has omitted piezoelectric and anisotropic effects. The analysis of Harvey and Tupholme[24] uses multiple scale techniques with the usual requirement that secular terms should be excluded. However, since the analysis treats interactions only between signals at frequencies $\omega_1$, $\omega_2$, $2\omega_1$, $2\omega_2$, $\omega_1 + \omega_2$ and $\omega_1 - \omega_2$ (taking $\omega_2 < \omega_1$), it may be developed from the continuous spectrum treatment leading to (1.24) rather like the treatment of periodic waveforms.

Then, after writing $k \equiv \omega_1/c$, $r \equiv \omega_2/\omega_1$ $(0 < r < 1)$ and seeking waveforms in the form of a double summation

$$u_j = \sum_{p=0}^{\infty}\sum_{q=0}^{\infty} C_{pq}(X)A_j[(p+rq)kX_2]e^{i(p+rq)k\theta} + \sum_{p=1}^{\infty}\sum_{q=1}^{\infty} D_{pq}(X)A_j(|p-rq|kX_2)e^{i|p-rq|k\theta} + \text{c.c.} \tag{3.14}$$

with $C_{00}(X) \equiv 0$, it is found that the coefficients $C_{pq}(X)$ and $D_{pq}(X)$ satisfy a system of coupled ordinary differential equations which generalize the coupled amplitude equations of Section 1.3. The interactions which contribute to $dC_{pq}/dX$ involve the products of pairs of terms from the right-hand side of (3.14) which involve the exponential factor $e^{i(p+rq)k\theta}$. The truncated system obtained by considering only the coefficients $C_{10}$, $C_{01}$, $C_{20}$, $C_{02}$, $C_{11}$ and $D_{11}$ is

$$\begin{aligned}
iJC'_{10}(X) &= 2K(2)C^*_{10}C_{20} + 2r(1+r)K(1+r)C^*_{01}C_{11} - r(1-r)K(1-r)C_{01}D_{11}\,,\\
iJC'_{01}(X) &= 2r^2K(2)C^*_{01}C_{02} + (1+r)K(1+r^{-1})C^*_{10}C_{11} + (1-r)K(r^{-1}-1)C_{10}D^*_{11}\,,\\
iJC'_{20}(X) &= -K(\tfrac{1}{2})C^2_{10} - (1-r^2)K\left(\frac{1-r}{2}\right)C_{11}D_{11}\,,\\
iJC'_{02}(X) &= -r^2K(\tfrac{1}{2})C^2_{01} + (1-r^2)K\left(\frac{1-r}{2r}\right)C_{11}D^*_{11}\,,\\
iJC'_{11}(X) &= -rK\left(\frac{r}{1+r}\right)C_{10}C_{01} - 2(1-r)K\left(\frac{1-r}{1+r}\right)\{rC_{02}D_{11} - C_{20}D^*_{11}\}\,,\\
iJD'_{11}(X) &= rK\left(\frac{1}{1-r}\right)C_{10}C^*_{01} + 2(1-r)K\left(\frac{2}{1-r}\right)C_{20}C^*_{11} + 2r(1+r)K\left(\frac{1+r}{1-r}\right)C^*_{02}C_{11}\,,
\end{aligned} \tag{3.15}$$

where the constant J and kernel $K(\cdot)$ may correspond to either Case E or Case O, or be replaced by $\tilde{J}$ and $\tilde{K}(\cdot)$ for Case F. Numerical and analytical predictions from these equations are used in Ref. 24 to indicate conditions under which the *inter-modulational distortion* between fundamental frequencies $\omega_1$ and $\omega_2$ is minimized. Results are given for lithium niobate (strongly piezoelectric) and magnesium oxide (cubically anisotropic elastic). For the former, the open circuit boundary condition (Case O) was applied. [It may be noted that for two quasi-monochromatic inputs, the ordinary derivatives in (3.15) may be interpreted as derivatives along $\epsilon\theta = X - \epsilon ct = \text{constant}$.]

Surface acoustic wave convolvers have much the same configuration as surface acoustic wave filters. They depend on material nonlinearity, but since they involve two counter-propa-

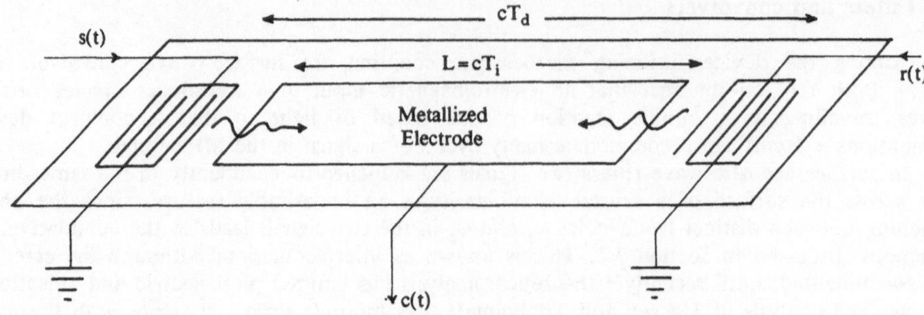


Fig. 11 Configuration of the reference and signal transducers and the output metallized electrode in a simplified surface acoustic wave convolver.

gating beams, interactions between the two input signals are negligible. The configuration is indicated in Fig. 11.

The reference wave r(t) and signal s(t) are launched by two interdigital transducers and the convolution output is measured from a metallized electrode of length $L(\equiv cT_i)$, where c is the wave speed. Since the *interaction time* is typically many thousands of periods of the two signals, it can be shown that, to a good approximation, the output voltage at the collector electrode $-\frac{1}{2}L < X < \frac{1}{2}L$ is

$$c[\tfrac{1}{2}(t+T_d)] = C_{NL}\int_{(t-T_i)/2}^{(t+T_i)/2} s(t-\tau)r(\tau)d\tau \,. \tag{3.16}$$

The nonlinear coefficient $C_{NL}$ incorporates both the linear modulus connecting the amplitudes of the potentials in the two excited surface acoustic waves to the input voltages r(t) and s(t) and also the nonlinear coupling K(2) from any frequency $\omega$ to its harmonic $2\omega$. Expression (3.16) differs from a true convolution of r(t) and s(t) by the finiteness of the integration range and by the compression of the time scale by the factor 2. Accordingly, all frequencies in the output spectrum are doubled. The efficiency of a convolver is essentially measured by $C_{NL}$. Recent surveys of the technology of surface acoustic wave convolvers may be found in Ref. 25.

### 3.4 Lateral spreading of surface acoustic waves

Most treatments of nonlinear effects in surface acoustic waves omit any dependence of the fields on the transverse coordinate $X_3$. By using ideas analogous to those used elsewhere in this volume to analyse wave modulation, we indicate in this section how allowance for gradual dependence of the fields on a transverse coordinate adds a diffractive term to Eq. (1.24). For simplicity, only elastic surface acoustic waves are treated here - further details, and an extension to piezoelectric surface acoustic waves may be found in Ref. 26.

In Section 1.2, displacements were allowed to depend on $\theta \equiv X_1 - ct$, $X_2$ and $X \equiv \epsilon X_1$. We now allow $u_i$ to depend additionally on $Z \equiv \epsilon^{1/2}(X_3 - gX_1)$. The term $gX_1$ is included so that, in the lowest approximation, signals propagate along 'surface rays' $dX_3/dX_1 = g$ which need not be parallel to the wave normal. The scaling $\epsilon^{1/2}$ is chosen so that diffractive terms are comparable in magnitude to nonlinear terms. Allowing $u_i$ to depend on the four coordinates $\theta$, $X_2$, X *and* Z in (1.1) - (1.4) then replaces (1.15) by

$$\begin{aligned}\rho c^2 u_{j,11} - c_{j\alpha m\beta}u_{m,\alpha\beta} &\doteq \epsilon^{1/2} g_N (c_{j\alpha mN} + c_{m\alpha jN})u_{m,\alpha Z} + \epsilon g_N g_L c_{jNmL} u_{m,ZZ} \\ &+ \epsilon(c_{j\alpha m1} + c_{m\alpha j1})u_{m,\alpha X} + \epsilon c_{j\alpha m\beta n\gamma}(u_{m,\beta}u_{n,\gamma})_{,\alpha} + O(\epsilon) \qquad \text{in } X_2 < 0\,, \end{aligned} \tag{3.17}$$

where, to simplify subsequent analysis, we have written $g_1 = -g$, $g_2 = 0$, $g_3 = 1$. As in Section 1.2, Greek suffices range only over the values 1 and 2, while $_{,1}$ denotes $\partial/\partial\theta$. The boundary condition likewise acquires terms $O(\epsilon^{1/2})$ and derivatives $\partial/\partial Z$.

Again the general $O(1)$ solution is readily identified. It is a linearized surface acoustic wave, with displacements $U_j$ which are given by (1.17) when $C(k,X)$ is generalized as $C(k,X,Z)$. Seeking $O(\epsilon^{1/2})$ perturbations yields an inhomogeneous linear problem. The linear operators are identical to those in (1.18) and (1.19) [and which have complementary function of the form (1.17) with C replaced by some new arbitrary function $F(k,X,Z)$]. Again there is a solvability condition for this problem, of the form (1.23). Since all the 'forcing' terms are linear in derivatives of $U_{j,Z}$, condition (1.23) involves the multiplicative factor $C_{,Z}$, which must not be zero. Thus, the condition does not impose a restriction on $C(k,X,Z)$ - instead it merely selects the parameter g. This is hardly surprising, since the $O(\epsilon^{1/2})$ problem is a linear, geometric acoustics problem for a beam of waves having wave normal parallel to the $X_1$-axis. It is well-known from crystal optics and crystal acoustics that the amplitude (and energy) travel at the group velocity. Consequently, it is readily shown that the (surface) vector having components $(1,0,g)$ is a vector which is normal at the point $(\omega,k,0)$ to the slowness curve $D(\omega,k_1,k_3) = 0$ for surface waves having specified frequency $\omega$. Moreover, the necessary procedure, which is to analyse (linearized) surface acoustic waves with displacements

$$u_j = V_j(X_2;k)e^{i(k_1X_1+k_3X_3-\omega t)} \tag{3.18}$$

and (surface) wave vector $\mathbf{k} = (k_1,0,k_3)$, yields a convenient formula for the $O(\epsilon^{1/2})$ correction to $U_j$. This correction is required before the $O(\epsilon)$ terms on the right-hand side of (3.17) may be evaluated and inserted into the solvability condition (1.23) once more.

The depth dependence $V_j$ is governed by a set of ordinary differential equations possessing solutions of the form $a_j^{(p)}\exp(iks^{(p)}X_2)$. Here, both $a_j^{(p)}$ and $s^{(p)}$ depend only on $k_1/\omega$ and $k_3/\omega$, where $k \equiv |\mathbf{k}|$. The boundary condition ($\tau_{2j} = 0$ at $X_2 = 0$) then allows the depth dependence in (3.18) to be constructed by superposing these partial waves

$$V_j(X_2;\mathbf{k}) = \sum_{p=1}^{3} B^{(p)}a_j^{(p)}\exp(iks^{(p)}X_2) \ , \qquad V_2(0;\mathbf{k}) = 1 \ , \qquad k > 0 \ .$$

The corresponding compatibility condition, ensuring that the boundary matrix equation may be solved for the multipliers $B^{(p)}$, is the *dispersion relation*

$$D(\omega,k_1,k_3) = 0 \ .$$

In $(\omega,k_1,k_3)$ space this defines a cone, so that at each frequency $\omega$ the phase speed $\omega/|\mathbf{k}| = \omega/k$ depends only on the direction, not the magnitude of the wave normal. The cross-section at each constant value of $\omega$ is a magnification of the *slowness curve* defined by $D(1,k_1,k_3) = 0$, as in Fig. 12. Then it is readily shown that, at each value $\omega$,

$$\frac{\partial D}{\partial k_3} = g\frac{\partial D}{\partial k_1}$$

so that the *ray direction* $dX_3/dX_1 = g$ is the normal to the slowness curve corresponding to the wave normal $\mathbf{k}$. Also, it may be checked that the $O(\epsilon^{1/2})$ correction to $U_j$ has the form

$$-\epsilon^{1/2}iC_{,Z}(k,X,Z)g_N\frac{\partial V_j}{\partial k_N}(X_2;k,0) \ . \tag{3.19}$$

When expression (3.19) is incorporated into (3.17) and the consequent terms $\overline{P}_j$ and $\overline{Q}_j$ in the solvability condition (1.23) for the $O(\epsilon)$ problem, it is found that all terms involving Z-derivatives are multiples of $C_{,ZZ}$. Consequently, the evolution equation is simply Eq. (1.24) with one additional term, namely:

$$k^{-1}M\frac{\partial^2 C}{\partial Z^2} + iJ\frac{\partial C}{\partial X} + \int_0^\infty K\left(\frac{\ell}{k}\right)(k-\ell)\ell C(k-\ell,X,Z)C(\ell,X,Z)d\ell = 0 \ . \tag{3.20}$$

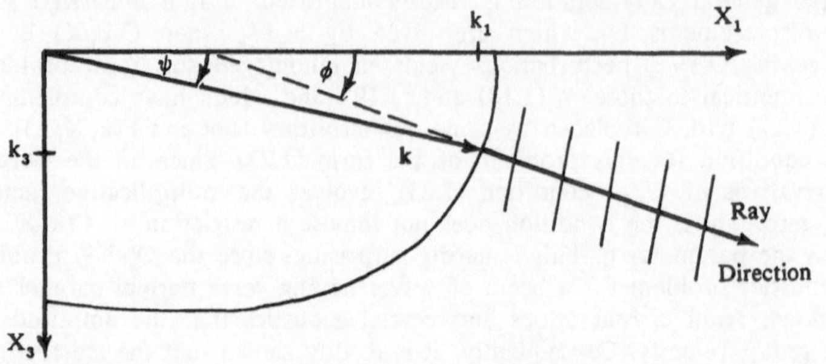


Fig. 12 Slowness curve $D(1,k_1,k_3) = 0$ (at frequency $\omega = 1$), showing the relation of the ray angle $\phi$ to the angle $\psi$ defining the wave normal.

Moreover, the additional coefficient M is related to J through a simple property of geometric acoustics, in the form

$$\frac{M}{J} = \frac{1}{2}\sec^2\phi \left.\frac{d\phi}{d\psi}\right|_{\psi=0},$$

where $\psi$ and $\phi$ are the inclinations of the wave normal and ray vector to the $X_1$-axis. This is just a diffraction coefficient, and has the value $\frac{1}{2}$ on materials for which surface waves behave isotropically. Clearly, it can be negative if the $X_1$-axis lies in a concave part of the slowness curve.

Equation (3.20) has not yet been investigated numerically. Although its linear terms are familiar from diffraction theory and are the same as in the nonlinear Schrödinger equation, there is one essential difference. Because *all* Fourier components may be present in the signal, the quadratic nonlinearity mixes components at all wavenumbers k. Except near a monochromatic source, we cannot concentrate on one dominant (central) frequency, as in the nonlinear Schrödinger equation.

## REFERENCES


1. D. F. Parker, Int. J. Engng. Sci. **26**, 59 (1988).
2. G. W. Farnell and E. L. Adler, in *Physical Acoustics IX*, W. P. Mason and R. N. Thurston, eds. (Academic Press, New York, 1972).
3. D. M. Barnett and J. Lothe, Physica Norvegica **3**, 13 (1973);
   D. M. Barnett and J. Lothe, J. Phys. F **4**, 671 (1974);
   P. Chadwick and G. D. Smith, Advances in Appl. Mech. **17**, 303 (1977);
   V. I. Al'shits and J. Lothe, Soviet Phys. Crystallogr. **23**, 509 (1978);
   P. Chadwick, Proc. Roy. Soc. Lond. A **401**, 203 (1985);
   P. Chadwick and T.C.T. Ting, Quart. Appl. Math. **45**, 419 (1987).
4. A. N. Stroh, J. Math. Phys. **41**, 77 (1962).
5. J. D. Achenbach, *Wave Propagation in Elastic Solids* (North-Holland, Amsterdam, 1973).
6. A. A. Maradudin and A. P. Mayer, these proceedings.
7. R. W. Lardner, Int. J. Engng. Sci. **21**, 1331 (1983).
8. A. P. Harvey, R. E. Craine and S. Syngellakis, in *Recent Developments in Surface Acoustic Waves*, D. F. Parker and G. A. Maugin, eds. (Springer-Verlag, Berlin-Heidelberg, 1988)
9. G. B. Whitham, Proc. Roy. Soc. Lond. A **283**, 238 (1965);
   J. C. Luke, Proc. Roy. Soc. Lond. A **292**, 403 (1966);
   G. B. Whitham, *Linear and Nonlinear Waves* (Wiley, New York, 1974).
10. N. Zabusky and M. D. Kruskal, Phys. Rev. Lett. **15**, 240 (1965);
    C. S. Gardner, J. M. Greene, M. D. Kruskal and R. M. Miura, Phys. Rev. Lett. **19**, 1095 (1967);

R. K. Dodd, J. C. Eilbeck J. D. Gibbon and P. Morris, *Solitons and Nonlinear Wave Equations* (Academic, London, 1982).
11. D. F. Parker and F. M. Talbot, in *Nonlinear Deformation Waves*, U. Nigul and J. Engelbrecht, eds. (Springer, Berlin-Heidelberg, 1983).
12. D. F. Parker and F. M. Talbot, J. Elast. **15**, 389 (1985).
13. E. A. David and D. F. Parker, in *Recent Developments in Surface Acoustic Waves*, D. F. Parker and G. A. Maugin, eds. (Springer-Verlag, Berlin-Heidelberg, 1988).
14. D. F. Parker, Physica **16D**, 385 (1985);
D. F. Parker, in *Waves and Stability in Continuous Media*, M. Maiellaro and L. Palese, eds. (Atti del III Meeting, Bari, 1989).
15. E. A. David and D. F. Parker, to appear in Wave Motion.
16. J. K. Hunter, in *Current Progress in Hyperbolic Systems, Riemann Problems and Computations*, W. B. Lindquist, ed. (Am. Math. Soc., 1989)
17. S. Fubini-Ghiron, Alta Frequenza **4**, 530 (1935);
see e.g. R. T. Beyer, in *Physical Acoustics* Vol IIB, W. P. Mason, ed. (Academic, New York, 1965)
18. G. Bleustein, Appl. Phys. Lett. **13**, 412 (1968);
Yu. V. Gulyaev, Soviet Phys. JETP Lett. **9**, 37 (1969).
19. G. A. Maugin, *Nonlinear Electromechanical Effects and Applications* (World Scientific, Singapore, 1985).
20. D. F. Parker and E. A. David, Int. J. Engng. Sci. **27**, 565 (1989).
21. H. F. Tiersten and J. C. Baumhauer, J. Appl. Phys. **45**, 4272 (1974).
H. F. Tiersten and J. C. Baumhauer, J. Appl. Phys. **58**, 1867 (1985).
22. A. N. Abd-Alla and G. A. Maugin, J. Acoust. Soc. Am. **82**, 1746 (1987).
23. G. A. Maugin and A. Hakmi, J. Acoust. Soc. Am. **76**, 826 (1984);
A. N. Abd-Alla and G. A. Maugin, in *Recent Developments in Surface Acoustic Waves*, D. F. Parker and G. A. Maugin, eds. (Springer-Verlag, Berlin-Heidelberg, 1988);
A. N. Abd-Alla, Ph.D. Thesis, University of Assiut, Egypt (1988).
24. A. P. Harvey and G. E. Tupholme, submitted to Int. J. Engng. Sci.
A. P. Harvey, Ph.D. Thesis, University of Southampton (1990).
25. C. Maerfeld, in *Rayleigh Wave Theory and Applications*, E. A. Ash and E. G. S. Paige, eds. (Springer-Verlag, Berlin-Heidelberg, 1985);
H. P. Grassl, L. Reindl and H.-W. Worz, in *Recent Developments in Surface Acoustic Waves*, D. F. Parker and G. A. Maugin, eds. (Springer-Verlag, Berlin-Heidelberg, 1988).
26. D. F. Parker and E. A. David, in *Elastic Wave Propagation*, M.F. McCarthy and M. A. Hayes, eds. (Elsevier, Amsterdam, 1989);
N. Kalyanasundaram, D. F. Parker and E. A. David, to appear in J. Elasticity.

# NONLINEAR THEORY OF ACOUSTOELECTRONIC WAVES IN SEMICONDUCTORS

Yu. V. Gulyaev, S. N. Ivanov, G. D. Mansfeld, P. E. Zilberman

Institute of Radioengineering and Electronics,
USSR Academy of Sciences,
Marx Avenue 18, 103907 Moscow, Centre, USSR

## 1. INTRODUCTION

This chapter presents the results of an investigation of the attenuation and amplification of acoustic waves by electrons in semiconductors. Emphasis is put on the criteria of applicability of the different existing theories to various regions of acoustic wave frequencies, semiconductor properties and external conditions. We consider the drag of free electrons by acoustic waves, which leads to the acoustoelectric effect, acoustomagnetoelectric effect, acoustothermal effect, etc. The nonlinear effects of acoustoelectronic interactions, such as the dependence of the acoustic wave attenuation coefficient and velocity on the amplitude of the acoustic waves, the non-ohmic behaviour of the current-voltage curve of the specimen, and the generation of harmonics, are also considered. The specific features of acoustoelectronic interactions for surface acoustic waves are pointed out. A comparison of the theory of acoustoelectronic interactions with the experimental data is briefly given.

## 2. GENERAL LINEAR THEORY OF ATTENUATION AND AMPLIFICATION OF ACOUSTIC WAVES BY ELECTRONS IN SEMICONDUCTORS

There are two main mechanisms for the interaction of acoustic waves with free electrons. In semiconductors the acoustoelectronic interaction occurs via the deformation potential and in piezoelectric semiconductors via the piezoelectric effect (see, for example, Refs. 1-2). The interaction due to the deformation potential is valid for all solids and is related to the change of energy of the electrons in the conduction band due to the deformation of the crystal lattice by the acoustic wave:

$$\delta\epsilon = \Lambda_{ij} \frac{\partial u_j}{\partial x_i} \tag{2.1}$$

Here $\Lambda_{ij}$ are the components of the deformation potential tensor and $u_j$ are the components of mechanical displacement in the acoustic wave. The square of the corresponding coupling constant, $K_d^2$, is proportional to $f_{AW}^2$, where $f_{AW}$ is the frequency of the acoustic wave. For a one-dimensional problem

$$K_d^2 = \frac{\epsilon q^2}{4\pi\rho v_s^2}\left(\frac{\Lambda}{e}\right)^2 \sim f_{AW}^2 \tag{2.2}$$

where $\epsilon$ and $\rho$ are the dielectric constant and density of the crystal respectively, $q = 2\pi f_{AW}/v_s$ is the wave vector of the acoustic wave, $v_s$ its velocity, e the charge of the electron and $\Lambda$ the deformation potential constant.

The interaction due to the piezoelectric effect is related to the electric field (electric potential) produced by the acoustic wave in piezoelectric semiconductors. It is described by the electric field contribution to the strain tensor $T_{ij}$ and the deformation contribution to the electric displacement $D_i$:

$$T_{ij} = C_{ijkl} u_{kl} - \beta_{kij} \frac{\partial \phi}{\partial x_k} \tag{2.3}$$

$$D_i = -\epsilon \frac{\partial \phi}{\partial x_i} - 4\pi\beta_{ijk} \frac{\partial u_j}{\partial x_k} . \tag{2.4}$$

Here $C_{ijkl}$ is the elastic modulus tensor, $\beta_{kij}$ the piezoelectric tensor, $u_{kl}$ the stress tensor and $\phi$ the electric potential of the acoustic wave. The corresponding coupling constant in this case is

$$K_p^2 = \frac{4\pi\beta^2}{\epsilon C} \tag{2.5}$$

A comparison of the coupling strengths for the two kinds of acoustoelectronic interaction for typical semiconductors shows that for acoustic wave frequencies

$$f_{AW} \leq 10 \text{ GHz} \qquad K_d^2 < K_p^2 \tag{2.6}$$

Thus, the deformation potential interaction usually becomes important only for the high frequencies of ultrasound, $f_{AW} > 10$ GHz. Due to high viscous attenuation of the acoustic waves at these frequencies existing acoustoelectronic devices are operated at frequencies below 2 GHz where the deformation potential interaction is negligibly small. From now on we shall mainly restrict our discussion to the case of piezoelectric semiconductors, such as CdS, GaAs and InSb, the generalization to nonpiezoelectric semiconductors with the deformation potential interaction being made, if necessary, very easily.

We shall consider the case when the acoustic wave wavelength is much bigger than the lattice constant, $\lambda_s >> d$, so that from the acoustic point of view the crystal is considered as a continuous medium.

From the physical point of view there are two different cases:

(i) The classical case:

$$\lambda_s >> \lambda_e \qquad \text{i.e. } k >> q \tag{2.7}$$

where $\lambda_e$ is the De Broglie wavelength of the electron, and k and q are the wavevectors of the electron and phonon respectively. In this case the electron may be considered as a point particle moving in the classical field of the acoustic wave. This case is usually realised in semiconductors at high temperatures, high values of the electron effective mass $m^*$ and low acoustic frequencies. For example, for InSb at 77 K, $m^* = 0.015\ m_0$ and for the acoustic wave frequency f = 2 GHz one has $\lambda_e = 10^{-5}$ cm and $\lambda_s \simeq 10^{-4}$ cm, so $\lambda_e << \lambda_s$ - the classical case.

(ii) The quantum case:

$$\lambda_s \leq \lambda_e \qquad \text{i.e. } k \leq q \tag{2.8}$$

In this case the interaction of electrons with an acoustic wave must be considered through a quantum mechanical approach as the diffraction of the electron De Broglie wave by the acoustic wave. In some works, the acoustic wave is considered as a flux of coherent phonons for this kind of interaction with electrons. For the same InSb crystal at T = 4.2 K and for f = 10 GHz we have $\lambda_e \simeq 5\times10^{-5}$ cm and $\lambda_s \simeq 2\times10^{-5}$ cm, i.e. $\lambda_e \geq \lambda_s$ - a typical quantum case.

Here we restrict ourselves to the classical case, since the basic conclusions turn out to be the same in both cases. The classical case, in turn, may be divided into 3 subcases:

(i)

$$\lambda_s >> L_e \qquad \omega_s << \tau_e^{-1} \tag{2.9}$$

where $L_e$ and $\tau_e$ are the characteristic length and time required to reach the local quasi-equilibrium Fermi distribution in the electron gas. This is the so-called "hydrodynamic" case (see Ref. 1), where the electron gas is characterized by local equilibrium kinetic coefficients, i.e. the mobility $\mu$, the diffusion coefficient $\mathscr{D}$, the heat conductivity $\kappa_e$ etc., which in principle are variable in space and time due to the propagation of the acoustic wave.

(ii)

$$\lambda_s << \ell \qquad \omega_s >> \tau^{-1} \qquad q\ell >> 1 \tag{2.10}$$

where $\ell$ and $\tau$ are the electron mean free path and momentum relaxation time respectively. This is the so-called "collisionless" case (see Ref. 2), where electron collisions are not taken into account.

(iii)

$$L_e >> \lambda_s \gtrless \ell \qquad \tau_e^{-1} << \omega_s << \tau^{-1} \tag{2.11}$$

the "intermediate frequency" case (see Refs. 3-4), where electron collisions must be explicitly taken into account.

The methods for calculating these three cases are quite different, case (iii) being, of course, the most difficult one.

It is useful to note here that at room temperatures and acoustic wave frequencies, $f_s < 0.5$ GHz, the hydrodynamic approach is valid for all semiconductors.

## 2.1 Hydrodynamic case

Under the conditions $\lambda_s >> L_e$, $\omega_s << \tau_e^{-1}$, the symmetrical part of the electron distibution function can be chosen to be of the form of the local quasi-equilibrium Fermi distribution at each point in space and time:

$$f_{el} = \frac{1}{\exp\left(\frac{\epsilon - \xi(x,t)}{k_B T}\right) + 1} \tag{2.12}$$

The chemical potential $\xi(x,t)$ is given by

$$\xi(x,t) = \xi_0 + \xi_1(x,t) \tag{2.13}$$

$$\xi_1(x,t) \sim e^{i(\omega t - qx)} \tag{2.14}$$

where $\omega$ and q are the frequency and wavevector of the acoustic wave, which propagates in the OX direction.

What are the quantities $L_e$ and $\tau_e$? Let us first determine them from qualitative considerations. Suppose the electron gas is brought away from thermal equilibrium. In the simplest case its relaxation to equilibrium is characterized by two times: $\tau$ (the momentum relaxation time) and $\tau_\epsilon$ (the energy relaxation time). For quasi-elastic collisions $\tau_\epsilon >> \tau$, so the relaxation of the electron gas to equilibrium is effectively described by the longest time $\tau_\epsilon$. $\tau_\epsilon$ is of the order of magnitude of the $\tau_e$ introduced above. There are several characteristic lengths which describe electron gas relaxation in space. The mean free path $\ell$ characterizes the electron momentum relaxation. The energy relaxation diffusion length $L_\epsilon = \sqrt{\mathscr{D}\tau_\epsilon}$ is the distance through which the electron moves in Brownian-like motion during the energy relaxation time. Usually $L_\epsilon >> \ell$. If there is a directed drift of electrons there appears one more length, $L_d$ =

$v_d\tau_\epsilon$ (where $v_d$ is the mean drift velocity of electrons), which characterizes the distance that electrons travel due to their drift velocity during the energy relaxation time $\tau_\epsilon$. The characteristic length $L_e$ introduced above is the longest of the three lengths $L_\epsilon$, $L_d$ and $\ell$,

$$L_e = max\{\ell, \sqrt{\mathscr{D}\tau_\epsilon}, v_d\tau_\epsilon\} \tag{2.15}$$

Thus, the conditions that must be satisfied for the hydrodynamic approach to be valid are:

$$T >> \tau_\epsilon \qquad \lambda_s >> max\{\ell, \sqrt{\mathscr{D}\tau_\epsilon}, v_d\tau_\epsilon\} \tag{2.16}$$

where T is the period of the acoustic wave, or

$$\omega_s\tau_\epsilon << 1 \qquad max\{q\ell, q\sqrt{\mathscr{D}\tau_\epsilon}, qv_d\tau_\epsilon\} << 1 \tag{2.17}$$

Conditions (2.16)-(2.17) follow, of course, from the strict kinetic theory of acoustic wave interaction with electrons (see below). Here it is necessary to note that often only the condition $q\ell << 1$ is considered as necessary for the hydrodynamic approach. In the majority of cases this is incorrect. Let us, for instance, take a typical example: n-InSb, T = 77K, $\mu = 10^6$ cm²/Vs, $\mathscr{D} = 7\times10^3$ cm²/s, $\tau = 10^{-12}$ s, $\tau_\epsilon = 10^{-10}$ s, $v_T = 10^7$ cm/s, $v_d = 10^6$ cm/s. From these data it follows that

$$\sqrt{\mathscr{D}\tau_\epsilon} = 10^{-3}\ \text{cm} \qquad v_d\tau_\epsilon = 10^{-4}\ \text{cm} \qquad \ell \approx 10^{-5}\ \text{cm} \tag{2.18}$$

so $\sqrt{\mathscr{D}\tau_\epsilon} >> \ell$, $v_d\tau_\epsilon >> \ell$ and condition (2.17) by no means can be reduced to $q\ell << 1$.

We shall consider further the simplest case of piezoelectric semiconductors with one type of carrier (electrons), not taking into account trapping or heating of electrons. The basic system of equations is now[1-4]

$$\rho \frac{\partial^2 u_i}{\partial t^2} = \frac{\partial T_{ij}}{\partial x_j} \tag{2.19}$$

$$\nabla\cdot\mathbf{D} = 4\pi e(N - N_d) \tag{2.20}$$

$$\nabla\cdot j = e\frac{\partial N}{\partial t} \tag{2.21}$$

$$T_{ij} = C_{ijkl}\frac{\partial u_k}{\partial x_l} - \beta_{kij}\frac{\partial \phi}{\partial x_k} + \Lambda_{ij}N \tag{2.22}$$

$$D_i = -\epsilon_{ij}\frac{\partial \phi}{\partial x_j} - 4\pi\beta_{ijk}\frac{\partial u_j}{\partial x_k} \tag{2.23}$$

$$j_i = \sigma_{ik}\left[E_{0k} - \frac{\partial \phi}{\partial x_k} - \frac{\Lambda_{ij}}{e}\frac{\partial u_{ij}}{\partial x_k}\right] - e\mathscr{D}_{ik}\frac{\partial n}{\partial x_k} \tag{2.24}$$

Here N and $N_d$ are the concentrations of free electrons and compensating positive charges.

In the one-dimensional case for $\mathbf{q} \parallel \mathbf{E}_0 \parallel OX$ the system of equations (2.19)-(2.24) reduces to the following:

$$\rho\frac{\partial^2 u}{\partial t^2} = C\frac{\partial^2 u}{\partial x^2} - \beta\frac{\partial^2 \phi}{\partial x^2} + \Lambda\frac{\partial N}{\partial x} \tag{2.25}$$

$$\epsilon\frac{\partial^2 \phi}{\partial x^2} + 4\pi\beta\frac{\partial^2 u}{\partial x^2} = -4\pi e(N - N_d) \tag{2.26}$$

$$e\frac{\partial N}{\partial t} + \frac{\partial j}{\partial x} = 0 \tag{2.27}$$

$$j = \sigma(N)\left[E_0 - \frac{\partial\phi}{\partial x} - \frac{\Lambda}{e}\frac{\partial^2 u}{\partial x^2}\right] - e\mathcal{D}(N)\frac{\partial N}{\partial x} \tag{2.28}$$

The kinetic coefficients $\sigma$ and $\mathcal{D}$ depend, generally speaking, on the electron concentration N and on the electron scattering mechanisms.

We now perform the linearization procedure by putting

$$\begin{aligned} N &= N_0 + n & \quad |n| &<< N_0 = N_d \\ j &= j_0 + j_{\sim} & \quad |j_{\sim}| &<< j_0 \end{aligned} \tag{2.29}$$

$$\left|\frac{\partial\phi}{\partial x}\right| << E_0 \qquad u, \phi, n, j_{\sim} \sim e^{i(\omega t - qx)} \tag{2.30}$$

We then have

$$\sigma(N) = \sigma_0(N_0) + \frac{\partial\sigma_0}{\partial N_0}n \qquad \mathcal{D}(N) = \mathcal{D}_0(N_0) + \frac{\partial\mathcal{D}_0}{\partial N_0}n \tag{2.31}$$

Since $\sigma(N) = eN\mu(N)$, where $\mu$ is the electron mobility

$$\frac{\partial\sigma_0}{\partial N_0} = e\mu_0(N_0)\left[1 + \frac{\partial ln\mu_0}{\partial lnN_0}\right] \tag{2.32}$$

For example, for degenerate semiconductors we have $\mu = e\tau/m^*$, where $\tau = \tau(\xi)$ is the momentum relaxation time, $m^*$ the effective mass, $\xi$ the Fermi energy, $\mu = \mu(\xi)$, $\xi = (3\pi^2)^{2/3}(h^2N^{2/3})/2m^*$, so we have $\mu = \mu(N)$.

After linearization we obtain:

$$j = j_0 + j_{\sim} = eN_0\mu_0E_0 + en\frac{1}{e}\frac{\partial\sigma_0}{\partial N_0}E_0 - eN_0\mu_0\frac{\partial\phi}{\partial x} - e\mathcal{D}_0\frac{\partial n}{\partial x} - N_0\mu_0\Lambda\frac{\partial^2 u}{\partial x^2} \tag{2.33}$$

The second term on the right hand side of Eq. (2.33) describes the motion of space charge bunches n under the action of a d.c. electric field $E_0$, with a drift velocity

$$v_d^* = \frac{1}{e}\frac{\partial\sigma_0}{\partial N_0}E_0 \equiv v_d r \tag{2.34}$$

that is different from the average electron drift velocity $v_d = \mu_0 E_0$ by a factor

$$r = \frac{d ln\sigma_0}{d lnN_0} \equiv 1 + \frac{d ln\mu_0}{d lnN_0} \tag{2.35}$$

where r is, generally speaking, not equal to 1.

Neglecting at this point the deformation potential coupling we have from Eqs. (2.27) and (2.33):

$$n = -\frac{e\phi}{k_B T}\frac{N_0}{\left[1 - \frac{i\omega}{q^2\mathcal{D}}\left(1 - \frac{v_d^*}{v}\right)\right]} \tag{2.36}$$

Thus one sees that the phase shift between n and $\phi$ depends on the sign of the factor $(1 - v_d^*/v)$: it changes from positive to negative at the point where the drift velocity of the electron bunches $v_d^*$ (not the average electron drift velocity $v_d$!) becomes larger than the velocity of sound $v_s$.[5]

After linearization the dispersion equation for the system of equations (2.25)-(2.28) is found in the usual way. The approximate solution of this dispersion equation is easily found if one puts

$$q = q_0 + \Delta q + i\alpha_e \qquad |\Delta q|,\ |\alpha_e| << q_0 = \frac{\omega}{v_{s0}} \tag{2.37}$$

$$v_{s0} = \sqrt{\frac{c}{\rho}} \qquad \frac{\Delta q}{q_0} = -\frac{\Delta v_s}{v_{s0}}$$

Here $\alpha_e$ is the electronic attenuation (amplification) coefficient of the acoustic wave, and $\Delta v_s/v_{s0}$ is the relative electron correction to the velocity of sound.

One obtains

$$\alpha_e = \left(\frac{K_p^2 + K_d^2}{2}\right) \frac{q_0\omega\tau_M\left(1 - \frac{v_d^*}{v_s}\right)}{\omega^2\tau_M^2\left(1 - \frac{v_d^*}{v_s}\right)^2 + \left(1 + q^2r_D^2\right)^2} \tag{2.38}$$

$$\frac{\Delta v_s}{v_{s0}} = \frac{1}{2}\,\frac{K_p^2\left[\omega^2\tau_M^2\left(1 - \frac{v_d^*}{v_s}\right)^2 + q^2r_D^2(1 + q^2r_D^2)\right] - K_d^2(1 + q^2r_D^2)}{\omega^2\tau_M^2\left(1 - \frac{v_d^*}{v_s}\right)^2 + \left(1 + q^2r_D^2\right)^2} \tag{2.39}$$

Here $r_D$ is the Debye length:

$$r_D = \begin{cases} \left(\dfrac{\epsilon k_B T}{4\pi e^2 N_0}\right)^{1/2} & \text{nondegenerate case} \quad (2.40) \\[2ex] \left(\dfrac{\epsilon \frac{2}{3}\xi_0}{4\pi e^2 N_0}\right)^{1/2} & \text{degenerate case} \quad (2.41) \end{cases}$$

$$\tau_M = \frac{\epsilon}{4\pi\sigma_0} \qquad \text{Maxwellian relaxation time} \tag{2.42}$$

For the nondegenerate case $d\mu_0/dN_0 = 0$ and $r = 1$, and if one puts $K_d^2 = 0$ then Eq. (2.38) is the well known expression for $\alpha_e$ in White's theory.[6] For the degenerate case

$$r = 1 + \frac{2}{3}\frac{d\ln\tau(\xi)}{d\ln\xi}$$

where $\tau(\xi)$ is the momentum relaxation time. If $\tau(\xi) \sim \xi^\nu$, then

$$r = 1 + \frac{2}{3}\nu \qquad \text{gain condition } v_d > v_s/r \tag{2.43}$$

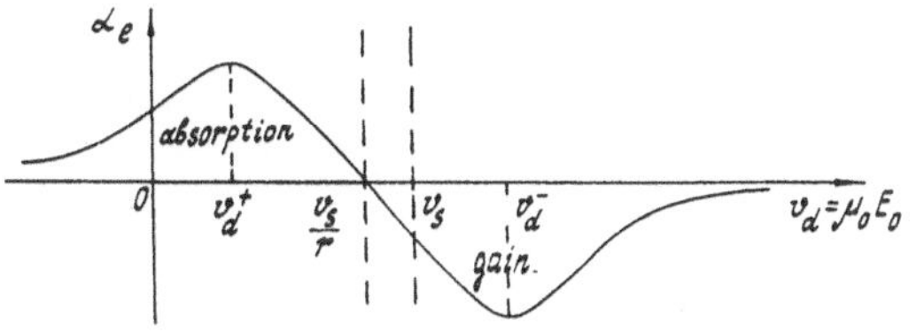


Fig. 1. The variation of the electronic attenuation coefficient $\alpha_e$ with drift electric field $E_0$ in the hydrodynamic case.

For different electron scattering mechanisms one has:

$$\nu = -1/2 \text{ (deformation potential scattering)}, \; r = 2/3, \text{ gain condition } v_d > 1.5v_s \tag{2.44}$$

$$\nu = +1/2 \text{ (piezoelectric scattering)}, \; r = 4/3, \text{ gain condition } v_d > 0.75v_s \tag{2.45}$$

$$\nu = +3/2 \text{ (impurity scattering)}, \; r = 2, \text{ gain condition } v_d > 0.5v_s \tag{2.46}$$

So, for example, for impurity scattering the amplification of acoustic waves starts already at the electron drift velocity $v_d$, which is equal to only half of the acoustic wave velocity. Exactly at that point, as already stated, the phase shift between the electric potential $\phi$ in the acoustic wave and the electron concentration wave n changes sign (the electron bunches begin to "overtake" sound). The dependence of the electronic attenuation (amplification) coefficient of sound, $\alpha_e$, and of the sound velocity $v_s$ on the drift electric field $E_0$ is shown in Figs. 1 and 2. The physical explanation of why the electronic attenuation of sound turns into amplification at the point $v_d = v_s/r$ where the phase shift between $\phi$ and n changes sign is illustrated by Fig. 3.

## 2.2 Collisionless case (classical): $q\ell \gg 1$

In this case, between collisions the electrons travel through distances much greater than the acoustic wave wavelength (see Fig. 4), so the local relation between electron density and the electric potential of the acoustic waves is no longer valid. Thus, in order to calculate the acoustoelectronic interaction, one must use the Boltzmann kinetic equation.

The basic system of equations is now:

$$\rho \frac{\partial^2 u}{\partial t^2} = C \frac{\partial^2 u}{\partial x^2} - \beta \frac{\partial^2 \phi}{\partial x^2} \tag{2.47}$$

$$\epsilon \frac{\partial^2 \phi}{\partial x^2} + 4\pi\beta \frac{\partial^2 u}{\partial x^2} = -4\pi e n \tag{2.48}$$

$$\frac{\partial f_p}{\partial t} + v\nabla_r f_p + \left(eE + \frac{e}{c}[v\times H]\right)\nabla_p f_p = \hat{J}\{f_p\} \tag{2.49}$$

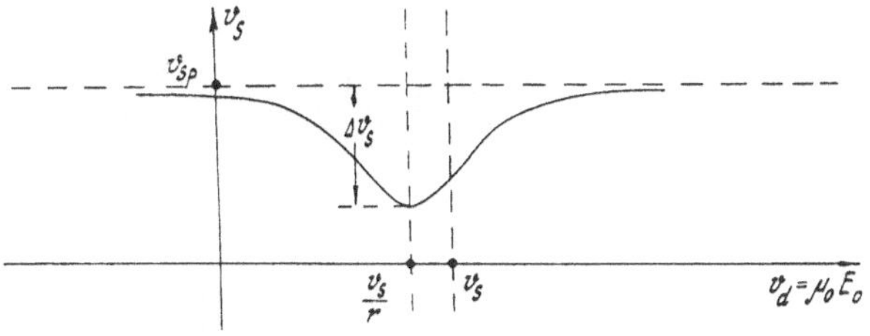

Fig. 2. The dependence of sound velocity $v_s$ on the drift electric field $E_0$ in the hydrodynamic case. $v_{sp} = [c(1 + K_p^2)/\rho]^{1/2}$, $K_p^2 = 4\pi\beta^2/(\epsilon c)$, $\Delta v_s/v_s \le K_p^2$.

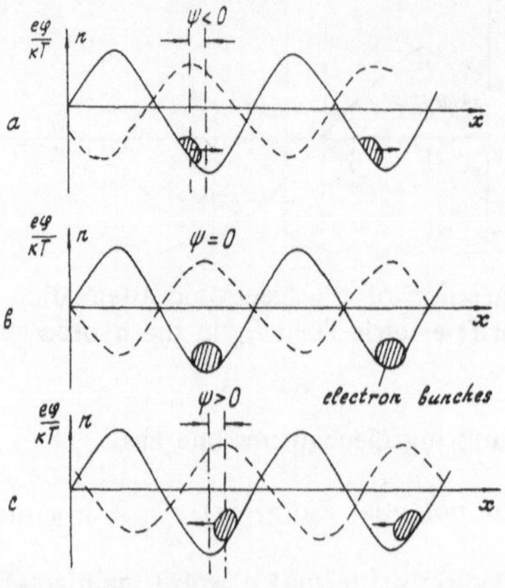


Fig. 3. Illustration of the mechanism of electronic sound attenuation/ amplification. The solid curve is $e\phi/(kT)$, the dashed line is n, $\psi$ is the phase shift. (a) $E_0 = 0$, (b) $\mu_0 E_0 = v_s/r$, (c) $\mu_0 E_0 > v_s/r$.

Here $f_p$ is the electron distribution function, **H** is the magnetic field (if any) and $\hat{\int}\{f_p\}$ is the collision integral.

The relation between the amplitudes of n and $\phi$ can now be found from the solution of the kinetic equation (2.49). In the linear case being considered here it has the general form:

$$n(\mathbf{q},\omega) = K(\mathbf{q},\omega)\phi(\mathbf{q},\omega) \tag{2.50}$$

$$K(\mathbf{q},\omega) = \frac{iq\sigma(\mathbf{q},\omega)}{e^2 v_s} \tag{2.51}$$

Here the nonlocal character of the interaction of electrons with acoustic waves is described by the dependence of the complex conductivity $\sigma(\mathbf{q},\omega)$ on the wave vector **q** and frequency $\omega$.

By putting $K = K' + iK''$ and solving the linearized equations (2.47) and (2.48) with the use of Eq. (2.50) in a usual manner, we obtain for the electronic attenuation coefficient of acoustic waves the expression

$$\alpha_e = \frac{K_p^2 q}{2} \frac{\dfrac{4\pi e^2}{\epsilon_0 q^2} K''(\mathbf{q},\omega)}{\left[1 - \dfrac{4\pi e^2 K'(\mathbf{q},\omega)}{\epsilon_0 q^2}\right]^2 + \left[\dfrac{4\pi e^2 K''(\mathbf{q},\omega)}{\epsilon_0 q^2}\right]^2} \tag{2.52}$$

Now the problem is reduced to finding the complex conductivity $\sigma(\mathbf{q},\omega)$ of the specimen, with spatial and temporal dispersion. In the collisionless case $\hat{\int}\{f_p\} = 0$. So for the one-dimensional case $\mathbf{q} \parallel \mathbf{E}_0 \parallel OX$, putting $f = f_0 + f_1$ (where $f_0$ is the electron distribution function in the d.c. electric field) we have from Eq. (2.49):

$$f_1 = \frac{e}{m^*}\left(\frac{1}{\omega - qv_x}\right)\frac{\partial f_0}{\partial v_x} q\phi \tag{2.53}$$

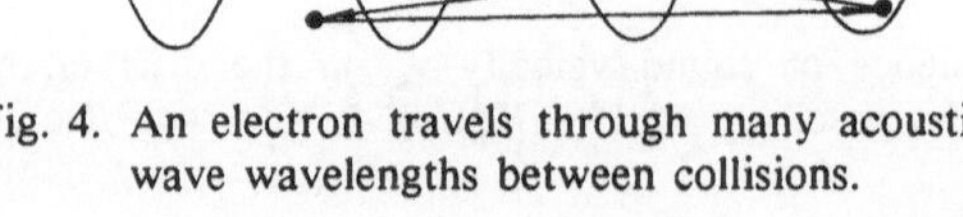

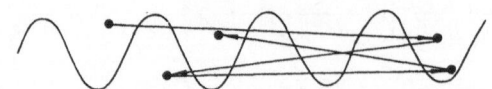

Fig. 4. An electron travels through many acoustic wave wavelengths between collisions.

$$n = \frac{2}{(2\pi\hbar)^2}\int f_1(\mathbf{p})d^3\mathbf{p} \equiv K(\mathbf{q},\omega)\phi \tag{2.54}$$

where

$$K(\mathbf{q},\omega) = \frac{2e}{(2\pi\hbar)^3}\int d^3\mathbf{v}\,\frac{q\dfrac{\partial f_0}{\partial v_x}}{(\omega - qv_x)} \tag{2.55}$$

Performing the integration around the pole at $v_x = \omega/q$, in the same manner as in the theory of Landau damping of plasma waves, we have:

$$K(\mathbf{q},\omega) = -\frac{eN_0}{k_B T}\left[1 + i\sqrt{\pi}\,\frac{m^*\omega(1 - \mu E_0/v_s)}{q\sqrt{2m^* k_B T}}\right] \tag{2.56}$$

One immediately obtains from (2.52) the electronic attenuation coefficient $\alpha_e$:

$$\alpha_e = \frac{K_p^2\omega\sqrt{\pi}\left(1 - \dfrac{\mu E_0}{v_s}\right)}{q^2 r_D^2\left(1 + \dfrac{1}{q^2 r_D^2}\right)^2\sqrt{\dfrac{2k_B T}{m^*}}} \tag{2.57}$$

This is the well known formula of Spector theory.[7]

The main contribution to the integral in Eq. (2.55) comes from the pole region at $v_x = \omega/q$, i.e. from electrons whose thermal velocity $\mathbf{v}_T$, projected onto the sound propagation direction, equals the sound velocity $v_s$. Since $v_T$ is usually much greater than $v_s$, this means that only electrons moving almost perpendicular to the acoustic wave direction contribute to collisionless acoustic wave attenuation or amplification. The dependence of $\alpha_e$ on the drift electric field $E_0$ is shown in Fig. 5.

### 2.3 Intermediate frequency case

Here $L_e >> \lambda_s \gtrless \ell$ and $\tau_e^{-1} << \omega_s < \tau^{-1}$, which includes the point $\lambda_s \simeq \ell$, i.e. $q\ell \simeq 1$. Usually at this point $\omega\tau \simeq q\ell(v_s/v) << 1$.

In this case electron collisions must be taken into account explicitly and for calculations one must use the kinetic equation approach (2.47)-(2.49) with the collision integral $\hat{J}\{f_p\} \neq 0$. Let us for simplicity put $\mathbf{H} = 0$ (i.e. no d.c. magnetic field). Then the kinetic equation (2.49) is:

$$\frac{\partial f_\mathbf{p}}{\partial t} + \mathbf{v}\nabla_\mathbf{r} f_\mathbf{p} + e\mathbf{E}\nabla_\mathbf{p} f_\mathbf{p} = \hat{J}\{f_\mathbf{p}\} \tag{2.58}$$

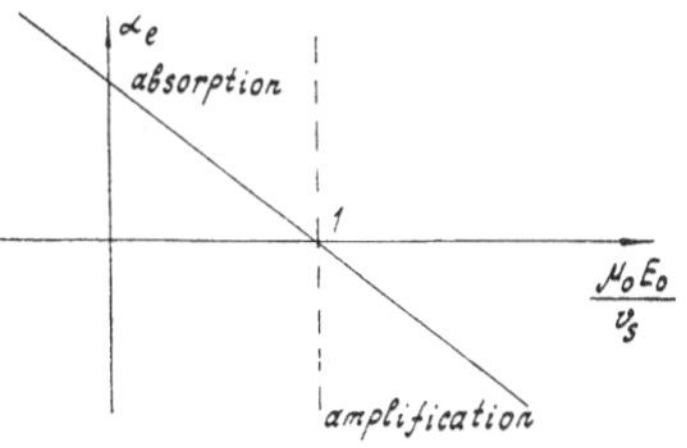


Fig. 5. The variation of the electronic attenuation coefficient $\alpha_e$ with drift electric field $E_0$ in the collisionless case.

Let us put

$$f = F + f^h + \psi \tag{2.59}$$

where

$$F = f_0 - eE_0 v\tau \frac{\partial f_0}{\partial \epsilon} \tag{2.60}$$

is the electron distribution function in the d.c. drift electric field $E_0$. $f_0$ is the quasiequilibrium Fermi distribution function in the presence of the acoustic wave

$$f_0 = \frac{1}{e^{(\epsilon-\xi)/(k_B T)} + 1} \qquad \xi = \xi_0 + \xi_1(x,t) \tag{2.61}$$

By $f^h$ we denote that part of the electron distribution function which leads to the results of the hydrodynamic White's theory. So if $f = F + f^h$, then from the kinetic equation we obtain $K^h(\mathbf{q},\omega)$, which, when substituted into Eq. (2.52), gives the well known White's theory expression for the electronic attenuation coefficient $\alpha_e$. $\psi$ is that part of the electron distribution function that gives the deviation from White's theory.

By solving the kinetic equation (2.58) in the hydrodynamic case one easily obtains $f^h$ in the form

$$f^h = f^h_+ + f^h_- \tag{2.62}$$

where the symmetric and antisymmetric parts are:

$$f^h_+ = -\xi^{(1)} \frac{\partial f_0}{\partial \epsilon} \tag{2.63}$$

$$f^h_- = -iqv\tau \left\{ (\xi^{(1)} + e\phi) \frac{\partial f_0}{\partial \epsilon} + eEv\tau\xi^{(1)} \frac{\partial^2 f_0}{\partial \epsilon^2} \right\} \tag{2.64}$$

Here

$$\xi^{(1)} = -\frac{q^2 r_D^2}{(q^2 r_D^2 - i\omega'\tau_M)} e\phi \qquad \omega' = \omega - qv_d \tag{2.65}$$

As already pointed out, by using $f = F + f^h$ from Eqs. (2.60) and (2.63)-(2.64) one obtains the results of White's theory. Now the question is, under what conditions is the contribution of $\psi$ (which is, generally speaking, not equal to zero!) negligibly small, so that one can use the formulae of White's theory?

Let us also divide $\psi$ into symmetric and antisymmetric parts:

$$\psi = \psi_+ + \psi_- \tag{2.66}$$

Then for $\psi_+$ and $\psi_-$ we obtain from the kinetic equation the following equations:

$$-i\omega\psi_+ + iqv\psi_- + eE\frac{\partial \psi_-}{\partial p} + \hat{J}\{\psi_+\} = \Pi_+ \tag{2.67}$$

$$-i\omega\psi_- + iqv\psi_+ + eE\frac{\partial \psi_+}{\partial p} + \hat{J}\{\psi_-\} = \Pi_- \tag{2.68}$$

where

$$\Pi_+ = i\omega f_+^h - i\mathbf{q}\mathbf{v}f_-^h + \left\{\frac{\xi^{(1)}}{k_B T} + 2\frac{E_-}{E_0} + \frac{E_-}{E_0}\frac{\xi^{(1)}}{e\phi}\right\}\hat{\int}\{F\} \tag{2.69}$$

$$\Pi_- = i\omega f_-^h \tag{2.70}$$

$$E_- = -\nabla\phi$$

By using the approximations $\partial/\partial p \simeq 1/p$, $\hat{\int}\{\psi_-\} \simeq \psi_-/\tau$ and $\hat{\int}\{\psi_+\} \simeq \psi_+/\tau_\epsilon$ we have from Eqs. (2.67)-(2.70):

$$\psi_+ \simeq \frac{\Pi_+\tau_\epsilon - (iq\ell + v_d/v_T)\Pi_-\tau}{1 + q^2\mathcal{D}\tau_\epsilon - i\omega'\tau_\epsilon} \tag{2.71}$$

and a similar expression for $\psi_-$. Further, by putting

$$K(\mathbf{q},\omega) = K^h(\mathbf{q},\omega) + K^\psi(\mathbf{q},\omega) \tag{2.72}$$

we have

$$K^\psi(\mathbf{q},\omega) \simeq \frac{\left\{i\omega'\tau_\epsilon q^2 r_D^2 - q^2 r_D^2\left(\frac{v_d}{v_T}\right)^2\frac{\tau_\epsilon}{\tau} - qv_d\tau_\epsilon\omega'\tau_M\right\}\frac{n_0}{k_B T}}{(1 + q^2\mathcal{D}\tau_\epsilon - i\omega\tau_\epsilon)(q^2 r_D^2 - i\omega'\tau_M)} \tag{2.73}$$

Now the conditions of applicability of White's theory must obviously be

$$|Im K^\psi| << |Im K^h| \qquad |Re K^\psi| << |Re K^h| \tag{2.74}$$

or, after evaluation,

$$q\sqrt{\mathcal{D}\tau_\epsilon} << 1 \qquad qv_d\tau_\epsilon << 1 \qquad \omega\tau_\epsilon << 1 \tag{2.75}$$

which coincides with conditions (2.17), obtained from intuitive considerations. How should the kinetic equation (2.58) be solved when $\hat{\int}\{f_\mathbf{p}\}$ is not small and conditions (2.75) are not satisfied?

For $q\ell \geq 1$, $f_\mathbf{p}$ has a pole structure

$$f_\mathbf{p} \simeq \frac{1}{\left[\left(\cos\theta - \frac{mv_s}{\hbar q}\right) - \frac{i}{q\ell}\right]} \qquad \theta = \widehat{\mathbf{p},\mathbf{q}} \tag{2.76}$$

and the approximation of the collision integral,

$$\hat{\int}\{f_\mathbf{p}\} \simeq -\frac{1}{\tau}[f_\mathbf{p} - f_{\mathbf{p}_0}] \tag{2.77}$$

is too inaccurate. One must take into account all higher harmonics of $f_\mathbf{p}$ since they are not small:

$$f_\mathbf{p} = f_s(\epsilon_\mathbf{p}) + \sum_{n=1}^{\infty} f_n(\epsilon_\mathbf{p})P_n(\cos\theta) \qquad \cos\theta = \frac{p_x}{p} \qquad \mathbf{q} \parallel OX \tag{2.78}$$

Thus for the collision integral one has, generally speaking,

$$\hat{J}\{f_{\mathbf{p}}\} = \hat{J}_s\{f_s(\epsilon_{\mathbf{p}})\} - \sum_{n=1}^{\infty} \frac{f_n(\epsilon_{\mathbf{p}})P_n(\cos\theta)}{\tau_n(\epsilon_{\mathbf{p}})} \tag{2.79}$$

where each harmonic $f_n(\epsilon_{\mathbf{p}})$ has its own relaxation time $\tau_n(\epsilon_{\mathbf{p}})$ which is expressed in terms of the scattering transition probability $W_n(\epsilon_{\mathbf{p}})$;

$$W(|\mathbf{p} - \mathbf{p}'|)_{p=p'} = \sum_n W_n(\epsilon_{\mathbf{p}})\cdot P_n(\cos[\hat{\mathbf{p},\mathbf{p}'}]) \tag{2.80}$$

Bearing these considerations in mind, we write

$$f_{\mathbf{p}} = F + f_{\mathbf{p}}^{(1)} \tag{2.81}$$

and linearise the kinetic equation (2.58) in $f_{\mathbf{p}}^{(1)}$. This is possible if the conditions opposite to those necessary for the applicability of White's theory (2.75) are satisfied:

$$q\sqrt{\mathscr{D}\tau_\epsilon} >> 1 \qquad qv_d\tau_\epsilon >> 1 \qquad q^2\mathscr{D}\tau_\epsilon >> \omega\tau_\epsilon \tag{2.82}$$

Under these conditions we have:

$$\frac{\partial f_{\mathbf{p}}^{(1)}}{\partial t} + v_x \frac{\partial f_{\mathbf{p}}^{(1)}}{\partial x} - \hat{J}\{f_{\mathbf{p}}^{(1)}\} = e\nabla\phi \frac{\partial F}{\partial \mathbf{p}} \tag{2.83}$$

where

$$\hat{J}\{f_{\mathbf{p}}^{(1)}\} = -\frac{f_{\mathbf{p}}^{(1)}}{\tau_0} + \sum_{\mathbf{p}'} W(\mathbf{p},\mathbf{p}')f_{\mathbf{p}'}^{(1)} \tag{2.84}$$

$$\tau_0^{-1} = \sum_{\mathbf{p}'} W(\mathbf{p},\mathbf{p}') \tag{2.85}$$

$$F = f_0 - e\mathbf{E}_0\mathbf{v}\tau \frac{\partial f_0}{\partial \epsilon} \tag{2.60 bis}$$

For elastic and isotropic scattering one obtains

$$\tau_0 \sum_{\mathbf{p}'} W(\mathbf{p},\mathbf{p}')f_{\mathbf{p}'}^{(1)} = \frac{\int dO' S(\theta',\phi';\theta,\phi) f_{\mathbf{p}}^{(1)}(\theta')}{\int dO' S(\theta',\phi';\theta,\phi)} \tag{2.86}$$

where $S(\theta',\phi';\theta,\phi)$ is the corresponding scattering cross-section, angles $\theta$ and $\phi$ are related to $\mathbf{p}$, $\theta'$ and $\phi'$ to $\mathbf{p}'$, and we have used $f_{\mathbf{p}} \sim e^{i(qx-\omega t)}$. The kinetic equation (2.83) then reduces to the linear Fredholm integral equation

$$f_{\mathbf{p}}^{(1)}(z) = \lambda \int_{-1}^{1} d\xi K(z,\xi) f_{\mathbf{p}}^{(1)}(\xi) + \Phi(z) \tag{2.87}$$

Here $z = \cos\theta$, $\xi = \cos\theta'$,

$$K(z,\xi) = -\left(\frac{1}{W_0 - z}\right)\tilde{S}(z,\xi) \tag{2.88}$$

$$\tilde{S}(z,\xi) = \tilde{S}(\cos\psi) = \frac{1}{2\pi}\int_0^{2\pi} d\phi \int_0^{2\pi} d\phi' S(\theta',\phi';\theta,\phi) \tag{2.89}$$

$$\psi = (\hat{\mathbf{p},\mathbf{p}'})$$

$$\Phi(z) = -\left(\frac{e\phi}{W_0 - z}\right)\left[\left(z - \frac{v_d^\epsilon}{v}\right)\frac{\partial f_0}{\partial \epsilon} - m v_d^\epsilon v z \frac{\partial^2 f_0}{\partial \epsilon^2} - \nu \frac{m v_d^\epsilon v}{\epsilon} z^2 \frac{\partial f_0}{\partial \epsilon}\right] \tag{2.90}$$

$$W = \frac{i\omega\tau - 1}{iq\ell} \qquad W_0 = \frac{i\omega\tau_0 - 1}{iq\ell_0} \qquad \ell = v\tau \qquad \ell_0 = v\tau_0 \tag{2.91}$$

$$v_d^\epsilon \equiv v_d(\epsilon) = \frac{eE_0\tau(\epsilon)}{m^*} \qquad \tau(\epsilon) \sim \epsilon^\nu \tag{2.92}$$

$$\lambda = \frac{2\pi}{iq\ell_0 \int dO' S(\theta',\phi';\theta,\phi)} \tag{2.93}$$

We have already suggested that for typical scattering mechanisms $\tau(\epsilon) \sim \epsilon^\nu$. For this case Eq. (2.87) can be solved explicitly (analytically). Here

$$\tilde{S}(\psi) = \frac{(1 - \cos\psi)^{3/2-\nu}}{[1 + \kappa(\epsilon) - \cos\psi]^2} \tag{2.94}$$

where

$$\kappa(\epsilon) = \frac{\hbar^2}{4m^* \epsilon r_D^2}$$

Substituting into Eqs. (2.88) and (2.89) one obtains:

$$n = K(\mathbf{q},\omega)e\phi \tag{2.95}$$

$$K(\mathbf{q},\omega) = \frac{3N_0}{2}\frac{(2m^*)^{3/2}}{3\pi^2\hbar^3 N_0}\int_0^\infty d\epsilon\sqrt{\epsilon}\,\frac{\partial f_0}{\partial \epsilon} \times \left[1 - \frac{Q_0(W_0)i\omega'\tau_0}{Q_0(W_0) + iq\ell_0 + h(\epsilon)} + \frac{v_d^\epsilon}{v}\frac{1}{iq\ell}\left(\nu - \frac{1}{2}\right)\right] \tag{2.96}$$

Here

$$h(\epsilon) = iq\ell_0[-1 + Q_0(W_0)(W_0 + B)] \tag{2.97}$$

where $Q_0(W_0)$ is a Legendre function of the second kind and B the infinite continued fraction:

$$B = \cfrac{1}{\beta_{11} - \cfrac{2^2}{\beta_{22} - \cfrac{3^2}{\beta_{33} - \dots}}} \tag{2.98}$$

$$\beta_{KK} = -(2K + 1)W_K \qquad W_K = \frac{i\omega\tau_K - 1}{iq\ell_K} \tag{2.99}$$

$$\tau_K^{-1} = \tau_0^{-1}\left(1 - \frac{\int_{-1}^{1} d\xi\, S(\xi)P_K(\xi)}{\int_{-1}^{1} d\xi\, S(\xi)}\right) \equiv \tau_0^{-1}(1 - \lambda_K) \tag{2.100}$$

$$\ell_K = v\tau_K \qquad \tau_1 = \tau_\epsilon \tag{2.101}$$

Finally, with the help of Eqs. (2.95)-(2.101) one obtains:

$$\alpha_e = K^2 q \frac{q^2 r_D^2 \langle \frac{1}{\epsilon}\rangle^{-1} \langle \frac{1}{\epsilon}\beta'' \left[1 - \frac{v_d^\epsilon}{v_s}\left(1 - \frac{v_s}{v}\frac{(\nu - \frac{1}{2})}{q\ell\beta''}\right)\right]\rangle}{\left(\langle \frac{1}{\epsilon}\rangle^{-1}\langle \frac{1}{\epsilon}(1 - \beta')\rangle + q^2 r_D^2\right)^2 + \left(\langle \frac{1}{\epsilon}\rangle^{-1} \langle \frac{1}{\epsilon}\beta'' \left[1 - \frac{v_d^\epsilon}{v_s}\left(1 - \frac{v_d^\epsilon}{v_s}\frac{(\nu - \frac{1}{2})}{q\ell\beta''}\right)\right]\rangle\right)^2} \tag{2.102}$$

where

$$\beta = \beta' + i\beta'' = \frac{1}{1 + \frac{v}{v_s}B} \tag{2.103}$$

$$< \dots > = \frac{(2m)^{3/2}}{3\pi^2\hbar^3 N_0}\int_0^\infty ( \dots ) \frac{\partial f_0}{\partial \epsilon}\epsilon^{3/2} d\epsilon \tag{2.104}$$

The criterion for the amplification of acoustic waves by electron drift takes the form

$$v_d > \frac{v_s}{r} \tag{2.105}$$

where

$$r = \frac{\langle \frac{1}{\epsilon}\tau(\epsilon)\beta''(\epsilon)\left(1 - \frac{v_s}{v}\frac{(\nu - \frac{1}{2})}{q\ell(\epsilon)\beta''(\epsilon)}\right)\rangle}{\langle \tau(\epsilon)\rangle\langle \frac{1}{\epsilon}\beta''(\epsilon)\rangle} \tag{2.106}$$

Physically, $r \neq 1$ means that at different frequencies and for different scattering mechanisms the main contribution to electronic attenuation (amplification) of acoustic waves comes from energetically different groups of electrons. Their "partial" drift velocities

$$v_d^\epsilon \equiv v_d(\epsilon) = \frac{e\tau(\epsilon)}{m^*}E_0 \tag{2.107}$$

are, generally speaking, different from one another and from the mean electron drift velocity $v_d$.

Formulae (2.102)-(2.107) solve the problem in a general way. For each particular electron scattering mechanism the infinite fraction $B(\epsilon)$ can be summarized. For example, for optical phonon scattering below the Debye temperature ($\nu = 0$) one obtains

$$B(\epsilon) = \cfrac{1}{x + \cfrac{2^2}{2x + \cfrac{3^2}{3x + \ldots}}}$$

$$= \frac{i}{2\sqrt{1 + x^2}\cdot F(1, -p, 1-p, -1/n)} \tag{2.108}$$

where

$$x = \frac{1}{q\ell_0} - iO \tag{2.109}$$

$$p = p(x) = -\frac{1}{2} + \frac{1}{2}\frac{x}{\sqrt{1 + x^2}} \tag{2.110}$$

$$n = \left(x + \sqrt{1 + x^2}\right)^2 \tag{2.111}$$

and F is the Gauss hyper-geometrical function,

$$1 \geq F \geq \frac{\pi}{4} \tag{2.112}$$

Thus, one obtains from Eq. (2.102) the following expression for $\alpha_e$:

$$\alpha_e = K^2 q \frac{2q^2 r_D^2}{(1 + q^2 r_D^2)^2} \langle \frac{1}{\epsilon} \rangle^{-1}$$

$$\times \langle \frac{1}{\epsilon} \frac{v_s}{v} \sqrt{1 + \frac{1}{q^2\ell_0^2}} \left(1 - \frac{v_d^\epsilon}{v_s}\right) F\left(1; \frac{\sqrt{1 + q^2\ell_0^2} - 1}{2\sqrt{1 + q^2\ell_0^2}}; \frac{3\sqrt{1 + q^2\ell_0^2} - 1}{2\sqrt{1 + q^2\ell_0^2}}; \frac{q^2\ell_0^2}{1 + \sqrt{1 + q^2\ell_0^2}}\right) \rangle \tag{2.113}$$

Other scattering mechanisms (e.g. $\nu = -1/2$ corresponding to acoustic phonon scattering due to the deformation potential; $\nu = 1/2$ corresponding to optical phonon scattering above the Debye temperature, or acoustic phonon scattering due to the piezoelectric potential; $\nu = 3/2$ corresponding to ionized impurity scattering, etc.) can be considered in a similar manner.

This theory describes satisfactorily all known experimental data concerning electronic acoustic wave attenuation and amplification for intermediate (and high!) frequencies $q\ell \lessgtr 1$, including the region $q\ell \sim 1$.

### 2.4 Surface acoustic wave case

In the uniform piezoelectric semiconductor, the interaction of surface acoustic waves with electrons can be treated in a manner very similar to that used in the previous sections, so we shall not cover this topic explicitly. A much more interesting case is that of a layered piezoelectric-semiconductor structure (see Fig. 6), which is of some practical importance. Here piezoelectric and semiconductor media are separated and so can be optimized more effectively.

To find the electronic surface acoustic wave attenuation (amplification) coefficient $\alpha_e$, one must write the equations corresponding to (2.19)-(2.24) for each medium and then use the relevant boundary conditions.

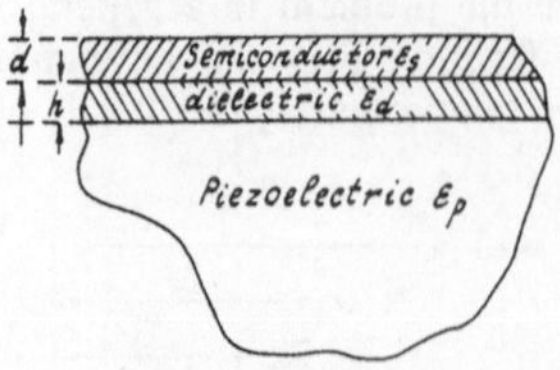


Fig. 6. A layered structure.

For the hydrodynamic case and a thin dielectric layer

$$h \ll \lambda_{SAW} \tag{2.114}$$

one obtains:

$$\alpha_e = -\frac{K^2 q}{2}\left(\frac{\epsilon_p}{\epsilon_p + 1}\right)\frac{\gamma \frac{\omega_c}{\omega}\left(\frac{\epsilon_s}{\epsilon_p + 1}\right)\tanh q_R d}{\gamma^2 + \left[F + \frac{\omega_c}{\omega}\frac{\epsilon_s}{(\epsilon_p + 1)}\tanh q_R d\right]^2} \tag{2.115}$$

Here $\gamma = (1 - v_d/v_s)$ is a drift parameter, $v_R$ and $q_R$ are the velocity and wavevector of the Rayleigh surface acoustic waves respectively, $F = F(q_R, d, \sigma)$ is a diffusion factor, $\omega_c = \sigma/(\epsilon_0\epsilon_s)$, where $\sigma$ is the conductivity of the semiconductor, $v_d$ is the electron drift velocity,

$$K^2 = 2\left(\frac{v_0 - v_\infty}{v_0}\right)F_1(q_R, h, d, \epsilon_p, \epsilon_d, \epsilon_s)$$

is the effective electromechanical coupling constant, $v_0$ and $v_\infty$ are the surface acoustic wave velocities in the piezoelectric medium under consideration along the free and metallized surfaces respectively, and $F_1$ is a factor of the order of unity. The excellent agreement between the experimental data for a $LiNbO_3$-$SiO_x$-InSb surface acoustic wave amplifier (Fig. 7) and the theoretical calculations of formula (2.115) shows the relevance of the theory (see Figs. 8 and 9).

To conclude this section, we can say that the theory of electronic attenuation and amplification of acoustic waves in solids is well developed for different structures and types of waves, different material parameters and different frequencies and that it agrees reasonably well with all known experimental results.

## 3. EFFECTS RELATED TO THE DRAG OF ELECTRONS BY ACOUSTIC WAVES IN SEMICONDUCTORS

An acoustic wave propagating in a semiconductor and loosing its energy due to electronic attenuation can also loose some of its momentum. This momentum is transferred to electrons thus producing a d.c. "drag" force on the electrons. This drag force may, in principle, produce d.c. kinetic phenomena similar to those produced by a temperature gradient $\nabla T$ or an electron concentration gradient $\nabla n$. Some examples are: a d.c. longitudinal current (or voltage in the open circuit case), which is called the acoustoelectric effect; a d.c. transverse current (or voltage) in a transverse d.c. magnetic field, which could be called the acoustomagnetoelectric effect; etc. This chapter is devoted to a consideration of some of these effects.

### 3.1 Acoustoelectric effect

The magnitude of the acoustoelectric current or voltage in the simplest one dimensional case of an isotropic semiconductor (see Fig. 10) can be obtained from the momentum conser-

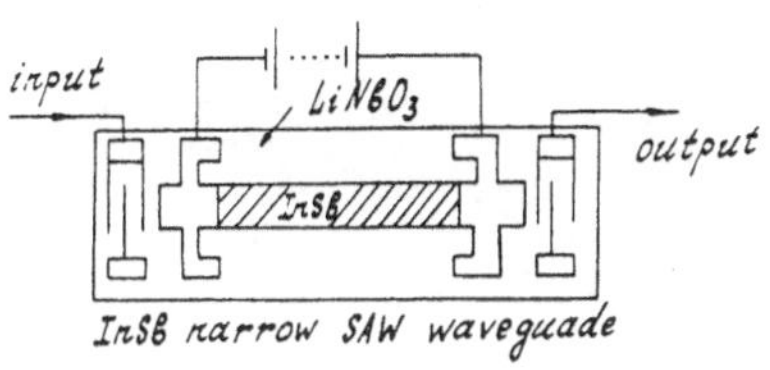

Fig. 7. The experimental configuration for cw operation.

vation law. Let W be the acoustic wave intensity (energy flux), measured in W/cm². The energy and momentum conservation laws for the interaction of acoustic waves with electrons then give

$$\frac{1}{v_s}\frac{\partial W}{\partial x} = -\alpha_e \frac{W}{v_s} \tag{3.1}$$

$$-\frac{1}{v_s}\alpha_e W = -m^* v_d \nu N_0 \tag{3.2}$$

Here $\nu$ is the frequency of electron collisions, $\nu = \tau^{-1}$. Introducing the mobility of the electrons $\mu = e\tau/m^* = e/(m^*\nu)$, we have from Eq. (3.2)

$$-\frac{1}{v_s}\alpha_e W = -\frac{e v_d N_0}{\mu} \equiv \frac{j_{ae}}{\mu} \tag{3.3}$$

From here we immediately obtain the acoustoelectric current $j_{ae}$

$$j_{ae} = \frac{\mu}{v_s}\alpha_e W \tag{3.4}$$

which is the well known Weinreich relation for the short-circuit case.[8]

For the open-circuit case by putting $j_{ae} = \sigma E_{ae} = e\mu N_0 E_{ae}$, we obtain the well known Weinreich relation for the acoustoelectric field,

$$E_{ae} = \frac{\alpha_e W}{e N_0 v_s} \tag{3.5}$$

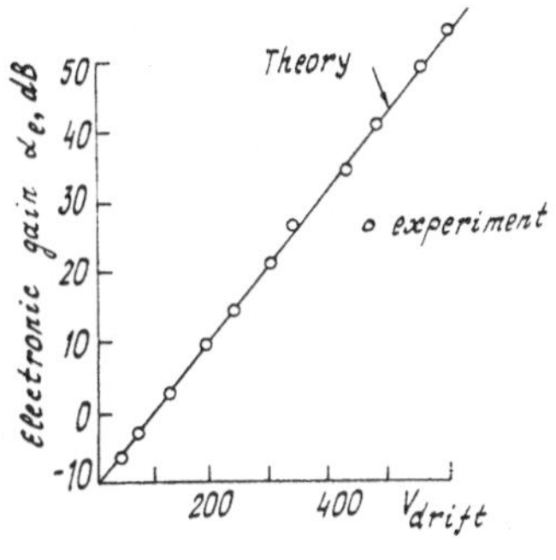

Fig. 8. Electronic gain versus drift voltage for a $LiNbO_3$-$SiO_x$-InSb structure. The frequency is 200 MHz, mobility 3000 cm²/Vs, conductivity $4.4\times10^{-6}$ ohm$^{-1}$ cm$^{-1}$ and length 1 cm.

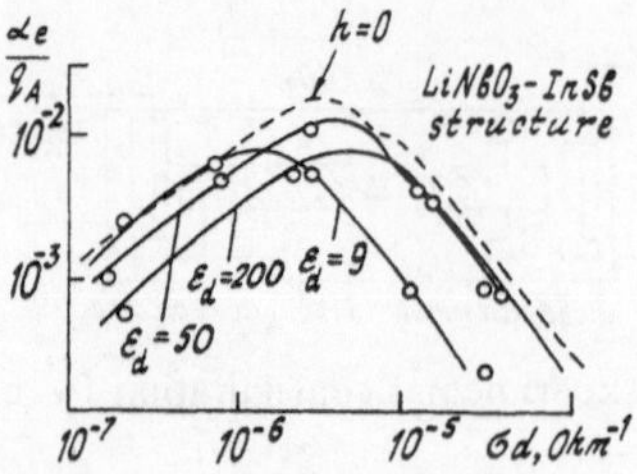


Fig. 9. Electronic attenuation of surface acoustic waves as a function of conductivity for a $LiNbO_3$-$SiO_x$-InSb structure. $q_R h = 0.07$, $K^2 = 0.048$, $\epsilon_p = 50$.

or, in the case of weak acoustic wave attenuation, for the acoustoelectric voltage,

$$V_{ae} = \frac{\alpha_e W}{eN_0 v_s} L \tag{3.6}$$

Usually such simple Weinreich relations are valid only in particular cases. Indeed, for example in the hydrodynamic case, acoustic waves interact not with individual electrons but with electron bunches and, hence, drag these bunches. However, the drift velocity of the electron bunches, which is used in the derivation of the Weinreich relations (3.3)-(3.4), is different, as was already shown above, from the average drift velocity of the electrons by the factor r [see Eqs. (2.34)-(2.35)]. Therefore, strictly speaking, one must substitute $v_d^* = rv_d$ instead of $v_d$ in the Weinreich relations (3.4)-(3.6), r being the same quantity as in the amplification condition $v_d r > v_s$, Eq. (2.43). So the correct Weinreich relations are

$$j_{ae} = \frac{\mu}{v_s} \alpha_e W r \tag{3.7}$$

$$E_{ae} = \frac{\alpha_e W}{eN_0 v_s} r \tag{3.8}$$

$$V_{ae} = \frac{\alpha_e W}{eN_0 v_s} L r \tag{3.9}$$

where

$$r = \frac{\partial ln\sigma}{\partial lnN_0} = 1 + \frac{d ln\mu_0}{d lnN_0}$$

and, for a degenerate semiconductor

$$r = 1 + \frac{2}{3}\frac{d ln\tau(\xi)}{d ln\xi}$$

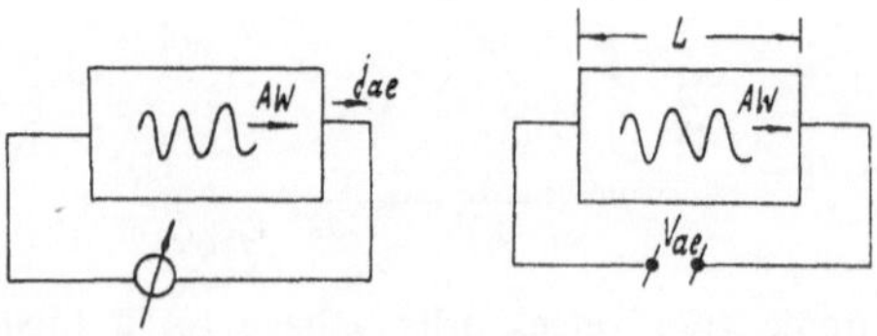


Fig. 10. Measurement of the acoustoelectric current $j_{ae}$ and acoustoelectric voltage $V_{ae}$ of an isotropic semiconductor.

The relations (3.7)-(3.9) follow, of course, from the strict theory (see below). Let us consider first the phenomenological theory for the hydrodynamic case in a one dimensional isotropic configuration.[9]

The current j in the presence of acoustic waves is given by the expression

$$j_- = \sigma\left(-\frac{\partial\phi}{\partial x}\right) - e\mathcal{D}\frac{\partial n}{\partial x} \tag{3.10}$$

[compare with Eq. (2.28) when $\Lambda = 0$, $E_0 = 0$].

$$\phi, n \sim u_0 e^{i(\omega t - qx)} \tag{3.11}$$

where $u_0$ is the acoustic wave amplitude.

$$\sigma = \sigma_0 + \frac{\partial\sigma_0}{\partial N_0}n \qquad \sigma_0 = e\mu_0(N_0)N_0$$

$$\frac{\partial\sigma_0}{\partial N_0} = \frac{\sigma_0}{N_0}\frac{\partial ln\sigma}{\partial lnN_0} \equiv \frac{\sigma_0}{N_0}r \tag{3.12}$$

So, from Eq. (3.10) one has:

$$j_- = -\sigma_0\frac{\partial\phi}{\partial x} - \frac{\sigma_0}{N_0}rn\frac{\partial\phi}{\partial x} - e\mathcal{D}_0\frac{\partial n}{\partial x} \tag{3.13}$$

We are interested in the d.c. part of this current which may only come from the nonlinear term

$$-\frac{\sigma_0}{N_0}rn\frac{\partial\phi}{\partial x}$$

So averaging over the period of the acoustic waves we have

$$\overline{j_-} \equiv j_{ae} = -r\frac{\sigma_0}{N_0}\overline{n\frac{\partial\phi}{\partial x}} = -r\frac{\sigma_0}{N_0}\frac{1}{2}Re\left[n\left(\frac{\partial\phi^*}{\partial x}\right)\right] \tag{3.14}$$

From the basic equations (2.25)-(2.28), one has:

$$\frac{\partial\phi}{\partial x} = \frac{4\pi\beta}{\epsilon}(iqu)\frac{q^2r_D^2 - i\omega\tau_M\left(1 - \frac{v_d r}{v_s}\right)}{\left[(1 + q^2r_D^2) - i\omega\tau_M\left(1 - \frac{v_d r}{v_s}\right)\right]} \tag{3.15}$$

$$\eta = \frac{\beta q^2}{e}n\frac{1}{\left[(1 + q^2r_D^2) - i\omega\tau_M\left(1 - \frac{v_d r}{v_s}\right)\right]} \tag{3.16}$$

Introducing the acoustic wave intensity,

$$W = \frac{c\omega}{q}\overline{\left(\frac{\partial u}{\partial x}\right)^2} \tag{3.17}$$

from Eqs. (3.14)-(3.16) we obtain

$$j_{ae} = \frac{\mu_0}{v_s}\alpha_e W r \tag{3.18}$$

where

$$r = \frac{d\ln\sigma_0}{d\ln N_0} = \frac{1}{e\mu_0}\frac{\partial\sigma_0}{\partial N_0} \tag{3.19}$$

Formula (3.18) is just the "corrected" Weinreich relation (3.7) which we derived above from qualitative momentum conservation considerations. The other Weinreich relations, for example

$$E_{ae} = \frac{\alpha_e W}{eN_0 v_s} r$$

follow immediately.

Let us give some numerical estimates. From (3.8), we have the following order of magnitude approximation:

$$E_{ae} \simeq \frac{K^2 q W}{eN_0 v_s} \tag{3.20}$$

For InSb with $N_0 = 10^{12}$ cm$^{-3}$, $v_s = 3\times10^5$ cm/s, $K^2 = 10^{-2}$ and $C = 10^{12}$ dyn/cm$^2$ and acoustic waves with $W = 1$ mW, $f = 200$ MHz ($q = 4.2\times10^3$ cm$^{-1}$, $\omega = 1.26\times10^9$ s$^{-1}$), we have $qu_0 = \{W/Cv_s\}^{1/2} = 1.8\times10^{-7}$ and hence from Eq. (3.20) $E_{ae} \simeq 1$ V/cm.

## 3.2 Acoustomagnetoelectric, acoustothermal and other drag effects

Let us first look at Fig. 11. As already shown for this geometry, there will either be an acoustoelectric current (in the short-circuit case) or an acoustoelectric voltage (in the open-circuit case) between contacts 1 and 2. In addition, acoustic waves, by exerting a sort of dragging force on the electrons, may produce other "drag" effects. We now consider some of these.

(i) *Acoustoelectric Hall effect*. This effect appears when contacts 1 and 2 are short-circuited and the acoustoelectric current, $j_{ae}$ flows in the X-direction. Hence in a transverse magnetic field **H** there will obviously appear a Hall voltage between contacts 3 and 4 (or a Hall current will flow if contacts 3 and 4 are short-circuited). The expression for the "acoustoelectric Hall effect" may be easily obtained in the simplest case from the usual Hall effect formulae where the ordinary electric current must be replaced by the acoustoelectric current, $j_{ae}$.

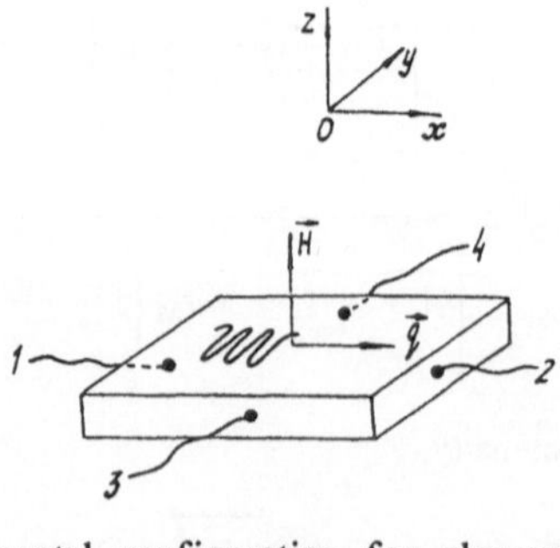


Fig. 11. Experimental configuration for observation of acousto-magnetoelectric, acoustothermal and other drag forces.

(ii) *Acoustomagnetoelectric effect.* By definition this effect is the appearance of a transverse voltage between contacts 3 and 4 in the presence of a transverse magnetic field **H**, when the contacts 1 and 2 are open-circuited so that there can be no longitudinal current. In order to obtain the magnitude of the acoustomagnetoelectric effect, let us consider the phenomenological expressions for the electric current averaged over the period of the acoustic waves in the presence of both longitudinal electric and transverse magnetic fields [compare Eq. (3.14)]:[10]

$$j_{ae}^{(x)} = \overline{\sigma_{xx} E_x} + \overline{\sigma_{xy} E_y} - \overline{e\mathscr{D}_{xx} \frac{\partial n}{\partial x}} = \sigma_{xx}^{(0)} E_{ae}^{(x)} + \sigma_{xy}^{(0)} E_{ae}^{(y)} + \frac{\partial \sigma_{xx}^{(0)}}{\partial N_0} \overline{n\left(-\frac{\partial \phi}{\partial x}\right)} \tag{3.21}$$

$$j_{ae}^{(y)} = -\overline{\sigma_{xy} E_x} + \overline{\sigma_{xx} E_y} + \overline{e\mathscr{D}_{xy} \frac{\partial n}{\partial x}} = -\sigma_{xy}^{(0)} E_{ae}^{(x)} + \sigma_{xx}^{(0)} E_{ae}^{(y)} - \frac{\partial \sigma_{xy}^{(0)}}{\partial N_0} \overline{n\left(-\frac{\partial \phi}{\partial x}\right)} \tag{3.22}$$

Remembering that [see Eqs. (3.14)-(3.18)]

$$-\overline{n\frac{\partial \phi}{\partial x}} = -\frac{1}{2} Re\left[n\left(\frac{\partial \phi}{\partial x}\right)^*\right] = -\frac{\alpha_e W}{e v_s} \tag{3.23}$$

we have

$$j_{ae}^{(x)} = \sigma_{xx}^{(0)} E_{ae}^{(x)} + \sigma_{xy}^{(0)} E_{ae}^{(y)} - \frac{\partial \sigma_{xx}^{(0)}}{\partial N_0} \frac{\alpha_e W}{e v_s} \tag{3.24}$$

$$j_{ae}^{(y)} = -\sigma_{xy}^{(0)} E_{ae}^{(y)} + \sigma_{xx}^{(0)} E_{ae}^{(x)} + \frac{\partial \sigma_{xy}^{(0)}}{\partial N_0} \frac{\alpha_e W}{e v_s} \tag{3.25}$$

The components of the conductivity tensor $\sigma_{ik}$ are well known and are given by:

$$\sigma_{xx}^{(0)} = \frac{e^2 N_0}{m^*}\left\langle \frac{\tau}{1 + \omega_H^2 \tau^2} \right\rangle \qquad \sigma_{xy}^{(0)} = \frac{e^2 N_0}{m^*}\left\langle \frac{\omega_H \tau^2}{1 + \omega_H^2 \tau^2} \right\rangle \tag{3.26}$$

where the symbol < ... > was given by Eq. (2.104).

For an open-circuited specimen

$$j_{ae}^{(x)} = 0 \qquad j_{ae}^{(y)} = 0 \tag{3.27}$$

By using here Eqs. (3.24) and (3.25) we immediately have:

$$E_{ae}^{(x)} \equiv E_{ae} = \frac{\alpha_e W}{e N_0 v_s} r_{\|} \tag{3.28}$$

$$E_{ae}^{(y)} \equiv E_{AME} = \frac{\alpha_e W}{e N_0 v_s} r_{\perp} \tag{3.29}$$

where the factors $r_{\|}$ and $r_{\perp}$ are:

$$r_{\|} = \frac{1}{2} \frac{\partial ln(\sigma_{xx}^{(0)2} + \sigma_{xy}^{(0)2})}{\partial ln N_0} \tag{3.30}$$

$$r_{\perp} = -\frac{\sigma_{xx}^{(0)} \sigma_{xy}^{(0)}}{(\sigma_{xx}^{(0)2} + \sigma_{xy}^{(0)2})} \frac{\partial ln(\sigma_{xy}^{(0)}/\sigma_{xx}^{(0)})}{\partial ln N_0} \tag{3.31}$$

One can easily see that in the absence of a magnetic field ($\sigma_{xy}^{(0)} = 0$, $\sigma_{xx}^{(0)} = \sigma_0$)

$$r_{\parallel} = \frac{\partial ln\sigma_0}{\partial lnN_0} \equiv r$$

introduced above, and $r_{\perp} = 0$, so $E_{AME} = 0$ and $E_{ae}$ is given by the Weinreich relation (3.8).

For a degenerate electron gas, putting $\tau(\epsilon) \sim \epsilon^{\nu}$ we have for the acoustomagnetoelectric field:

$$E_{AME} = \frac{\alpha_e W}{eN_0 v_s} \frac{2}{3}\nu \frac{\mu_0 H/c}{[1 + (\mu_0 H/c)^2]} \tag{3.32}$$

where c is the velocity of light.

It can be seen that for different electron scattering mechanisms, the acoustomagnetoelectric effect may have a different sign (for example, $\nu = 1/2$ for acoustic phonon scattering and $\nu = -3/2$ for impurity scattering). So the acoustomagnetoelectric effect may be a very good tool for determining electron scattering mechanisms in semiconductors, which is very important in semiconductor electronics.

The physical reason for the appearance of the acoustomagnetoelectric effect is the following. The "drag forces" on electrons of different energies, caused by acoustic waves and the compensating acoustoelectric field $E_{ae}^{(x)}$, are, generally speaking, different due to bunching of the electrons and the dependence of their "partial" mobility $\mu(\epsilon) = e\tau(\epsilon)/m^*$ on their energy. The forces do not compensate one another exactly for individual electrons. So, in spite of the absence of a net longitudinal electric current in the X-direction, there exist two currents equal in value and flowing in opposite directions: the acoustoelectric current $j_{ae}$ and a compensating electric current $\sigma_{xx}^{(0)} E_{ae}$. Both currents depend, of course, on the magnetic field. They consist, generally speaking, of electrons of different average energy and with different average mobilities, so the Hall effects on these currents do not compensate each other. The difference in the Hall voltages produced by these two currents gives the acoustomagnetoelectric effect.

To calculate the acoustomagnetoelectric (as well as the acoustoelectric) effect for the collisionless case $q\ell >> 1$, it is necessary to use the kinetic equation approach. Calculations similar to those described above for the electronic attenuation coefficient, give for this case (non-degenerate semiconductor, weak magnetic field):

$$E_{AME} = \frac{\alpha_e W}{eN_0 v_s} \frac{\mu H}{c} \left[ \frac{\Gamma\left(\frac{5}{2}\right)}{\Gamma\left(\frac{5}{2} + \nu\right)} \right]^2 \left\{ \Gamma(1 + 2\nu) - \frac{\Gamma\left(\frac{5}{2} + 2\nu\right)}{\Gamma\left(\frac{5}{2} + \nu\right)} \Gamma(1 + \nu) \right\} \tag{3.33}$$

where $\Gamma$ is the well known $\Gamma$-function and $\tau(\epsilon) \sim \epsilon^{\nu}$. It can be seen that in this case the acoustomagnetoelectric effect can be used to determine electron scattering mechanisms.

(iii) *Longitudinal acoustothermal effect.* As already pointed out, two currents which consist of electrons of different average energies (namely the acoustoelectric current and the current due to the compensating electric field) flow in opposite directions. Under these conditions there exists in the specimen a net heat flux produced by electrons interacting with acoustic waves. If one makes the ends of the specimen thermo-isolated, there appears a compensating longitudinal temperature gradient which we call the longitudinal acoustothermal effect. This effect can be calculated in a similar way to the acoustoelectric and acoustomagnetoelectric effects.[11] For example, for the hydrodynamic case in a degenerate electron gas, one obtains

$$\nabla T = \frac{\alpha_e W}{k_B N_0 v_s} \frac{2}{\pi^2} \frac{\xi_0}{T} \left(1 + \frac{\kappa_L}{\kappa_e}\right)^{-1} \tag{3.34}$$

and for the collisionless case in a non-degenerate electron gas

$$\nabla T = -\frac{\alpha_e W}{k_B N_0 v_s}\frac{g\sqrt{\pi}}{s}\frac{\Gamma(1+\nu)}{\Gamma\left(\frac{7}{2}+\nu\right)}\left(1+\frac{\kappa_L}{\kappa_e}\right)^{-1} \tag{3.35}$$

where $\kappa_L$ and $\kappa_e$ are the lattice and electron heat conductivities, $k_B$ is the Boltzmann constant and $\xi_0$ the Fermi energy.

One may imagine that, due to the existence of the "acoustic wave dragging force" on electrons in semiconductors (or in any other conducting material), there will appear a whole set of kinetic phenomena that are similar to those produced by electric fields, temperature gradients, etc. All these effects have been calculated and they have some very interesting features. For example, Nernst-Ettinghausen and Peltier effects due to acoustoelectric currents are a factor of $(\xi_0/k_B T)^2$ times greater than the same effects when caused by ordinary electric currents. Hence, for a degenerate electron gas they may be very large. Also, in quasilinear theory, the acoustoelectric Peltier coefficient, $\Pi_{ae}$, does not go to zero when $T \rightarrow 0$ as is the case for ordinary electric currents. This means that the Peltier effect due to acoustoelectric currents may produce much deeper cooling than the ordinary Peltier effect.

### 3.3. Acoustoelectric effect on surface acoustic waves

Surface acoustic waves propagating along the surface of a conducting body (semiconductor) will drag electrons, as one may imagine, along the surface (see Fig. 12a). This will lead to the appearance of circular acoustoelectric currents (and hence to a d.c. magnetic momentum of the specimen, induced by the surface acoustic wave). This further leads to the appearance of a transverse acoustoelectric effect via the presence of an alternating transverse electric field. This transverse field is associated with the surface acoustic waves in the piezoelectric material and is comparable in magnitude with the longitudinal field ($E_{z-} = \kappa E_{x-}/iq$, where $\kappa$ is the decay coefficient of the surface acoustic waves in the Z-direction, see Fig. 12a). Of course, there will also be an ordinary longitudinal acoustoelectric effect.

To calculate these effects one must first find the "driving" currents

$$j_{ae}^{\|} = e\mu\overline{nE_{x-}} \tag{3.36}$$

and

$$j_{ae}^{\perp} = e\mu\overline{nE_{z-}} \tag{3.37}$$

These are calculated in a similar manner to that used above. Thereafter, one must find the d.c. current distribution inside the specimen using the boundary conditions. For example, in the hydrodynamic case we have for the transverse acoustoelectric voltage:[12]

$$V_{ae}^{\perp} = \frac{\alpha_e W}{e N_0 v_s} d \frac{F_1}{\omega' \tau_M}\left\{1 + \frac{1}{2\left(1 + 2\sqrt{1+\omega'^2\tau_M^2}\right)} - 2\kappa r_D \frac{1+\sqrt{1+\omega'^2\tau_M^2}}{1+2\sqrt{1+\omega'^2\tau_M^2}}\right\} \tag{3.38}$$

where $\omega' = \omega - qv_d$ and $F_1$ is a factor of the order of unity. It can be seen that the transverse acoustoelectric effect is of the same order of magnitude as the longitudinal one and can change sign depending on the frequency, the parameters of the material and the drift velocity.

There are two physical reasons for the transverse acoustoelectric effect: the bunching of electrons by the longitudinal alternating electric field associated with the surface acoustic wave; and the presence of the surface acoustic wave transverse alternating electric field, which is shifted in phase with respect to the electron bunches so that the transverse electric current averaged over a period of the wave is not equal to zero.

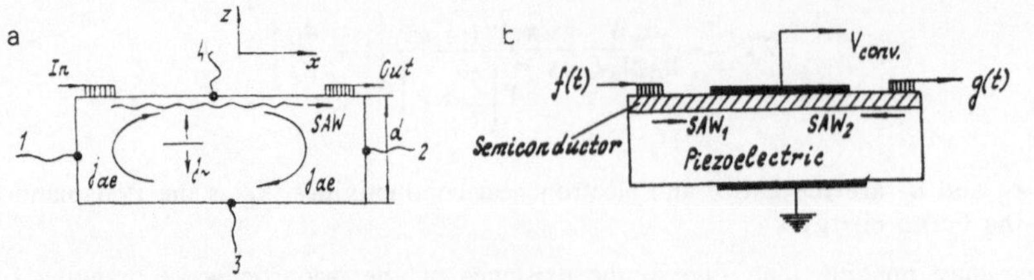


Fig. 12. (a) Surface acoustic waves propagating along the surface of a conducting body will drag electrons along that surface. (b) Convolution of surface acoustic wave signals.

### 3.4 Convolution of surface acoustic wave signals

If there are two surface acoustic waves propagating in opposite directions (see Fig. 12b) the acoustoelectric current will be produced by both waves jointly:

$$j_{ae}^{\perp} = e\mu\overline{(n_1 + n_2)(E_1 + E_2)} = e\mu(\overline{n_1E_1} + \overline{n_2E_2} + \overline{n_1E_2} + \overline{n_2E_1})$$

$$= j_{ae}^{\perp(1)} + j_{ae}^{\perp(2)} + j_{ae}^{conv} \tag{3.39}$$

Here the mixing term, the acoustoelectric convolution current, is given by

$$j_{ae}^{conv} = e\mu(\overline{n_1E_2} + \overline{n_2E_1}) \tag{3.40}$$

where $n_{1,2}$ and $E_{1,2}$ are the alternating electron concentration and alternating electric field in both waves.

$$j_{ae}^{conv} = \frac{e\mu}{2} Re(n_1E_2^* + n_2E_1^*) \tag{3.41}$$

Let us put

$$n_1 = n_{10}e^{i\omega t - iqx} \qquad E_1 = E_{10}e^{i\omega t - iqx} \tag{3.42}$$

$$n_2 = n_{20}e^{-i\omega t - iqx} \qquad E_2 = E_{20}e^{-i\omega t - iqx} \tag{3.43}$$

where $n_1$, $E_1$ ~ f; $n_2$, $E_2$ ~ g; and f and g are the amplitudes of the two surface acoustic waves propagating in opposite directions (see Fig. 12b).

Then one has

$$j_{ae}^{conv} = \frac{e\mu}{2} Re[n_{10}E_{20}^* e^{2i\omega t} + n_{20}E_{10}^* e^{-2i\omega t}] \tag{3.44}$$

Thus the convolution current is uniform along the surface and has a second harmonic time variation. Calculating in a similar way the amplitudes $n_{10}$, $E_{10}$, $n_{20}$ and $E_{20}$, we obtain for the amplitude of the convolution voltage:

$$V_{\perp}^{conv} = \frac{K^2q^3Cd}{eN_0} F_2(qr_D, \omega\tau_M) \frac{1}{T} \int_0^T f(\tau)g(t-2\tau)d\tau \tag{3.45}$$

where $F_2(qr_D, \omega\tau_M)$ is a factor of the order of unity and T is the surface acoustic wave propagation transit time in the specimen. Thus, one sees that the second harmonic transverse acoustoelectric voltage is indeed the convolution of the two surface acoustic wave signals f(t) and g(t), which justifies the term we have already used for $j_{ae}^{conv}$ and $V_{\perp}^{conv}$.

Let us now estimate the loss due to acoustoelectronic convolution. The convolution acoustoelectric field is of the order of

$$E_{\perp}^{conv} \simeq \frac{K^2 q^3 C}{eN_0} fg \qquad (3.46)$$

and

$$E_1 \simeq \beta q f \qquad (3.47)$$

$$\frac{E_{\perp}^{conv}}{E_1} \simeq \frac{\beta q^2}{eN_0} g$$

So the convolution loss (in dB) is given by the expression

$$\Gamma = 20 log \frac{E_{\perp}^{conv}}{E_1} = 20 log \left( \frac{\beta q^2 g}{eN_0} \right) \text{dB} \qquad (3.48)$$

Thus this loss depends on the amplitude of the other wave g. Usually, the intensity of the second wave is taken to be equal to 1 mW, and the loss expressed in units called "dbm".

The amplitude of the surface acoustic waves, u, is related to their intensity W by the approximate formula $qu \sim \sqrt{W/(Cv_s)}$ (u = f, g). So by putting $W_2$ = 1 mW, we have $qg = \sqrt{1mW/Cv_s}$. Typically, $C = 10^{12}$ dyn/cm$^2$, $v_s = 3\times10^5$ cm/s, so

$$qg \simeq 10^{-7} \qquad \text{at } W = 1 \text{ mW} \qquad (3.49)$$

For a surface acoustic wave frequency of 200 MHz, $q \simeq 10^3$ cm$^{-1}$, $N_0 = 10^{14}$ cm$^{-3}$ and $\beta = 3\times10^7$ V/cm. From Eq. (3.48) the loss is then approximately given by

$$\Gamma \simeq -60 \text{ dbm} \qquad (3.50)$$

which is in reasonable agreement with the experimental results.

It is important to note that the loss in an acoustoelectronic convolver is much less than in convolvers based on the nonlinearity of the elastic properties of a material. Acoustoelectronic convolvers of different types are used in practice for signal processing.

## 4. NONLINEAR THEORY OF ACOUSTIC WAVE PROPAGATION, ATTENUATION AND AMPLIFICATION IN SEMICONDUCTORS

When the amplitude of an acoustic wave in a semiconductor is big enough, the nonlinear effects become noticeable. The main nonlinear effects are:

(i) Dependence of the attenuation (amplification) coefficient of the acoustic wave and the acoustic wave velocity on its amplitude.
(ii) The non-ohmic behaviour of the current-voltage curve of the specimen.
(iii) The generation of harmonics.
(iv) Current oscillations due to the formation and motion of acoustoelectronic domains.
(v) Parametric amplification of subharmonics.

On the one hand, these effects restrict the dynamic range of various acoustoelectronic devices but, on the other hand, they may give rise to the creation of new devices based on acoustoelectronic nonlinearities. Here we shall give a brief outline of the nonlinear analytical theory of acoustoelectronic phenomena published by one of the authors in 1970.[13]

Let us consider a piezoelectric semiconductor so oriented that the propagation of acoustic waves in it can be described in terms of a one dimensional problem. We shall consider the hydrodynamic case and neglect electron trapping and heating. The whole theory is made under the assumption that the amplitude and the velocity of acoustic waves change little within a wavelength or during a period of the wave, which is usually the case.

The whole problem is divided into two parts. The "quick" part consists of:

(i) Solving the problem of the propagation of acoustic waves in a small space-time interval where the amplitude and velocity of the wave do not change noticeably.
(ii) Deriving the averaged equation describing the small variations of the wave amplitude in space and time - the continuity equation for the sound flux.

The "slow" part consists of:

(i) Solving the equation for the amplitude of the wave together with the Poisson equation and the continuity equation for the current.
(ii) Taking into account the boundary conditions and finding the distributions of the acoustic wave intensity, electric field and current inside the specimen.

Thus the aims of the calculation are to find the laws governing the variation in space and time of the intensity, velocity and shape of the acoustic waves, the distribution of the average electric field in the specimen and its current-voltage curve.

The basic equations for the "quick" problem are as follows:

$$\rho \frac{\partial^2 u}{\partial t^2} - C \frac{\partial^2 u}{\partial x^2} = -\beta \frac{\partial^2 \phi}{\partial x^2} + \chi \frac{\partial^3 u}{\partial x^2 \partial t} \tag{4.1}$$

$$\frac{\partial^2 \phi}{\partial x^2} + \frac{4\pi\beta}{\epsilon} \frac{\partial^2 u}{\partial x^2} = -\frac{4\pi e}{\epsilon}(N - N_0) \tag{4.2}$$

$$\frac{\partial N}{\partial t} + \frac{1}{e} \frac{\partial j}{\partial x} = 0 \tag{4.3}$$

$$j = eN\mu \left( E_0 - \frac{\partial \phi}{\partial x} \right) - e\mathscr{D} \frac{\partial u}{\partial x} \tag{4.4}$$

where $\chi$ is the viscosity coefficient of the crystal. The solutions to the first part of the "quick" problem, the nonlinear electronic and lattice acoustic wave attenuation coefficients, $\alpha_e$ and $\alpha_L$ respectively, and the shape of the wave, are as follows:

$$\alpha_e = -\eta q \left( \frac{p - \gamma}{a^2 W} \right) \tag{4.5}$$

$$\alpha_L = \kappa q \left( \frac{\gamma}{p} + \frac{3\sqrt{3}}{2\pi} \sqrt{\omega} \left( \frac{p - \gamma}{p + \gamma\sqrt{W}} \right) \right) \tag{4.6}$$

where

$$\eta = \frac{4\pi\beta^2}{\epsilon C} \equiv K^2 \qquad p = \frac{\mu E_0 - v_s}{\mathscr{D} q} \qquad \gamma = \frac{v_d - v_s}{\mathscr{D} q}$$
$$\kappa = \frac{\chi\omega}{C} \qquad a = q r_D \qquad W = \frac{W}{W_c} \qquad W_c = q^2 r_D^2 \frac{k_B T N_0 v_s}{\eta} \tag{4.7}$$

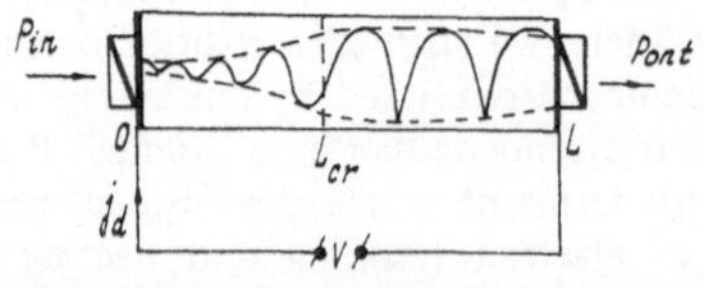


Fig. 13. The shape of the amplifying wave becomes non-sinusoidal due to the effective generation of harmonics.

The shape of the wave is non-sinusoidal due to the effective generation of harmonics, as shown in Fig. 13.

Let us consider now the "slow" part of the problem to find the stationary amplification of acoustic waves of very large amplitude in a finite crystal. The basic equations are:

$$\frac{\partial W}{\partial t} + v_s \frac{\partial W}{\partial x} = -[\alpha_e(E_0, W) + \alpha_L(E_0, W)]v_s W \tag{4.8}$$

$$I = eN_0\mu E_0 - e\mathcal{D}\frac{\partial N_0}{\partial x} + \frac{\mu}{v_s}\alpha_e(E_0, W)W \tag{4.9}$$

$$\frac{\partial N_0}{\partial t} + \frac{1}{e}\frac{\partial I}{\partial x} = 0 \tag{4.10}$$

$$\alpha_e = -\eta q\left(\frac{p - \gamma}{a^2 W}\right) \tag{4.11}$$

$$\frac{\partial E_0}{\partial x} = \frac{4\pi e}{\epsilon}(N_0 - N_i) \tag{4.12}$$

$$\alpha_L = \kappa q\left(\frac{\gamma}{p} + \frac{3\sqrt{3}}{2\pi}\sqrt{W}\left(\frac{p - \gamma}{p + \gamma\sqrt{W}}\right)\right) \tag{4.13}$$

Here $E_0$ is the d.c. electric field inside the specimen. The boundary conditions are

$$W(x)|_{x=0} = W_0 \tag{4.14}$$

$$\int_0^L E_0(x)dx = V \tag{4.15}$$

where L is the length of the specimen and V the applied voltage.

The results of the calculation are presented in Figs. 14 and 15. One can see that there is good agreement between the theoretical and experimentally measured current-voltage curves of the specimen for similar initial acoustic wave intensities. Also, there is good agreement between theory and experiment for the shape of the wave and acoustic wave intensity, and the electric field distribution along the specimen, as obtained in Ref. 14 and other works.

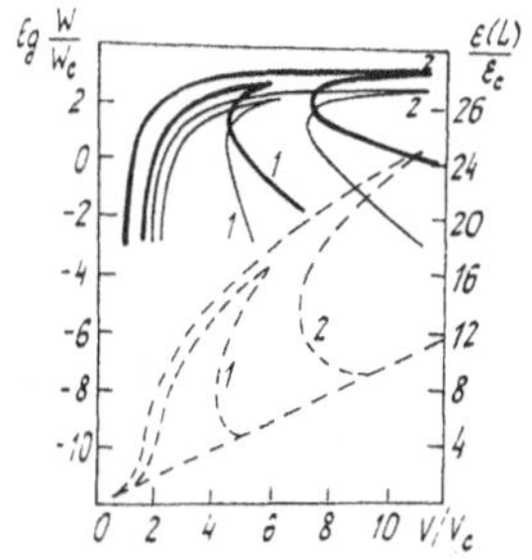


Fig. 14. The variation of the acoustic wave first and second harmonic output intensity ($W^{(1)}$ and $W^{(2)}$ respectively) and of the electric field near the anode [$\epsilon(L)$] with applied voltage V. $\rho$ = 750 Ωcm, $\mu$ = 200 cm²/Vs, $v_s$ = 1.75×10⁵ cm/s, $\eta$ = 3.7×10⁻², kT = 0.026 eV, f = 500 MHz, $qr_D$ = 1, $W_c$ = 0.84 W/cm², $V_c$ = 1.17 A/cm², $\epsilon_c$ = 875 V/cm, L = 2 mm. The initial intensities are $W_{01} = 10^{-7}$ W/cm² (curves 1) and $W_{02} = 10^{-3}$ W/cm² (curves 2). The heavy solid line is $log(W^{(1)}/W_c)$, the light solid line $log(W^{(2)}/W_c)$, and the dashed line $log[\epsilon(L)/\epsilon_c]$.

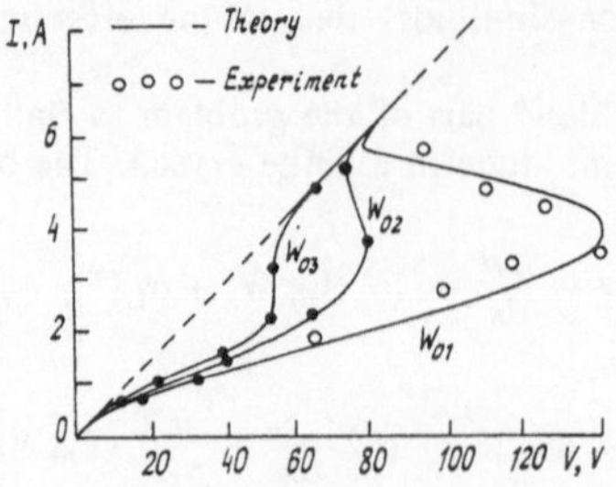


Fig. 15. Current voltage curves[13,14] for ⟨110⟩, ⟨001⟩ shear waves in n-InSb. T = 77K, n = $3.2\times10^{14}$ $cm^{-3}$, $\mu = 6\times10^5$ $cm^2/Vs$, H = 3 kOe, $\mu H/c >> 1$. The initial intensities are $W_{01} = 10^{-3}$ $W/cm^2$, $W_{02} = 10^{-5}$ $W/cm^2$ and $W_{03} = 10^{-7}$ $W/cm^2$.

It is interesting to note that this S-type current-voltage curve was derived theoretically for the first time in Ref. 15 in connection with so-called stationary acoustic waves under the sound amplification condition. In Ref. 15, it was necessary to take into account lattice sound attenuation for the restriction of the sound amplitude, while in Ref. 13 this attenuation was not necessary (the amplitude restriction being due to the finite length L of the crystal).

## 5. NONLINEAR EFFECTS OF THE ACOUSTOELECTRONIC INTERACTION IN A COLLISIONLESS REGIME

### 5.1 Momentum nonlinearity

With increasing frequency of the acoustic wave, the wavelength becomes smaller than the mean free path of the electrons and the condition $q\ell > 1$ is satisfied. Within a classical approach $\hbar q < p_T$ or $p_F$, where $\hbar q$ is the momentum of the acoustic phonons, $p_T$ the mean thermal electron momentum for non-degenerate statistics, and $p_F$ the momentum of the electron in the Fermi level for a degenerate electron gas. It then follows that for the acoustoelectronic interaction in a collisionless regime, only the electrons whose momentum projected onto the direction of propagation is equal (or close) to $m^* v_s$ can take part in the energy exchange with the acoustic wave.[1] When $\hbar q \sim p_T$, the quantum approach explains the acoustoelectronic interaction in terms of transitions of electrons belonging to the "interaction space", a certain volume in the momentum space near the planes ($m^* \mathbf{v}_s \pm \hbar \mathbf{q}/2$) separated by $\hbar \mathbf{q}$ (the momentum of emitted or absorbed acoustic phonons) (see Fig 16a).[17,18] Due to electron transitions induced by intense acoustic waves (phonon flux) the electron distribution in the interaction space can essentially be distorted as compared with the linear interaction case. This intensity dependent population change in the interaction space governs the probabilities of the electron transitions. As a result, the acoustoelectronic interaction depends on the intensity and is nonlinear. We shall describe this type of nonlinearity as "momentum" nonlinearity.

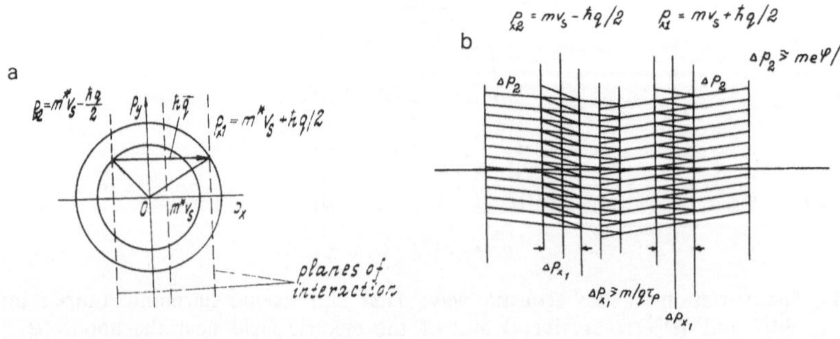


Fig. 16. Illustration of (a) the transitions involved in acoustoelectronic interactions, (b) the regions in momentum space occupied by free electrons interacting with acoustic waves.

The possibility of the existence of such an effect was pointed out in Ref. 19 for the case of a broad acoustic wave spectrum. The problem was theoretically treated using classical[20] and ultra-quantum[21] approaches. A detailed theoretical study of the collisionless nonlinear interaction problem was given in Refs. 22-25. Experimentally, the phenomenon of the momentum nonlinearity was observed and investigated for the first time in Refs. 26 and 27. Later the effect of the breakdown of this kind of nonlinearity by strong electric and by weak magnetic fields was experimentally observed in Refs. 28 and 29, and theoretically analyzed in Refs. 30 and 31.

If the intensities of the acoustic waves are such that $e\phi_{-}/k_BT \geq 1$, the capture of electrons by the potential wells of the acoustic wave plays a dominant role in the interaction and a "concentration" nonlinearity occurs for $q\ell >> 1$.[32,33]

### 5.2 Analysis of the phenomenon of momentum nonlinearity

Let us discuss the problem of the distortion of the electron distribution function by acoustic waves. The rate of acoustoelectronic interaction is governed by the energy and momentum conservation laws. In the simplest one-dimensional case of acoustic waves propagating in the X direction ($q = q_x$) these laws may be expressed in the form

$$\frac{(p_x \pm \hbar q)^2}{2m^*} - \frac{p_x^2}{2m^*} \mp \hbar q v_s = 0 \tag{5.1}$$

The positive sign corresponds to phonon absorption, the negative to phonon emission. Using Eq. (5.1), the absolute value of the electron momentum, projected onto the direction of propagation of the acoustic wave, which obeys the conservation laws is given by

$$p_x = m^* v_s \pm \frac{\hbar q}{2} \tag{5.2}$$

These conditions define two planes in momentum space: electrons which can emit phonons are located in the plane $p_{x1} = m^* v_s + \hbar q/2$, and those which can absorb phonons in the plane $p_{x2} = m^* v_s - \hbar q/2$.

Such planes describe the idealized collisionless situation. Due to collisions, there is some uncertainty in momentum and energy, and these idealised planes in momentum space become somewhat broadened. For electron relaxation times $\tau$

$$\Delta\left\{\frac{(p_x \pm \hbar q)^2}{2m^*} - \frac{p_x^2}{2m^*} \mp \hbar q v_s\right\} \geq \frac{\hbar}{\tau} \tag{5.3}$$

From Eq. (5.3) the "broadening" can easily be evaluated:

$$\Delta p_x \geq \frac{m^*}{q\tau} \tag{5.4}$$

The uncertainty given by Eq. (5.3) is the intrinsic broadening because it does not depend on the intensity of the acoustic waves. The ultra-quantum approach to the problem of electron phonon interaction, developed in Refs. 19, 21 and 22, is valid if this uncertainty is much less than the momentum of the acoustic phonon:

$$\Delta p_x \simeq \frac{m^*}{q\tau} << \hbar q \tag{5.5}$$

or, equivalently,

$$\frac{\hbar^2 q^2}{2m^*} >> \frac{\hbar}{\tau}$$

The uncertainty in $p_{x1}$ is illustrated in Fig. 16b as a broadening of the planes $p_x = m^* v_s \pm \hbar q/2$. It is evident that the strongest influence of the intense acoustic wave on the electron distribution function is to be expected exactly in this region $\Delta p_x$ of the momentum space. In other words, the nonequilibrium part of the distribution function will be noticeably changed by the strong externally excited acoustic waves in the regions of momentum interaction space $\Delta p_x$.

In the quantum description, acoustic wave amplification takes place if the number of radiative transitions from the plane $p_{x1} = m^* v_s + \hbar q/2$ to the plane $p_{x2} = m^* v_s - \hbar q/2$ induced by the acoustic waves is greater than the number of absorptions from $p_{x2}$ to $p_{x1}$. This happens if the electronic states in the plane $p_{x1}$ are more highly populated than those at $p_{x2}$. Such a population inversion can exist only if $v_d > v_s$. When $v_d = 0$ or $v_d < v_s$ the transitions from $p_{x2}$ to $p_{x1}$ are dominant and this situation corresponds to absorption. If $v_d = v_s$ there is neither absorption nor amplification.

For amplification of acoustic waves at low intensities the number of electrons leaving the space near the plane $p_{x1}$ and the number of electrons entering the region near $p_{x2}$ are negligible compared with the equilibrium concentrations in these regions of the momentum space. As a result the amplification process remains linear; the amplification coefficient does not depend noticeably on the acoustic wave intensity. In the case of amplification of acoustic waves of greater amplitude, however, the concentration of the "excess" electrons near $p_{x2}$ and the "deficiency" of electrons near $p_{x1}$ may be comparable with the equilibrium concentrations in these regions. This is possible only if the relaxation processes are unable to keep up with the electron transitions. The acoustoelectronic interaction, which depends on the number of transitions between the empty and occupied states near $p_{x1}$ and $p_{x2}$ in the momentum space, becomes sensitive to the wave intensity in this case. The typical characteristic time that is necessary for a change in the population due to the acoustic wave is of the order of the quantum transition time $\hbar/e\phi_-$, where $\phi_-$ is the electric potential of the wave. If this time is smaller than the electron relaxation time

$$\frac{\hbar}{e\phi_-} < \tau \tag{5.6}$$

then the changes in population will dominate the relaxation process.

When the condition (5.6) is fulfilled the value of the acoustoelectronic amplification coefficient will decrease and nonlinear acoustoelectronic amplification will take place. In order to evaluate the threshold intensity for such a nonlinearity, we rewrite expression (5.6) in the following form:

$$\frac{e\phi_-\tau}{\hbar} = \frac{e\phi_-}{kT}\left(\frac{kT\tau}{\hbar}\right) = \left(\frac{e\phi_-}{kT}\right)\frac{p}{\hbar q}q\ell >> 1 \tag{5.7}$$

Because of the fact that $q\ell >> 1$ and $p/\hbar q >> 1$ the expression (5.7) may be easily fulfilled even if $e\phi/kT << 1$, i.e. this kind of nonlinearity occurs at much lower intensities than the concentration one.

A more careful analysis[21,22] results in the following threshold value of the intensity of sound for the onset of the momentum nonlinearity:

$$W_c^{(m)} = \frac{\epsilon\omega^2}{4\pi e^2 v_s K^2(1 + q^{-2}r_D^{-2})^2}\left(\frac{\hbar}{\tau}\right)^2 \tag{5.8}$$

It is interesting to compare this expression with the formula of the critical intensity for the onset of the concentration nonlinearity.

$$W_c^{(c)} = \frac{\epsilon\omega^2}{4\pi e^2 v_s K^2(1 + q^{-2}r_D^{-2})^2}(kT)^2 \tag{5.9}$$

The two expressions are identical except for the last factor. Due to the fact that $kT >> \hbar/\tau$ the threshold for the momentum nonlinearity is much less than that for the concentration nonlinearity. For InSb at liquid nitrogen temperatures they differ by a factor of $5\times10^3$.

The quantum approach gives the following relation between the linear and nonlinear coefficients of attenuation, $\alpha_e^{lin}$ and $\alpha_e$ respectively:

$$\alpha_e = \alpha_e^{lin}\left(1 + \frac{W}{W_c^{(m)}}\right)^{1/2} \tag{5.10}$$

The classical approach gives:

$$\alpha_e = \alpha_e^{lin}\left(1 + \frac{W}{W_c^{(m)}}\right)^{1/4} \tag{5.11}$$

A rigorous theoretical treatment of the momentum nonlinearity shows that the nonlinear parameter (5.8) depends not on the transport relaxation time but only on that part of the relaxation time which describes the transitions from the initial state. We shall call this the "departure" time. It can be understood in the following way. The collision integral in the Boltzmann equation may be written as

$$-\frac{f - f_0}{\tau} = \int_{k'} [W_{kk'}f_k(1 - f_{k'}) - W_{k'k}f_{k'}(1 - f_k)]dk' \tag{5.12}$$

Here $W_{kk'}$ are the transition probabilities and $f_k$ is the electron distribution function.

The first term in the integral describes the transitions ("departure") of the electrons from a given state k in momentum space to other states k′. The second term describes the rate at which electrons arrive at a given state from the rest of the states in momentum space. The net relaxation time may be written as

$$\frac{1}{\tau} = \frac{1}{\tau_{dep}} - \frac{1}{\tau_{arr}} \tag{5.13}$$

Here $\tau_{dep}$ and $\tau_{arr}$ are the "departure" and "arrival" times respectively. When a change in electron distribution is induced at some point in momentum space the relaxation process is dramatically modified. For example, if there is a deficiency of electrons at some point (for instance in the vicinity of plane 1 in Fig. 16), the rate of arrival of electrons at this point is unchanged, because the distribution of electrons in the rest of momentum space is not disturbed. However, the rate of departure of electrons from the given point changes, essentially because the number of electrons there has decreased ($f_k$ has decreased). As a result, only the "departure" processes determine the stationary population of the electrons at the point in momentum space under consideration. Similar arguments can be applied to the situation where a region has an excess number of electrons (region 2 in Fig. 16b).

Hence, experiments with nonlinear electron-phonon interactions can give information about a very important physical parameter of the electronic system, i.e. the "departure relaxation time". It is clear that $\tau_{dep}$ depends on the scattering mechanisms of the electrons.

To conclude this section, we discuss the application of this theory to a real experimental situation. As pointed out before, the best material for the observation and study of the momentum nonlinearity is indium antimonide (n-InSb, T = 77 K, $\mu = 6.7\times10^5$ cm²/Vs, f ~ 2 GHz). For this material, the parameter $\hbar\tau q^2/m \simeq 2$, that is it lies just in between the ranges of validity of the quantum and classical approaches. Moreover, instead of Eq. (5.3), a rigorous analysis gives the expression:

$$\Delta\left\{\frac{(p_x \pm \hbar q)^2}{2m^*} - \frac{p_x^2}{2m^*} \mp \hbar q v_s\right\} \geq max\left(\frac{\hbar}{\tau}, e\phi_-\right) \tag{5.14}$$

It follows from Eq. (5.14) that the width of the region of interaction in momentum space is

$$\Delta p_{x2} \geq m^* \frac{e\phi_-}{\hbar q} \tag{5.15}$$

and, in the case of intense waves, the regions $\Delta p_{x1,2}$ intersect. In this case only the classical approach is applicable. One can expect that, in a real experimental situation, the onset of the momentum nonlinearity will be best described by quantum theory, but as the sound intensity is increased further the classical model becomes more applicable.

## 5.3 Experimental investigation of the momentum nonlinearity

We use the well known "echo-pulse method" for the detection and measurement of the propagation parameters of acoustic signals. When an electric field pulse is applied to the crystal, the drift of electrons changes the electronic part of the attenuation. If the drift velocities of electrons exceed the velocity of sound, amplification is observed. The quantity that is measured in our experiments is the change of the electronic attenuation in the "active" part $L_{act}$ of the crystal to which the electric field was applied:

$$G = \int_0^{L_{act}} \alpha_e(E)dx - \int_0^{L_{act}} \alpha_{e_0} dx \tag{5.16}$$

Due to the fact that in real experiments $v_d/v_s >> 1$ and $|\alpha_e(E)| >> |\alpha_{e_0}|$, in many cases the measured quantity is considered to be an amplification.

The electronic amplification G in the nonlinear case, when $W \geq W_c^{(m)}$, depends on the sound intensity. The intensity $W_{aw}(x)$ is itself a function of the input sound intensity and the lattice attenuation constant $\alpha_L$. The latter quantity was measured in additional experiments. The measurements of G and $\alpha_L$ were made to within an accuracy of 10%. The input acoustic wave intensity was evaluated after measurement of the efficiency of the electroacoustic transducer $\eta$. The error in the input level was ±3 dB. The input intensity of the acoustic wave was controlled by changing the output level of the microwave generators.

In some experiments, the acoustoelectric current $I_{ae}$ was also measured. The experiments were performed in almost short circuit conditions for the acoustoelectric current. [In the circuit R(sample) ≃ 10 Ω, R(pulse generator) = 0.5 Ω]. Measurement of the two quantities $G = G(W/W_c^{(m)})$ and $I_{ae} = I_{ae}(W/W_c^{(m)})$ as a function of acoustic intensity level gives detailed information about the nonlinear interactions.

As mentioned above, n-InSb is a very useful material for studying nonlinear acoustoelectronic gain in a collisionless regime. In the microwave region it is quite easy to reach $q\ell > 10$.

We have studied nonlinear acoustoelectronic interactions for piezoactive shear acoustic wave propagation along the ⟨110⟩ axis in InSb crystals. The samples were cut in the shape of rectangular bars oriented along the ⟨110⟩ axis. The lengths of the samples were between 7 - 12 mm and their cross-sections in the range 1 - 2 mm². To prevent heating of the samples by the drift current, the samples were put in a liquid nitrogen bath.

Experimental investigation of the amplification showed that it depends strongly on the input intensities. The greater the input power, the smaller the amplification. Examples of the dependence of the gain on the acoustic wave intensity, as a function of the drift velocity and the frequency of the acoustic waves, are shown in Figs. 17 and 18. As the drift velocity increases the decrease in gain becomes more pronounced. The qualitative explanation for this is quite simple: for bigger drift velocities, the acoustic flux in the crystal reaches high values within shorter distances.

The stronger influence of input intensity on the gain at high frequencies, and at fixed drift velocity, qualitatively has a similar explanation: for the material parameters used in the experiments the linear amplification coefficient increases with the frequency. We can evaluate the threshold intensity for the momentum nonlinearity with the aid of the curves in Figs. 17 and 18. For a quantitative comparison of the experimental results with the theory, it is necessary to solve the continuity equation for the acoustic flux

$$\frac{\partial W}{\partial x} = -[\alpha_L + \alpha_e(W)]W \tag{5.17}$$

by using the theoretical expression for the acoustoelectronic attenuation coefficient. Following Ref. 21,

$$\alpha_e = \alpha_e^{lin}\left[1 + \frac{W}{W_c^{(m)}}\right]^{1/2} \tag{5.18}$$

The result can be obtained by the numerical solution of the algebraic system

$$-\alpha_L L_{act} = F_\gamma\left(\frac{W_{out}}{W_c^{(m)}}\right) - F_\gamma\left(\frac{W_{in}}{W_c^{(m)}}\right)$$

$$F_\gamma\left(\frac{W}{W_c^{(m)}}\right) = \frac{2\gamma^2}{(\gamma^2 - 1)} ln\left|\sqrt{1 + \frac{W}{W_c^{(m)}}} - \gamma\right| + \frac{1}{(\gamma + 1)} ln\left|\sqrt{1 + \frac{W}{W_c^{(m)}}} + 1\right| + \frac{1}{(\gamma - 1)} ln\left|\sqrt{1 + \frac{W}{W_c^{(m)}}} + 1\right| \tag{5.19}$$

$$\gamma = \frac{\alpha_{e_0}^{lin}}{\alpha_L}\left(1 - \frac{v_d}{v_s}\right) \qquad G = \frac{W_{out}}{W_{in}}$$

The results of this calculation are shown in Figs. 17 and 18 by solid lines and illustrate good agreement between theory and experiment.

The dependence of the acoustoelectric current on the acoustic wave intensity for different values of $v_d/v_s$ is shown in Fig. 19. Instead of the linear dependence that is to be expected for linear interaction, the experimental curve obeys the law $I_a \sim \sqrt{W_{in}}$. The measured values of the current are almost two orders of magnitude lower than those predicted by linear theory.

The theoretical expression for the acoustoelectric current was obtained using the equation for current density $j = env_d + \mu\alpha_e W/v_s$ together with Eq. (5.17). The result is:

$$I_{ae} = S_0 \frac{\mu\gamma W_c^{(m)}}{v_s L_{act}}\left[\Phi\left(\frac{W_{out}}{W_c^{(m)}}\right) - \Phi\left(\frac{W_{in}}{W_c^{(m)}}\right)\right]$$

where (5.20)

$$\Phi\left(\frac{W}{W_c^{(m)}}\right) = 2\left[\left(1 + \frac{W}{W_c^{(m)}}\right)^{1/2} + \gamma ln\left|\sqrt{1 + \frac{W}{W_c^{(m)}}} - \gamma\right|\right]$$

and $S_0$ is the cross-section of the specimen. The results of the calculations are shown in Fig. 19 by solid lines. The value of $I_{ae}$ depends upon the ratios $W_{out}/W_c^{(m)}$, $W_{in}/W_c^{(m)}$ and on $W_c^{(m)}$. The ratio $W_{in}/W_c^{(m)}$ that was found from the amplification data was used for the calculation of the curves of the acoustoelectric current, and the best fit between experiment and theory was obtained by a suitable choice of $W_c^{(m)}$. The best agreement was obtained when $W_c^{(m)} \simeq 2\times10^{-1}$ W/cm² (for f = 1.9 GHz).

Having compared the experimental data to results obtained from the quantum theory, it is also interesting to make this comparison for the classical approach. At high acoustic wave intensities, the classical approach gives a better description of the experimental data than the quantum theory (see Fig. 20). The fitting parameter is $\tau_{dep}$ in both theories. In the classical approach, the best fit with the experimental results is obtained for $\tau_{dep} \simeq 0.5\tau$. The quantum theory best agrees with experiment when $\tau_{dep} \simeq 0.3\tau$.

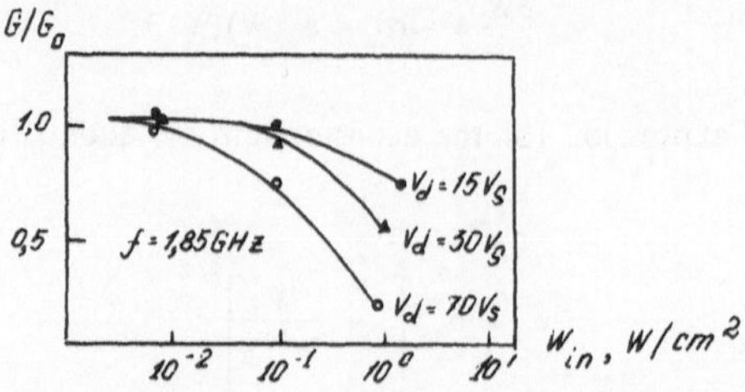


Fig. 17. The dependence of the normalised electronic amplification coefficient on the acoustic wave input intensity for various fixed electron drift velocities.

The decrease in electronic gain with the increase of acoustic wave intensity results in an interesting phenomenon: "stationary wave" formation. The results of a study of this effect are shown in Fig. 21. The effect manifests itself through the fact that the output intensity of the acoustic wave in the amplifying sample remains constant when the input intensity is varied through a wide range. Stationary wave formation can be understood qualitatively in the following way: when the intensity of the wave increases, the modulus of the amplification coefficient decreases and becomes close to the lattice attenuation coefficient. As a result, the acoustic wave will propagate without change of amplitude. Any decrease in the intensity will increase the amplification and *vice versa*.

### 5.4 Breakdown of the momentum nonlinearity in strong electric and weak magnetic fields

Momentum nonlinearity arises as a result of a strong distortion of the electron distribution function in the interaction space. The nonlinear effects turn out to be quite noticeable if the characteristic time for "capture" of electrons by the acoustic wave potential

$$\tau_1 \sim \begin{cases} \sqrt{m^*}/q^2e\phi_- & \text{in classical theory} \\ \hbar/e\phi_- & \text{in quantum theory} \end{cases}$$

is less than the relaxation time $\tau$. In order to restore the linear interaction regime, it is necessary to sweep the captured electrons away from the potential wells in the acoustic wave field.

In principle, this can be done by an external electric field. The characteristic time required for the excess electrons to escape from the potential well is, in accordance with Ref. 30, $\tau_2 = \Delta p_x/eE$, where (in the strong field approximation $v_d q\tau > 1$) the momentum uncertainty is $\Delta p_x \simeq (eEm^*/q)^{1/2}$. In a strong electric field $\tau$ can become less than $\tau_1$ and the nonlinearity must vanish.

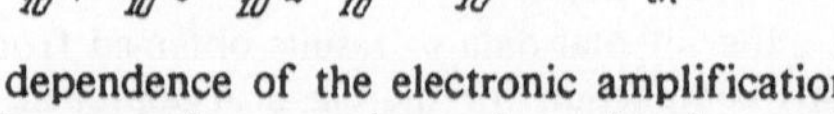

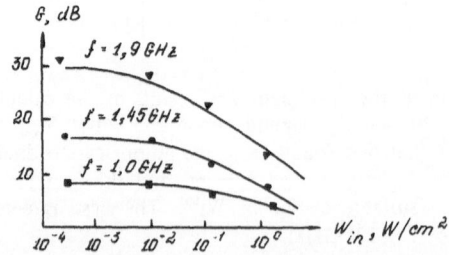

Fig. 18. The dependence of the electronic amplification coefficient on the acoustic wave input intensity for various acoustic wave frequencies. The electron drift velocity is fixed at $v_d = 30v_s$ and the electron concentration $n = 4\times10^{14}$ cm$^{-3}$. The solid curves correspond to the theory.

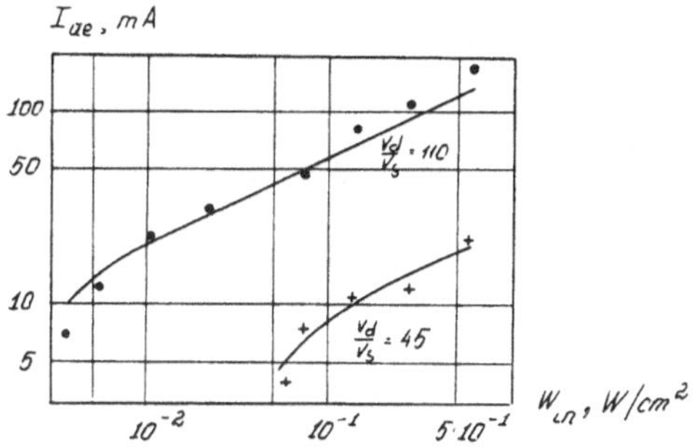


Fig. 19. The dependence of the acoustoelectric current on the acoustic wave input intensity $W_{in}$ for fixed electron drift velocities. The frequency f = 1.85 GHz and electron concentration n = $1.84\times10^{14}$ $cm^{-3}$. The solid curves correspond to the theory.

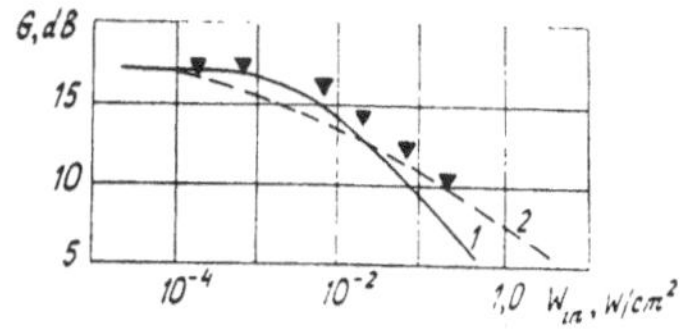


Fig. 20. The variation of the electronic amplification coefficient with the acoustic wave input intensity. The solid curve shows results obtained using a quantum approach, the dashed curve a classical approach and the points give the experimental data.

The experiments (in which the effect was observed for the first time) were carried out at 2 GHz in a field E ~ 200 V/cm (where $qv_d\tau \simeq 4$). The input sound intensity was $W_{in} \sim 10^{-1}$ W/cm². The characteristic times were: $\tau \simeq 6\times10^{-12}$ s, $\tau_1 \simeq 1.5\times10^{-12}$ s, $\tau_2 \simeq 0.8\times10^{-12}$ s and hence $\tau > \tau_1 > \tau_2$. The dependence of the total electronic sound amplification at high electron velocities was measured as a function of the input intensity.

The theory gives a local coefficient for electronic amplification. In order to find this local gain $\alpha_e$ at a point x = L we used the graphical procedure developed in Ref. 37. The resulting values of $\alpha_e(L)$ are plotted in Fig. 22 as a function of normalised drift velocity. It can be seen that when $v_d/v_s > 120$ the gain increases and the nonlinearity breaks down. Using the data in Fig. 22, one can evaluate $\tau_{dep}$. In accordance with the theory,[30] the minima on the curves correspond to the condition $eEq\tau_{dep}/m^* \simeq 1$. The value of $\tau_{dep}$ obtained from this relation is $\tau_{dep} \simeq 0.3\tau$ and is in good agreement with the data obtained from threshold measurement experiments.

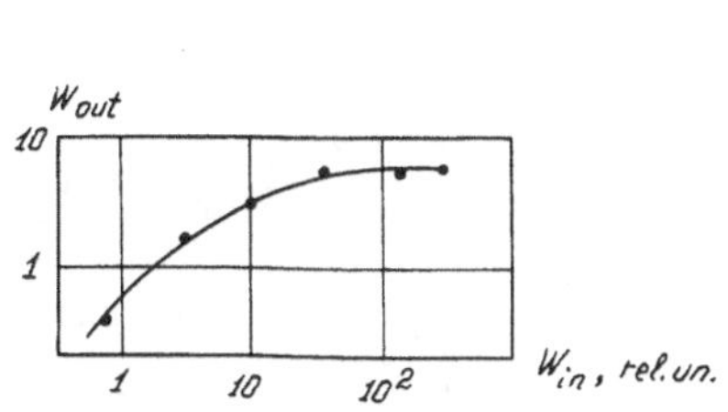


Fig. 21. The change in the acoustic wave output intensity as a function of the input intensity. f = 2.3 GHz and n = $4\times10^{14}$ $cm^{-3}$.

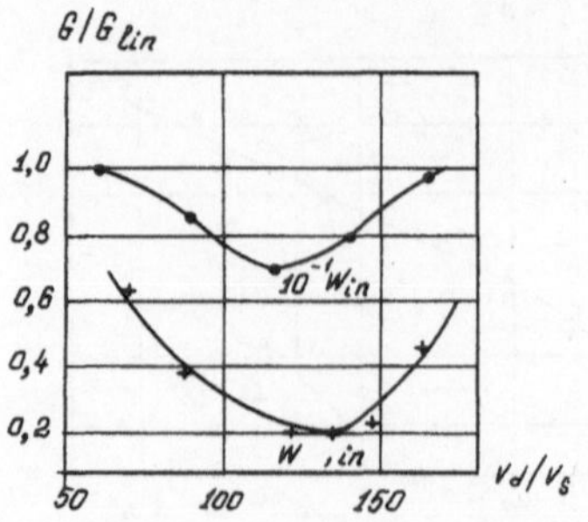


Fig. 22. The local acoustoelectronic amplification factor as a function of the drift velocity for various acoustic wave input intensities.

There is another possible way to break down the nonlinear effects. A weak magnetic field can restore the electron distribution function in the regions of momentum space that are responsible for acoustoelectronic wave interaction. This is possible if the Lorentz force becomes comparable to the electric field forces of the acoustic wave acting on the captured electrons:

$$\frac{e}{c}\langle v\rangle H \geq qe\phi \tag{5.21}$$

The analysis of this physical situation showed[35] that the nonlinear sound amplification coefficient in a weak magnetic field can be written in the form

$$\alpha = a\alpha_e^{lin} + b\alpha_{e_0}^{lin} \tag{5.22}$$

where $\alpha_e^{lin}$ is the electronic amplification coefficient given by the linear theory. The condition for nonlinearity when $q\ell >> 1$ is

$$a^{-1} = \sqrt{\frac{e\phi_{-}}{E_{el}} q\ell} > 1 \tag{5.23}$$

where $E_{el}$ is the energy of the electrons and $b^{-1} = (e\phi/E_{el})qR$ takes into account the influence of the magnetic field on the amplification, R being the cyclotron radius.

The condition for breakdown of the nonlinearity is $b \geq a$ or

$$\frac{e\phi}{E_{el}} \leq \frac{\ell^2}{R^2} \tag{5.24}$$

Using Eqs. (5.21) and (5.23) this can be rewritten as

$$1 \leq q^2\ell^2 \frac{e\phi}{E_{el}} \leq min \begin{cases} \frac{\ell}{R} q\ell \\ \frac{\ell^2}{R^2}(q^2\ell^2) \end{cases} \tag{5.25}$$

The above condition is rather difficult to satisfy experimentally and the experiment must be carefully arranged for the effect to be observed. Estimates were made for InSb with an electron concentration $n = 2\times10^{14}$ cm$^{-3}$ and mobility $\mu = 5\times10^5$ cm$^2$/Vs; at the frequency f = 1.5 GHz the parameter $q\ell \simeq 10$.

The requirement of a weak magnetic field means that the distortion of the electron trajectory in the magnetic field must be small, and the mean free path of the electrons $\ell$ should be less than R. This condition, together with Eq. (5.24), restricts the values of the magnetic

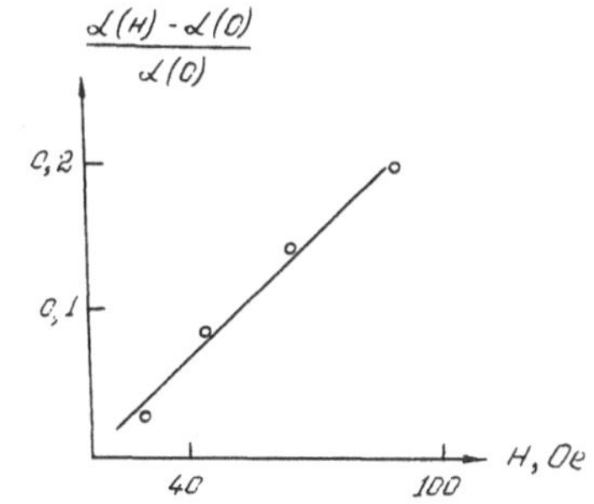


Fig. 23. The normalised change in gain as a function of the transverse magnetic field.

field which can be used in the experiments to $20 < H < 100$ Oe. It is then necessary to evaluate the maximum allowed electric field intensity to avoid the breakdown of the nonlinearity by the electric rather than the magnetic field. It is quite easy to estimate this electric field intensity by assuming the force acting on the electron due to the electric field to be smaller than the Lorentz force, $E < \langle v \rangle H/e$. From this relation, it follows that $v_d \leq 20v_s$. This agrees with the results of our previous experiments, where the breakdown of the nonlinearity by the electric field was not observed at such drift velocities. To observe the breakdown of the nonlinearity due to a weak magnetic field, $H < 100$ Oe, the intensity of the acoustic waves must not exceed 1 W/cm$^2$. Hence it is possible to observe the effect using the moderate intensities $10^{-2}$ - $10^{-1}$ W/cm$^2$, which are large enough for the onset of nonlinear amplification. In all the experiments described here, the input intensity did not exceed 0.5 W/cm$^2$. For a magnetic field normal to the propagation direction, the electronic amplification coefficient increases towards its linear value. It follows from Eq. (5.22) that the change in $\alpha$ must depend linearly on the magnetic fields. The experimental measurements of the changes in $[\alpha(H) - \alpha_0]/\alpha_0$ as a function of the magnetic field are shown in Fig. 23. In accordance with the theoretical conclusion, this dependence is linear with a slope of $0.5\times10^{-2}$ Oe$^{-1}$. The theoretical value is $10^{-2}$ Oe$^{-1}$. Such an agreement seems to be satisfactory if one takes into account the fact that the formula (5.22) was the asymptotic one and was obtained for strong nonlinearity, whilst in the experiments the intensity was restricted.

Thus the dynamic range of the acoustoelectronic amplifier, which is limited by the effects of the nonlinearity, can be restored by putting the crystal into a strong electric or weak magnetic field. The experimental results confirm that the physical picture of the phenomenon of momentum acoustoelectronic nonlinearity is correct both qualitatively and quantitatively.

### 5.5 Electron concentration nonlinearity in the collisionless regime of acoustoelectronic interaction

This section is devoted to the problem of acoustoelectronic interactions with acoustic wave input intensities higher than those used in the experiments on momentum nonlinearity. At such high acoustic wave intensities the potential field of the acoustic wave becomes so deep that the condition $e\phi_-/kT > 1$ is fulfilled together with $q\ell > 1$ for the majority of electrons. The experimental problem is how to excite acoustic waves with such high intensities in the sample. The use of conventional transducers is restricted by an electrical breakdown of the piezodielectric films. The momentum nonlinearity makes it impossible to obtain high intensities by means of acoustic wave amplification. In our experiments we used multipass amplification of acoustoelectronic fluctuations[36] in piezosemiconductors within alternating electric drift fields, and the transformation of the wide amplified spectrum of noise into a narrow spectral line by the nonlinear interaction of the fluctuations.

To obtain the effect of multipass amplification of fluctuations, the sample was put into an alternating electric field whose period was equal to twice the transit time of the acoustic wave through the sample. The acoustoelectronic fluctuations are amplified for one half of the period of the field and, on reflection, change their direction of propagation simultaneously with the change in sign of the electric field, thereby being amplified again. This reflection

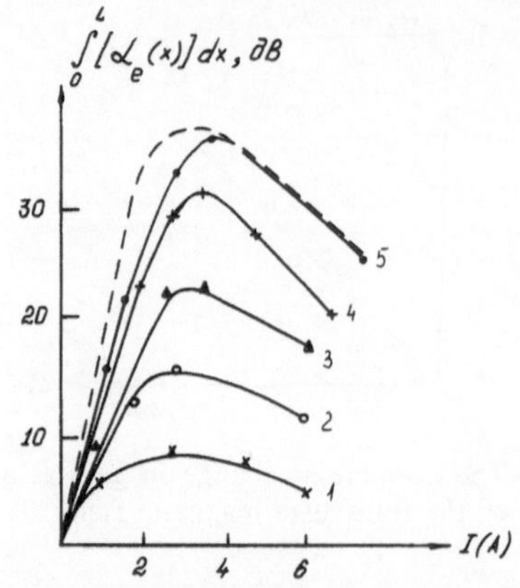


Fig. 24. The dependence of the acoustoelectronic amplification factor on the electron drift current for various acoustic wave input intensities: 1. $W_{in}$ = 45 W/cm², 2. $W_{in}$ = 4.5 W/cm², 3. $W_{in}$ = 0.45 W/cm², 4. $W_{in} = 4.5\times10^{-2}$ W/cm², 5. $W_{in} = 4.5\times10^{-3}$ W/cm².

and amplification process repeats itself many times so that the integrated intensity of the acoustic flux reaches a few hundred W/cm². The growth in the intensity is accompanied by a narrowing of the spectrum. The process was found experimentally to produce a very narrow spectral line acoustic flux with an intensity of 50 W/cm² at 1050 MHz in n-InSb crystals with a carrier concentration of $6\times10^{13}$ cm$^{-3}$.[34] The "input" intensity of sound in such experiments can be changed step by step by a suitable choice of starting point. Without a drift field, the acoustic wave intensity changes by $\exp(-\alpha_L L)$ after each successive pass through the specimen.

The experimentally measured dependence of the electronic amplification on the drift current for various acoustic wave input intensities is shown in Fig. 24. The dashed line in this figure gives the result of linear theory†. It is interesting to compare this curve with the experimental data for low input intensity (curve 5 in Fig. 24). When the drift velocity is relatively small, the experimental values are less than those given by the linear theory. Evidently, this is due to the momentum nonlinearity which decreases the amplification. At high drift velocities, this nonlinear mechanism is broken down and the experimental data coincide with the theory. When the input intensity increases (curves 4, 3, 2 and 1 in Fig. 24) the electron gain decreases. This may be due to the onset of another nonlinear mechanism, the concentration nonlinearity. In order to prove this suggestion, it is convenient to plot the electronic amplification as a function of the acoustic wave input intensity at various fixed drift currents, as shown in Fig. 25. The vertical dashed line corresponds to the condition $e\phi_-/kT \simeq 1$. It can be seen that at high intensities, when $e\phi_-/kT > 1$, the slope of the curves closely follows a $W^{-1}$ law. Such behaviour was predicted for the case of concentration nonlinearity.[32,33] In accordance with the theory developed for the case $q\ell \gg 1$, the local electronic attenuation coefficient is:

$$\alpha_e(W) = \frac{3\pi^{3/2}}{4}\frac{m^* n_0 v_s}{\tau W}\left(v_s - \frac{3\pi}{16}v_d\right) = \frac{\tilde{\alpha}_0(v_d)}{W} \tag{5.26}$$

In order to take into account the dependence of the sound intensity on position, the continuity equation for acoustic flux was used:

$$\frac{dW}{dx} = -[\alpha_e(W) - \alpha_L]W \tag{5.27}$$

†The theoretical curve is calculated for electron temperatures equal to the lattice temperature. Numerical estimates show that for maximum values of the electric field the electron temperature does not exceed 100 K. It is important to note that if the effects of heating are taken into account, the curves in Figs. 24 and 25 go higher than in the absence of heating.

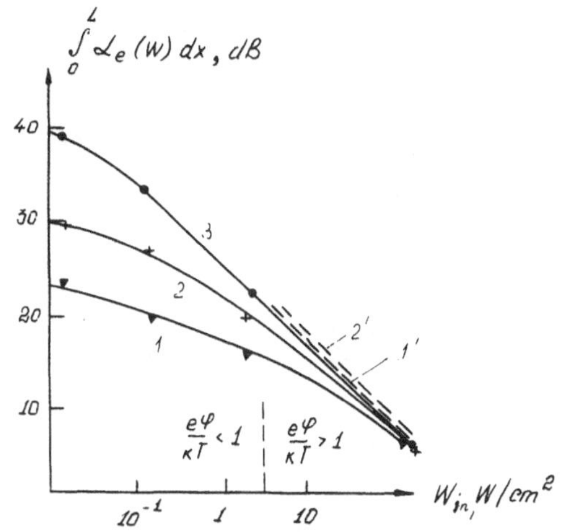


Fig. 25. The electronic amplification factor as a function of the acoustic wave input intensity at three different drift currents: 1. I = 1.5 A, 2. I = 2.0 A and 3. I = 3.0 A. The solid lines give the experimental results and the theoretical results for the first two drift currents are given by the dashed lines.

With the theoretical expression for $\alpha_e(W)$, Eq. (5.26), this equation was solved. As a result, the relation between input and output acoustic wave intensities was obtained. The measured gain is then

$$\int_0^L \alpha[W(x)]dx = 10\log\frac{W_{out}}{W_{in}} = 10\log\left[1 + \frac{\tilde{\alpha}_0}{\alpha_L W_{in}}\,\frac{(1 - e^{-\alpha_L L})}{e^{-\alpha_L L}}\right]e^{-\alpha_L L}\ \mathrm{dB} \qquad (5.28)$$

The results obtained from Eq. (5.28) are also shown in Fig. 25 for various drift currents. It can be seen that the experimental results are in good agreement with the theory describing the concentration nonlinearity in the collisionless case of acoustoelectronic interaction.

## 6. CONCLUSION

The results of the investigation of a wide variety of nonlinear acoustoelectronic phenomena are presented in this chapter. These results give a complete physical picture of the processes responsible for manifestations of the nonlinear acoustoelectronic effects.

## REFERENCES


1. V L Gurevich, FTP **2**, 1557 (1968)
2. V I Pustovoit, UFN **97**, 257 (1969)
3. P E Zilberman, A G Mishin, JETP **65**, 1474 (1973)
4. Yu V Gulyaev, A G Kozorezov, FTT **18**, 2048 (1976)
5. Yu V Gulyaev, Phys. Lett. **29A**, 187 (1969)
6. D L White, J. Appl. Phys. **33**, 2547 (1962)
7. H N Spector, Appl. Phys. Lett. **7**, 82 (1965)
8. G Weinreich, Phys. Rev. **107**, 317 (1957)
9. Yu V Gulyaev, FTT **8**, 3366 (1966)
10. E M Epstein, Yu V Gulyaev, FTT **9**, 376 (1967)
11. Yu V Gulyaev, E M Epstein, FTT **9**, 864 (1967)

12. Yu V Gulyaev, A Yu Karabanov, A M Kmita, A V Medved, Sh S Tursunov, FTT **12**, 2595 (1970)
13. Yu V Gulyaev, IEEE Trans. Sonics and Ultrasonics **SU-17**, 111 (1970); FTT **12**, 415 (1970)
14. Yu V Gulyaev, G D Mansfeld, A A Rubtsov, Acta Phys. Slov. **32**, 99 (1982); FTP **15**, 319 (1981)
15. P E Zilberman, FTT **9**, 309 (1967)
16. L D Landau, JETP **16**, 574 (1946)
17. A B Pippard, Phil. Mag. **46**, 1104 (1955)
18. A I Ahiezer, M I Kaganov, G Yu Lubarsky, JETP **32**, 837 (1955)
19. P E Zilberman, FTT **12**, 1014 (1970)
20. Yu M Galperin, V D Kagan, JETP **59**, 321 (1970); ibid **59**, 1657 (1970)
21. P E Zilberman, JETP **60**, 1943 (1971)
22. P E Zilberman, FTP **5**, 1940 (1971)
23. Yu M Galperin, V D Kagan, V I Kozub, JETP **62**, 1521 (1972)
24. Yu M Galperin, P E Zilberman, S N Ivanov, V D Kagan, G D Mansfeld, Usp. Fiz. Nauk. **105**, 774 (1971)
25. Yu M Galperin, V L Gurevich, V I Kozub, Usp. Fiz. Nauk. **128**, 107 (1979)
26. S N Ivanov, I M Kotelyansky, G D Mansfeld, E N Khazanov, JETP Lett. **13**, 283 (1971)
27. P E Zilberman, S N Ivanov, I M Kotelyansky, G D Mansfeld, E N Khazanov JETP **63**, 1745 (1972)
28. S N Ivanov, G D Mansfeld, E N Khazanov, FTT **15**, 282 (1973)
29. S N Ivanov, G D Mansfeld, E N Khazanov, FTT **15**, 2972 (1973)
30. P E Zilberman, A G Mishin, FTT **14**, 902 (1973); Radiotech. and Electron. **17**, 2384 (1972)
31. Yu M Galperin, V I Kozub, FTT **17**, 2222 (1975)
32. V D Kagan, FTT **16**, 1766 (1974)
33. B D Laikhtman, Yu V Pogorelsky, JETP **75**, 1892 (1978)

# NONLINEAR WAVES IN FERROMAGNETIC FILMS

A D Boardman

Department of Physics,
University of Salford
Salford, M5 4WT, UK

Yu. V Gulyaev, S A Nikitov

Institute of Radioengineering and Electronics
USSR Academy of Sciences
Marx Avenue 18, 103907 Moscow, Centre, USSR

Wang Qi

Department of Physics
Shanghai University of Science and Technology
Shanghai, China

## 1. MAGNETIC MATERIALS

Typically a magnetic material[1] may consist of elements from a transition group of the periodic table and involve atoms of Fe, Co, Mn and Ni, for example. The atoms of such elements have unfilled electron shells so that an uncompensated magnetic moment can arise from the spin of the electrons. If the permeability of a substance is less than unity, so that its magnetic susceptibility is negative, then it is said to be diamagnetic. The opposite situation of positive susceptibility is of interest here, since this feature characterises paramagnetic and other very interesting magnetic media. Broadly speaking, if only a small concentration of magnetic ions exists then the magnetic materials are known as paramagnets. Ordinary paramagnetic substances have permeabilities close to unity but the transition elements give rise to substantial values. In paramagnetic materials, however, the magnetic ions are practically independent of each other. The orientation of the magnetic moments of these ions is chaotic, with the average moment of the whole volume being equal to zero.

In magnetic materials that are classed as ferromagnetic the magnetic ions are close together. Because of this there is a strong exchange interaction between them that produces ordering and a possible spontaneous magnetic moment in zero applied magnetic field, below the Curie point. This is essentially a quantum effect connected with the intersection of the wave functions of the magnetic ions. In the simplest cases the magnetic moments either all point in the same direction or, for a simple antiferromagnet, lie on two anti-parallel sub-lattices. If the total magnetic moment is not zero then this is known as the ferromagnetic case. Examples of such materials are Fe, Co, Mn and Ni. Materials with more than one magnetic sublattice, but with a zero magnetic moment, are in the antiferromagnetic class. Examples of the latter are $MnF_2$, MnO and NiO. In materials that contain several magnetic sublattices a ferromagnetic state does not have to consist of a parallel alignment of magnetic moments provided that there exists a non-zero net saturation magnetisation. This would be a ferrite-type ferromagnetic spin order termed ferrimagnetic. The classical example is the insulator YIG which is $Y_3I_5O_{12}$

(yttrium-iron-garnet) and consists of 20 sublattices. It is a cubic ferrimagnetic insulator. The oscillations of these sublattices relative to one another will occur at frequencies greater than $10^3$ GHz. At lower frequencies the sublattices move together so that YIG behaves simply as a ferromagnet with total saturated magnetic moment $M_0$. For this reason we shall now only consider the ferromagnetic state as being characteristic of the kind of materials most commonly used for devices.

Since the mid seventies YIG films have become available through liquid-phase-epitaxy. They are deposited onto substrates of gallium-gadolinium-garnet (GGG). It is possible to generate them with thicknesses from 0.2 $\mu$m up to 100 $\mu$m and with large planar dimensions (up to 7-8 cm). Such films are characterised by a high level of structural perfection and very small damping of any spin waves that may propagate in them. All of this is very important from the standpoint of using YIG films for the construction of microwave devices exploiting certain kinds of waves launched at GHz frequencies. These waves can be thought of as a disturbance to the magnetic moments and hence the spin system. They are, as a consequence, called spin waves or, more commonly, magnetostatic waves in a sense to be defined later. In YIG films the only magnetic ions are ferric ions that have zero orbital momentum. This means that YIG has very narrow linewidths during ferromagnetic resonance. This chapter will call, synonymously, magnetic materials ferrites or YIG and will assume ferromagnetic behaviour. Although the Curie point of YIG is well above room temperature, in the absence of an applied magnetic field a finite sample of ferrite degenerates into small domains[1,2] associated with alternating magnetisation. The boundaries between these domains consume power so it is necessary to apply a strong enough applied magnetic field to remove this domain structure in order to prevent this loss corrupting any launched spin waves. In the rest of this chapter it is assumed that the medium has reached magnetic saturation and that the saturation magnetisation is constant. It is interesting that spin waves in YIG can be easily excited by coupling in microwave power through short-circuited or open-circuited metallic micro-stripline transducers placed onto the film. In the simplest experimental configurations the losses due to the transformation of electromagnetic waves to magnetostatic waves are about 3 dB but can be quite close to zero with careful transducer design.

## 2. MAGNETOSTATIC (DIPOLE) WAVES

The total energy of a ferromagnetic medium is

$$W(\mathbf{r}) = -\mathbf{H}_e\cdot\mathbf{M} + W_d(\mathbf{r},\mathbf{M}) + W_{ex}(\mathbf{r},\mathbf{M}) + W_a(\mathbf{M}) \tag{2.1}$$

where $\mathbf{M} \equiv \mathbf{M}(\mathbf{r})$ is the magnetisation at the point $\mathbf{r}$. The first term in Eq. (2.1) is the Zeeman energy and $W_d$ is the dipole-dipole energy; both of these energies are created by all the magnetic dipole moments. $W_{ex}$ is the exchange energy of the moment $\mathbf{M}$ at the point $\mathbf{r}$ with the magnetic moments existing at all the other points. $W_a$ is the anisotropy energy but YIG and certain other ferromagnets have only weak anisotropy so that $W_a$ is often much less than the other terms in Eq. (2.1). In the absence of an external magnetic field, as has already been stated, a real piece of ferromagnetic material has a domain structure. Even though the magnetic moment is uniform inside a single domain there are changes from one domain to another. If a strong enough external bias magnetic field $\mathbf{H}_e$ is applied to a sample under investigation, the magnetisation saturates to a value $\mathbf{M}_0$ and an internal field $\mathbf{H}_i$ is established that is parallel to $\mathbf{M}_0$. For example, in YIG at room temperature, $4\pi M_0$ is equal to 1750 G and it should be noted that it is very common in the literature to define the saturation magnetisation in non-SI units and with a factor of $4\pi$ included.

The torque exerted by a magnetic field $\mathbf{H}$ on a magnetic moment $\mathbf{M}$ is $\mathbf{M}\times\mathbf{H}$ and this torque, being due to the rate of change of the total spin angular momentum, is proportional to $\partial\mathbf{M}/\partial t$. Hence a precession of the magnetisation about the applied magnetic field occurs that is determined by the Landau-Lifshitz equation[2-5]

$$\frac{\partial \mathbf{M}}{\partial t} = \gamma[\mathbf{M}\times\mathbf{H}_{eff}] - \left(\frac{\omega_r}{M_0^2}\right)\mathbf{M}\times[\mathbf{M}\times\mathbf{H}_{eff}] \tag{2.2}$$

where $\gamma$, the electron gyromagnetic ratio, is equal to $1.76\times10^7$ rad$s^{-1}$Oe, $\omega_r$ is a relaxation parameter and $\mathbf{H}_{eff}$ is called the effective magnetic field. It is the *applied* (real) magnetic field that will appear in Maxwell's equations, together with fields arising from various interactions where the latter add up to form what can be called the *material* field.[6] Hence the total field is $\mathbf{H}_{eff} = \mathbf{H}_e + \mathbf{H}_d + \mathbf{H}_{ex} + \mathbf{H}_a$. Not all of these terms will be of interest here since the aim is to expose some of the most interesting ideas on nonlinear magnetic materials. $\mathbf{H}_e$ is obviously the externally applied field but $\mathbf{H}_d$ is a demagnetising field that will be present, for example, when a magnetic field is applied perpendicular to the plane of a thin film. The last two fields in $\mathbf{H}_{eff}$ are due to exchange and magnetic anisotropy and, although potentially important, will not concern us any further in this chapter. It is also assumed here that the magnetic material is inelastic so that the medium is rigid in the sense that the spacing between the magnetic moments is constant.[6] The spins can communicate with each other, however, by changing direction and permitting a wave of spin to flow through the material. In a typical system high frequency magnetic fields will be propagated and at a particular frequency resonance can occur. This appears as a resonant peak in the imaginary part of the diagonal and off-diagonal components of the permeability tensor. This ferromagnetic resonance, during which energy from the electromagnetic field is strongly absorbed, is observed as a resonant absorption line that is additional to other sources of loss. It is characterised by a quantity $\Delta H(\omega)$ that is the half-width of this resonant line. For YIG this is rather small and particularly good films have a linewidth of only $2\Delta H(\omega) \simeq 0.3$ Oe. Equation (2.2) has both uniform and non-uniform solutions describing uniform and non-uniform ferromagnetic resonance. The frequency of uniform resonance is $\omega_H = \gamma H_e$ if the applied field lies in the plane of the ferromagnetic film but it has the value $\omega_H = \gamma(H_e - 4\pi M_0)$ if the field is perpendicular to the surfaces of the film. Apart from these types of solution Eq. (2.2) can also have travelling wave solutions. In a travelling wave energy must be able to transfer from one point to another and only dipole-dipole and exchange energies can be responsible for such a transformation. Dipole interactions are long-range and are essential only at long wavelengths (small wavenumbers). Exchange interactions are short-range and are essential only for short wavelengths (large wavenumbers). For waves with wavenumbers, q, in the range $10\ \mathrm{cm}^{-1} \leq q \leq 10^3\ \mathrm{cm}^{-1}$ dipole interactions are predominant, whilst for waves with wavenumbers in the range $10^4\ \mathrm{cm}^{-1} \leq q \leq 10^6\ \mathrm{cm}^{-1}$ exchange interactions are predominant. Waves associated with dipole interactions are often called dipole spin waves while waves associated with exchange interactions are usually called exchange spin waves. Waves for which dipole and exchange interactions are of equal importance go by the name of dipole-exchange waves.[2-5]

The main restriction on the presentation here is that *only* dipole spin waves in ferromagnetic films will be considered. This is a reasonable assumption since many device applications are based upon films that are in this regime and, in any case, the concepts to be discussed here are best exposed, initially, without the complicating presence of exchange forces. Since, for much of the wavenumber range, the phase velocity of dipole-spin waves is much less than the velocity of light in a magnetic crystal, electromagnetic retardation can be neglected in Maxwell's equations thus producing so-called magnetostatic wave equations.

The solution of Eq. (2.2) must be coupled to the following magnetostatic equations[2,3]

$$\nabla\times\mathbf{H} = 0 \qquad \nabla\cdot\mathbf{B} = 0 \tag{2.3}$$

where $\mathbf{B}$ is the magnetic induction. Together with the boundary conditions this will give a dispersion law $\omega(\mathbf{q})$ where $\omega(\mathbf{q})$ is the angular frequency of a wave with wavevector $\mathbf{q}$. This dispersion equation depends not only upon the value of the applied field but also, quite critically, upon its orientation. It will also depend upon the dimensions of the magnetic film and the nature of the surrounding media.

There are three main types of magnetostatic waves that occur in a thin ferromagnetic film

(1) *Surface magnetostatic waves.* These, as the name implies, propagate along a surface. The externally applied magnetic field is parallel to the surface and the direction of propagation is perpendicular to this field, or at some angle to it. Usually these waves are discussed for thin magnetic films for which two coupled surfaces are involved. The surface modes are members of a class of waves now known as Damon-Eshbach waves. Their dispersion relation (or spectrum), shown in Fig. 1a, is[2,3,4,6,7]

$$\omega^2 = \omega_H^2 + \omega_H\omega_M + \frac{\omega_M^2}{4}(1 - e^{-2qd}) \tag{2.4}$$

where d is the thickness of the magnetic film, $\omega$ is the angular frequency of the wave, q is the wavenumber, $\omega_M = 4\pi\gamma M_0$, $\omega_H = \gamma H_i$ and $H_i$ is the net internal field of the film. For the surface wave case $H_i$ is equal to the applied field since the demagnetising factor is zero. The group velocity is $\frac{1}{4}(\omega_M^2/\omega)\, d \exp(-2|qd|)$, which is essentially proportional[7] to $d/\omega$ for $qd \ll 1$.

Magnetostatic surface waves have a frequency-wavenumber plot that has only one dispersion branch so that they are single mode waves. They are also nonreciprocal in the sense that the surface of the film on which the wave propagates depends upon the direction of the vectors $\mathbf{H}_e$ and $\mathbf{q}$. Quite simply, reversing the direction of propagation of the wave will cause it to propagate along the opposite surface. The waves are actually reciprocal with respect to reversing the sign of q. Changing the sign of q causes the energy distribution to change so that forward and backward waves are associated with different surfaces.

(2) *Backward volume magnetostatic waves.* These waves propagate along an external magnetic field that is also lying in the plane of the film. They have an infinite set of dispersion branches and it is possible to excite, at a particular frequency, a lot of modes. The group and phase velocities of backward volume magnetostatic waves have opposite signs, as shown in Fig. 1b, and this is the reason for their name. If $q \to 0$ then only the lowest mode has a non-zero group velocity.[2-4,6]

(3) *Forward volume magnetostatic waves.* These waves propagate in normally magnetized magnetic films.[2-4,6] Their spectrum is symmetric with respect to the direction of wave propagation in the plane of the film. The group and phase velocities are sketched in Fig. 1c to show that they have the same sign and that the waves are also multimode in character. The modes, however, are readily distinguished by their field structure within the film.

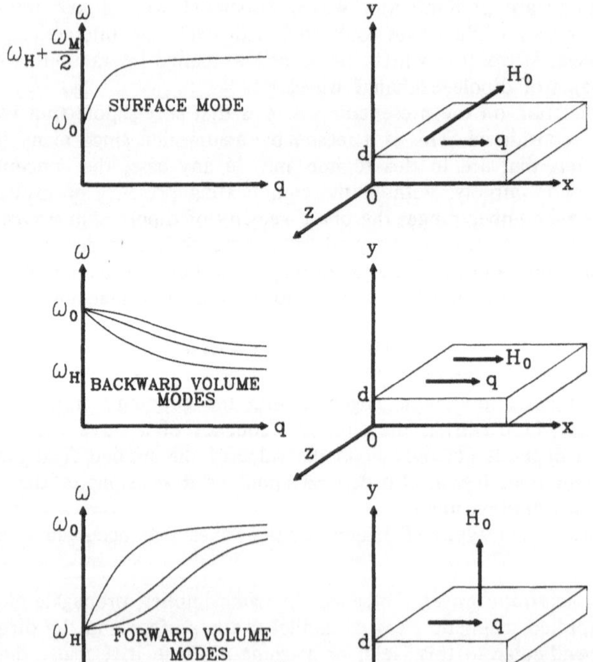


Fig. 1 Sketches of typical dispersion curves for linear magnetostatic waves in a thin ferromagnetic film. The film configurations shown are for (a) surface waves (b) backward volume modes (c) forward volume modes. $\omega_M$, $\omega_H$ and $\omega_0$ are defined in the text.

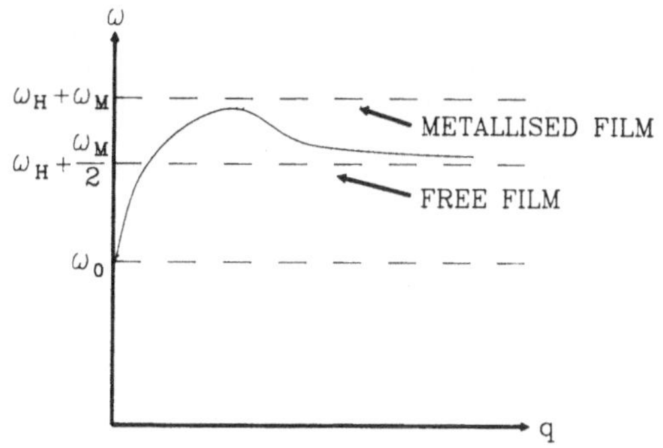


Fig. 2 Sketch of the influence of a metal coating or the near presence of a metal film on the linear Damon-Eshbach surface modes. The Damon-Eshbach modes are the dotted curve labelled free film. The coated (metallised) film frequencies are raised but the near presence of a metal plate creates a peak in the dispersion curve.

Although it is traditional now to refer to these waves as volume modes, they are, in fact, *guided* modes. A more modern terminology, and one that would produce an accord with optics, would restrict the word 'volume' to bulk waves in infinite unbounded media.

Magnetostatic surface waves and backward volume magnetostatic waves do not possess isotropic properties. Also it is very interesting that magnetostatic surface waves exist only if the angle between the direction of the propagation vector **q** and the Ox axis of Fig. 1a is less than a critical value $\phi$, where $\cos^2\phi = \omega_M/(\omega_H + \omega_M)$. As the angle $\phi$ increases from zero, the frequency spectrum of the magnetostatic surface waves narrows until it vanishes upon entering the guided wave region.

Magnetostatic surface wave spectra are strongly dependent on the boundary conditions of the film. For example, if one of the surfaces of the film is metallized,[8,9] then the upper boundary of the frequency spectrum will grow. If there is a gap between the metal and the surface of the film, then the spectrum of a magnetostatic surface wave changes dramatically and part of the spectrum will acquire a negative group velocity, as shown in Fig. 2.

All of these magnetostatic wave types occur within a dipole approximation. If $W_{ex}$ is taken into account in the total energy of a ferromagnetic substance, the spectrum of the spin waves changes in a non-trivial way. In particular, it is necessary to consider exchange boundary conditions that cause short-wavelength exchange waves to propagate perpendicular to the

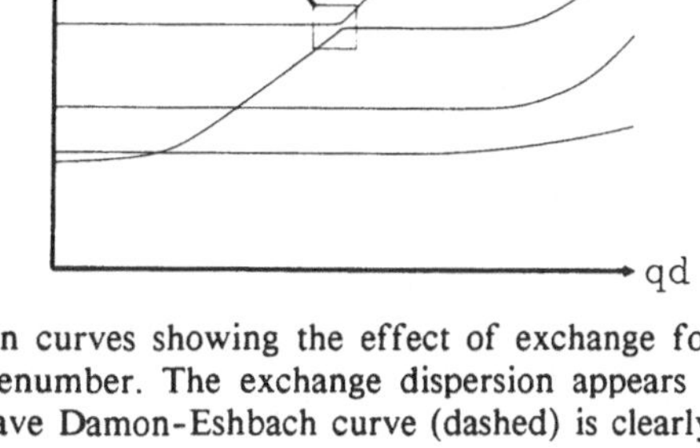

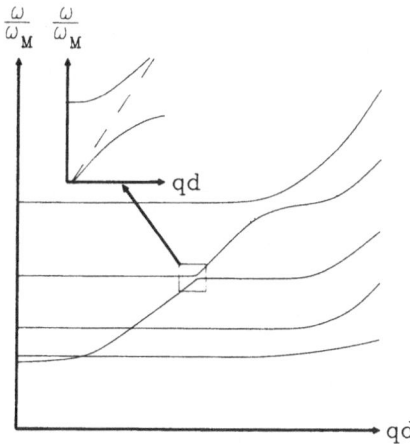

Fig. 3 Typical dispersion curves showing the effect of exchange for zero or small z-component of the wavenumber. The exchange dispersion appears as straight lines. Splitting of the surface wave Damon-Eshbach curve (dashed) is clearly shown.

surfaces of the film.[7,10] This propagation between the film surfaces creates standing spin waves. The dispersion relation of these exchange-dominated spin waves depends only on the wavenumber component perpendicular to the film surfaces and consists of a set of lines parallel to the propagation wavenumber axis. They can be introduced by letting them cross the dipole-spin Damon-Eshbach curves causing them to split,[7,10] as shown in Fig. 3. Such splitting will not be encountered in the nonlinear work to be described here because exchange has not been included. As emphasised before, this is justified on the grounds that attention will be restricted to wavenumbers that are low enough to avoid this complication.

## 3. NONLINEAR MAGNETOSTATIC WAVES

The damping term[4,5] in Eq. (2.2) will now be omitted. It can be included, if necessary, but the experimental situations envisaged here will depend upon such narrow linewidths that the inclusion of damping at this stage will only be a distraction. The existence of magnetostatic waves means that the magnetisation in the film is the sum of the saturation magnetisation $\mathbf{M}_0$ and a time-dependent magnetisation $\mathbf{m}(t)$, where $\mathbf{M} = \mathbf{M}_0 + \mathbf{m}(t)$. Also the magnetic field is the effective internal field ($\mathbf{H}_{eff} \equiv \mathbf{H}_i$) plus a time-dependent quantity $\mathbf{h}(t)$, where $\mathbf{H} = \mathbf{H}_i + \mathbf{h}(t)$. Usually, Eq. (2.2) is linearised but many higher *nonlinear* terms can, in principle, be retained. In general the nonlinearity depends in an interesting way upon the amplitude of the spin waves, and strong deviations from linear behaviour appear. These deviations can be divided into two groups.

The first includes processes in which new waves are created as the principal wave propagates through the material. For example, a wave with wavevector $\mathbf{q} \simeq 0$ and angular frequency $\omega$ may decay[5,11] to two waves with frequency $\omega/2$ having wavevectors $\mathbf{q}_1$ and $\mathbf{q}_2$, where $\mathbf{q}_1 = -\mathbf{q}_2$. In this case $\Delta q = |\mathbf{q} - \mathbf{q}_{1,2}| >> q$ and a large momentum exchange occurs. Three waves appear in the interaction, or, using quantum language, three magnons. The magnon is the quantum unit of energy of a spin wave and clearly a three-magnon event is a *second-order* nonlinear effect. Another possibility is the mutual interaction of four waves, or four magnons, in which two magnons at frequency $\omega$ combine to produce magnons at frequencies $\omega_1$ and $\omega_2$. Again, $q \simeq 0$ and $\mathbf{q}_1 = -\mathbf{q}_2$, so that $\Delta q >> q$ and is not small. The presence of four magnons in the interaction shows that it is a *third-order* nonlinear effect. The theoretical framework that has been used for this first group is macroscopic magnon theory, characterised by conservation laws[12,13] for momentum and energy. It will be returned to in section 3.1.

The second group of nonlinear effects is characterised[12] by the condition $\Delta q = |\mathbf{q} - \mathbf{q}_{1,2}| << q$. For four-magnon interactions this condition characterises all self-interaction processes, such as self-phase modulation leading to modulational instability and solitons. Modulational instability is the exponential *growth* in time of a single perturbation wave created by a third-order nonlinearity coupled to anomalous dispersion (see other parts of this book). It is the mixing of four waves (magnons) that achieve phase-matching through the self-phase modulation of the waves induced by the third-order nonlinearity, operating at the fundamental frequency. In this case the dispersion equation takes on the form $\omega(\mathbf{q},|A|)$, from which the nonlinear coefficient determining possible soliton behaviour can be derived. Also the instability develops over a finite length L so that an indeterminacy $\Delta q$ in the wavenumber sets in. Thus, for this process, the conservation laws that are such an important feature of the first group of nonlinearities are not applicable.

For self-phase modulation to be the dominant process in a thin magnetic film it must be possible to avoid any complications due to second-order nonlinear effects. This is especially important here since, unlike the isotropic dielectric media so familiar in high-frequency optics, ferromagnetic media exhibit all orders of nonlinearity. Hence, in principle, second-harmonic generation or second-order dc fields could occur, even though only third-order-based solitonic phenomena are being sought.

Another theoretical method can be used, that is not based upon Holstein-Primakoff transformations. It involves the classical development of the appropriate nonlinear terms of the Landau-Lifshitz equation using perturbation theory. It is a continuum theory that works entirely with the classical magnetisation so that no magnon concepts are really necessary. If it is remembered that, in this second group, $\Delta q << q$ then these are momentum changes that do not span very much of the magnon Brillouin zone. A classical, continuum approach, analogous to the use of the theory of sound for low q phonon problems, ought to be enough. This type of perturbation theory involves the expansion of the magnetisation into linear, second-order

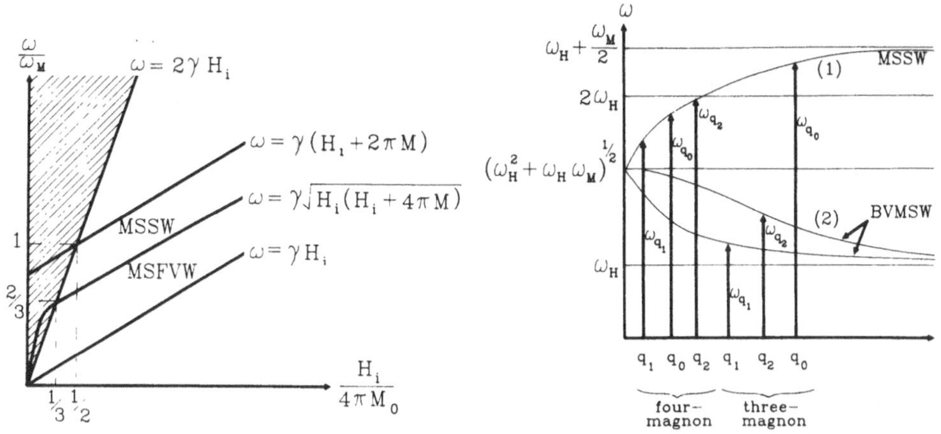


Fig. 4 (a) Limiting regions of the linear dispersion of magnetostatic waves on a thin ferromagnetic film. $H_i$ is the internal field and the shaded region is where nonlinear effects will set in. (b) Illustation of some possible energy conservation laws for three- and four-magnon processes. All disperson curves are plotted on the same plane against modulus of wavenumber.

and third-order terms. The development is stopped at third-order since that will embrace all the interesting nonlinear effects. The second-order fields actually vanish for magnetostatic surface waves that are directed perpendicular to the applied field but not for other directions or for the forward and backward guided waves. Nevertheless, even though second-order fields exist in principle, frequencies that initiate within the narrow magnetostatic wave bands of thin films are usually non-resonant and lie outside the original linear band. These are likely to be of zero amplitude or, at best, weakly driven modes. If $H_i = 10^3$ a typical band covers the range 2.8 GHz to 4.5 GHz so that for all frequencies $\omega$ initiated within this magnetostatic wave region the second-harmonic $2\omega$ lies above the linear eigenfrequency band and can be neglected.

As discussed above, there are two groups of effects that will be referred to here as decay (processes) instabilities (three-magnon and four-magnon effects) and self-action processes (modulational instability, soliton generation). They will generally appear above certain power thresholds and will be addressed below. For bulk (infinite with no boundaries except at infinity) materials the first group of nonlinearities are very well-known and are traditionally called first- and second-order Suhl instabilities.[14,15] For those readers approaching this subject from optics it should be pointed out that a first-order Suhl instability is a manifestation of a second-order magnetic nonlinearity which in turn involves three magnons.[11] The second-order Suhl instability arises from a third-order nonlinearity and involves four magnons.[11]

### 3.1 Three-magnon effects

The word instability will be used here to mean that the wave, after reaching a certain threshold power, will change in some way. Modulational instability[16,17] is one such instability that is also well-known in optics. It is, however, a four-magnon process characterised by the breakup or decay of a steady state cw wave into a stream of pulses. This case will not be discussed in this subsection, nor will soliton formation that is also a four-magnon process.

Mathematically, instability can be proved if, initially, the wave suffers unchecked exponential growth in time. The onset of instability heralds breakup of the wave. For a three-magnon process in which $\Delta q = |\mathbf{q} - \mathbf{q}_{1,2}| \sim q$ the outcome of this breakup is a decay of the wave into two other waves. The proof of this takes the form of finding the conditions under which a perturbation to the magnon system will begin to grow, as opposed to die away. If instability above a threshold power is established then the calculation ceases with the statement that breakup will occur and that the only kind of breakup contemplated for $\Delta q = |\mathbf{q} - \mathbf{q}_{1,2}| \gg q$ is the creation of new waves.[18,19] This physical process is seen experimentally as the creation of satellite frequencies (waves) in addition to that of the initiating, principal, wave.

Suppose now that the thin film is supporting a wave with wavevector $\mathbf{q}_0$ and frequency $\omega_{\mathbf{q}_0}$. Under a three-magnon process with $\Delta q \sim q$ a pair of secondary waves with frequency and wavevector $(\omega_{\mathbf{q}_1},\mathbf{q}_1)$ and $(\omega_{\mathbf{q}_2},\mathbf{q}_2)$ can be created above a threshold power. It has been shown that this threshold is lower than that for the four-magnon decay to be described below. Hence three-magnon decay, unless it is forbidden, will occur first.

For the three-magnon process both energy (frequency) and momentum (wavevector) will be conserved so that the matching conditions are[11,18,20]

$$\omega_{\mathbf{q}_0} = \omega_{\mathbf{q}_1} + \omega_{\mathbf{q}_2} \qquad \mathbf{q}_0 = \mathbf{q}_1 + \mathbf{q}_2 \tag{3.1.1}$$

In Fig. 4a the linear dependence of the frequency on the internal magnetic field $H_{eff} \equiv H_i$ and the saturation magnetisation $M_0$ of the film are introduced. This diagram is well-known[21] and is of considerable help in deciding which frequency regions will sustain three-magnon processes. The loci $\omega_H = \gamma H_i$ and $\omega = \gamma\sqrt{H_i(H_i + 4\pi M_0)}$ bound the spectrum of the linear volume (guided) magnetostatic waves. The loci $\omega = \gamma\sqrt{H_i(H_i + 4\pi M_0)}$ and $\omega = \gamma(H_i + 2\pi M_0)$ bound the linear magnetostatic surface wave spectrum. The energy conservation law shows that, since the minimum frequency that a volume magnetostatic wave can have is $\omega = \gamma H_i$, a surface wave that is going to decay into a volume wave must have a frequency at least equal to $2\gamma H_i$. All waves, then, that decay through a three-magnon process originate with frequencies above the locus $\omega = 2\gamma H_i$ shown in Fig. 4a. The point is emphasised further in Fig. 4b where a possible decay scheme for surface waves into backward volume waves is shown. Another feature of Fig. 4a is that the locus $\omega = 2\gamma H_i$ confines the three-magnon decay of surface waves to quite small internal magnetic fields in the region where the locus $\omega = \gamma\sqrt{H_i(H_i + 4\pi M_0)}$ has more curvature. It is interesting that even the region above the upper surface wave locus, and above the locus $\omega = 2\gamma H_i$, can take part in three-magnon decay provided that external pumping at these frequencies occurs. The whole shaded part of Fig. 4a is known as the *coincidence limiting* region,[21] since it is where the internal field equals the threshold field for Suhl instability.

Magnetostatic surface waves have been investigated experimentally for this kind of instability and the theoretically expected frequency-satellites have been observed in the spectrum of the transmitted signals of thin YIG films. Very simply, high power intensive magnetostatic surface waves with frequency $\omega$ decay to volume magnetostatic waves with frequencies $\omega_1 = \omega_2 = \omega/2$, propagating in mutually opposite directions and parallel to the magnetic field. Because of the multimode character of backward or forward magnetostatic waves, the surface waves can generate volume waves with different mode numbers. After these volume waves are created by the original decay process they can mix to produce magnetostatic surface waves again, but with shifted frequencies. These frequencies can also be readily observed.

The two theoretical techniques to be used in this chapter are both classical in origin but one uses a Hamiltonian formalism[5,11] and the other a wave mixing perturbation method.[22] For the analysis of three-magnon and, subsequently, four-magnon decay, a macroscopic magnon theory has been developed. This type of theory[23] permits the use of the magnetisation rather than the spins on the atoms. It is valid only for wavenumbers much less than the reciprocal of the lattice spacing and is therefore to be used well away from the magnon zone boundaries.[23] The emphasis is not on the atoms and their spins, so continuum concepts and phenomenological constants can be used, and the equations can be reduced to canonical (i.e. accepted orthodox) form.

An important representation for a magnon system can be obtained through the introduction of Holstein-Primakoff field variables.[5,11] The Holstein-Primakoff method transforms the Hamiltonian of the spin system to harmonic oscillator form. This action permits the magnetisation and the Hamiltonian to be expanded in terms of magnon creation and annihilation operators. Products of three or more operators characterise transitions between magnon states.

Creation and annihilation operators first appeared in the quantum theory of the simple harmonic oscillator, where the oscillator has a quantum number n labelling a stationary state containing n quanta. This state has eigenvector $|n\rangle$ and a transition in which one quantum of energy equal to $\hbar\omega$ is absorbed *raises* the vector to $|n+1\rangle$. Similarly a transition with the emission of one quantum of energy *lowers* the vector to $|n-1\rangle$. The raising and lowering operators are called creation and destruction (annihilation) operators, respectively defined as $\alpha$ and $\alpha^*$. Since the spin system, like the phonon field and the electromagnetic field, can be represented by a Hamiltonian function that describes a set of harmonic oscillators, a representation of the magnon system in terms of creation and annihilation operators, as opposed to spin operators, is completely justified.

Briefly, some important mathematical properties of $\alpha$ and $\alpha^*$ are[24]

$$\alpha^* |n\rangle = \sqrt{(n+1)}|n+1\rangle \qquad \alpha|n\rangle = \sqrt{n}|n-1\rangle \tag{3.1.2}$$

$$\alpha^*\alpha|n-1\rangle = (n-1)|n-1\rangle \qquad \alpha^*\alpha|n+1\rangle = (n+1)|n+1\rangle \qquad \alpha^*\alpha|n\rangle = n|n\rangle \tag{3.1.3}$$

Suppose now that a magnetic field is applied along the z-axis. The $i^{th}$ spin vector of the set of spins that constitutes the magnetic material will have a z-component $S_{iz}$ along this field direction. $S_{iz}$ will have quantum values ranging from -S through to S. It is therefore logical to introduce a spin-deviation occupation number $n = S - S_{iz}$, where n is the reduction in the z-component of the $i^{th}$ spin, in the whole collection of spins that make up the magnetic material, from its maximum possible value S. Hence creation and annihilation operators $\alpha \equiv \alpha_i$ and $\alpha^* \equiv \alpha_i^*$ that change n by ±1 can now be introduced and they will cause $S_{iz}$ to change by plus or minus one spin-deviation number. As is the case in the quantum theory[24] of angular momentum and spin it is more convenient to work with $S_i^{\pm} = S_{ix} \pm S_{iy}$ where

$$S_i^{\pm}|S_{iz}\rangle = [S(S+1) - S_{iz}(S_{iz} \pm 1)]^{1/2}|S_{iz} + 1\rangle \tag{3.1.4}$$

The form for the eigenvalue given in Eq. (3.1.4) is quite easy to prove from the quantum theory of angular momentum and spin, provided that it is remembered that the magnitude of the spin vector **S** is $\sqrt{S(S+1)}$ and not simply S. In other words the eigenvalue of the spin operator $S^2$ is $S(S + 1)$.

The eigenvalues in the [ ] brackets evaluate to[5]

$$S(S+1) - S_{iz}(S_{iz} \pm 1) = \begin{cases} 2S\left[1 - \dfrac{n-1}{2S}\right]n \\ 2S(n+1)\left[1 - \dfrac{n}{2S}\right] \end{cases} \tag{3.1.5}$$

where n, n + 1 and n - 1 will now be treated as operators. Their location in the factorisation is important as will now be seen. A comparison of Eqs. (3.1.5) and (3.1.4) with the creation and annihilation equations (3.1.2) and (3.1.3) suggests the following form for the operators $S_i^+$, $S_i^-$ and $S_{iz}$:

$$S_i^+ = (2S)^{1/2}[1 - \alpha_i^*\alpha_i]^{1/2}\alpha_i \tag{3.1.6}$$

$$S_i^- = (2S)^{1/2}\alpha_i^*[1 - \alpha_i^*\alpha_i]^{1/2} \tag{3.1.7}$$

$$S_{iz} = S - \alpha_i^*\alpha_i \tag{3.1.8}$$

The location of the $\alpha^*$, $\alpha$ operators, after expanding the square roots in Eqs. (3.1.6) and (3.1.7), is justified in the following way. The lowest terms, after the linear terms, will be $\alpha_i^*\alpha_i\alpha_i$ and $\alpha_i^*\alpha_i^*\alpha_i$, if the order that has been selected is correct. These operators act upon $|n\rangle$ to give

$$\begin{aligned} \alpha_i^*\alpha_i\alpha_i|n\rangle &= \sqrt{n}(n-1)|n-1\rangle \\ \alpha_i^*\alpha_i^*\alpha_i|n\rangle &= n\sqrt{n+1}|n+1\rangle \end{aligned} \tag{3.1.9}$$

The Holstein-Primakoff transformations of Eqs. (3.1.6) to (3.1.8) are nonlinear and it must always be understood that the square root must be expanded. The neglect of higher order terms is justified at low temperatures since it is the average value of $\alpha_i^*\alpha_i$ that has to be neglected as the series is cut-off. Nevertheless, higher products of $\alpha_i^*\alpha_i$ can always be included and it would appear that going as high as products of four operators gives an adequate explanation of the observed results for thin magnetic films. The operators also satisfy the usual commutation relations. However, since this development is based upon the spin system it is one stage removed from the classical, macroscopic, concept of magnetisation. The latter is the physical phenomenological quantity that would appear naturally in Maxwell's equations. It is therefore more useful, if it can be justified, to work in terms of magnetisation so this formulation will now be introduced.

Classically, the initial equations required are the nonlinear equation (2.2) and the magnetostatic equations (2.3). The magnetisation is $\mathbf{M}_0 + \mathbf{m}$, where the high frequency, alternating, magnetization $\mathbf{m}$ has components $m_x$, $m_y$ that will be assumed to have a magnitude much less than $M_0$. The total magnetisation satisfies Eq. (2.2) and since the second term on the right-hand side is going to be neglected here[2,4]

$$\mathbf{M}\cdot\left(\frac{d\mathbf{M}}{dt}\right) = \gamma\mathbf{M}\cdot[\mathbf{M}\times\mathbf{H}] = 0 \tag{3.1.10}$$

so that

$$\frac{d(|\mathbf{M}|^2)}{dt} = 0 \tag{3.1.11}$$

The integration constant can only be $M_0^2$ where $M_0^2 = M_x^2 + M_y^2 + M_z^2$ so that the total magnetisation is always conserved. This result, for $\mathbf{M} = \mathbf{z}M_0 + \mathbf{m}$, becomes

$$m_x^2 + m_y^2 + m_z^2 + M_0^2 + 2M_0 m_z = M_0^2 \tag{3.1.12}$$

Therefore,

$$M_z \simeq M_0\left[1 - \frac{m_x^2 + m_y^2}{2M_0^2}\right] \qquad \frac{m_x^2 + m_y^2}{2M_0^2} \ll 1 \tag{3.1.13}$$

As indicated above it is more physical in an investigation of relaxation processes to work with the magnetisation. Furthermore, since an analysis based upon the actual spins exists in terms of creation and annihilation operators, such concepts should be introduced here. Instead of the creation and annihilation operators of quantum mechanics, however, field variables that are semi-classical analogues of the magnon creation and annihilation operators are introduced. These are defined as $\alpha(\mathbf{r})$ and $\alpha^*(\mathbf{r})$ where $\mathbf{r} = (x, y, z)$ is a position vector of the space in which the magnetisation exists. They satisfy the commutation relationship

$$[\alpha(\mathbf{r}), \alpha^*(\mathbf{r}')] = \delta(\mathbf{r} - \mathbf{r}') \tag{3.1.14}$$

The magnetisation can be expressed in terms of these operators by using Eqs. (3.1.6) to (3.1.8) and the fact that the magnetisation $\mathbf{M}$ is related to the spin momentum vector $\mathbf{S}$ through $\mathbf{M} = \gamma\mathbf{S}$, where $\gamma$ is the gyromagnetic ratio and the spin is measured in units of $\hbar$. In terms of these operators, the combinations $M_\pm = M_x \pm iM_y$ become[5]

$$M_+ = M_x + iM_y = \sqrt{2\gamma M_0}\left[1 - \frac{\gamma\alpha^*(\mathbf{r})\alpha(\mathbf{r})}{2M_0}\right]^{1/2}\alpha(\mathbf{r}) \tag{3.1.15}$$

$$M_- = M_x - iM_y = \sqrt{2\gamma M_0}\,\alpha^*(\mathbf{r})\left[1 - \frac{\gamma\alpha^*(\mathbf{r})\alpha(\mathbf{r})}{2M_0}\right]^{1/2} \tag{3.1.16}$$

and

$$M_z = M_0 - \gamma\alpha^*(\mathbf{r})\alpha(\mathbf{r}) \tag{3.1.17}$$

Once again it must be emphasised that the square roots have a formal significance only, i.e. they are shorthand for an infinite series of operators created from the $\alpha^*\alpha$ terms.

$M_x$ and $M_y$, after expanding the square root and retaining terms up to $(\gamma\alpha^*\alpha)/(4M_0)$, are

$$M_x = \frac{M_+ + M_-}{2} \simeq \frac{1}{2}\sqrt{2\gamma M_0}\left[\alpha(\mathbf{r}) + \alpha^*(\mathbf{r}) - \frac{\gamma}{4M_0}\alpha^*(\mathbf{r})\alpha(\mathbf{r})\alpha(\mathbf{r}) - \frac{\gamma}{4M_0}\alpha^*(\mathbf{r})\alpha^*(\mathbf{r})\alpha(\mathbf{r})\right] \tag{3.1.18}$$

$$M_y = -\frac{i(M_+ - M_-)}{2} \simeq -\frac{i}{2}\sqrt{2\gamma M_0}\left[\alpha(\mathbf{r}) - \alpha^*(\mathbf{r}) - \frac{\gamma}{4M_0}\alpha^*(\mathbf{r})\alpha(\mathbf{r})\alpha(\mathbf{r}) + \frac{\gamma}{4M_0}\alpha^*(\mathbf{r})\alpha^*(\mathbf{r})\alpha(\mathbf{r})\right] \tag{3.1.19}$$

The nonlinear terms, compared to the linear $\alpha$ terms, are actually second-order in size. It is easily appreciated that the presence of a product of three $\alpha$-type operators means that three magnons are involved. Hence Eqs. (3.1.18) and (3.1.19) will give an adequate description of the confluence and splitting processes involving three magnons. In either process both momentum and energy are conserved. The combinations $\alpha^*(\mathbf{r})\alpha(\mathbf{r})\alpha(\mathbf{r})$ and $\alpha^*(\mathbf{r})\alpha^*(\mathbf{r})\alpha(\mathbf{r})$ are interesting. In the first one we imagine two magnons being annihilated to create a third one, while in the second combination we imagine one magnon being annihilated to create two more. The first process is a confluence and the second process is a three-magnon splitting or decay. For bulk materials such a nonlinear process is known as the first-order Suhl instability in which, for example, two oppositely travelling magnons are created from a single magnon near to zero wavenumber. Since energy is conserved in the process then if the original magnon that is *annihilated* has a frequency of $\omega$, the final *created* magnons[5,11] must have a frequency of $\omega/2$.

In the foregoing discussion the basic formalism for discussing three-magnon decay has been established. It can be seen that a formalism based upon creation and annihilation operators is very useful in this context and it remains now to show, briefly, how the full calculations can be done. Apart from stating that we now need to invoke Hamilton's equations and to effect a further transformation into momentum space through the use of Fourier transforms, the rest of the presentation is a matter of detail. The discussion, therefore will be confined to showing the principal results.

If W is the total energy (the Hamiltonian) of the magnetic material then *functional derivatives*[26] (denoted by the appearance of $\delta$) of W with respect to $\alpha$ and $\alpha^*$ can be introduced to give[12,20]

$$\frac{\partial\alpha^*(\mathbf{r})}{\partial t} = i\frac{\delta W}{\delta\alpha(\mathbf{r})} \tag{3.1.20a}$$

$$\frac{\partial\alpha(\mathbf{r})}{\partial t} = -i\frac{\delta W}{\delta\alpha^*(\mathbf{r})} \tag{3.1.20b}$$

Equations (3.1.20) closely resemble Hamilton's equations of classical dynamics. Obviously, such equations are not confined to a description of magnetic media and are a general description of any nonlinear media that can be expressed in this canonical form.

In the representation of the Hamiltonian in terms of the Holstein-Primakoff field variables $\alpha^*(\mathbf{r})$ and $\alpha(\mathbf{r})$, it is not sufficient to work in $\mathbf{r}$ space. As is seen quite often in the literature, it is much more revealing when working with coupled systems to work with Fourier components ($\mathbf{q}$ space). At the level of development that works with the spins on atomic sites, the Fourier transform changes the representation from operators at individual sites to collective operators expressed as Fourier amplitudes. This advantage is also apparent in the present case since we can change from the Holstein-Primakoff type of field variables to magnon field variables, defined from Fourier expansions.[5,20,23]

The formulation of these problems goes back fifty years[25] but until recently attention was focused upon the bulk and, if thin films were considered, three-magnon effects were not studied specifically in relation to the guiding properties of the film. More recent work[18-20,26,27] concentrates on these issues and very interesting results are emerging.

Guiding within a thin film can be introduced at the Fourier transform stage by defining a wavevector $\mathbf{q} = (q_y, q_z)$ that is parallel to the (y,z) plane and, for the purpose of this chapter, parallel to the film surfaces. The transverse x-dependence, representing the presence of modal fields, is to be included through a function $\theta_{\mathbf{q}}(x,\mathbf{q})$. The spatial Fourier transforms of $\alpha(\mathbf{r})$, $M(\mathbf{r})$ and the magnetostatic potential $\psi(\mathbf{r})$ are then defined as

$$\alpha(\mathbf{r}, x) = \int \alpha_{\mathbf{q}} e^{i\mathbf{q}\cdot\mathbf{r}} \theta_{\mathbf{q}}(x,\mathbf{q}) d\mathbf{q}$$

$$\mathbf{M}(\mathbf{r}, x) = \int \mathbf{M}_{\mathbf{q}} e^{i\mathbf{q}\cdot\mathbf{r}} \theta_{\mathbf{q}}(x,\mathbf{q}) d\mathbf{q} \tag{3.1.21}$$

$$\psi(\mathbf{r}, x) = \int \psi_{\mathbf{q}} e^{i\mathbf{q}\cdot\mathbf{r}} \theta_{\mathbf{q}}(x,\mathbf{q}) d\mathbf{q}$$

where $\mathbf{r}$ is now a spatial vector lying in the plane of the film (y,z plane), $\alpha_{\mathbf{q}}$, $\mathbf{M}_{\mathbf{q}}$ and $\psi_{\mathbf{q}}$ are Fourier amplitudes and the factor $\theta_{\mathbf{q}}(x,\mathbf{q})$, representing the spatial variation of the modal field across the thin film, is given by

$$\theta_{\mathbf{q}} = e^{px} + \lambda e^{-px} \tag{3.1.22}$$

where $p^2 = q_y^2 + q_z^2/(1 + \chi_1)$, $\lambda = [\mu_1 + \mu_2\cos(\phi) - 1]/[\mu_1 - \mu_2\cos(\phi) + 1]$, $\mu_1(\omega)$ and $\mu_2(\omega)$ are the familiar linear components of the Polder form of the magnetic permeability tensor, $\chi_1$ is the magnetic susceptibility associated with $\mu_1$, the internal magnetic field is along the z-axis and $\phi$ is the angle between the propagation direction along $\mathbf{q}$ and the y-axis. All the Fourier amplitudes of $\mathbf{M}$ and $\psi$ can be expressed in terms of $\alpha_{\mathbf{q}}$ and $\alpha^*_{\mathbf{q}}$, although it can be a fairly long and tedious process. The easiest results to obtain are expressions for the magnetisation components. For example[20]

$$M_x = \frac{1}{2}\sqrt{2\gamma M_0}\int [\alpha_{\mathbf{q}} + \alpha^*_{-\mathbf{q}}] e^{i\mathbf{q}\cdot\mathbf{r}} \theta_{\mathbf{q}}(x,\mathbf{q}) d\mathbf{q}$$

$$- \frac{\gamma}{8M_0}\int\int\int [\alpha^*_{-\mathbf{q}}\alpha_{\mathbf{q}'}\alpha_{\mathbf{q}''} + \alpha^*_{-\mathbf{q}}\alpha^*_{-\mathbf{q}'}\alpha_{\mathbf{q}''}] e^{i(\mathbf{q}\cdot\mathbf{r}+\mathbf{q}'\cdot\mathbf{r}+\mathbf{q}''\cdot\mathbf{r})} \theta_{\mathbf{q}}(x,\mathbf{q}) d\mathbf{q}\, \theta_{\mathbf{q}'}(x,\mathbf{q}') d\mathbf{q}'\, \theta_{\mathbf{q}''}(x,\mathbf{q}'') d\mathbf{q}'' \tag{3.1.23}$$

Similar expressions are obtainable for the other magnetisation components and the negative wavenumber subscripts arise from changing the variable from $\mathbf{q}$ to $-\mathbf{q}$ whenever a complex conjugate is used.

The Hamiltonian of the magnetic system assumes the series form

$$W = W_2 + W_3 + .... \tag{3.1.24}$$

where $W_i$ (i = 2, 3) denotes two- and three-magnon interactions. The application of Hamilton's equations and, for this section, restricting attention to three-magnon decay processes would, formally, seem to provide a good basis for a solution of the problem being addressed. Nevertheless, yet another transformation, popularly known as a Bogoliubov transformation, is normally required to diagonalise the Hamiltonian. It will have been noticed in Eq. (3.1.23), for example, that states $\mathbf{q}$ and $-\mathbf{q}$ appear and this means in effect that the Hamiltonian is not yet diagonal. If it is not diagonal then the correct form of the normal modes and the eigenvalues have not been obtained. Hence it is an essential, yet normal, step in eigenvalue problems to diagonalise the representation of the Hamiltonian. In other words, if we can diagonalise the Hamiltonian by constructing some special matrix or unitary transformation to perform this task then the eigenvalues will in practice have been found. In classical wave propagation language, determining the eigenvalues means finding the dependence of the frequency of propagation upon the wavenumber.

It is often the case that the off-diagonal terms of the Hamiltonian matrix are quite small compared to the diagonal ones so the process of diagonalising the Hamiltonian need not be carried out. In this case, the operators $\alpha_{\mathbf{q}}$, $\alpha^*_{\mathbf{q}}$ are the ones to use. If the final diagonalisation stage is necessary then the $\alpha_{\mathbf{q}}$ operators are converted to new operators $b_{\mathbf{q}}$ by a unitary transformation.[5,11] This mathematical procedure is not difficult but the details are rather too complicated for any real pedagogic value to be gained by giving them here. Only the results will be given, therefore. The operators $b_{\mathbf{q}}$ are constructed to diagonalise the linear part of the Hamiltonian so that, with the inclusion of three-magnon processes,[20]

$$W = \int \omega_{\mathbf{q}} b_{\mathbf{q}} b^*_{\mathbf{q}} d\mathbf{q} + \iiint [V_{\mathbf{q},-\mathbf{q}',\mathbf{q}''} b_{\mathbf{q}} b_{\mathbf{q}'} b^*_{\mathbf{q}''} + \text{c.c.}]\delta(\mathbf{q} - \mathbf{q}' - \mathbf{q}'')d\mathbf{q}\,d\mathbf{q}'d\mathbf{q}'' \tag{3.1.25}$$

where $\omega_{\mathbf{q}}$ is the linear dispersion law of magnetostatic waves and the new variables $b_{\mathbf{q}}$ are usually simply related to $\alpha_{\mathbf{q}}$ through a proportionality coefficient defined in terms of $\omega_H$ and $\omega_M$. If the matrix elements V, corresponding to three-magnon effects, are zero then W has the form that is expected for a set of *independent* harmonic oscillators. This is, of course, correct since the collective magnon system, in the absence of nonlinear effects, is precisely represented by such a set of oscillators. Also if $W = \Sigma W_{\mathbf{q}}$, where $W_{\mathbf{q}} = \omega_{\mathbf{q}} b_{\mathbf{q}} b^*_{\mathbf{q}}$, then the Heisenberg equations of motion[23] give

$$i\frac{db_{\mathbf{q}}}{dt} = \omega_{\mathbf{q}} b_{\mathbf{q}} \tag{3.1.26a}$$

This is interesting because, provided that normal absorption losses are not dominant, $b_{\mathbf{q}}$ will have a simple $e^{i\omega t}$ time dependence. The influence of the three-magnon interaction terms of Eq. (3.1.25) adds another term to this equation to make it look like

$$\frac{db_{\mathbf{q}}}{dt} + i\omega_{\mathbf{q}} b_{\mathbf{q}} = \text{three-magnon perturbation} \tag{3.1.26b}$$

Two questions then arise. The first concerns the possibility that this extra term will stimulate purely exponential decay or growth in the form of $b_{\mathbf{q}}$, which is what is meant by instability. The second concerns the possibility that such instability can only proceed above a certain power threshold. Both of these questions will now be answered. For a thin magnetic film $b_{\mathbf{q}} \simeq \alpha_{\mathbf{q}}$ so the equation from which the decay instability due to three-magnon processes can be estimated is[12,20]

$$\frac{\partial\alpha_{\mathbf{q}}}{\partial t} + i\omega_{\mathbf{q}}\alpha_{\mathbf{q}} = -2i\iint V_{\mathbf{q},-\mathbf{q}',\mathbf{q}''}\alpha_{\mathbf{q}'}\alpha^*_{\mathbf{q}''}\delta(\mathbf{q} - \mathbf{q}' + \mathbf{q}'')d\mathbf{q}'d\mathbf{q}'' \tag{3.1.27}$$

where $V_{\mathbf{q},-\mathbf{q}',\mathbf{q}''}$ is the three-magnon interaction matrix element and $\delta(\mathbf{q} - \mathbf{q}' + \mathbf{q}'')$ is the usual Dirac $\delta$-function. The classical first-order Suhl instability could, for example, involve the decay of a wave with wavenumber $\mathbf{q} = 0$ into two counter-propagating waves with wavenumbers $\mathbf{q}'$ and $-\mathbf{q}''$. This mechanism will also be considered here but this time the decay of a magnetostatic surface wave with momentum $\mathbf{q} \simeq 0$ into two backward volume magnetostatic waves propagating in opposite directions is considered. In the limit $qd \ll 1$, for magnetic films of thickness d, the matrix elements in Eq. (3.1.27) become[20]

$$V_{\mathbf{q},-\mathbf{q}',\mathbf{q}''} = -\frac{i\omega_M}{2\pi}\left[\frac{g}{2M_0}\right]^{1/2}\beta^2[(1 + \lambda) + 2qd](1 + qd)d \tag{3.1.28}$$

Since the magnetostatic surface waves decay approximately into waves with equal and opposite wavenumbers, $|\mathbf{q}'| = |\mathbf{q}''|$, $\beta = [2\Delta\omega/\omega_M]^{1/2}$, $\Delta\omega = \omega_{\mathbf{q},\mathbf{q}''} - \omega_H$ and $\Delta\omega$ is the difference between the frequency of the backward volume magnetostatic waves and the bottom edge of the backward volume magnetostatic wave spectrum. In the thin film case $\lambda \simeq \lambda_1 \simeq -1 + Gqd$, where G is a constant.

The stability of the exact harmonic solution of Eq. (3.1.27) can now be found by introducing a small perturbation. The total perturbed solution will be written as

$$\alpha_{\mathbf{q}} = \alpha_0 \delta(\mathbf{q} - \mathbf{q}_0) e^{-i\omega_{\mathbf{q}} t} + \xi_{\mathbf{q}}(t) e^{-i\omega_{\mathbf{q}} t} \tag{3.1.29}$$

where $|\xi_{\mathbf{q}}(t)| << |\alpha_0|$, $\mathbf{q}_0$ is the unperturbed wavenumber and $\alpha_0$ is the unperturbed amplitude. If Eq. (3.1.29) is substituted into Eq. (3.1.27) then the only term on the left-hand side of the equation that survives is $\partial\xi_{\mathbf{q}}(t)/\partial t \times e^{-i\omega_{\mathbf{q}} t}$. The final form of Eq. (3.1.27) is therefore

$$\frac{\partial \xi_{\mathbf{q}}(t)}{\partial t} = -2iV_{\mathbf{q},-\mathbf{q}_0,\mathbf{q}_0} |\alpha_0|^2 \delta(\mathbf{q}) e^{-i\omega_{\mathbf{q}} t} - 2iV_{\mathbf{q},\mathbf{q}+\mathbf{q}_0,\mathbf{q}_0} \alpha_0^* \xi_{\mathbf{q}+\mathbf{q}_0} e^{i(\omega_{\mathbf{q}_0} - \omega_{\mathbf{q}+\mathbf{q}_0} + \omega_{\mathbf{q}})t}$$
$$- 2iV_{\mathbf{q},-\mathbf{q}_0,\mathbf{q}_0-\mathbf{q}} \alpha_0 \xi^*_{\mathbf{q}_0-\mathbf{q}} e^{i(\omega_{\mathbf{q}_0-\mathbf{q}} - \omega_{\mathbf{q}_0} + \omega_{\mathbf{q}})t} \tag{3.1.30}$$

In (3.1.30) only the last term on the right-hand side needs to be retained because only three-magnon processes are of interest. Any three-magnon process satisfies a conservation of momentum law that permits $\mathbf{q} = 0$ to split into $\mathbf{q}$ and $-\mathbf{q}$, for example. This will certainly not be true for the first term, involving the delta function $\delta(\mathbf{q})$, and it is therefore rejected. For the second and third terms the three-magnon energy conservation law can be used to check if the argument of the exponential functions is zero. The point about this condition is that the integration of Eq. (3.1.30) leads to functions of the form $\sin(\Delta\omega t)/\Delta\omega$ which, provided that $\Delta\omega \simeq 0$, will be significant. The arguments can only be zero provided that

$$\omega_{\mathbf{q}_0} + \omega_{\mathbf{q}} = \omega_{\mathbf{q}+\mathbf{q}_0} \qquad \omega_{\mathbf{q}_0-\mathbf{q}} + \omega_{\mathbf{q}} = \omega_{\mathbf{q}_0} \tag{3.1.31}$$

The second relationship of (3.1.31), provided that $\mathbf{q}_1 = \mathbf{q}_0 - \mathbf{q}$ and $\mathbf{q}_2 = \mathbf{q}$, is the desired conservation law while the first relationship violates it. Only the third term of (3.1.30) is retained, therefore, to give

$$\frac{\partial \xi_{\mathbf{q}}(t)}{\partial t} = -2iV_{\mathbf{q},-\mathbf{q}_0,\mathbf{q}_0-\mathbf{q}} \alpha_0 \xi^*_{\mathbf{q}_0-\mathbf{q}} \tag{3.1.32}$$

The time derivative of this equation and its complex conjugate give

$$\frac{\partial^2 \xi_{\mathbf{q}}}{\partial t^2} = -2iV_{\mathbf{q},-\mathbf{q}_0,\mathbf{q}_0-\mathbf{q}} \alpha_0 \frac{\partial \xi^*_{\mathbf{q}_0-\mathbf{q}}}{\partial t} = 4|V_{\mathbf{q},-\mathbf{q}_0,\mathbf{q}_0-\mathbf{q}}|^2 |\alpha_0|^2 \xi_{\mathbf{q}} \tag{3.1.33}$$

The solution of Eq. (3.1.33) is

$$\xi_{\mathbf{q}} = Ce^{i\sigma t} \tag{3.1.34}$$

where C is a constant,

$$\sigma = \pm \sqrt{4|V|^2 |\alpha_0|^2} \tag{3.1.35}$$

and $|V|^2 = |V_{\mathbf{q},-\mathbf{q}_0,\mathbf{q}_0-\mathbf{q}}|^2$.

This equation shows that, in the absence of damping, three-magnon decay can occur for arbitrarily small values of power since there are no restrictions upon $|\alpha_0|^2$. Under experimental conditions, however, there will always be some damping and this can be introduced into the equations phenomenologically through the transformation $\sigma \to \sigma - i\Gamma$, where $\Gamma$ is a damping parameter. In fact $\Gamma$ is simply $\gamma\Delta H$, where $\Delta H$ is the natural linewidth of the resonance peak in the real and imaginary parts of the susceptibility.

The introduction of damping converts Eq. (3.1.35) to

$$(\sigma - i\Gamma)^2 + 4|V|^2 |\alpha_0|^2 = 0 \tag{3.1.36}$$

and this has the solutions

$$\sigma_{1,2} = i\left(\Gamma \pm \sqrt{4|V|^2|\alpha_0|^2}\right) \tag{3.1.37}$$

This equation shows that, because of damping, a three-magnon decay process is only possible provided that the propagated power of the wave exceeds a threshold defined by

$$|\alpha_0|^2 > \frac{(\gamma\Delta H)^2}{4|V|^2} \tag{3.1.38}$$

where V is given by Eq. (3.1.28).

The power carried by *linear* magnetostatic surface waves on a thin ferromagnetic film is

$$P = \frac{Lq^2d^2\omega_0^2|A|^2}{\pi\omega_H} \tag{3.1.39}$$

where L is the width of the film, i.e. the aperture of the magnetostatic surface waves, and $|A|$ is the amplitude of the magnetostatic potential which is connected to $\alpha_0$. The threshold condition can be substituted directly into Eq. (3.1.39) to obtain an estimate of the threshold power. The use of Eq. (3.1.39) is justified by the fact that the power level is low enough to avoid distortion of the modal fields by the nonlinearity.

For three-magnon decay the power threshold is[20]

$$P_{th} \simeq \frac{Ld^2M_0^2\omega_0^2\omega_H(\gamma\Delta H)^2}{4\pi q^2d^2(\Delta\omega)^2(\omega_0 - \omega_H)^2} \tag{3.1.40}$$

### 3.2 Four-magnon effects

Three-magnon processes will be suppressed when, for example, the angular frequency is less than $2\omega_H$ but this does not mean that nonlinear effects will not occur. Nonlinear phenomena will then appear as four-magnon processes[12,19,20,27,28] for which

$$2\omega_{\mathbf{q}_0} = \omega_{\mathbf{q}_1} + \omega_{\mathbf{q}_2} \qquad 2\mathbf{q}_0 = \mathbf{q}_1 + \mathbf{q}_2 \tag{3.2.1}$$

Equations (3.2.1) are the four-magnon conservation laws for a process describing the decay of a degenerate pair of magnons $(q_0, \omega_{q_0})$ into another pair of magnons $(q_1, \omega_{q_1})$ and $(q_2, \omega_{q_2})$. Once again a magnetostatic wave can be unstable due to four-magnon decay processes analogous to the three-magnon processes described in the previous section. Four degenerate magnons can, however, take part in self-action phenomena such as self-phase modulation associated with modulational instability and pulse envelope solitons or self-focusing associated with spatial solitons.

A description of four-magnon events can also be developed from the Hamiltonian using Holstein-Primakoff transformations. As was discussed in detail in the earlier section several transformations are needed to bring the Hamiltonian into collective operator and diagonal form. In this case certain triple products of the operators, although three-magnon processes, will contribute to the four-magnon decay matrix elements as the diagonalisation in terms of the $b_{\mathbf{q}}$ operators is effected. The equation of motion for $b_{\mathbf{q}}$ that accounts for only four-magnon nonlinearity is[20]

$$\frac{\partial b_{\mathbf{q}}}{\partial t} + (i\omega_q + \gamma_q)b_q = -i\iiint T_{q,q',q'',q'''}b_{q'}^*b_{q''}b_{q'''}\delta(\mathbf{q} + \mathbf{q}' - \mathbf{q}'' - \mathbf{q}''')d\mathbf{q}'d\mathbf{q}''d\mathbf{q}''' \tag{3.2.2}$$

where $T_{q,q',q'',q'''}$ are matrix elements of the four-magnon interaction. This is an effect arising from third-order nonlinearity as opposed to the second-order nonlinearity producing the three-magnon matrix elements of the previous section. If the magnons in the interaction are practically degenerate, then $\mathbf{q} \simeq \mathbf{q}' \simeq \mathbf{q}'' \simeq \mathbf{q}''' \simeq \mathbf{q}_0$ and the matrix elements are simply

$T_{q_0,q_0,q_0,q_0}$. Usually an analytical form for $T_{q_0,q_0,q_0,q_0}$ can be found under this assumption for magnetostatic waves decaying into other magnetostatic waves, or forward and backward volume magnetostatic waves decaying into themselves. In all cases of decay the conservation laws must be satisfied.

Equation (3.2.2) for $b_\mathbf{q}$ has the simple solution

$$b_\mathbf{q} = b_0 e^{-i\omega_0 t}\delta(\mathbf{q} - \mathbf{q}_0) \tag{3.2.3}$$

where

$$\omega_0 = \omega_{\mathbf{q}_0} + T_0|b_0|^2 \tag{3.2.4}$$

and $T_0 \equiv T_{q_0,q_0,q_0,q_0}$.

This solution can now be investigated, just as was done for three-magnon processes, to see how stable it is against perturbations. In an earlier discussion in this chapter it was stated that two types of instability can be generated by four-magnon interactions. They are distinguished by the magnitude of $\Delta q$, the change in the wavenumber, involved in each type. If $\Delta q \simeq q$ then the magnetisation and the magnetostatic potential are superpositions of all the waves taking part in the process and their amplitudes will vary slowly. For $\Delta q \simeq q$ the instability is four-magnon decay and the satisfaction of the momentum and energy conservation laws means that it will, in principle, appear only above a threshold power. The second type of instability is characterised by $\Delta q << q$. It appears for this case that the strict form of the conservation rules need not be satisfied. This type of instability is an example of the wave interacting with itself. In fact the instability develops over a finite propagation distance L which, in turn, leads to an indeterminacy of the wavenumber equal to $\Delta q \simeq 1/L$.

The stability investigation begins with the introduction of a perturbation $\delta b_\mathbf{q}$ of the form

$$\delta b_\mathbf{q} = \alpha(x)e^{-i\omega_1(\mathbf{q}_0-\boldsymbol{\kappa})t}\delta(\mathbf{q} - \mathbf{q}_0 + \boldsymbol{\kappa}) + \beta(x)e^{-i\omega_2(\mathbf{q}_0+\boldsymbol{\kappa})t}\delta(\mathbf{q} - \mathbf{q}_0 - \boldsymbol{\kappa}) \tag{3.2.5}$$

where propagation of the magnetostatic waves is in the x-direction and $\alpha(x)$ and $\beta(x)$ are slowly varying amplitudes. It does not matter, at this stage, what kind of magnetostatic wave is being investigated since that is determined by the matrix T and the influence of the boundary conditions on the power flow.

In Eq. (3.2.5) $\mathbf{q}_0$ is the initial wavenumber and $\mathbf{q}_0 \pm \boldsymbol{\kappa}$ are the new wavevectors created by the four-magnon interaction. Experimentally these new wavevectors, since they create new frequencies $\omega_{1,2}(\mathbf{q}_0 \pm \boldsymbol{\kappa})$, can be seen on a spectrum analyser as so-called satellite frequencies to the original input frequency of the wave. The precise appearance of these satellites depends upon the experiment but they have been observed for thin films, magnetised normally, and it is now clear that they require four-magnon interactions for their explanation. Another important experimental point is that the films are nonlinear for any power above a threshold value and the satellites do not disappear as the power is increased. It is observed that they increase in number and eventually combine into a noise peak. These experimental features are sketched in Fig. 5. The satellites appear on the left of the signal frequency $f_0$. This is simply because for the experiment reported in Fig. 5 the region $f_0 + \Delta f$, where $\Delta f$ is the frequency shift caused by the four-magnon processes, is inaccessible. Relatively small changes in the applied magnetic field, however, will permit both $f_0 + \Delta f$ and $f_0 - \Delta f$ to be observed. The threshold power, as was seen for three-magnon processes, is inversely proportional to $|T|^2$ and proportional to $\gamma\Delta H$, where $\Delta H$ is the ferromagnetic resonance linewidth.

Modulational instability arises when

$$\Delta q = |\mathbf{q} - \mathbf{q}_{1,2}| << q \qquad \text{and} \qquad \Delta\omega = |\omega_\mathbf{q} - \omega_{\mathbf{q}_{1,2}}| << \omega_\mathbf{q} \tag{3.2.6}$$

In this case the magnetization and potential is identified only with one wave whose amplitude is slowly varying compared to the carrier wave characteristics. Furthermore, the Lighthill criterion[16,17,29]

$$\frac{\partial\omega}{\partial(|A|^2)}\frac{\partial^2\omega}{\partial q^2} < 0 \tag{3.2.7}$$

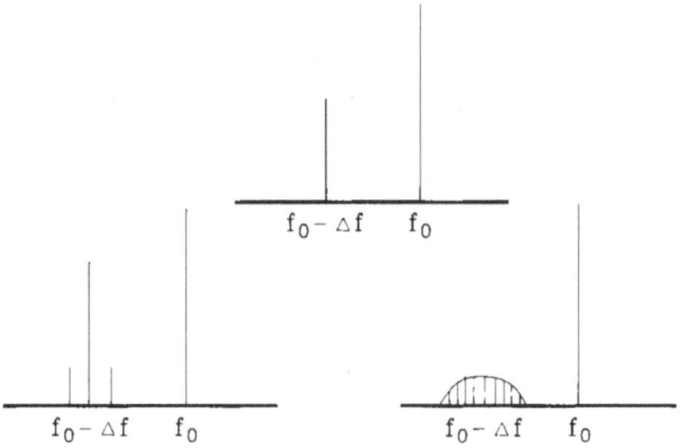


Fig. 5 Sketch of the appearance of satellite frequencies due to four-magnon decay processes.

must be satisfied. This is a general relationship that modulational instability in any physical system must obey. It is, perhaps, more familiar as a necessary condition for the generation of bright envelope solitons. Hence the onset of modulational instability is a part of the self-interaction,[20,22,27-37] four-magnon, regime of nonlinearity. It sets in motion initially unstable waves that can be expected to have stable solitons as their asymptotic state. A modulational instability means that experimentally an envelope with frequency $\Omega$ and wavenumber $\kappa$ will appear, for which $\Omega << \omega_0$ and $\kappa << q_0$, where $\omega_0$ and $q_0$ are the initial frequency and wavenumber. The wavenumber shift $\kappa$ must reach a threshold value for the instability to ensue and this has been shown to be[32]

$$\kappa_0 = \left[\frac{2|\omega_r|}{|\omega''|}\right]^{1/2} \tag{3.2.8}$$

where $\omega_r$ is the relaxation frequency of the material and essentially measures the damping, and $\omega'' = \partial^2\omega/\partial q^2$ is the linear second derivative of the frequency with respect to the wavenumber q.

Equation (3.2.8) shows that dispersion curves with large $\omega''$ are required if $\kappa << q$ and these can only be found by introducing retardation effects. This is an interesting discovery that has recently been confirmed experimentally[32] for forward volume magnetostatic waves. The observations were made close to the edge of the absorption band at a threshold power of only 3 $\mu$W. This is a fascinating discovery since it is several orders of magnitude lower than previously reported thresholds. A qualification is that the sample must not be too small and the exciting antenna must be quite wide. There is clearly much more work to be done in this area and the stimulus to study solitons and solitonic behaviour is very great.

## 4. MAGNETOSTATIC SURFACE WAVES: ROLE OF SELF-PHASE MODULATION

In this section magnetostatic wave instability arising only from self-interaction is briefly reviewed. Self-action includes self-phase modulation and self-focusing as possibilities and is a degenerate four-magnon process. As explained earlier, any theoretical results that claim to be about only self-action and its associated modulational instability apply to experimental systems for which three-magnon processes are excluded. This is fairly easy to arrange since three-magnon processes are possible only above a certain frequency in the magnetostatic surface wave band. A more serious difficulty, experimentally, is guaranteeing a clear separation between self-action four-magnon processes and the kind of four-magnon *decay* process discussed in the previous section. It does not seem to be obvious from the literature how this has been done experimentally but, nevertheless, there does seem to be strong evidence for solitonic behaviour in thin magnetic films. It is now certain that nonlinear (solitonic!) pulse evolution has been observed,[33-36] and hence some form of self-phase modulation. The experimental evidence also supports the assumption that harmonic generation, theoretically the consequence of very weak driving sources appearing in the nonlinear Schrödinger equation, may also be legitimately neglected. A very rough way of supporting that assumption would be to note that harmonic generation, whether it is due to second-order or third-order nonlinearity, attempts to produce states that are outside the linear frequency band of the waves or are only weakly coupled to the linear driving fields. For example, in the case of magnetostatic surface waves, in the absence of exchange forces, the linear frequency band lies on the $\omega$ interval $(\omega_H^2 + \omega_H\omega_M)^{1/2} < \omega < \omega_H + \frac{1}{2}\omega_M$. Hence any attempt by the nonlinearity to drive $2\omega$, $3\omega$ or dc terms will be an attempt to drive states outside the original band. Intuitively we can see that this will be a weak effect.

The basic assumptions here, therefore, are that exchange will be neglected, harmonic and dc generation will be neglected and four-magnon decay will be ignored. The nonlinearities stimulating the self-action processes are third-order in magnitude and, in the terminology of the previous discussions, $\Delta q << q$ so that the wave interacts with *itself*. In less classical language the fact that the conditions $\Delta q \simeq q$ and $\Delta q << q$ separate self-action from decay processes means that self-action is associated with the diagonal terms of the Hamilitonian representation whereas decay processes arise from the off-diagonal terms. This concept will not be pursued any further here. Instead, results from a classical perturbation theory will be summarised. Since both self-action and four-magnon decay involve wavevectors that are fairly close to the centre of the magnon Brillouin zone a purely classical electromagnetic theory should give a complete description of all the effects.

If only self-action is considered then a wave with a carrier frequency $\omega$ is mixed, by a degenerate four-wave third-order nonlinear process, to produce a wave that still has the same frequency. If a periodic wavetrain is set up in such a nonlinear medium then it can be probed with a small perturbation. If small perturbations grow with time then the wavetrain is said to be *unstable*. It is usually easy, both theoretically and experimentally, to distinguish between stability with respect to transverse and longitudinal perturbations. The terminology derives from whether the perturbations are transverse or parallel to the direction of propagation of the unperturbed wave. If the amplitude of a wave is A, its angular frequency is $\omega$ and its wavenumber is q, then for longitudinal perturbations propagating along the direction of the wave it will become unstable when[16,17]

$$\left[\frac{\partial^2\omega}{\partial q^2}\right]\Bigg|_0 < 0 \qquad \left[\frac{\partial\omega}{\partial(A^2)}\right]\Bigg|_0 > 0 \tag{4.1}$$

where the subscript denotes the equilibrium state. The first term is called the group-velocity dispersion and the second term obviously gives the nonlinear shift in the frequency. Equation (4.1) is the famous Lighthill criterion that caused such a sensation when it predicted that cw waves on deep water are, in fact, unstable. CW waves that are unstable with respect to transverse perturbations have $[\partial\omega/\partial(A^2)]|_0 > 0$. If, for example, an initial plane wave is unstable with respect to longitudinal modulations then perturbations will cause it to decay into pulse envelope solitons that have been discussed elsewhere in this book. Waves that are unstable with respect to transverse perturbations break into self-focused channels that can be called spatial solitons. It is only the former that will be discussed here since a lot of work remains to be done on self-focusing possibilities.

As opposed to the creation-annihilation operator formalism used in the previous section, a perturbation theory, developed from the Landau-Lifshitz material equation and Maxwell's equations, can be used. This kind of theory leads to an equation for the nonlinear magnetostatic potential function and then, through the boundary conditions, to a nonlinear cw dispersion equation. The nonlinear shift in the frequency or wavenumber, and the group velocity dispersion, can be calculated to see if solitons or forms of modulational instability are possible. The use of the Lighthill criterion is a remarkably simple recipe. Its use requires only two pieces of information. The group-velocity dispersion is obtained by first finding the linear dispersion equation and the nonlinear coefficient is obtained by finding the nonlinear cw dispersion equation.

In the absence of a formal theory an estimate of the possibilities for soliton formation can easily be obtained. This is done by considering the z-component of the magnetisation projected along the applied field direction. As the power of the wave is increased, and hence the nonlinear regime is reached, the precession of the magnetisation vector increases and the projection of the vector along z, and hence $\omega_M$, decreases in size. Reference to the frequencies of the edges of the initially linear band will show, from the occurrence of $\omega_M$ in the formula for the band edge, whether the nonlinearity is controlled by a positive or negative nonlinear coefficient. A quick sketch of the linear dispersion curves will indicate the sign of the group-velocity dispersion. These two pieces of information,[29] from the Lighthill criterion, give an early indication of whether solitons are going to be possible, in principle, or not.[37] Naturally, this is not the whole story because a full assessment at all wave numbers is required, the influence of exchange must be considered, and the steepness of the dispersion curves must be checked to see if the Schrödinger equation is really applicable. The latter point will be especially important when retardation effects, ignored throughout this chapter, have to be considered. If solitons are not found to be possible then it should be realised that the criterion above applies to *bright solitons*. These are pulses that propagate against a dark background. Another solution of the Schrödinger equation exists called a *dark soliton* and these are "black holes" propagating against a bright uniform background. Dark solitons are possible for positive nonlinear coefficients and positive group-velocity dispersion and, in view of recent developments in optics, are far from purely theoretical entities.

Only magnetostatic surface waves on a ferromagnetic material, propagating perpendicular to the applied magnetic field, will be considered here. For these, using perturbation theory, it is possible to derive the following inhomogeneous partial differential equation[22] for the nonlinear magnetostatic potential $\psi^{NL}$

$$\frac{\partial^2\psi^{NL}}{\partial x^2} - q^2\psi^{NL} = \frac{4q^4A|A|^2(\chi_1^2+\chi_2^2)}{\mu_1 M_0^2}\left\{(\chi_1+\chi_2)e^{3qx} + \alpha_1^2\alpha_2(\chi_1-\chi_2)e^{-3qx} + \alpha_1\chi_1^2\frac{\chi_2-\chi_1}{\chi_1^2+\chi_2^2}e^{qx} - \alpha_1\alpha_2\chi_1^2\frac{\chi_1+\chi_2}{\chi_1^2+\chi_2^2}e^{-qx}\right\} \tag{4.2}$$

where $\omega$ is the angular frequency, q is the wavenumber and the inhomogeneous part has been generated, within the spirit of perturbation theory, from the linear solutions. In the limit of small power, A is the amplitude of the linear wave. The inhomogeneous equation (4.2) will now be treated as an equation for generating the nonlinear magnetostatic potential in terms of A, and the wave number and frequency will be assumed to be the nonlinear values. In view of the way in which Eq. (4.2) has been derived it may appear to be internally inconsistent to treat it in this way. Nevertheless, this will be done here, in the absence of a more enlightened theory.

The other quantities in Eq. (4.2) are defined as

$$\alpha = \frac{\mu_1+\mu_2-1}{\mu_1-\mu_2+1} \qquad \alpha_1 = \frac{\mu_1-\mu_2-1}{\mu_1+\mu_2+1} \tag{4.3}$$

$$\mu_1 = 1 + 4\pi\chi_1 \qquad \mu_2 = 4\pi\chi_2 \qquad \chi_1 = \frac{\omega_H\omega_M}{\omega_H^2-\omega^2} \qquad \chi_2 = \frac{\omega_M\omega}{\omega_H^2-\omega^2} \tag{4.4}$$

where $\mu_1$ and $\mu_2$ are the usual elements of the linear Polder permeability tensor for a ferromagnetic material.

The complete nonlinear potential function, constructed from the particular integral and the complementary solution of Eq. (4.2), is

$$\psi^{NL} = A[(1+L_1qx)e^{qx} + \alpha_1(1+L_2qx)e^{-qx} + L_3e^{-3qx} + L_4e^{3qx}]e^{i(qy-\omega t)} + \text{c.c.} \tag{4.5}$$

where

$$\left.\begin{aligned} L_1 &= -\frac{4\chi_1^2(\chi_1-\chi_2)\alpha_1}{\mu_1}\frac{q^2|A|^2}{M_0^2} \\ L_2 &= \frac{4\chi_1^2(\chi_1+\chi_2)\alpha_2}{\mu_1}\frac{q^2|A|^2}{M_0^2} \end{aligned}\right\} \tag{4.6}$$

$$L_3 = \frac{(\chi_1 - \chi_2)(\chi_1^2 + \chi_2^2)}{2\mu_1}\alpha_2\alpha_1^2\frac{q^2|A|^2}{M_0^2}$$

$$L_4 = \frac{(\chi_1 + \chi_2)(\chi_1^2 + \chi_2^2)}{2\mu_1}\frac{q^2|A|^2}{M_0^2}$$

**The dispersion equation for cw magnetostatic surface wave propagation along a thin ferromagnetic film of thickness d can now be found by applying standard boundary conditions. This dispersion equation is**

$$\omega^2 - \left[\omega_H^2 + \omega_M\omega_H + \frac{\omega_M^2}{4}(1 - e^{-2qd})\right] = G \tag{4.7}$$

**where**

$$G = \frac{\omega_H^2 - \omega^2}{4}e^{-qd}\left[F_1(\mu_1-\mu_2+1)e^{qd} - F_2(\mu_1+\mu_2-1)e^{-qd} - F_3(\mu_1-\mu_2-1) + F_4(\mu_1+\mu_2+1)\right] \tag{4.8}$$

**and**

$$F_1 = \mu_1L_1 - (3\mu_1 - \mu_2 - 1)L_3 + (3\mu_1 + \mu_2 + 1)L_4 - L_5(0)$$

$$F_2 = -\mu_1L_2$$

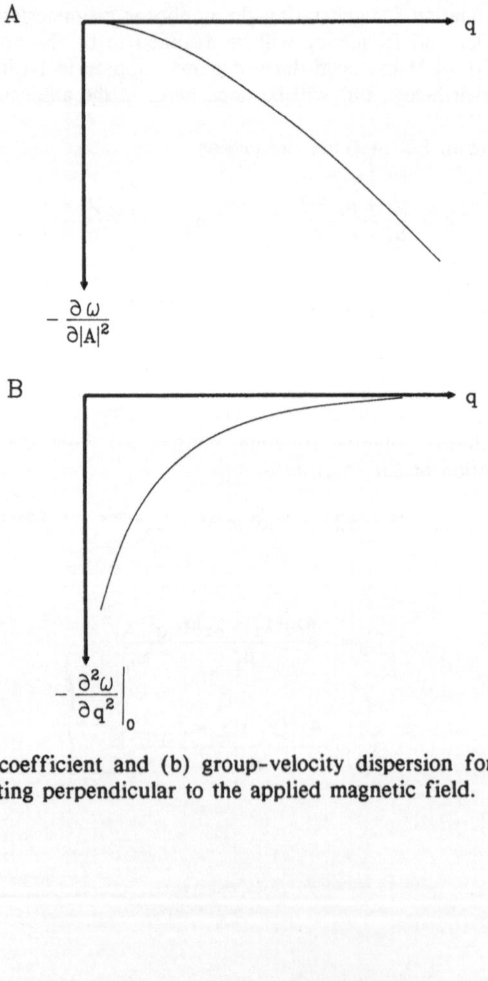


**Fig. 6 (a) Nonlinear coefficient and (b) group-velocity dispersion for surface magnetostatic waves propagating perpendicular to the applied magnetic field.**

$$F_3 = [\mu_1 - (\mu_1 + \mu_2 + 1)qd]L_1e^{-qd} - (3\mu_1 - \mu_2 + 1)L_3e^{3qd} + (3\mu_1 + \mu_2 - 1)L_4e^{-3qd} - L_5(-d)$$

$$F_4 = -[\mu_1 - (\mu_1 - \mu_2 + 1)qd]L_2e^{qd} \tag{4.9}$$

$$L_5(-d) = \frac{2q^2|A|^2}{M_0^2}[-2\alpha_1\chi_1^2(\chi_1 - \chi_2)e^{-qd} + 2\alpha_1\alpha_2\chi_1^2(\chi_1 + \chi_2)e^{qd} - \alpha_1^2\alpha_2(\chi_1 - \chi_2)(\chi_1^2 + \chi_2^2)e^{3qd} + (\chi_1 + \chi_2)(\chi_1^2 + \chi_2^2)e^{-3qd}]$$

The nonlinear coefficient and the group-velocity dispersion obtained from these equations has been calculated numerically and the results are shown in Fig. 6. They show that both the nonlinear coefficient and the group-velocity dispersion for magnetostatic surface waves are negative for all wavenumbers. This means that pulse envelope solitons are not possible for this film and applied magnetic field configuration. Also, since the nonlinear coefficient is negative, self-focusing, which is the asymptotic state of instability with respect to transverse perturbations, should occur. Self-focusing leads to spatial solitons which may soon attract a lot of attention.

## 5. BISTABILITY AND MULTISTABILITY OF MAGNETOSTATIC SURFACE WAVES

The previous results can be used to investigate magnetostatic surface wave propagation in a ferromagnetic film that has a periodic structure.[38] Suppose that the top surface of a thin ferromagnetic film has a periodically corrugated part and that there is propagation along the y-axis. The corrugated part can be defined as

$$x = d[1 + \xi(y)] = d(1 + \epsilon e^{iQy} + \epsilon^* e^{-iQy}) \tag{5.1}$$

where $|\epsilon|d$ is the amplitude of the corrugation, $Q = 2\pi/\Lambda$, $\Lambda$ is the period and $\epsilon^*$ is the complex conjugate of $\epsilon$. Simplification, without loss of substance, is obtained by setting $|\epsilon| << 1$, $|\epsilon|Qd << 1$ and the period of the corrugation equal to $\lambda/2$, where $\lambda$ is the wavelength of the magnetostatic surface wave. The last assumption is the usual Bragg reflection condition.

This problem can be solved using a straightforward coupled wave method. In this method the magnetostatic wave potential in the periodic region of the waveguide is written as a Bloch function but only two terms, corresponding to waves with phase velocities parallel and antiparallel to the y-axis, are dominant and it is these that are retained. The magnetostatic potential can therefore be written as

$$\Psi = A_+[e^{q_+x} + \alpha_1e^{-q_+x} + L_1xq_+e^{q_+x} + \alpha_1L_2xq_+e^{-q_+x} + L_3e^{-3q_+x} + L_4e^{3q_+x}]e^{i(q_+y-\omega t)}$$
$$+ A_-[e^{q_-x} + \alpha_2e^{-q_-x} + L_1'xq_-e^{q_-x} + \alpha_2L_2'xq_-e^{-q_-x} + L_3'e^{-3q_-x} + L_4'e^{3q_-x}]e^{i(q_-y-\omega t)} \tag{5.2}$$

where

$$\alpha_1 = \frac{\mu_1 + \mu_2 + 1}{\mu_1 - \mu_2 - 1} \qquad \alpha_2 = \frac{\mu_1 - \mu_2 + 1}{\mu_1 + \mu_2 - 1}$$

$\mu_1$ and $\mu_2$ are the familiar elements of the magnetic permeability tensor for ferromagnetic materials, $q_+$ is the forward wavenumber, $q_-$ is the backward wavenumber, $\omega$ is the angular frequency, $|q_+| \simeq |q_-| \simeq Q/2$ and $q_\pm = Q/2 \pm \Delta q$ where $\Delta q$ is small wavenumber change. The coefficients $L_i$ are

$$\left.\begin{aligned} L_1 &= -\frac{4\chi_1^2(\chi_1 - \chi_2)\alpha_1}{\mu_1}\frac{q_+^2|A_+|^2}{M_0^2} \\ L_2 &= \frac{4\chi_1^2(\chi_1 + \chi_2)\alpha_2}{\mu_1}\frac{q_+^2|A_+|^2}{M_0^2} \end{aligned}\right\} \tag{5.3}$$

$$
\left.\begin{aligned}
L_3 &= \frac{(\chi_1 - \chi_2)(\chi_1^2 + \chi_2^2)}{2\mu_1}\alpha_2\alpha_1^2 \frac{q_+^2|A_+|^2}{M_0^2} \\
L_4 &= \frac{(\chi_1 + \chi_2)(\chi_1^2 + \chi_2^2)}{2\mu_1}\frac{q_+^2|A_+|^2}{M_0^2}
\end{aligned}\right\}
$$

The equations for $L_i'$ are similar with $q_+$ replaced by $q_-$ and $A_+$ replaced by $A_-$.

The boundary conditions can be "transformed" from the corrugated surface back to the surface $x = d$ so that they become

$$
\Psi^{(1)} + \frac{\partial\Psi^{(1)}}{\partial x}\xi = \Psi^{(2)} + \frac{\partial\Psi^{(2)}}{\partial x}\xi \tag{5.4}
$$

$$
B_x^{(1)} + \frac{\partial B_x^{(1)}}{\partial x}\xi - B_y^{(1)}\xi_y' = B_x^{(2)} + \frac{\partial B_x^{(2)}}{\partial x}\xi - B_y^{(2)}\xi_y' \tag{5.5}
$$

The superscript (1) refers to a vacuum region ($x > d$) above the film, superscript (2) labels the thin ferromagnetic film occupying the region $0 < x < d$ and $\xi_y'$ is the derivative of the function $\xi$ with respect to y.

From Eqs. (5.2) and (5.5) the following coupled equations can be established

$$
A_+[D_+ + \eta_1] = A_-G_- \tag{5.6}
$$

$$
A_-[D_- + \eta_2] = A_+G_+ \tag{5.7}
$$

where

$$
D_\pm = \omega_H^2 + \omega_H\omega_M - \omega^2 + \frac{\omega_M^2}{4}[1 - e^{-2q_\pm d}] \tag{5.8}
$$

$$
G_+ = \frac{Qd\epsilon}{8\alpha_2}e^{-Qd/2}\,(\omega_H^2 - \omega^2)(\mu_1 - \mu_2 - 1)\left[(\mu_1 + \mu_2 - 3)e^{Qd/2} + \alpha_1(\mu_1 - \mu_2 - 1)e^{-Qd/2}\right] \tag{5.9}
$$

$$
G_- = \frac{Qd\epsilon^*}{8\alpha_1}e^{-Qd/2}\,(\omega_H^2 - \omega^2)(\mu_1 + \mu_2 - 1)\left[(\mu_1 - \mu_2 - 3)e^{Qd/2} + \alpha_2(\mu_1 + \mu_2 - 1)e^{-Qd/2}\right] \tag{5.10}
$$

$$
\begin{aligned}
\eta_1 = \frac{\omega_H^2 - \omega^2}{4}e^{-Qd/2}\Bigg\{&(\mu_1 - \mu_2 + 1)\left[\eta_+(d) + \frac{1}{q_+}\frac{\partial}{\partial x}\eta_+(d) + N_+(d)\right] \\
&+ (\mu_1 - \mu_2 - 1)\left[\eta_+(0) - \frac{1}{q_+}\frac{\partial}{\partial x}\eta_+(0) - N_+(0)\right]\Bigg\}
\end{aligned} \tag{5.11}
$$

$$
\begin{aligned}
\eta_2 = \frac{\omega_H^2 - \omega^2}{4}e^{-Qd/2}\Bigg\{&(\mu_1 + \mu_2 + 1)\left[\eta_-(d) + \frac{1}{q_-}\frac{\partial}{\partial x}\eta_-(d) + N_-(d)\right] \\
&+ (\mu_1 + \mu_2 - 1)\left[\eta_-(0) - \frac{1}{q_-}\frac{\partial}{\partial x}\eta_-(0) - N_-(0)\right]\Bigg\}
\end{aligned} \tag{5.12}
$$

and, for example, some nonlinear terms are

$$
\eta_+(d) = L_1q_+xe^{q_+x} + \alpha_1L_2q_+xe^{-q_+x} + L_3e^{-3q_+x} + L_4e^{3q_+x} \tag{5.13}
$$

$N_+$ and $N_-$ are also nonlinear terms.

If $\Omega$ is the frequency for which the unperturbed wavenumber is Q/2 then for a perfectly smooth, linear film $D(Q/2,\Omega) = 0$. A good estimate of the influence of both the corrugation and the nonlinearity can therefore be obtained by expanding the D factors in Eqs. (5.6) and (5.7) around the point (Q/2,Ω). This expansion samples the neighbourhood of Q/2 by an amount $\Delta q \ll Q$ and the neighbourhood of Ω by an amount $\Delta\omega \ll \Omega$. Hence we write

$$D_{\pm} = \pm \frac{\partial D}{\partial q}\Delta q + \frac{\partial D}{\partial \omega}\Delta\omega = \pm \frac{\partial D}{\partial q}\delta + \frac{\partial D}{\partial \omega}\Delta\omega \tag{5.14}$$

The application of Eq. (5.14) to Eqs. (5.6) and (5.7) yields

$$\delta_{1,2} = \frac{e^{Qd}}{d\omega_M^2}\left[\eta_2 - \eta_1 \pm \left\{(\eta_2 - \eta_1)^2 + 4[4\Omega^2(\Delta\omega)^2 - 2\Omega\Delta\omega(\eta_1 + \eta_2) - G_+G_-]\right\}^{1/2}\right] \tag{5.15}$$

These equations are now ready to be solved parametrically and they will show, as expected, that this waveguide is bistable. Indeed, for some parameters the guide will be multistable.[39] This is an area of great interest since it holds out the promise of novel microwave signal processing. Furthermore, there may not be so many problems in finding realistic materials for devices in this frequency range as there seem to be in the high-frequency optical window.

## REFERENCES


1. C. Kittel, *Introduction to Solid State Physics* (John Wiley and Sons, 1976).
2. M. S. Sodha and N. C. Srivastrava, *Microwave Propagation in Ferrimagnetics* (Plenum Press, 1981).
3. R. W. Damon and J. R. Eshbach, J. Phys. Chem. Solid **19**, 308 (1961).
4. A. G. Gurevich, *Ferrites at Microwave Frequencies* (Heywood and Company, London, 1963).
5. M. Sparks, *Ferromagnetic Relaxation Theory* (McGraw-Hill, 1964).
6. F. R. Morgenthaler, Proc. IEEE **76**, 138 (1988)
7. W. Schilz, Philips Research Reports **28**, 50 (1973).
8. T. Yukawa, J. Ikenoue, S. Yamada and K. Abe, J. Appl. Phys. **16**, 2187 (1977); J. Appl. Phys. **49**, 346 (1978).
9. R. E. De Wames and T. Wolfram, J. Appl. Phys. **41**, 5243 (1976).
10. M. Sparks, Phys. Rev. B **1**, 3831 (1970).
11. R. M. White, *Quantum Theory of Magnetism* (McGraw-Hill, 1970).
12. V. E. Zakharov, V. S. Lvov and S. S. Starobinets, Sov. Phys. Solid State **11**, 2368 (1970).
13. V. E. Zakharov, Sov. Phys. JETP **24**, 740 (1967).
14. H. Suhl, J. Phys. Chem. Solids **1**, 209 (1957).
15. H. Suhl, Proc. IRE **44**, 1270 (1956).
16. G. B. Whitham, *Linear and Nonlinear Waves* (John Wiley and Sons, New York, 1974).
17. V. I. Karpman, *Nonlinear Waves in Dispersive Media* (Pergamon Press, Oxford, 1974).
18. A. G. Temiryazev, Sov. Phys. Solid State **29**, 179 (1987).
19. P. E. Zilberman, S. A. Nikitov and A. G. Temiryazev, JETP Lett. **42**, 110 (1985).
20. A. D. Boardman and S. A. Nikitov, Phys. Rev. B **38**, 1144 (1988).
21. W. S. Ishak, Proc. IEEE **76**, 171 (1988).
22. A. D. Boardman, S. A. Nikitov, T. P. Shen and Wang Qi, Phys. Rev. (to be published).
23. C. Kittel, *Quantum Theory of Solids* (John Wiley and Sons, New York, 1963).
24. J. L. Powell and B. Crasemann, *Quantum Mechanics* (Addison-Wesley, London, 1961).
25. T. Holstein and H. Primakoff, Phys. Rev. **58**, 1098 (1940).
26. B. L. Moiseiwitsch, *Variational Principles* (Interscience Publishers, 1966).
27. A. M. Mednikov, Sov. Phys. Solid State **23**, 136 (1981).
28. Yu. V. Gulyaev, P. E. Zilberman, S. A. Nikitov and A. G. Temiryazev, Sov. Phys. Solid State **28**, 1553 (1986); ibid. **29**, 1031 (1987).
29. A. D. Boardman, G. S. Cooper, A. A. Maradudin and T. P. Shen, Phys. Rev. B **34**, 8273 (1986).
30. A. D. Boardman, Yu. V. Gulyaev and S. A. Nikitov, Sov. Phys. JETP, **95**, 2150 (1989).
31. A. D. Boardman and S. A. Nikitov, Sov. Phys. Solid State **31**, 568 (1989).
32. A. G. Temiryazev, JETP Lett. **50**, 228 (1989).
33. P. De Gasperis, R. Marcelli and G. Miccoli, Phys. Rev. Lett. **59**, 481 (1987).
34. B. A. Kalinikos, N. G. Kovshikov and A. N. Slavin, Sov. Tech. Phys. Lett **10**, 392 (1984).

35. B. A. Kalinikos, N. G. Kovshikov and A. N. Slavin, JETP Lett. **38**, 414 (1983); Sov. Phys. Solid State 27, 135 (1985); JETP **67**, 138 (1988).
36. B. A. Kalinikos and A. N. Slavin, Sov. Phys. Solid State **26**, 2077 (1984).
37. A. K. Zvezdin and A. F. Popkov, Sov. Phys. JETP **84**, 350 (1982).
38. Yu. V. Gulyaev, S. A. Nikitov, V. P. Plessky, Sov. Phys. Solid State, **23**, 1321 (1981).
39. A. D. Boardman, S. A. Nikitov and Wang Qi, Phys. Rev. (to be published).

# THE LAGRANGIAN APPROACH TO NONLINEAR WAVE PROPAGATION

J M Arnold

Department of Electronics and Electrical Engineering
University of Glasgow
Glasgow, Scotland

## 1. INTRODUCTION

The increasing importance of nonlinear phenomena in optics makes it essential that a theoretical framework should be developed having sufficient generality to encompass the full range of nonlinear phenomena under investigation, yet which is sufficently flexible to enable analytical calculation of the basic phenomena to be realistic. Nonlinear optics, by definition, involves nonlinear partial differential equations. In particular, Maxwell's equations are coupled to a Schrödinger equation *via* a nonlinear polarisation. The wide range of Hamiltonians which can appear in the Schrödinger equation dictates the generality required of the theoretical methods.

There exist many approaches to the solution of the nonlinear partial differential equations encountered in optics. The purpose of this chapter is to describe only one, derived from a formulation of the physics as a variational problem. In this approach, which mimics Lagrangian classical mechanics, a 'principle of least action' is determined; the Euler-Lagrange equations of this variational problem are the nonlinear partial differential equations which govern the dynamical behaviour of the system. There are many advantages associated with the formulation of nonlinear wave systems as Lagrangian variational problems. First, there is the fact that Lagrangians are directly related to the energies of various parts of the system, and this connection to physical quantities has benefits in understanding the behaviour of the wave types supported by the system. Secondly, the Lagrangian (or, more generally, the action-functional) is well-defined even when the fields do not themselves satisfy the Euler-Lagrange equations. Thirdly, the process of 'averaging' over various quantities is perfectly natural in the Lagrangian approach. Averaging appears in many contexts in nonlinear optics - averaging over fast space-time scales, averaging over microscopic states to obtain a macroscopic variable, averaging over the transverse field in an optical fibre, etc - and is generally used to effect a significant reduction in the number of dynamical variables needed to specify the problem. Fourthly, the multiple-scales approximation method does not encounter secular terms when applied to Lagrangians, whereas these always appear when applying multiple-scales to the Euler-Lagrange partial differential equations.

The theory of Lagrangians in nonlinear field systems has two distinct aspects. First there is the question of how to describe a given system by means of a Lagrangian (or action-functional); secondly, there is the problem of obtaining the solution of the variational problem. It is often thought (and indeed stated in many textbooks) that Lagrangians have to be arrived at by trial and error. Our first task in this chapter will be to show that there is a systematic procedure for constructing a variational principle for an electromagnetic field coupled to an arbitrary quantum-mechanical system. To be precise, what is obtained is not a Lagrangian but a more general 'action functional', but the distinction is not very significant in practice. Having formulated a completely general variational principle in Section 2, we go on to consider approximate methods of analysing the variational problem in Section 3. In particular we consider Whitham's method,[1] and a refinement due to Kawahara[2] which leads more efficiently to the multiple-scales expansion method. Whitham's ideas have been enormously influential in the theory of nonlinear fluid mechanics but have yet to make much impact on nonlinear optics.[3,4] This is mainly because the principal physical motivation for the recent interest in nonlinear optics derives from integrable phenomena such as solitons, and it has seemed more natural to develop theory from this viewpoint. However, the physical systems of practical interest are not truly integrable, and the need has now been recognised for a perturbation theory for 'nearly integrable' systems, which must be of sufficient generality. The recent work of Anderson, Lisak and Bondeson[5-7] shows clearly the advantages of variational methods of formulating this question, and the Whitham-Kawahara method provides a natural and general framework for new developments in the theory of nonlinear optics.

## 2. LAGRANGIAN FORMULATIONS OF NONLINEAR WAVE EQUATIONS

### 2.1 Nonlinear Klein-Gordon equation

We begin by studying purely dispersive nonlinear wave systems. Probably the simplest such system is described by the nonlinear Klein-Gordon equation

$$\frac{\partial^2\phi}{\partial x^2} - \frac{1}{c^2}\frac{\partial^2\phi}{\partial t^2} = \mu\phi + \lambda\phi^3 \tag{2.1}$$

for a real field $\phi$, with constant parameters $\mu$ and $\lambda$. It is easy to demonstrate that Eq. (2.1) is the Euler-Lagrange equation for the optimisation of the functional $L$, given by

$$\mathcal{L} = \iint \mathrm{L}\, dx\, dt \tag{2.2}$$

The Lagrangian density, L, is given in terms of the field $\phi$ and its derivatives as

$$\mathrm{L} = -\frac{1}{2}\left\{\left(\frac{\partial\phi}{\partial x}\right)^2 - \frac{1}{c^2}\left(\frac{\partial\phi}{\partial t}\right)^2\right\} - \frac{\mu}{2}\phi^2 - \frac{\lambda}{4}\phi^4 \tag{2.3}$$

L does not depend on x or t explicitly, only through $\phi$ and its derivatives.

(Notational note: in the following, L, H and S are functions of field variables, whereas their italicised forms *L, H, S* are functionals obtained by integration of the respective functions over space-time variables.)

To prove stationarity of *L* with respect to small variations in $\phi$, let us replace $\phi$ in (2.3) by $\phi + \epsilon$, where $\epsilon$ is an arbitrary function of x and t. The variation in L, to first order in $\epsilon$, is

$$\delta\mathrm{L} = -\left\{\frac{\partial\phi}{\partial x}\frac{\partial\epsilon}{\partial x} - \frac{1}{c^2}\frac{\partial\phi}{\partial t}\frac{\partial\epsilon}{\partial t}\right\} - \mu\phi\epsilon - \lambda\phi^3\epsilon \tag{2.4}$$

Hence, the variation in *L* is

$$\delta\mathcal{L} = \iint \delta\mathrm{L}\, dx\, dt$$

$$= -\iint\left\{\frac{\partial\phi}{\partial x}\frac{\partial\epsilon}{\partial x} - \frac{1}{c^2}\frac{\partial\phi}{\partial t}\frac{\partial\epsilon}{\partial t} + \mu\phi\epsilon + \lambda\phi^3\epsilon\right\} dx\, dt \tag{2.5}$$

Integration by parts removes the derivatives of $\epsilon$; if we assume that $\epsilon = 0$ at the boundaries of the integration domain, then

$$\delta L = \iint \left\{ \frac{\partial^2\phi}{\partial x^2} - \frac{1}{c^2}\frac{\partial^2\phi}{\partial t^2} - \mu\phi - \lambda\phi^3 \right\} \epsilon \, dx \, dt \tag{2.6}$$

In order that the first-order variation $\delta L = 0$ for arbitrary $\epsilon$, it is necessary that

$$\frac{\partial^2\phi}{\partial x^2} - \frac{1}{c^2}\frac{\partial^2\phi}{\partial t^2} - \mu\phi - \lambda\phi^3 = 0 \tag{2.7}$$

This equation is obviously identical with (2.1), so we have proved that (2.1) is the Euler-Lagrange equation for stationarity of the Lagrangian $L$. This example demonstrates in a simple manner the essential features of the Lagrangian formulation.

For dispersive waves (i.e. nondissipative), one intuitively expects some sort of energy conservation relation to hold, and therefore that the Lagrangian is linked to a Hamiltonian formulation. In general, to establish a Hamiltonian one would have to find canonically conjugate variables $(\pi, \psi)$ and a functional $H$ such that the system

$$\frac{\delta H}{\delta \pi} = \frac{\partial \psi}{\partial t} \tag{2.8a}$$

(Hamilton's equations)

$$\frac{\delta H}{\delta \psi} = -\frac{\partial \pi}{\partial t} \tag{2.8b}$$

was totally equivalent to the Euler-Lagrange equations of the given wave system. In many cases, these can be obtained by analogy with classical mechanics. In the simple case under consideration here, this procedure works completely. Let

$$\pi = \frac{\partial \mathrm{L}}{\partial \phi_t}, \qquad \phi = \psi \tag{2.9}$$

where $\phi_t$ denotes $\partial\phi/\partial t$. Then, from (2.3)

$$\pi = \frac{1}{c^2}\phi_t \tag{2.10}$$

Next, apply the Legendre transformation

$$\mathrm{H} = \pi\phi_t - \mathrm{L} \tag{2.11}$$

and replace all $\phi_t$'s by $\pi$'s according to (2.10). This gives

$$\mathrm{H} = c^2\pi^2 + \frac{1}{2}\left\{ \left(\frac{\partial\phi}{\partial x}\right)^2 + \mu\phi^2 + \frac{\lambda}{2}\phi^4 \right\} - \frac{c^2\pi^2}{2}$$

$$= \frac{c^2\pi^2}{2} + \frac{1}{2}\left\{ \phi_x^2 + \mu\phi^2 + \frac{\lambda}{2}\phi^4 \right\} \tag{2.12}$$

The required Hamiltonian is

$$H = \int \mathrm{H} \, dx \tag{2.13}$$

Essentially, we are regarding $\phi$ as the 'displacement' of a particle placed at x, and the term in brackets in (2.12) as the 'potential' energy density of the particles. The first term, $c^2\pi^2/2$, is then the 'kinetic' energy density, and $\phi_t/c^2$ is the 'momentum' of the particles. Then the steps indicated to obtain $H$ are exactly as prescribed by classical mechanics. Hamilton's equations (2.8) turn out to be

$$c^2\pi = \frac{\partial \phi}{\partial t} \tag{2.14a}$$

$$-\frac{\partial^2 \phi}{\partial x^2} + \mu\phi + \lambda\phi^3 = -\frac{\partial \pi}{\partial t} \tag{2.14b}$$

$$= -\frac{1}{c^2}\frac{\partial^2 \phi}{\partial t^2} \tag{2.15}$$

Thus, with (2.13) to define $H$ along with (2.12), clearly Hamilton's equations are equivalent to the given equation (2.1). The link between Lagrangian and Hamiltonian formulations will turn out to be very useful in studying more complicated systems in Sections 2.2 and 2.3.

The advantage of the Lagrangian formulation is that the Lagrangian is well-defined for a wider class of functions $\phi$ than can satisfy the Euler-Lagrange equations. This freedom to 'probe around' a neighbourhood of an exact solution in some suitable function space provides very powerful techniques for the approximate determination of solutions, as we shall see later on.

## 2.2 Electromagnetic field coupled to a classical electron oscillator

To progress towards more realistic models of nonlinear optics, we next investigate a system consisting of simplified Maxwell equations to describe the electromagnetic field in 1-dimension + time, interacting with an oscillator through a polarisation, P.[1]

If we assume that the atoms producing the polarisation are very dilute, Maxwell's equations can be approximated by

$$\frac{\partial E}{\partial x} = \mu_0 \frac{\partial H}{\partial t} \tag{2.16a}$$

$$\frac{\partial H}{\partial x} = \epsilon_0 \frac{\partial E}{\partial t} + \frac{\partial P}{\partial t} \tag{2.16b}$$

where it is assumed that the electric and magnetic fields E and H are transverse to the direction of propagation (x), and are mutually orthogonal; P represents a macroscopic polarisation due to the local spatial average of the polarisation induced by E in the atomic dipoles.

Equation (2.16a) implies the existence of a potential $\phi$, such that

$$\mu_0 H = -\frac{\partial \phi}{\partial x} \tag{2.17a}$$

$$E = -\frac{\partial \phi}{\partial t} \tag{2.17b}$$

Substituting for E and H in terms of the potential $\phi$ in (2.16b) gives

$$\frac{\partial^2 \phi}{\partial x^2} - \frac{1}{c^2}\frac{\partial^2 \phi}{\partial t^2} = -\frac{1}{c^2\epsilon_0}\frac{\partial P}{\partial t} \tag{2.18}$$

Differentiating (2.18) with respect to t and using (2.17b) also yields

$$\frac{\partial^2 E}{\partial x^2} - \frac{1}{c^2}\frac{\partial^2 E}{\partial t^2} = \frac{1}{c^2\epsilon_0}\frac{\partial^2 P}{\partial t^2} \tag{2.19}$$

The polarisation P is modelled by a collection of atoms, each consisting of one electron moving through a displacement q along the transverse plane in a potential well U(q). The dynamical equations are then

$$P = Neq \tag{2.20}$$

$$m\frac{\partial^2 q}{\partial t^2} + \frac{\partial U}{\partial q} = eE = -e\frac{\partial \phi}{\partial t} \tag{2.21}$$

with m and e representing the mass and charge of the electron respectively and N the density of atoms.

Substituting (2.20) in (2.18) gives

$$\frac{\partial^2 \phi}{\partial x^2} - \frac{1}{c^2}\frac{\partial^2 \phi}{\partial t^2} = -\frac{Ne}{\epsilon_0 c^2}\frac{\partial q}{\partial t} \tag{2.22}$$

Equations (2.21) and (2.22) form a coupled system for the dynamical field variables $\phi$ and q.

It can be proved[1] that Eqs. (2.21) and (2.22) are the Euler-Lagrange equations, under independent variations of $\phi$ and q, of the Lagrangian $L$ with Lagrange density

$$L = \frac{1}{2}\epsilon_0(\phi_t^2 - c^2\phi_x^2) - Ne\phi_t q + N\left\{\frac{1}{2}mq_t^2 - U(q)\right\} \tag{2.23}$$

and

$$\mathcal{L} = \iint L \, dx \, dt \tag{2.24}$$

While the demonstration of this fact is an exercise in formal manipulation, the physical significance of L is more important at this stage, so we shall spend some time investigating this aspect of our development. It is, in fact, very instructive to find the Hamiltonian density H corresponding to L. Following the classical mechanical analogy used in Section 2.1, we first define the canonical momenta

$$\pi = \frac{\partial L}{\partial \phi_t} = \epsilon_0\phi_t - Neq \tag{2.25a}$$

$$p = \frac{\partial L}{\partial q_t} = mNq_t \tag{2.25b}$$

Performing the Legendre transformation

$$H = pq_t + \pi\phi_t - L \tag{2.26}$$

and substituting for $q_t$ and $\phi_t$ in terms of p and $\pi$ gives

$$H = \frac{c^2}{2}\left\{\epsilon_0\phi_x^2 + \mu_0(\pi + Neq)^2\right\} + \frac{p^2}{2mN} + NU(q) \tag{2.27}$$

The electron charge e plays the role of a coupling constant; when e = 0, the Hamiltonian density H splits into 'field' and 'atomic' parts

$$H_f = \frac{c^2}{2}(\epsilon_0\phi_x^2 + \mu_0\pi^2) \tag{2.28a}$$

$$H_a = \frac{p^2}{2mN} + NU(q) \tag{2.28b}$$

representing the energy densities of the uncoupled (i.e. noninteracting) field and atoms respectively. The Hamiltonian density H for the coupled field atom system is obtained by forming

$$H = H_f + H_a \tag{2.29}$$

and then making the replacement

$$\pi \to \pi + Neq \tag{2.30}$$

It is left as an exercise to show that Hamilton's equations

$$\frac{\delta H}{\delta \pi} = \frac{\partial \phi}{\partial t} \qquad \frac{\delta H}{\delta p} = \frac{\partial q}{\partial t}$$

$$\frac{\delta H}{\delta \phi} = -\frac{\partial \pi}{\partial t} \qquad \frac{\delta H}{\delta q} = -\frac{\partial p}{\partial t} \tag{2.31}$$

are equivalent to the original equations (2.21) and (2.22).

The above model of field + classical oscillator has been thoroughly studied by Whitham.[1] When U(q), the atomic potential, is quadratic

$$U(q) = \lambda_1 q^2 + \lambda_2$$

then the derived equations (2.21), (2.22) form a linear system, leading to the classical Lorentz theory of linear dielectric dispersion. When U(q) contains higher powers of q, the resulting system is nonlinear.

## 2.3 Maxwell equations coupled to a 2-level quantum system

In any reasonable model of nonlinear optics, we would expect the atoms to be described by quantum mechanics. Therefore, the problem arises as to how to find a Lagrangian formulation for Maxwell's equations coupled to a quantum-mechanical system. The procedure is a modification of the method of the last section.

Consider first a 2-level atom, described by the Schrödinger equation

$$i\hbar \frac{\partial \nu}{\partial t} = \hat{H}_0 \nu \tag{2.32}$$

where

$$\nu = \begin{pmatrix} \nu_1 \\ \nu_2 \end{pmatrix} \qquad \hat{H}_0 = \frac{\hbar\omega_0}{2}\begin{pmatrix} -1 & 0 \\ 0 & 1 \end{pmatrix} \tag{2.33}$$

The expectation value of the energy of the atom, which is a classical observable, is

$$H_0 = \nu^+ \hat{H}_0 \nu \tag{2.34}$$

The Schrödinger equation is obtained formally from the relations

$$i\hbar \frac{\partial \nu}{\partial t} = \frac{\partial H_0}{\partial \nu^+}$$

$$-i\hbar \frac{\partial \nu^+}{\partial t} = \frac{\partial H_0}{\partial \nu} \tag{2.35}$$

which are clearly similar to Hamilton's equations for a classical Hamiltonian with conjugate variables ($\nu^+$, $\nu$). The Schrödinger equation is also obtained by minimising the action-functional *S* obtained from the Legendre-like relation

$$S = i\hbar\nu^{+}\frac{\partial\nu}{\partial t} - \nu^{+}\hat{H}_{0}\nu \tag{2.36a}$$

$$\mathcal{S} = \int S\, dt \tag{2.36b}$$

with respect to $\nu^{+}$:

$$\frac{\delta\mathcal{S}}{\delta\nu^{+}} = \frac{\partial S}{\partial\nu^{+}} = i\hbar\frac{\partial\nu}{\partial t} - \hat{H}_{0}\nu \tag{2.37}$$

and stationarity of $\mathcal{S}$ with respect to variations in $\nu^{+}$ implies (2.32). If we impose the unitary condition

$$\nu^{+}\nu = 1 \tag{2.38}$$

then, by differentiation,

$$\nu^{+}\frac{\partial\nu}{\partial t} + \frac{\partial\nu^{+}}{\partial t}\nu = 0 \tag{2.39}$$

so S can also be expressed as

$$S = -i\hbar\frac{\partial\nu^{+}}{\partial t}\nu - \nu^{+}\hat{H}_{0}\nu \tag{2.40}$$

Taking the variation of $\mathcal{S}$ with respect to $\nu$ now gives

$$\frac{\partial S}{\partial\nu} = -i\hbar\frac{\partial\nu^{+}}{\partial t} - \nu^{+}\hat{H}_{0} \tag{2.41}$$

Since $\hat{H}_{0}$ is hermitian, (2.41) is the hermitian conjugate of (2.37).

Thus the action functional $\mathcal{S}$, obtained from the Hamiltonian $H_0$ by the transformation (2.36), plays the role of the Lagrangian for a single atom.

For K noninteracting atoms at sites $x = x_k$, the total energy is

$$\sum_{k=1}^{K} \nu_k^{+}\hat{H}_{ok}\nu_k \tag{2.42}$$

and the appropriate action functional is

$$S = \sum_{k=1}^{K}\left\{i\hbar\nu_k^{+}\frac{\partial\nu_k}{\partial t} - \nu_k^{+}\hat{H}_{ok}\nu_k\right\} \tag{2.43}$$

where the subscript k represents the particular atomic site. In the limit of small spacing between the atoms, the sum can be replaced by an integral and the discrete index by a continuous one. Thus

$$S = N\int\left\{i\hbar\nu^{+}\frac{\partial\nu}{\partial t} - \nu^{+}\hat{H}_{0}\nu\right\}dx \tag{2.44}$$

$$H_a = N\nu^+\hat{H}_0\nu \tag{2.45}$$

where N is the density of atoms and $H_a$ is the macroscopic energy density of the atomic ensemble.

The total energy density of an uncoupled system of electromagnetic field and 2-level atomic ensemble is the sum of $H_f$ and $H_a$, as before:

$$\begin{aligned} H(e{=}0) &= H_f + H_a \\ &= \frac{c^2}{2}(\epsilon_0\phi_x^2 + \mu_0\pi^2) + N\nu^+\hat{H}_0\nu \end{aligned} \tag{2.46}$$

Finally, the fully interacting Hamiltonian is obtained by the replacement (2.30), where now q is the expectation value of the electron displacement. This is well known from elementary atomic quantum mechanics to be

$$q = q_0\nu^+\sigma_1\nu \tag{2.47}$$

where $\sigma_1$ is the Pauli matrix

$$\sigma_1 = \begin{pmatrix} 0 & 1 \\ 1 & 0 \end{pmatrix} \tag{2.48}$$

and $q_0$ is a constant with dimensions of length. Thus we have a final expression for the Hamiltonian density H identical to (2.27), but with q given by (2.47) and $H_a$ by (2.45).

Having obtained the Hamiltonian density, an action-functional $S$ is defined by combining Legendre transforms on $(\pi, \phi)$ and $(\nu^+, \nu)$, to give

$$S = \pi\phi_t + \frac{i\hbar}{2}(\nu^+\nu_t - \nu_t^+\nu) - H \tag{2.49}$$

$$\mathcal{S} = \iint S \, dx \, dt \tag{2.50}$$

where the bracketed term in (2.49) is a symmetrised form obtainable by partial integration over t.

The canonical field momentum $\pi$ can be eliminated from S, using (2.25a), to obtain

$$S = -\frac{\epsilon_0}{2}(c^2\phi_x^2 - \phi_t^2) + N\left\{-\alpha\phi_t(\nu^+\sigma_1\nu) - \frac{\hbar\omega_0}{2}(\nu^+\sigma_3\nu) + \frac{i\hbar}{2}(\nu^+\nu_t - \nu_t^+\nu)\right\} \tag{2.51}$$

Stationarity of $\mathcal{S}$ with respect to variations in $\phi$ and $\nu^+$ leads to

$$\frac{\partial^2\phi}{\partial x^2} - \frac{1}{c^2}\frac{\partial^2\phi}{\partial t^2} = -\frac{Ne}{\epsilon_0 c^2}\frac{\partial q}{\partial t} \tag{2.52}$$

and

$$i\hbar\frac{\partial\nu}{\partial t} = \left\{\frac{\hbar\omega_0}{2}\sigma_3 + \alpha\phi_t\sigma_1\right\}\nu \tag{2.53}$$

where the dipole moment $\alpha$ is given by $\alpha = eq_0$. Equation (2.52) is identical to (2.22) for the classical oscillator case, and (2.53) is exactly the correct Schrödinger equation for the 2-level system driven by an electric field $E = -\phi_t$.

Thus, we may conclude that the equations of nonlinear optics in the semiclassical approximation can be formulated as the condition for stationarity of the action-functional

$$\mathcal{S} = \iint \mathrm{S}\, dx\, dt$$

with the density S given by (2.49).

### 2.4 Arbitrary quantum systems

Clearly, the foregoing analysis can be extended to include the case of an arbitrary quantum system, which may be composed of multi-level noninteracting atoms, atoms interacting through a weak potential, extended electrons in a semiconductor, or atoms connected to a reservoir, so long as a Schrödinger equation of the form

$$i\hbar \frac{\partial \nu_k}{\partial t} = \hat{H}_{ok} \nu_k \qquad (k = 1, \ldots, K) \qquad (2.54)$$

can be found for each of K noninteracting subsystems. The action-functional

$$\mathcal{S} = \int \left[ \int -\frac{\epsilon_0}{2}(c^2\phi_x^2 - \phi_t^2)\, dx + \sum_{k=1}^{K} \left\{ -eq_k \phi_t - \nu_k^+ \hat{H}_{ok} \nu_k + i\hbar \nu_k^+ \frac{\partial \nu_k}{\partial t} \right\} \right] dt \qquad (2.55)$$

where the displacement is

$$q_k = \nu_k^+ \hat{q}_k \nu_k \qquad (2.56)$$

will act as a suitable functional whose minimisation with respect to variations in all the dynamical variables will be equivalent to Maxwell's equations coupled to the quantum system through a polarisation

$$P = e \sum_k q_k \qquad (2.57)$$

The quantum system may have K = 1 and a large number of dimensions for $\nu$ (e.g. semiconductors) or, at the other extreme, a very large K and a 2-dimensional $\nu$ (e.g. noninteracting 2-level atoms). All intermediate possibilities can also occur. In addition various other classical wave types can be coupled in the same way (for example acoustic waves in a crystal lattice, or pressure waves in a gas). The operator quantities required are $\hat{H}_{ok}$ and $\hat{q}_k$; all other quantities remain classical.

## 3. APPROXIMATE SOLUTION METHODS

### 3.1 Introduction

In Section 2 we saw that the problem of the electromagnetic field coupled to a 2-level atomic system could be represented by the equivalent problem of optimising the functional $\mathcal{S}$ with respect to variations in the electromagnetic potential $\phi$ and the atomic state vectors ($\nu$, $\nu^+$), where

$$\mathcal{S} = \iint \mathrm{S}\, dx\, dt \qquad (3.1)$$

$$S = -\frac{\epsilon_0}{2}(c^2\phi_x^2 - \phi_t^2) + N\left\{-\alpha\phi_t(\nu^+\sigma_1\nu) - \frac{\hbar\omega_0}{2}(\nu^+\sigma_3\nu) + \frac{i\hbar}{2}(\nu^+\nu_t - \nu_t^+\nu)\right\} \tag{3.2}$$

This equivalent formulation is the source of many approximation methods obtained by taking for $\phi$ and $\nu$ trial functions of a definite form, but containing a small number of parameters which are themselves subjected to variation; that choice of parameters which optimises $S$ then gives the trial functions which best approximate the true solutions $\phi$ and $\nu$. There are many such approximation methods, depending on the physical content which it is required that the leading order approximation should display. We shall study approximations which separate 'slow' and 'fast' scales in the functions $\phi$ and $\nu$, concepts which will become more definite as we proceed.

Generally we expect that at least two distinct time scales will appear in nonlinear optics: the period of 'carrier' oscillations (~ $10^{-14}$ seconds), and the time-scale of amplitude or phase modulations impressed on the carrier (> $10^{-12}$ seconds, typically). Similarly, spatial scales will be separated, by the scale of the wavelength, into 'short' (~ $10^{-6}$ m) and 'long' (~ $10^{-4}$ m) scales in the same proportion as the two time-scales. Two methods will be considered here which exploit this disparity between scales: Whitham's method,[1] and the multiple-scales expansion.[2] Although logically the former method is derived from the latter, nevertheless Whitham's method has secured a place in the theory in its own right because it is richer in physical content than the general multiple-scales method, which tends to be rather formal.

### 3.2 Whitham's method

The basic idea of Whitham's method, applied to the Lagrangian density

$$L = L(\phi, \phi_x, \phi_t) \tag{3.3}$$

for the single scalar field $\phi$, is to use as trial functions for $\phi$ certain *periodic travelling wave* solutions $\Phi$ of the Euler-Lagrange equations. These are themselves obtained by postulating that $\Phi$ is a function only of a *single* variable

$$\xi = \beta x - \omega t \tag{3.4}$$

where $\beta$ is the wavenumber of the travelling wave and $\omega$ is its frequency. Clearly, since

$$\frac{\partial\Phi}{\partial x} = \beta\frac{d\Phi}{d\xi} \tag{3.5a}$$

$$\frac{\partial\Phi}{\partial t} = -\omega\frac{d\Phi}{d\xi} \tag{3.5b}$$

partial differential equations in $\phi(x,t)$ become ordinary differential equations in $\Phi(\xi)$; they optimise the *average Lagrangian*.

$$\overline{L} = \oint L(\Phi, \beta\Phi_\xi, -\omega\Phi_\xi)\, d\xi \tag{3.6}$$

obtained directly from (3.3) by integration of L over a $\xi$-period of $\Phi$. By standard procedures one shows that the Euler-Lagrange equation for this variational principle is

$$\frac{\partial L}{\partial\Phi} - \frac{\partial}{\partial x}\frac{\partial L}{\partial\Phi_x} - \frac{\partial}{\partial t}\frac{\partial L}{\partial\Phi_t} = 0 \tag{3.7}$$

$$\Phi_x = \beta\Phi_\xi \tag{3.8a}$$

$$\Phi_t = -\omega\Phi_\xi \tag{3.8b}$$

Since $\partial/\partial x = \beta d/d\xi$, $\partial/\partial t = -\omega d/d\xi$, we have

$$\frac{\partial L}{\partial \Phi} - \frac{d}{d\xi}\left\{\beta \frac{\partial L}{\partial \Phi_x} - \omega \frac{\partial L}{\partial \Phi_t}\right\} \tag{3.9}$$

This *ordinary* differential equation can often be solved explicitly, with the introduction of a constant of integration A which appears as a first integral of (3.9). As a simple example, consider the Klein-Gordon Lagrangian (2.3). This reduces, following (3.4) to (3.9), to

$$(\beta^2 - k^2)\Phi_{\xi\xi} = \mu\Phi + \lambda\Phi^3 \tag{3.10}$$

where $k = \omega/c$ [(3.10) is, of course, (2.1) with (3.5)]. Equation (3.10) can be integrated once (after multiplying by $\Phi_\xi$) to obtain

$$\frac{1}{2}(\beta^2 - k^2)\Phi_\xi^2 = -A + \frac{1}{2}\mu\Phi^2 + \frac{\lambda}{4}\Phi^4 \tag{3.11}$$

where $-A$ is an integration constant. Equation (3.11) can be integrated in terms of elliptic integrals. The maximum value of $\Phi$ occurs when $\Phi_\xi = 0$; hence

$$\Phi_{max}^2 = -\frac{\mu}{\lambda} \pm \sqrt{\left(\frac{\mu}{\lambda}\right)^2 + \frac{4A}{\lambda}} \tag{3.12}$$

Thus, given $\mu$ and $\lambda$, the peak value of the field $\Phi$ is determined by the integration constant A.

This is a general feature of periodic travelling wave solutions of nonlinear wave equations: a first integral A determines the amplitude of the oscillations which form the travelling wave. We are also free to specify that the period of $\Phi$ in $\xi$ should be $2\pi$. This gives a second relation, between $\beta$, $\omega$ and A. In the case of the Klein-Gordon equation considered above, assuming $A > 0$ and $\mu, \lambda > 0$, we get

$$\frac{1}{\sqrt{2}}(k^2 - \beta^2)^{1/2} \int_0^{\Phi_{max}} \frac{d\Phi}{\sqrt{A - \mu\Phi^2/2 - \lambda\Phi^4/4}} = \frac{\pi}{2} \tag{3.13}$$

as this relation. Since it expresses a relationship between $\beta$, $\omega$ and $\Phi_{max}$ (*via* A) it is a *nonlinear dispersion relation.* Note that, if $\lambda = 0$ (which removes the nonlinear term from the Klein-Gordon equation), the nonlinear dispersion relation (3.13) reduces to

$$\frac{1}{\sqrt{2}}(k^2 - \beta^2)^{1/2}\left(\frac{2}{\mu}\right)^{1/2}\frac{\pi}{2} = \frac{\pi}{2}$$

giving

$$k^2 - \beta^2 = \mu \tag{3.14}$$

which is the correct *linear* dispersion relation for (3.10) with $\lambda = 0$. Equation (3.14) is independent of the wave amplitude $\Phi_{max}$, as expected for linear waves; in the nonlinear case, however, the right-hand side of (3.14) becomes dependent on $\Phi_{max}$.

In this way, we obtain a 1-parameter family of periodic travelling wave solutions

$$\Phi = \Phi(\xi; A) \qquad \xi = \beta x - \omega t$$

The next step in Whitham's procedure is based on the following argument. If the modulation impressed on the carrier wave in the fully nonlinear, generally space-time dependent case, is varying sufficiently slowly, then a general solution $\phi$ should look locally like a periodic travelling wave $\Phi(\xi; A)$; however, the values of $\beta$, $\omega$ and A may acquire slow (x, t) dependences to reflect the modulation. This suggests that we use as a trial function

$$\phi = \Phi\,(\xi;\, A) \tag{3.15}$$

in the Lagrangian (3.3), with $\xi$ and A (x, t)-dependent. Specifically, we choose

$$\beta = \frac{\partial \xi}{\partial x} \tag{3.16a}$$

$$\omega = -\frac{\partial \xi}{\partial t} \tag{3.16b}$$

as the (x, t)-dependent wave number and frequency respectively, which must also satisfy the consistency condition

$$\frac{\partial \beta}{\partial t} = -\frac{\partial \omega}{\partial x} \tag{3.17}$$

The variational problem for $\phi$ can now be reduced to one for $\xi$ and A. Substituting for $\phi$ from (3.15) in the Lagrangian (3.3) gives

$$L = L(\Phi,\, \beta\Phi_\xi + \Phi_A A_x,\, -\omega\Phi_\xi + \Phi_A A_t) \tag{3.18}$$

The crucial step in the Whitham procedure involves decoupling $\xi$ from (x, t) and treating $\xi$ as a third *independent* variable; the coupling to (x, t) is restored by adding an extra integration over $\xi$ in forming the final Lagrangian

$$\mathcal{L} = \frac{1}{2\pi}\iint\oint_0^{2\pi} L\; d\xi\; dx\; dt \tag{3.19}$$

The innermost integral is

$$\overline{L} = \frac{1}{2\pi}\oint_0^{2\pi} L\; d\xi \tag{3.20}$$

so

$$\mathcal{L} = \iint \overline{L}\; dx\; dt \tag{3.21}$$

It is desirable to recognise $\xi$ in these equations as a dummy integration variable decoupled from (x, t) by replacing it with $\theta$ in the integrations; thus,

$$\overline{L} = \frac{1}{2\pi}\oint_0^{2\pi} L\,(\Phi,\, \beta\Phi_\theta + \Phi_A A_x,\, -\omega\Phi_\theta + \Phi_A A_t)\; d\theta \tag{3.22}$$

and $\Phi = \Phi(\theta;\, A)$ is a member of the 1-parameter family of periodic travelling-wave solutions. In this way, $\xi$ can be retained as having its original meaning, as the 'potential' which determines $\beta$ and $\omega$ through (3.16) and (3.17). Note here that when A is a constant, $A_x = 0$ and $A_t = 0$, and (3.20) reduces to (3.6), from which A appeared as an integration constant.

Whitham has shown (Ref. 1, p. 495) that (3.22) is indeed the complete and *exact* variational principle for $\phi = \Phi(\xi; x, t)$. Variations in $\Phi(\Phi \rightarrow \Phi + \delta\Phi)$ lead to

$$\frac{\partial}{\partial\theta}\frac{\partial L}{\partial\Phi_\theta} + \frac{\partial}{\partial t}\frac{\partial L}{\partial\Phi_t} + \frac{\partial}{\partial x}\frac{\partial L}{\partial\Phi_x} - \frac{\partial L}{\partial\Phi} = 0 \tag{3.23}$$

Variations in $\xi$ $(\xi \rightarrow \xi + \delta\xi)$ lead to

$$-\frac{\partial}{\partial t}\frac{\partial \overline{L}}{\partial\omega} + \frac{\partial}{\partial x}\frac{\partial \overline{L}}{\partial\beta} = 0 \tag{3.24}$$

with

$$\overline{L} = \frac{1}{2\pi}\oint_0^{2\pi} L\,(\Phi,\ \omega\Phi_\theta + \Phi_t,\ \beta\Phi_\theta + \Phi_x)\ d\theta \tag{3.25}$$

These *exact* equations become *approximate* equations if it is assumed that $\Phi_t$ and $\Phi_x$ are negligible. Then (3.23) - (3.25) become

$$\frac{\partial}{\partial\theta}\frac{\partial L}{\partial\Phi_\theta} - \frac{\partial L}{\partial\Phi} = 0 \tag{3.26}$$

$$-\frac{\partial}{\partial t}\frac{\partial \overline{L}}{\partial\omega} + \frac{\partial}{\partial x}\frac{\partial \overline{L}}{\partial\beta} = 0 \tag{3.27}$$

$$\overline{L} = \frac{1}{2\pi}\oint_0^{2\pi} L\,(\Phi,\ \omega\Phi_\theta,\ \beta\Phi_\theta)\ d\theta \tag{3.28}$$

Equation (3.26) is, essentially, Eq. (3.9) which defines $\Phi$ as a periodic travelling-wave, with an integration 'constant' A. A may, in fact, now depend slowly on (x, t) because these variables are decoupled from $\theta$. The periodic travelling-wave solution $\Phi = \Phi[\theta, A(x,t)]$ may now be substituted into (3.28), and the averaging carried out, to yield

$$\overline{L} = \overline{L}(\omega, \beta, A) = \overline{L}(-\xi_t, \xi_x, A) \tag{3.29}$$

Finally

$$\mathcal{L} = \iint \overline{L}\ dx\ dt \tag{3.30}$$

is optimised with respect to variations in $\xi$ and A to obtain Euler-Lagrange equations for $\xi$ and A which are on the *slow* scales, all fast variations having been removed by the averaging.

The averaged Lagrangian has many useful properties. Firstly, since no derivatives of A appear in L in (3.28), variation with respect to A leads to an *algebraic* equation linking $\omega$, $\beta$ and A, which turns out to be the nonlinear dispersion relation referred to earlier. Secondly, the Euler-Lagrange equations for $\omega$ and $\beta$, along with the consistency relation (3.17), turn out to be a first-order quasi-linear system which can be studied further using the method of characteristics. Thirdly, it is often not necessary to know the full nonlinear periodic travelling wave solution $\Phi$. In many cases, essentially those which are weakly nonlinear, it suffices to use the leading order Fourier component.

$$\Phi \sim A \cos\theta \tag{3.31}$$

In such cases, the calculation of the average Lagrangian from L is almost trivial, and enormous simplification results. This method is called Stokes' approximation.

### 3.3 Multiple-scales expansion

Whitham's method, as described in the previous Section, suffers one major defect. It is known from quite general arguments (e.g. Ref. 8, pp 501-508) that the slow-scale behaviour of a weakly nonlinear dispersive wave equation should be governed by the nonlinear Schrödinger equation for the amplitude of the wave,

$$i\frac{\partial A}{\partial x} = k_1\frac{\partial^2 A}{\partial t^2} + k_2|A|^2 A \tag{3.32}$$

where $k_1$ and $k_2$ are constants representing the strengths of dispersion and nonlinearity respectively. The nonlinear Schrödinger equation is a differential equation for A, but the average Lagrangian in (3.28) does not contain any derivatives of A, so the resulting Euler-Lagrange equation for A is algebraic. Moreover, the nonlinear Schrödinger equation is for a *complex* A, whereas our A is *real*. Clearly, the basic Whitham procedure is too approximate to include the effects of dispersion fully, and a higher-order version must be sought. This difficulty was fully appreciated by Whitham, and studied extensively by Newell.[9] The systematic method for dealing with this problem is the multiple-scales method. This method can appear rather formal and quite tedious when applied directly to differential equations; when applied to Lagrangians it is a very efficient and flexible tool.

Suppose we have a single real-valued field $\phi$, and a Lagrangian density $L(\phi, \phi_x, \phi_t)$. We assume that L is quadratic in $\phi_x$ and $\phi_t$, but nonquadratic in $\phi$. We are interested in the *near-linear* case, when the peak amplitude of $\phi$ is small. The basic strategy is to introduce the field expansion

$$\phi = \epsilon\phi_1 + \epsilon^2\phi_2 + \epsilon^3\phi_3 \ldots \tag{3.33}$$

where $\epsilon$ is a small parameter. The higher-order terms involving powers of $\epsilon$ greater than 1 will not contribute to the analysis we shall carry out, and we neglect them from the start to simplify the notation. We take as a trial function for $\phi$, in the Stokes approximation,

$$\phi = \epsilon\phi_1 = \epsilon(\Phi e^{i\theta} + \Phi^* e^{-i\theta}) \tag{3.34}$$

Here $\Phi$ is a *slowly varying* complex function of (x, t). The trial function is then substituted in the *exact* average Lagrangian (3.25) to obtain

$$\overline{L} = \frac{1}{2\pi}\oint_0^{2\pi} L[\epsilon\phi_1, \epsilon(\phi_{1t} - \omega\phi_{1\theta}), \epsilon(\phi_{1x} + \beta\phi_{1\theta})]\, d\theta \tag{3.35}$$

Next, we introduce the multiple length and time scales

$$X_n = \epsilon^n x \tag{3.36a}$$

$$T_n = \epsilon^n t \tag{3.36b}$$

and the associated derivative expansions

$$\frac{\partial}{\partial x} = \epsilon\frac{\partial}{\partial X_1} + \epsilon^2\frac{\partial}{\partial X_2} \ldots \tag{3.37a}$$

$$\frac{\partial}{\partial t} = \epsilon\frac{\partial}{\partial T_1} + \epsilon^2\frac{\partial}{\partial T_2} \ldots \tag{3.37b}$$

The variational problem is then expressed by

$$\delta L = 0 \tag{3.38a}$$

$$L = \iint \overline{L}\, dX_1\, dT_1\, dX_2\, dT_2 \ldots \tag{3.38b}$$

The average Lagrangian is formally expanded in powers of $\epsilon$, each power of $\epsilon$ in the expansion involving only derivatives with respect to $(X_n, T_n)$ up to a certain order in n, and this permits a recursive scheme for the solution of the Euler-Lagrange equations.

As an example, we consider the Klein-Gordon equation (2.1), having Lagrange density (2.3). The average Lagrangian, after substituting the trial function (3.34), is

$$\overline{L} = -\epsilon^2\left\{\beta^2|\Phi|^2 + |\Phi_x|^2 + i\beta(\Phi\Phi_x^* - \Phi^*\Phi_x)\right\} + \frac{\epsilon^2}{c^2}\left\{\omega^2|\Phi|^2 + |\Phi_t|^2 - i\omega(\Phi\Phi_t^* - \Phi^*\Phi_t)\right\} - \mu\epsilon^2|\Phi|^2 - \frac{\lambda}{2}\epsilon^4|\Phi|^4 \tag{3.39}$$

Introducing the derivative expansions (3.37), and expanding, we obtain

$$\overline{L} = \epsilon^2\{L_0 + \epsilon L_1 + \epsilon^2 L_2 + \ldots\} \tag{3.40}$$

with

$$L_0 = (k^2 - \beta^2 - \mu)|\Phi|^2 \tag{3.41a}$$

$$L_1 = -i\beta\left(\Phi\Phi_{X_1}^* - \Phi^*\Phi_{X_1}\right) - \frac{i\omega}{c^2}\left(\Phi\Phi_{T_1}^* - \Phi^*\Phi_{T_1}\right) \tag{3.41b}$$

$$L_2 = -|\Phi_{X_1}|^2 + \frac{1}{c^2}|\Phi_{T_1}|^2 - \frac{\lambda}{2}|\Phi|^4 - i\beta\left(\Phi\Phi_{X_2}^* - \Phi^*\Phi_{X_2}\right) - \frac{i\omega}{c^2}\left(\Phi\Phi_{T_2}^* - \Phi^*\Phi_{T_2}\right) \tag{3.41c}$$

At this stage, we choose

$$k^2 - \beta^2 - \mu = 0 \tag{3.42}$$

(i.e. the linear dispersion relation). This choice is possible here because the complex representation (3.34) has an arbitrary phase, which can be fixed by (3.42). Finally we carry out the variation of $L$ with respect to the field $\Phi$ and its conjugate, and equate to zero separately all the orders in $\epsilon$. This gives

$$i\beta\Phi_{X_1} + \frac{i\omega}{c^2}\Phi_{T_1} = 0 \tag{3.43a}$$

$$2i\beta\Phi_{X_2} + 2i\frac{\omega}{c^2}\Phi_{T_2} + \Phi_{X_1X_1} - \frac{1}{c^2}\Phi_{T_1T_1} - \lambda|\Phi|^2\Phi = 0 \tag{3.43b}$$

From (3.42) we find that the group velocity is

$$\nu = \frac{d\omega}{d\beta} = c\frac{dk}{d\beta} = c\frac{\beta}{k} \tag{3.44}$$

and therefore (3.43a) is

$$\Phi_{X_1} + \frac{1}{\nu}\Phi_{T_1} = 0 \tag{3.45}$$

which implies that $\Phi$ is an arbitrary function of

$$\xi_1 = X_1 - \nu T_1 \tag{3.46}$$

Equation (3.45) can be used in (3.43b) to obtain

$$2i\beta\left(\Phi_{X_2} + \frac{1}{\nu}\Phi_{T_2}\right) + \Phi_{\xi_1\xi_1}\left(1 - \frac{\nu^2}{c^2}\right) - \lambda|\Phi|^2\Phi = 0 \tag{3.47}$$

This final equation is essentially the nonlinear Schrödinger equation.

Identical procedures can be carried out for multicomponent fields such as those described by the atom-field Lagrangians in Section 2.3. Of course, the multiple-scales method in this form does not actually *solve* the coupled problems of nonlinear optics; it reduces them to simpler problems. In the case of the nonlinear Schrödinger equation derived above, large classes of exact solutions are obtainable by the celebrated Inverse Scattering Method (ISM), which then form approximate solutions of the given nonlinear system whose multiple-scales approximation at second-order is the nonlinear Schrödinger equation. However, the multiple-scales method can also be applied directly to the *determination* of explicit solutions. Examples of this technique can be found in the work of Anderson, Lisak and Bondeson.[5,7]

## REFERENCES


1. G. B. Whitham, *Linear and nonlinear waves* (Academic, 1974).
2. T. Kawahara, Plasma Phys. **18**, 305-316 (1977).
3. R. Bullough, P. M. Jack, P. W. Kitchenside and R. Saunders, Phys. Scr. **20**, 364-381 (1979).
4. P. M. Jack, Ph.D. thesis, UMIST.
5. D. Anderson, Phys. Rev. A **27**, 3135-3145 (1983).
6. D. Anderson and M. Lisak, Phys. Rev. A **27**, 1393-1398 (1983).
7. A. Bondeson, D. Anderson and M. Lisak, Phys. Scr. **20**, 479 (1979).
8. R. K. Dodd, J. C. Eilbeck, J. D. Gibbon, H. C. Morris, *Solitons and nonlinear wave equations* (Academic, 1982).
9. A. C. Newell, *Solitons in mathematics and physics* (SIAM, Philadelphia, 1985).

# THE OPTICAL KERR EFFECT IN FIBERS

Bruno Crosignani and Paolo Di Porto

Dipartimento di Fisica and Fondazione Ugo Bordoni
Universita' dell'Aquila Roma
L'Aquila, Italy Italy

Emanuele Caglioti

Dipartimento di Fisica
Universita' di Roma
Roma, Italy

## 1. INTRODUCTION

It has become evident in the last ten years that a single-mode fiber constitutes the ideal medium for observing nonlinear optical effects in silica, in that it is able to provide an almost lossless diffraction-free long interaction medium over a wide range of wavelengths. This circumstance, besides allowing one to perform with relative ease nonlinear optics experiments which would have otherwise required, in bulk media, the use of high-intensity laser sources, has offered to people working in the field of telecommunications the possibility of conceiving new kinds of optical devices (and has been, sometimes, a source of detrimental side effects).

Among the various nonlinear effects, a prominent role is played by the so-called *optical Kerr effect*, a third-order effect often modelled, as we will see below, through a self-induced intensity dependent contribution to the refractive index of the propagation medium. Its popularity is mainly due, in addition to the host of ultra-fast physical effects it generates, to the formal simplicity and generality of its analytical description. This has in turn attracted the interest of researchers working in different fields, who were able to recognize analogies and similarities with other fundamental problems of nonlinear physics.

The main aim of this chapter is to deduce, in a simple and yet rigorous way, the set of equations describing nonlinear propagation in an optical fiber in the presence of the optical Kerr effect. In the second part of the chapter, the authors try to give the reader a bird's-eye view of some of the problems which admit simple analytical descriptions, in order to familiarize him/her with the various applications associated with the optical Kerr effect. In the third part, a Hamiltonian approach to nonlinear propagation is presented which consists of the description of the propagating field in terms of a finite number of suitable parameters possessing a direct physical meaning, and thus offering the reader a particularly intuitive view of the problem.

## 2. MACROSCOPIC MAXWELL'S EQUATIONS

In the absence of charges and currents, the macroscopic Maxwell equations take, in a nonmagnetic material like silica for which $\mathbf{B} = \mu_0\mathbf{H}$, the form

$$\nabla\cdot\mathbf{D} = 0$$

*Nonlinear Waves in Solid State Physics*
Edited by A.D. Boardman *et al.*, Plenum Press, New York, 1990

$$\nabla \cdot \mathbf{H} = 0$$

$$\nabla \times \mathbf{E} = -\mu_0 \frac{\partial \mathbf{H}}{\partial t} \tag{1}$$

$$\nabla \times \mathbf{H} = \frac{\partial \mathbf{D}}{\partial t}$$

the influence of the material medium being contained in the "additive relation"

$$\mathbf{D} = \epsilon_0 \mathbf{E} + \mathbf{P} \tag{2}$$

where **P** is called the electric polarization. The relation between **P** and **E** is in general a very complicated one, possibly nonlinear and nonlocal in space and time, and it is most often introduced in a phenomenological way. In a linear spatially nondispersive isotropic medium it takes the simple form

$$\mathbf{P}(\mathbf{r}, t) = \epsilon_0 \int_0^\infty \chi(\mathbf{r}, t')\mathbf{E}(\mathbf{r}, t-t')dt' \tag{3}$$

which allows one to obtain in a straightforward way the well-known "constitutive relation"

$$\mathbf{D}_\omega(\mathbf{r}) = \epsilon_0[1 + \chi_\omega(\mathbf{r})]\mathbf{E}_\omega(\mathbf{r}) = \epsilon_\omega(\mathbf{r})\mathbf{E}_\omega(\mathbf{r}) \tag{4}$$

connecting the time Fourier transform $\mathbf{D}_\omega$ of the electric induction to the time Fourier transform $\mathbf{E}_\omega$ of the electric field through the dielectric constant $\epsilon_\omega(\mathbf{r})$ [or the refractive index $n(\omega,\mathbf{r}) = (\epsilon_\omega/\epsilon_0)^{1/2}$]. The set of equations (1), (2), and (3) [or (4)] is the starting point for developing the theory of *linear* propagation in dielectric waveguides.

In general, the phenomenological expression of the polarization **P** can be written as the sum of a linear and of *nonlinear* contributions of increasing order in the form (adopting the convention of summation over repeated indices and dropping, for simplicity, the dependence on the spatial variable **r**)

$$P_i(t) = \epsilon_0 \int_{-\infty}^{t} dt' \chi_{ij}^{(1)}(t-t')E_j(t') + \epsilon_0 \int_{-\infty}^{t} dt' \int_{-\infty}^{t} dt'' \chi_{ijk}^{(2)}(t-t', t-t'')E_j(t')E_k(t'')$$

$$+ \epsilon_0 \int_{-\infty}^{t} dt' \int_{-\infty}^{t} dt'' \int_{-\infty}^{t} dt''' \chi_{ijkl}^{(3)}(t-t', t-t'', t-t''')E_j(t')E_k(t'')E_l(t''') + \ldots \tag{5}$$

the tensorial character of the linear and nonlinear response tensors $\chi_{ijkl\ldots}^{(n)}$ allowing for the description of anisotropic media. Once **P** is expressed, according to Eq. (5), as the sum of a linear contribution $\mathbf{P}^{(L)}$ and a nonlinear one $\mathbf{P}^{(NL)}$,

$$\mathbf{P}(t) = \mathbf{P}^{(L)}(t) + \mathbf{P}^{(NL)}(t) \tag{6}$$

the set of equations (1), (2) and (6) completely describes, in principle, nonlinear propagation in a material medium.

## 3. SOLVING MAXWELL'S EQUATIONS

When faced with the formidable task of finding a solution of the set of equations (1), (2) and (6) with the appropriate boundary conditions, two main approaches are usually adopted.

## 3.1 The wave equation approach

In this approach, Maxwell's equations, together with the additive and constitutive relations, are manipulated to give the wave equation

$$\nabla^2\mathbf{E} - \frac{1}{c^2}\frac{\partial^2}{\partial t^2}\mathbf{E} - \nabla(\nabla\cdot\mathbf{E}) = \mu_0 \frac{\partial^2}{\partial t^2}\mathbf{P} \tag{7}$$

where c is the velocity of light *in vacuo*. By taking the time Fourier transform of Eq. (7) one obtains, after some straightforward algebra in which use is made of Eqs. (1), (2), (4) and (6),

$$\nabla^2\mathbf{E}_\omega + k^2n^2(\omega,\mathbf{r})\mathbf{E}_\omega + 2\nabla[\mathbf{E}_\omega\cdot\nabla \ln n(\omega,\mathbf{r})] = -\mu_0\omega^2\mathbf{P}_\omega^{(NL)} - \nabla\left[\frac{1}{\epsilon_\omega}\nabla\cdot\mathbf{P}_\omega^{(NL)}\right] \tag{8}$$

where $n(\omega)$ and $\epsilon_\omega$ are the refractive index and the dielectric constant of the linear medium. At this point, the standard procedure hinges upon expanding $\mathbf{E}_\omega(\mathbf{r})$ in terms of the modes of the linear medium and looking for the evolution of the expansion coefficients. In particular, as far as this chapter is concerned, these modes are assumed to be the guided modes of an optical fiber possessing a symmetry axis z and a refractive index independent of z, that is $n(\omega,\mathbf{r}) = n(\omega, x, y)$, so that, adopting the complex representation of the field,

$$\mathbf{E}(\mathbf{r}, t) = \sum_m c_m(z,\omega)\mathbf{E}_m(x,y)e^{-i\beta_m(\omega)z+i\omega t} \tag{9}$$

where $\mathbf{E}_m(x,y)$ and $\beta_m(\omega)$ represent respectively the modal configuration and the associated propagation constant. The z-dependence of the $c_m$'s is looked for by inserting Eq. (9) into Eq. (8) (which is second order in d/dz) and then by reducing it to a first-order equation by introducing the slowly-varying approximation, which amounts to neglecting second-order z-derivatives of the modal amplitudes under the assumption

$$\left|\frac{d^2c_m}{dz^2}\right| << \beta_m\left|\frac{dc_m}{dz}\right| \tag{10}$$

This allows us, after exploiting the definition of propagation modes and their orthogonality, to reduce Eq. (8) to a set of coupled differential equations whose linear part contains only terms of the kind $dc_m/dz$ and $c_m$. Actually, as has been recently shown,[1] the slowly varying approximation is actually redundant and its (artificial) necessity arises because one has deduced from Maxwell's equations, which are inherently first-order, a second-order equation (the wave equation). It is thus more appropriate to rely on an alternative approach, the coupled-mode theory, which deals directly with first-order equations for the $c_m$'s (obtained by inserting the field into Maxwell's equations) without any approximation.

## 3.2 The coupled-mode approach

In this approach, the field is again expanded in terms of the guided modes of the linear waveguide [see Eq. (9)], but now it is directly inserted into Maxwell's equations, thus obviously giving equations containing only the first-order derivatives with respect to z of the $c_m$'s. More precisely, the electric field, written in a more general form to that appearing in Eq. (9) in order to include forward (+) and backward (-) propagation, reads

$$\mathbf{E}(x,y,z,t) = \sum_m\sum_\sigma c_m^{(\sigma)}(z,\omega)\mathbf{E}_m^{(\sigma)}(x,y)e^{-i\beta_m^{(\sigma)}(\omega)z+i\omega t} \tag{11}$$

where $\sigma$ = +,- and $\beta_m^{(+)} = -\beta_m^{(-)} = \beta_m > 0$. It is then possible to deduce, with full generality and without any approximation,[2] the following set of equations for the $c_m^{(\sigma)}$'s

$$e^{-i\beta_m z}\frac{dc_m^{(+)}}{dz} + \frac{i\omega}{4p_0}\int_{-\infty}^{+\infty} dx \int_{-\infty}^{+\infty} dy(\mathbf{P}_{\omega,T}^{(NL)}\cdot \mathbf{E}_{m,T} - P_{\omega,z}^{(NL)} E_{m,z}) = 0 \tag{12}$$

$$e^{i\beta_m z}\frac{dc_m^{(-)}}{dz} - \frac{i\omega}{4p_0}\int_{-\infty}^{+\infty} dx \int_{-\infty}^{+\infty} dy(\mathbf{P}_{\omega,T}^{(NL)}\cdot \mathbf{E}_{m,T} + P_{\omega,z}^{(NL)} E_{m,z}) = 0 \tag{13}$$

where $\mathbf{P}_{\omega,T}^{(NL)}$ is the complex amplitude of the corresponding induced polarization vector and the subscripts T and z stand for "transverse" and "longitudinal" respectively. We have taken advantage of the relations $\mathbf{E}_{m,T}^{(+)} = \mathbf{E}_{m,T}^{(-)} = \mathbf{E}_{m,T}$ and $E_{m,z}^{(+)} = -E_{m,z}^{(-)} = -E_{m,z}$ and of the orthogonality condition among guided modes

$$\int_{-\infty}^{+\infty} dx \int_{-\infty}^{+\infty} dy(\mathbf{E}_{m,T}^{(+)}\times \mathbf{H}_{n,T}^{*(+)})\cdot\hat{\mathbf{z}} = 2p_0\delta_{mn} \tag{14}$$

$p_0$ being a normalization constant and $\mathbf{H}_{n,T}$ the transverse magnetic modal configurations.

In many cases of interest, the longitudinal parts of the modal configurations are small with respect to the transverse ones (weakly-guiding hypothesis) and the terms containing $P_{\omega,z}^{(NL)}$ can be neglected in Eqs. (12) and (13).

## 4. THIRD-ORDER NONLINEAR POLARIZATION IN SILICA

The lowest order significant nonlinear contribution in silica is that associated with third-order polarization since $\chi_{ijk}^{(2)}$ is zero in material media possessing inversion symmetry. The reader particularly interested in the origin and behavior of the physical mechanisms contributing to $\chi_{ijkl}^{(3)}$ is referred to Ref. 3 for an excellent and exhaustive discussion on the subject. For the purpose of this chapter, it will suffice to recall that $\mathbf{P}^{(NL)} = \mathbf{P}^{(3)}$ can be written in the form

$$\mathbf{P}^{(3)}(t) = \epsilon_0\chi^{(3)}\mathbf{E}(t)\cdot\mathbf{E}(t)\mathbf{E}(t) + \mathbf{E}(t)\int a(t-\tau)\mathbf{E}(\tau)\cdot\mathbf{E}(\tau)d\tau + \mathbf{E}(t)\int b(t-\tau)\mathbf{E}(\tau)\cdot\mathbf{E}(\tau)d\tau \tag{15}$$

where, in order to preserve causality, one has to assume $a(t-\tau) = b(t-\tau) = 0$ for $\tau > t$. In Eq. (15), the first term represents the fast responding electronic contributions arising from a direct distortion of the electronic clouds from their region of linear response under the influence of the propagating field. This term can be easily worked out by assuming a $\delta$-function behavior for $\chi_{ijkl}^{(3)}(t-t', t-t'', t-t''')$ and by recalling that, for isotropic media, the nonlinear susceptibility $\chi_{ijkl}^{(3)}$ has to be invariant under all spatial symmetry transformations. The last two terms model slower nuclear nonlinearities which are a consequence of electric field induced changes in the motion of the nuclei. In particular, contributions of this latter kind can be associated with

(i) *molecular reorientation* mechanisms (unlikely, however, to be of particular relevance in solids) exhibiting a simple relaxational response and
(ii) *Raman type nonlinearities* resulting from the modulation of the electronic polarizability by the (Raman) vibrational modes driven by the incident field. These exhibit a response function possessing a resonance frequency in the infrared region.

## 5. THE OPTICAL KERR EFFECT IN SILICA

The *optical Kerr effect* is associated with the contributions to the third-order polarization

$\mathbf{P}^{(3)}(t)$ which vibrate at (approximately) the same frequency as the propagating field and are thus able to act as a source for the field itself.

Let us consider for simplicity the case of a linearly polarized field and let us neglect, as a first approximation, the non-instantaneous response of the medium represented by the last two terms in Eq. (15) (they turn out to be important only for very short pulses). In this case, Eq. (15) takes the simple scalar form

$$P^{(3)}(t) = \epsilon_0 \chi^{(3)} E^3(t) \tag{16}$$

from which, after introducing the signal E(t) such that

$$E(t) = \frac{\hat{E}(t) + \hat{E}^*(t)}{2} \tag{17}$$

it follows that the contribution vibrating at approximately the same frequency as the field has the form

$$P^{(3)}(t) = \frac{3}{4}\epsilon_0 \chi^{(3)} \left|\hat{E}(t)\right|^2 E(t) \tag{18}$$

where $|\hat{E}(t)|^2$ is the so-called "instantaneous optical intensity".

A nonlinear polarization of the form described by Eq. (18) can be approximately interpreted, as we shall see below, in terms of a nonlinear refractive index, thus allowing for an intuitive representation of the optical Kerr effect. To see this, let us introduce Eq. (18), together with the linear part of the polarization, into Eq. (2). After taking the time Fourier transform of both sides, this gives

$$D_\omega = \epsilon_\omega E_\omega + P_\omega^{(3)} = \epsilon_\omega E_\omega + \frac{3}{4}\epsilon_0 \chi^{(3)} \left|\hat{E}(t)\right|^2 E_\omega \tag{19}$$

Note that, in performing the Fourier-transform, we have tacitly assumed the existence of two different time scales in $P^{(3)}(t)$, namely $1/\omega_0$ ($\omega_0$ being the average frequency of the field) and that associated with $|\hat{E}(t)|^2$ (typically the pulse duration). The Fourier-transform operates only on the "fast" time scale (that is $1/\omega_0$) and thus one obtains a dielectric constant

$$\tilde{\epsilon}_\omega(t) = \epsilon_\omega + \frac{3}{4}\epsilon_0 \chi^{(3)} \left|\hat{E}(t)\right|^2 \tag{20}$$

which, besides depending on $\omega$, exhibits a "slow" time-scale dependence. Equation (20) allows one to write

$$\tilde{n}(\omega) = \sqrt{\frac{\tilde{\epsilon}_\omega}{\epsilon_0}} = \sqrt{n_1^2(\omega) + \frac{3}{4}\chi^{(3)} |\hat{E}(t)|^2} \simeq n_1(\omega) + n_2 \left|\hat{E}(t)\right|^2 \tag{21}$$

where $n_1(\omega)$ is the linear refractive index and $n_2 = (3/4)\chi^{(3)}/n_1$ is the so-called nonlinear refractive index coefficient ($n_2 \simeq 10^{-22}\ m^2V^{-2}$ for silica) or, equivalently,

$$\tilde{n}(\omega) \simeq n_1(\omega) + N_2 I(t) \tag{22}$$

where I(t) is the optical intensity (Watt/cm$^2$) and $N_2 \simeq 10^{-16}$ cm$^2$/Watt.

Equation (18) represents the optical Kerr effect in its simplest form, that is for linearly polarized light and neglecting the non-instantaneous response of the medium. In order to include the contributions which can be obtained by dropping the last hypothesis, we can extract from Eq. (15) the terms vibrating at approximately the same frequency as the field by observing that

$$E(t)\int_{-\infty}^{t} d\tau\, a(t-\tau)E(\tau)E(\tau) = E(t)\int_{0}^{+\infty} d\tau\, a(\tau)E(t-\tau)E(t-\tau)$$

$$\simeq \left\{\left[\int_0^{+\infty} d\tau a(\tau)\right]\frac{\hat{E}(t)\hat{E}^*(t)}{2} - \frac{1}{2}\left[\int_0^{+\infty} d\tau \tau a(\tau)\right]\frac{d}{dt}[\hat{E}(t)\hat{E}^*(t)] + \frac{1}{4}\left[\int_0^{+\infty} d\tau \tau^2 a(\tau)\right]\frac{d^2}{dt^2}[\hat{E}(t)\hat{E}^*(t)] + \ldots\right\}E(t) \tag{23}$$

which, inserted in Eq. (15), allows us to write the nonlinear polarization as

$$P^{(3)}(t) = \left\{\left[\frac{3}{4}\epsilon_0\chi^{(3)} + b_0\right]|\hat{E}(t)|^2 - b_1\frac{d}{dt}\left[|\hat{E}(t)|^2\right] + \frac{1}{2}b_2\frac{d^2}{dt^2}\left[|\hat{E}(t)|^2\right]\right\}E(t) + \ldots = S(t)E(t) \tag{24}$$

where S(t) contains the slowly-varying contributions and we have set

$$b_n = \frac{1}{2}\int_0^{\infty} d\tau\, \tau^n a(\tau) \tag{25}$$

The expression of the nonlinear polarization furnished by Eq. (24), or its generalization to a field (not necessarily linearly polarized) resulting from the superposition of the guided modes of an optical waveguide, can now be inserted into the set of coupled equations (12) and (13) in order to obtain the evolution of the modal amplitudes.

## 6. THE NONLINEAR COUPLED EQUATIONS

We wish to consider explicitly the case of a mono-mode fiber, propagating just one linearly polarized mode, since it allows us to understand the basic features of the procedure leading to the set of nonlinear differential equations describing the evolution of the modal amplitudes. To this end, let us first generalize the expression of the monochromatic field appearing in Eq. (11) to a nonstationary case by writing

$$\hat{E}(\rho,z,t) = 2\sum_m\sum_\sigma E_m^{(\sigma)}(\rho)\int_0^{\infty} d\omega \tilde{c}_m^{(\sigma)}(z,\omega)e^{-i\beta_m^{(\sigma)}(\omega)z+i\omega t}$$

$$= 2\sum_m\sum_\sigma E_m^{(\sigma)}(\rho)e^{i\omega_0 t-i\beta_m^{(\sigma)}(\omega_0)z}\int_0^{\infty} d\omega \tilde{c}_m^{(\sigma)}(z,\omega)e^{i(\omega-\omega_0)t-i[\beta_m^{(\sigma)}(\omega)-\beta_m^{(\sigma)}(\omega_0)]z}$$

$$= \sum_m\sum_\sigma E_m(\rho)e^{i\omega_0 t-i\beta_m^{(\sigma)}(\omega_0)z}\Phi_m^{(\sigma)}(z,t) \tag{26}$$

where $\rho = (x,y)$ and the $\Phi_m^{(\sigma)}(z,t)$'s represent modal amplitudes varying on time and space scales that are long compared with $1/\omega_0$ and $1/k_0$ ($k_0 = \omega_0/c$) respectively. For a mono-mode fiber, considering a single forward propagating mode and neglecting the longitudinal part of the field,

$$\hat{E}(\rho,z,t) = E(\rho)\hat{x}\Phi(z,t)e^{i\omega_0 t-i\beta(\omega_0)z} \tag{27}$$

where, according to Eq. (12),

$$\frac{d\tilde{c}^{(+)}(z,\omega)}{dz} = -\frac{i\omega}{4p_0}\int_{-\infty}^{+\infty}dx\int_{-\infty}^{+\infty}dy\; e^{i\beta(\omega)z}E(\rho)\mathbf{P}^{(NL)}_{\omega,T}\cdot\hat{\mathbf{x}} \tag{28}$$

Let us now multiply both sides of Eq. (28) by the factor

$$e^{i(\omega-\omega_0)t-i[\beta(\omega)-\beta(\omega_0)]z}$$

and integrate over the positive values of $\omega$. By recalling the definition of $\Phi(z,t)$, we can immediately show that

$$2\int_0^{\infty}d\omega\,\frac{d\tilde{c}^{(+)}(z,\omega)}{dz}e^{i(\omega-\omega_0)t-i[\beta(\omega)-\beta(\omega_0)]z} = \left(\frac{\partial}{\partial z}+\frac{1}{V}\frac{\partial}{\partial t}-\frac{i}{2A}\frac{\partial^2}{\partial t^2}-\frac{1}{3!B}\frac{\partial^3}{\partial t^3}+\dots\right)\Phi \tag{29}$$

where

$$V=\left[\left(\frac{d\beta}{d\omega}\right)_{\omega_0}\right]^{-1}\qquad A=\left[\left(\frac{d^2\beta}{d\omega^2}\right)_{\omega_0}\right]^{-1}\qquad B=\left[\left(\frac{d^3\beta}{d\omega^3}\right)_{\omega_0}\right]^{-1} \tag{30}$$

represent respectively the group-velocity, the second-order group dispersion, the third-order group dispersion and so on. On the right-hand side of Eq. (28) we have

$$-\frac{2i}{4p_0}\int_{-\infty}^{+\infty}dx\int_{-\infty}^{+\infty}dy\,E(\rho)\int_0^{\infty}d\omega e^{i(\omega-\omega_0)t-i[\beta(\omega)-\beta(\omega_0)]z}\,\omega e^{i\beta(\omega)z}\,P^{(3)}_{\omega}$$

$$= -\frac{1}{4p_0}e^{i\beta(\omega_0)z-i\omega_0 t}\int_{-\infty}^{+\infty}dx\int_{-\infty}^{+\infty}dyE(\rho)\,\frac{\partial}{\partial t}\hat{P}^{(3)}(t) \tag{31}$$

so that, recalling Eqs. (24), (27), and (29), we can finally write

$$\left(\frac{\partial}{\partial z}+\frac{1}{V}\frac{\partial}{\partial t}-\frac{i}{2A}\frac{\partial^2}{\partial t^2}+\dots\right)\Phi(z,t)$$

$$= -\frac{i\omega_0}{4p_0}\int_{-\infty}^{+\infty}dx\int_{-\infty}^{+\infty}dyE^2(\rho)S(t)\Phi(z,t) - \frac{1}{4p_0}\int_{-\infty}^{+\infty}dx\int_{-\infty}^{+\infty}dyE^2(\rho)\frac{\partial}{\partial t}[S(t)\Phi(z,t)] \tag{32}$$

The normalization constant $p_0$ can be expressed through Eq. (14) which, after taking advantage of the approximate relation

$$\hat{\mathbf{z}}\times\mathbf{H}_{m,T} = \sqrt{\frac{\epsilon_0}{\mu_0}}\,n_1\mathbf{E}_{m,T} \tag{33}$$

yields

$$2p_0 = n_1 \sqrt{\frac{\epsilon_0}{\mu_0}} \int_{-\infty}^{+\infty} dx \int_{-\infty}^{+\infty} dy E^2(\rho) \tag{34}$$

If we consider, for example, the case described by Eq. (18), that is

$$\hat{P}^{(3)}(t) = \frac{3}{4}\epsilon_0 \chi^{(3)} e^{i\omega_0 t - i\beta(\omega_0)z} E^3(\rho) \left| \Phi(z,t) \right|^2 \Phi(z,t) \tag{35}$$

then Eq. (32) furnishes the nonlinear differential equation obeyed by $\Phi(z,t)$ in the form

$$\left( \frac{\partial}{\partial z} + \frac{1}{V}\frac{\partial}{\partial t} - \frac{i}{2A}\frac{\partial^2}{\partial t^2} + \ldots \right) \Phi(z,t) = -iR \left| \Phi \right|^2 \Phi - \frac{R}{\omega_0}\frac{\partial}{\partial t}\left( \left| \Phi \right|^2 \Phi \right) \tag{36}$$

where

$$R = k_0 n_2 \frac{\int_{-\infty}^{+\infty} dx \int_{-\infty}^{+\infty} dy E^4(\rho)}{\int_{-\infty}^{+\infty} dx \int_{-\infty}^{+\infty} dy E^2(\rho)} \tag{37}$$

It is worthwhile to note that if the modal transverse configuration is normalized to one, that is

$$\int_{-\infty}^{+\infty} dx \int_{-\infty}^{+\infty} dy E^2(\rho) = 1 \tag{38}$$

then

$$\frac{1}{\sigma} = \int_{-\infty}^{+\infty} dx \int_{-\infty}^{+\infty} dy E^4(\rho) \tag{39}$$

has the dimension of the inverse of an area and $\sigma$ can be interpreted as the "effective" area of the mode.

Equation (36) can be easily generalized to include the Raman type nonlinearities, associated with the contributions to $\hat{S}(t)$ containing the time derivative of the instantaneous optical intensity $|\hat{E}(t)|^2$. By introducing these on the right-hand side of Eq. (32), new terms are added on the right-hand side of Eq. (36) which becomes

$$\left( \frac{\partial}{\partial z} + \frac{1}{V}\frac{\partial}{\partial t} - \frac{i}{2A}\frac{\partial^2}{\partial t^2} + \ldots \right) \Phi$$

$$= -iR \left| \Phi \right|^2 \Phi - \frac{R}{\omega_0}\frac{\partial}{\partial t}\left( \left| \Phi \right|^2 \Phi \right) - iRb_1 \Phi \frac{\partial}{\partial t}\left( \left| \Phi \right|^2 \right) - \frac{Rb_1}{\omega_0}\frac{\partial}{\partial t}\left( \Phi \frac{\partial}{\partial t} \left| \Phi \right|^2 \right) + \ldots \tag{40}$$

It is convenient at this point to remember that the optical Kerr effect in silica can also affect the transverse configuration of the propagating field. This effect (known as "self-focusing") is completely missing in our treatment since we have developed the field in terms of guided modes alone. Its description would require the introduction of the radiation modes of the waveguide but its role is not relevant at the low optical powers usually employed in connection with telecommunications fibers.

The preceding derivation can be generalized in a straightforward way to the case of a fiber supporting an arbitrary number of modes. The only novel feature of this is associated with the fact that, if the field consists of the superposition of two orthogonal linearly polarized states, the relation between $\mathbf{P}^{(3)}$ and $\mathbf{E}$ acquires a tensorial character as can easily be realized by inspecting Eq. (15) and noting that terms of the kind $E_y E_x^*$ do oscillate at the same frequency as the field. More precisely, considering only transverse fields, the direct generalization of Eq. (18) is[4]

$$\mathbf{P}_T^{(3)}(t) = \mathbf{X}{:}\mathbf{E}_T(t) \tag{41}$$

where the tensor $\mathbf{X}$ is given by the two-by-two matrix

$$\mathbf{X} = \frac{\epsilon_0 \chi^{(3)}}{2} \begin{bmatrix} \hat{E}_T \cdot \hat{E}_T^* + \frac{1}{2}|\hat{E}_x|^2 & \frac{1}{2}\hat{E}_y \hat{E}_x^* \\ \frac{1}{2}\hat{E}_y^* \hat{E}_x & \hat{E}_T \cdot \hat{E}_T^* + \frac{1}{2}|\hat{E}_y|^2 \end{bmatrix} \tag{42}$$

The procedure adopted for studying propagation in a mono-mode fiber can now be exactly followed to derive the set of nonlinear equations describing propagation in the presence of the optical Kerr effect in a multimode fiber. In practice, however, the relevant situations are those pertaining to a single-mode fiber supporting two polarization states.

### 6.1 Nonlinear propagation in a straight single-mode fiber

The transverse part of the signal reads in this case

$$\begin{aligned} \hat{E}_T(\rho, z, t) = {} & E_1(\rho)\hat{x}[e^{i\omega_0 t - i\beta_1(\omega_0)z}\Phi_1^+(z,t) + e^{i\omega_0 t + i\beta_1(\omega_0)z}\Phi_1^-(z,t)] \\ & + E_2(\rho)\hat{y}[e^{i\omega_0 t - i\beta_2(\omega_0)z}\Phi_2^+(z,t) + e^{i\omega_0 t + i\beta_2(\omega_0)z}\Phi_2^-(z,t)] \end{aligned} \tag{43}$$

where the $\Phi_i^+$'s and the $\Phi_i^-$'s (i = 1,2) obey the following set of nonlinear equations[5]

$$\begin{aligned} L_1^\pm \Phi_1^\pm = {} & \mp i R_{11}\left(|\Phi_1^\pm|^2 + 2|\Phi_1^\mp|^2\right)\Phi_1^\pm \mp \frac{2i}{3}R_{12}\left(|\Phi_2^+|^2 + |\Phi_2^-|^2\right)\Phi_1^\pm \mp \frac{2i}{3}R_{12}\Phi_1^{\mp *}\Phi_2^+\Phi_2^- \\ & \mp \frac{2i}{3}R_{12}e^{\pm 2i\delta\beta z}\Phi_1^\mp \Phi_2^{\mp *}\Phi_2^\pm \mp \frac{i}{3}R_{12}e^{\pm 2i\delta\beta z}\Phi_1^{\pm *}\Phi_2^{\pm 2} \end{aligned} \tag{44a}$$

$$\begin{aligned} L_2^\pm \Phi_2^\pm = {} & \mp i R_{22}\left(|\Phi_2^\pm|^2 + 2|\Phi_2^\mp|^2\right)\Phi_2^\pm \mp \frac{2i}{3}R_{21}\left(|\Phi_1^+|^2 + |\Phi_1^-|^2\right)\Phi_2^\pm \mp \frac{2i}{3}R_{21}\Phi_2^{\mp *}\Phi_1^+\Phi_1^- \\ & \mp \frac{2i}{3}R_{21}e^{\pm 2i\delta\beta z}\Phi_2^\mp \Phi_1^{\mp *}\Phi_1^\pm \mp \frac{i}{3}R_{21}e^{\pm 2i\delta\beta z}\Phi_2^{\pm *}\Phi_1^{\pm 2} \end{aligned} \tag{44b}$$

where either the upper or the lower sign applies, $\delta\beta = \beta_1 - \beta_2$ and

$$L_i^\pm = \frac{\partial}{\partial z} \pm \frac{1}{V_i}\frac{\partial}{\partial t} \mp \frac{i}{2A_i}\frac{\partial^2}{\partial t^2} + \ldots \qquad i = 1,2 \tag{45}$$

$$R_{ij} = k_0 n_2 \int_{-\infty}^{+\infty} dx \int_{-\infty}^{+\infty} dy E_i^2(\rho) E_j^2(\rho) \qquad i,j = 1,2 \tag{46}$$

$V_i$ and $A_i$ are the group velocity and the group-velocity dispersion of the i-th mode respectively. Note that in the two limiting situations of *high-birefringence* and *low-birefringence fibers*, one can respectively neglect the rapidly oscillating terms containing $\exp(\pm 2i\delta\beta z)$ or set $\delta\beta = 0$ in Eqs. (44). Also, due to the structure of the system of Eqs. (44), backward (forward) traveling modes have to be present at the fiber input (output) in order to be excited, so that the propagation problem can in many cases be reduced to that of forward or backward propagation alone.

### 6.2 Nonlinear propagation in highly-twisted or optically active single-mode fibers

In the preceding developments the fibers were considered ideal, that is without any imperfection in their geometry or structure but for the presence of the nonlinearity associated with the optical Kerr effect. Most often this is not the case and the equations of propagation contain coupling terms, responsible for the electromagnetic power exchange among different modes, having their physical origin in fiber imperfections. This coupling mechanism, besides often being detrimental in itself to fiber-optic telecommunications, is sometimes capable of concealing the nonlinear effects we are investigating. Formally, its presence is accounted for by adding to the right-hand side of Eqs. (44) a term $\pm iK(z)\Phi^{\pm}$, where K(z) is a (real) coupling coefficient resulting from random fiber imperfections.

Let us now suppose that the single-mode fiber is either linearly twisted around its symmetry axis at a uniform twist rate $\tau$ or that a magnetic field H is applied along the direction of propagation producing a gyroelectric perturbation proportional to H through the Verdet constant.[6] In both cases, these operations result in an additional source of coupling which can be formalized by adding to $\pm iK(z)$ a suitable coefficient. In the case of the *twisted fiber*, the propagating field is more suitably described in a local system of coordinates $\hat{x}(z)$, $\hat{y}(z)$ which rotates with the twist, so that in Eq. (43) one has to substitute the fixed axes $\hat{x}$, $\hat{y}$ with the local birefringence axes $\hat{x}(z)$, $\hat{y}(z)$. In the case of the *optically active fiber* the expression for the field remains unchanged. The set of equations (44) becomes, for the twisted fiber,[5]

$$L_1^{\pm}\phi_1^{\pm} = \pm[iK(z) + \tau]e^{\pm i\delta\beta z}\phi_2^{\pm} + NLT$$
$$L_2^{\pm}\phi_2^{\pm} = \pm[iK(z) - \tau]e^{\mp i\delta\beta z}\phi_1^{\pm} + NLT \qquad (47)$$

where the symbol NLT stands for the nonlinear terms appearing on the right-hand sides of Eqs. (44), while for the active fiber it reads

$$L_1^{\pm}\phi_1^{\pm} = \pm[iK(z) \pm \alpha]e^{\pm i\delta\beta z}\phi_2^{\pm} + NLT$$
$$L_2^{\pm}\phi_2^{\pm} = \pm[iK(z) \mp \alpha]e^{\mp i\delta\beta z}\phi_1^{\pm} + NLT \qquad (48)$$

where $\alpha$ is a real constant accounting for the optical activity.

In order to reduce the set of equations (47) and (48) to a more amenable form, it is convenient to perform a linear-to-circular polarization transformation by introducing the right and left circularly polarized eigenmodes $E(\rho)\hat{e}_r$ and $E(\rho)\hat{e}_l$ [we assume, for simplicity, $E_1(\rho) = E_2(\rho) = E(\rho)$], where

$$\hat{e}_r = \frac{1}{\sqrt{2}}(\hat{x} + i\hat{y}) \qquad \hat{e}_l = \frac{1}{\sqrt{2}}(\hat{x} - i\hat{y}) \qquad (49)$$

After introducing the "a priori" unknown propagation constants $\beta_r$ and $\beta_l$ relative to the right and left circularly-polarized eigenmodes, we write the electric field as

$$\mathbf{E}(\rho, z, t) = E(\rho)e^{i\omega_0 t}\left[e^{-i\beta_r z}\phi_r^+(z,t) + e^{i\beta_r z}\phi_r^-(z,t)\right]\hat{e}_r + E(\rho)e^{i\omega_0 t}\left[e^{-i\beta_l z}\phi_l^+(z,t) + e^{i\beta_l z}\phi_l^-(z,t)\right]\hat{e}_l \qquad (50)$$

and

$$\mathbf{E}(\rho, z, t) = E(\rho)e^{i\omega_0 t}\left[e^{-i\beta_r z}\phi_r^+(z,t) + e^{i\beta_l z}\phi_r^-(z,t)\right]\hat{e}_r + E(\rho)e^{i\omega_0 t}\left[e^{-i\beta_l z}\phi_l^+(z,t) + e^{i\beta_r z}\phi_l^-(z,t)\right]\hat{e}_l \qquad (51)$$

respectively for the case of the twisted and active fibers. In fact, the twisted fiber possesses an intrinsic right or left "handedness" in the sense that a circularly polarized mode, either forward or backward traveling, always "sees" the fiber twisted in the same sense so that one has to attribute to $\phi_r^-$ and $\phi_l^-$ a propagation constant $-\beta_r$ and $-\beta_l$ respectively. This is no longer true in an active fiber where the direction of the magnetic field introduces a preferential orientation, so that the backward traveling right (left) circularly polarized mode has the propagation constant of the forward traveling left (right) circularly polarized mode.

By neglecting chromatic dispersion ($1/A = 0$) it is possible, in the limit of large $\tau$ or $\alpha$, to obtain simple propagation equations for the $\phi_{r,l}$ which in the two cases considered read

$$\left(\frac{\partial}{\partial z} \pm \frac{1}{V}\frac{\partial}{\partial t}\right)\phi_r^\pm = \mp\frac{2i}{3}R\left(|\phi_r^\pm|^2 + 2|\phi_l^\pm|^2 + 2|\phi_r^{\bar{\mp}}|^2 + 2|\phi_l^{\bar{\mp}}|^2\right)\phi_r^\pm$$

$$\left(\frac{\partial}{\partial z} \pm \frac{1}{V}\frac{\partial}{\partial t}\right)\phi_l^\pm = \mp\frac{2i}{3}R\left(|\phi_l^\pm|^2 + 2|\phi_r^\pm|^2 + 2|\phi_r^{\bar{\mp}}|^2 + 2|\phi_l^{\bar{\mp}}|^2\right)\phi_l^\pm \tag{52}$$

where

$$\frac{1}{V} = \frac{1}{2}\left(\frac{1}{V_1} + \frac{1}{V_2}\right) \tag{53}$$

the $\beta$'s appearing in Eq. (50) being given by

$$\beta_r = \frac{\beta_1 + \beta_2}{2} - \tau \qquad \beta_l = \frac{\beta_1 + \beta_2}{2} + \tau \tag{54}$$

and

$$\left(\frac{\partial}{\partial z} \pm \frac{1}{V}\frac{\partial}{\partial t}\right)\phi_r^\pm = \mp\frac{2i}{3}R\left(\left|\phi_r^\pm\right|^2 + 2\left|\phi_l^\pm\right|^2 + 2\left|\phi_r^{\bar{\mp}}\right|^2 + 2\left|\phi_l^{\bar{\mp}}\right|^2\right)\phi_r^\pm \mp \frac{4i}{3}R\phi_r^{\bar{\mp}}\,\phi_l^\pm\phi_l^{\bar{\mp}*}$$

$$\left(\frac{\partial}{\partial z} \pm \frac{1}{V}\frac{\partial}{\partial t}\right)\phi_l^\pm = \mp\frac{2i}{3}R\left(\left|\phi_l^\pm\right|^2 + 2\left|\phi_r^\pm\right|^2 + 2\left|\phi_r^{\bar{\mp}}\right|^2 + 2\left|\phi_l^{\bar{\mp}}\right|^2\right)\phi_l^\pm \mp \frac{4i}{3}R\phi_l^{\bar{\mp}}\,\phi_r^\pm\phi_r^{\bar{\mp}*} \tag{55}$$

the $\beta$'s appearing in Eq. (51) being

$$\beta_r = \frac{\beta_1 + \beta_2}{2} - \alpha \qquad \beta_l = \frac{\beta_1 + \beta_2}{2} + \alpha \tag{56}$$

Note that the (random) coupling coefficient K(z) does not appear any more in the set of equations (52) and (55) and, accordingly, nonlinear propagation is not affected by fiber imperfections in the case of high twist-rate or optical activity.

## 7. AVAILABLE SIMPLE ANALYTICAL SOLUTIONS

The nonlinear differential equations introduced in the previous sections, formidable as they may seem, in some cases admit manageable analytical solutions. We wish to present in this section a few of them with the purpose of acquainting the reader with some typical types of behaviour that can be expected when dealing with nonlinear propagation.

## 7.1 The mono-mode fiber

The evolution of a single forward traveling mode is described by Eq. (40) which can be, in many situations of interest where the envelope $\phi(z,t)$ is not varying too fast, approximated by

$$\left(\frac{\partial}{\partial z} + \frac{1}{V}\frac{\partial}{\partial t} - \frac{i}{2A}\frac{\partial^2}{\partial t^2}\right)\Phi = -iR\left|\Phi\right|^2\Phi \tag{57}$$

In a frame of reference moving with velocity V, and after introducing suitable dimensionless variables $\zeta = z/(T_0^2|A|)$, $\tau = (t-z/V)/T_0$ and $u = (RT_0^2|A|)^{1/2}\Phi$, where $T_0$ is the width of the incident pulse, Eq. (57) can be put in the compact form

$$i\frac{\partial}{\partial\zeta}u - \text{sgn}(A)\frac{1}{2}\frac{\partial^2}{\partial\tau^2}u + \left|u\right|^2 u = 0 \tag{58}$$

where sgn is the sign function. Equation (58) is often called the *nonlinear Schrödinger equation.* For a wide class of well-behaved initial value problems, Eq. (58) can be solved exactly by means of the method of the inverse scattering transform.[7] We limit ourselves here to reporting some particular solutions possessing an immediate physical meaning. Thus, for example, if we neglect chromatic dispersion ($1/A = 0$), Eq. (57) admits the solution

$$\Phi(z,t) = e^{-iR\left|\Phi(z=0,t-z/V)\right|^2 z}\,\Phi(z=0,t-z/V) \tag{59}$$

which clearly exhibits the *self-modulation* process induced on the field by the optical Kerr effect. If, on the contrary, chromatic dispersion is present (and group-velocity dispersion is negative, $1/A < 0$), a particular solution of Eq. (57) is the one (bright) *soliton* solution (for the general form of this solution see Ref. 7)

$$\Phi(z,t) = \Phi_0 e^{iz/2A\sigma^2}\,\text{sech}\left[\frac{t - z/V}{\sigma}\right] \tag{60}$$

with

$$-\frac{1}{A\sigma^2} = R\left|\Phi_0\right|^2 \tag{61}$$

which shows how the exact balance between the influences of chromatic dispersion and nonlinearity can give rise to distortionless propagation. Equation (60) can be compared with the dark soliton solution

$$\Phi(z,t) = \Phi_0 e^{-iz/A\sigma^2}\,\tanh\left[\frac{t - z/V}{\sigma}\right] \tag{62}$$

whose existence requires positive group-velocity dispersion ($1/A > 0$) in order to satisfy the relation

$$\frac{1}{A\sigma^2} = R\left|\Phi_0\right|^2 \tag{63}$$

It is also worth mentioning that the equation

$$\left(\frac{\partial}{\partial z} + \frac{1}{V}\frac{\partial}{\partial t} - \frac{i}{2A}\frac{\partial^2}{\partial t^2} - \frac{1}{3!B}\frac{\partial^3}{\partial t^3}\right)\Phi = -iR\left|\Phi\right|^2\Phi - \frac{R}{\omega_0}\frac{\partial}{\partial t}\left(\left|\Phi\right|^2\Phi\right) \tag{64}$$

is completely integrable by means of the inverse scattering transform and that an explicit one soliton solution can be worked out.[7]

### 7.2 The two-mode fiber

Let us now consider the case of a single-mode fiber supporting two polarization states. The set of equations (44) takes, in the case of forward propagation, the form

$$L_1\Phi_1 = -iR_{11}\left|\Phi_1\right|^2\Phi_1 - \frac{2i}{3}R_{12}\left|\Phi_2\right|^2\Phi_1 - \frac{i}{3}R_{12}e^{2i\delta\beta z}\Phi_1^*\Phi_2^2$$
$$L_2\Phi_2 = -iR_{22}\left|\Phi_2\right|^2\Phi_2 - \frac{2i}{3}R_{12}\left|\Phi_1\right|^2\Phi_2 - \frac{i}{3}R_{12}e^{-2i\delta\beta z}\Phi_2^*\Phi_1^2 \tag{65}$$

In particular, for a *high-birefringence* fiber ($\delta\beta$ large), the last term appearing on the right-hand side of Eqs. (65) can be dropped due to its rapidly oscillating behavior and the set of equations (65) becomes

$$L_1\Phi_1 = -iR\left(\left|\Phi_1\right|^2 + \frac{2}{3}\left|\Phi_2\right|^2\right)\Phi_1$$
$$L_2\Phi_2 = -iR\left(\left|\Phi_2\right|^2 + \frac{2}{3}\left|\Phi_1\right|^2\right)\Phi_2 \tag{66}$$

having assumed, for the sake of simplicity, $R_{11} = R_{12} = R$. Again, if we neglect chromatic dispersion ($1/A = 0$) and higher-order group velocity dispersions, we can immediately write the explicit solution of the set of equations (65) as

$$\Phi_1(z,t) = \exp\left[-\frac{2i}{3}R\int_0^z dz'\left|\Phi_2\left(z{=}0, t - \frac{z'}{V_2} + \frac{z-z'}{V_1}\right)\right|^2 - iR\left|\Phi_1\left(z{=}0, t - \frac{z}{V_1}\right)\right|^2 z\right]\Phi_1\left(z{=}0, t - \frac{z}{V_1}\right)$$
$$\Phi_2(z,t) = \exp\left[-\frac{2i}{3}R\int_0^z dz'\left|\Phi_1\left(z{=}0, t - \frac{z'}{V_1} + \frac{z-z'}{V_2}\right)\right|^2 - iR\left|\Phi_2\left(z{=}0, t - \frac{z}{V_2}\right)\right|^2 z\right]\Phi_2\left(z{=}0, t - \frac{z}{V_2}\right) \tag{67}$$

In the same limit of zero chromatic dispersion, it is possible to solve the set of equations (65) in the case of a low-birefringence fiber ($\delta\beta = 0$, $\beta_1 = \beta_2 = \beta$, $V_1 = V_2 = V$). To this end, it is expedient to use the circularly polarized eigenmodes $E(\rho)\hat{e}_r$ and $E(\rho)\hat{e}_l$ introduced in section 6.2 [see Eq. (49)] and to write the field in the form

$$\mathbf{E}(\rho,z,t) = E(\rho)e^{i\omega_0 t - i\beta(\omega_0)z}[\Phi_r(z,t)\hat{e}_r + \Phi_l(z,t)\hat{e}_l] \tag{68}$$

where

$$\Phi_r = \frac{1}{\sqrt{2}}(\Phi_1 - i\Phi_2) \qquad \Phi_l = \frac{1}{\sqrt{2}}(\Phi_1 + i\Phi_2) \tag{69}$$

and

$$\Phi_1 = \frac{1}{\sqrt{2}}(\Phi_r + \Phi_l) \qquad \Phi_2 = \frac{i}{\sqrt{2}}(\Phi_r - \Phi_l) \tag{70}$$

Written in terms of the $\Phi_r$, $\Phi_l$ variables, the set of equations (65) (where we put $\delta\beta = 0$) becomes

$$\left(\frac{\partial}{\partial z} + \frac{1}{V}\frac{\partial}{\partial t}\right)\Phi_r = -\frac{2i}{3}R\left(\left|\Phi_r\right|^2 + 2\left|\Phi_l\right|^2\right)\Phi_r$$

$$\left(\frac{\partial}{\partial z} + \frac{1}{V}\frac{\partial}{\partial t}\right)\Phi_l = -\frac{2i}{3}R\left(\left|\Phi_l\right|^2 + 2\left|\Phi_r\right|^2\right)\Phi_l \qquad (71)$$

which is of the form of the set of equations (66) and thus can be easily solved [see Eqs. (67)], the expressions for $\phi_1$ and $\phi_2$ being then found through Eq. (70).

## 8. FINITE DIMENSIONAL HAMILTONIAN DESCRIPTION OF NONLINEAR PROPAGATION

The task of finding general manageable analytical solutions to the set of equations describing nonlinear propagation is a rather difficult one and most often people have to resort to time consuming cumbersome computational procedures to obtain numerical solutions. Besides, the very nature of the problem, and its description in terms of nonlinear partial differential equations, does not lend itself much to physical intuition. To partially satisfy the requirement of a simpler analytical model of nonlinear propagation, with associated simpler numerical analysis, we have introduced a formalism[8] where each modal field amplitude $\Phi_i(z,t)$ is represented by a finite number of parameters possessing a direct physical meaning (as, for example, the center-of-mass or the width of the pulse traveling on a particular mode), a circumstance which admittedly furnishes a limited description of a system possessing an infinite number of degrees of freedom. These parameters can be chosen in such a way as to be formally interpreted as conjugate variables of a dynamical Hamiltonian system and their evolution is described by a set of first-order ordinary differential equations which represent Hamilton's equations of motion. They lend themselves to numerical integration more easily than the set of nonlinear partial differential equations and offer a direct physical insight into the propagation problem.

In order to illustrate our approach, one has to recall the basic concepts of the *Lagrangian formulation* for continuous systems.[9] Within its framework, the field equations are deduced by means of a variational principle

$$\delta\int_{t_1}^{t_2} dtL = \delta\int_{t_1}^{t_2} dt\int dx\mathcal{L}\left(\eta, \frac{\partial\eta}{\partial t}, \frac{\partial\eta}{\partial x}\right) = 0 \qquad (72)$$

where L is the Lagrangian and $\mathcal{L}$ the Lagrangian density [we refer, for simplicity, to a unidimensional system described by the generalized coordinate $\eta(x,t)$, whose variation is assumed to vanish at $t_1$ and $t_2$ and also at the extremities of the space integration], which gives rise to the Euler-Lagrange equations of motion

$$\frac{\partial}{\partial t}\frac{\partial\mathcal{L}}{\partial(\partial\eta/\partial t)} + \frac{\partial}{\partial x}\frac{\partial\mathcal{L}}{\partial(\partial\eta/\partial x)} - \frac{\partial\mathcal{L}}{\partial\eta} = 0 \qquad (73)$$

Considering, for example, the specific case described by the set of equations (65), it is convenient to choose as generalized coordinates the modal amplitudes $\Phi_i(z,t)$ and their complex conjugate $\Phi_i^*(z,t)$, so that

$$\mathcal{L} = \mathcal{L}[\{\Phi_i\},\{\Phi_i^*\},\{\Phi_{i,t}\},\{\Phi_{i,t}^*\},\{\Phi_{i,z}\},\{\Phi_{i,z}^*\}]$$

where $\{\Phi_i\}$ stands for $\{\Phi_1,\Phi_2\}$ and $\{\Phi_{i,t}\} = \{\partial\Phi_i/\partial t\}$, $\{\Phi_{i,z}\} = \{\partial\Phi_i/\partial z\}$, and the Euler-Lagrange equations read

$$\frac{\partial}{\partial t}\frac{\partial L}{\partial \Phi_{i,t}} + \frac{\partial}{\partial z}\frac{\partial L}{\partial \Phi_{i,z}} - \frac{\partial L}{\partial \Phi_i} = 0$$

$$\frac{\partial}{\partial t}\frac{\partial L}{\partial \Phi^*_{i,t}} + \frac{\partial}{\partial z}\frac{\partial L}{\partial \Phi^*_{i,z}} - \frac{\partial L}{\partial \Phi^*_i} = 0 \qquad i = 1,2 \qquad (74)$$

By performing the change of dependent variables

$$u_1(z,t) = \Phi_1(z,t)e^{-i\delta\beta z/2} \qquad u_2(z,t) = \Phi_2(z,t)e^{i\delta\beta z/2} \qquad (75)$$

the set of equations (65) becomes, by keeping only the second-order group-velocity dispersion and assuming $R_{11} = R_{22} = R$,

$$\left(\frac{\partial}{\partial z} + \frac{1}{V_1}\frac{\partial}{\partial t} - \frac{i}{2A}\frac{\partial^2}{\partial t^2}\right)u_1 + i\frac{\delta\beta}{2}u_1 = -iR\left(\left|u_1\right|^2 + \frac{2}{3}\left|u_2\right|^2\right)u_1 - \frac{iR}{3}u_1^* u_2^2$$

$$\left(\frac{\partial}{\partial z} + \frac{1}{V_2}\frac{\partial}{\partial t} - \frac{i}{2A}\frac{\partial^2}{\partial t^2}\right)u_2 - i\frac{\delta\beta}{2}u_2 = -iR\left(\left|u_2\right|^2 + \frac{2}{3}\left|u_1\right|^2\right)u_2 - \frac{iR}{3}u_2^* u_1^2 \qquad (76)$$

which is generated by the field Lagrangian density

$$L = Im\left\{\sum_{j=1}^{2}\left(u_j^* u_{j,z} + \frac{1}{V_j}u_j^* u_{j,t}\right)\right\} + \frac{1}{2}\sum_{j=1}^{2}\frac{1}{A_j}\left|u_{j,t}\right|^2 + \frac{\delta\beta}{2}\left(\left|u_1\right|^2 - \left|u_2\right|^2\right)$$

$$+ \frac{R}{6}\left(\left|u_1\right|^4 + \left|u_2\right|^4\right) + \frac{R}{3}\left(\left|u_1\right|^2 + \left|u_2\right|^2\right)^2 + \frac{R}{6}(u_2^2 u_1^{*2} + u_2^{*2} u_1^2) \qquad (77)$$

Let us suppose that the field equations are to be generated by a given Lagrangian density and let us postulate, as in Ref. 10, a prescribed functional dependence of the i-th field amplitude on a finite number N of z-dependent parameters, that is

$$\Phi_i(z,t) = F_i(\{b_n^{(i)}(z)\}, t) \qquad n = 1,2,...,N \qquad (78)$$

where the $F_i$'s are arbitrary functions. If Eqs. (78) are inserted into the general expression of the Lagrangian density, we obtain a particular Lagrangian density $\tilde{L}$ such that the equations of evolution of the *generalized coordinates* $\{b_n^{(i)}(z)\}$ can be obtained as Euler-Lagrange equations of the reduced variational principle[10] (note the exchange of roles between t and z)

$$\delta\int dz\langle\tilde{L}\rangle = 0 \qquad (79)$$

where

$$\langle\tilde{L}\rangle = \int_{-\infty}^{+\infty} dt\tilde{L} \qquad (80)$$

In general, one can pass from the Lagrangian to the Hamiltonian formalism by introducing the canonical momentum $\pi$ conjugate to $\eta$,

$$\pi = \frac{\partial L}{\partial(\partial\eta/\partial t)} \qquad (81)$$

and the Hamiltonian density

$$H = \pi \frac{\partial \eta}{\partial t} - L \tag{82}$$

a procedure which can, of course, be followed also in the case of the Lagrangian $\langle L \rangle$. Most often, however, it is not possible to explicitly express $L$, and thus $H$, in terms of the conjugate momenta $\pi$ and thus the Hamiltonian formalism does not possess a particular usefulness. Basically, the essence of our approach[8] consists of a "smart" choice of the generalized coordinates $\{b_n^{(i)}\}$ which allows us to express in a very simple way the conjugate momenta and thus to take full advantage of the Hamiltonian formalism. More precisely, we characterize each modal amplitude

$$\Phi_i(z,t) = |\Phi_i(z,t)| e^{i\Psi_i(z,t)} \qquad i = 1,2 \tag{83}$$

in terms of a finite number of z-dependent parameters

$$\{b_n^{(j)}\} = \{a_n^{(j)}(z), M_n^{(j)}(z)\} \qquad n = 1,2,...,N \tag{84}$$

where the $a_n^{(j)}(z)$ are defined through the relations

$$\Psi_j(z,t) = \sum_{n=0}^{N} a_n^{(j)} t^n \tag{85}$$

In general, we can write

$$\tilde{L} = \tilde{L}_0 + \tilde{L}_1 \tag{86}$$

where

$$\tilde{L}_0 = \frac{1}{2i} \sum_{j=1}^{2} \left( \Phi_j^* \frac{\partial}{\partial z} \Phi_j - \Phi_j \frac{\partial}{\partial z} \Phi_j^* \right) \tag{87}$$

is the Lagrangian density corresponding to the operator $\partial/\partial z$, so that

$$\langle \tilde{L}_0 \rangle = \int_{-\infty}^{+\infty} dt \tilde{L}_0 = \frac{1}{2i} \sum_{j=1}^{2} \int_{-\infty}^{+\infty} dt \left( \Phi_j^* \frac{\partial}{\partial z} \Phi_j - \Phi_j \frac{\partial}{\partial z} \Phi_j^* \right) = \sum_{j=1}^{2} \int_{-\infty}^{+\infty} dt |\Phi_j|^2 \frac{\partial \Psi_j}{\partial z} = \sum_{j=1}^{2} \sum_{n=0}^{N} \dot{a}_n^{(j)} M_n^{(j)} \tag{88}$$

where the dot stands for derivation with respect to z and

$$M_n^{(j)} = \int_{-\infty}^{+\infty} dt \, t^n \left| \Phi_n^{(j)} \right|^2 \tag{89}$$

After writing

$$\langle \tilde{L}_1 \rangle = \langle \tilde{L} \rangle - \langle \tilde{L}_0 \rangle \tag{90}$$

it is possible to write down the Euler-Lagrange equations associated with the reduced variational principle

$$\frac{\partial}{\partial z}\frac{\partial\langle\tilde{L}\rangle}{\partial\dot{a}_n^{(j)}} - \frac{\partial\langle\tilde{L}\rangle}{\partial a_n^{(j)}} = 0$$

$$\frac{\partial}{\partial z}\frac{\partial\langle\tilde{L}\rangle}{\partial\dot{M}_n^{(j)}} - \frac{\partial\langle\tilde{L}\rangle}{\partial M_n^{(j)}} = 0 \qquad j = 1,2;\ n = 1,2,\ldots,N \quad (91)$$

By assuming $\langle\tilde{L}_1\rangle$ to be not explicitly dependent on the $\{\dot{a}_n, \dot{M}_n\}$ [this is, for example, actually the case for the situation described by Eq. (77) and it is valid for a quite general class of nonlinear interaction], we have

$$\frac{\partial\langle\tilde{L}\rangle}{\partial\dot{a}_n^{(j)}} = \frac{\partial\langle\tilde{L}_0\rangle}{\partial\dot{a}_n^{(j)}} = M_n^{(j)} \tag{92}$$

so that the moments $M_n^{(j)}$ represent the *conjugate momenta* of the variable $a_n^{(j)}$. The set of equations (91) then reduces to [see Eq. (88)]

$$\frac{\partial}{\partial z}M_n^{(j)} = \frac{\partial\langle\tilde{L}\rangle}{\partial a_n^{(j)}} = \frac{\partial\langle\tilde{L}_1\rangle}{\partial a_n^{(j)}}$$

$$-\frac{\partial\langle\tilde{L}\rangle}{\partial M_n^{(j)}} = -\frac{\partial\langle\tilde{L}_0\rangle}{\partial M_n^{(j)}} - \frac{\partial\langle\tilde{L}_1\rangle}{\partial M_n^{(j)}} = -\dot{a}_n^{(j)} - \frac{\partial\langle\tilde{L}_1\rangle}{\partial M_n^{(j)}} = 0 \qquad (93)$$

After setting

$$\langle\tilde{H}\rangle = -\langle\tilde{L}_1\rangle \tag{94}$$

we get

$$\frac{\partial}{\partial z}M_n^{(j)} = -\frac{\partial\langle\tilde{H}\rangle}{\partial a_n^{(j)}}$$

$$\frac{\partial}{\partial z}a_n^{(j)} = \frac{\partial\langle\tilde{H}\rangle}{\partial M_n^{(j)}} \qquad (95)$$

which can be interpreted as the Hamilton equations described by the set of conjugate variables $\{M_n^{(j)}, a_n^{(j)}\}$ with the Hamiltonian

$$\langle\tilde{H}\rangle = \sum_{j=1}^{2}\sum_{n=0}^{N} M_n^{(j)}\dot{a}_n^{(j)} - \langle\tilde{L}\rangle = -\langle\tilde{L}_1\rangle \tag{96}$$

Note that our approach requires the assumption of a particular form of the $|\Phi_j(z,t)|^2$, that is

$$|\Phi_j(z,t)|^2 = G_j(M_0^{(j)}(z), M_1^{(j)}(z), \ldots, M_N^{(j)}(z), t) \tag{97}$$

where $G_j$ is an arbitrary function which satisfies Eq. (89). For example, for N = 2 we can write

$$|\Phi_j(z,t)|^2 = \frac{M_0^{(j)}}{\sigma_j} f_j\left(\frac{t - \tau_j}{\sigma_j}\right) \tag{98}$$

where $f_j$ is an arbitrary function normalized to one with mean value zero and variance equal to one, $\tau_j$ and $\sigma_j$ being related to $M_0^{(j)}$, $M_1^{(j)}$, and $M_2^{(j)}$ through the relations

$$\tau_j = \frac{M_1^{(j)}}{M_0^{(j)}} \qquad \sigma_j^2 = \frac{M_2^{(j)}}{M_0^{(j)}} - \tau_j^2 \tag{99}$$

## 8.1 Propagation in a monomode fiber

In the case of a single-mode fiber supporting one polarization state, Eq. (77) reduces to

$$L = |\Phi|^2 \frac{\partial \Psi}{\partial z} + \frac{1}{V}|\Phi|^2 \frac{\partial \Psi}{\partial t} + \frac{1}{2A}\left[\left(\frac{\partial |\Phi|}{\partial t}\right)^2 + |\Phi|^2\left(\frac{\partial \Psi}{\partial t}\right)^2\right] + \frac{1}{2}R|\Phi|^4 \tag{100}$$

where we have set

$$\Phi(z,t) = |\Phi(z,t)|e^{i\Psi(z,t)} \tag{101}$$

so that, taking into account Eqs. (98) and (99),

$$\langle \tilde{L} \rangle = \langle \tilde{L}_0 \rangle + \langle \tilde{L}_1 \rangle$$

$$= \sum_{n=0}^{N} \dot{a}_n M_n + \frac{1}{V}\sum_{n=1}^{N} n a_n M_{n-1} + \frac{1}{2A}\sum_{n=1}^{N}\sum_{m=1}^{N} nm a_n a_m M_{m+n-2} + \frac{g}{8A}\left(\frac{M_0}{\sigma^2}\right) + \frac{Rh}{2}\left(\frac{M_0^2}{\sigma}\right) \tag{102}$$

with

$$g = \int_{-\infty}^{+\infty} dx \frac{1}{f}\left(\frac{df}{dx}\right)^2 \qquad h = \int_{-\infty}^{+\infty} dx f^2(x) \tag{103}$$

If we limit ourselves to the case N = 2, it is convenient to define the $a_n$'s and the $M_n$'s in a slightly different way through the relations [see Eqs. (85) and (89)]

$$\Psi(z,t) = a_0(z) + a_1(z)t + a_2(z)\left(t - \frac{M_1}{M_0}\right)^2 \tag{104}$$

and

$$M_0 = \int_{-\infty}^{+\infty} dt |\Phi|^2 \qquad M_1 = \int_{-\infty}^{+\infty} dt |\Phi|^2 t \tag{105}$$

$$M_2 = \int_{-\infty}^{+\infty} dt |\Phi|^2 \left(t - \frac{M_1}{M_0}\right)^2 \tag{106}$$

It can be easily checked that the new $a_n$ and $M_n$ are still conjugate Hamiltonian variables and that

$$\langle \tilde{L} \rangle = \langle \tilde{L}_0 \rangle - \langle \tilde{H} \rangle = \langle \tilde{L}_0 \rangle + \frac{a_1 M_0}{V} + \frac{1}{2A}(M_0 a_1^2 + 4a_2^2 M_2) + \frac{g}{8A}\frac{M_0^2}{M_2} + \frac{Rh}{2}\frac{M_0^{5/2}}{M_2^{1/2}} \qquad (107)$$

From Hamilton's equations (95) it immediately follows that

$$\frac{dM_0}{dz} = -\frac{\partial\langle \tilde{H}\rangle}{\partial a_0} = 0$$

$$\frac{da_1}{dz} = \frac{\partial\langle \tilde{H}\rangle}{\partial M_1} = 0 \qquad (108)$$

which implies $M_0$ = const = $\mathcal{M}_0$ (energy conservation) and $a_1 = \alpha_1$ . In particular, by choosing $\alpha_1 = 0$, one has

$$\frac{dM_1}{dz} = -\frac{\partial\langle \tilde{H}\rangle}{\partial a_1} = \frac{\mathcal{M}_0}{V}$$

$$\frac{da_0}{dz} = \frac{\partial\langle \tilde{H}\rangle}{\partial M_0} = -\frac{g}{4A}\frac{\mathcal{M}_0}{M_2} - \frac{5Rh}{4}\frac{\mathcal{M}_0^{3/2}}{M_2^{1/2}} \qquad (109)$$

so that $M_1 = (\mathcal{M}_0/V)z$ while $a_0(z)$ can be determined by direct integration once $M_2(z)$ is known. The two remaining Hamilton equations yield

$$\frac{dM_2}{dz} = -\frac{\partial\langle \tilde{H}\rangle}{\partial a_2} = \frac{4a_2M_2}{A}$$

$$\frac{da_2}{dz} = \frac{\partial\langle \tilde{H}\rangle}{\partial M_2} = -\frac{2a_2^2}{A} + \frac{g}{8A}\frac{\mathcal{M}_0^2}{M_2^2} + \frac{Rh}{4}\frac{\mathcal{M}_0^{5/2}}{M_2^{3/2}} \qquad (110)$$

which can be deduced in terms of the reduced Hamiltonian

$$H_0 = -2a_2^2\frac{M_2}{A} - \frac{g}{8A}\frac{M_0^2}{M_2} - \frac{Rh}{2}\frac{M_0^{5/2}}{M_2^{1/2}} \equiv -2a_2^2\frac{M_2}{A} + V(M_2) \qquad (111)$$

Obviously, $H_0$ is a conserved function of z so that

$$H_0 = E \qquad (112)$$

is a first integral of motion, its minimum value $E_m$ being achieved for $a_2 = 0$ and $M_2 = \mathcal{M}_2$ where

$$\mathcal{M}_2^{1/2} = \frac{g}{2h}\frac{1}{\mathcal{M}_0^{1/2}|A|R} \qquad (113)$$

is the value at which the potential $V(M_2)$ has a minimum (see Fig. 1).

A particular solution of the propagation problem is given by [see Eqs. (108), (109) and (110)]

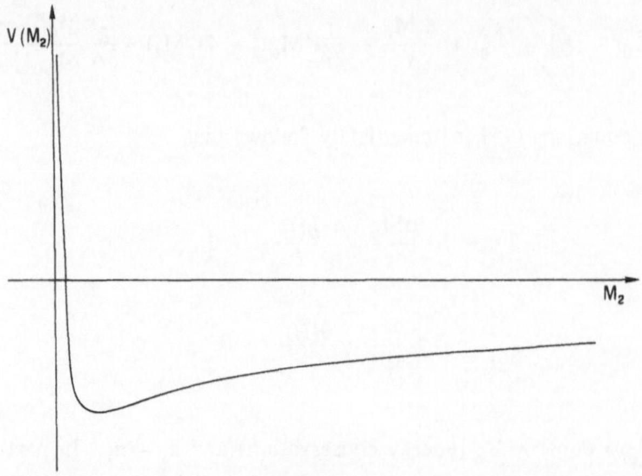


Fig. 1. Qualitative behavior of the potential $V(M_2)$ as a function of $M_2$ for $A < 0$ (for $A > 0$ the potential does not exhibit a minimum).

$$a_0 = -\left(\frac{g}{4A}\frac{\mathcal{M}_0}{\mathcal{M}_2} + \frac{5Rh}{4}\frac{\mathcal{M}_0^{3/2}}{\mathcal{M}_2^{1/2}}\right)z \qquad M_0 = \mathcal{M}_0$$

$$a_1 = 0 \qquad M_1 = \frac{\mathcal{M}_0}{V}z \tag{114}$$

$$a_2 = 0 \qquad M_2 = \mathcal{M}_2$$

If we now choose

$$f(x) = \frac{\pi}{4\sqrt{3}}\,\mathrm{sech}^2\left(\frac{\pi x}{2\sqrt{3}}\right) \tag{115}$$

one has $g = \pi^2/9$, $h = \pi\sqrt{3}/18$ and Eq. (113) exactly reproduces Eq. (61), with $|\Phi_0|^2 = (\pi/4\sqrt{3})(\mathcal{M}_0^{3/2}/\mathcal{M}_2^{1/2})$ and $\tau = (2\sqrt{3}/\pi)(\mathcal{M}_2^{1/2}/\mathcal{M}_0^{1/2})$ . Besides, by taking advantage of Eq. (113), one has $a_0 = z/2A\tau^2$ so that, by using Eqs. (98), (101) and (104), we can see that our solution exactly reproduces, at least in this case, the fundamental soliton solution (60).

## REFERENCES


1. B. Crosignani, P. Di Porto, and A. Yariv, "Coupled-mode theory and slowly varying approximation in guided-wave optics", submitted for publication in Opt. Commun.
2. H. Kogelnik, "Theory of dielectric waveguides" in *Integrated Optics*, T. Tamir ed. (Springer-Verlag, Berlin, 1979)
3. A. Owyoung, "The origins of the nonlinear refractive indices of liquids and glasses", Ph.D. Thesis (California Institute of Technology, 1971, unpublished)
4. B. Crosignani, A. Cutolo, and P. Di Porto, "Coupled-mode theory of nonlinear propagation in multimode and single-mode fibers: envelope solitons and self-confinement", J. Opt. Soc. Am. **72**, 1136 (1982)
5. B. Crosignani and A. Yariv, "Nonlinear interaction of copropagating and counterpropagating waves in straight and highly twisted single-mode fibers", J. Opt. Soc. Am. B **5**, 507 (1988).
6. R. Ulrich and A. Simon, "Polarization optics of twisted single-mode fibers", Appl. Opt. **18**, 2241 (1979).

7. Y. Kodama and A. Hasegawa, "Nonlinear pulse propagation in a monomode dielectric guide", IEEE J. Quantum Electron. **QE-23**, 510 (1987);
G. P. Agrawal, *Nonlinear Fiber Optics* (Academic Press, New York, 1989)
8. E. Caglioti, B. Crosignani, and P. Di Porto, "Hamiltonian description of nonlinear propagation in optical fibers", Phys. Rev. A **38**, 4036 (1988);
E. Caglioti. B. Crosignani, P. Di Porto, S. Trillo, and S. Wabnitz, "Finite dimensional description of nonlinear pulse propagation in optical fibers with applications to soliton switching", J. Opt. Soc. Am. B 7, 374 (1990)
9. See, e.g., H. Goldstein, *Classical Mechanics* (2nd edition, Addison-Wesley, 1980)
10. D. Anderson, "Variational approach to nonlinear pulse propagation in optical fibers", Phys. Rev. A 27, 3135 (1983).

## SECOND HARMONIC GENERATION IN OPTICAL FIBERS

R. H. Stolen

AT&T Bell Laboratories
Holmdel, New Jersey, USA

## 1. GENERAL SURVEY OF THE PROBLEM

### 1.1 Introduction

Second harmonic generation was not expected to play an important role in fiber nonlinear optics. In contrast to bulk and planar guided-wave nonlinear optics, where so-called $\chi^{(2)}$ phenomena such as second harmonic generation and three-photon parametric amplification have dominated the field, fibers usually demonstrate $\chi^{(3)}$ phenomena. These $\chi^{(3)}$ nonlinear effects have been extensively studied in fibers and include spectral broadening, stimulated Raman and Brillouin scattering, nonlinear switching, and parametric four-photon mixing.[1] The $\chi^{(2)}$ phenomena are expected to be absent in glass fibers because of inversion symmetry in the fiber core.

Of course, there are higher order terms such as electric quadrupole, magnetic dipole, and surface terms which can produce a weak second harmonic - even in a centrosymmetric fiber. Such a weak second harmonic has been seen as a faint green output from several hundred meters of multimode or single mode fiber when intense mode-locked Q-switched Nd:YAG laser pulses were coupled into the fiber.[2] The green output is near the limit of visibility using powers sufficient to blow the input end off the fiber. Third harmonic light has also been seen in fibers.[3]

The report of efficient second harmonic generation in short lengths of fiber thus came as a tremendous surprise. Conversion efficiencies of 5% were reported in only 50 cm of fiber and the output used to pump a dye laser.[4] Even though the peak input power was around 20 kW at 1.064 $\mu$m, this result was impressive enough to set people dreaming of efficient and inexpensive frequency doublers that could be used with diode lasers. So far this goal is a long way off and the present record is a 13% conversion efficiency with a peak input power at 1.064 $\mu$m of 900 W.[5]

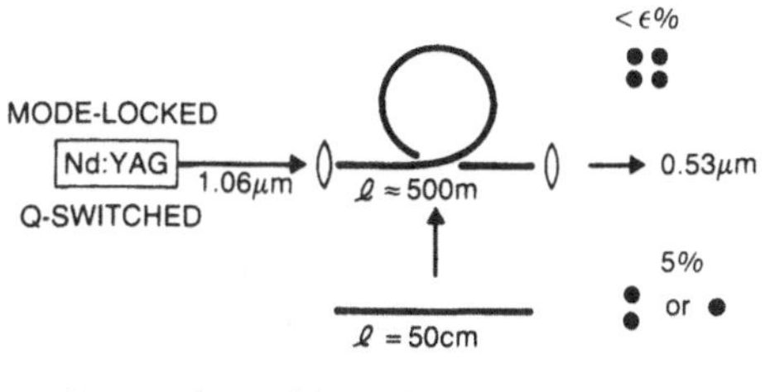


Fig. 1. Observation of second harmonic generation in long and short fibers (Fujii *et al* 1980, Ohmori and Sasaki 1981, Österberg and Margulis 1986).

$$E_1 \longrightarrow \qquad \begin{array}{c} \longrightarrow E_2 \\ \longrightarrow E_1 \end{array}$$

$$\frac{dE_2}{dz} \sim \chi_2 E_1^2 e^{i(2k_1 - k_2)z}$$

$$k = \frac{n\omega}{c}$$
$$n_1 \neq n_2$$

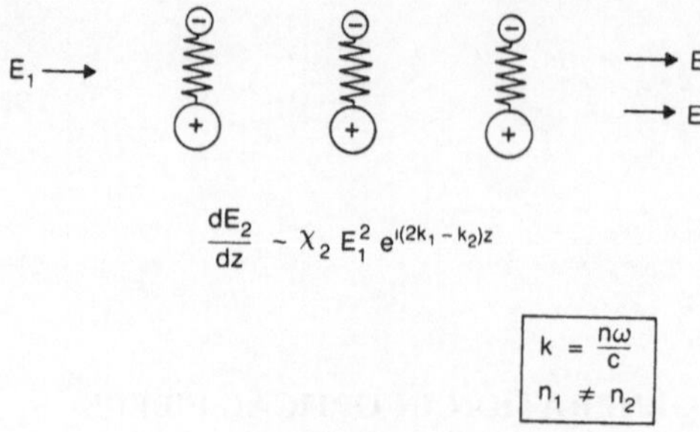

Fig. 2. Phase mismatch problem.

## 1.2 Defects

It appears that the strong second harmonic generation arises from something in the glass which is produced or moved around in just such a way as to produce a non-centrosymmetric core. There are two reasons for this belief. First, the short fibers did not initially produce second harmonic generation. The green harmonic output grew slowly over several hours as the fiber was pumped with mode-locked Q-switched laser pulses with peak powers in the 5-20 kW range. The fibers themselves were doped with both germanium and phosphorus and no second harmonic generation was seen in fibers with pure silica cores.[4]

The other important observation deals with the fiber output mode in the green. As will be discussed in more detail in section 3, the higher order nonlinearities as well as surface contributions from the fiber's core-cladding interface require, by symmetry, a conversion from the fiber's lowest-order transverse mode to a mode of odd symmetry. This type of output was seen in the weak output of long fibers and was usually seen in the strong second harmonic generation in short fibers as well. However, occasionally the green light appeared in the fiber's lowest order transverse mode. The only way both fundamental and harmonic waves can be in the same fiber mode is for the symmetry of the fiber core material to actually change. A bulk $\chi^{(2)}$ has thus appeared where there was none before!

## 1.3 Phasematching

A noncentrosymmetric core is not sufficient for strong second harmonic generation. As illustrated in Fig. 2 the polarization wave radiating at the harmonic frequency typically gets out of phase with the free running wave after a short distance because the refractive index is not the same at the fundamental and harmonic frequencies. Methods of matching the phase velocities of the fundamental and harmonic waves are called phasematching.

It appears that the efficient second harmonic generation fibers are phasematched by alternating the sign of the $\chi^{(2)}$ nonlinearity with exactly the right periodicity to compensate for

$$\chi_2 = |\chi_2| e^{ikz}$$

$$\frac{dE_2}{dz} \sim \chi_2 E_1^2 e^{i(2k_1 - k_2)z}$$

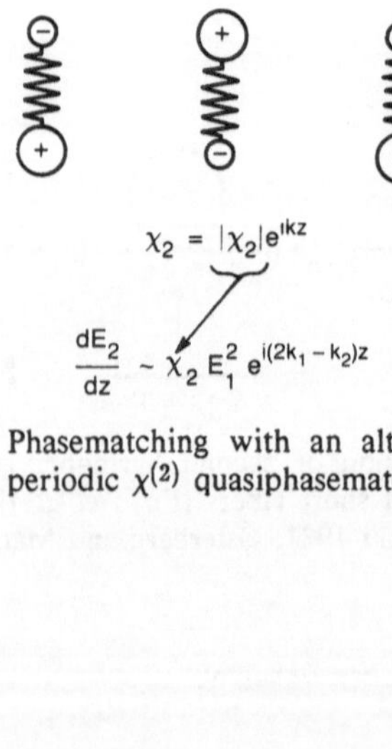

Fig. 3. Phasematching with an alternating or periodic $\chi^{(2)}$ quasiphasematching.

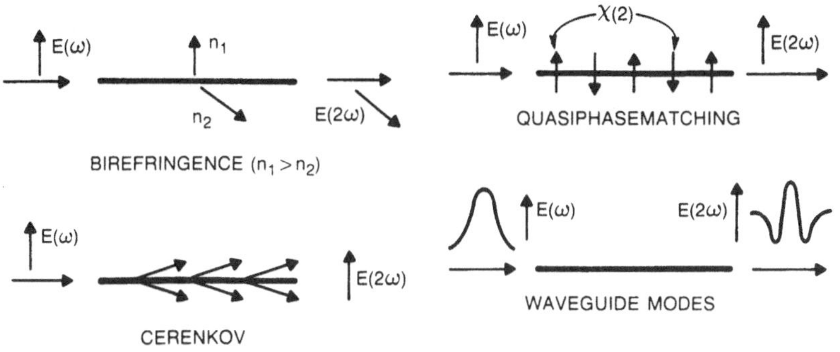


Fig. 4. Various solutions to phasematching.

the mismatch in wavevectors.[6] This is called quasiphasematching and was one of the first proposed methods for phasematching second harmonic generation.[7] The $\chi^{(2)}$ grating only has to be periodic and any combination of an alternating plus a uniform $\chi^{(2)}$ would have the same effect.[8]

Four phasematching schemes are illustrated schematically in Fig. 4. The first and most common method uses birefringence to match the refractive indices at the fundamental and harmonic wavelengths. This is particularly useful in planar waveguides where both birefringence intrinsic to a crystal and birefringence from guide geometry can be used. This is not applicable to glass fibers and, even in crystals, the largest nonlinear coefficient corresponds to parallel rather than orthogonal fundamental and harmonic polarizations.

A phasematching scheme which is possible in fibers and with parallel polarizations utilizes the different wave velocities of different transverse waveguide modes. This is seldom the most desirable approach, however, because cancellations and poor mode overlap reduce efficiency. Also the intensity is highest in small guides which support only a single transverse mode. Phasematched second harmonic generation by waveguide modes has been seen in fibers and is treated in section 3.

In another approach which permits the use of parallel polarizations, second harmonic generation is emitted at an angle out of the waveguide - that is Čerenkov radiation.[9] The problem, however, is that the radiation emerges in the form of an arc which cannot all be focused down to the diffraction limit.

Quasiphasematching is the most promising approach to phasematching. Recently, lithium niobate channel guides have been periodically poled to produce quasiphasematching.[10] It is predicted that mW's of blue light should be possible with 10's of mW's of near infrared power.

## 1.4 Phenomenological mixing model

One possible mechanism for second harmonic generation in fibers looks at localized defects in the glass. The obvious question is how can the fiber organize itself to produce not only defect centers with the right symmetry but also arrange those defects with just the right alternating period for phasematching? Right now there is no accepted microscopic model but there is a macroscopic phenomenological model which describes many of the observed features.

In the macroscopic model, intense fields at the pump and harmonic frequencies mix to orient the defects which produce the direct dipole-allowed response. In a macroscopic picture, two pump waves mix with the harmonic wave by a third-order nonlinear process to form a dc polarization. The dc polarization alternates in sign with exactly the right period for phasematching. The constant relating the macroscopic polarization to the pump and harmonic fields is $\chi^{(3)}$. By way of comparison, $\chi^{(3)}$ also produces third harmonic generation and parametric four photon mixing.

The assumption is then made that the $\chi^{(2)}$ grating which is written into the fiber is proportional in some way to the macroscopic dc polarization. It is important to note that this is not the same thing as saying that the macroscopic dc polarization has actually oriented the defects. The mixing model is, in fact, much more general than a specific model of orientated

$$\left.\begin{array}{c}E_3\\E_a \cdot E_s\\E_0\end{array}\right\} \sim \chi_3 E^3$$

$$E_0 \sim \chi_3 E_1^* E_1^* E_2 \, e^{i(k_2 - 2k_1)z} \qquad \boxed{\chi_2 = \gamma \, E_0}$$

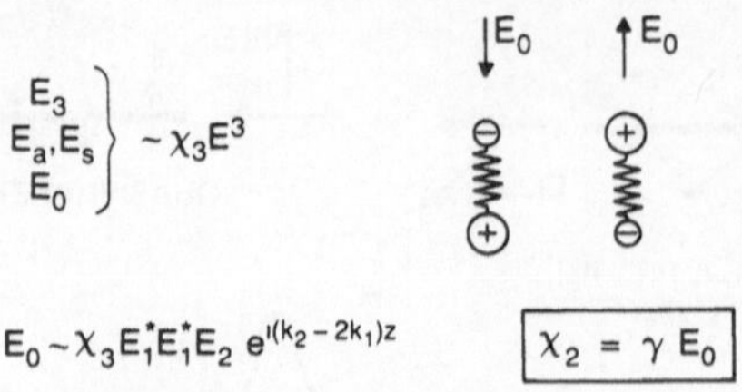

Fig. 5. Orientation of defects with the same periodicity
as the macroscopic polarization.

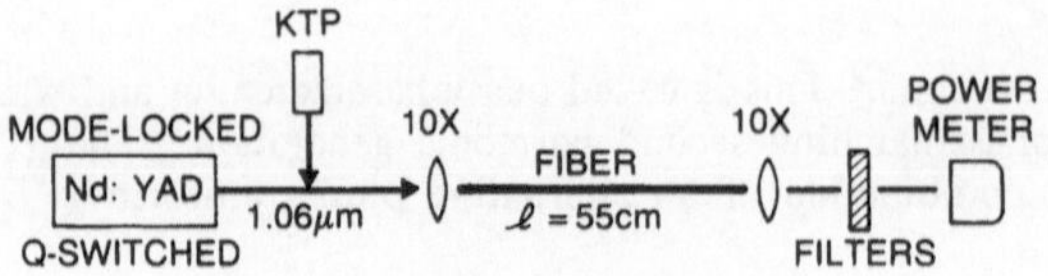

Fig. 6. Seeding experiment.

defects. For example, the same macroscopic model for an alternating $\chi^{(2)}$ grating was arrived at by looking at the sum of one- and two-photon absorption.[11] Thus the phenomenological $\chi^{(3)}$ does not necessarily have the value related to the nonlinear index of silica.

The most important implication of the mixing model is that the process requires the presence of harmonic light. This implies that long preparation times are required only when the process has to start from noise. If the harmonic light were injected into the fiber along with the fundamental, the grating should be formed quickly. Indeed when such an experiment was done the grating was formed in a few minutes.[6] The fiber in that experiment was drawn to be single-mode at both 1.064 and 0.532 $\mu$m in order to rule out such mechanisms as a bulk quadrupole or core-cladding surface contributions to the harmonic output. These coupling mechanisms have odd symmetry and would require one wave to be in a fiber mode of odd symmetry.[12]

Injection of the harmonic along with the fundamental is now called seeding and is the preferred method for making $\chi^{(2)}$ gratings in fibers. This still leaves open the question of the source of the seed when the process starts from noise. There is no answer yet to this question but some of the issues are discussed at the end of section 3.

### 1.5 Bandwidth considerations

Besides leading to the seeding technique, the mixing model also predicts the effective length of the $\chi^{(2)}$ gratings formed in the fibers.[13] This is actually two problems. The first problem is the fundamental limit imposed on the effective coherence length by the spectral bandwidths of the input pulses. The second problem is competing nonlinear effects.

The basic mixing model treats spectrally narrow input fields and assumes no saturation on

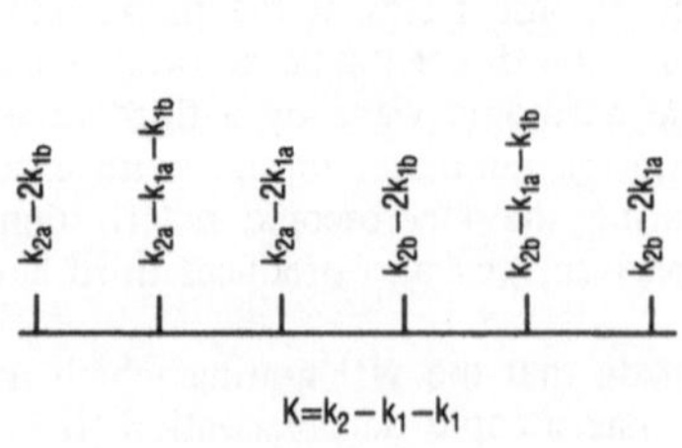

Fig. 7. Components of grating formed by mixing
two pump and two seed wavelengths.

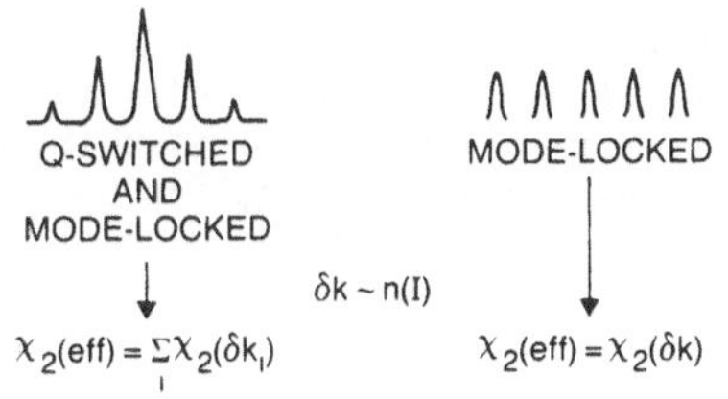


Fig. 8. Writing a grating with Q-switched mode-locked pulses as opposed to just using mode-locked pulses.

the induced nonlinearity. If we relax the requirement of spectrally narrow input fields by choosing just two pump and two harmonic frequencies the dc polarization can be a sum of polarizations with six different periods. The $\chi^{(2)}$ grating will be washed out after some effective coherence length which is inversely proportional to the initial bandwidths.

Competing nonlinear processes produce a broader range of grating periods which further reduces the effective coherence length. For example, self-and cross-phase modulation increase the bandwidths of the pump and harmonic seed beams as the pulse travels along the fiber. Also, a mode-locked Q-switched laser produces a burst of intense mode-locked pulses of different peak power. The intensity is high enough for the grating period produced by each subpulse to be different because of the intensity-dependent refractive index.

A measure of the effective coherence length can be obtained by measuring the second harmonic intensity as the fundamental wavelength is tuned. Figure 9 compares three such measurements.[8,14,13] The first two measurements used mode-locked Q-switched pulses and the third used only mode-locked pulses. The first measurement actually tuned the wavelength of a Raman-shifted pulsed dye laser. The other two measurements both used 1.064 $\mu$m pulses which were first chirped by passing through a separate 400 meter fiber. It is clear that the bandwidth is an order of magnitude narrower using the mode-locked pulses.

An effective coherence length can be obtained from the data using the usual definition of $L_{coh} = 2\pi/\Delta k$. A bandwidth of 2.4 Å corresponds to a coherence length of 3.3 cm. To relate $\delta\omega$ to $\delta k$ we let $k(\omega)$ and $k(2\omega)$ vary linearly about $\omega_0$ with $a = dk(\omega)/d\omega|_{\omega_0}$ and $b = dk(\omega)/d\omega|_{2\omega_0}$ so that $\delta k = 2(a-b)\delta\omega$. The intensity of phasematched second harmonic generation goes as $I(2\omega) = \sin^2(\delta kL/2)/(\delta kL/2)^2$ and the coherence length corresponds to the full width in $\delta k$ at about 0.40 of the maximum intensity.

The mixing model will be treated in detail in section 2. Some of the interesting features include questions of saturation and the relative phases of the seed and generated harmonic waves.

The wavelength detuning measurements give information about the effective grating length but they don't actually prove the existence of a $\chi^{(2)}$ grating since almost any phasematching scheme would produce a similar curve. It is certain that there is actually phasematching because the second harmonic generation in the fiber is coherent with second harmonic generation from a crystal pumped by the same laser.[14]

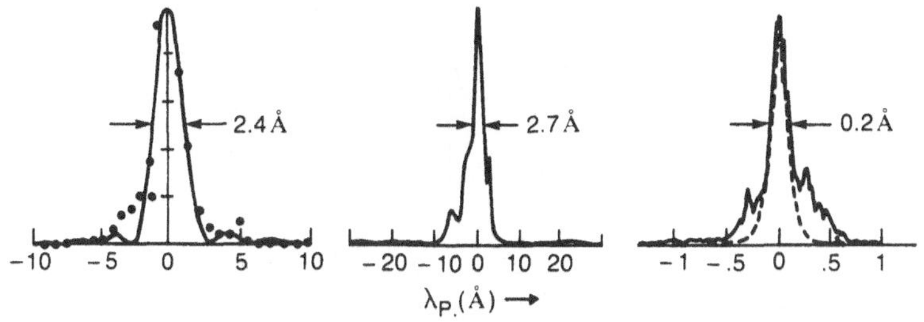


Fig. 9. Second harmonic intensity versus fundamental wavelength detuning.

There are, however, experiments which prove that a $\chi^{(2)}$ grating has been formed in the fiber. In one experiment, a grating was written in the fiber and the fiber then stretched. The input wavelength for maximum second harmonic generation shifted to longer wavelengths by exactly the increase in grating period from stretching.[15] In the second experiment, a birefringent fiber was written with orthogonal fundamental and harmonic polarizations. The grating was then read with the polarization of fundamental field rotated by 90°. The shift in wavelength for maximum second harmonic generation was exactly predicted by the fiber birefringence.[15]

### 1.6 Microscopic picture

There is no general agreement concerning a microscopic model for the self-organized $\chi^{(2)}$ gratings in fibers although a picture is emerging for the fibers in which the effect has so far been observed.

There are several tools for investigating microscopic models. First different dopants and wavelengths can be tried. These experiments point to a germanium defect as the primary source of the dipole-allowed nonlinearity.[16] The original experiments used fibers doped with both germanium and phosphorus.[4] Ge-doped fibers have shown self-seeded second harmonic generation starting with red light from an eximer laser although the output of the fiber was extremely unstable.[17] Gratings have also been written by seeding starting with a 1.3 $\mu$m laser pulse although the effect was much weaker than with 1.064 $\mu$m.[18]

Excitation experiments with blue light[19] and spin resonance studies[20] suggest that an electron is removed from a germanium defect by blue or green light. The germanium defect then relaxes to a metastable state which lacks inversion symmetry.[19] In the presence of an external dc field, the excited defects are lined up along the field while their orientation is random in the absence of any orienting fields.[19]

Experiments in which the $\chi^{(2)}$ grating is destroyed by annealing[21] or bleaching[22] are obviously of interest. So far the annealing and bleaching experiments are consistent with a picture of excited Ge defects which are then organized into the $\chi^{(2)}$ grating. These experiments will be discussed in more detail later. Annealing some fibers before writing increases the conversion efficiency by 10 times.[5]

The mixing of the pump and harmonic fields in some way controls the relaxation of the excited defects so they are aligned with the alternating dc polarization. The problem is that the macroscopic field from the dc polarization is of the order of a few V/cm so it is hard to imagine that this is actually providing the orientational force. Most of the effort is going into generating models which can deal in some way with this basic fact and some of these general models will be further treated in section 4.

### 1.7 Other self-organized systems

This type of self-organized behavior is not unique to second harmonic generation in fibers. A self-organized refractive index grating has also been observed in fibers.[23] These "Hill gratings" appear when incident and reflected narrow-band green light interferes in a Ge-

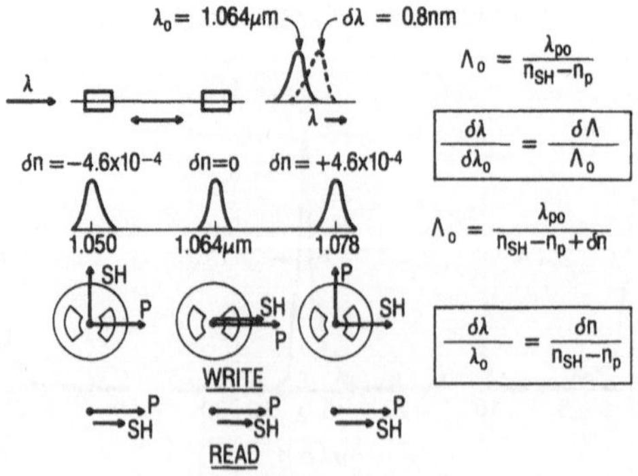


Fig. 10. Experiments demonstrating the existence of a $\chi^{(2)}$ grating.[15]

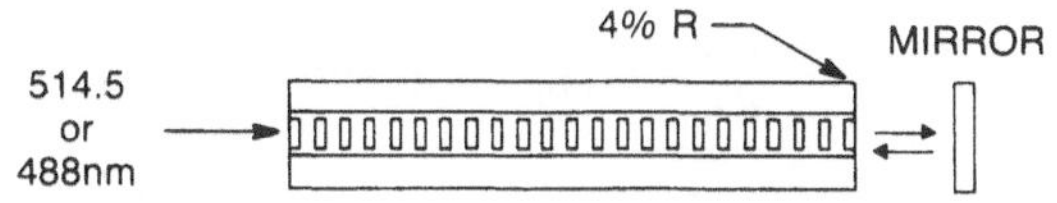


Fig. 11. Formation of a Hill grating in a fiber.

doped fiber. There has been speculation that the same germanium defect contributes to both the index grating and to second harmonic generation. This speculation is based on the observation that the index grating can be written from outside the fiber by ultraviolet light matched to the 245 nm absorption band of Ge-doped silica.[24]

In ferroelectric crystals there is a long history of self-organized index gratings.[25] In these crystals it is the intrinsic lack of inversion symmetry that is important. The light generates carriers which then drift creating large internal fields. The index changes are produced by the fields by way of the large linear electrooptic coefficient of these materials. The term "photorefractive" has been restricted to these specific processes involving large internal electric fields although it is tempting to generalize its meaning so that photorefractive refers to any permanent index changes induced by light.

We can go further afield and look at general self-organized systems. Self organization appears in many physical and biological systems and is often related to bifurcation and chaos. In one particular class of such systems a weak perturbing force biases the system to chose a preferred orientation in the presence of much larger random forces.[26]

## 2. MIXING MODEL

### 2.1 Basic plane-wave mixing model

The simplest form of the mixing model starts by assuming the presence of some harmonic light along with the pump. These waves mix to produce a dc polarization in the same direction as the field polarization of the waves. For simplicity we assume that both pump and harmonic waves have the same linear polarization. The dc polarization $P_{dc}$ is given by:

$$P_{dc} = \frac{3}{4}\chi^{(3)}\left|E^*(2\omega)E(\omega)E(\omega)\right|\cos(\Delta kz) \tag{1}$$

$$\Delta k = k(2\omega) - 2k(\omega)$$

where $\Delta k$ is the wavevector mismatch.

The model then assumes that a periodic dipole-allowed response is produced proportional to the alternating $P_{dc}$.

$$\chi^{(2)} \sim P_{dc} \tag{2}$$

The general statement of the model then becomes:

$$\chi^{(2)} = \gamma\left|E^*(2\omega)E^2(\omega)\right|\cos(\Delta kz) \tag{3}$$

We now have a $\chi^{(2)}$ to use in the nonlinear wave equation. The constant $\gamma$ is a phenomenological constant which is a kind of $\chi^{(3)}$. Since we don't have a microscopic model to calculate $\gamma$ we will try to simplify the equations by not worrying about the constants. There is not much point in worrying about factors of 2, n and $\pi$ when the major factor is totally unknown. This has the added advantage of avoiding long discussions concerning whose convention is followed for $\chi^{(3)}$.

In setting up the formalism it is often convenient to think in terms of a simple model in which the macroscopic dc field orients bipolar defects. The mixing model, however, is given by Eq. (3) and makes no statement about microscopic models.

These assumptions lead to the coupled equations:

$$\frac{dE_2}{dz} = i\alpha\left[E_2 + E_2^* e^{-2i\Delta kz}\right]$$

$$\frac{dE_2^*}{dz} = -i\alpha\left[E_2 e^{2i\Delta kz} + E_2^*\right] \qquad (4)$$

$$\alpha \sim \gamma\omega_2\left|E_{10}\right|^4$$

1 and 2 refer to pump and second harmonic and it is assumed that the pump is unchanged along the fiber so that $E_1(z) = E_1(0)$.

These coupled equations come from the wave equation with a second-order nonlinear source term and complex slowly varying amplitudes are defined in the usual way.

$$\nabla^2\epsilon - \frac{n^2}{c^2}\frac{\partial^2}{\partial t^2}\epsilon = \frac{4\pi\chi^{(2)}}{c^2}\frac{\partial^2}{\partial t^2}\epsilon^2$$

$$\epsilon = \epsilon_1(z) + \epsilon_2(z) \qquad (5)$$

$$\epsilon_i(z) = \frac{1}{2}E_i e^{i\phi(z)} e^{i(k_i z - \omega_i t)} + \text{c.c.}$$

$$= \frac{E_i e^{i(\ )} + E_i^* e^{-i(\ )}}{2}$$

The coupled equations of (4) can be simplified with a substitution.

$$E_2 = F e^{i\alpha z}$$

$$\frac{dF}{dz} = i\alpha F^* e^{-i\kappa z}$$

$$\frac{dF^*}{dz} = -i\alpha F e^{i\kappa z} \qquad (6)$$

$$\kappa = 2(\Delta k + \alpha)$$

If $F \sim \exp(gz)$ then:

$$g = -i(\Delta k + \alpha) \pm iG$$

$$G = \sqrt{\Delta k^2 + 2\alpha\Delta k} \qquad (7)$$

The general solution for $2\alpha < \Delta k$ is:

$$E_2(z) = e^{-i\Delta kz}\left[A e^{iGz} + B e^{-iGz}\right]$$

$$= E_{20}(2\omega) e^{-i\Delta kz}\left[\cos(Gz) + i\,\frac{\Delta k + 2\alpha}{G}\sin(Gz)\right] \qquad (8)$$

If second harmonic generation is not initially phasematched then $\Delta k >> 2\alpha$.

$$\frac{\Delta k + 2\alpha}{G} \approx 1 \qquad (9)$$

$$E_2(z) = E_2(0)e^{i\alpha z}$$

Thus the second harmonic field does not grow in amplitude but only shifts slightly in phase along the fiber. The $\chi^{(2)}$ grating is then:

$$\chi^{(2)} = \gamma E_{10}^2 |E_{20}| \cos[(\Delta k+\alpha)z] \tag{10}$$

where we refer to the writing field amplitudes as 10 and 20.

If this grating is then read with a low power pump $E_1$,

$$\frac{dE_2}{dz} = i\xi\chi^{(2)}(z)E_1^2 e^{-i\Delta kz}$$

$$= \frac{i\gamma}{2}\xi E_{10}^2 |E_{20}| E_1^2 e^{i\alpha z}$$

$$E_2(L) = \int_0^L \left(\frac{dE_2}{dz}\right) dz \tag{11}$$

$$= \frac{i\gamma\xi}{\alpha_1} |E_{20}| \frac{E_1^2}{E_{10}^2} e^{i\alpha L/2} \sin\left(\frac{\alpha}{2}L\right)$$

$$\alpha = \alpha_1 |E_{10}|^4$$

where we have removed the pump power from alpha of Eq. (4).

The constant $\xi$ of Eq. (11) can be established by turning off the seeding field in Eq. (9). This corresponds to reading with a pump power equal to the writing pump power.

$$E_2(L) = E_{20}(E^{i\alpha L} - 1)$$

$$= 2iE_{20}e^{i\alpha L/2} \sin\left(\frac{\alpha}{2}L\right) \tag{12}$$

Therefore $\gamma\xi = 2\alpha_1$ and the second harmonic intensity is:

$$|E_2(L)|^2 = 4|E_{20}|^2 \frac{|E_1|^4}{|E_{10}|^4} \sin^2\left(\frac{\alpha}{2}L\right) \tag{13}$$

Note that it also works out this way if the problem is done carefully keeping track of all the factors. The somewhat unrealistic feature of this result is that the seed power clamps the amplitude of the $\chi^{(2)}$ grating. A small seeding field produces a weaker grating. This question will be discussed further in the next section.

## 2.2 Relative phase of the $\chi^{(2)}$ grating

The simplest form of the mixing model assumes that the $\chi^{(2)}$ grating is in phase with the alternating dc polarization. It has been pointed out that this is unnecessarily restrictive and some phase shift should be allowed between the $\chi^{(2)}$ grating and the dc polarization.[27]

It is simple to add a phase shift to the mixing problem. The term $\gamma$ in Eq. (3) and hence the term $\alpha$ of Eq. (4) become complex. Since $\gamma$ is a phenomenological constant anyway, why not a complex phenomenological constant? There is some basis for such a complex relation because the depletion models will have a different phase relation than the force models.[27,11] Any sort of resonance in $\chi^{(3)}$ will also lead to a phase shift.

The complex $\alpha$ changes the equations in (6) and (7) so that:

$$\alpha \rightarrow \alpha' + i\alpha''$$

$$\frac{d^2F}{dz^2} = \alpha\alpha^* F - i\kappa' \frac{dF}{dz}$$

$$\kappa' = 2(\Delta k + \alpha') \tag{14}$$

$$g = -i(\Delta k + \alpha') \pm iG$$

$$G = \sqrt{\Delta k^2 + 2\alpha'\Delta k - \alpha''^2}$$

Now in the same limit of $\Delta k >> \alpha'$ and $\alpha''$ Eq. (9) becomes:

$$E_2(z) = E_{20} e^{-\alpha'' z} e^{i\alpha' z} \tag{15}$$

If we had picked $-\alpha''$ to start with we would now have gain for $E_2(z)$.

There have been some experiments to investigate this question.[28] The experiments seeded a fiber with various seeding powers while maintaining a constant mode-locked, Q-switched 1.06 $\mu$m power. The seed power was then removed and the grating along the fiber investigated by looking at the second harmonic power out as the fiber was cut back from the output end.

The essential result was that the maximum second harmonic output power was almost independent of seed green power for seed powers ranging from 1.0 $\mu$W to 1000 $\mu$W. Maximum second harmonic output power was about 300 $\mu$W. At high seed powers, however, the $\chi^{(2)}$ grating started out at its maximum value right at the beginning of the fiber while at low seed powers the grating grows in amplitude along the fiber. Second harmonic from a constant grating shows the expected quadratic length dependence while second harmonic from an increasing grating shows an increase as a very high power of length. The length of maximum second harmonic also moves further into the fiber as seed power is decreased reaching a limit of about 20 cm which looks similar to self-seeded fibers.[28]

If these results are interpreted in terms of the relative phase between $\chi^{(2)}$ and $P_{dc}$ it would appear that the phase shift depends in some way on the seeding power so that there is gain only for low seed powers and the grating and the dc polarization are in phase for high seed powers. A constant grating may also imply saturation which is an open question at present.

An experiment has been reported which looks for a phase shift in the $\chi^{(2)}$ grating by looking for a phase shift in the second harmonic generation output.[29] A fiber is first prepared by seeding. The seed power is then reduced so the green beam becomes a probe. The output of the fiber is then the sum of the probe and the second harmonic generation produced in the fiber. By shifting the phase of the probe, the power output traces out a series of maxima and minima as shown in Fig. 12.

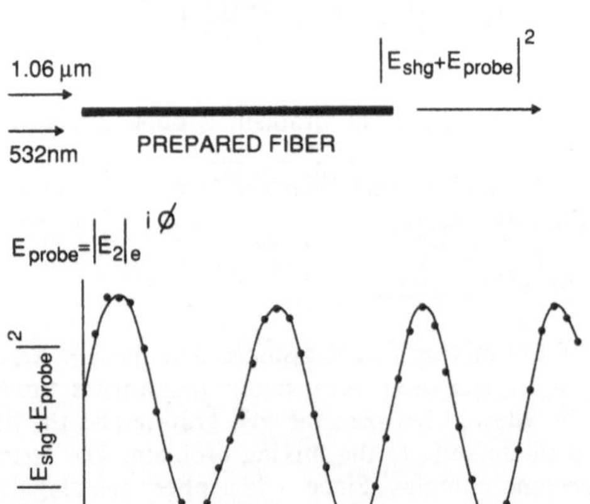


Fig. 12. Experiment probing the phase of the $\chi^{(2)}$ grating.[19]

The plot of intensity vs. phase shows what the phase difference was at zero phase shift. The experimental result is within experimental accuracy of $\pi/2$ which is the result predicted by the steady state mixing model with no phase shift of the grating. This implies that either there is no phase shift in the steady state limit (which would agree with the previous cutback measurements), or that the phase shift is less than the experimental uncertainty of about 10%.

### 2.3 Basic mixing model in a fiber

While all of the treatment has dealt with plane waves we are really talking about $\chi^{(2)}$ gratings in fibers and it is necessary to include the effect of the waveguide modes. The assumption of linear polarizations is still fine as long as a polarization preserving fiber is used. Even in an ordinary fiber, linear polarization can be maintained over the relatively short lengths of the $\chi^{(2)}$ gratings.

Equation (13) can then be written in terms of fundamental and second harmonic powers.

$$p(2\omega) = 4p_0(2\omega)\frac{p^2(\omega)}{p_0^2(\omega)}\sin^2(\alpha_f z/2)$$

$$\alpha_f = \frac{192\pi^4\omega\gamma\chi^{(3)}}{\epsilon_0 n(2\omega)n^2(\omega)c^3A_{eff}^2}p_0^2(\omega) \tag{16}$$

The effective area is an integral over the mode fields $\Psi_i$.

$$\frac{1}{A_{eff}^2} = \frac{\langle\Psi_1^4\Psi_2^2\rangle}{\langle\Psi_2^2\rangle\langle\Psi_1^2\rangle^2}$$

$$\langle\Psi_i\rangle = \int_0^{2\pi}\int_0^{\infty}\Psi_i\, r\,dr\,d\theta \tag{17}$$

The field in the fiber becomes:

$$\epsilon = A\Psi(r)\cos(kz-\omega t)$$

$$p = |A|^2N^2 \tag{18}$$

$$N^2 = \frac{cn}{8\pi}\int_0^{\infty}\int_0^{2\pi}\Psi^2(r,\theta)\, r\,dr\,d\theta$$

To get Eq. (16) it doesn't work to just convert the intensities of Eq. (13) to power divided by an effective area. It is necessary to start at the beginning and work the problem through in terms of the mode fields.

The reason that the effective area is large corresponds to the poor overlap between the fundamental and second harmonic modes in a typical fiber. This is illustrated schematically in Fig. 13. If both the pump and second harmonic are in the fiber's lowest order $LP_{01}$ mode, the mode radius for the fundamental will be much larger than that of the harmonic. The radial dependence of $\chi^{(2)}$ will be very much like that of the harmonic.

$$\chi^{(2)}(r) \sim \Psi_1^2(r)\Psi_2(r) \tag{19}$$

### 2.4 Four-wave mixing picture

One way of looking at the optical rectification from $\chi^{(3)}$ is as a limit of four photon mixing. In this limit, the frequency difference between the pump and anti-Stokes frequency becomes the same as the pump frequency. Thus the frequency of the Stokes wave goes to zero.

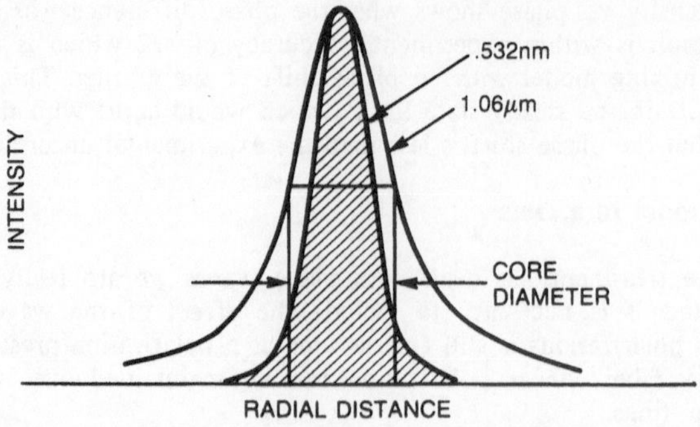


Fig. 13. Overlap of fundamental and second harmonic modes.

This argument can be turned around to produce a slightly different version of the mixing model which may give some insight as to how the process can start without an external seed. We assume that defects in the fiber have been excited so that there is some random distribution of polarizations. This random field can act as a nonpropagating Stokes wave. Only that component of the random field will grow which has a period equal to the phasematching period.

Beyond such general statements this is not an easy problem to do. The nonpropagating Stokes wave is a kind of polariton which is hard to fit into the usual coupled-wave formalism of nonlinear optics. This type of picture does suggest that the simple mixing model is a limiting case of a more general macroscopic model.

The mixing model looks very much like a parametric interaction based on a fifth-order nonlinearity.[30]

$$P(2\omega) = \chi^{(5)} E^*(2\omega) E^4(\omega) \tag{20}$$

A possible source of the initial green light in the absence of an external seed may be parametric amplification of noise by way of the macroscopic $\chi^{(5)}$.[30]

### 2.5 Saturation

The simple mixing model assumes that $\chi^{(2)}$ is linearly proportional to $P_{dc}$. In general, $\chi^{(2)}$ need not be linearly proportional to $P_{dc}$ and will thus have higher-order terms in cos(n$\Delta$kz) where n is an integer. These higher order terms will not result in phase-matched second harmonic although they do correspond to saturation of the nonlinearity. Experimentally there is evidence both for and against saturation.

For example, the absolute efficiency of conversion at some given peak power leads to a figure of merit for a fiber of $\chi_{eff} L_{coh}$ where $\chi_{eff}$ is the peak amplitude of the alternating $\chi^{(2)}$ along z. This value for a seeding experiment with 230 W mode locked 1.06 μm peak power

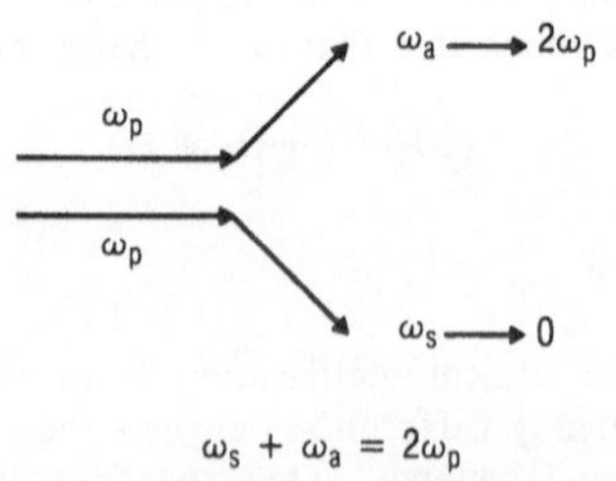


Fig. 14. Parametric mixing.

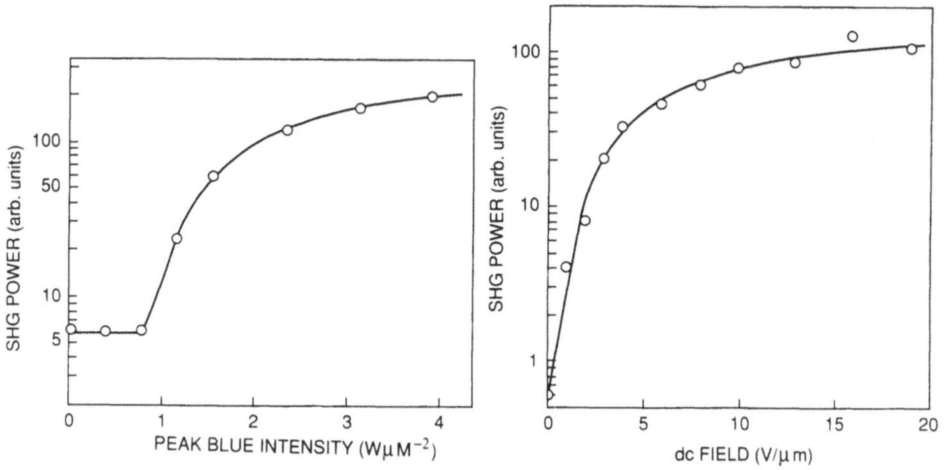


Fig. 15. Saturated second harmonic-power at constant poling field of 4V/μm vs. peak blue intensity in a Ge-doped fiber.[19]

was $1.86\times10^{-10}$ esu. The figure of merit for a similar fiber prepared with Q-switched mode-locked pulses was $4.2\times10^{-11}$ esu. Taking $L_{coh}$ to be 35 cm in the mode-locked prepared fiber and 3.5 cm for the fiber prepared with mode-locked Q-switched pulses, $\chi_{eff}$ becomes $5.3\times10^{-12}$ esu and $1.2\times10^{-11}$ esu. It is remarkable that $\chi_{eff}$ comes out within a factor of two even though the input powers used to write the gratings were very different. This suggests saturation.[13]

Evidence for saturation also comes from measurements of second harmonic power from fibers prepared with blue light with a constant dc poling field.[19] These studies will be discussed further in section 4. The important feature here is the suggestion of saturation as seen in Fig. 15. There is some question however whether the saturation is quite so apparent if the points are replotted on a linear scale.[31]

Evidence against saturation comes from birefringent fibers seeded with green polarization along or orthogonal to the pump polarization. If saturation occurred the gratings should have reached the same $\chi_{eff}$. Actually, the parallel-seeded grating was three times larger than the orthogonal-seeded grating.[15]

## 2.6 Growth of the $\chi^{(2)}$ grating

The simple mixing model says nothing about the dynamics of the $\chi^{(2)}$ grating. The mixing model just treats the system after it has reached steady state. A more general approach must treat the growth of the grating itself and include time constants and saturation. There are experiments on the relative preparation rate as a function of green seed power.[32] The growth of second harmonic generation was initially exponential and saturated at longer time. The exponent of initial growth increases as the square root of the green seed power. In these experiments care was taken to ensure that the probing infra-red pulses did not temporally overlap with the preparing pulses within the fiber.[32]

There is also an optical rectification from $\chi^{(2)}$ itself which affects the dynamics of grating formation.

$$
\begin{gathered}
P_{dc} \sim \chi^{(2)} E(\omega) E^*(\omega) \\
\chi^{(2)} = |\gamma_1| e^{i\phi_1} \chi^{(3)} E^*(2\omega) E^2(\omega) + |\gamma_2| e^{i\phi_2} \chi^{(2)} E(\omega) E^*(\omega)
\end{gathered}
\tag{21}
$$

There are thus two contributions to $P_{dc}$ which don't necessarily have to be in phase. This turns the simple mixing problem into a dynamical problem in which the $\chi^{(2)}$ grating changes with $P_{dc}$ and $P_{dc}$ changes with the $\chi^{(2)}$ until everything reaches some steady state. The limit in which the seeding intensity is much greater than any second harmonic produced in the fiber is thus a particularly simple limit.

### 2.7 Bandwidth and grating length

The relation of spectral bandwidth to $\chi^{(2)}$ grating length was discussed earlier. Here we elaborate somewhat on the formalism involved. The basic mixing model treats only cw waves leading to a $\chi^{(2)}$ of:

$$\chi^{(2)} = \gamma E^*(\omega_1 + \omega_2)E(\omega_1)E(\omega_2)\cos(\Delta kz) \tag{22}$$

where we now allow for pump waves of different frequency. As discussed in section 1, the total grating is then the sum of all the separate gratings formed by mixing all possible combinations of pump and second harmonic waves.

$$\chi^{(2)} = \gamma E^*(2\omega_0)E^2(\omega_i)\int d\Delta k\ u(\Delta k)\cos(\Delta kz) \tag{23}$$

In Eq. (23) $\omega_0$ is the center frequency of the fundamental pulse spectrum and the weighting function $u(\Delta k)$ takes into account the sum of all frequencies within the fundamental and second harmonic pulse spectra that have the same value of $\Delta k$. If we assume Gaussian frequency distributions for the pump and second harmonic pulses of $1/e$ halfwidths of $\delta(\omega)$ and $\delta(2\omega) = \sqrt{2}\delta(\omega)$, the integral becomes:

$$\chi^{(2)} = \gamma E^*(2\omega_0)E^2(\omega_0)\cos(\Delta k_0 z)e^{-(z/L_{coh})^2} \tag{24}$$

where the coherence length $L_{coh}$ is given by:

$$\left[L_{coh}\right]^{-1} = \frac{1}{2}(a - b)\delta(\omega) \tag{25}$$

To get a and b we let $k(\omega)$ and $k(2\omega)$ vary linearly about $\omega_0$ and $2\omega_0$. Numbers for silica are given later.

The second harmonic output power is thus a sum of the squares of a cosine and sine transform of an exponential.

$$E(2\omega) \sim \int_0^L \cos(\Delta k_0 z)e^{i\Delta kz}e^{-(z/L_{coh})^2}dz \tag{26}$$

Figure 16 illustrates the second harmonic power for the case where $L = L_{coh}$ and $L >> L_{coh}$. To see the contribution of the cosine and sine transforms, the cosine transform alone is shown as a dotted line.

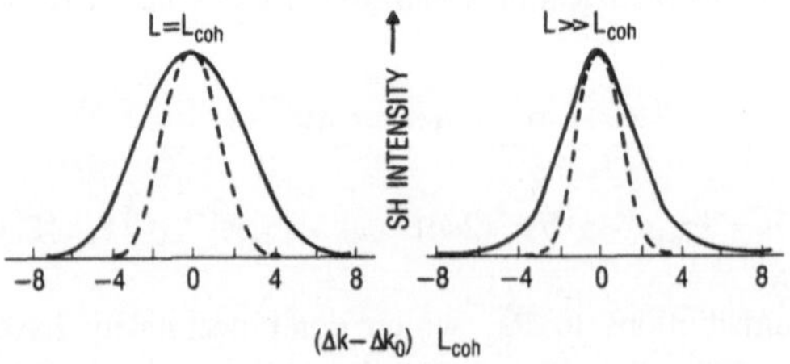


Fig. 16. Second harmonic power vs. $\Delta k$-$\Delta k_0$ for Gaussian pulses.

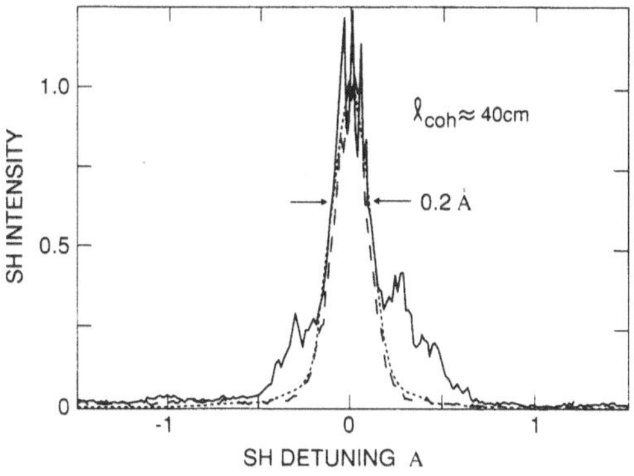


Fig. 17. Second harmonic intensity vs. wavelength detuning.

The preceding approach assumes the mixing limit in which the second harmonic light from each grating component doesn't rewrite the grating. An experiment was done in this mixing limit by seeding a 55 cm Ge-P-doped fiber with mode-locked pulses rather than the usual mode-locked Q-switched pulses.[13] The advantage is that the power is low enough that competing nonlinear effects are negligible. When the grating is read by removing the green seed beam, the bandwidth of the second harmonic light produced by the grating is then determined strictly by the bandwidths of the input pulses. Figure 17 shows the second harmonic output as a function of input frequency. The dashed line is the spectrum of the seed second harmonic pulse propagated through the fiber. The spectral FWHM at the second harmonic frequency is 0.213 Å. The second harmonic seed pulse was not cross-phase modulated by the pump as the spectrum was the same with or without the pump. Also, the mode-locked Nd:YAG 1.064 $\mu$m pulses are not transform limited (for which the FWHM = 0.11 Å).

The peak writing powers were 230 W at 1.064 $\mu$m and 3.3 W at 532 nm in pulses of approximately 80 ps duration. Seeding took three hours at these low powers. The fiber core size was chosen to support two modes at 532 nm and one mode at 1.064 $\mu$m to optimize the overlap integral. The 532 nm light actually propagates in the $LP_{01}$ mode. The spectral broadening of the reading 1.064 $\mu$m beam was accomplished by passing the pulse through about 400 meters of a separate fiber. The 400 m fiber also chirps the pulse so that the second harmonic produced at any given time during the pulse may be considered to be produced by monochromatic radiation.[33]

The relation between $\delta\omega$ and $(\Delta k - \Delta k_0)$ uses the same (a-b) of Eq. (25). Estimates of (a-b) and $k_0$ are obtained using index data for fused silica.[34]

$$a = \left.\frac{dk}{d\omega}\right|_{\omega_0} = 1.463/c$$

$$b = \left.\frac{dk}{d\omega}\right|_{2\omega_0} = 1.485/c \qquad (27)$$

The coherence length in the fiber comes from the refractive index at the pump and second harmonic.

$[SiO_2]$

$$n(1.064) = 1.4493$$

$$n(0.532) = 1.4604$$

$$\Delta n = 0.0111 \qquad (28)$$

$$\Delta k = \frac{\Delta n \omega_2}{c} = 1310 \text{ cm}^{-1}$$

$$\delta k \approx 0.0015 \frac{\omega_2}{c} = 180 \text{ cm}^{-1}$$

Adding in the waveguiding $\delta k$:

$$\Delta k_0 = \Delta k + \delta k = 1490 \text{ cm}^{-1} \tag{29}$$

The material contribution $\Delta k$ will probably be slightly larger than the present estimate because Ge and P doping increase the dispersion somewhat.

## 3. PHASEMATCHING AND QUADRUPOLE EFFECTS

### 3.1 Phasematching

Waveguide modes can be used to phasematch second harmonic generation in fibers. The only problem is that it is hard to get enough difference between the effective indices of the waveguide modes to compensate for the material dispersion. The basic idea is illustrated by the plot of effective refractive index vs. normalized frequency (Fig. 18).[35] The wavevector mismatch appears because $n_2 > n_1$ so we can get phasematching by putting the pump in the lowest order mode and the second harmonic in a higher order mode. The problem is that the index difference between 1.064 and 0.532 $\mu$m is 0.011 which is difficult to make up with waveguide modes. For example, with the pump in the $LP_{01}$ and the second harmonic in the $LP_{02}$ mode the core cladding index difference would have to be at least 0.024 which is large for a fiber. Standard communications fibers use a core-cladding index difference of around 0.005. Recently fibers have been made with index differences of 0.03 and used for waveguide phasematching of second harmonic generation.[36,37]

Another difficulty with the use of waveguide modes for phasematching is that variations in fiber diameter limit the coherence length to distances of the order of a cm. We can treat a wandering phasematching frequency using an approach developed for dealing with the effect of optical inhomogeneities in lithium niobate.[38] A linear change in phasematching frequency produces a characteristic asymmetry in a measurement of second harmonic generation intensity as the pump wavelength is tuned. The integrated second harmonic power as the pump wavelength is tuned will still increase linearly with fiber length. However, because this quantity does not require perfect phasematching, the integral of second harmonic generation over frequency is a constant.

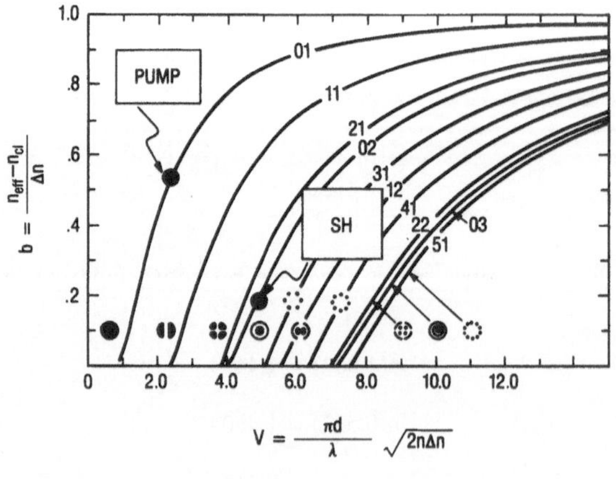


Fig. 18. Effective refractive index vs. normalized frequency V.

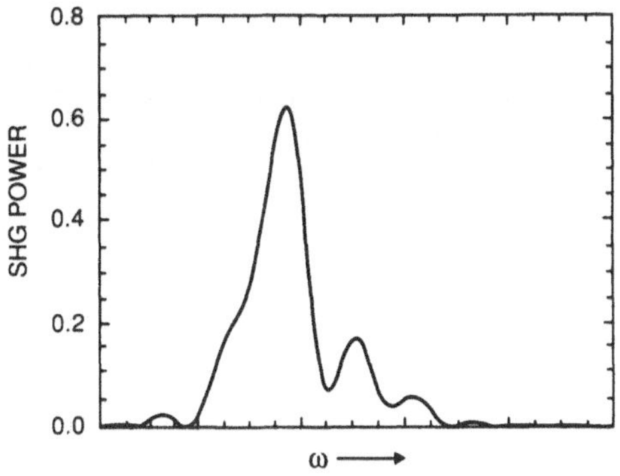


Fig. 19. Second harmonic generation output for a linear variation in phasematching wavelength.[38]

$$\frac{dE_2}{dz} \sim -\frac{i}{k}E_1^2 e^{i\Delta\beta z} \rightarrow E_2(L) = C\int_0^L e^{i\Delta\beta z}dz$$

Perfect phasematching: $\Delta\beta = a(\nu - \nu_0)$

$$E_2(L) = \frac{1}{ia(\nu - \nu_0)}\left[e^{ia(\nu-\nu_0)L} - 1\right]$$

$$EE^* = \frac{\sin^2\left(\frac{a}{2}(\nu-\nu_0)L\right)}{\left[\frac{a}{2}(\nu-\nu_0)\right]^2} = \frac{L^2\sin^2(x)}{x^2} \qquad (30)$$

$$\int_0^\infty EE^* d\nu = \frac{2L}{a}\int \frac{\sin^2(x)}{x^2}dx = \frac{2\pi L}{a}$$

Wandering phasematching: $\Delta\beta = a(\nu-\nu_0) - f(z)$

$$EE^* = \int_0^L dz\int_0^L dz' e^{ia(\nu-\nu_0)(z-z')}e^{-af(z)z}e^{af(z')z'}$$

$$\int_0^\infty EE^* d\nu = \frac{2\pi}{a}\int_0^L dz\int_0^L dz' e^{-a[f(z)z-f(z')z']}\delta(z-z') \qquad (31)$$

$$\int_0^\infty EE^* d\nu = \frac{2\pi}{a}\int_0^L dz = \frac{2\pi L}{a}$$

## 3.2 Overlap integrals

Phasematching with waveguide modes is not as efficient as might be hoped because problems with mode overlap reduce the conversion efficiency. The mode overlap shows up in the overlap integral which we derive leaving out most of the constants. Also phasematching is assumed.

$$\frac{dE_2}{dz} = i\chi^{(2)} E_1^2$$

$$E = A\Psi \tag{32}$$

$$\frac{dA_2}{dz}\Psi_2 = i\chi^{(2)} A_1^2 \Psi_1^2$$

Multiply by $\Psi_2$ and integrate over r and $\theta$.

$$\frac{dA_2}{dz}\langle\Psi_2^2\rangle = i\chi^{(2)} A_1^2 \langle\Psi_2\Psi_1^2\rangle$$

$$\langle\Psi\rangle \equiv \int_0^\infty \int_0^{2\pi} \cdot \Psi r dr d\theta \tag{33}$$

The amplitudes and spot patterns of the four lowest-order fiber modes are illustrated in Fig. 20 along with their LP mode designations. The patterns look very much like laser modes.

It is clear that symmetry will rule out most mode combinations in the overlap integral. The physical meaning of such cancellation is that dipoles on one side of the core are radiating out of phase with dipoles on the other side. For example, the integral will be zero for pump $LP_{01}$ and second harmonic $LP_{11}$. The integral will not be zero for pump $LP_{01}$ with second harmonic $LP_{02}$ but even here there will be some cancellation because the outer part of the $LP_{02}$ mode is out of phase with the inner part. The pump could be in the $LP_{11}$ mode and the second harmonic in the even $LP_{01}$ mode because the square of the odd mode is even. Usually phasematching in a fiber requires the pump to be in the lower-order mode.

## 3.3 Quadrupole and surface contributions

The simplest way to visualize the quadrupole and surface contributions is to imagine little oscillators pinned to the surface formed by the core-cladding interface (Fig. 21). By symmetry all the dipoles will be oriented the same way with respect to the surface. A linearly polarized $LP_{01}$ mode in a fiber will drive the oscillators on opposite sides of the core in such a way that the second harmonic generation fields will be exactly out of phase. Thus a lowest-order $LP_{01}$ mode cannot be excited at the second harmonic frequency. If the fiber will support a mode of odd symmetry, this mode can be excited.

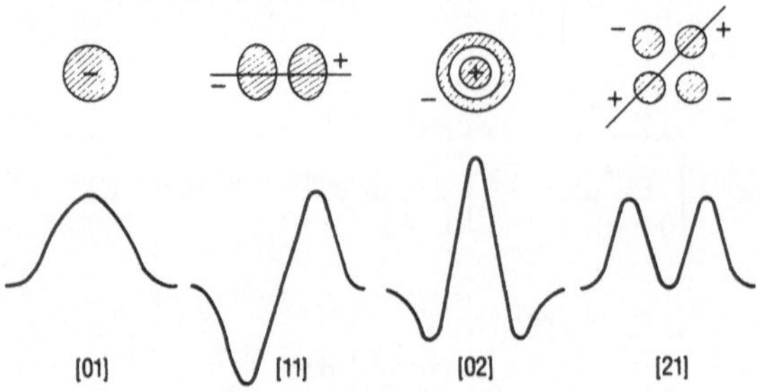


Fig. 20. Amplitudes of the lowest-order fiber modes.

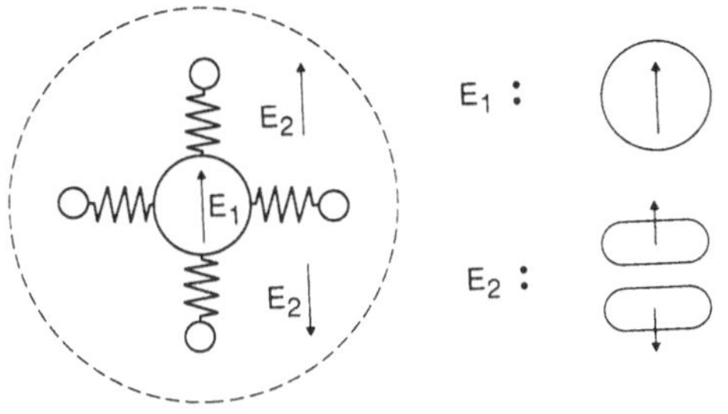


Fig. 21. Symmetry of surface and quadrupole contributions to fiber second harmonic generation.

The quadrupole terms all involve some sort of $(\nabla E)$ or $(\nabla \cdot E)$.

$$\mathbf{P}(2\omega) \sim C_1 \nabla E^2 + C_2(\mathbf{E}\cdot\nabla)\mathbf{E} + C_2\mathbf{E}(\nabla\cdot\mathbf{E})$$

for

$$\mathbf{E}(\omega) = \hat{x}E_x(\omega) \tag{34}$$

$$\mathbf{P}(2\omega) \sim C_1 \nabla E_x^2 + \hat{x}(C_2 + C_3)E_x \frac{d}{dx}E_x$$

If the pump wave is in the $LP_{01}$ mode, the second harmonic field will again have the same odd symmetry as the surface contribution.

### 3.4 Second harmonic generation in long fibers

The first question with respect to the surface and quadrupole contributions is whether this offers an explanation of the weak second harmonic generation observed in long fibers.[2] Experiments showed that the output was in a higher-order mode so it was assumed that there was phasematching even though the fiber was most likely a communications type fiber with nowhere near the indices necessary for phasematching.[39]

The understanding of this question is still open and the general history goes something like this. First it was shown that quadrupole contributions are sufficient to produce the weak second harmonic generation seen in long fibers.[40] Then it was argued that if all contributions are included, cancellations occur and the net effect is much smaller than would be obtained by choosing a few selected terms.[12] The reply was that the cancellation does not occur far off phasematching which has to be the case in both the weak second harmonic generation in long fibers and the startup of the $\chi^{(2)}$ gratings in self-seeded fibers.[41]

It is not easy to be sure that all the terms are indeed included. Quadrupole and surface terms can be formally transformed into each other which makes it a little harder to be sure some terms are not included twice. Terhune and Weinberger handled the wandering phasematching problem by breaking the fiber up into short sections each of which had a different phasematching frequency.[12] The approach did assume that the fundamental frequency was somewhere near the phasematching frequency.

There has been some sloppiness in just what is meant by "quadrupole". Some authors use the term for anything that has the right symmetry while others restrict the term to radiation from adjacent oppositely-directed dipoles. So far this question has not been sorted out. The primary reason is that there are more than enough questions related to the seeded second harmonic generation to keep all the parties involved busy.

### 3.5 Startup of the $\chi^{(2)}$ grating

So far there has been a fairly general belief that the self-seeded $\chi^{(2)}$ gratings start with quadrupole second harmonic generation. Most authors have been on both sides of the question at one time or another. For example, the estimates of the magnitude of the quadrupole second harmonic generation are advanced as a starting field for the mixing process.[41] The initial experiments on external seeding used a fiber which was single-mode at both pump and second harmonic wavelengths. This fiber did not form a grating in many hours with pump alone even though the grating was formed in only 5 minutes with the external seed. A fiber from the same preform with a diameter large enough to support a mode of odd symmetry did self-seed and always produced the second harmonic in the higher order mode as would be expected from quadrupole second harmonic generation.[13]

There are also some observations which run counter to the idea that the initial seed comes from a quadrupole contribution. If the green seeding light is coupled into a higher-order mode or some combination of modes, that pattern will be observed when the grating is subsequently read out by 1.06 $\mu$m pump light alone.[16] It appears that the grating accurately records the mode combination used in seeding. Thus it is hard to believe that the process could start with a second harmonic mode of odd symmetry and then switch over to produce an even symmetry mode. Occasionally the second harmonic output does indeed appear in the fiber's lowest-order mode and it seems to have started out in that mode with no observed change in symmetry.[14] Thus there is some evidence that the self-seeding process starts from some dipoles which are initially present in the fiber.

In either case there are some real difficulties with the self-seeding process. If the fields produced by mixing with an external seeding beam are too weak to provide enough force for orientation of defects then how can the process start from noise?

## 4. MICROSCOPIC MODELS

### 4.1 Dopants and defect models

Self-seeded second harmonic generation in fibers was first observed in fibers doped with both germanium and phosphorus.[4] These fibers used germanium to raise the core index to form the guide. The ideal cladding would be pure silica but at the deposit and melting temperature of silica, the silica tubes used in the MCVD preform making process often sagged or collapsed. Phosphorus was added to lower the processing temperature slightly at the expense of a small increase in cladding index. Subsequently, phosphorus was implicated in loss increases from radiation damage and conversion of $H_2$ to OH. It was natural to assume that phosphorus was the dominant defect in self-organized second harmonic generation.

Once fibers with different doping were prepared using the seeding technique it was discovered that fibers doped only with germanium could work almost as well as Ge-P-doped fibers.[16] It was also found that fibers doped only with phosphorus could either not be seeded to produce second harmonic generation at all[16] or would produce only very weak second harmonic generation.[42]

The best conversion efficiency yet reported uses both germanium and phosphorus doping.[5] It is also interesting to note, that fibers doped only with germanium do not self-seed at 1.064 $\mu$m[32] although self-seeded second harmonic generation has been observed using 647

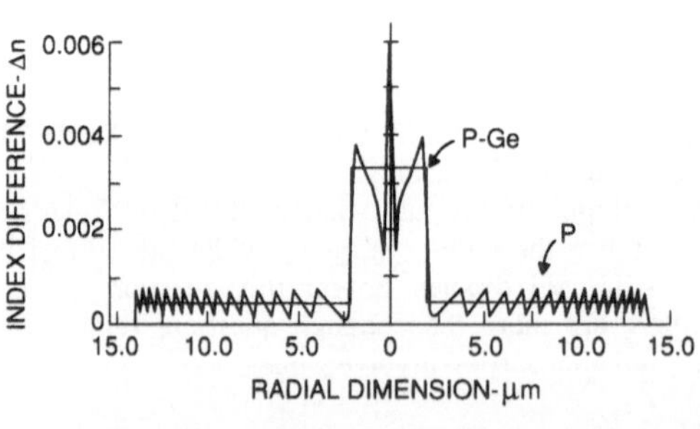


Fig. 22. Refractive index profile of a preform doped with Ge and P. This is not a particularly good preform because of the large index excursions caused by a poor balance between dopant burnoff and compensation during preform collapse. Fibers from this preform were used in the first seeding experiments.[6]

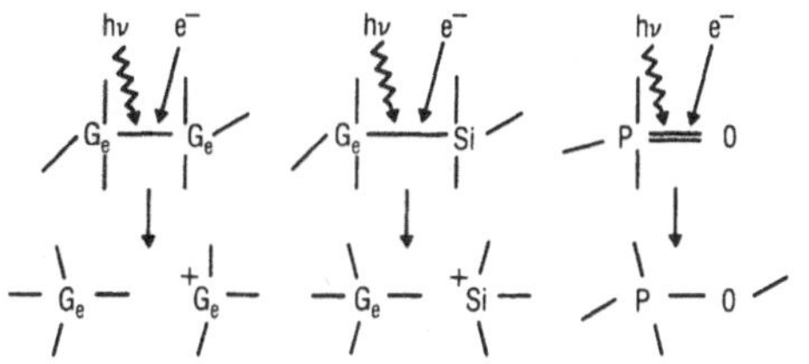


Fig. 23. Three idealized defect structures in silica.

nm.[17] Just to keep things confusing, in early experiments Tom and Stolen observed self-seeding in a phosphorus-doped silica fiber with a borosilicate cladding.[43] Fibers with other dopants have also been prepared to produce externally-seeded second harmonic generation. Among these are aluminum and various rare earths but the effect was much weaker than in Ge-doped fibers.[42]

It is clear that there may be many different defect centers which can be organized to produce $\chi^{(2)}$ gratings in fibers. It is also clear that all observations to date of strong self-organized second harmonic generation have involved germanium. It is very reasonable to concentrate development of a microscopic model on fibers doped only with germanium.

### 4.2 General models

We will concentrate on three idealized defect models because they provide a focus both for general pictures of the self-organization process and for various experimental studies.

The first defect is the Ge E′ center formed by removing an electron from two adjacent germanium atoms. Such a defect would occur in an oxygen deficient germanium-doped glass. Right now, this particular picture is favored by several different groups.[44] The second defect would also produce a Ge E′ center which requires only a germanium ion next to an oxygen vacancy.[45] The third defect is a phosphorus impurity which contains one nonbonding orbital because of the extra electron contributed by the pentavalent phosphorus. This is one of the favorite defects in $P{:}SiO_2$ and it is a useful example for the present discussion.[46]

Specific models can involve the orientation of defects already present in the glass or the orientation of defects created by multiphoton absorption. Models also separate into those which force an excited defect into the desired direction or those in which only certain select orientations are excited by the light. It has also been argued that the impurity is only indirectly involved and that trapping of electrons or holes at defect sites is the direct source of the dipole induced susceptibility.[16]

The direction of the nonbonding orbital of the phosphorus impurity could be changed by raising one of the electrons to a higher state in the presence of a strong dc field. The Ge-Ge bond of the second example could lose an electron by multiphoton absorption after which there is a configurational relaxation in which the Ge-Ge distance increases. The remaining electron has to choose one of the germaniums which could be forced by an external field. In the middle Ge-Si defect, removal of the electron would permit a choice of final orientation only if the configurational changes involved were more general. In both the P and Ge-Si defects a selective absorption mechanism could produce a $\chi^{(2)}$ grating.

Every specific microscopic model can be picked apart fairly quickly. As will be discussed later, experiments show that the dipole active centers seem to lose an electron which rules out tunneling of neutral dangling oxygens like the phosphorus defect. All of the force models suffer from the lack of a mixing field strong enough to orient anything. The selective depletion picture can't explain why excitation of defects by a secondary source enhances the effect. Models of trapped charges have to face the question of why such charges are not excited by a much lower-order process than that which created them. So far the value of these specific models has been to clarify thinking on certain experiments and to lead to concepts which are more general than any specific microscopic model.

### 4.3 Excitation of defects

Most of the possible mechanisms for creating $\chi^{(2)}$ gratings require some sort of absorption by electronic states. This is obviously necessary in models which involve the Ge E′

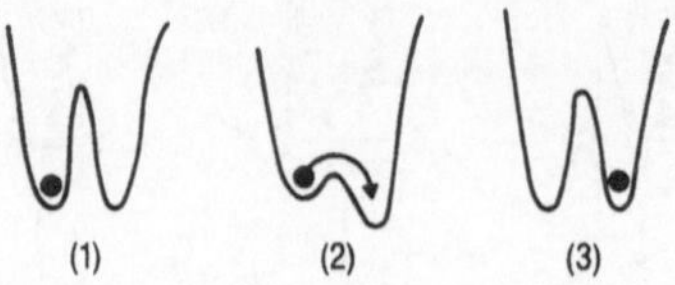

Fig. 24. Particle moving between two equivalent wells.

center. To change the direction of the dangling oxygen connected to the phosphorus ion it is also first necessary to excite an electron from the P-O double bond. The whole process can be viewed as a ball which can move between two equivalent potential wells in an external field by first lowering the barrier height by multiphoton absorption. Once the particle has changed position the fields are removed and the particle is trapped in its new position.

There are several studies showing that multiphoton absorption is essential to the writing process. In one experiment a $\chi^{(2)}$ grating was written in a Ge-P-doped fiber by seeding with 1.319 $\mu$m light and its harmonic. The absorption spectrum from 450 to 800 nm was almost identical in the second harmonic generation prepared fiber and in an unprepared fiber irradiated with only 457 nm light. The implication is that third harmonic light (measured to be about 27 $\mu$W) created the defects which are then aligned to produce the $\chi^{(2)}$ grating.[18]

We will be deliberately vague about multiphoton absorption and include all such possibilities as direct multiphoton absorption by the defect and generation of harmonic light which subsequently excites defects.

Electron spin resonance experiments on prepared fibers showed a signal characteristic of a germanium E′ center.[20] The fractional increase in Ge E′ center concentration correlated with second harmonic generation conversion efficiency in three different fibers. The E′ centers were distributed uniformly over 2m of fiber as compared with the $\chi^{(2)}$ grating which only extended over the first few cm. The implication is that the E′ centers are formed by multiphoton absorption along the whole length of the fiber and are properly arranged in a $\chi^{(2)}$ grating over the distance given by the bandwidths of the fundamental and seed beams. Also the most Ge E′ centers were seen in a fiber containing phosphorus leading to speculation that the phosphorus provides traps for the free electrons.

Annealing experiments on prepared fibers showed that the second harmonic conversion efficiency stayed constant up to 100° C and then deteriorated so that it was down to about 10% at 400° C. Also some deterioration in second harmonic efficiency was found over several months even at room temperature. When the second harmonic efficiency decreased the induced loss was not simultaneously bleached out. This suggests that the aligned defect centers are just redistributed on heating but they haven't disappeared.[21]

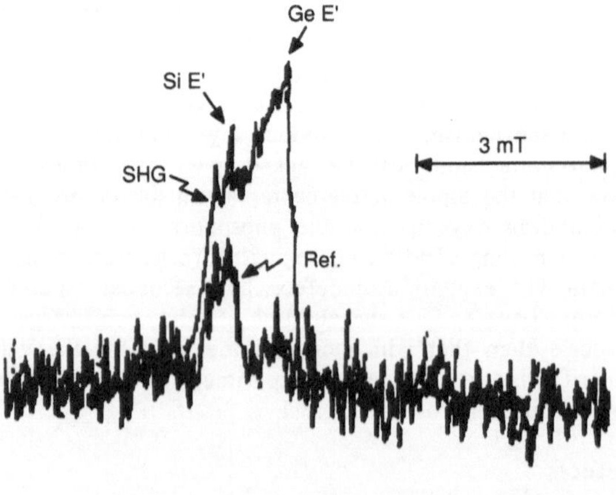

Fig. 25. ESR derivative signal of Si E′ and Ge E′ centers in a Ge-P doped fiber before and after preparation for second harmonic generation.[20]

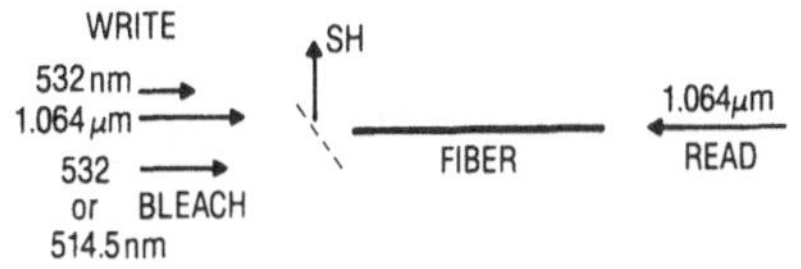


Fig. 26. Bleaching Experiment.

Bleaching experiments lead to similar conclusions as the annealing studies. In a bleaching experiment, a grating is first prepared in a fiber by seeding with 1.064 μm and 532 nm. Then the second harmonic output is monitored as the grating is bleached with green light. The green light can be either 532 nm from the doubled YAG or it can be 514.5 nm from an Argon laser and is directed backwards through the fiber to avoid any reseeding of the fiber.[22] Seeding a bleached fiber again produces a $\chi^{(2)}$ grating which suggests that the bleaching merely rearranges the defects. There is another interesting result of the bleaching measurements which is probably extremely important - but no-one knows what to make of it. The decay rate of the $\chi^{(2)}$ grating amplitude A goes as:

$$\frac{dA}{dt} \sim -P^4(2\omega)A^2 \tag{35}$$

The $A^2$ dependence on bleaching power is not expected. Bleaching of independent defects would be expected to produce a linear dependence in A. The $A^2$ decay rate was found in a Ge-doped fiber but similar results are obtained from bleaching measurements in a Ge-P-doped fiber.[32]

Hydrogen appears to increase the efficiency of second harmonic generation in prepared fibers. If fibers are heated in a hydrogen atmosphere and then seeded to form $\chi^{(2)}$ gratings, the efficiency of second harmonic generation is increased about four times.[47] Just what the hydrogen has to do with the detailed mechanism is not known but heating induces reactions between $H_2$ and defects in the glass.[47] Studies by Raman scattering show that in fibers which can be prepared for efficient second harmonic generation the intensity of defect lines at 490 and 603 $cm^{-1}$ is much less than non-preparable fibers.[48] Speculation is that the defects associated with these Raman lines provide the dipole-allowed nonlinear response. It is also possible that these measurements are evidence of hydrogen enhancement. It is known that these defect lines are reduced in intensity in high OH silicas.[49]

## 4.4 Electric field studies

An electric field will induce an effective $\chi^{(2)}$ through $\chi^{(3)}$ and a measure of $\chi^{(3)}$ can be obtained in this way.[50] The enhancement factor from oriented defects in seeded second harmonic generation experiments is two orders of magnitude greater than this intrinsic dc-induced second harmonic generation. It is clearly of interest to examine whether an external electric field will also orient defects in the glass.

In the first such experiment, 10 kV were applied across 1 cm of Ge-P-doped silica from the preform used to make efficient second harmonic generation fiber.[51] If 1.06 μm light was present, electric field-induced second harmonic generation from $\chi^{(2)}_{eff} = \chi^{(3)} E_{dc}$ was observed. Because there was no phase matching the second harmonic generation signal was small and came only from the last coherence length (about 30 μm). In a few minutes the second harmonic generation signal dropped indicating that defects were oriented to produce second harmonic generation out of phase with the dc field-induced second harmonic generation. When the dc field was removed the oriented defects produced a second harmonic generation signal which decayed in a few minutes.[51]

The essential results are that first it is possible to orient defects with an external field but that a field about $10^4$ times the internal mixing field is insufficient to generate a large permanent $\chi^{(2)}$ in the glass.

The dc field can also be imposed across the fiber itself.[52] It is possible to produce much higher fields in the fibers themselves, by using internal electrodes.[19]

These internal poling techniques have not only induced large $\chi^{(2)}$'s in the fiber core but also produced large linear electrooptic coefficients. It is the subject of some debate whether these two effects are one and the same or only indirectly related.[53]

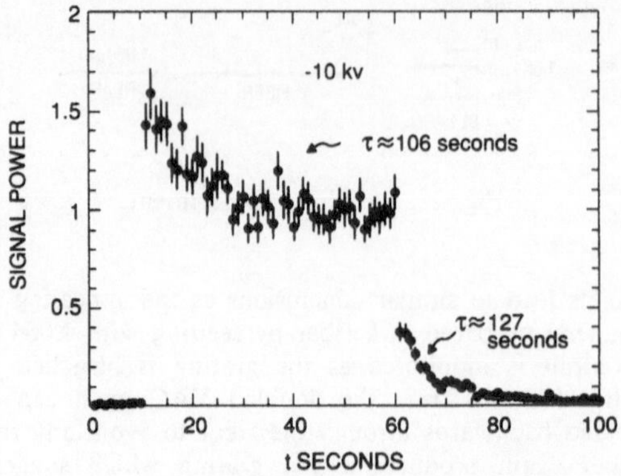


Fig. 27. Electric field-induced second harmonic generation in a bulk Ge-P-doped silica sample.[51]

DC-poled second harmonic generation in a single-mode fiber suffers from lack of phase matching. If the fiber supports several modes, phasematching can come from the waveguide modes as discussed earlier. Another approach to phasematching the dc-induced second harmonic generation is to use an alternating dc field imposed at the phasematching period. This has been done by polishing a short section of fiber and using a periodic electrode grating structure. Phasematching was performed by rotating the grating slightly. Not only was phasematching of the optimal pump and second harmonic $LP_{01}$ modes accomplished but other modes of the fiber could also be observed by using the grating to make up the phase mismatch.[54]

### 4.5 The force problem

The $E_{dc}$ from the macroscopic $\chi^{(3)}$ is only ~ 1 V/cm for fibers written with mode-locked Q-switched laser pulses. It gets even weaker when seeding with just mode-locked pulses. Also, $E_{dc}$ is even weaker still when starting from noise.

Could the effective $\chi^{(3)}$ be much larger? For example, in an orientational model it is not necessarily the macroscopic dc field that has to interact with the dipole. Mixing can actually take place in the orbitals of the defect itself. After all, the macroscopic polarization is just the sum of the polarizations from all the bonds in the glass. If the hyperpolarizability ($\gamma$) of the defect itself is much larger than $\gamma$ for the Si-O bonds then the interaction energy between the induced dipole and the intrinsic dipole could be much larger than the interaction energy between the intrinsic dipole and the macroscopic dc field. It is known that molecular systems with conjugated bonds such as long chain organic molecules have enhanced hyperpolarizabilities. However, we can extrapolate from existing data on $\gamma$ vs. chain length and find an enhancement of only one to two orders of magnitude.[55]

One intriguing possibility is that of a discontinuity in the hyperpolarizabiltiy, $\gamma$, as a configurational relaxation occurs. Such a discontinuity has been observed in liquid crystal phase

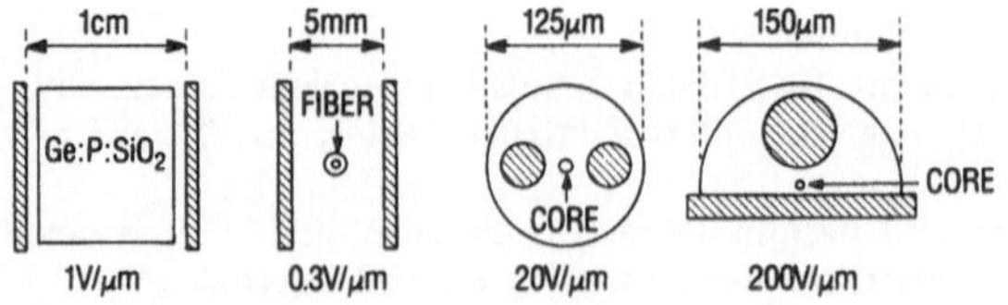


Fig. 28. Methods of applying large dc fields for poling experiments.

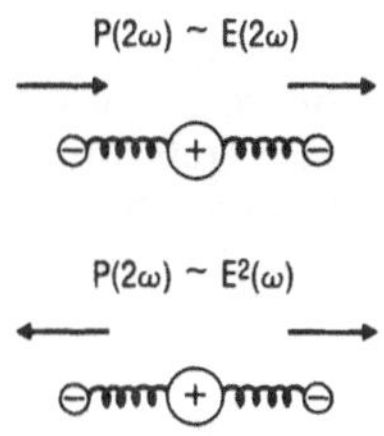


Fig. 29. Simple picture of selective absorption by sum of one and two photon processes.

transitions.[56] Similar behavior might be expected any time a soft mode is involved in a phase transition leading to bifurcation. There is one denominator in expansions of $\chi^{(3)}$ that has the frequency of the soft mode so that this term leads to the discontinuity.

A relatively weak field can produce stable orientation if the centers are organized as domains. The problem with dipole type defects in glass is that they will align themselves anti-parallel which will not produce a net $\chi^{(2)}$. Something like the exchange force in ferromagnetism is needed to produce domains with a net $\chi^{(2)}$.

Another model has been proposed based on orientation of GeO molecules trapped in the $SiO_2$ network.[57]

One proposal for a selective absorption mechanism uses simultaneous one- and two-photon absorption.[11] This can be understood in terms of a simple model of charges on springs. One photon absorption drives the charges on each side in the same direction while two photon absorption drives the charges in opposite directions.

The displacement of the charges on one side is enhanced while the displacement on the other side is reduced by the 2-photon contribution. The enhanced motion corresponds to a larger induced optical dipole moment and increased absorption coefficient. This enhancement will switch back and forth with a period given by $(k_2 - 2k_1)$ which is just the period required for phasematching of second harmonic generation.

## 5. CONCLUSION

It should be apparent that the subject of second harmonic generation in fibers is in a fairly early stage of development. By the time a field is well enough organized to be the subject of a series of lectures, there has been enough work that things start looking reasonably tidy. In contrast to such an ideal, second harmonic generation in fibers is a mess of loose ends and contradictory results. Perhaps this is a unique opportunity to see the workings of science as it really is.

The various approaches to self-organized second harmonic generation in fibers can be divided into phenomenological macroscopic models, general microscopic models, and specific microscopic models. Most of the progress has been made in the phenomenological model but it should be clear that even here there is a long way to go before we understand the dynamics of the grating formation.

The general microscopic models can be divided up into some specific classifications as pictured in Fig. 30. One division is into force and selective absorption models. In force

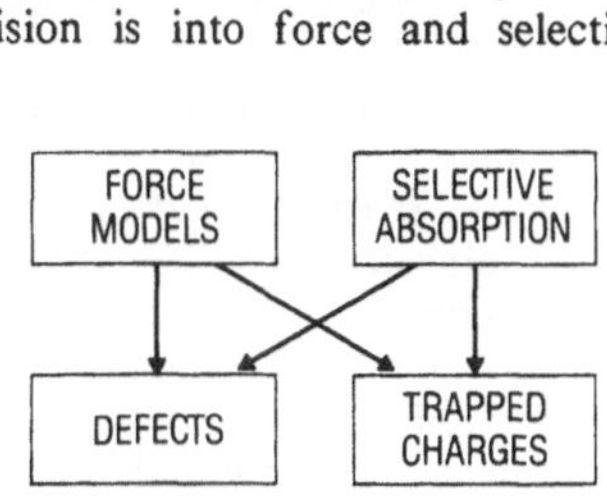


Fig. 30. Classification of general microscopic models.

models, the defects are aligned in some way by the mixing field. In selective absorption models only the proper defects are excited to produce the $\chi^{(2)}$ grating. The second division is whether the dipole-allowed response comes from the actual defect or whether it comes from trapped charges. As a general rule people with a background in color centers favor the defects as the source of the nonlinearity while people from a photorefractive background like models with trapped charges.

The defect models can be further broken down into those that get the dipole response from excited or unexcited defects. The trapped charge models can get the dipole response directly from the charges in the traps or indirectly through a large internal electric field.[58,59] The large internal dc field induces second harmonic generation through the macroscopic $\chi^{(3)}$.

It is well not to neglect the other possibilities even if they don't explain the microscopic details of Ge-doped fibers. It may well turn out that some of the other possibilities are just what is needed to make a stable device or have a huge non-saturable $\chi^{(2)}$.

Obviously the goal of most researchers is to make an efficient doubler. It is well to keep in mind, however, that there is a lot of interesting science in these self-organized effects in fibers. This is a rich and variegated field which can provide new insights into the structure and defects of glasses. The $\chi^{(2)}$ and index gratings in fibers may well be some of the simpler self-organized systems and prove to be an important system for study of this branch of science.

## REFERENCES


1. R. H. Stolen, "Nonlinear properties of optical fibers", in *Optical Fiber Telecommunications*, S. E. Miller and A. G. Chenowyth eds. (Academic Press, New York, 1979) Ch. 5; G. P. Agrawal, *Nonlinear Optical Fibers* (Academic Press, San Diego, 1989)
2. Y. Fujii, B. S. Kawasaki, K. O. Hill, and D. C. Johnson, "Sum frequency generation in optical fibers", Opt. Lett. **5**, 48 (1980); Y. Ohmori and Y. Sasaki, "Phasematched sum-frequency light generation in optical fibers", Appl. Phys. Lett. **39**, 466 (1981).
3. J. M. Gabriagues, "Third-harmonic and three-wave sum-frequency light generation in an elliptical-core optical fiber", Opt. Lett. **8**, 183 (1983).
4. U. Österberg and W. Margulis, "Efficient second harmonic generation in an optical fiber", XIV Internat. Quantum Electron. Conf., San Fransisco (1986) paper WBB1; and "Dye laser pumped by Nd:YAG laser pulses frequency doubled in an optical fiber", Opt. Lett. **11**, 516 (1986).
5. M. C. Farries, "Efficient second harmonic generation in an optical fibre", Proc. Colloquium on Non-linear Optical Waveguides (London, IEE, 1988); and Laser Focus **24**, 12 (1988).
6. R. H. Stolen and H. W. K. Tom, "Self-organized phase-matched harmonic generation in optical fibers", Digest of Conference on Lasers and Electrooptics (Optical Society of America, 1987), paper ThL2; Opt. Lett. **12**, 585 (1987).
7. J. A. Armstrong, N. Bloembergen, J. Ducuing, and P. S. Pershan, "Interactions between light waves in a nonlinear dielectric", Phys. Rev. **127**, 1918 (1962).
8. M. C. Farries, P. St. J. Russell, M. E. Fermann, and D. N. Payne, "Second-harmonic generation in an optical fiber by self-written $\chi^{(2)}$ grating", Electron. Lett. **23**, 322 (1987).
9. N. A. Sanford and W. C. Robinson, "Direct measurement of effective indices of guided modes in $LiNbO_3$ waveguides using the Čerenkov second harmonic", Opt. Lett. **12**, 445 (1987); T. Taniuchi and K. Yamamoto, "Second harmonic generation using proton-exchanged $LiNbO_3$ waveguide", Optoelectronics-Devices and Technologies 2, 53 (1987).
10. E. J. Lim, M. M. Fejer, and R. L. Byer, "Second harmonic generation of green light in periodically poled planar lithium niobate waveguide", Electron. Lett. **25**, 174 (1989); J. Webjorn, F. Laurell, and G. Arvidsson, "Periodically domain-inverted lithium niobate channel waveguides for second harmonic generation", Digest of Topical Meeting on Nonlinear Guided Wave Phenomena (Optical Society of America, 1989), p. 6.
11. N. B. Baranova and B. Ya. Zeldovich, "Extension of holography to multifrequency fields", Pisma ZhETF **45**, 562 (1987) [JETP Lett. **45**, 717 (1987)].
12. R. W. Terhune and D. A. Weinberger, "Second-harmonic generation in fibers", J. Opt. Soc. Am. B **4**, 661 (1987).
13. H. W. K. Tom, R. H. Stolen, G. D. Aumiller, and W. Pleibel, "Preparation of long-coher-

ence-length second-harmonic-generating optical fibers by using mode-locked pulses", Opt. Lett. **13**, 512 (1988).
14. W. Margulis and U. Österberg, "Second-harmonic generation in optical glass fibers", J. Opt. Soc. B **5**, 312 (1988)
15. M. C. Farries, M. E. Fermann, P. St. J. Russell, and D. N. Payne, "Tunable second-order susceptibility gratings for harmonic generation in optical fibers", Digest of the Optical Fiber Communications Conf. '88 (Optical Society of America/IEEE, New Orleans, LA, Jan. 1988), paper THE2;
M. E. Fermann, M. C. Farries, P. St. J. Russell, and L. Poyntz-Wright, "Tunable holographic second-harmonic generators in high-birefringence optical fibers", Opt. Lett. **13**, 282 (1988).
16. M. A. Saifi and M. J. Andrejco, "Second-harmonic generation in single-mode and multimode fibers", Opt. Lett. **13**, 773 (1988);
E. Snitzer, "Rare earth fiber lasers", J. Less-Common Metals **148**, 45 (1989).
17. B. Valk, E. M. Kim, and M. M. Salour, "Second harmonic generation in Ge-doped fibers with a mode-locked Kr+ laser", Appl. Phys. Lett. **51**, 722 (1987).
18. M. C. Farries and M. E. Fermann, "Frequency doubling of 1.319 $\mu$m radiation in an optical fibre by optically written $\chi^{(2)}$ grating", Electron. Lett. **24**, 294 (1987).
19. M. V. Bergot, M. C. Farries, M. E. Fermann, L. Li, L. J. Poyntz-Wright, P. St. J. Russell, and A. Smithson, "Generation of permanent optically induced second-order nonlinearities in optical fibres by poling", Opt. Lett. **13**, 592 (1988).
20. T. E. Tsai, E. J. Friebele, D. L. Griscom, M. A. Saifi, and U. Österberg, "Correlation of defect centers with second harmonic generation in Ge and G-P-doped silica core single mode fibers", Topical Meeting on Nonlinear Guided-Wave Phenomena: Physics and Applications, 1989 Technical Digital Series, Vol. 2 (Optical Society of America, Washington, DC 1989), p. 250.
21. M. E. Fermann, "Characterisation techniques for special optical fibres", Ph. D. Thesis, University of Southamption (1988).
22. F. Ouellette, K. O. Hill, and D. C. Johnson, "Light-induced erasure of self-organized $\chi^{(2)}$ gratings in optical fibers", Opt. Lett. **13**, 515 (1988).
23. K. O. Hill, Y. Fujii, D. C. Johnson, and B. S. Kawasaki, "Photosensitivity in optical fibre waveguides: application to reflection filter fabrication", Appl. Phys. Lett. **32**, 647 (1978);
J. Stone, "Photorefractivity in GeO2-doped silica fibers", J. Appl. Phys. **62**, 4371 (1987);
C. P Kuo, U. Österberg, C. T. Seaton, G. I. Stegeman, and K. O. Hill, "Optical fibers with negative group-velocity dispersion in the visible", Opt. Lett. **13**, 1032 (1988).
24. G. Meltz, W. W. Morey, and W. H. Glenn, "Formation of Bragg gratings in optical fibers by a tranverse holographic method", Opt. Lett. **14**, 823 (1989).
25. A. M. Glass, "The photorefractive effect", Optical Engineering **17**, 470 (1978);
J. Feinberg and R. W. Hellwarth, Opt. Lett. **5**, 519 (1980);
B. Fischer, M. Cronin-Golomb, J. O. White, A. Yariv, and R. Neurgaonkar, "Amplifying continuous wave phase conjugate mirror with strontium barium niobate", Appl. Phys. Lett. **40**, 863 (1983).
26. D. K. Kondepudi, F. Moss, and P. V. E. McClintock, "Observation of symmetry breaking, state selection and sensitivity in a noisy electronic system", Physica **21D**, 296 (1986).
27. B. Ya. Zeldovich, private communication
28. B. Batdorf, C. Krautschik, U. Österberg, G. Stegeman, and T. F. Morse, "Length dependence of second harmonic generation in optical fibers prepared with 1.064 $\mu$m and 532 nm light", Topical Meeting on Nonlinear Guided-Wave Phenomena: Physics and Applications, 1989 Technical Digest Series, Vol. 2 (Optical Society of America, Washington, D. C. 1989), p. 259.
29. W. Margulis, I. C. S. Carvalho, and J. P. von der Weid, "Phase measurement in frequency-doubling fibers", Opt. Lett. **14**, 700 (1989).
30. P. Chmela, "Second-harmonic generation from quantum noise owing to fifth-order nonlinearity", Opt. Lett. **13**, 669 (1988).
31. V. Mizrahi and J. E. Sipe, "Generation of permanent optically induced second-order nonlinearities in optical fibers by poling: comment", Appl. Opt. **28**, 1976 (1989).
32. A. Krotkus and W. Margulis, "Investigations of the preparation process for efficient second-harmonic generation in optical fibers", Appl. Phys. Lett. **52**, 1942 (1988).
33. W. J. Tomlinson, R. H. Stolen, and C. V. Shank, "Compression of optical pulses chirped by self-phase modulation in fibers", J. Opt. Soc. Am. B **1**, 139 (1984).

34. Index data for fused $SiO_2$ from *American Institute of Physics Handbook*, B. E. Grey ed. (McGraw-Hill, N. Y., 1963), p. 6:25
35. D. Gloge, "Weakly guiding fibers", Appl. Opt. **10**, 2252 (1971).
36. R. Kashyap, "Photo-induced enhancement of second harmonic generation in optical fibres", Topical Meeting on Nonlinear Guided-Wave Phenomena, 1989 Technical Digest Series, Vol. 2 (Optical Society of America, Washington, D. C. 1989), p. 255.
37. M. E. Fermann, L. Li, M. C. Farries, and D. N. Payne, "Frequency-doubling by modal phase matching in poled optical fibres", Electron. Lett. **24**, 894 (1988).
38. F. R. Nash, G. D. Boyd, M. Sargent III, and P. M. Bridenbaugh, "Effect of optical inhomogeneities on phase matching in nonlinear crystals", J. Appl. Phys. **41**, 2564 (1970).
39. Y. Ohmori and Y. Sasaki, "Two-wave sum-frequency light generation in optical fibers", IEEE J. Quantum Electron. **QE-18**, 758 (1982).
40. F. P. Payne, "Frequency doubling in single-mode optical fibres", 4th Int. Symp. on Optical and Optoelectronic Applied Science and Engineering, The Hague (SPIE, 1987).
41. F. P. Payne, "Second-harmonic generation in single-mode optical fibres", Electron. Lett. **23**, 1215 (1987).
42. M. C. Farries, M. E. Fermann, and P. St. J. Russell, "Second harmonic generation in optical fibres", Topical Meeting on Nonlinear Guided-Wave Phenomena: Physics and Applications, 1989 Technical Digest Series, Vol. 2 (Optical Society of America, Washington, D. C. 1989), p. 246.
43. Unpublished work
44. L. J. Poyntz-Wright and P. St. J. Russell, "Spontaneous relaxation processes in irradiated germanosilicate optical fibres", Electron. Lett. **25**, 478 (1989).
45. Y. Watanabe, H. Kawazoe, K. Shibuya, and K. Muta, "Structure and mechanism of formation of drawing- or radiation-induced defects in $SiO_2$:$GeO_2$ optical fiber", Jpn. J. Appl. Phys. **25**, 425 (1986).
46. D. L. Griscom, E. J. Friebele, K. J. Long, and J. W. Fleming, "Fundamental defect centers in glass: Electron spin resonance and optical absorption studies of irradiated phosphorus-doped silica glass and optical fibers", J. Appl. Phys. **54**, 3743 (1983).
47. F. Ouellette, K. O. Hill, and D. C. Johnson, "Enhancement of second-harmonic generation in optical fibers by a hydrogen and heat treatment", Appl. Phys. Lett. **54**, 1086 (1989).
48. J. M Gabriagues and H. Fevrier, "Analysis of frequency-doubling process in optical fibers using Raman spectroscopy", Opt. Lett. **12**, 720 (1987).
49. R. H. Stolen and G. E. Walrafen "Water and its relation to broken bond defects in fused silica", J. Chem. Phys. **64**, 2623 (1976).
50. C. G. Bethea, "Electric field induced second harmonic generation in glass", Appl. Opt. **14**, 2435 (1975).
51. V. Mizrahi, U. Österberg, J. E. Sipe, and G. I. Stegeman, "Test of a model of efficient second-harmonic generation in glass optical fibers", Opt. Lett. **13**, 279 (1988).
52. V. Mizrahi, U. Österberg, C. Krautschik, G. I. Stegeman, J. E. Sipe, and T. F. Morse, "Direct test of a model of efficient second-harmonic generation in glass optical fibers", Appl. Phys. Lett. **53**, 557 (1988).
53. M. E. Fermann, L. Li, M. C. Farries, D. N. Payne, and P. St. J. Russell, "Second-harmonic generation in poled optical fibres: dynamics and phasematching techniques", Topical Meeting on Nonlinear Guided-Wave Phenomena: Physics and Applications, 1989 Technical Digest Series, Vol. 2 (Optical Society of America, Washington, D. C. 1989), paper PD6.
54. R. Kashyap, "Phase-matched electric-field-induced second-harmonic generation in optical fibers", J. Opt. Soc. Am. B **6**, 313 (1989).
55. J. P. Herman and J. Ducuing, "Third-order polarizabilities of long-chain molecules", J. Appl. Phys. **45**, 5100 (1974).
56. S-H. Chen, C-L. Kuo, and M-C. Lee, "Quasi-static electric-field-enhanced degenerate four-wave mixing in a nematic liquid-crystal film", Opt. Lett. **14**, 122 (1989).
57. N. M. Lawandy, "Mechanism for efficient second harmonic generation in Ge and P-doped optical fibers", Optical Society of America 1988 Annual Meeting (Santa Clara, CA) paper FS4.
58. D. Z. Anderson, "Efficient second-harmonic generation in glass fibers: The possible role of photo-induced charge redistribution", SPIE proc., Vol. 1148 (1989).
59. E. M. Dianov, P. G. Kazansky, and D. Yu Stepanov, "On the problem of photoinduced second harmonic generation in optical fibers", Sov. J. Quantum Electron. **16**, 887 (1989).

# SOLITONS IN OPTICAL FIBRES

K J Blow and N J Doran

British Telecom Research Laboratories
Martlesham Heath
Ipswich, IP5 7RE
England

## 1. THEORY OF OPTICAL PROPAGATION

### 1.1 Introduction

This chapter aims to present a review of some central properties of nonlinear short pulse effects in conventional silica optical fibres. Solitons are the key feature of such propagation and will form a central theme for the chapter. In particular we shall examine soliton generation and interaction with linear and nonlinear effects. There will be an underlying drive towards the potential exploitation of soliton effects but the emphasis will be on the fundamental physics.

Why should we be interested in nonlinear optics in fibres at all? The important parameter which determines the strength of the nonlinear optical effects is the product of the optical intensity and the interaction length in the medium. In bulk nonlinear optics a focused geometry is used. The optical intensity is then $I = P/\pi a^2$ where $P$ is the power and the interaction length is limited, because of the beam waist ($2a$) due to diffraction, to be $L = 2\pi a^2/\lambda$. In an optical fibre with the same field width as the diffraction limited beam waist the intensity would be the same as for the focused case but the interaction length is only limited by the absorption length of the fibre ($\Gamma^{-1}$). The enhancement in the intensity length product is approximately $\lambda/2\pi a^2\Gamma$ which can be as large as $10^9$ in a typical optical fibre. This enhancement of the optical nonlinear effects has led to the observation of many phenomena at relatively low power levels, some of which will be discussed during the course of this chapter.

### 1.2 Linear pulse propagation

Nonlinear properties are best understood from a firm foundation in the relevant linear properties. Therefore, we shall first review the essential linear pulse propagation effects and their mathematical description.

In a weakly guiding optical fibre where the refractive index variations are small, we can use the scalar wave equation [1] for the transverse field component $E$

$$(\nabla^2 - \frac{\varepsilon(r)}{c^2}\frac{\partial^2}{\partial t^2})E = 0. \tag{1.1}$$

The electric field $E$ can be separated into a radial part, $R(r,\theta)$, and a longitudinal part, $V(z,t)$. The radial part is easily shown to satisfy the following wave equation which generates the mode fields as well as the propagation constants through the eigenfunctions, R, and eigenvalues, $\beta^2$, respectively

$$(\nabla_T^2 + (n^2k^2 - \beta^2))R = 0. \tag{1.2}$$

The eigenvalues $\beta^2$ of equation (1.2) are real and if there is just one discrete eigenvalue at wavelength $\lambda$ then the fibre is said to be single mode at that wavelength. Strictly speaking 'single mode' fibres have two orthogonally polarised modes but we will be concerned with cylindrically symmetric fibres where these two modes are degenerate. The eigenvalue $\beta$ is frequency dependent through the frequency dependence of the refractive index, n, as well as through k, the free space wavevector which acts as a scaling term. These dependencies are described as chromatic and waveguide dispersion respectively. The distinction between these two 'types' of dispersion is somewhat artificial since it is only the resultant which matters in terms of the properties of the mode. It is convenient to expand the propagation constant around some fixed, though arbitrary, frequency $\omega_0$. Thus

$$\beta = \beta_0 + \beta'(\omega-\omega_0) + \frac{\beta''}{2}(\omega-\omega_0)^2 + \frac{\beta'''}{3}(\omega-\omega_0)^3 + ... \tag{1.3}$$

where $\beta'$ will be shown to be the reciprocal of the group velocity and $\beta''$ the dispersion coefficient.

The longitudinal equation contains the eigenvalue $\beta^2(\omega)$ as obtained from equation (1.2). Transforming to frequency space we write

$$\frac{\partial^2 V(\omega,z)}{\partial z^2} + \beta^2(\omega)V(\omega,z) = 0. \tag{1.4}$$

We are interested in the propagation of a pulse envelope which in turn can be considered to have a well defined packet of frequencies. That is we take

$$V(t,z) = u(t,z)\exp[i(\omega_0 t - \beta_0 z)] + c.c. \tag{1.5}$$

Substituting the Fourier transform of (1.5) into (1.4) and taking just the terms in $exp[i(\omega_0 t - \beta_0 z)]$ results in a wave equation

$$\frac{\partial^2 u}{\partial z^2} - 2i\beta_0\frac{\partial u}{\partial z} + (\beta^2 - \beta_0^2)u = 0 \tag{1.6}$$

which can easily be factorised to obtain the following simple equation for the forward travelling wave.

$$i\frac{\partial u}{\partial z} = (\beta - \beta_0)u \tag{1.7}$$

Equation (1.7) describes the frequency packet $u(\omega)$. However, it is conventional to write the evolution equation in the time domain. To do this we use the expansion of the propagation constant, equation (1.3), and Fourier transform to obtain

$$i(\frac{\partial u}{\partial z} + \beta'\frac{\partial u}{\partial t}) + \frac{\beta''}{2}\frac{\partial^2 u}{\partial t^2} = 0. \tag{1.8}$$

Equation (1.8) generates the evolution of a pulse envelope $u(z,t)$ under the influence of dispersion. In a standard single mode fibre the coefficient $\beta''$ changes sign at about $1.3\mu m$ and near this wavelength of zero second order dispersion it would be necessary to include the third order term in equation (1.3).

We now transform the variables so that we are in a frame moving at the group velocity *i.e.* $T = t - \beta' z$, $Z = z$ and write the evolution equation in a simpler form

$$i\frac{\partial u}{\partial Z} + \frac{\beta''}{2}\frac{\partial^2 u}{\partial T^2} = 0. \tag{1.9}$$

This equation can be solved by Fourier transforming from the time domain to the frequency domain, solving for the evolution of the spectral components of $u(Z,\omega)$ and finally taking the inverse Fourier transform to recover $u(Z,T)$. The solution for an initially gaussian pulse $\exp[-\frac{T^2}{4\tau^2}]$ is

$$u(Z,T) = (1+\frac{i\beta''Z}{2\tau^2})^{-\frac{1}{2}}\exp\left[-\frac{T^2}{4\tau^2(1+\frac{i\beta''Z}{2\tau^2})}\right]. \qquad (1.10)$$

Equation (1.10) shows that the pulse broadens as it propagates, this property is also usually referred to as dispersion, but the broadening does not depend on the sign of $\beta''$. If we write $u(Z,T) = |u(Z,T)|\exp^{i\phi(Z,T)}$ then $\phi$ represents the phase of the pulse. This phase factor is quadratic in time near the peak and the sign of its curvature is the same as the sign of $\beta''$. The derivative of the pulse phase $\frac{d\phi}{dT}(Z,T)$ can be used to define a 'local' or 'instantaneous' frequency within the pulse. Since the pulse phase is quadratic in time, near the peak, the local frequency will be a linear function of time and this is known as a chirp. More generally, pulses whose bandwidths are not limited by the Fourier transform of their intensity are referred to as chirped. The linear chirp acquired by a pulse propagating in an optical fibre depends on the sign of the dispersion which can be either positive or negative depending on the wavelength.

### 1.3 The nonlinear response of silica fibres

So far we have discussed only the linear propagation properties of optical fibres. Now we consider briefly the description of the nonlinearity which is to be included in the evolution equation. In any material which is weakly nonlinear the nonlinear polarisation can be expanded as a Taylor series in the electric field. This expansion can be written schematically as

$$P_{NL} = \chi_1 E + \chi_2 : E.E + \chi_3 : E.E.E. \qquad (1.11a)$$

The first term represents the linear response of the medium where the induced polarisation must be at the same frequency as the driving field. The second term on the right hand side involves two fields and therefore has the possibility of an induced field at a new frequency. This term is the origin of second harmonic generation and will not be considered here; in a centrosymmetric material $\chi_2$ is identically zero. The third term is our main concern. Including this alone we can write a more explicit form of equation (1.11a) where the frequencies of the waves are included (note we will assume that the susceptibilities are frequency independent).

$$P_{NL}(\omega_4) = \chi_1 E(\omega_4) + \chi_3 E(\omega_1)E(\omega_2)E(\omega_3) \qquad (1.11b)$$

There are four waves involved; three mixing waves and one induced wave. Energy and momentum conservation requires the following equations to be valid.

$$\omega_4 = \omega_1 + \omega_2 + \omega_3 \;\; ; \;\; k_4 = k_1 + k_2 + k_3 \qquad (1.12)$$

where both the frequencies and momenta can be of either sign. Here we consider the case where all the frequencies have the same magnitude, this is the appropriate regime for the propagation of a single wave packet. Then the polarisation may be written simply as

$$P = (\chi_1 + \chi_3 |E|^2)E. \qquad (1.13)$$

Then, using the usual relationship between the refractive index and the susceptibility ($\chi = n_0^2 - 1$) the nonlinear refractive index can be expressed as

$$n_{NL} = n_0 + \chi_3 \frac{n_0}{2}|E|^2 \qquad (1.14)$$

where $n_0$ is the linear refractive index. The coefficient in front of the $|E|^2$ term is often replaced by $n_2$ and identified as the Kerr coefficient. In the case of two copropagating waves, $E_A$ and $E_B$, there is a cross term which results in a more general definition for nonlinear refractive index

$$n_A = n_0 + n_2(|E_A|^2 + 2|E_B|^2). \qquad (1.15)$$

In silica $n_2$ is small $2\times10^{-22}(m/V)^2$ so that even for intense fields we can assume that the nonlinear index change is always much less than any other index in the problem. It is therefore a good approximation to include the nonlinear term as a perturbation in equation (1.2) and to consider only changes in the eigenvalue $\beta^2$ and not in the fields. After integrating out the radial field dependence to give an effective area $A_{eff}$ and redefining $n_2$ to be in terms of the intensity then the expansion for the propagation constant equation (1.3) is written in the nonlinear case as

$$\beta_{NL} = \beta_L + n_2 k_0 I = \beta_L + \frac{n_2 k_0}{A_{eff}} |u|^2 \tag{1.16}$$

where $\beta_L$ is the linear propagation constant as in equation (1.3). Including this nonlinear term in the wave equation (1.7) results in a nonlinear wave equation for the envelope $u$ which is now in units of $W^{1/2}$ [2].

$$i\frac{\partial u}{\partial Z} = (\beta_L - \beta_0)u + \frac{n_2 k_0}{A_{eff}} |u|^2 u \tag{1.17}$$

Here the Kerr coefficient is $n_2 = 3.2\times10^{-16}(cm^2/W)$ in silica. The full evolution equation for the pulse envelope $u$ analogous to (1.9) becomes

$$i\frac{\partial u}{\partial Z} + \frac{\beta''}{2}\frac{\partial^2 u}{\partial T^2} + \frac{k_0 n_2}{A_{eff}} |u|^2 u = 0. \tag{1.18}$$

The coefficients in equation (1.18) can be removed by the following simple transformations (note we introduce the arbitrary timescale $\tau$ which is generally convenient to take as the pulse width)

$$z = \frac{\beta'' Z}{\tau^2}, \; t = \frac{T}{\tau}, \; U = \tau\left(\frac{n_2 k_0}{\beta'' A_{eff}}\right)^{1/2} u \tag{1.19}$$

to generate the Nonlinear Schrodinger Equation (NLS)

$$i\frac{\partial U}{\partial z} \pm \frac{1}{2}\frac{\partial^2 U}{\partial t^2} + |U|^2 U = 0 \tag{1.20}$$

where the sign of the second derivative takes the sign of the dispersion parameter $\beta''$. This equation is central to our description of nonlinear pulse propagation in optical fibres and will form the kernel of much of the analysis in subsequent sections. We have

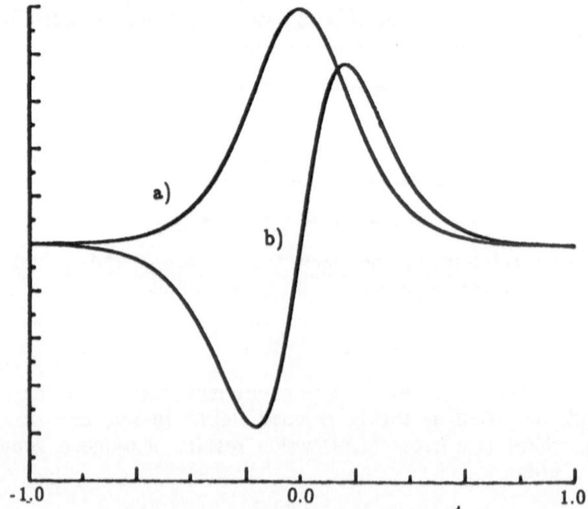


Fig. 1.1 a) The pulse intensity for a sech shaped pulse, b) the instantaneous frequency shift acquired as a result of self phase modulation.

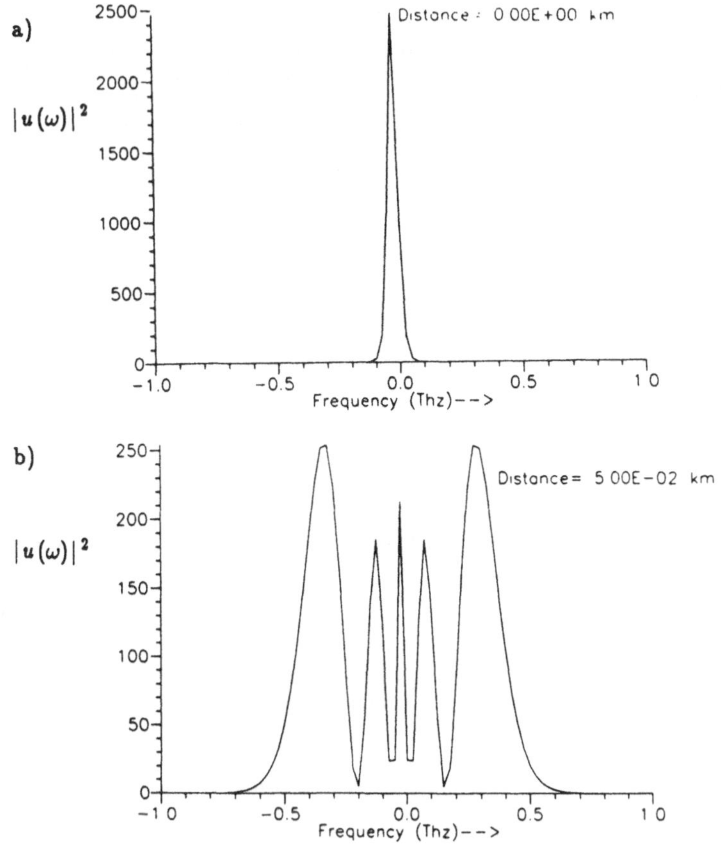


Fig. 1.2 a) Power spectrum of a sech shaped pulse and b) the power spectrum which results from the effect of self phase modulation.

already discussed the solution of equation (1.20) in the absence of the nonlinearity, let us now discuss the solution in the absence of dispersion. The simple nonlinear evolution equation for the pulse $u(T,Z)$ is now

$$i\frac{\partial u}{\partial Z} = \frac{n_2 k_0}{A_{eff}}|u|^2 u. \tag{1.21}$$

This equation can be solved for an arbitrary initial pulse since there is no time dependence. We obtain

$$u(Z,T) = u_0(T)\, exp[i(\frac{n_2 k_0}{A_{eff}}|u_0|^2 z)] \tag{1.22}$$

where $u_0(T)$ is the initial pulse. The nonlinear phase, $\Phi$, acquired during propagation is given by

$$\Phi = \frac{n_2 k_0}{A_{eff}} L |u|^2. \tag{1.23}$$

The local phase is simply proportional to the local intensity which results in a frequency (derivative of the phase) chirp across the pulse. This effect is called self-phase-modulation (SPM) and is illustrated in figure (1.1) where we show schematically the pulse ($Asech(t)$ is used as an example) and the nonlinear 'chirp'. In figure (1.2) the associated real power spectrum for a pulse ($7psFWHM, 100Wpeak$) both before and after propagation in $50m$ of (zero dispersion) optical fibre is shown. It should be noted that the power spectrum broadens as well as acquiring structure during propagation. The nonlinear pulse propagation thus generates new frequencies; this is the very real effect of SPM.

Having discussed the linear and nonlinear parts of equation (1.20) we return now to the full NLS. This equation is not so easily integrated as equation (1.21), however considerable progress can still be made on the solutions. The important observation is that for the + sign there is an exact solution

$$U(t,z) = \exp(i\frac{z}{2})\ sech(t), \tag{1.24}$$

which can easily be verified by direct substitution in equation (1.20). Equation (1.24) represents the simplest **SOLITON** solution of the NLS.

Solitons are nonlinear waves which propagate without changing form (or a change which is at most periodic) and do not mutually interact. The first property is clear from (1.24) since $|U|$ is not z dependent. The latter (non interacting) property is not obvious and indeed not necessary for the suppression of dispersion, this property will be discussed further in the next section. Solitons can be thought of as the nonlinear eigenmodes of the NLS.

Since the equation is nonlinear there is, therefore, a certain amplitude required in order to create a soliton of a given width. From the transformations (1.19) the peak intensity in Watts is given by

$$P(Watt) = \frac{1}{\tau^2}(\frac{\beta'' A_{eff}}{n_2 k_0}). \tag{1.25}$$

This is one of the two important scales which need to be considered in optical fibre soliton propagation. The other scale is the soliton period which is the length scale on which the soliton phase evolves and is related to the dispersion length i.e. the length over which a linear pulse doubles in width (see equation (1.10)). The soliton period in normalised units is $z = \frac{\pi}{2}$. For a standard fibre with a dispersion of $15ps/nm/km$ for pulses of sech width $\tau$ in picoseconds the formulae are

$$P(Watts) = \frac{6.9}{\tau^2}\ ,\ E(pJ) = \frac{13.6}{\tau}\ ,\ z(km) = 0.009\tau^2. \tag{1.26}$$

Thus soliton powers should be available from modern high power lasers producing pulses of a few picoseconds and the effects of dispersion suppression readily observable over propagation distances of order a kilometre. Such observations were first made in 1980 [3] and are now being observed with semiconductor laser sources coupled to optical fibre amplifiers to increase the power [4].

The solution (1.24) can be formally obtained by separating the $z$ and $t$ dependence to obtain an ordinary differential equation for the pulse shape. However, the single soliton is the only solution which can be obtained by this method. The way to make further progress was identified by Zakarov and Shabat [5]. They used the method of inverse scattering theory to solve the initial value problem for the NLS. The procedure is a nonlinear analogue of the way that Fourier transforms are used to solve the linear dispersion problem. The NLS is, in some sense, transformed to obtain a description of the input pulse in terms of the eigenvalues and eigenfunctions of a scattering problem. The evolution of these eigenvalues is described by a linear system of equations. Finally, the output pulse can be reconstructed from the inverse scattering equations. In particular we can obtain a great deal of information about the initial value problem by using the direct scattering transform. We calculate the eigenvalues $\varsigma_j$ of the following equations

$$i\frac{\partial V_1}{\partial t} + U(t)V_2 = \varsigma V_1 \tag{1.27a}$$

$$i\frac{\partial V_2}{\partial t} + U^*(t)V_1 = -\varsigma V_2. \tag{1.27b}$$

Zakarov and Shabat showed that the imaginary part of the discrete eigenvalues gives the energies of the solitons which are contained within the pulse and the real part of the

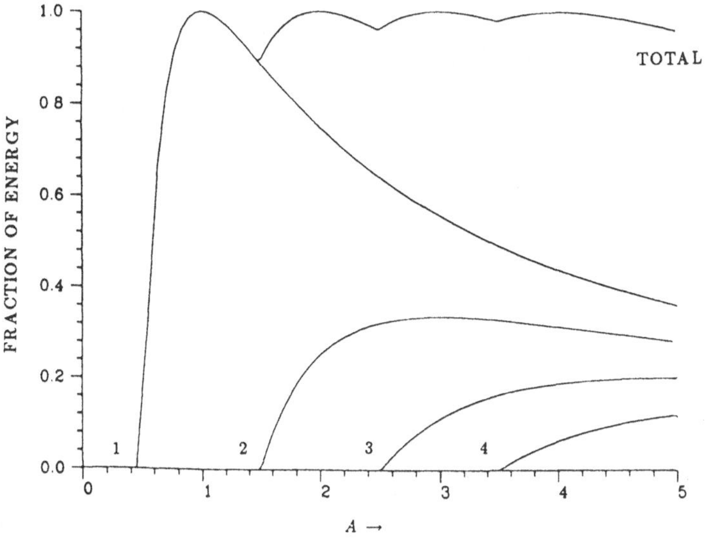


Fig. 1.3 Proportion of energy contained in the lowest four eigenvalues for a pulse $A\operatorname{sech}(t)$ as a function of $A$. Also shown is the total energy contained in the solitons.

eigenvalue gives the velocity of the soliton. The soliton energy will be less than or equal to the total pulse energy and indeed can be zero if no solitons are created. The eigenvalues are the constants of the evolutionary motion and are analogous to the frequencies in the linear dispersion problem. The number of discrete eigenvalues indicates the number of solitons in the initial pulse and we can think of any pulse as a superposition of solitons with each soliton corresponding to one of the eigenvalues together with some 'radiation energy' which disperses during propagation. Inverse scattering allows us to invert the process and determine the potential (i.e pulse shape) from which a particular eigenvalue can be obtained. Thus soliton solutions can be obtained for any arbitrary energy. Satsuma and Yajima [6] studied equations (1.27) in detail and derived some exact solutions for the eigenvalues and eigenfunctions. This is a simple scattering problem to solve numerically with $U(t)$ the potential and $\varsigma$ the eigenvalues and indeed is an extremely useful method of determining the soliton content of a pulse. As an example of the use of equation (1.27) we consider the initial pulse $A\operatorname{sech}(t)$ [6]. When $A = 1$ we have a single soliton and indeed at all integer values the pulse is made up of that number of solitons. At arbitrary values of $A$ the eigenvalues are purely imaginary and given by [6]

$$\varsigma_j = i(A - j + 1/2) \qquad j > 0 \ , \ \varsigma > 0. \tag{1.28}$$

In figure (1.3) we plot the proportion of pulse energy in each eigenvalue (soliton) and the total energy. It can be seen that for large $A$ a high proportion of the energy is contained in the solitons and little 'radiation' (non-soliton) remains.

We have seen that shape maintaining solutions exist to the simple nonlinear pulse evolution equation for optical fibres in the anomalous dispersion regime. Any arbitrary pulse consists of a number of solitons, the fraction of the pulse energy contained by these solitons indicates the fraction of the pulse energy that will perform soliton type propagation. The scattering problem allows us to determine both the number and velocities of the solitons. In fact for bright pulse propagation in optical fibres all solitons travel with the linear group velocity appropriate to their frequency. These properties, and many other solutions of the NLS, will be discussed in the next section.

Table (2.1)

| | $\alpha = +1$ | $\alpha = -1$ |
|---|---|---|
| $\lvert U \rvert \to 0$<br>$\lvert t \rvert \to \infty$ | Bright Solitons | Enhanced<br>Pulse Broadening |
| $\lvert U \rvert \to 1$<br>$\lvert t \rvert \to \infty$ | Modulation Instability<br>Bright Solitons | Dark Solitons |

## 2. SOLUTIONS OF THE NLS AND THE EFFECT OF PERTURBATIONS

### 2.1 Introduction

In the first section we discussed the physics of linear dispersion resulting from the frequency dependence of the fundamental mode propagation constant and self phase modulation which arises from the intensity dependence of the refractive index. We also showed that when these two effects occur together the propagation of light in an ideal single mode optical fibre is described by the Nonlinear Schrodinger Equation

$$i\frac{\partial U}{\partial z} + \frac{\gamma}{2}\frac{\partial^2 U}{\partial t^2} + |U|^2 U = 0 \tag{2.1}$$

where the coefficient $\gamma$ is +1 for negative group velocity dispersion and -1 for positive group velocity dispersion. In order to solve this equation we treat it as an evolution equation in the variable $z$ with the appropriate boundary conditions in time and we will consider two choices. The first is $|U| \to 0$ , $t \to \infty$ which is used for pulses of light and the second is $|U| \to 1$ , $t \to \infty$ which is used for modulated continuous wave (CW) beams. In table (2.1) we show, in matrix form, the four possible combinations of the sign of $\alpha$ and the choice of boundary condition.

The range of effects which are covered include solitons, instabilities, and enhanced nonlinear pulse broadening. In this section we will concentrate on these effects and give some examples of their observation. Finally we will consider the effect of the real world and discuss the perturbations which affect the NLS in practice and the influence they have on the soliton solutions.

### 2.2 Bright solitons

Bright solitons were mentioned in the first section as a simple solution of the NLS when $\alpha = +1$ and we have the boundary condition $|U(t)| \to 0$, $|t| \to \infty$. The simplest soliton (equation 1.24) is a single pulse which propagates without change in shape along the fibre. In general, equation (1.24) is one of a continuous family of single solitons which are given by

$$U(z,t) = \beta e^{i\beta^2\frac{z}{2}} sech(\beta t) \tag{2.2}$$

where the parameter beta is the inverse pulse width and is connected with the scaling properties of the NLS as discussed in the first section. The total normalised energy, $E$, of a pulse $U(t)$ is

$$E = \int |U|^2 dt \tag{2.3}$$

which for the single soliton, equation (2.2), is $2\beta$. This is a surprising result, at first sight, since it predicts that for longer pulses (small $\beta$) the soliton threshold is lower. If this is the case why is it so difficult to observe them in practice? The answer to this is contained in the phase factor in equation (2.2) which reveals that the length scale on which the soliton is observed is proportional to $\beta^{-2}$ which becomes increasingly large as we go to

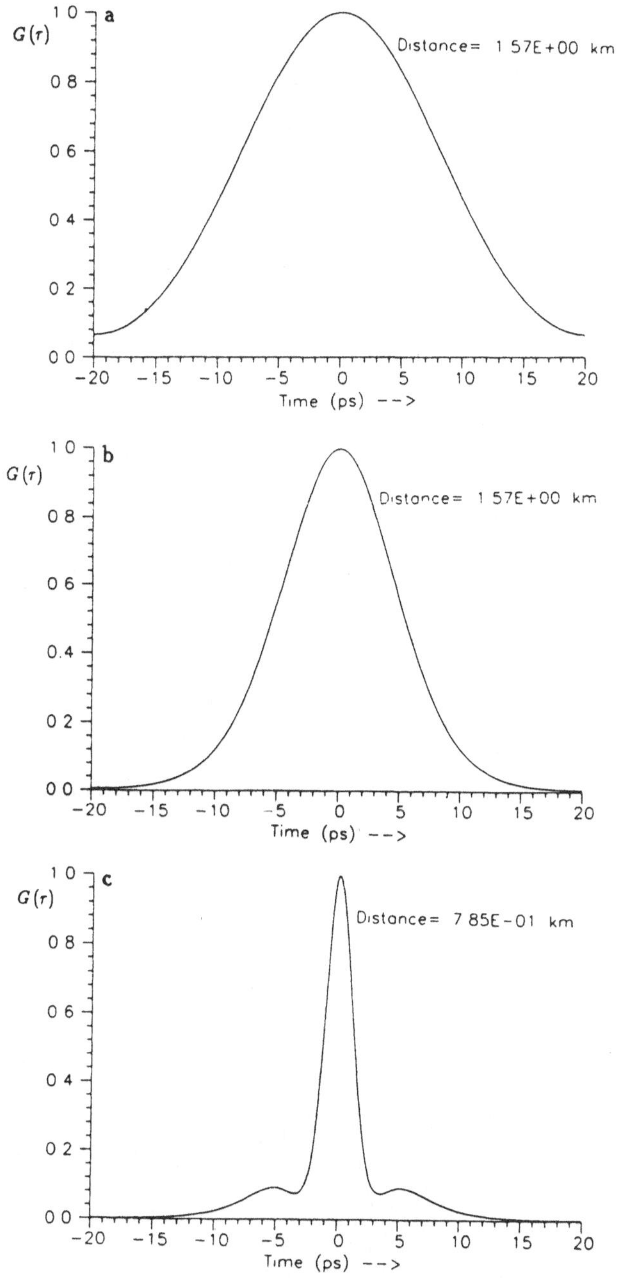


Fig. 2.1 Autocorrelation functions for a) a pulse whose power is below the soliton power, b) a single soliton and c) an $N = 2$ soliton at half the soliton period.

smaller $\beta$. In order to see this more clearly consider the following example: for a pulse with a FWHM of $7ps$ in a standard communication grade optical fibre the peak power of the single soliton at $1.55\mu m$ is $1W$ and the soliton length scale is about $1km$. If we had used $70ps$ pulses, which are much easier to produce, the soliton power would be $10mW$ but the length scale for observing the soliton would be $100km$ which is much longer than the absorption length, ignored so far, of the optical fibre. Hence, although the soliton would be easier to produce we would not be able to distinguish it from a linear pulse.

The first experimental observation of optical solitons was reported by Mollenauer, Stolen and Gordon in 1980 [3] in which they used a Tl:KCl colour centre laser. This laser produced pulses with a FWHM of $7ps$ and a peak power of $1W$ at $1.5\mu m$ which is a wavelength at which standard fibres have negative group velocity dispersion. Since it is difficult to measure pulses of this duration directly, say with a photodiode, an experimental technique known as autocorrelation was used which measures the following quantity

$$G(\tau) = \int_{-\infty}^{\infty} U(t)\, U(t + \tau)dt \tag{2.4}$$

and it is this quantity which is measured in the experiments. Figure (2.1a) shows the theoretical autocorrelation trace at the output when the input pulse is launched into the fibre below the power of the single soliton. The output pulse has broadened due to the effects of linear dispersion. When the input power is increased so that the pulse has a peak power of about $1W$ the output pulse has the same width as the input pulse as can be seen in figure (2.1b). At higher launched powers the output pulse contains further structure and to understand this we must return to the properties of soliton solutions of the NLS.

The single bright soliton, given by equation (2.2), is the simplest of the stationary soliton solutions of the NLS. The most general single soliton has a non-zero velocity in the chosen co-moving frame of reference used to derive equation (2.1). The velocity is parameterised by a coefficient $\varsigma$ and the full equation for the single soliton is given by

$$U(z,t) = \beta e^{i(\beta^2 - \varsigma^2)\frac{z}{2} + i\varsigma t + i\phi} sech(\beta(t - t_0 + \varsigma z)) \tag{2.5}$$

and we have also included an arbitrary phase $\phi$ and initial position $t_0$. The parameters $\varsigma$ and $\beta$ are related to the real and imaginary parts of the Zakharov-Shabat eigenvalue as

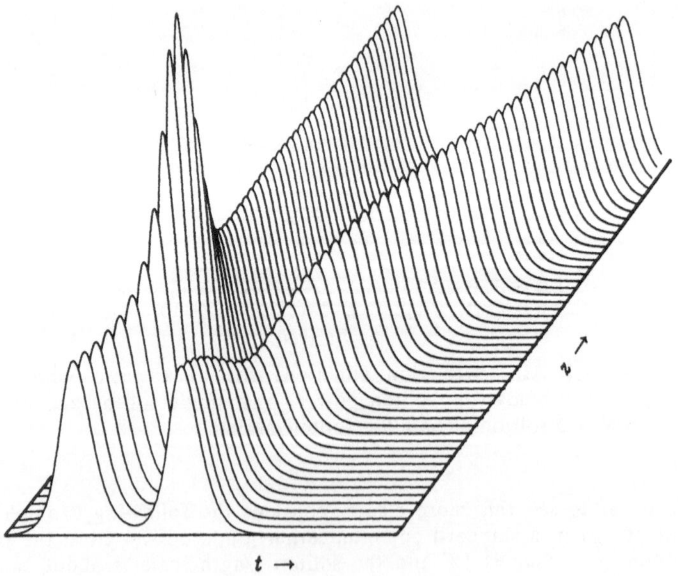


Fig. 2.2 Evolution of two solitons with different frequencies. The two pulses pass through each other with some distortion during the collision but emerge unscathed.

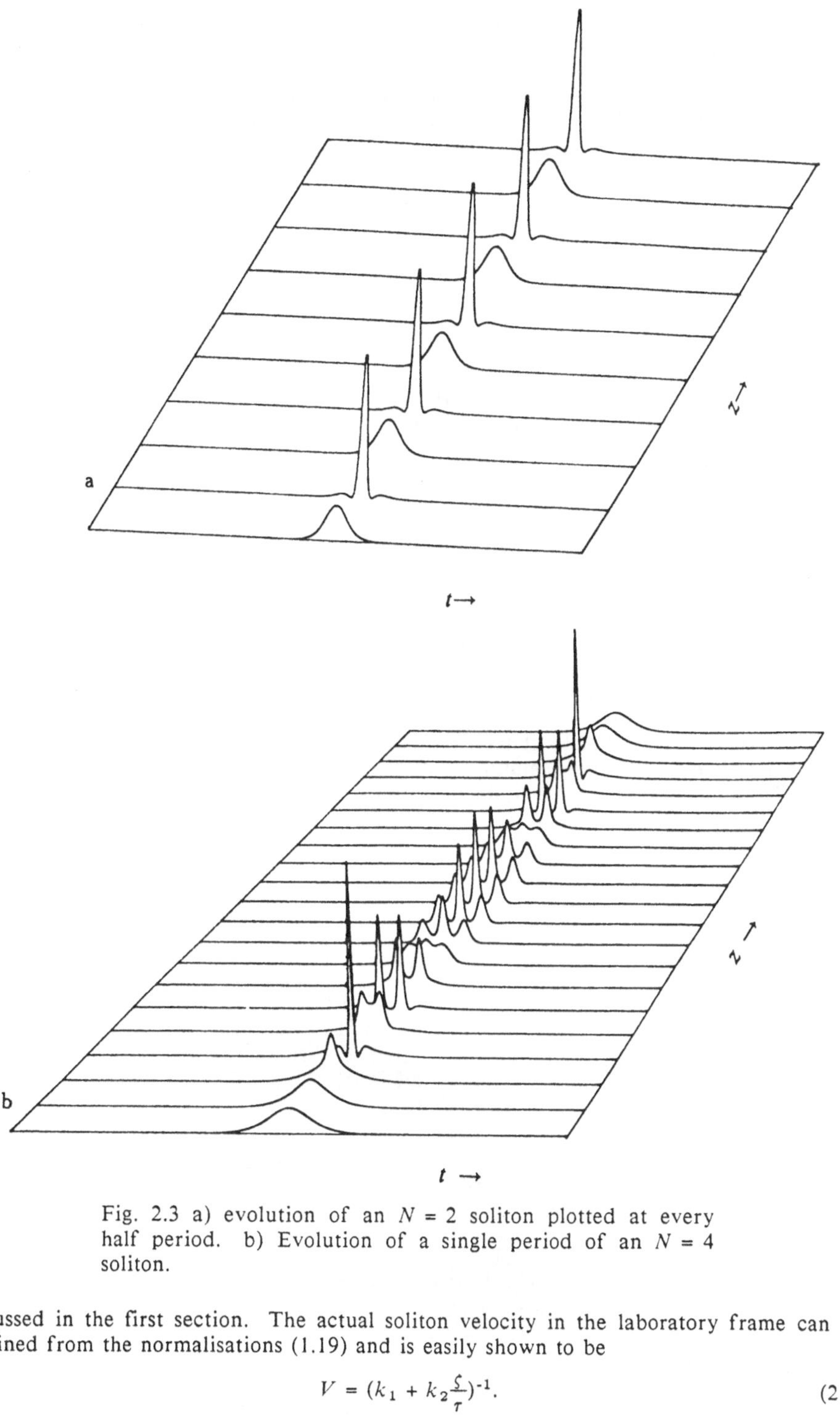


Fig. 2.3 a) evolution of an $N = 2$ soliton plotted at every half period. b) Evolution of a single period of an $N = 4$ soliton.

discussed in the first section. The actual soliton velocity in the laboratory frame can be obtained from the normalisations (1.19) and is easily shown to be

$$V = (k_1 + k_2\frac{\zeta}{\tau})^{-1}. \tag{2.6}$$

We can see from this expression that $\frac{\zeta}{\tau}$ is the difference between the frequency of the pulse and the chosen centre frequency used in the slowly varying envelope approximation used in the first section. Thus, this new velocity is exactly what one would expect for a linear pulse and is not a special feature of the NLS.

We can now look at the properties of collections of single solitons, initially well separated, by using the parameter $t_0$ in equation (2.5). Let us look at the evolution of two single solitons of the same width ($\beta = 1$) but travelling with different velocities in the co-moving frame. In figure (2.2) we show the collision of two such solitons with $t_0 = 5$ and $t_0 = 0$. Although the interaction is complicated, with strong pulse shaping occurring, the pulses pass through each other and emerge unchanged. This is one of the most important properties of solitons and is a central result of inverse scattering theory which was mentioned briefly in the first section. The only change which has occurred as a result of the collision is a sideways displacement of the pulse trajectory. In general we can construct superpositions of many solitons providing they are not placed too close to each other initially.

So far we have discussed the evolution of solitons when they have different velocities. In this case the interactions between the solitons are finite in number as the asymptotic state will be an ordered sequence of the solitons with the fastest appearing first and the slowest last with no further interaction. The final class of soliton solutions arises from superpositions of solitons travelling with the same velocity which we arbitrarily take to be zero. Consider the following initial condition

$$U(z = 0,t) = N sech(t) \tag{2.7}$$

which is the superposition of $N$ single solitons. In figure (2.3) we show the evolutions for $N = 2$ (2.3a) and $N = 4$ (2.3b), both evolutions are periodic with a period $\frac{\pi}{2}$. These soliton states show strong pulse structure with increasing complexity as $N$ is increased. This can be seen quite dramatically in figure (2.4) where we show the evolution of an $N = 10$ soliton for a single period. At the mid point of the evolution the pulse has split into nine sub pulses but these eventually perfectly recombine to reconstruct the initial pulse shape. In fact this reconstruction occurs for all integer values of $N$ and these solutions are known as bound states of single solitons. In principle we can obtain the analytic solution for the pure multiple soliton state from inverse scattering theory; the simple $N = 2$ bound state has been obtained [6] and is given by

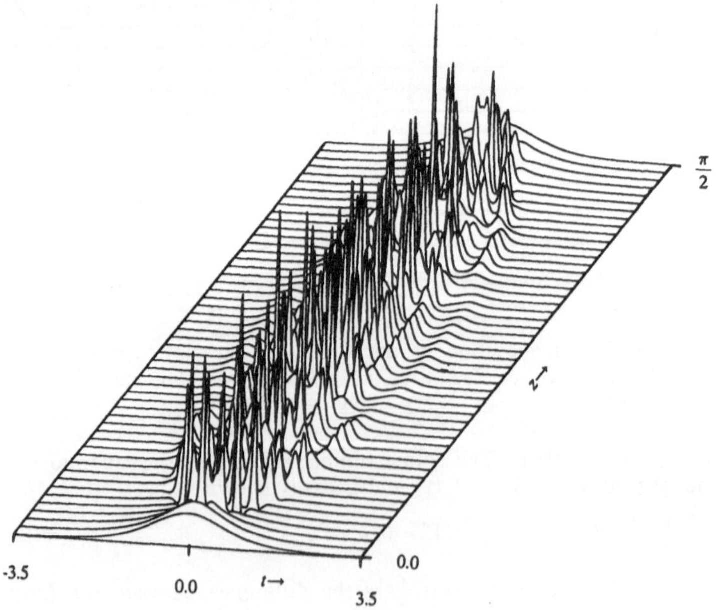


Fig. 2.4 Evolution an an $N = 10$ soliton for a single period.

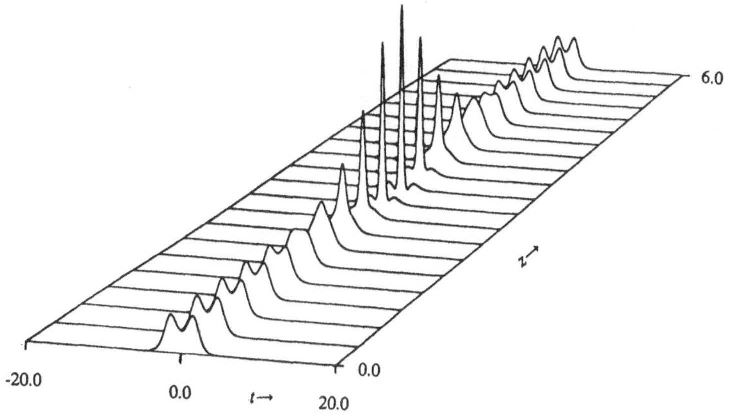


Fig. 2.5 Evolution of two pulses with the same frequency over one cycle of the periodic motion.

$$U(z,t) = \frac{4e^{-i\frac{z}{2}}[cosh(3t) + 3e^{-4iz}cosh(t)]}{cosh(4t) + 4cosh(2t) + 3cos(4z)} \tag{2.8}$$

It was this higher order solution of the NLS that Mollenauer et al observed and its autocorrelation function at $z = \frac{\pi}{4}$ is shown in figure (2.1c).

The bound states which evolve from the initial condition (2.7) are superpositions of solitons at the same time. Let us now consider the superposition of solitons at different times but still with identical velocities. Such states are described by an initial condition of the form

$$U(z = 0,t) = sech(t - \Delta) + sech(t + \Delta) \tag{2.9}$$

which is only an exact soliton bound state in the limits $\Delta = 0$, where we have an $N = 2$ bound soliton, and $\Delta \rightarrow \infty$, where we have two independent single solitons. The evolution from equation (2.9) is shown in figure (2.5) for $\Delta = 1.5$ and consists of a repetitive sequence of collapse followed by separation. This mode of oscillation is often referred to as a breather soliton and the period of the breathing motion is approximately equal to $\frac{\pi}{2}e^{\Delta}$ [7]. Although equation (2.9) does not give an exact two soliton solution, it has been shown [8] that the difference between equation (2.9) and the exact soliton initial-state is extremely small.

The properties of bright solitons we have so far discussed can also be expressed in terms of the direct scattering eigenvalues covered in the first section, equation (1.28). The single soliton, equation (2.5) has an eigenvalue of $(\varsigma + i\beta)/2$ so that the speed is related to the real part and the energy is related to the imaginary part. The $N$ soliton bound state has $N$ imaginary eigenvalues in the sequence $\frac{1}{2}, \frac{3}{2}, \ldots \frac{2N-1}{2}$. We can now calculate the soliton energy in two ways, first from equation (2.3) we obtain

$$E = \int N^2 sech^2(t)dt = 2N^2, \tag{2.10}$$

second the expression in terms of the eigenvalues of equation (1.27) is

$$E = 4\sum_i \text{Im}\ \varsigma_i = 4\sum_{i=1}^{N} \frac{2i-1}{2} = 2N^2 \tag{2.11}$$

which gives the same value showing that all the energy is contained in the solitons.

## 2.3 Dark solitons

Dark solitons are pulses of darkness in a beam of constant intensity. They occur as solutions of the NLS when the dispersion has the opposite sign to that for the bright solitons, $\alpha = -1$, and we use the boundary condition $|U(t)| \to 1$, $|t| \to \infty$. The simplest dark soliton is given by

$$U(z,t) = e^{iz} tanh(t) \tag{2.12}$$

which has the same property of the single bright soliton of being a self maintaining pulse which does not change its intensity distribution as it propagates. The principle difference between the two types of soliton is that the dark soliton changes sign across the pulse or, equivalently, the pulse contains an abrupt $\pi$ phase change. When we consider the general single soliton the situation is slightly more complicated than for the bright solitons, the equation is

$$U(z,t) = \frac{(\varsigma - i\nu)^2 + e^{2\nu(t - t_0 - \varsigma z)}}{1 + e^{2\nu(t - t_0 - \varsigma z)}} \tag{2.13}$$

where $\varsigma^2 + \nu^2 = 1$ and $0 < \varsigma < 1$. Thus, there is now a relationship between the dark pulse width parameter $\nu$ and the velocity parameter $\varsigma$ which were independent quantities for the bright solitons. Moreover, there are no solutions analogous to the multiple bright soliton bound states. For non-zero $\varsigma$ the pulse has a more gradual phase change which is smaller than $\pi$ and the intensity does not go to zero at the centre of the soliton.

Inverse scattering theory also gives a surprising result which is that no two dark solitons can have the same scattering eigenvalue $\varsigma$. Therefore, if we try to create many dark soliton states they will necessarily have different velocities, not particularly useful for communications purposes. It is possible, as for the bright solitons, to calculate exactly, in principle, the many soliton states [9] although these are no less complex for the dark solitons. In figure (2.6) we have plotted the evolution of a pair of dark solitons with equal and opposite velocities. The solitons again pass through each other but the interaction is much simpler than for the bright solitons. These pulses of darkness have been observed experimentally [10] using a spectral pulse shaping technique [11] to obtain the appropriate intensity distribution and phase shift. At low powers the dark pulse broadens due to linear dispersion in exactly the same way as the bright pulse would (use the superposition principle which is valid in the linear regime). As the power is increased the pulse dispersion is reduced until the balance point is reached.

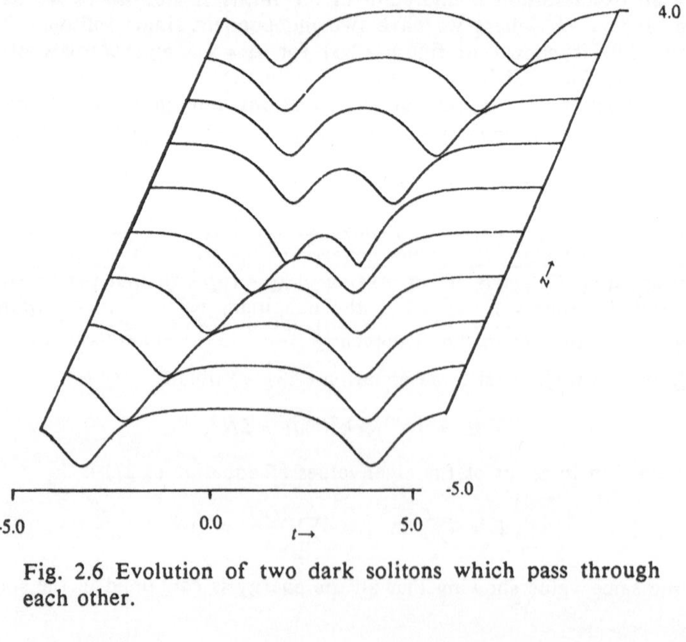


Fig. 2.6 Evolution of two dark solitons which pass through each other.

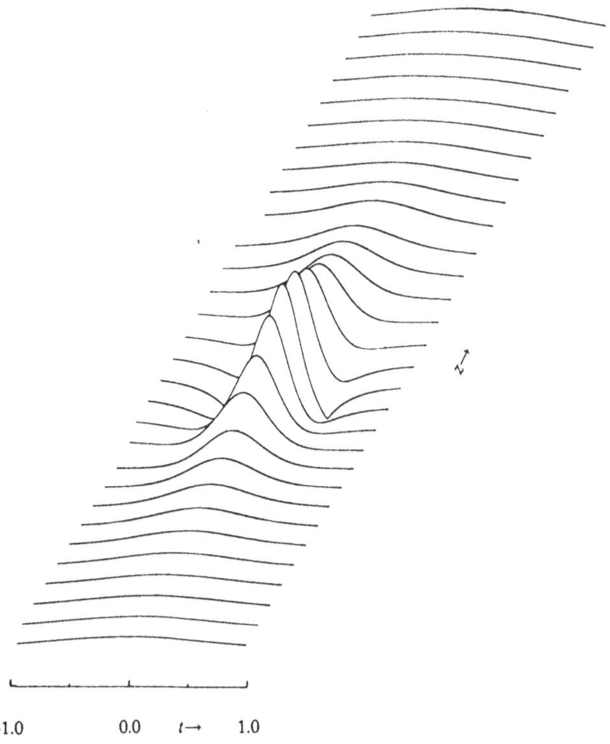


Fig. 2.7 Development of the modulation instability on a weakly modulated CW signal. The return to a weakly modulated signal is an example of Fermi-Pasta-Ulam recurrence.

## 2.4 Modulation instability

If we return to the bright soliton dispersion regime but change the boundary condition to $|U(t)| \to constant$, $|t| \to \infty$ then we obtain a different type of evolution. It turns out that bright solitons can also be obtained in this case [12] but these are similar to those already discussed and will not be considered further. We are interested in an effect known as a modulation instability which was first discussed for the NLS by Benjamin and Feir [13] in a study of deep water waves. We begin by weakly modulating the plane wave (self phase modulation) solution of the NLS as follows

$$U(z,t) = Ae^{i|A|^2 z}(1 + \varepsilon(z)e^{i\delta t} + \mu(z)e^{-i\delta t}) \tag{2.14}$$

$\delta$ being the frequency of the modulation. Substituting this into the NLS and linearising in the small quantities $\varepsilon$ and $\mu$ we obtain

$$i\frac{d\varepsilon}{dz} = (\frac{\delta^2}{2} - |A|^2)\varepsilon - |A|^2\mu^* \tag{2.15a}$$

$$i\frac{d\mu^*}{dz} = |A|^2\varepsilon - (\frac{\delta^2}{2} - |A|^2)\mu^* \tag{2.15b}$$

which are the evolution equations for the modulation amplitudes. Equations (2.15) are a two dimensional eigenvalue equation so we substitute $e^{i\Lambda z}$ for $(\varepsilon, \mu^*)$ and obtain the following equation for the eigenvalues

$$\Lambda = \frac{\delta}{2}\sqrt{\delta^2 - 4|A|^2}. \qquad (2.16)$$

When $\Lambda$ is imaginary the sideband amplitudes are unstable. The condition for this instability is $0<|\delta|<2|A|$ and the maximum growth rate (gain) occurs for $\delta = \sqrt{2}|A|$ where it takes the value $|A|^2$. This effect may be important in certain forms of optical transmission [14] where the data is sent as a modulated carrier wave. We can use the physical parameters already mentioned in this section to estimate the scale of frequencies involved in this effect. The single soliton data give a peak power of $1W$ for $|A| = 1$ when the time scale is $4ps$. Thus the peak of the gain occurs at a frequency of $\frac{\sqrt{2}}{8\pi}THz$ which is about $50GHz$ and the gain length would be about $1km$. In figure (2.7) we show

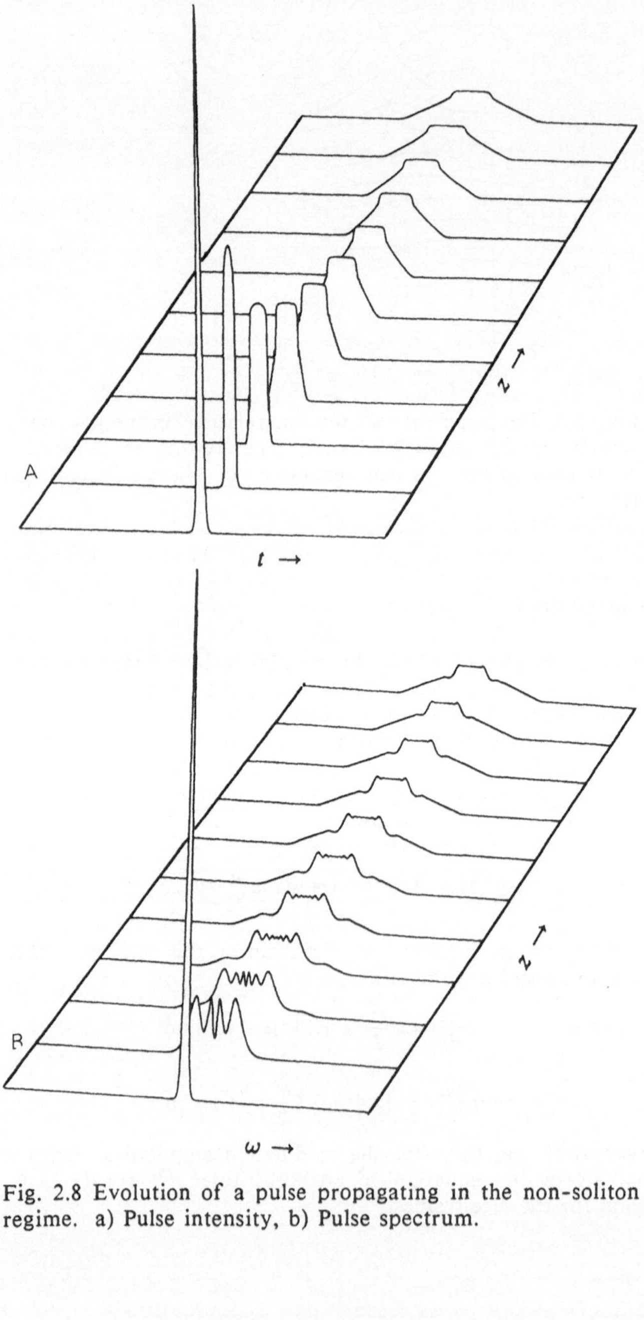


Fig. 2.8 Evolution of a pulse propagating in the non-soliton regime. a) Pulse intensity, b) Pulse spectrum.

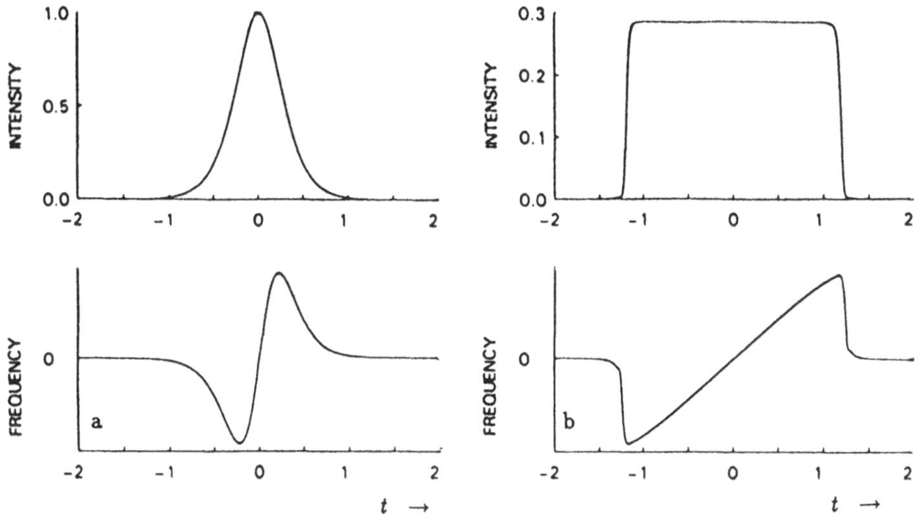


Fig. 2.9 Comparison of pulse shapes and chirps produced by a) self phase modulation alone and b) propagation in the non-soliton regime.

the evolution of a plane wave weakly modulated at the frequency of the maximum instability obtained by exact integration of the NLS. The modulation grows exponentially at first but when the modulation amplitude is comparable with the plane wave amplitude the instability is suppressed and the field eventually returns to its weakly modulated state. Of course, the linearised equations (2.15) only describe the initial stages of the evolution and do not account for the effects of pump depletion which gives rise to the saturation seen here. The return to the initial state is an example of a phenomenon known as Fermi-Pasta-Ulam recurrence [15].

Finally we note that the modulation instability is related to the very high order soliton evolution at short distances. If we consider a long pulse so that the soliton period is large compared to the propagation distance then no nonlinear effects will be seen until the pulse intensity is sufficiently high that soliton structure is seen at a small fraction of the soliton period, see, for example, the $N = 10$ evolution in figure (2.4) where the first structure appears at about one tenth of the soliton period. This pulse may also be considered as quasi-CW and the structure interpreted as the modulation instability occurring near the centre of the pulse. In general the modulation instability is observed using high peak power, long pulses where the effective soliton number may be several hundred.

### 2.5 Enhanced pulse broadening

Let us return to the dispersion regime where it is possible to observe the dark solitons and find out what happens to pulses of light. We have already pointed out at the start of this section that the chirp induced by dispersion and self modulation are of the same sign in this regime. Thus we would expect the two effects to reinforce each other as indeed is the case. In figure (2.8) we show the temporal and spectral evolution of the pulse, both of which show a marked broadening. In the early stage of the evolution the broadening occurs predominantly in the spectrum as a result of self phase modulation. As the propagation progresses the effect of dispersion becomes more important as the spectral width increases and eventually most of the broadening occurs in the time domain. These nonlinear propagation effects are characteristic of this dispersion regime for all values of $|A|$ providing a sufficiently large distance is available. Let us now compare the chirp acquired by the pulse with that due to self phase modulation alone. This is shown in figure (2.9), the combined effects lead to a more linear chirp across the pulse. The linear chirp can be used to compress the pulse by propagation in a linearly dispersive medium with the opposite sign of dispersion. The nonlinearity enables us to achieve an overall compression of the original input pulse and this is shown in figure (2.10). In practice pulse compression ratios of greater than 100 are readily achievable.

## 2.6 Perturbations

So far we have discussed the range of solutions and effects which can be obtained for the ideal NLS. Although equation (2.1) is a very good description of nonlinear pulse propagation in optical fibres, it is incomplete. In a real fibre we have to account for the effects of loss, higher order dispersion, stimulated Raman scattering and the effects of non-ideal sources of light. The simplest example of a perturbation which we will briefly describe is absorption. In a real fibre the energy of an optical pulse decreases as it propagates due to various effects including, Rayleigh scattering from the intrinsic density fluctuations in the glass, phonon scattering in the Urbach tail and residual water vibrations. A good optical fibre has a loss of about $0.2dBkm^{-1}$ at $1.5\mu m$ which corresponds to an absorption length of about $15km$. The NLS can be modified to include this effect and the equation becomes

$$i\frac{\partial U}{\partial z} + \frac{\alpha}{2}\frac{\partial^2 U}{\partial t^2} + |U|^2 U = -i\frac{\Gamma}{2}U. \tag{2.17}$$

As an example we will consider the effect of the absorption on the propagation of bright solitons. To do this we must resort to a numerical solution of equation (2.17) as we no longer have any results from inverse scattering theory to use.

At this point we need to make some mention of numerical solutions of the NLS and related equations. The simplest technique is to use a finite difference approximation to the derivative terms in equation (2.17). Thus we replace

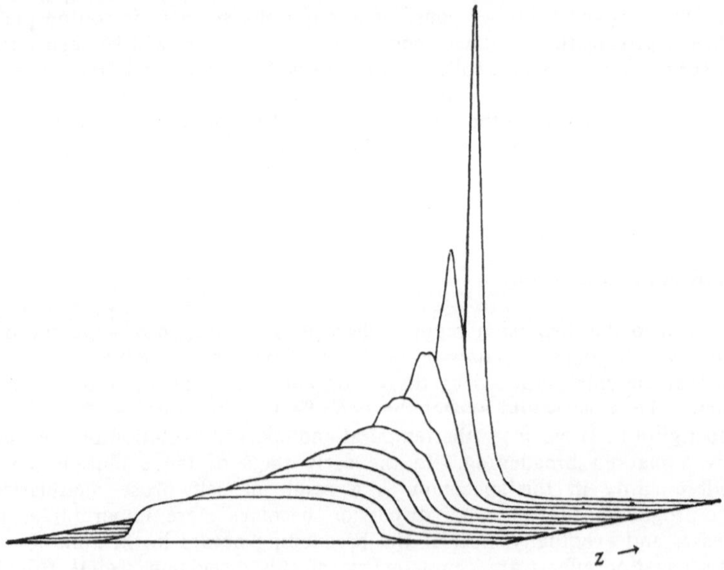


Fig. 2.10 Pulse compression of the pulse in figure 2.9b by propagation in a linear dispersive medium.

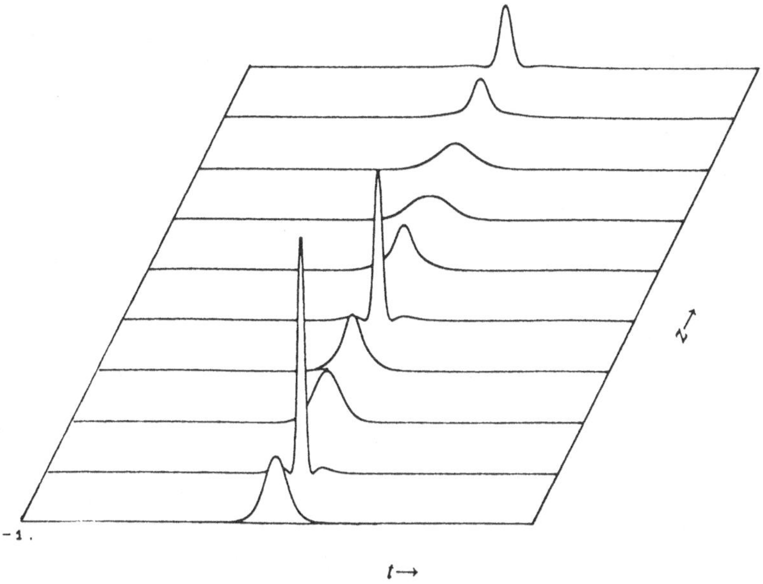


Fig. 2.11 Evolution of a single soliton in the presence of loss.

$$\frac{\partial U}{\partial z} \rightarrow \frac{U(z + dz) - U(z)}{dz} \tag{2.18a}$$

$$\frac{\partial^2 U}{\partial t^2} \rightarrow \frac{U(t + dt) - 2U(t) + U(t - dt)}{dt^2} \tag{2.18b}$$

to obtain a discrete representation of the field $U$ at times separated by the grid spacing $dt$ and at subsequent points down the fibre separated by the integration step $dz$. Although this method is very simple to use, a computationally better technique is the split step fourier method. In this procedure equation (2.17) is replaced by two equations which are integrated alternately by the same step length, these two equations are

$$i\frac{\partial U}{\partial z} + \frac{\alpha}{2}\frac{\partial^2 U}{\partial t^2} + i\frac{\Gamma}{2}U = 0 \tag{2.19a}$$

$$i\frac{\partial U}{\partial z} + |U|^2 U = 0 \tag{2.19b}$$

where we now use the fact that both equations (2.19) can be integrated exactly in frequency space (2.19a) and in real space (2.19b) respectively. The transfer from the time to the frequency domain being achieved by using fast fourier transforms which are extremely accurate and efficient computational procedures. In practice we use combinations of step lengths to increase the order of accuracy of the integration and this is described in more detail elsewhere [16].

The evolution of the $N = 2$ soliton in the presence of loss is shown in figure (2.11), the value of $\Gamma$ used in this calculation was 0.02. The oscillatory motion of the $N = 2$ soliton (see figure (2.3a)) is seen in the early stages of the evolution. However, the oscillation slows down and eventually freezes out at large distances. In general this is observed for all soliton states, so can we find some general argument to enable us to understand this result. To do this we need two features of the NLS which we have already discussed, the conservation laws in the presence of perturbations and the scaling properties. The pulse energy, conserved by the pure NLS, can easily be shown to evolve according to the following equation

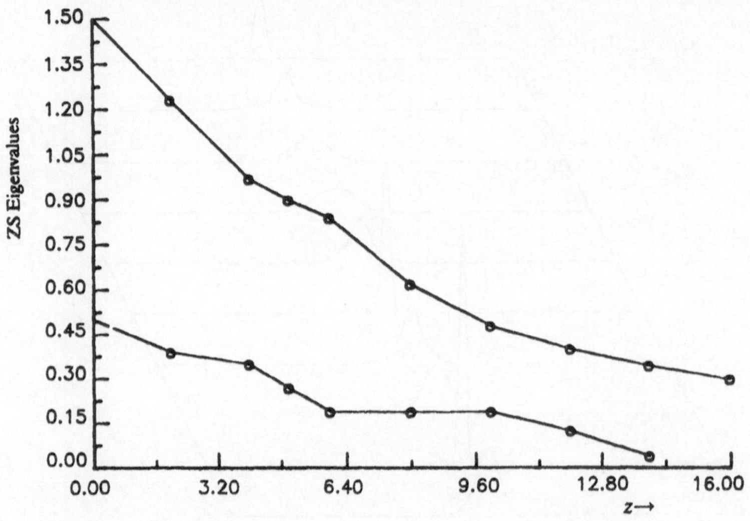


Fig. 2.12 Motion of the Zakharov-Shabat eigenvalues in the presence of loss.

$$\frac{dE}{dz} = -\Gamma E \tag{2.20}$$

which has the solution

$$E = E_0 \, e^{-\Gamma z}. \tag{2.21}$$

We also note that the single soliton given in equation (2.2) has energy $2\beta$ and that all soliton solutions are subject to the scaling relations (1.19) which would yield a similar linear dependence on the scale parameter $\beta$. Thus, the soliton solution with $\beta = 1$ could evolve by adiabatically adjusting $\beta$ to satisfy the energy equation (2.21). If we denote the $N = 2$ soliton solutions given in equation (2.8) by $U_2(z,t)$ then the adiabatically connected solution would be $\beta U_2(\beta^2 z, \beta t)$ with energy $8\beta$. Our adiabatically evolving solution in the presence of loss is finally obtained by replacing $\beta$ by $e^{-\Gamma z}$. Since the pulse width is $\beta^{-1}$ we can see that the adiabatic solution would have a width which increases exponentially. This broadening effect can be seen clearly in the early stages of propagation in figure (2.11) but it is clear that this situation cannot be maintained as the pulse energy would have to move outward with exponentially increasing speed. We can also see that the period of oscillation of the adiabatic solution is given by $\frac{\pi}{2}\beta^{-2}$ which increases with propagation distance, in other words the oscillation in pulse shape must slow down and presumably freezes out when the new period becomes longer than the absorption length. This is indeed what is observed in figure (2.11). Thus we have been able to use our knowledge of the properties and solutions of the pure NLS together with the energy equation to understand the general behaviour of solitons in the presence of loss.

If we return to the single soliton we can make one further step by using a WKB form for the total pulse phase

$$\Phi_{tot} = \int_0^z \phi(z')dz' \tag{2.22}$$

where

$$\phi(z) = -\frac{e^{-2\Gamma z}}{2} \tag{2.23}$$

and we obtain

$$\Phi_{tot} = \frac{(1 - e^{-2\Gamma z})}{4\Gamma}. \tag{2.24}$$

Using this result and the evolution of $\beta$ we can obtain an adiabatically evolving single soliton

$$U(z,t) = e^{-\Gamma z} e^{i\frac{(1 - e^{-2\Gamma z})}{4\Gamma}} sech(e^{-\Gamma z}t) \tag{2.25}$$

which is a result that can also be derived by applying perturbation theory to the direct scattering equations [17].

One further technique which is useful in studying the perturbed NLS is to follow the Zakharov Shabat eigenvalues during the evolution. This is easily done by solving equation (1.27) with the numerically calculated pulse $U(z,t)$. The results enable us to interpret the full solution in terms of the motion of the soliton eigenvalues. A typical evolution is shown in figure (2.12) for the $N = 2$ eigenvalues in the presence of loss $\Gamma = 0.1$. The eigenvalues decrease in magnitude, characteristic of the increase in the length scale of the evolution and broadening of the pulse. At about $z = 14$ the lowest eigenvalue ceases to exist and at this point the pulse evolution loses the characteristic $N = 2$ structure. This technique has been used in a number of problems to give extra insight into the complicated pulse dynamics that are observed.

## 3. STIMULATED RAMAN SCATTERING AND SOLITON RAMAN GENERATION

### 3.1 Introduction

There are many ways in which light can interact with matter. Perhaps the simplest is the coupling of the electromagnetic field to the electronic states. This coupling leads to many phenomena but for our purposes can broadly be split into two types.

1) Real transitions involve rearranging the electron distribution in a permanent way; for example, the electron may be raised into an excited state. Here, we are interested in silica based materials where the photon energy in the near infrared is small compared to the electron band gap so that there are no energetically allowed transitions available.

2) Virtual transitions involve states which are energetically forbidden. These transitions can be thought of as occurring by borrowing the extra energy $\Delta E$ required to make the transition from the uncertainty principle but only for a short time, of order $\hbar/\Delta E$ which is sub-femtosecond for silica, before returning to the initial state. These sorts of transition are responsible for the Kerr effect in silica in which the refractive index is a function of the optical intensity.

In this section we will be mainly concerned with the interaction between the electromagnetic field and the nuclear motion of the medium. The nuclear motion is described through the phonon modes with the acoustic and optic phonons giving rise to separate nonlinear optical effects. The acoustic phonon modes are responsible for Brillouin scattering which is mostly relevant to narrow linewidth sources [18]. Raman scattering, the subject of this section, is caused by the interaction of the light with the optic phonon modes. The scattered light is lower in frequency, as a result of emitting the phonon, and is mainly observed in the forward direction. In silica the high optical phonon frequencies lead to the observed frequency shift of the scattered light of $13.2THz$ which is extremely large and generally leads to non-phase matched stimulated Raman scattering (SRS). Finally we note that the phonon energy is much smaller than the optical energy which means that the optical energy is approximately conserved in the Raman interaction.

### 3.2 Stimulated Raman scattering

SRS is normally formulated for plane waves where one has to take into account the coupling to waves with k-vectors in different directions. The situation in single mode optical fibres is somewhat simpler since the k-vectors of the propagating modes are all parallel. Thus for our purposes we need only deal with the scalar amplitude of the transverse modes. We begin our discussion of SRS in single mode fibres with an heuristic derivation of the classical equations describing the process. We proceed from the wave equations for the optical and phonon fields and add nonlinear interaction terms of the appropriate frequency. The conservation of optical energy is then used to obtain a relationship between the nonlinear coefficients. The forward travelling wave equations including a generalised nonlinear interaction can be written as

$$i\frac{\partial E_L}{\partial z} = \beta_L E_L + G_L(\omega_L;E_L,E_S,Q) \tag{3.1a}$$

$$i\frac{\partial E_S}{\partial z} = \beta_S E_S + G_S(\omega_S;E_L,E_S,Q) \tag{3.1b}$$

$$i\frac{\partial Q}{\partial z} = \kappa Q + G_Q(\omega_Q;E_L,E_S,Q) \tag{3.1c}$$

where the functions $G$ can be calculated exactly from the free energy of the wave interaction. Here, we simply assume that the functions are quadratic in the fields and that the frequency of the interaction is the same as that of the left hand side of equations (3.1) by using $\omega_L - \omega_S = \Omega_Q$. This gives the following form for the interaction functions

$$G_L = g_L E_S Q \tag{3.2a}$$

$$G_S = g_S E_L Q^* \tag{3.2b}$$

$$G_Q = g_Q E_L E_S^* \tag{3.2c}$$

where, for example, we remember that the field $Q^*$ has frequency $-\Omega_Q$. Substituting equations (3.2) into equations (3.1) we obtain the following nonlinear evolution equations

$$i\frac{\partial E_L}{\partial z} = \beta_L E_L + g_L E_S Q \tag{3.3a}$$

$$i\frac{\partial E_S}{\partial z} = \beta_S E_S + g_S E_L Q^* \tag{3.3b}$$

$$i\frac{\partial Q}{\partial z} = \kappa Q + g_Q E_L E_S^* \tag{3.3c}$$

which describe the interaction of the pump, Stokes and optical phonon fields. At this point we can reduce the problem by working in the steady state regime where both the space and time dependence (implicit in the frequency dependence of $\kappa$) of the phonon motion are dropped. When this is done we obtain the simple relation for the phonon field from equation (3.3c)

$$Q = -\frac{g_Q}{\kappa} E_L E_S^* \tag{3.4}$$

where $\kappa$ is a constant evaluated at the phonon frequency. Thus, we can now use equation (3.4) to eliminate the phonon field from equation (3.3) to obtain our nonlinear evolution equations for the optical fields.

$$i\frac{\partial E_L}{\partial z} = \beta_L E_L - \frac{g_L g_Q}{\kappa}|E_S|^2 E_L \tag{3.5a}$$

$$i\frac{\partial E_S}{\partial z} = \beta_S E_S - \frac{g_S g_Q^*}{\kappa^*}|E_L|^2 E_S. \tag{3.5b}$$

We now have the equations for our optical fields which involve two nonlinear coefficients. In order to proceed we need to calculate the quantity conserved by equations (3.5) and compare this with the total optical energy. Multiplying equation (3.5a) by $E_L^*$ and adding this equation to its complex conjugate (similarly for equation (3.5b)) we obtain the evolution equations for the optical intensity

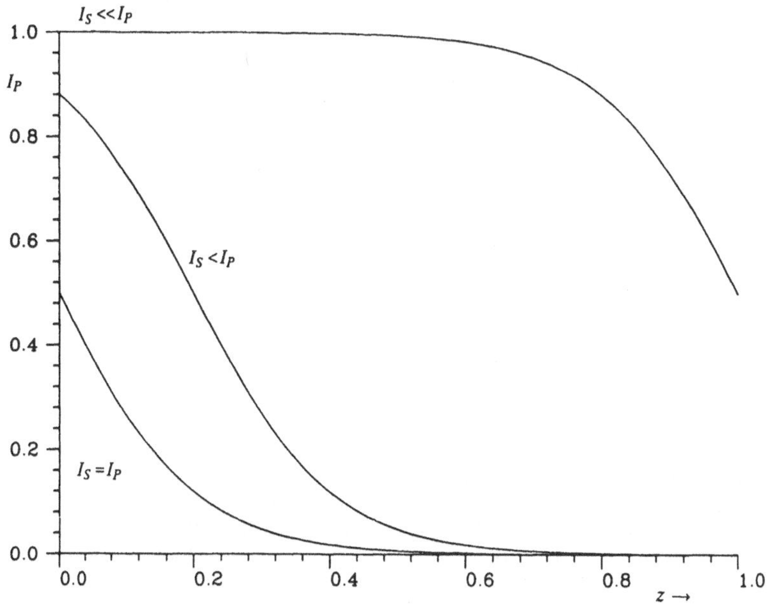


Fig. 3.1 Evolution of the pump intensity for various values of the initial Stokes intensity.

$$\frac{dI_L}{dz} = -\chi_L I_S I_L \tag{3.6a}$$

$$\frac{dI_S}{dz} = -\chi_S I_S I_L \tag{3.6b}$$

where $\chi_S = \text{Im}(g_S g_Q^*/\kappa^*)$ and $\chi_L = \text{Im}(g_L g_Q/\kappa)$. Multiplying equation (3.6a) by $\chi_S$ and (3.6b) by $\chi_L$ and subtracting we obtain the conserved quantity

$$\chi_L I_S - \chi_S I_L = constant. \tag{3.7}$$

We can now obtain the necessary relation between the nonlinear coefficients by imposing the condition that the total photon number is conserved. Photon number is defined as the energy in the optical wave divided by the photon energy $\hbar\omega$ and this condition can be expressed as

$$\frac{I_L}{\omega_L} + \frac{I_S}{\omega_S} = constant. \tag{3.8}$$

Comparing equations (3.7) and (3.8) we obtain the relation

$$\omega_L \chi_S = -\omega_S \chi_L \equiv -\omega_S \omega_L \chi_R. \tag{3.9}$$

Thus, we can finally write our equations describing CW SRS in the steady state regime

$$\frac{dI_L}{dz} = -\omega_L \chi_R I_S I_L \tag{3.10a}$$

$$\frac{dI_S}{dz} = \omega_S \chi_R I_S I_L. \tag{3.10b}$$

The solution of equations (3.10) is readily achieved by changing variables to

$$S = \frac{I_L}{\omega_L} + \frac{I_S}{\omega_S} \tag{3.11a}$$

$$D = \frac{I_L}{\omega_L} - \frac{I_S}{\omega_S} \tag{3.11b}$$

where $S$ is the total energy and is a constant and $D$ is the difference between the pump and Stokes energies. The equation for the evolution of the energy difference is

$$\frac{dD}{dz} = -\omega_S\omega_L\frac{\chi_R}{2}(S^2 - D^2) \tag{3.12}$$

which can easily be integrated to obtain the solution

$$D = S\tanh(-\frac{\omega_S\omega_R\chi_R S}{2}(z - z_0)) \tag{3.13}$$

where the parameter $z_0$ is defined by $D_0 = S\ tanh(\omega_S\omega_L\chi_R S z_0/2)$ and the fibre propagation begins at $z = 0$. In figure (3.1) we show some solutions for the pump and Stokes power for various values of the ratio $D_0/S$ with a constant value for $S$. When $I_S \ll I_P$ we observe exponential gain over the whole fibre length shown here. As $I_S$ is increased we observe linear gain (for the Stokes radiation) at small distances and complete transfer of the pump power at larger distances.

### 3.3 Dispersion and loss

So far, we have assumed that SRS can be described as an interaction between a CW pump and Stokes wave. In fibres SRS couples the pump to many Stokes waves as can be seen from the small signal gain spectrum shown in figure (3.2) [19]. Since the large signal gain is exponential, as we have shown, only the Stokes radiation near the peak of the spectrum is observed when we observe amplified spontaneous light. In practice SRS is not generated using a CW pump as the power levels required are too high. This problem can be overcome by using a pulsed source such as a Q-switched and modelocked laser. Figure (3.3) shows a typical spectrum of the light emanating from a fibre in the presence of SRS. There are many peaks corresponding to the higher order Stokes lines (i.e when the first

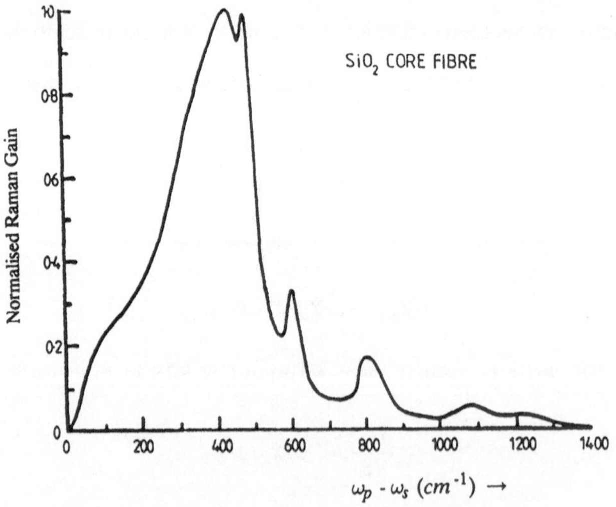


Fig. 3.2 Gain spectrum for stimulated Raman scattering as measured in a Silica core fibre.

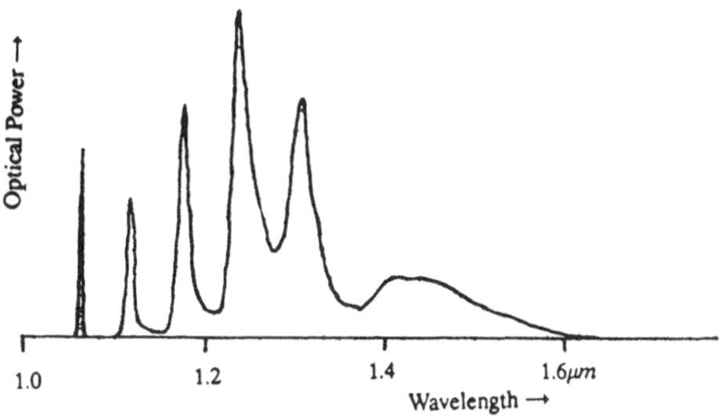


Fig. 3.3 Stimulated Raman scattering in an optical fibre well above the first Stokes threshold. Five cascaded orders can be seen.

Stokes line becomes sufficiently intense to pump its own Stokes band) with significant radiation between the peaks reflecting the width of the gain spectrum.

If we examine the solution of the SRS equations (3.13) we see that whatever the initial value of the pump and Stokes intensities we can always achieve 100% conversion to the Stokes frequency providing we have a long enough fibre. This is not observed in practice, and the explanation is contained in the linear term in equation (3.10). The linear term $\beta E$ contains the information about the fibre loss and dispersion which we have already discussed in the soliton sections. In the presence of group velocity dispersion different frequencies of light in the fibre travel with different speeds. Thus, the Stokes radiation will travel slightly faster or slower, depending on the sign of the dispersion, than the pump. Since the light source is pulsed we can see that interaction will only occur when the pump and Stokes light overlap. This effect is usually referred to as 'walk off' and we can estimate the length of fibre in which the interaction will occur as

$$L_{wo} = \frac{\tau}{D\ \Delta\lambda_{ps}} \tag{3.14}$$

where $\tau$ is the pump pulse width, $D$ is the fibre dispersion and $\Delta\lambda_{ps}$ is the Stokes shift. In a typical fibre at $1.064\mu m$ with a modelocked Nd:YAG laser producing $100ps$ pulses we obtain a walk off length $L_{wo}$ of $40m$ ($D = -40ps/nm/km$ and $\Delta\lambda_{ps} = 60nm$). As mentioned in the first section the fibre dispersion can be reduced to zero in the $1 - 2\mu m$ wavelength range and is typically zero at about $1.3\mu m$ in communication fibres. If the wavelength of zero dispersion falls between the pump and Stokes wavelengths then the SRS interaction can occur over lengths of many kilometres [20].

Finally we will briefly consider the effect of absorption on SRS in optical fibres. It is not difficult to include the effect of a uniform (in frequency) loss in the simple equations (3.10) describing SRS. We expect that at low powers, where the SRS gain length is large compared to the absorption length, no SRS will be seen since the pump power is being absorbed faster than Stokes radiation is generated. In the opposite limit absorption will have little effect as the pump light is rapidly converted to Stokes light. In the situation where the absorption length and gain length are comparable we can approximate the effect of loss by using an effective fibre length given by

$$L_{abs} = \frac{(1 - e^{-\Gamma Z})}{\Gamma}. \tag{3.15}$$

In fact, equation (3.15) can be used under all circumstances as it reduces to $L_{abs} = Z$ in the low loss limit and $L_{abs} = \Gamma^{-1}$ in the high loss limit.

### 3.4 Soliton self frequency shift

So far, in this section, we have considered the more traditional aspects of SRS such as Stokes gain, anti-Stokes loss, pulse walk off, absorption and pump depletion. Optical fibres are unusual in that they support solitons, this, combined with SRS leads to a subtle phenomenon known as the soliton self frequency shift. This effect was first observed experimentally [21], [22] as a change in the central frequency of the soliton wave packet during propagation. In our description of SRS so far, we have mainly been concerned with the peak of the Raman gain spectrum. In order to understand this new effect we must turn our attention to the low frequency part of the gain spectrum.

The Raman line in silica has a width which is comparable to its displacement from the peak, leading to Raman gain and loss at frequencies close to the pump frequency. Thus, when the bandwidth of the pump pulse is taken into account we can see that there is a possibility of the peak of the pulse spectrum self pumping the wings. If this were to occur the energy on the high frequency (anti-Stokes) side would decrease and the energy on the low frequency (Stokes) side would increase. The net effect would be for the average pulse frequency to decrease as observed in the experiments. Since the low frequency gain is small and the propagation is dominated by the soliton properties we can treat the Raman gain as a perturbation of the NLS [23]. We have already shown that the Raman gain at the Stokes frequency is described by a term like $\chi_R(\delta\omega)|E_p|^2E_s$, so that for small Stokes shifts $\delta\omega$ we can expand $\chi_R$ to first order in $\delta\omega$. Once this is done we can transform back to the time domain and obtain a term proportional to $E\,\dfrac{\partial|E|^2}{\partial t}$ thus giving a modified NLS

$$i\frac{\partial U}{\partial z} + \frac{\alpha}{2}\frac{\partial^2 U}{\partial t^2} + |U|^2U = c\,U\,\frac{\partial|U|^2}{\partial t} \tag{3.16}$$

where the coefficient $c$ is related to the nonlinear susceptibility [23]. We can now use the equations of motion for the conserved quantities, in a similar manner to that discussed in section 2, to calculate the effect of the perturbation on the solitons [23]. The quantity of interest is the mean pulse frequency and its equation of motion is

$$\frac{d\omega_0}{dz} = -\frac{4c\beta''}{15\tau^4}. \tag{3.17}$$

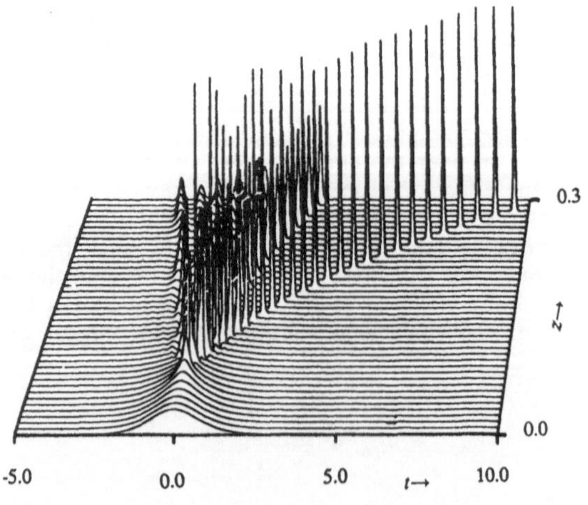


Fig. 3.4 Break up of a high order soliton in the presence of the soliton self frequency shift.

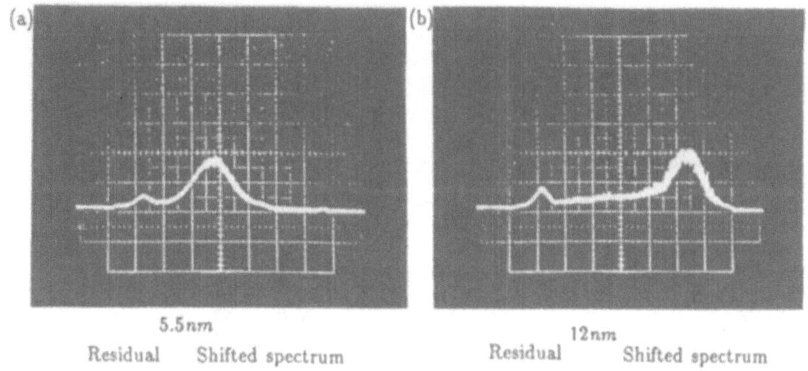


Fig. 3.5 Soliton self frequency shift observed in an $8km$ length of optical fibre.

Thus, the frequency decreases linearly with propagation distance but the overall effect varies as the fourth power of the pulse width. This strong dependence on pulse width means that the effect can only be observed when the pulse width is less than about $1ps$. In figure (3.4) we show the evolution of a high order soliton obtained from a numerical integration of equation (3.16). The pulse breaks up and a single soliton accelerates away from the residual pulse. The time displacement of this soliton is proportional to the integral of $\omega_0$ and this increases as $z^2$. Finally, in figure (3.5) we show an experimental result obtained with an $8km$ length of fibre. The pulse wavelength has shifted by about $15nm$. Some residual energy can be seen at the source wavelength which is partly a result of not launching a perfect soliton and partly due to the effects of absorption.

### 3.5 Soliton Raman generation

We have mentioned the effect of dispersion on SRS in optical fibres and noted, in particular, that group velocity matching can lead to very long interaction lengths. The

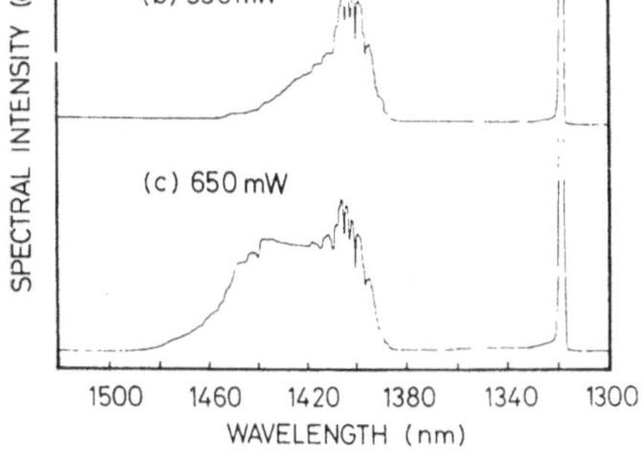

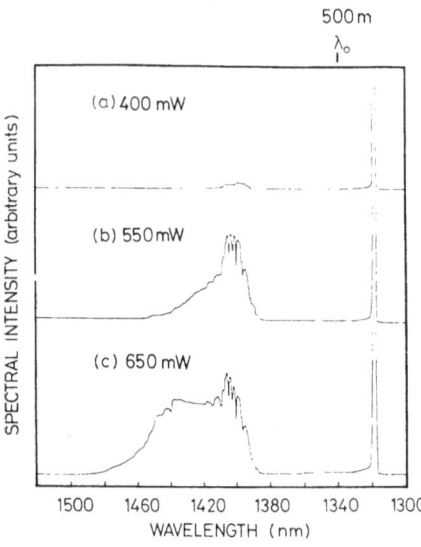

Fig 3.6 Output spectrum from $500m$ of low dispersion fibre showing a) onset of SRS, b) first Stokes band and c) broadening of the Stokes band. (Courtesy of Dr J.R.Taylor)

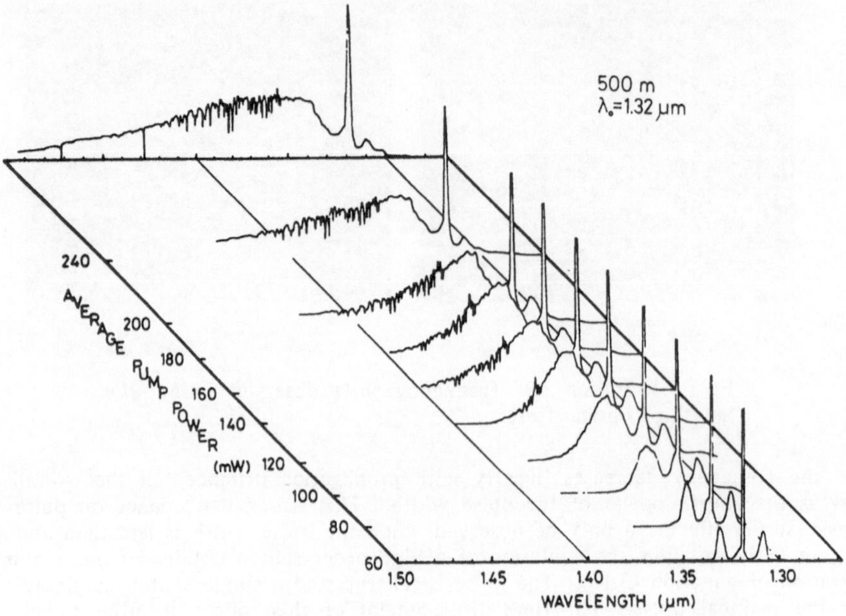


Fig. 3.7 Output spectra, as a function of input power, from an optical fibre where the pump is in the soliton regime. The modulation instability sidebands develop at low power and act as a seed for the Stokes band at higher powers. (Courtesy of Dr J.R.Taylor)

combination of this and the possibility of solitons at the Stokes wavelength leads to a process known as soliton Raman generation. Experimentally it has been observed that under these conditions solitons can be spontaneously produced at the Stokes wavelength [24], [25], [21]. The solitons produced have a width as low as $100fs$ which is the response time of the SRS process.

In figure (3.6) we show a sequence of spectra taken for different launched powers of a $1.32\mu m$ pump in a $500m$ length of optical fibre where the pump is in the normal dispersion regime. As the SRS threshold is reached radiation appears at the Stokes wavelength and as the power is increased the width of the Stokes band increases. If we observe the temporal profiles of the Stokes band we see a narrow pulse appears which gradually grows and eventually dominates. A slightly different sequence is observed if the pump wavelength is also in the soliton regime. At powers below the onset of SRS the sidebands characteristic of the modulational instability are observed, see figure (3.7). As the input power is increased the sidebands become asymmetric and eventually a wide Stokes band is created. The temporal sequence for this situation is simply the modulation of the broad pulse at low powers followed by a growth of the central spike until virtually all of the Stokes energy is contained in the central ($100fs$) part of the autocorrelation trace. The threshold for this process is lower than that for the pump in the non-soliton regime as a result of the 'seeding' from the modulational instability.

Although these two experiments result in similar Stokes solitons they arise in a rather different way. The second situation can be readily understood in terms of the modulation instability which causes short soliton-like structures to appear (see section 2, figure (2.7)). These pulses are then subject to the self frequency shift which causes the asymmetric broadening of the Stokes band. However, when the pump is in a region of normal group velocity dispersion there is no modulational instability to initiate the soliton formation. It turns out [26], [27] that the gain from the SRS is sufficient to stabilise and compress the solitons even if there is only noise present initially. In section 2 we discussed the broadening effect of loss on the soliton pulse width. If we change the sign of $\Gamma$ the equations describe gain and we see that exponential compression will occur. Of course, this cannot continue indefinitely and the process saturates when the bandwidth of the soliton is comparable with the bandwidth of the Raman gain spectrum [27]. This leads to a predicted soliton pulse width of about 100*fs* which is consistent with the observations.

### 3.6 Single wave formulation

The simple formulation of SRS outlined in section (3.2) is adequate when the bandwidth of the light at the pump or Stokes lines is small compared to their separation. Similarly the modified NLS described in section (3.4) is adequate when the soliton bandwidth is small compared to the Stokes shift. In the previous section we discussed some experimental results in which both of these assumptions is violated. In figure (3.7) the radiation fills the spectrum from the pump to the Stokes wavelength with no clear distinction between the two. Under these conditions we have strong coupling between the soliton and Raman phenomena and a more exact equation is necessary.

The fundamental assumption which can no longer be made is that the time dependence of the optic phonon polarisation wave can be neglected. Mathematically, this means we must return to equation (3.3c) to include the frequency dependence of the phonon propagation constant $\kappa$. We expand $\kappa$ to first order in $\Omega$ to obtain

$$\kappa = \kappa_0 - i\gamma + \kappa'_0\Omega \tag{3.18}$$

which can now be substituted into equation (3.3c). Using the properties of the optical phonon dispersion curve we can neglect the derivative with respect to $z$ and the small constant term $\kappa_0$ to give the following equation for the time dependence of the phonon polarisation

$$i\kappa'_0\frac{\partial Q}{\partial t} + i\gamma Q = g_Q E_L E_S^*. \tag{3.19}$$

This equation can be solved for $Q$ once the optical fields are known and we obtain

$$Q(t) = -i\frac{g_Q}{\kappa'_0}\int_{-\infty}^{t} e^{-\frac{\gamma}{\kappa'_0}(t-t')} E_s^*(t')E_p(t')dt' \tag{3.20}$$

which has the form of a time delayed response to the optical field. In general a frequency dependent susceptibility has a counterpart in the time domain of a response function. Under normal conditions where the fields are quasi-CW the frequency domain description is more convenient. However, in the strongly transient regime we must return to the time domain in order to obtain a simple picture [16]. The third order nonlinear polarisation is modelled as follows

$$P_{NL}(t) = \chi_3 E(t)\int_{-\infty}^{t} f(t-t')E(t')^2dt' \tag{3.21}$$

where the delayed response is applied to the optical intensity. The response function $f(t-t')$ can be obtained from the Raman gain spectrum by first applying a Kramers-Kronig transform to obtain the real part of the susceptibility and then Fourier transforming [28]. Following through the analysis similar to that presented in the first section we can derive a generalised NLS for the nonlinear response given in equation (3.21).

$$i\frac{\partial U}{\partial z} = \{\beta(\omega_0 - i\frac{\partial}{\partial t}) - \beta_0\}\{U\} + \sigma[1 - i\tau_{shock}\frac{\partial}{\partial t}]\{U\int f(t-t')|U(t')|^2dt'\} \tag{3.22}$$

where

$$\sigma = \frac{3N\chi^3\omega_0}{n_{eff}c}$$

and

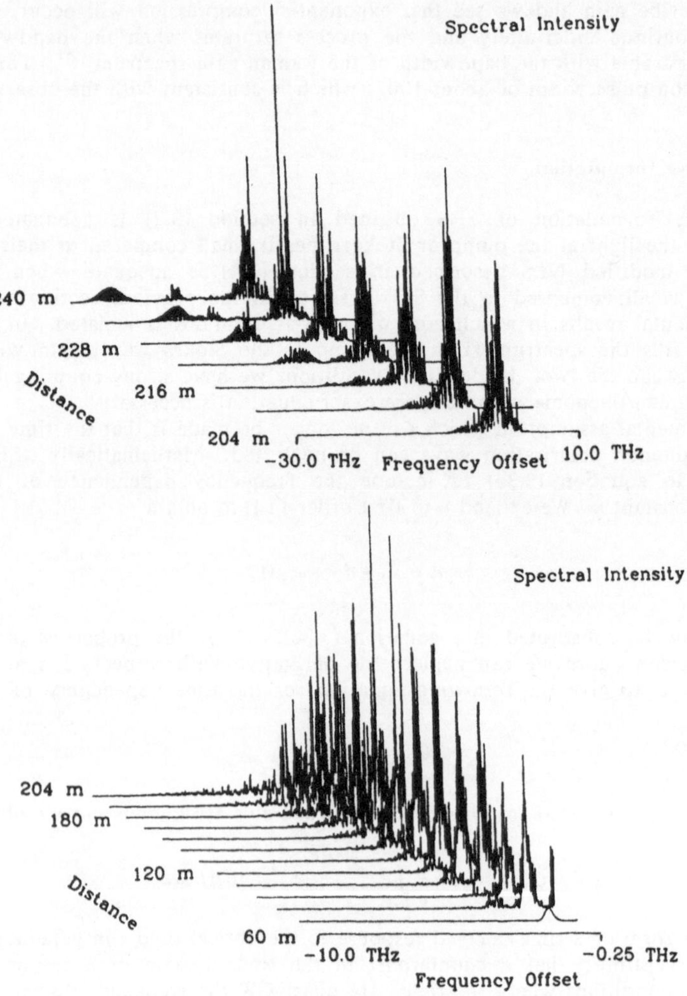


Fig. 3.8 Theoretical evolution of a pump wave with noise. At short distances the pump is depleted by the modulation instability. At longer distance we see the simultaneous onset of SRS and soliton formation. The lowest frequency feature at 240$m$ is a pair of solitons which are both experiencing a strong self frequency shift.

$$\tau_{shock} = \frac{1}{\omega_0} + \frac{\partial}{\partial\omega}\log(N/n_{eff})$$

$\beta(\omega_0 - i\frac{\partial}{\partial t}) - \beta_0$ is the dispersion operator and is usually expanded about $\omega_0$ to give the familiar dispersion terms. The cubic nonlinear term in the NLS is now replaced by a time delayed response to the intensity. The coefficient $\tau_{shock}$ is small and is really only needed to distinguish between energy and photon number, both being conserved if $\tau_{shock} = 0$ whereas only the photon number is conserved in general. In figure (3.8) we show a numerical integration of equation (3.22) for a CW pump beam together with a white noise seed. In the final frame there are two distinct features. A band of radiation at the Stokes frequency due to SRS and a band at a greater separation from the pump with strong

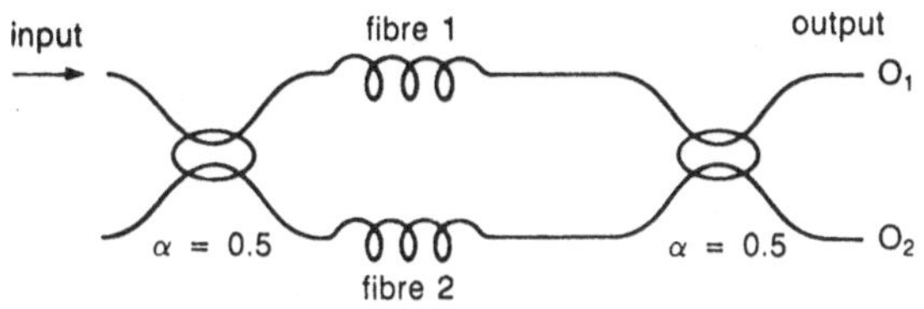


Fig. 4.1 Schematic diagram of an all fibre Mach-Zehnder.

interference fringes which has split off from the pump and moved down in frequency. This latter feature is actually a pair of solitons undergoing a self frequency shift.

## 4. NONLINEAR FIBRE DEVICES AND SOLITON SWITCHING

### 4.1 Introduction

We have seen, in section 1, that the effect of the Kerr nonlinearity, to first order, is to induce a nonlinear change in phase velocity of the mode with a resultant nonlinear phase shift as in equation (1.23)

$$\phi = k_0 n_2 L\ I. \tag{4.1}$$

This phase shift is proportional to the length, intensity and nonlinear coefficient. In any two mode configuration, where the fields are mixed, a phase sensitive nonlinear response can be obtained if any of these three variables are different for the two paths involved. This is the principle of operation of 'weakly' nonlinear fibre devices. There are two main examples; (1) nonlinear interferometers, (2) nonlinear couplers. In the former the coupling occurs at discrete points whereas in the latter it occurs continuously. However, it has been shown [29], that nonlinear couplers can be reliably modelled as concatenated nonlinear interferometers with both devices showing the same nonlinear characteristics.

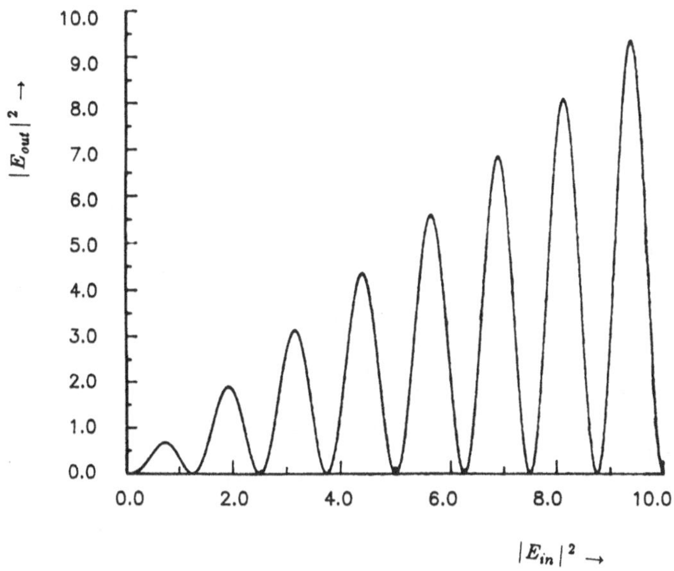


Fig. 4.2 Output power as a function of input power for a balanced Mach-Zehnder interferometer with different fibres in the two arms.

Interferometers will be taken as our example here but it should be borne in mind that the conclusions (for example on the use of solitons) apply to the general class of device.

### 4.2 Nonlinear interferometers

As an example consider an all-fibre Mach-Zehnder shown in figure (4.1). If the power splitting ratio of the couplers is $\alpha$ :(1-$\alpha$) then the coupling equations with inputs $E_1$ and $E_2$ and outputs $E_3$ and $E_4$ are

$$E_3 = \alpha^{1/2}E_1 + i(1-\alpha)^{1/2}E_2 \tag{4.2a}$$

$$E_4 = i(1-\alpha)^{1/2}E_1 + \alpha^{1/2}E_2. \tag{4.2b}$$

The configuration consists of coupling followed by propagation and finally coupling as in equation (4.2). If $\alpha = 0.5$ then the device is symmetric. The simplest (conceptually at least) way of breaking the symmetry of the two paths, and thus permitting a nonlinear response, is to have fibres with different core areas in the two arms. This would lead to different intensities in the two arms even for symmetric couplers where the power would be the same in both arms.

The output from $O_1$ with input $E_{IN}$ as indicated in figure (4.1) is

$$|E_{O}^2| = \frac{|E_{IN}|^2}{2}(1 - \cos[n_2 k_0 L|E_{IN}|^2(\frac{1}{A_1} - \frac{1}{A_2})]) \tag{4.3}$$

where the $A_i$ are the mode field areas in the two arms. Equation (4.3) can be written in a simplified form

$$|E_{out}|^2 = \frac{|E_{in}|^2}{2}(1 - \cos\Delta\phi) \tag{4.4}$$

where $\Delta\phi$ is just the difference in the nonlinear phases as given by equation 4.1. This standard sinusoidal nonlinear response of these devices is shown in figure (4.2).

The most practically straightforward way of breaking the symmetry in a Mach-Zehnder is to have two identical, but unbalanced, couplers. For example if the power

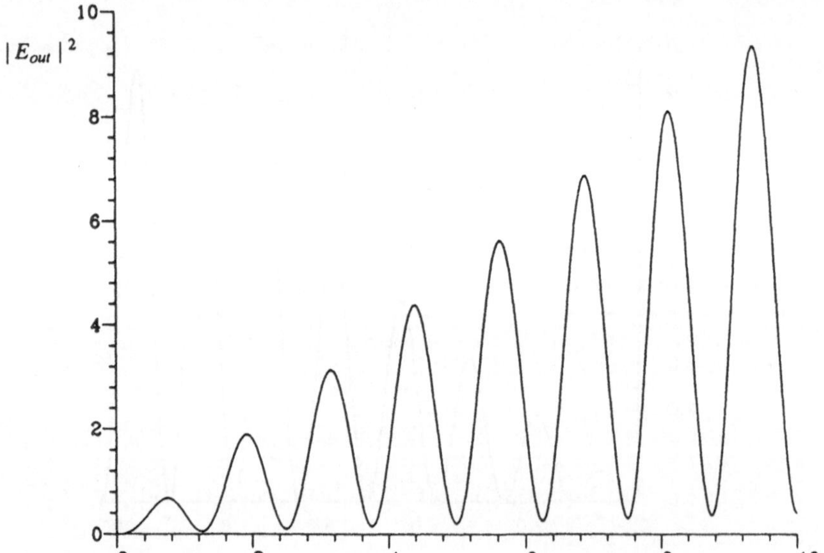


Fig. 4.3 Output power as a function of input power for an unbalanced Mach-Zehnder interferometer ($\alpha$ = 0.4).

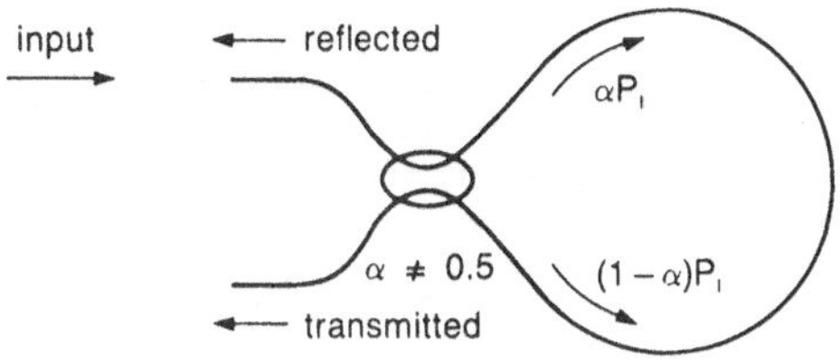


Fig. 4.4 Schematic diagram of an all fibre Sagnac interferometer.

splitting ratio of the couplers is $\alpha$:(1-$\alpha$) for both and identical fibres are used throughout then the nonlinear response is

$$|E_{out}|^2 = |E_{in}|^2(1 - 2\alpha(1-\alpha)(1 + \cos[(1-2\alpha)n_2k_0L|E_{in}|^2]))). \tag{4.5}$$

This formula shows the same periodic behaviour as equation (4.4). The minimum transmission is no longer zero but is now given by

$$1 - 4\alpha(1-\alpha), \tag{4.6}$$

for $\alpha = 0.4$ this response is shown in figure (4.3).

In standard silica fibres we can calculate the power required to induce a $\pi$ phase shift from equation (4.1). Taking a mode area of $50\mu m^2$ and a wavelength of $1.5\mu m$ gives

$$\phi = 2.7 \times P\ L \tag{4.7}$$

where $P$ is in kWatt and $L$ is in metres so that about 1kW is required for a single cycle in a one metre long interferometer (note it is actually a phase difference which is needed so that, in practice, the power would be a small multiple of this number). This very high power can only be significantly reduced if the device length is increased.

Long interferometers are susceptible to environmental fluctuations and would be difficult to fabricate with exactly matched lengths in the arms. In order to reduce the power requirements and increase the practical length an alternative interferometer may be considered. This is the all-fibre Sagnac interferometer and is illustrated in figure (4.4). This is a four port coupler with two of the ports spliced. In effect this is the same as the Mach-Zehnder but now the two paths are the two routes around the loop (clockwise or anticlockwise) and the two couplers are replaced by one which is traversed twice. The symmetry can be broken in the same way as above by selecting an unequal power splitting coupler. Thus the nonlinear response is as given in equation (4.5) where the field $E_3$ corresponds to the field transmitted. The lower power response (equation 4.6 ) shows that linearly the configuration acts as a mirror with a reflectivity determined by $\alpha$ (e.g. for $\alpha = 0.4$ the reflectivity is 94%). For this reason the device is often called the nonlinear optical loop mirror NOLM. This device has proved to be excellent for experimental study and all the following results were obtained with this device.

### 4.3 Pulse response of nonlinear fibre devices

The ultimate use of nonlinear devices is to perform ultrafast all-optical processing and since the nonlinear response in silica is of order femtoseconds or less silica based devices seem to offer this possibility. We will therefore discuss the response of the devices to short pulses and take the nonlinearity itself as instantaneous.

Pulses will acquire a nonlinear phase shift which will vary throughout the varying intensity of the pulse (see equation (1.22)). The device response, equations (4.4) or (4.5), applies to the instantaneous intensity and since the pulses arrive in coincidence the device response will be different for different parts of the pulse. This will result in pulse shaping effects in the output pulses. Thus the output pulse shape will be altered as a result of the intensity dependent transmission; this phenomenon is illustrated in figure (4.5). Not only does this pulse shaping destroy the pulses but it also seriously degrades the depth of contrast between high and low transmission of the device. If we calculate the total output energy, as would be measured by a detector, then the response ( for the NOLM with $\alpha$=0.4 ) is as shown in figure (4.6) for $sech^2$ pulses. The figure shows a 'staircase' response for the pulse energy and results in a device with almost no contrast. The maximum transmission now instead of being 100% is much reduced. The exact details of figure (4.6) depend on the pulse shape and indeed if the pulses were 'square' ( i.e. instantaneous rise and fall times) we would recover the response of equation (4.3) for the total energy.

We should note that pulse shaping is common to all nonlinear devices with an instantaneous response and has been observed in nonlinear coupler experiments [30].

### 4.4 Solitons in nonlinear fibre devices

A solution to the problem of the above pulse shape effects and subsequent performance degradation in device performance has recently been proposed. This is to use soliton

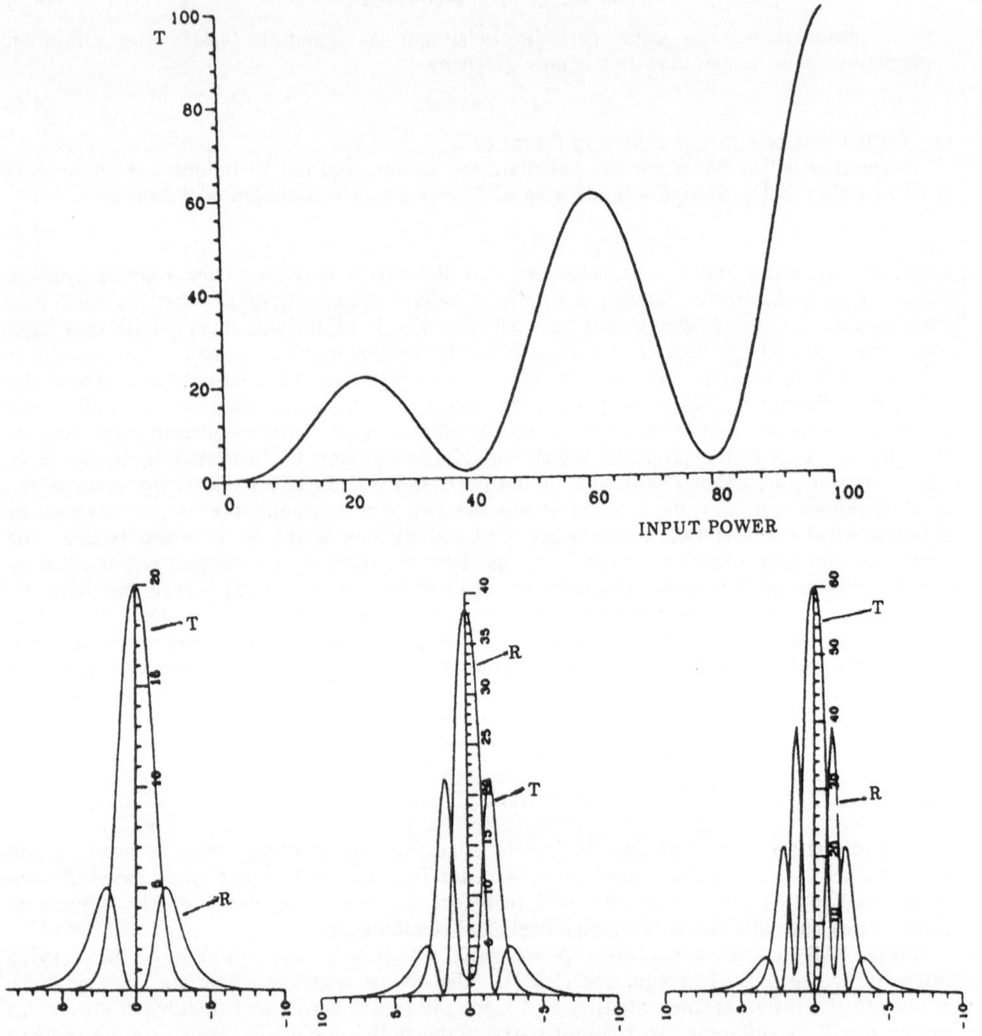


Fig. 4.5 Transmission coefficient for a fibre loop mirror as a function of input power. Also shown are the reflected and transmitted pulse shapes for various values of the input peak power.

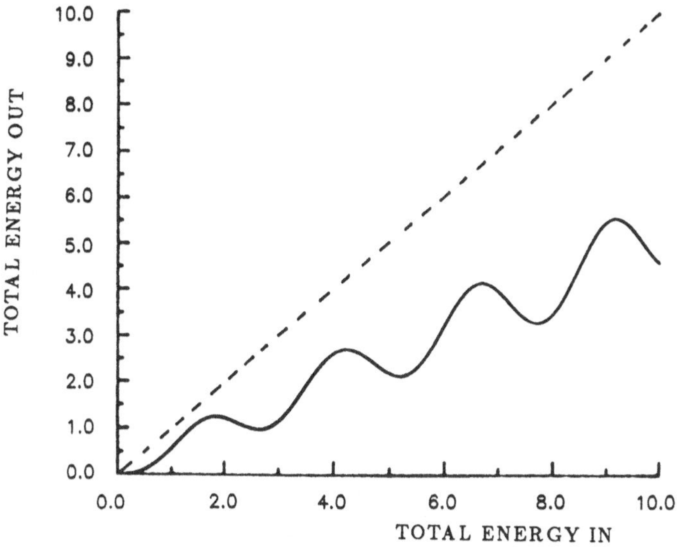


Fig. 4.6 Total energy transmission for an input pulse *Asech*($t$). The dotted line shows perfect transmission.

effects in the device. So far the device response has included only the effect of SPM and we have ignored dispersion effects. Dispersion will be significant when the propagation lengths in the devices are close to the dispersion lengths. This will be particularly significant for short pulses and long devices.

Consider now the response of phase sensitive devices to solitons. Recall from equation (1.24) that the single soliton solution is

$$u(t,z) = e^{\frac{iz}{2}} sech(t) \tag{4.8}$$

this shows that there is a phase proportional to the length which applies to the whole pulse. This is a remarkable property indicating the integrity of the soliton. What we require is such a uniform phase which applies not just to the exact single soliton but to a general pulse in the soliton regime.

Consider the general pulse *Asech*($t$) ; we need to know the response of the device as a function of $A$ with the expectation that the existence of solitons will result in a uniform phase which is not only a function of distance but also of $A$. We shall calculate an average phase for this general pulse as a function of distance. The first problem is to derive a consistent definition of such a phase. As an example we shall take the following definition

$$\Phi = \tan^{-1}(\int \mathrm{Im}\ u|u|^2 , \int \mathrm{Re}\ u|u|^2 ). \tag{4.9}$$

This definition is not unique but does give a measure of an integrated phase. In figure (4.7) we plot $\Phi$ as a function of distance for various values of $A$ which is obtained from numerical integration of the NLS. The linearity of the phase distance dependence is remarkable as is the clear dependence on $A$.

In order to understand this phase property in more detail we will now construct a simple model. We assume that the soliton features dominate and that the uniform phase is determined by the largest eigenvalue of the ZS scattering problem, see equation (1.27). Recalling equation (1.28) for the eigenvalue of *Asech*($t$) and that the motion of the solitons is determined by exponentials of twice the eigenvalue squared [5] then we assert that the soliton phase of the pulse *Asech*($t$) is

$$\phi_s = 2(A - 1/2)^2\ z. \tag{4.10}$$

This is the linear phase plotted in figure (4.7). The agreement with the exact calculation is excellent and leads us to expect that this model will apply generally. It is indeed

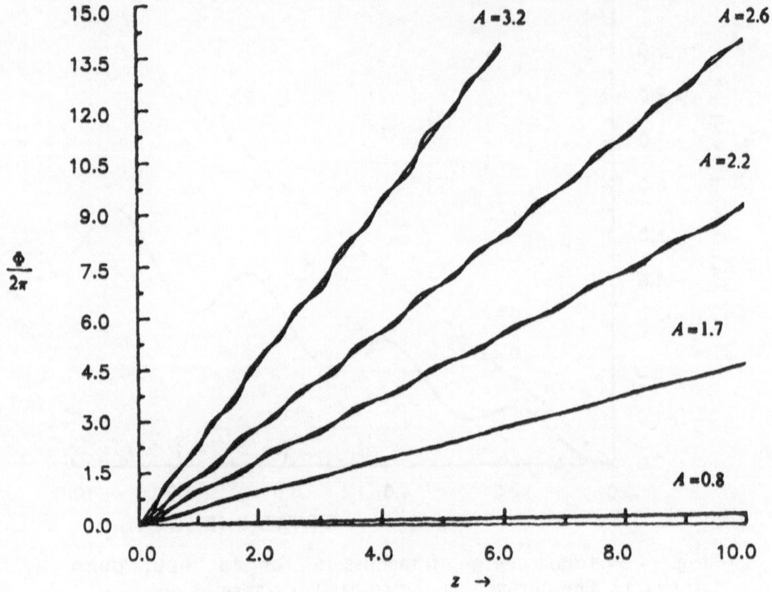


Fig. 4.7 Evolution of the average pulse phase for an input pulse $A sech(t)$ propagating in the soliton regime.

remarkable that such a simple model is so accurately followed and further illustrates the power of the scattering transform concept. In passing we should note that equation (4.7) is, of course, exact for the single soliton case $A = 1$ .

Using equation (4.10) it is straightforward to derive simple response functions for the nonlinear interferometers; the only thing to note is that we must integrate over the pulse. The energy transmitted $E_T$ of a NOLM for $A sech(t)$ pulses is

$$E_T = E_{IN}(1 - 2\alpha(1-\alpha)(1 + \cos 2z((1-2\alpha)A^2 + (\alpha^{1/2} - (1-\alpha)^{1/2})A)). \qquad (4.11)$$

In figure (4.8) we take as an example a loop of length $4\pi$ in soliton units and display the results of an exact NLS integration for varying input pulse energy compared with the soliton phase model (4.10). The comparison shows how reliable the simple model is. Indeed the only variations occur for large launched energies where the pulse shaping

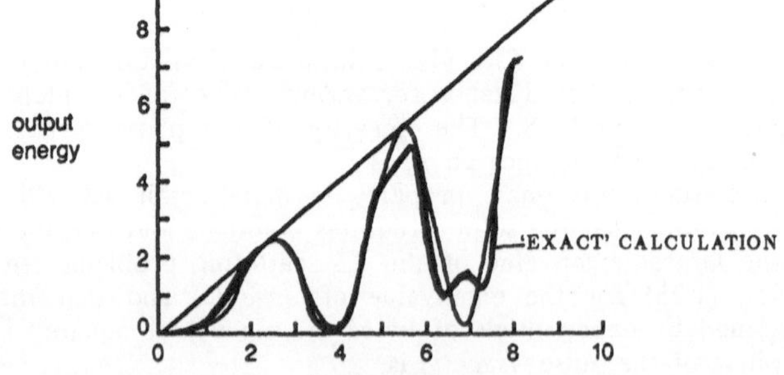


Fig. 4.8 Comparison between the exact calculation and the soliton phase model.

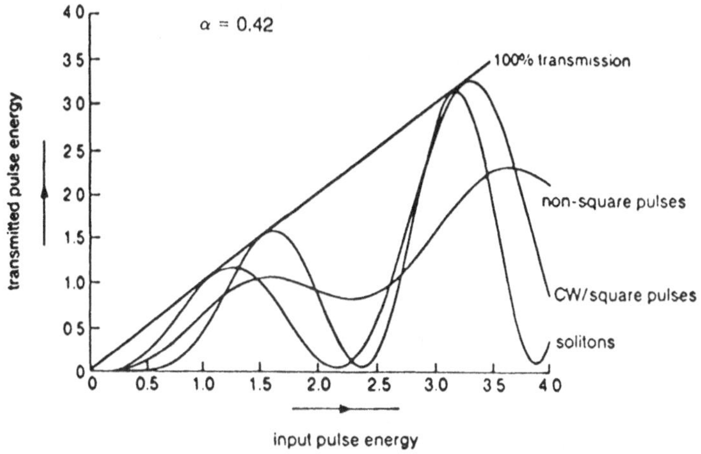


Fig. 4.9 Comparison of total energy transmission obtained when operating the nonlinear fibre loop mirror under different conditions.

effects of the NLS, which we have ignored, are important. In figure (4.9) we compare the soliton with CW (or square pulse) and non-soliton response of the NOLM. The restoration of the deep contrast, CW response obtained from incorporating soliton effects is clearly visible.

### 4.5 Soliton switching experiments

The above concepts have recently been demonstrated in experiments involving loop mirrors [31], [32]. The problem to overcome is to have a source of pulses which are short enough that the loop length is of order a soliton period or preferably more. For loop lengths of order 100m this means that pulses of order a picosecond or less are required.

In the first experiment [31] the loop length was 100m and pulses of ~$700fs$ FWHM or $400fs$ sech width were used with the source a 'soliton laser' operating at ~$1.5\mu m$. From equations (1.26) this gives a loop length of $2.9\pi$ or about six soliton periods. The coupler had an $\alpha$=0.42 so that the parameters corresponded to those illustrated in figure (4.8). It should be emphasised that the model and numerical simulations indicated that there is no need for the loop length to be an integer number of soliton periods and indeed although this appears quite close to such a situation the difference is sufficient to illustrate this point.

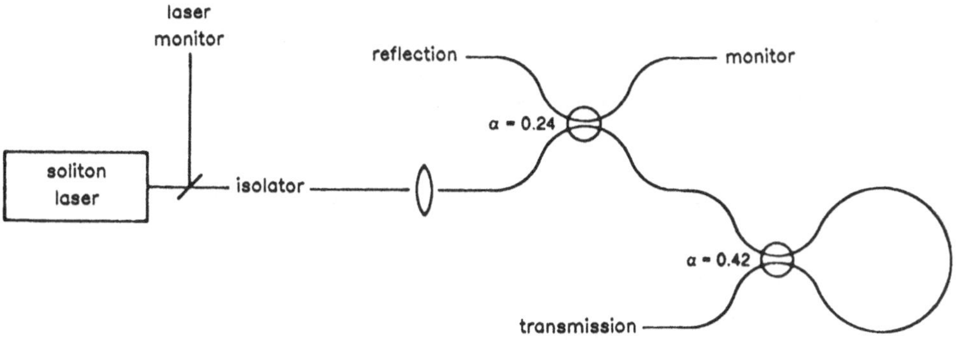


Fig. 4.10 Schematic diagram of the experimental arrangement used to investigate soliton self switching.

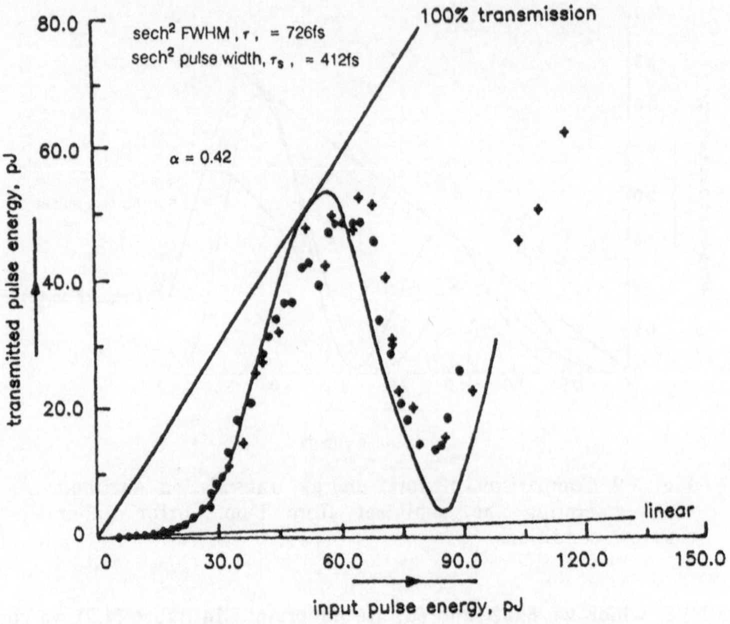


Fig. 4.11 Comparison between experimental results and theoretical prediction for soliton self switching in a nonlinear fibre loop mirror.

The experimental setup is shown in figure (4.10) and the results of the experiment displayed in figure (4.11). The continuous lines are exact calculations using a numerical simulation of the NLS although the model calculations would have been almost indistinguishable from this. The points represent the experimental measurements. Excellent agreement can be seen and indeed very good transmittance ( ~93% ) near the maximum predicted is observed. The excellent characteristics of the device obtained confirm the potential of solitons in all-optical processing.

A number of details are worth noting. The pulse energy is calculated to be 46pJ at the maximum transmittance. This represents a particularly low value for all-optical switching. This could be reduced for longer length loops and for longer pulses. The potential speed of the configuration is faster than that illustrated here although for short pulses the self-Raman effects can degrade the characteristics [32]. In the experiment shown here no such effects were observed. The pulse shape at the maximum transmittance was unaltered from that launched indicating the possibility of concatenating the devices.

## 5. NONLINEAR EXTERNAL CAVITY LASER MODELOCKING

### 5.1 Introduction

In this section we will discuss a rather different application of nonlinear effects in fibres to enhance the performance of modelocked lasers. In recent years an exciting new field in laser modelocking has opened up in which nonlinear optical response is the key element. A modelocked laser is usually constructed from a long cavity which supports many longitudinal modes and one transverse mode. If such a laser is CW pumped then these modes all lase with an energy distribution which reflects the gain profile of the lasing medium and the cavity losses. The laser output, in time, is then given by

$$U(t) = \sum_n A_n e^{i\phi_n} e^{in\omega t} \tag{5.1}$$

where $A_n$ is the amplitude of the longitudinal mode, $\phi_n$ its phase and $\omega$ is the mode spacing. If these modes have no particular phase relationship, $\phi_n$ random, then the output is very noisy with large amplitude spikes. If, however, we contrive to make all the phases the same, $\phi_n = constant$, then it is easy to show from equation (5.1) that the output is a train of pulses with separation $\omega^{-1}$ and a shape given by the Fourier transform of $A$. The central task of laser modelocking is therefore to achieve the situation in which the longitudinal laser modes all have the same phase.

The two principle methods for achieving this state are known as active and passive modelocking. Active modelocking consists of applying some external modulation to the cavity at frequency $\omega$ so that adjacent modes are coupled to each other and the phase information can be transferred. Passive modelocking is usually achieved by including a saturable absorber in the laser cavity so that the cavity loss is only low for high intensity pulses. The fundamental mechanism for passive modelocking is a nonlinear coupling of the cavity modes which also has the effect of allowing phase information to pass between the modes. The recent innovation to occur was to observe enhanced modelocking in an external cavity constructed from an optical fibre in which the pulses of light travelled as solitons.

In the next section we review the properties of the colour centre laser used in this experiment and follow this with an exposition of the theoretical model used to study this system.

### 5.2 The F centre laser

The F-centre laser (FCL) has an extremely large bandwidth of available laser modes under the gain curve. Modelocking is achieved by a technique known as synchronous pumping. The laser is pumped with another modelocked laser with the cavities matched to the same round trip time. As it is not possible to lock the phases of all of the FCL modes by the synchronous pumping, then the actual bandwidth of the laser gain must be restricted. This is achieved by adjusting a set of birefringent plates. For small bandwidths the laser produces transform limited pulses. As the bandwidth is increased the pulse width decreases and remains transform limited until a minimum width of $8ps$ is reached. Any further increase in bandwidth beyond this only leads to noisier pulses, with some structure on smaller time scales, rather than to narrower ones.

### 5.3 The soliton laser

Figure (5.1) is a schematic diagram of the essential elements of the apparatus used in Mollenauer and Stolen's experiment. The laser cavity is formed by two mirrors that contain the lasing medium, in this case $Tl$ doped $KCl$, and birefringence plates which act

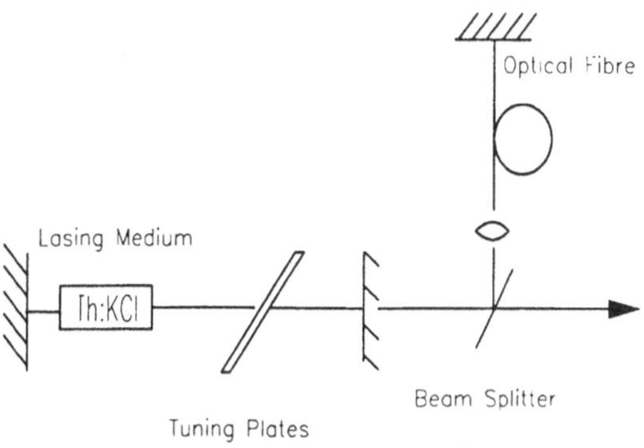


Fig. 5.1 Schematic diagram of the soliton laser.

as a tuning element and band pass filter. The laser crystal is axially pumped with 1.06$\mu$m radiation synchronously with the cavity. The output from the main laser cavity is taken through a beam splitter and launched into a monomode optical fibre, which is formed into an external cavity by a third mirror thus providing feedback to the laser cavity.

By this device of using an optical fibre in an external cavity, the output pulse width could be reduced from $8ps$ to as low as $100fs$. This output pulse width depended on the length of the optical fibre and is consistent with the width of an $N = 2$ soliton whose period is twice the length of the fibre which is given by

$$\tau = (0.46D\lambda^2 L)^{1/2} \tag{5.1}$$

where $L$ is the length of the fibre in km, $D$ is the dispersion in $ps/nm/km$ and $\lambda$ is the wavelength in $\mu$m [33].

Since soliton propagation is a nonlinear phenomenon it should depend critically on the intensity of the optical beam launched into the fibre. Initially, operation was found to occur close to the $N = 2$ soliton power. However, later studies found stable operation over a wide range of powers [34], [35].

## 5.4 Laser models

We will treat the two processes of mode amplification and mode locking independently, and present simple phenomenological models of each process which retain the essential qualitative features observed in the experiments. Since the operation of the soliton laser is dominated by the properties of the optical fibre we would expect the explanation to be essentially independent of the choice of laser model. We will outline two simple models of lasers, one of an inhomogeneously broadened laser and one of a homogeneously broadened laser. The amplification of a single mode on successive passes through the lasing medium can be described as an inhomogeneous saturable amplifier on the modes:

$$E_{cav}(k) = \left[\frac{f(k)}{1+\varepsilon|E'_{cav}(k)|^2}\right]E'_{cav}(k) \tag{5.2a}$$

where

$$f(k) = \frac{\alpha}{1+(\frac{k}{\beta})^2}. \tag{5.2b}$$

Here, $E_{cav}(k)$ is the complex amplitude of the laser mode displaced by $k$ from the centre frequency after passing through the crystal and tuner plates. $f(k)$ is the effective small signal gain of the laser; $\alpha$ is the small signal gain at the operating frequency, and $\beta$ describes the effective width of the gain curve. Both $\alpha$ and $\beta$ are adjustable parameters: $\alpha$ by altering the pump power: and $\beta$ through the birefringence plates. The saturation, $\varepsilon$ is a property of the lasing medium and is not easily adjusted. After many passes through the lasing medium, the mode amplitude tends to the steady state, given by

$$\begin{aligned} |E(k)| &= \frac{f(k)}{\varepsilon^{1/2}} && |k| < \beta(\alpha-1)^{1/2} \\ &= 0 && |k| > \beta(\alpha-1)^{1/2}. \end{aligned} \tag{5.3}$$

If all the phases of these modes were locked to some reference phase, then the output from the laser would be simply the inverse Fourier transform of (5.3).

The mechanism of synchronous pumping assists the laser to mode lock by exciting sidebands which resonate with the laser modes. Those modes close to the operating frequency have their phases gradually locked to the reference phase of the pumping by the build up of stimulated emission generated by the sidebands. Away from the centre frequency, the amplitude of the sidebands decays rapidly and their effect is swamped by the spontaneous noise. The bandwidth of effective mode locking is determined by the pump modulation and the spontaneous noise floor, and is almost independent of the mode amplification process. We can use a simple model for the mode locking in which the

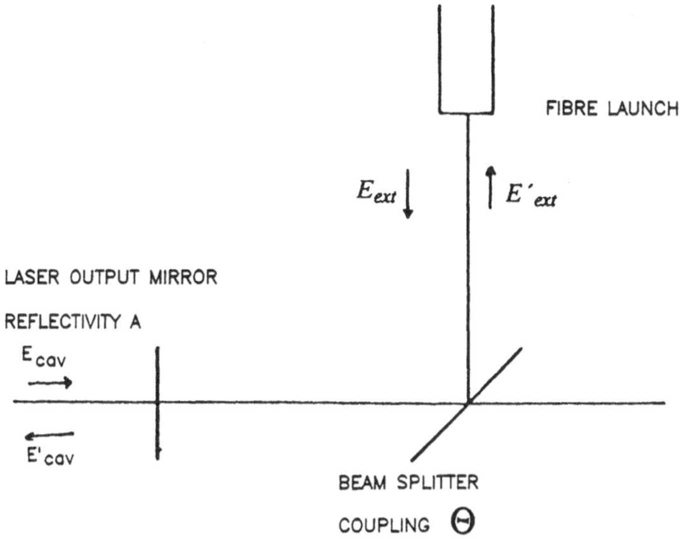


Fig. 5.2 Expanded diagram of the cavity coupling showing the notation used.

phases of modes within a finite bandwidth are gradually contracted to zero, and those outside the band are unaltered

$$\phi(k) = \left[1-\frac{1}{5}\left(\frac{k_0^2-k^2}{k_0^2}\right)\right]\phi'(k) \qquad |k|<k_0 \tag{5.4}$$
$$= \phi'(k) \qquad |k|>k_0$$

where $\phi(k)$ is the phase of the Fourier mode $E(k)$.

The homogeneous model is based on the active mode locking theory of Haus [36]. The evolution of the laser field, $E$, within the active medium is described by the following equation

$$E_{cav} = \frac{\alpha E'_{cav}}{(1+I)(1+\beta^2\omega^2)} + \Delta\frac{d^2E'_{cav}}{d\omega^2} + N \tag{5.5}$$

where $I = \int |E'_{cav}|^2 dt$. The steady state pulse width results from the competition between the compression induced by the modelocking, represented by the parameter $\Delta$, and the finite bandwidth of the equivalent filter, $\beta$ (i.e the effective bandwidth of any intracavity filter and the gain bandwidth). The modelocking term describes the effect of preferential amplification at the peak of the gain modulation. $N$ is an additive noise source which serves to destabilise the weakly modelocked states. In the steady state and the absence of noise, this equation has bound state solutions with $I$ an eigenvalue. In the small $\beta$ limit, equation (5.5) can be related to the quantum mechanical Harmonic oscillator whose solutions are the Gaussian-Hermite polynomials [36]. The pulse width is then given by $\Delta^{-1/4}(2\beta)^{1/2}$, showing the balance between modelocking and filtering effects.

### 5.5 Cavity coupling

The soliton laser is formed by adding an external cavity, which contains an optical fibre, to the F-centre laser. At the output mirror of the laser the electric fields in the fibre and the laser cavities are mixed, and we need to discuss in some detail this mixing. In figure (5.2) we show the output mirror, beam splitter and fibre launch which defines our notation. The input and output fields on either side of the mirror are connected by the following equation

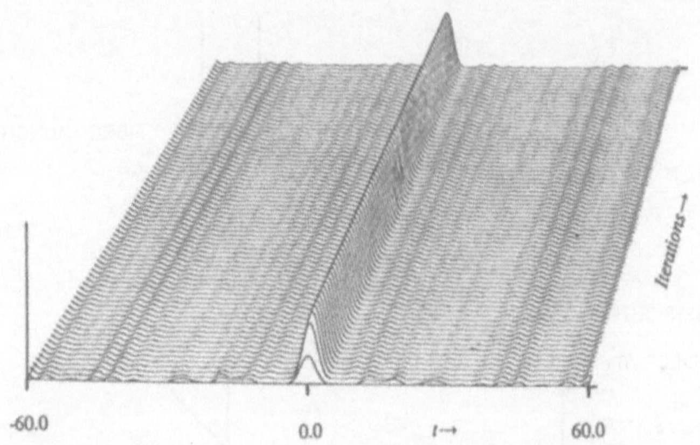


Fig. 5.3 Evolution of the laser pulse in the absence of the external cavity.

$$E'_{ext} = -i\Theta(1-R^2)^{1/2}E_{cav} \tag{5.6a}$$

$$E'_{cav} = RE_{cav} + \Theta(1-R^2)^{1/2}e^{iP\pi}E_{ext}. \tag{5.6b}$$

Here, the output mirror has a field reflection coefficient $R$. The beam splitter parameter and the launch losses into the optical fibre have been combined into one coupling coefficient, $\Theta$. We have neglected the 'ringing' effect of back reflections from the output mirror being relaunched into the fibre which is $O(\Theta^2)$. The phase shift, $P$, accounts for the possibility of an interferometric mismatch between the two cavities. $P$ turns out to be one of the most important parameters in the model, for it determines whether the two fields mix constructively or destructively. A change in the optical path through the fibre arm of a single wavelength would take $P$ through its full range 0-2$\pi$.

## 5.6 Nonlinear external cavity: 1. optical fibre

The propagation of light in the optical fibre is described by the NLS as we have

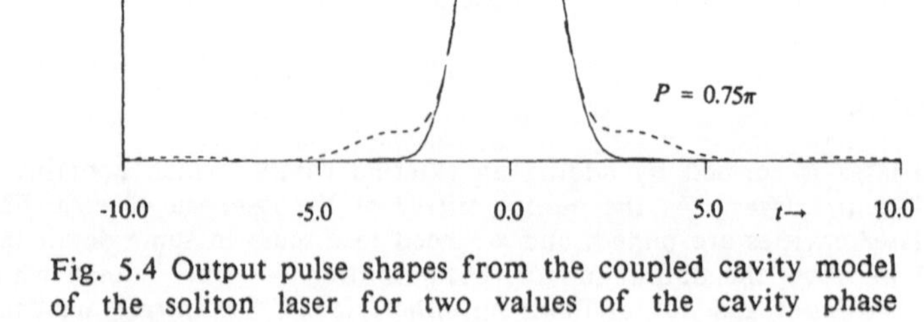

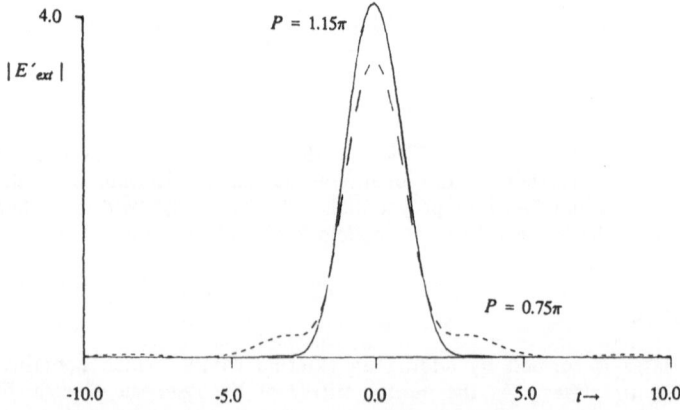

Fig. 5.4 Output pulse shapes from the coupled cavity model of the soliton laser for two values of the cavity phase mismatch.

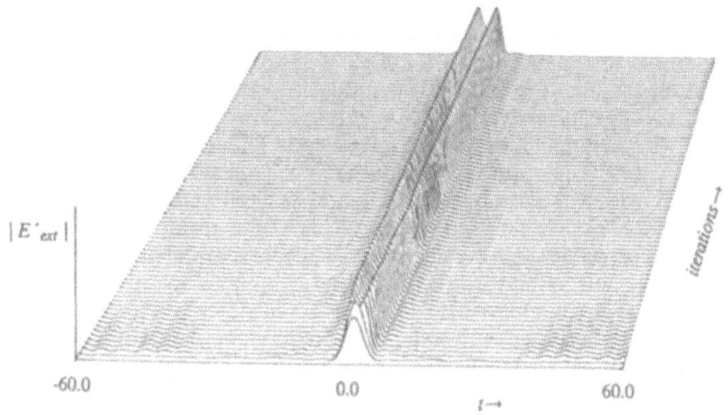


Fig. 5.5 Double pulse solution of the coupled cavity equations.

already discussed. Combining this with the inhomogeneous laser model (5.2) and the mirror equations (5.6) we have a complete set of equations describing the coupled cavity laser. In figure (5.3) we show the output of the laser in the absence of the external cavity. The steady state pulse is broad (FWHM ~4) with some additive noise due to the unlocked portion of the gain spectrum.

The $N = 2$ soliton is thought to be a simple operating point of the soliton laser and in figure (5.4) we show two steady state pulses obtained when the system was started close to the $N = 2$ state but with different values of the interferometric phase mismatch $P$. Both pulses are narrower than that in figure (5.3) and are approximately sech shaped. Indeed, analysis of the ZS eigenvalues shows that the pulses are a two soliton bound state. Other nonlinear stable states were also found by starting the laser from an initial state consisting of small amplitude noise. An example is shown in figure (5.5) where the output has evolved into a double humped solution. This solution is an unbound state of two solitons.

In general these theoretical results show that there are many stable solutions consisting of both bound and unbound double solitons.

### 5.7 Nonlinear external cavity: 2. discrete elements

In the introduction to this section we described modelocking in terms of a nonlinear coupling between the longitudinal modes of the laser. In the soliton laser it is the nonlinear soliton propagation which provides this coupling. However, it may be possible to provide this coupling through other, more general, nonlinear elements. In this section we will describe some theoretical calculations based on discrete nonlinear elements.

The first model we use is that of a saturable absorber described by

$$E_{ext}(t) = \frac{|E'_{ext}|^2}{1+|E'_{ext}|^2} E'_{ext}(t) \tag{5.7}$$

which gives rise to pulse compression and can be used to passively modelock a laser when used intracavity.

The second model is that of a saturable amplifier described by

$$E_{ext}(t) = \frac{1+|E'_{ext}|^2}{1+2|E'_{ext}|^2} E'_{ext}(t) \tag{5.8}$$

which gives rise to pulse broadening.

The transmission functions are defined in time rather than in frequency space. This ensures that coupling will occur between the modes of different frequency in the fundamental laser cavity. These models can now be combined with the homogeneous laser model (5.5) and the mirror equations to provide another complete set of coupled cavity equations. In figure (5.6) we show the evolution of the *RMS* pulse width with the curve (5.6a) corresponding to the uncoupled cavity. Curves (5.6b) and (5.6c) show the effect of the saturable absorber and the saturable amplifier respectively with $\Theta = 0.8$ in each case. The values of the interferometric phase mismatch $P$ were chosen to be in the middle of the range over which modelocking was achieved. The external cavity has changed the situation dramatically. The transient region has decreased and the final state has an *RMS* pulse width of 0.6 which is largely determined by the filter bandwidth. The modelocking of the laser is dominated by the external cavity. These results show very little difference between the effects of pulse compression and pulse broadening in the external cavity. It is clear that the important effect is the coupling of modes of different frequency in the fundamental cavity even though different values of $P$ were used in each case.

These models represent two extreme transmission characteristics. The differences are reflected in different values of the cavity phase mismatch required to achieve modelocking. The mirror equations (5.6) show that the returning wave will be in phase with the cavity when $P = 1$ and out of phase when $P = 0$. The saturable absorber is optimum for $P = 1$ as this adds preferentially to the high intensity parts of the laser cavity field. By contrast, the saturable amplifier is optimum for $P = 0$ as the external cavity field now subtracts preferentially from the low intensity parts of the laser field.

### 5.8 The non-soliton laser

The results of the previous section showed that improved modelocking could be obtained with quite diverse nonlinear elements by choosing the interferometric phase to give pulse compression in the main cavity. This theoretical prediction has recently been confirmed [35]. We discussed, in the first section, the waveguide contribution to the dispersion. This can be used to move the wavelength of zero dispersion. In the experiment [35] a fibre was chosen with a dispersion zero at a longer wavelength than $1.6\mu m$ so that (bright) soliton solutions do not occur at the FCL wavelength ($1.5\mu m$). This fibre was shown to lead to nonlinear pulse broadening and was thus qualitatively similar to the saturable amplifier considered in section (5.7). The fibre was used in a standard soliton laser coupled cavity configuration and the results obtained are shown in figure (5.7). When the external cavity is included (5.7b) the laser output pulse is significantly shorter than the laser produces alone (5.7a). This result clearly demonstrates that the crucial property of the external cavity is its nonlinear response. The soliton laser is one example of a much wider class of nonlinear external cavity modelocked lasers.

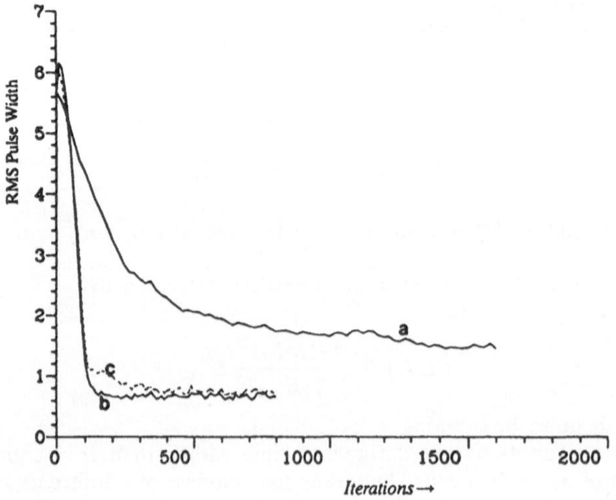


Fig. 5.6 Evolution of the laser RMS pulse width a) no external cavity, b) saturable amplifier and c) saturable absorber.

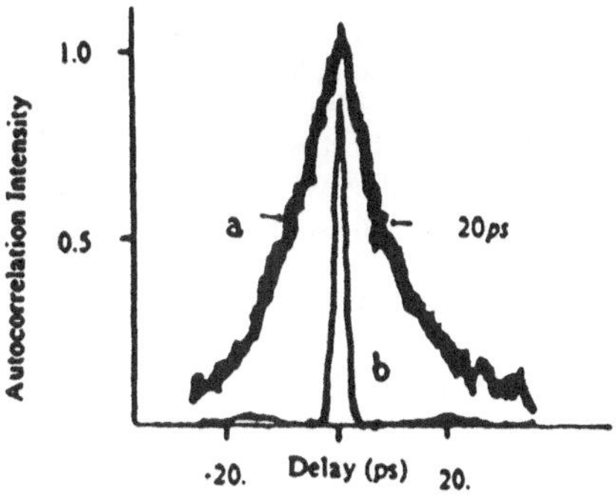


Fig. 5.7 Experimental results obtained with a non-soliton supporting fibre in an external cavity laser a) laser alone and b) with the external cavity.

Recently, more examples of nonlinear coupled cavity modelocking have been observed with various lasers and nonlinear elements. Colour centre lasers have been modelocked with a semiconductor laser in the cavity [37] which has a response similar to the saturable amplifier discussed here although with a much longer response time. More recent results [38] have been achieved with $Ti{:}Al_2O_3$ as the lasing medium and a (non-soliton) optical fibre in the external cavity.

## 6. SOLITONS: THE FUTURE

Finally we consider where soliton and short pulse effects in fibres are likely to lead.

Long distance non dispersing propagation has been demonstrated [39] although not yet for a random bit stream. Other nonlinear effects in particular Raman effects are seen to be the most serious limitation on such transmission and mean that it is only in the very long distance regime where solitons may prove to be of benefit.

Solitons and nonlinear fibre effects have been utilised in many short pulse generation schemes. The role of nonlinear fibre effects is central to most pulse compression techniques, Raman soliton generation and 'soliton laser' applications. We have seen here how the 'soliton laser' has been generalised to nonlinear external cavity lasers and we expect many examples of this phenomenon to be useful in the future.

Switching is a topic of current interest and a number of schemes based on soliton properties are being tried. We have discussed one example of 'whole pulse' switching using solitons but the general concept has much wider applicability.

## REFERENCES

1. A.W.Snyder and J.D.Love, Optical Waveguide Theory, Chapman and Hall, London (1983)
2. A.Hasegawa and F.Tappert, Transmission of Stationary Nonlinear Optical Pulses in Dispersive Dielectric Fibres 1. Anomalous Dispersion, Appl Phys Lett 23 142-144 (1973)
3. L.F.Mollenauer, R.H.Stolen and J.P.Gordon, Experimental Observation of Picosecond Pulse Narrowing and Solitons in Optical Fibres, Phys Rev Lett 45 1095-8 (1980)
4. M.Nakazawa, Y.Kimura and K.Suzuki, Soliton Amplification and Transmission with $Er^{3+}$ Doped Fibre Repeater Pumped by GaInAsP Laser Diode, Elect Lett 25 199-200 (1989)
5. V.E.Zakharov and A.B.Shabat, Exact Theory of Two Dimensional Self Focusing and One Dimensionsal Self Modulation of Nonlinear Waves in Nonlinear Media, Sov Phys JETP 34 62-69 (1972)

6. J.Satsuma and N.Yajima, Initial Value Problems of One Dimensional Self-Modulation of Nonlinear Waves in Dispersive Media, Prog Theor Phys Suppl 55 284-306 (1974)
7. K.J.Blow and N.J.Doran, Bandwidth Limits of Nonlinear (Soliton) Optical Communication Systems, Elect Lett 19 429-430 (1983)
8. C.Desem, Ph.D Thesis, Univeristy of New South Wales, Sydney, Australia (1988)
9. K.J.Blow and N.J.Doran, Multiple Dark Soliton Solutions of the Nonlinear Schrodinger Equation, Phys Lett A 107A 55 (1985)
10. A.M.Weiner, J.P.Heritage, R.J.Hawkins, R.N.Thurston, E.M.Kirschner, D.E.Laird and W.J.Tomlinson, Experimental Observation of the Fundamental Dark Soliton in Optical Fibres, Phys Rev Lett 61 2445-48 (1988)
11. C.Froely, B.Colombeau and M.Vampouille,, in Progress in Optics, Vol 10 edited by E.Wolf (North-Holland, Amsterdam, 1983), pp 115-121
12. Y.C.Ma, The Perturbed Plane Wave Solutions of the Cubic Schrodinger Equation, SIAM 60 43-48 (1979)
13. T.B.Benjamin and J.E.Feir, The Disintegration of Wave Trains in Deep Water Part 1. Theory, J Fluid Mech 27 417-430 (1967)
14. D.Anderson and M.Lisak, Modulational Instability of Coherent Optical Fibre Transmission Signals, Optics Lett 9 468 (1984)
15. E.Fermi, J.Pasta and S.Ulam, Collected Papers of Enrico Fermi, ed. E.Segre 2 978 (1965)
16. K.J. Blow and D. Wood, Theoretical Description of Transient Stimulated Raman Scattering in Optical Fibres, IEEE J Quantum Electronics 25 2665-73 (1989)
17. Elements of Soliton Theory, G.R.Lamb, Wiley Interscience, New York (1980)
18. D.Cotter, Stimulated Brillouin Scattering in Monomode Optical Fibre, J Optical Commun 4 10-19 (1983)
19. R.H. Stolen, C. Lee and R.K. Jain, Developement of the Stimulated Raman Spectrum in Single-Mode Silica Fibres, J Opt Soc Am B 1, 652-657 (1984)
20. K.J.Blow and B.P.Nelson, Observation of Stimulated Raman Scattering and Nonlinear Pulse Broadening at 1.32$\mu m$ in Monomode Optical Fibres,IEE PROC J 134 161-162 (1987)
21. E.M.Dianov, A.YA Karasik, P.V.Mamyshev A.M.Prokhorov, V.N.Serkin, M.F.Stel'makh and A.A.Fomichev, Stimulated Raman Conversion of Multisoliton Pulses in Quartz Optical Fibres, Pisma Zh Eksp Teor Fiz 41 242-244 (1985)
22. F.M.Mitschke and L.F.Mollenauer, Discovery of the Soliton Self Frequency Shift, Optics Lett 11 659 (1986)
23. J.P.Gordon, Theory of the Soliton Self Frequency Shift, Optics Lett 11 662 (1986)
24. A.S.Gouveia-Neto, A.S.L.Gomes and J.R.Taylor, High Efficiency Single Pass Soliton-Raman Compression in an Optical Fibre around 1.4$\mu$m, Optics Letts 12 1035-1037 (1987)
25. A.S.Gouveia-Neto, A.S.L.Gomes, J.R.Taylor, Femtosecond Soliton Raman Generation, IEEE J Quantum Electronics 24 332-40 (1988)
26. K.J.Blow, N.J.Doran and D.Wood, Trapping of Energy into Solitary Waves in Amplified Nonlinear Dispersive Systems, Optics Lett 12 1011-13 (1987)
27. K.J.Blow, N.J.Doran and D.Wood, Generation and Stabilisation of Short Soliton Pulses in the Amplified Nonlinear Schrodinger Equation, J Opt Soc Am B 5 381-3 90 (1988)
28. R.H.Stolen, J.P.Gordon, W.J.Tomlinson and H.A.Haus, Raman Response Function of Silica-Core Fibres, J Opt Soc Am B 6 1159-66 (1989)
29. N.J.Doran, Solitons in Optical Fibres and Nonlinear Fibre Devices, Paper WA6 IQEC (1988)
30. S.R.Friberg, A.M.Weiner, Y.Silberberg, B.G.Sfez and P.W.Smith, Femtosecond Switching in a Dual-Core Fibre Nonlinear Coupler, Optics Lett 13 904-6 (1988)
31. K.J.Blow, N.J.Doran and B.K.Nayar, Experimental Demonstration of Optical Soliton Switching in an All-Fibre Nonlinear Sagnac Interferometer, Optics Lett 14 754-6 (1989)
32. M.N.Islam, E.R Sunderman, R.H.Stolen, W.Pleibel and J.R.Simpson, Soliton Switching in a Fibre Nonlinear Loop Mirror, Optics Letts 14 811-3 (1989)
33. N.J.Doran and K.J.Blow, Solitons in Optical Communications, IEEE J Quantum Electronics 19 1883-88 (1983)

34. F.M.Mitschke and L.F.Mollenauer, Stabilizing the Soliton Laser, IEEE J Quantum Electronics 22 2242-2250 (1986)
35. K.J.Blow and B.P.Nelson, Improved Modelocking of an F-Centre Laser with a Nonlinear Nonsoliton External Cavity, Optics Lett 13 1026-29 (1988)
36. H.A.Haus, A Theory of Forced Mode Locking, IEEE J Quantum Electronics 11 323-330 (1975)
37. P.N.Kean, X.Zhu, D.W.Crust, R.S.Grant, N.Langford and W.Sibbett, Enhanced Modelocking of Colour Centre Lasers, Optics Lett 14 39-41 (1989)
38. J.Goodberlet, J.Wang, J.G.Fujimoto and P.A.Schulz, Femtosecond Passively Modelocked $Ti{:}Al_2O_3$ Laser with a Nonlinear External Cavity, Optics Lett 20 1125-7 (1989)
39. L.F.Mollenauer and K.Smith, Demonstration of Soliton Transmission over more than $4000km$ in Fibre with Loss Periodically Compensated by Raman Gain, Optics Lett 13 675-7 (1988)

## LINEAR AND NONLINEAR THEORY OF SURFACE POLARITON DIFFRACTION

A. I. Voronko, Yu. V. Gulyaev, G. N. Shkerdin

Institute of Radioengineering and Electronics
USSR Academy of Sciences
Marx Avenue 18, 103907 Moscow, Centre, USSR

### 1. INTRODUCTION

It is now well established that dielectric structures with interfaces can support surface and guided electromagnetic waves. This has brought about the emergence and rapid development of such fields as integrated optics[1-3] and surface acoustics.[4] The characteristic distances over which major processes at interfaces occur are, in fact, microdistances. This is important both from the point of view of investigating the interface microstructure,[5-6] using spectroscopic methods, and for the development and design of integrated-optical devices for information processing.[7] A complete description of surface wave interactions must include nonlinear effects and this aspect will be considered here.

The investigations of nonlinear surface electromagnetic waves that have been reported up to now[8-12] have opened up the field of energy controlled processing that may lead to the development, in principle, of new elements for optical signal processing.[13-14]

We will consider the propagation of surface electromagnetic waves whose optical wavelength range permits the use of a phenomenological approach in which the electrodynamic properties of the guiding media are characterized by tensor dielectric permittivities $\epsilon_{ij}$ and tensor magnetic permeabilities $\mu_{ij}$. Only the nonmagnetic case $\mu_{ij} = \mu_0\delta_{ij}$ will be considered, however.

One of the simplest cases, an interface between two media characterised by the dielectric tensors $\epsilon_{ij}^{(1)}$ and $\epsilon_{ij}^{(2)}$ (complex in the general case), is illustrated in Fig. 1. Even in this simple structure, surface electromagnetic wave propagation is possible, with many of the properties of the waves being typical of waves in more complicated structures. It is assumed that the structure is planar. Any dependence of the dielectric constant on y is neglected by setting $d/dy = 0$, and propagation is along the x axis. This approach, common for such investigations, is substantiated by the following. Firstly, structures whose width along the y axis greatly exceeds the wavelength $\lambda$ of the electromagnetic radiation (it is this circumstance that constitutes the condition for planarity) are often used. Secondly, TE- and TM-polarized waves are separable in this case and hence are independent. This is a simplifying move that still permits a clear explanation of many phenomena at interfaces. TE and TM modes are also known as s- and p-polarized modes respectively and they are defined by the field component groups ($E_y$, $H_x$, $H_z$) and ($H_y$, $E_x$, $E_z$).

A surface polariton is a localized p-polarized mode[5-6] propagating along the interface between two media in the case where the dielectric constant of one of the media is negative, i.e. $\epsilon_{ij}^{(2)} = \epsilon(\omega)\delta_{ij}$, $\epsilon(\omega) < 0$. These waves are called polaritons because the dielectric permittivity may be negative in a wavelength region where an intrinsic crystal resonance exists. In this case the light propagating in the crystal assumes a polaritonic (hybrid) character. Depending on the elementary excitations involved, the surface polaritons can be classified as surface-plasmons, -phonons, -excitons or -magnons. In other words, the terminology reflects the specific conditions necessary for the existence of the surface polariton but, in principle, these waves are all really of the type shown in Fig. 1.

*Nonlinear Waves in Solid State Physics*
Edited by A.D. Boardman *et al.*, Plenum Press, New York, 1990

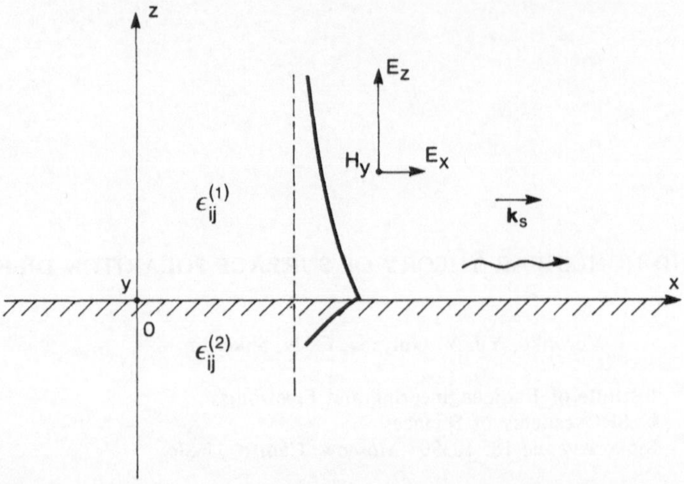


Fig. 1. Boundary between two semi-infinite dielectric media which can support the propagation of surface polaritons. The media are characterized by $\epsilon_{xx}$, $\epsilon_{zz}$ at $z > 0$ and $\epsilon(\omega)$ at $z < 0$; $\mathbf{k}_s$ is the wave vector of the surface polariton.

Anisotropy can be introduced into the problem through $\epsilon_{ij}^{(1)}$ or $\epsilon_{ij}^{(2)}$, but we shall concentrate on $\epsilon_{ij}^{(1)}$ which will be assumed to be a uniaxial crystal with $\epsilon_{11}^{(1)} = \epsilon_{22}^{(1)} = \epsilon_{xx}$ and $\epsilon_{33}^{(1)} = \epsilon_{zz}$. This situation is chosen by keeping in mind that, in the phonon region of the spectrum,[15] such a case is common.

Figure 1 shows the wave number and the transverse envelope of the field ($H_y$ component). In the case of plane electromagnetic wave propagation in a homogeneous medium the normalised wave number is also the refractive index of the medium in the x direction, and the transverse amplitude envelope is simply a plane perpendicular to the propagation direction x. This chapter will deal mainly with these two surface polariton characteristics. Indeed the initial investigation of surface waves in any structure, linear or not, reduces to the investigation of these quantities.

Surface polariton excitation by a bulk electromagnetic wave requires the projection of the bulk electromagnetic wave number onto the interface to be equal (phase-matched) to the surface polariton wave number. In a medium at $z > 0$, however, the bulk wave number ($k_v = k_0\sqrt{\epsilon_{xx}}$) is always too small[5-6] and the phase-matching condition is never fulfilled at an ideal interface. It can be achieved, however, by introducing a prism into the system, thus providing effective surface polariton excitation. A typical phase-matching geometry is shown in Fig. 2. This method is called attenuated total-internal-reflection or ATR.[16,17]

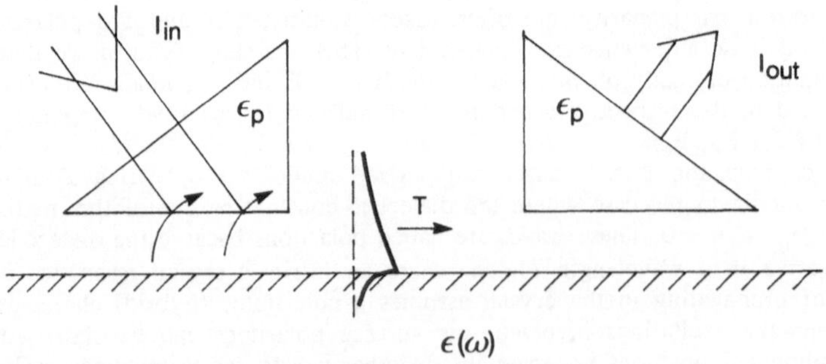


Fig. 2. Excitation and detection of a surface polariton by means of attenuated total-internal-reflection (phase method). T is the amplitude of the surface polariton.

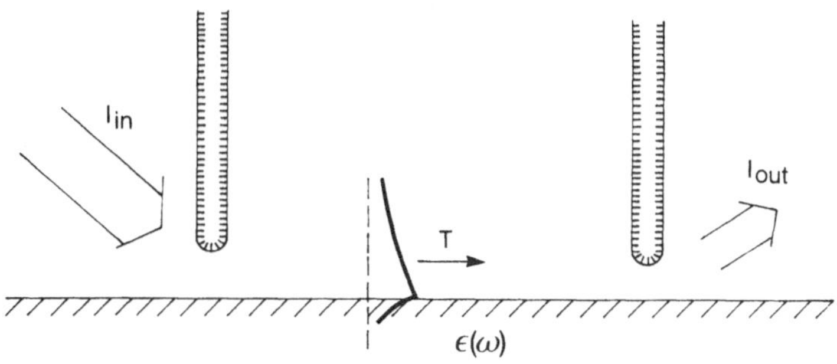


Fig. 3. Excitation and detection of a surface polariton by means of a diaphragm (integral method).

Another surface polariton excitation method has been developed in which the bulk and surface waves are matched by the transverse envelope. In other words, the bulk radiation is focused into a spot of the order of magnitude of the surface polariton field (the integral method). A typical example of the integral method is that realized at a razor blade (Fig. 3).[18] In this case the distance between the razor blade and interface is chosen so that the bulk radiation incident on this structure is localized in the surface polariton localization zone. This provides effective surface polariton excitation without fulfilment of the phase matching conditions.

Thus there exist, for smooth surfaces, two linear excitation methods, i.e phase-matching and integral. We will not discuss nonlinear excitation methods,[19-21] and nonlinearity will be considered only in cases where it may influence diffraction processes that have an analogue in the linear case. The integral surface polariton excitation method is one such linear diffraction process. The main feature of the integral method is the bulk-to-surface radiation energy conversion due to the presence of an inhomogeneity, introduced by the razor blade, at the interface. This is the opposite of surface polariton propagation over irregular surfaces where the surface radiation is transformed into bulk radiation.[22]

Let us take as an example surface polariton reflection from a dielectric barrier. The geometry of the process is shown in Fig. 4. By a barrier we mean a jump in the permittivity of

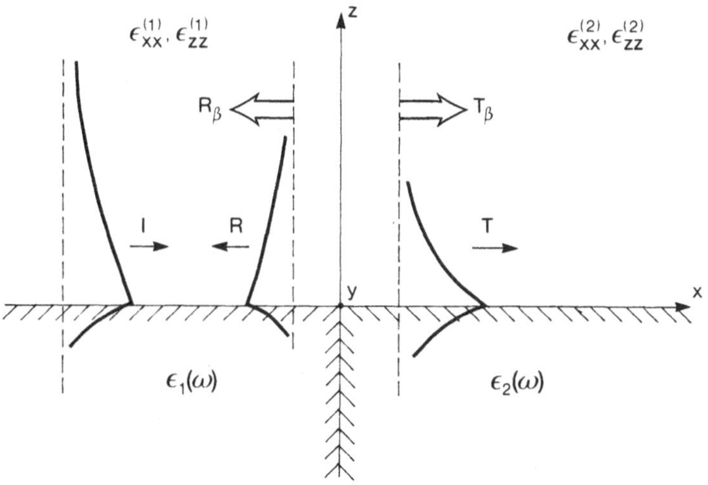


Fig. 4. Surface polariton reflection at a dielectric barrier. The dielectric permittivities to the left [$x < 0$, $\epsilon_{xx}^{(1)}$, $\epsilon_{zz}^{(1)}$ and $\epsilon_1(\omega)$] and to the right [$x > 0$, $\epsilon_{xx}^{(2)}$, $\epsilon_{zz}^{(2)}$ and $\epsilon_2(\omega)$] correspond to those in Fig. 1. R, $R_\beta$ and T, $T_\beta$ are the reflection and transmission coefficients of a surface polariton and a surface polariton radiation mode, respectively.

one or both of the media at the interface along which the polaritons propagate. Two properties of such barriers that will be of interest to us should be stressed. First, the dielectric barrier is vertical and, second, surface polaritons are possible on the left- and right-hand sides of the barrier. The latter requirement concerns the dielectric constant values on both sides of the barrier and may always be satisfied. This idea may easily be generalized to arbitrary dielectric barrier geometries.

If a surface polariton is incident on a dielectric barrier, as shown in Fig. 4, reflected and transmitted surface polaritons are excited. However, since the transverse surface polariton envelopes are different on the left- and right-hand sides of the barrier, satisfaction of the field continuity conditions at the barrier in the presence only of surface waves is impossible.

There is only one way to resolve this difficulty and that is to take into account the bulk radiation diffracted during the process of reflection. There are a number of approaches to the solution of this problem. In the case of the geometry of Fig. 4 the most expedient method would seem to be that developed in Ref. 23. However, in this approach the structure was screened with metal mirrors at $z = \pm d$ to form a metal waveguide, and the bulk field was written in the form of a sum of the normal modes of the metal-clad guide. A rigorous discussion of this approach will not be given here since we prefer to give a final comparison in term of numerical results.

We will now explain the mode-method of investigating surface diffraction problems. This method does not introduce additional conditions, such as structure screening and so on. In addition to the description of the formalism, its application will be illustrated. First of all, the problem of normal incidence of a surface polariton on a dielectric barrier will be considered. Then the process of nonlinear surface polariton diffraction at a thin resonant film will be discussed.

## 2. LINEAR THEORY OF SURFACE POLARITON DIFFRACTION AT A DIELECTRIC BARRIER

The investigation of both linear and nonlinear surface diffraction can be formulated within the framework of classical dispersion theory, taking into account the fact that we are dealing with surface modes. The surface polariton diffraction at a dielectric barrier will be described as the excitation of all possible normal electrodynamic modes of the structure. The excitation of these eigenmodes is controlled by perturbations in the translational symmetry along the surface polariton propagation direction. In other words, violation of momentum conservation results in the ability of a surface polariton, having a fixed momentum or wave number, to excite normal modes at the barrier with arbitrary wave numbers. The excitation process is described by the overlap integrals of the normal modes, as in the theory of quantum mechanics.[24]

### 2.1 General definitions

The surface polaritons considered here are solutions of Maxwell's equations, defined in the form

$$\mathrm{curl}\,\mathcal{E} = -\frac{\partial \mathcal{B}}{\partial t} \qquad \mathrm{curl}\,\mathcal{H} = \frac{\partial \mathcal{D}}{\partial t} \tag{2.1}$$

where the material relations

$$\mathcal{D}_i = \epsilon_0 \epsilon_{ij} \mathcal{E}_j \qquad \mathcal{B}_i = \mu_0 \mathcal{H}_i \tag{2.2}$$

will be assumed. In the coordinate system chosen here the fields have the following (x,z) dependence

$$\{\mathcal{E}(x,z),\ \mathcal{H}(x,z)\} = \{\mathbf{E}(z),\ \mathbf{H}(z)\} e^{i(k_x x - \omega t)} \tag{2.3}$$

where $k_x$ is a wave number. Since we are considering planar structures the electromagnetic waves can be separated into two independent polarizations (s and p). Furthermore, it is easy to show that the surface polaritons may only be p-polarized, having the components $\mathcal{H}_y$, $\mathcal{E}_z$ and $\mathcal{E}_x$. Since in a normal non-magnetic diffraction process no extra polarization is generated, it is necessary to consider only p-polarized modes. In addition, the wave number in Eq. (2.3) may be associated with both surface polariton and other modes in the system, thus making the formalism general.

The following relations for the field components of p-polarized modes may be derived

$$E_x(z) = -\frac{i}{\omega\epsilon_0}\frac{1}{\epsilon_{xx}}\frac{\partial H_y(z)}{\partial z} \qquad E_z(z) = \frac{i}{\omega\epsilon_0}\frac{1}{\epsilon_{zz}}\frac{\partial H_y(z)}{\partial z} \tag{2.4}$$

and the $H_y(z)$ component is the solution of

$$\frac{\partial^2 H_y(z)}{\partial z^2} + (k_0^2\epsilon_{zz} - k_x^2)\frac{\epsilon_{xx}}{\epsilon_{zz}}H_y(z) = 0 \tag{2.5}$$

The asymptotic behaviour of $H_y(z)$ as $z \to \infty$ must be fixed[25] and the interface boundary conditions and surface wave energy flow must be considered. The boundary conditions describe the behaviour of the electromagnetic radiation as jumps in the dielectric constant value occur.[26] They, of course, follow directly from Eqs. (2.1) and (2.2) and are widely used in the study of numerous processes in both bulk[26] and integrated optics.[1-3] The process that we are considering here is characterized by two types of boundary conditions, i.e. the continuity of the field components at $z = 0$ and the continuity of the field components at $x = 0$. The position of the dielectric barrier corresponds to the planes $z = 0$ and $x = 0$, and thus the solution of Eq. (2.5) may be sought independently in all of the four quadrants of Fig. 4. The same is true for the components $E_x$ and $E_z$. The solutions must satisfy the continuity of field components tangential to the media interfaces, i.e.

$$H_y(z{=}0_-) = H_y(z{=}0_+) \qquad E_x(z{=}0_-) = E_x(z{=}0_+) \tag{2.6a}$$

$$\mathcal{H}_y(x{=}0_-,z) = \mathcal{H}_y(x{=}0_+,z) \qquad \mathcal{E}_z(x{=}0_-,z) = \mathcal{E}_z(x{=}0_+,z) \tag{2.6b}$$

where the boundary conditions (2.6a) describe the behaviour of the components $H_y(z)$ and $E_x(z)$ of each of modes at $z = 0$ and the boundary conditions (2.6b) describe the behaviour of the components $\mathcal{H}_y(x,z)$ and $\mathcal{E}_z(x,z)$ of the total field in the system at $x = 0$.

This reasoning is abstract at this stage and is not related to the initial statement of the problem. This is that surface polaritons, whose existence ensures the continuity of their components at $z = 0$, are incident on the left-hand side of the dielectric barrier. It is only during the process of the transformation of the surface wave into all other modes of the system that the boundary conditions for $x = 0$ come into force. The continuity conditions, Eq. (2.6a), give the normal modes of the structure for $x < 0$ and $x > 0$ and then, from Eq. (2.6b), the amplitudes of these modes can be found. This separation of steps in the investigation of diffracted radiation is the principal feature of the mode-method.

Thus, the surface diffraction process can be investigated, in principle, in two-dimensions even though it involves bulk radiation as well. After the continuity at $z = 0$ is imposed and the eigenmodes are found, each of these two-dimensional modes is characterized by two parameters. These are the transverse envelope (dependence of $H_y$ on z) and the wave number $k_x$. Each wave propagates in the x-y plane and carries energy along only the x-direction. Therefore the x-component of the energy flow is an important characteristic of the wave. The energy flow density per unit length along the y-axis is the integral along the z-axis which from the Poynting vector is given by

$$P_x = \frac{1}{2}Re\int_{-\infty}^{+\infty}[\mathcal{E}\mathcal{H}^*]_x dz \tag{2.7}$$

There are two types of solution for Eq. (2.5) having finite values of $P_x$. For the first, each of the waves goes to zero at infinity. For the second, the integral over these waves satisfies the above mentioned condition, but each of the waves satisfies the less stringent condition $|H_y(\pm\infty)| < C$ at $|z| \to \infty$, where C is a positive constant. The first type of solution corresponds to localized modes (discrete spectrum), the second to radiation modes (continuous spectrum). A rigorous discussion leading to such a classification is given in Ref. 25.

## 2.2 Surface Polaritons

Localized interface modes have been investigated rather exhaustively,[5-6] but for completeness the main relationships will be given here. Surface polaritons decay as $|z| \rightarrow \infty$ so that

$$H_y^s(z) = \begin{cases} A_1 e^{-k_1 z} & z > 0 \\ A_2 e^{k_2 z} & z < 0 \end{cases} \tag{2.8}$$

where $k_1$ and $k_2$ are transverse wave numbers expressed in terms of the surface polariton wave number $k_s = k_x = k_0 n_f$ as

$$k_1^2 = \frac{\epsilon_{xx}}{\epsilon_{zz}}(n_f^2 - \epsilon_{zz})k_0^2 \qquad k_2^2 = [n_f^2 - \epsilon(\omega)]k_0^2 \tag{2.9}$$

Here $k_0 = \omega/c = 2\pi/\lambda$ is the wave number in vacuum or air and $n_f$ is the surface polariton propagation constant. Obviously, the surface polariton has a localized character if $k_1^2,\ k_2^2 > 0$. This is satisfied provided that $n_f^2 > \epsilon_{zz}$. Therefore a surface polariton is a slow wave with a phase velocity $v_s = c/n_f < c/\sqrt{\epsilon_{zz}}$, where c is the velocity of light *in vacuo*.

The continuity condition of Eq. (2.6a) leads, from Fig. 1, to the dispersion relation

$$\frac{k_1}{\epsilon_{xx}} = -\frac{k_2}{\epsilon(\omega)} \tag{2.10}$$

and $A_1 = A_2$. From Eq. (2.10) it follows that the dielectric function of one of the media must be negative if the equation is to have any localised solutions. The dispersion equation of a surface polariton is therefore

$$n_f^2 = \epsilon_{zz}\epsilon(\omega)\frac{\epsilon(\omega) + \epsilon_{zz}}{\epsilon^2(\omega) - \epsilon_{xx}\epsilon_{zz}} \tag{2.11}$$

If we model the frequency dependence of $\epsilon(\omega)$ on the optical phonon case[5] (Fig. 5) then the dispersion curve of the surface polariton, shown in Fig. 6, is obtained from Eq. (2.11). The curve is, in principle, similar to that obtained at plasma frequencies. It should be taken into account, however, that for plasma oscillations $\omega_T = 0$, so that the curve in Fig. 6 for plasmons would begin at the coordinate origin.

The above analysis has defined the surface polariton wave number and the field envelope, Eq. (2.8). The constant $A_1$ consists of a normalization constant $A_0$ and an energy parameter (amplitude) S, i.e. $A_1 = SA_0$. The value of S is determined by the excitation conditions (for example S = T in Figs. 2 and 3). The normalization constant $A_0$ remains undefined and will be derived later.

Equations (2.3) and (2.8) yield the following total form for the surface polariton components:

$$\{\mathcal{E}^s(x,z), \mathcal{H}^s(x,z)\} = S\{\mathbf{E}^s(z), \mathbf{H}^s(z)\}e^{i(k_s x - \omega t)} \tag{2.12}$$

where the transverse envelopes of the surface polariton are

$$H_y^s(z) = A_0 \begin{cases} e^{-k_1 z} & z > 0 \\ e^{k_2 z} & z < 0 \end{cases} \tag{2.13}$$

$$E_z^s(z) = -\frac{1}{\omega\epsilon_0}k_s A_0 \begin{cases} \dfrac{1}{\epsilon_{zz}}e^{-k_1 z} & z > 0 \\ \dfrac{1}{\epsilon(\omega)}e^{k_2 z} & z < 0 \end{cases} \tag{2.14}$$

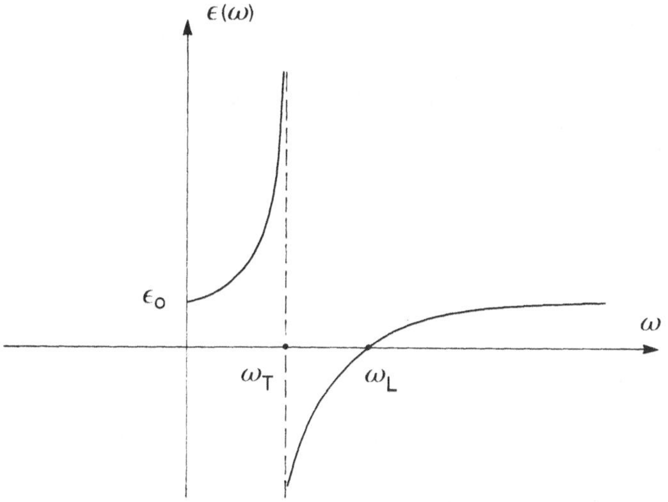


Fig. 5. Dependence of the dielectric permittivity of a surface-active crystal $\epsilon(\omega)$ upon frequency in the neighbourhood of a dipole oscillation. $\omega_L$ and $\omega_T$ are the longitudinal and transverse frequency of the oscillations, respectively.

$$E_x^s(z) = \frac{i}{\omega\epsilon_0} A_0 \begin{cases} \dfrac{k_1}{\epsilon_{xx}} e^{-k_1 z} & z > 0 \\ -\dfrac{k_2}{\epsilon(\omega)} e^{k_2 z} & z < 0 \end{cases} \tag{2.15}$$

## 2.3 Surface polariton radiation modes

As has already been noted a surface polariton radiation mode must be confined at $|z| \to \infty$. This is satisfied by the following solutions of Eq. (2.5)

$$H_y^\beta(z) = \begin{cases} B_1 e^{-i\beta z} + B_2 e^{i\beta z} & z > 0 \\ B e^{\rho z} & z < 0 \end{cases} \tag{2.16}$$

where $\beta$ and $\rho$ are called transverse surface polariton radiation mode wave numbers, and are defined through the equations

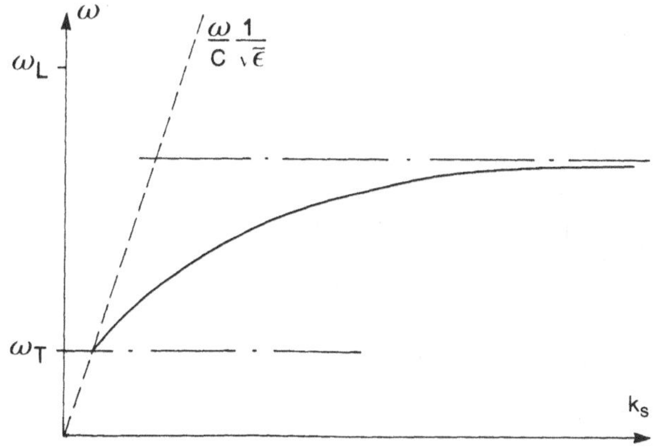


Fig. 6. Typical dispersion law of a surface polariton in the absence of damping ($Im\epsilon_{ij} = 0$).

$$k_x^2 + \beta^2 \frac{\epsilon_{zz}}{\epsilon_{xx}} = k_0^2 \epsilon_{zz} \qquad k_x^2 - \rho^2 = k_0^2 \epsilon(\omega) \tag{2.17}$$

From Eq. (2.6a) the following set of equations for the constants $B_1$, $B_2$ and B are obtained:

$$B_1 + B_2 = B \qquad B_1 - B_2 = i \frac{\rho}{\beta} \frac{\epsilon_{xx}}{\epsilon(\omega)} B$$

Thus we have two equations for three unknowns. It may easily be seen, however, that this is directly associated with the continuity in $\beta$-space of the surface polariton radiation mode spectrum, i.e. it is possible to find the solution of a surface polariton radiation mode for an arbitrary value of $\beta$. If the constant B is taken to be independent then, for the remaining constants, we obtain

$$B_1 = \frac{1}{2} B \left( 1 + i \frac{\rho}{\beta} \frac{\epsilon_{xx}}{\epsilon(\omega)} \right) \qquad B_2 = \frac{1}{2} B \left( 1 - i \frac{\rho}{\beta} \frac{\epsilon_{xx}}{\epsilon(\omega)} \right) \tag{2.18}$$

Equations (2.17)-(2.18) clearly show that the surface polariton radiation modes are continuous and contain both unattenuated waves ($k_x^2 > 0$), for which $0 < \beta^2 < k_0^2 \epsilon_{xx}$, and decaying waves ($k_x^2 < 0$), for which $k_0^2 \epsilon_{xx} < \beta^2 < m^2$. The limitation on $\beta$ in the latter relation is necessary for $\rho^2$ to be positive which, from Eq. (2.17), is

$$\rho^2 = k_0^2 [-\epsilon(\omega) + \epsilon_{zz}] - \frac{\epsilon_{zz}}{\epsilon_{xx}} \beta^2 \tag{2.19}$$

The value of m, derived from Eq. (2.19) for $\rho = 0$, is

$$m = k_0 \sqrt{\frac{\epsilon_{xx}}{\epsilon_{zz}} [-\epsilon(\omega) + \epsilon_{zz}]} \tag{2.20}$$

It should be noted that in principle there may exist waves with imaginary $\rho$ values ($\rho^2 < 0$). Such surface polariton radiation modes would have the same structure for both $z > 0$ and $z < 0$. However, it is likely that $\epsilon(\omega) = -|\epsilon(\omega)|$ would be large so, from Eq. (2.19), it follows that surface polariton radiation modes with imaginary $\rho$ values would correspond to large $\beta$ values. Since these waves oscillate rather fast, their contribution to a diffraction process would be small and they will be neglected.

As in the case of the surface polariton, the constant B consists of a surface polariton radiation mode normalization constant $B_\beta$ and its energy amplitude $S_\beta$. Therefore the form of the surface polariton radiation mode components is

$$\{\mathcal{E}^\beta(x,z), \mathcal{H}^\beta(x,z)\} = S_\beta \{\mathbf{E}^\beta(z), \mathbf{H}^\beta(z)\} e^{i(k_x x - \omega t)} \tag{2.21}$$

where the transverse envelopes are

$$H_y^\beta(z) = B_\beta \begin{cases} \cos(\beta z) + \frac{\rho}{\beta} \frac{\epsilon_{xx}}{\epsilon(\omega)} \sin(\beta z) & z > 0 \\ e^{\rho z} & z < 0 \end{cases} \tag{2.22}$$

$$E_z^\beta(z) = -\frac{1}{\omega \epsilon_0} k_x B_\beta \begin{cases} \frac{1}{\epsilon_{zz}} \left[ \cos(\beta z) + \frac{\rho}{\beta} \frac{\epsilon_{xx}}{\epsilon(\omega)} \sin(\beta z) \right] & z > 0 \\ \frac{1}{\epsilon(\omega)} e^{\rho z} & z < 0 \end{cases} \tag{2.23}$$

$$E_x^\beta(z) = \frac{i}{\omega\epsilon_0} B_\beta \begin{cases} \frac{\beta}{\epsilon_{xx}}\left[\sin(\beta z) - \frac{\rho}{\beta}\frac{\epsilon_{xx}}{\epsilon(\omega)}\cos(\beta z)\right] & z > 0 \\ -\frac{\rho}{\epsilon(\omega)} e^{\rho z} & z < 0 \end{cases} \tag{2.24}$$

## 2.4 The structure of a diffracted field

The treatment of the continuous spectrum of the surface polariton radiation modes and the discrete spectrum of the surface polaritons requires a knowledge of how these modes are interrelated. This requires an investigation of certain overlap integrals which define the normalization and orthogonality conditions of the modes. This problem is treated in almost any book on integrated optics.[1-3] The orthogonality condition of the surface polariton modes follows directly from Maxwell's equations and is valid for fields belonging to the same guiding system.

It should be remembered here that we are now independently investigating waves with boundary conditions (2.6a) on the left- and right-hand sides of the dielectric barrier, i.e. we are interested in eigenmodes of a structure which is infinite along x and has the boundary conditions of Fig. 1. In this case, if $H_y^s(z)$ and $E_z^s(z)$ are components of the surface polariton, and $H_y^\beta(z)$ and $E_z^\beta(z)$ are those of the surface polariton radiation mode, then, after some simple transformations, we obtain

$$\int_{-\infty}^{+\infty} H_y^{s*}(z)E_z^\beta(z)dz = \int_{-\infty}^{+\infty} H_y^{\beta*}(z)E_z^s(z)dz = 0 \tag{2.25}$$

The normalization conditions for surface polaritons and surface polariton radiation modes are chosen so that the energy amplitudes of the modes are clearly distinguished in the expressions for the fields. For surface polaritons the normalisation condition is

$$\int_{-\infty}^{+\infty} H_y^{s*}(z)E_z^s(z)dz = -\frac{1}{\omega\epsilon_0}k_s \tag{2.26}$$

and for surface polariton radiation modes the normalisation (orthogonality) condition is

$$\int_{-\infty}^{+\infty} H_y^{\beta*}(z)E_z^{\beta'}(z)dz = -\frac{1}{\omega\epsilon_0}k_x\delta(\beta - \beta') \tag{2.27}$$

where $\delta(\beta - \beta')$ is the usual delta function. Equations (2.26) and (2.27) lead to the following expressions for the normalization constants $A_0$ and $B_\beta$:

$$A_0 = \left(\frac{1}{2k_2\epsilon(\omega)} + \frac{1}{2k_1\epsilon_{zz}}\right)^{-1/2} \tag{2.28}$$

$$B_\beta = \left(\frac{2}{\pi}\epsilon_{zz}\frac{\beta^2}{\beta^2 + \rho^2[\epsilon_{xx}/\epsilon(\omega)]^2}\right)^{1/2} \tag{2.29}$$

The amplitudes S and $S_\beta$ can be derived from the initial conditions for the problem of surface polariton diffraction at a dielectric barrier. In order to find the values of S and $S_\beta$ it is necessary to satisfy the boundary conditions of Eq. (2.6b), i.e. the continuity of $\mathcal{H}_y(x,z)$ and $\mathcal{E}_z(x,z)$, the components of the total field in the system. From the equations for the orthogonality and normalization conditions for these components we may write

(1) for x > 0:

$$\mathcal{H}_y^\tau(x,z) = \left[ TH_y^\tau(z)e^{ik_s^\tau x} + \int_0^{m_2} T_\beta H_{2y}^\beta(z)e^{ik_x^{(2)}x}\,d\beta \right] e^{-i\omega t} \tag{2.30}$$

$$\mathcal{E}_z^\tau(x,z) = \left[ TE_z^\tau(z)e^{ik_s^\tau x} + \int_0^{m_2} S_\beta E_{1z}^\beta(z)e^{ik_x^{(2)}x}\,d\beta \right] e^{-i\omega t} \tag{2.31}$$

where $S = T$, $S_\beta = T_\beta$ and $k_s = k_s^\tau$. The expressions for the components of the transmitted surface polariton, $H_y^\tau(z)$ and $E_z^\tau(z)$, and for the transmitted surface polariton radiation mode, $H_{2y}^\beta(z)$ and $E_{2z}^\beta(z)$, are the same as in Eqs. (2.13)-(2.14) and Eqs. (2.22)-(2.23) with dielectric constants $\epsilon_{xx}^{(2)}$, $\epsilon_{zz}^{(2)}$ and $\epsilon_2(\omega)$. The value of $m_2$ is given by Eq. (2.20).

(2) for x < 0:

$$\mathcal{H}_y^r(x,z) = \left[ RH_y^r(z)e^{-ik_1^i x} + \int_0^{m_1} R_\beta H_{1y}^\beta(z)e^{-ik_x^{(1)}x}\,d\beta \right] e^{-i\omega t} \tag{2.32}$$

$$\mathcal{E}_z^r(x,z) = -\left[ RE_z^r(z)e^{-ik_1^i x} + \int_0^{m_1} R_\beta k_x^{(1)} E_{1z}^\beta(z)e^{-ik_x^{(1)}x}\,d\beta \right] e^{-i\omega t} \tag{2.33}$$

where $S = R$, $S_\beta = R_\beta$ and the wave number of the surface polariton is $k_s = k_s^i$. The components of the reflected surface polariton, $H_y^r(z)$ and $E_z^r(z)$, and the reflected surface polariton radiation mode, $H_{1y}^\beta(z)$ and $E_{1z}^\beta(z)$, are the same as in Eqs. (2.13)-(2.14) and Eqs. (2.22)-(2.23) with dielectric constants $\epsilon_{xx}^{(1)}$, $\epsilon_{zz}^{(1)}$ and $\epsilon_1(\omega)$ (see Fig. 4). For these dielectric constants $m_1$ is given by Eq. (2.20).

Note that the use of the orthogonality and normalization conditions of Eqs. (2.25)-(2.27), written in terms of the energy flow as defined in Eq. (2.7), gives the total energy flow along x as

x > 0 (the transmitted energy flow)

$$P_x^\tau = \frac{1}{\omega\epsilon_0} Re \left[ T^* T k_s^\tau + \int_0^{m_2} T_\beta T_\beta^* k_x^{(2)}\,d\beta \right] \tag{2.34}$$

x < 0 (the reflected energy flow)

$$P_x^r = -\frac{1}{\omega\epsilon_0} Re \left[ R^* R k_s^\tau + \int_0^{m_1} R_\beta R_\beta^* k_x^{(1)}\,d\beta \right] \tag{2.35}$$

where the first terms in Eqs. (2.34) and (2.35) correspond to the surface polariton energy flow and the last terms to the bulk field. In investigating the bulk field as part of the total diffracted radiation [corresponding to the final terms in Eqs. (2.30)-(2.31) and (2.32)-(2.33)] we are interested in the two limiting cases $x = 0$ and $x \to \infty$ (i.e. $x >> \lambda$). The first case is important for satisfying the boundary conditions of Eq. (2.6b) while the second is important

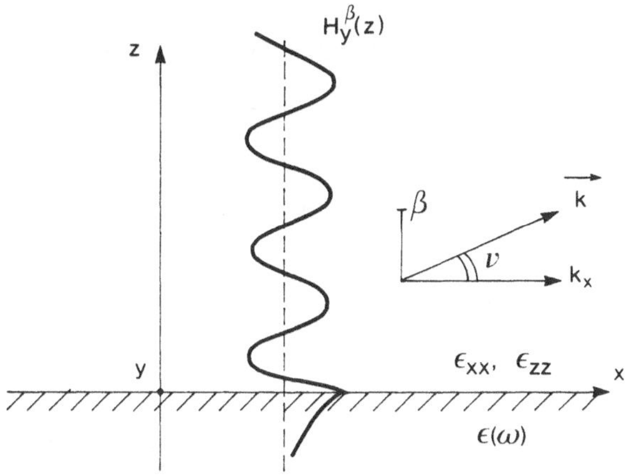


Fig. 7. The structure of surface polariton radiation modes. $k_x$ is the longitudinal and $\beta$ the transverse component of the wave vector **k** of a surface polariton radiation mode.

in the investigation of the far field radiation. This conclusion is based on an experimental observation. An important quantity is $P_x(\nu)$, the energy flow density of the bulk radiation as a function of the angle $\nu$ of Fig. 7. In Eqs. (2.34)-(2.35) the integration is carried out with respect to the surface polariton radiation mode wave number $\beta$. In order to find $P_x(\nu)$, the integration should be carried out with respect to $\nu$. This can be done rather simply by writing

$$k_x = k_0\sqrt{\epsilon_{zz}}\cos(\nu) \qquad \beta = k_0\sqrt{\epsilon_{xx}}\sin(\nu) \tag{2.36}$$

This gives $d\beta = k_0\sqrt{\epsilon_{xx}}\cos(\nu)d\nu$ and the following simple expression for $P_x(v)$:

$$P_x(\nu) = S_\beta(\nu)S_\beta^*(\nu)\sqrt{\epsilon_{xx}\epsilon_{zz}}\cos^2(\nu) \tag{2.37}$$

This expression is valid for $S_\beta = R_\beta$ and $S_\beta = T_\beta$ and will be important in the numerical investigation of the diffraction process.

## 2.5 The surface polariton diffraction process

Having found the fields and energy flows at the interface, we continue with an investigation of the diffraction process, i.e. the mixing of the fields on the left- and right-hand sides of the dielectric barrier. We wish to find the reflected mode amplitudes R and $R_\beta$ and the transmitted mode amplitudes T and $T_\beta$ for an incident surface polariton with amplitude I. The continuity conditions Eq. (2.6b) at the dielectric barrier are

$$\begin{aligned} Re\left[IH_y^i(z)e^{i(k_s^i x-\omega t)} + \mathcal{H}_y^r(x,z)\right]_{x=0} &= Re[\mathcal{H}_y^T(x,z)]_{x=0} \\ Re\left[IE_z^i(z)e^{i(k_s^i x-\omega t)} + \mathcal{E}_z^r(x,z)\right]_{x=0} &= Re[\mathcal{E}_z^T(x,z)]_{x=0} \end{aligned} \tag{2.38}$$

where the transverse envelopes of the incident surface polariton are the same as those of the reflected surface polariton, i.e. $H_y^i(z) = H_y^r(z)$ and $E_y^i(z) = -E_y^r(z)$. After substituting the expressions for the diffracted fields $\mathcal{H}_y^r(x,z)$, $\mathcal{E}_z^r(x,z)$ [Eqs. (2.30)-(2.31)] and $\mathcal{H}_y^T(x,z)$, $\mathcal{E}_z^r(x,z)$ [Eqs. (2.32)-(2.33)] into Eqs. (2.38) the following set of equations for the amplitudes of the system eigenmodes excited during diffraction[27] are obtained:

$$(I + R)H_y^i(z) + \int_0^m R_\beta H_{1y}^\beta(z)d\beta = TH_y^\tau(z) + \int_0^m T_\beta H_{2y}^\beta(z)d\beta \tag{2.39}$$

$$(I - R)E_z^i(z) - \int_0^m R_\beta E_{1z}^\beta(z)d\beta = TE_z^\tau(z) + \int_0^m T_\beta E_{2z}^\beta(z)d\beta \tag{2.40}$$

In order to simplify the theory it is assumed that $m = min(m_1, m_2)$. From Eq. (2.20) it follows that if $|\epsilon_1(\omega)|, |\epsilon_2(\omega)| >> \epsilon_{xx}^{(1)}, \epsilon_{zz}^{(1)}, \epsilon_{xx}^{(2)}, \epsilon_{zz}^{(2)}$, then $|m_1 - m_2| << m_1, m_2$. Therefore the contribution to the diffraction process made by the surface polariton radiation mode in the interval $\beta \propto |m_1 - m_2|$ is small and will be neglected.

Equations (2.39) and (2.40) are the first step in solving any class of surface diffraction problems using the mode-method. These problems are associated with finding the eigenmodes of the system, expanding the diffracted fields in terms of these modes and then "sewing together" the corresponding diffracted field components with the objective of determining the amplitudes of the eigenmodes. The initial conditions consist of defining the radiation that excites the structure. In our case it is the surface polariton (Fig. 4). Naturally, there is an alternative situation in which the structure is excited by the bulk field, i.e. a surface polariton radiation mode continuum. The excitation of the surface polariton at the dielectric barrier by the bulk field, which is the inverse of the process that we are investigating, may also be studied using the mode-method.[28]

As a second step in the study of a surface diffraction problem the set of equations (2.39)-(2.40) can be transformed into another form more convenient for analysis. For comparison it can be observed that in solving similar problems via a screening method[23] expressions like (2.39) and (2.40) are derived but instead of integrals with respect to the modes, sums of modes of the metal screening waveguide which describe the bulk radiation appear. However, further transformations of the set of equations (2.39) and (2.40) and the corresponding system in Ref. 23 are different. In the screening method a certain number of points on the z-axis are taken at which values of the fields are calculated. The problem of the fields at $z \to \infty$ is automatically obviated via screening. The mode-method described here uses a different transformation technique, so the fields at $z \to \infty$ are treated differently. Indeed, it is sufficient to use the orthogonality and orthonormalization conditions of the system eigenmodes, Eqs. (2.25)-(2.27). If we multiply Eq. (2.40) by the complex-conjugate of $H_y^\tau(z)$ [Eq. (2.13)], integrate with respect to z, and use Eqs. (2.25) and (2.26) for the amplitude of the transmitted surface polariton, then the following expression is obtained:

$$-\frac{1}{\omega\epsilon_0}k_z^\tau T = (I - R)\int_{-\infty}^{+\infty} E_z^i(z)H_y^{\tau*}(z)dz - \int_0^m R_\beta \left\{\int_{-\infty}^{+\infty} E_{1z}^\beta(z)H_y^{\tau*}(z)dz\right\}d\beta \tag{2.41}$$

If Eq. (2.40) is multiplied by $H_{2y}^{\beta'*}(z)$ [Eq. (2.22)] then, after performing a similar transformation, but taking into account Eq. (2.27) for the surface polariton radiation mode amplitude, we obtain

$$-\frac{1}{\omega\epsilon_0}k_x^{(2)'}T_{\beta'} = (I - R)\int_{-\infty}^{+\infty} E_z^i(z)H_{2y}^{\beta'*}(z)dz - \int_0^m R_\beta \left\{\int_{-\infty}^{+\infty} E_{1z}^\beta(z)H_{2y}^{\beta'*}(z)dz\right\}d\beta \tag{2.42}$$

After multiplying the complex-conjugate of Eq. (2.39) by $E_z^i(z)$ [Eq. (2.14)] and then by $E_{1z}^{\beta'}(z)$ [Eq. (2.23)], and integrating with respect to z, the amplitudes of the reflected surface polariton and surface polariton radiation modes become

$$-\frac{1}{\omega\epsilon_0}k_z^{i*}(I^* + R^*) = T^*\int_{-\infty}^{+\infty} H_y^{\tau*}(z)E_z^i(z)dz + \int_0^m T_\beta^* \left\{\int_{-\infty}^{+\infty} H_{2y}^{\beta*}(z)E_z^i(z)dz\right\}d\beta \tag{2.43}$$

$$- \frac{1}{\omega\epsilon_0} k_x^{(1)'*} R_{\beta'}^* = T^* \int_{-\infty}^{+\infty} H_y^{\tau *}(z) E_{1z}^{\beta'}(z) dz + \int_0^m T_\beta^* \left[ \int_{-\infty}^{+\infty} H_{2y}^{\beta *}(z) E_{1z}^{\beta'}(z) dz \right] d\beta \tag{2.44}$$

The set of equations (2.41)-(2.44) is far more convenient to use than Eqs. (2.39) and (2.40), since it does not contain the variable z. It also permits the derivation of the diffracted mode values (R, $R_\beta$) and (T, $T_\beta$). It is, however, still an integral set with respect to $R_\beta$ and $T_\beta$ and remains rather complicated even for numerical calculations. It is therefore desirable to use further transformations to process the various overlap integrals. We begin with the surface polariton radiation mode mutual overlap integrals on both sides ($x < 0$ and $x > 0$) of the dielectric barrier, namely

$$\int_{-\infty}^{+\infty} E_{1z}^{\beta}(z) H_{2y}^{\beta' *}(z) dz = - \frac{k_x^{(1)}}{\omega\epsilon_0} [(E_{1z}^{\beta}, H_{2y}^{\beta}) \delta(\beta - \beta') - (E_{1z}^{\beta}, H_{2y}^{\beta'})] \tag{2.45}$$

$$(E_{1z}^{\beta}, H_{2y}^{\beta}) = B_\beta^{(1)} B_\beta^{(2)} \frac{\pi}{2} \frac{1}{\epsilon_{zz}} \left[ 1 + \frac{\rho_1 \rho_2}{\beta^2} \frac{\epsilon_{xx}^{(1)}}{\epsilon_1(\omega)} \frac{\epsilon_{xx}^{(2)}}{\epsilon_2(\omega)} \right] \tag{2.46}$$

$$(E_{1z}^{\beta}, H_{2y}^{\beta'}) = B_\beta^{(1)} B_{\beta'}^{(2)} \frac{1}{\epsilon_{zz}} \frac{\rho_1 \dfrac{\epsilon_{xx}^{(1)}}{\epsilon_1(\omega)} - \rho_2 \dfrac{\epsilon_{xx}^{(2)}}{\epsilon_2(\omega)} + \dfrac{1}{\epsilon_1(\omega)} \dfrac{\beta^2 - \beta'^2}{\rho_1 + \rho_2}}{\beta^2 - \beta'^2} \tag{2.47}$$

The final results can be made more convenient by introducing the definitions

$$\int_{-\infty}^{+\infty} E_z^i(z) H_y^{\tau *}(z) dz = - \frac{1}{\omega\epsilon_0} k_s^i (E_z^i, H_y^{\tau})$$

$$\int_{-\infty}^{+\infty} E_z^i(z) H_{2y}^{\beta *}(z) dz = - \frac{1}{\omega\epsilon_0} k_s^i (E_z^i, H_y^{\beta})$$

$$\int_{-\infty}^{+\infty} E_{1z}^{\beta}(z) H_y^{\tau *}(z) dz = - \frac{1}{\omega\epsilon_0} k_x^{(1)} (E_{1z}^{\beta}, H_y^{\tau})$$

These overlap integrals appear in Eqs. (2.41) and (2.44). After a simple integration with respect to z we obtain

$$(H_y^{\tau}, E_z^i) = A_0^{(1)} A_0^{(2)} \left[ \frac{1}{\epsilon_{zz}^{(1)}} \frac{1}{k_1^{(1)} + k_2^{(2)}} + \frac{1}{\epsilon_1(\omega)} \frac{1}{k_2^{(1)} + k_2^{(2)}} \right] \tag{2.48}$$

$$(E_z^i, H_{2y}^{\beta}) = A_0^{(1)} B_\beta^{(2)} \left[ \frac{1}{\epsilon_{zz}^{(1)}} \frac{k_1^{(1)} + \rho_2 \dfrac{\epsilon_{xx}^{(2)}}{\epsilon_2(\omega)}}{\beta^2 + (k_1^{(1)})^2} + \frac{1}{\epsilon_1(\omega)} \frac{1}{k_2^{(1)} + \rho_2} \right] \tag{2.49}$$

$$(E_{1z}^{\beta}, H_y^{\tau}) = A_0^{(2)} B_\beta^{(1)} \left[ \frac{1}{\epsilon_{zz}^{(1)}} \frac{k_1^{(2)} + \rho_1 \dfrac{\epsilon_{xx}^{(1)}}{\epsilon_1(\omega)}}{\beta^2 + (k_1^{(2)})^2} + \frac{1}{\epsilon_1(\omega)} \frac{1}{k_2^{(2)} + \rho_1} \right] \tag{2.50}$$

A perturbation method for the solution of the set of equations (2.41) to (2.44) will now be developed. In order to do this we analyze Eq. (2.45). The mutual overlap integral of surface polariton radiation modes on both sides of the dielectric barrier contains a singular term [the first on the right-hand side of Eq. (2.45)], which describes the interaction between surface polariton radiation modes having equal transverse wave numbers $\beta' = \beta$ [Eq. (2.46)], and also a term [Eq. (2.47)] corresponding to nonsingular surface polariton radiation mode interaction ($\beta' \neq \beta$). The first step in the perturbation method for this two-dimensional dispersion problem is to neglect the nonsingular interaction Eq. (2.47). In comparison with standard methods for the investigation of dispersion problems in quantum mechanics,[25] which as a rule are restricted to two- or three-particle interactions, we do not exclude from consideration the surface polariton radiation mode continuum, but in this first step it is treated approximately. Thus, if only the singular term is left in Eq. (2.45), the set of equations (2.41)-(2.44) transforms, in accordance with the definitions given above, into

$$Tk_s^\tau = (I - R)k_s^i(E_z^i, H_y^\tau) - \int_0^m R_\beta k_x^{(1)}(E_{1z}^\beta, H_y^\tau)d\beta \tag{2.51}$$

$$T_\beta k_x^{(2)} = (I - R)k_s^i(E_z^i, H_{2y}^\beta) - R_\beta k_x^{(1)}(E_{1z}^\beta, H_{2y}^\beta) \tag{2.52}$$

$$I + R = T(H_y^\tau, E_z^i) + \int_0^m T_\beta(H_{2y}^\beta, E_z^i)d\beta \tag{2.53}$$

$$R_\beta = T(H_y^\tau, E_{1z}^\beta) + T_\beta(H_{2y}^\beta, E_{1z}^\beta) \tag{2.54}$$

where on the right-hand sides of Eqs. (2.52) and (2.54) there are now algebraic expressions instead of integrals. This results from the reduction of these integrals through the use of the delta function equation (2.45) in the corresponding equations (2.42) and (2.44). Thus, Eqs. (2.52) and (2.54) are purely algebraic equations and, after some elementary transformations, the surface polariton radiation mode amplitudes become

$$R_\beta = \frac{(I - R)k_s^i(E_z^i, H_{2y}^\beta)(H_{2y}^\beta, E_{1z}^\beta) + Tk_x^{(2)}(H_y^\tau, E_{1z}^\beta)}{k_x^{(2)} + k_x^{(1)}(E_{1z}^\beta, H_{2y}^\beta)(H_{2y}^\beta, E_{1z}^\beta)} \tag{2.55}$$

$$T_\beta = \frac{(I - R)k_s^i(E_z^i, H_{2y}^\beta) + Tk_x^{(2)}(H_y^\tau, E_{1z}^\beta)(E_{1z}^\beta, H_{2y}^\beta)}{k_x^{(2)} + k_x^{(1)}(E_{1z}^\beta, H_{2y}^\beta)(H_{2y}^\beta, E_{1z}^\beta)} \tag{2.56}$$

Further transformations can be obtained if Eq. (2.55) is substituted into the equation for the transmitted surface polariton amplitude [Eq. (2.51)]. If like terms are collected then the following expressions emerge

$$T = (I - R)T_0 \tag{2.57}$$

where $T_0 = k_s^i T_1/T_2$, and

$$T_1 = (H_y^\tau, E_z^i) - \int_0^m \frac{(E_z^i, H_{2y}^\beta)(H_{2y}^\beta, E_{1z}^\beta)(E_{1z}^\beta, H_y^\tau)k_x^{(1)}}{k_x^{(2)} + k_x^{(1)}(E_{1z}^\beta, H_{2y}^\beta)(H_{2y}^\beta, E_{1z}^\beta)}d\beta \tag{2.58}$$

$$T_2 = k_s^\tau + \int_0^m \frac{(H_y^\tau, E_{1z}^\beta)(E_{1z}^\beta, H_y^\tau)}{k_x^{(2)} + k_x^{(1)}(E_{1z}^\beta, H_{2y}^\beta)(H_{2y}^\beta, E_{1z}^\beta)}d\beta \tag{2.59}$$

Equation (2.57) is also an algebraic equation, since $T_1$ and $T_2$ are simply numbers for a particular problem that can easily be found by standard numerical methods for evaluating integrals.

The expression for T, in the form of Eq. (2.57), makes it possible to transform Eqs. (2.55) and (2.56) into the more convenient form:

$$R_\beta = (I - R)R_\beta^0 \qquad T_\beta = (I - R)T_\beta^0 \tag{2.60}$$

where

$$R_\beta^0 = \frac{k_s^i(E_z^i, H_{2y}^\beta)(H_{2y}^\beta, E_{1z}^\beta) + T_0 k_x^{(2)}(H_y^\tau, E_{1z}^\beta)}{k_x^{(2)} + k_x^{(1)}(E_{1z}^\beta, H_{2y}^\beta)(H_{2y}^\beta, E_{1z}^\beta)} \tag{2.61}$$

$$T_\beta^0 = \frac{k_s^i(E_z^i, H_{2y}^\beta) + T_0 k_x^{(2)}(H_y^\tau, E_{1z}^\beta)(E_{1z}^\beta, H_{2y}^\beta)}{k_x^{(2)} + k_x^{(1)}(E_{1z}^\beta, H_{2y}^\beta)(H_{2y}^\beta, E_{1z}^\beta)} \tag{2.62}$$

Using the fields in the form of Eqs. (2.60) is convenient in so far as the energy (I - R) and structural parameters, Eqs. (2.61) and (2.62), are independently separated. It may easily be seen that, after substitution of Eqs. (2.67) and (2.60) into Eq. (2.53),

$$I + R = (I - R)N_{ef} \tag{2.63}$$

where $N_{ef}$ is

$$N_{ef} = T_0(H_y^\tau, E_z^i) + \int_0^m T_\beta^0(H_{2y}^\beta, E_z^i)d\beta \tag{2.64}$$

Finally, the reflected surface polariton amplitude is given by

$$R = I\left(\frac{N_{ef} - 1}{N_{ef} + 1}\right) \tag{2.65}$$

The fundamental character of the latter expression is that in outward appearance it resembles the usual Fresnel law for the reflection of plane electromagnetic waves in bulk media.[26] The basic quantity in the generalized Fresnel law, Eq. (2.65), is the effective reflection factor of the dielectric barrier, namely $N_{ef}$. The expression for this quantity, Eq. (2.64), consists of two parts, of which the first corresponds to surface polariton interaction and the second describes the interaction of surface polaritons and surface polariton radiation modes normalized with respect to the surface polariton radiation mode interaction, i.e. the term $k_x^{(2)} + k_x^{(1)}(E_{1z}^\beta, H_{2y}^\beta)(H_{2y}^\beta, E_{1z}^\beta)$ in Eqs. (2.55)-(2.62).

The eigenmode amplitudes R, T, $R_\beta$ and $T_\beta$ of the structure, evaluated using the mode-method, should satisfy the energy conservation law but this must be proved. In accordance with the general expression for the energy flux at the interface, Eqs. (2.34)-(2.35), the energy conservation law during the diffraction process is

$$Re\left[II^* k_s^i - RR^* k_s^i - \int_0^m R_\beta R_\beta^* k_x^{(1)} d\beta - \int_0^m T_\beta T_\beta^* k_x^{(2)} d\beta - TT^* k_s^\tau\right] = 0 \tag{2.66}$$

This corresponds to the fact that the total energy flux of all excited fields must be equal to the energy flux of the incident surface polariton, $II^* k_1$. The validity of Eq. (2.66) can be established from the set of equations (2.51) to (2.54) from which the final expressions for the diffracted field amplitudes was derived. Keeping in mind the relation $Re(II^* - RR^*) = Re[(I - R)(I^* + R^*)]$ we rewrite Eq. (2.56) using Eq. (2.53)

$$Re\left[T[(I - R)(H_y^\tau, E_z^i) - Tk_s^\tau] + \int_0^m T_\beta[(I - R)(H_{2y}^\beta E_{1z}^\beta) - T_\beta k_x^{(2)}]d\beta - \int_0^m R_\beta R_\beta k_x^{(1)}\right] = 0 \tag{2.67}$$

After some simple transformations, and using Eqs. (2.51) and (2.52), Eq. (2.67) reduces to Eq. (2.54). Since the field amplitudes satisfy each of Eqs. (2.51) to (2.54), the energy conservation law for the process of diffraction is therefore proved. This is a very important theoretical fact since the quantities being evaluated are physical in nature and it supports the use of perturbation theory. This conclusion is also verified by numerical calculations.

## 2.6 Numerical results

The method developed here, which permits surface polariton reflection to be treated in both visible and submillimeter ranges, has been tested by numerical calculations on the structure considered in Ref. 23. The results for the reflection of surface plasmon-polaritons (visible range) are shown in Fig. 8 and for phonon-polaritons (submillimeter range) in Fig. 9. Comparing these results with those of Ref. 23, it can be seen that the values of R and T approximately coincide. This appears to be due to the fact that surface polariton radiation modes participate only weakly in the reflection process, the fraction of the surface polariton energy released by the volume radiation being only about 10%. In contrast, for the structure shown in Fig. 9, this fraction rises to 50% and higher. The distribution of the volume diffracted fields in Figs. 8 and 9 was calculated using expression (2.37).

In terms of clarity, the generalized Fresnel law for the reflectance of surface polaritons at a dielectric barrier, Eq. (2.65), is undoubtedly an advantage of the theory given here. However, this is not the most important factor since, as may easily be shown, a similar law could be obtained within a framework of cruder approximations than those made here. The main advantage of the method is in its self-consistency in taking account of all the possible fields in the process of diffraction. The energy conservation law, established as an analytical extension of the theory, reveals the reliability of the mode-method, far more than a simple comparison of the results obtained with those of bulk diffraction. Nevertheless the appearance of the Fresnel law, Eq. (2.65), is physically appealing, since a relatively long chain of definitions

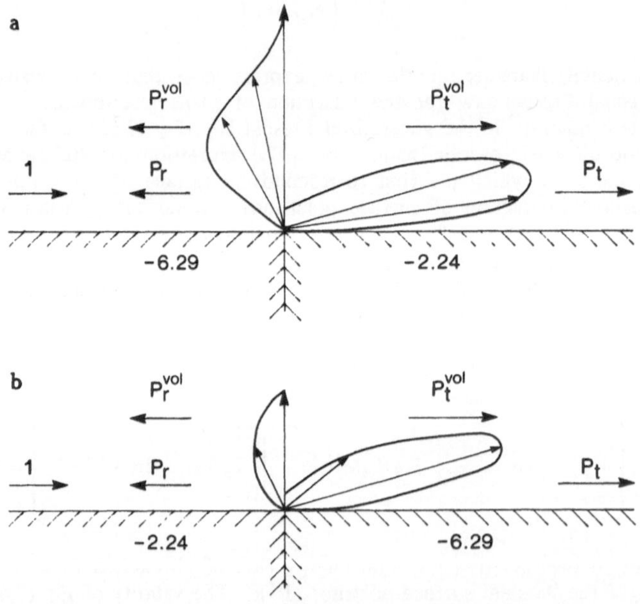


Fig. 8. Relative redistribution of the energies of surface and volume waves during the reflection of a surface plasmon-polariton [$\epsilon(\omega) = 1 - \omega_p^2/\omega^2$] for a system with $\epsilon_{xx}^{(1)} = \epsilon_{zz}^{(1)} = \epsilon_{xx}^{(2)} = \epsilon_{zz}^{(2)} = 1$, $\lambda = 1.06\ \mu$m. (a) The polariton goes through a dielectric barrier from a medium with $\omega_p/\omega = 2.7$ to a medium with $\omega_p/\omega = 1.8$. $P_r = 0.0027$, $P_t = 0.9053$, $P_r^{vol} = 0.0075$, $P_t^{vol} = 0.0789$. (b) Same as (a) but with the direction reversed. $P_r = 0.0036$, $P_t = 0.9157$, $P_r^{vol} = 0.0019$, $P_t^{vol} = 0.0788$.

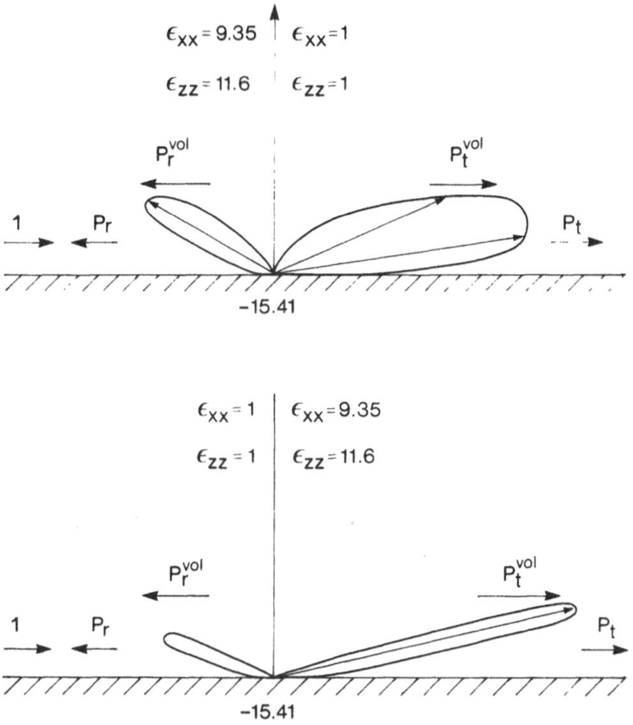


Fig. 9. Normal diffraction of surface phonon-polaritons $[\epsilon(\omega) = \epsilon_\infty + (\epsilon_0 - \epsilon_\infty)\omega^2/(\omega_T^2 - \omega^2)]$ for a system with a mutual surface-active crystal (GaAs, $\lambda$ = 222 $\mu$m): $\epsilon_\infty$ = 11.1, $\epsilon_0$ = 13.1, $\omega/\omega_T$ = 1.04. (a) The polariton goes through a dielectric barrier from a medium with $\epsilon_{xx}$ = 9.35, $\epsilon_{zz}$ = 11.6 to a medium with $\epsilon_{xx} = \epsilon_{zz}$ = 1. $P_r$ = 0.0036, $P_t$ = 0.3104, $P_r^{vol}$ = 0.1550, $P_t^{vol}$ = 0.5300. (b) Same as (a) but with the direction reversed. $P_r$ = 0.3135, $P_t$ = 0.3375, $P_r^{vol}$ = 0.0007, $P_t^{vol}$ = 0.3498.

and processing of the theory has, in the final analysis, been reduced to two process parameters, i.e. to $T_0$ and $N_{ef}$. These may be relatively easily calculated. The method opens up the possibility of a more exact definition of the quantities involved by using second and subsequent steps of the perturbation method. This will take into account the nonsingular surface polariton radiation mode interaction [Eq. (2.47)]. Such an investigation is well removed from the aims of this work, however.

## 3. BASIC THEORY OF DIFFRACTION OF NONLINEAR SURFACE WAVES

The geometry for nonlinear diffraction of a surface polariton is shown in Fig. 10. The structure consists of a nonlinear film ($\epsilon_{nl}$) situated on a surface-active crystal $\epsilon_2(\omega)$. The end of the film (x = 0) is the dielectric barrier in this case. In the x < 0 region there is a two-medium structure and in the x > 0 region there is a three-medium structure. The method of investigation of this nonlinear process is the same, in principle, as that used in the preceding section. First the eigenmodes of the structure must be found both in the x < 0 region and x > 0 region. Then the amplitudes of these modes must be determined from the corresponding boundary conditions. The eigenmodes in the x < 0 region have already been investigated in the preceding section (see sections 2.2 and 2.3). Our major task in this section is to find the eigenmodes of the x > 0 region containing the nonlinear film when the dielectric permittivity of the film has a thermal, Kerr-like, form

$$\epsilon_{nl}(\mathcal{E}) = \epsilon_1(\omega) + \alpha_e|\mathcal{E}^{(1)}(x,z)|^2 \tag{3.1}$$

where the linear part of the dielectric permittivity is associated with excitons of the form

$$\epsilon_l(\omega) = \epsilon_\infty \frac{\omega^2 - \omega_L^2}{\omega^2 - \omega_T^2} \tag{3.2}$$

The frequencies $\omega_L$ and $\omega_T$ are the longitudinal and transverse oscillation frequencies respectively and $\alpha_e$ is a nonlinear coefficient.

The field $\mathcal{E}^{(1)}(x,z)$ in the expression (3.1) is the total electric field in the film, including both a surface and a volume diffraction field. As was shown in the first part of this chapter, the dependence of a volume field on the x,z-coordinates is complicated [$\mathcal{E}^{(1)}(x,z)$ is, in principle, the same as that given by Eq. (2.31)]. Therefore the dependence of $\epsilon_{nl}$ on the x,z-coordinates will also be very complicated. This is a major problem in the study of nonlinear diffraction. Nevertheless, we will use Fourier and mode-methods for investigating the diffraction of a surface polariton on a nonlinear film.

### 3.1 The effective boundary conditions

It has already been mentioned that the x > 0 region is a three-medium structure. Generally speaking, the method of investigation of the eigenmodes consists of two steps. First, it is necessary to find all the solutions in the three layers, substrate [$-\infty < z < 0$, $\epsilon(x,z) = \epsilon_2(\omega)$], film [$0 < z < d$, $\epsilon(x,z) = \epsilon_{nl}(x,z)$] and vacuum [$d < z < +\infty$, $\epsilon(x,z) = 1$], of the following Maxwell equations:

$$[i\omega\mu_0\mathcal{H}_y(x,z)]e^{-i\omega t} = \frac{\partial\mathcal{E}_x(x,z)}{\partial z}e^{-i\omega t} - \frac{\partial\mathcal{E}_z(x,z)}{\partial x}e^{-i\omega t} \tag{3.3a}$$

$$[i\omega\epsilon_0\epsilon(x,z)\mathcal{E}_x(x,z)]e^{-i\omega t} = \frac{\partial\mathcal{H}_y(x,z)}{\partial z}e^{-i\omega t} \tag{3.3b}$$

$$[i\omega\epsilon_0\epsilon(x,z)\mathcal{E}_z(x,z)]e^{-i\omega t} = -\frac{\partial\mathcal{H}_y(x,z)}{\partial x}e^{-i\omega t} \tag{3.3c}$$

Second, these solutions must be matched at the interfaces using the following boundary conditions:

$$\mathcal{E}_x^{(2)}(x,0) = \mathcal{E}_x^{(1)}(x,0) \qquad \mathcal{H}_y^{(2)}(x,0) = \mathcal{H}_y^{(1)}(x,0) \tag{3.4a}$$

$$\mathcal{E}_x^{(v)}(x,d) = \mathcal{E}_x^{(1)}(x,d) \qquad \mathcal{H}_y^{(v)}(x,d) = \mathcal{H}_y^{(1)}(x,d) \tag{3.4b}$$

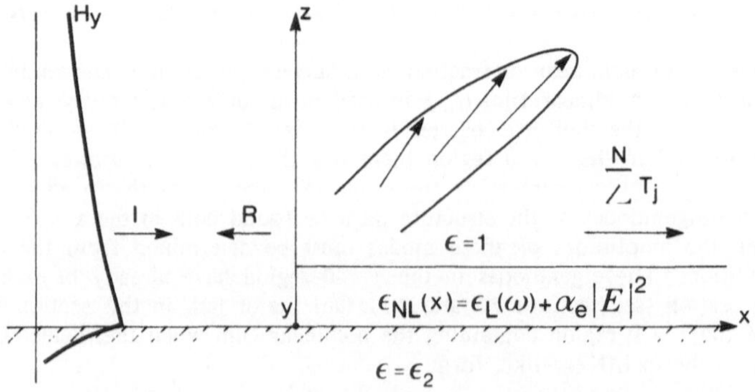


Fig. 10. Diffraction of surface polaritons by a thin nonlinear resonance film.

However, the x,z dependence of $\epsilon_{nl}$ leads in this case to huge difficulties when finding the solution of Eqs. (3.3). We will, however, work out an approximate way to solve this problem for the very interesting situation when the thickness of the film is quite small, i.e. $d << \lambda$, where $\lambda$ is the wavelength. For this case, effective boundary conditions can be obtained instead of separately investigating Eqs. (3.3) and (3.4). The condition that $d << \lambda$ enables the theory to be simplified by expanding the field components to

$$\mathcal{E}_x(x,d) = \mathcal{E}_x(x,0) + d\frac{\partial \mathcal{E}_x}{\partial z} + O\left(\frac{d^2}{\lambda^2}\right) \tag{3.5a}$$

$$\mathcal{H}_y(x,d) = \mathcal{H}_y(x,0) + d\frac{\partial \mathcal{H}_y}{\partial z} + O\left(\frac{d^2}{\lambda^2}\right) \tag{3.5b}$$

From Eqs. (3.3a) and (3.5a) we can derive

$$\mathcal{E}_x(x,d) = \mathcal{E}_x(x,0) + d\left(i\omega\mu_0\mathcal{H}_y(x,0) + \frac{\partial \mathcal{E}_z(x,0)}{\partial x}\right) + O\left(\frac{d^2}{\lambda^2}\right) \tag{3.6}$$

Equations (3.3b) and (3.5b) yield

$$\mathcal{H}_y(x,d) = \mathcal{H}_y(x,0) + d\left(i\omega\epsilon_0\epsilon(x,z)\mathcal{E}_x(x,0)\right) + O\left(\frac{d^2}{\lambda^2}\right) \tag{3.7}$$

Equation (3.6) can be rewritten using boundary conditions (3.4b) to give

$$\mathcal{E}_x^{(1)}(x,0) - \mathcal{E}_x^{(v)}(x,0) = d\left(\frac{\partial \mathcal{E}_z^{(v)}(x,0)}{\partial x} - \frac{\partial \mathcal{E}_z^{(1)}(x,0)}{\partial x}\right) + O\left(\frac{d^2}{\lambda^2}\right) \tag{3.8}$$

Similarly, rewriting Eq. (3.7) using boundary condition (3.4b) we get

$$\mathcal{H}_y^{(1)}(x,0) - \mathcal{H}_y^{(v)}(x,0) = i\omega\epsilon_0\left(\mathcal{E}_x^{(v)}(x,0) - \epsilon_{nl}(x,0)\mathcal{E}_x^{(1)}(x,d)\right)d + O\left(\frac{d^2}{\lambda^2}\right) \tag{3.9}$$

The continuity condition for the z-component of the induction, $\mathcal{D}_z^{(v)}(x,d) = \mathcal{D}_z^{(1)}(x,d)$, and the material relations (2.2), $\mathcal{D}_z^{(v)}(x,z) = \mathcal{E}_z^{(v)}(x,z)$ and $\mathcal{D}_z^{(1)}(x,z) = \epsilon_{nl}(x)\mathcal{E}_z^{(1)}(x,z)$, lead to

$$\frac{\partial \mathcal{E}_z^{(v)}(x,0)}{\partial x} - \frac{\partial \mathcal{E}_z^{(1)}(x,0)}{\partial x} = \frac{\partial}{\partial x}\left[\left(1 - \frac{1}{\epsilon_{nl}(x)}\right)\mathcal{E}_z^{(v)}(x,0)\right] + O\left(\frac{d}{\lambda}\right) \tag{3.10}$$

The main interest here is in the frequency region $\omega \simeq \omega_L$. The linear part of the dielectric permittivity of the film is then close to zero $[\epsilon_l(\omega) \simeq 0]$ and $\epsilon_{nl}$ has a relatively big nonlinearity shift $\alpha_e|\mathcal{E}|^2$ for this frequency region. The value of $1/\epsilon_{nl}(x)$ is much greater than the value of $\epsilon_{nl}(x)$ in this case. This enables us to obtain from Eqs. (3.8)-(3.10) and from the boundary conditions (3.4a) the following effective boundary conditions

$$\mathcal{E}_x^{(2)}(x,0) = \mathcal{E}_x^{(v)}(x,0) + d\frac{\partial}{\partial x}\left[\left(1-\frac{1}{\epsilon_{nl}(x)}\right)\mathcal{E}_z^{(v)}(x,0)\right] \tag{3.11a}$$

$$\mathcal{H}_y^{(2)}(x,0) = \mathcal{H}_y^{(v)}(x,0) \tag{3.11b}$$

where only the terms containing $(d/\lambda)\times[1/\epsilon_{nl}(x)]$ are taken into consideration and $\epsilon_{nl}(x)$ should be evaluated at $z = 0$.

It is important that only the fields outside the film are present in Eq. (3.11). Instead of solving Eqs. (3.3) in each layer we can now find the solution (3.3) in the two linear media z < 0 (metal) and x > d (vacuum) from the following wave equations

$$\frac{\partial^2 \mathcal{H}_y^{(v)}(x,z)}{\partial x^2}) + \frac{\partial^2 \mathcal{H}_y^{(v)}(x,z)}{\partial z^2} + k_0^2 \mathcal{H}_y^{(v)}(x,z) = 0$$

$$\frac{\partial^2 \mathcal{H}_y^{(2)}(x,z)}{\partial x^2} + \frac{\partial^2 \mathcal{H}_y^{(2)}(x,z)}{\partial z^2} + k_0^2 \epsilon_2(\omega) \mathcal{H}_y^{(2)}(x,z) = 0 \tag{3.12}$$

and include the nonlinear film only in the effective boundary conditions (3.11). The $\mathcal{E}_x^{(v,2)}(x,z)$ and $\mathcal{E}_z^{(v,2)}(x,z)$ components are obtained from Eqs. (3.3b) and (3.3c) respectively. The dependence of the $\mathcal{E}_z^{(1)}(x,z)$ component on the coordinates in the film implies a knowledge of the space dependence of the dielectric permittivity $\epsilon_{nl}(x)$.

Thus the form of the nonlinear eigenmodes can be found from Eqs. (3.11)-(3.12) and (3.3) but we should define more precisely the nonlinear permittivity [Eq. (3.1)] and the expression for the energy flow [Eq. (2.7)] in the nonlinear film. For TM waves propagating along the thin film

$$|\mathbf{E}^{(1)}(x,z)|^2 = |\mathcal{E}_x^{(1)}(x,z)|^2 + |\mathcal{E}_z^{(1)}(x,z)|^2$$

It can be shown that $\mathcal{E}_z^{(1)}(x,d) = \mathcal{E}_z^{(v)}(x,d)/\epsilon_{nl}$ and $\mathcal{E}_x^{(1)}(x,d) = \mathcal{E}_x^{(v)}(x,d)$. Usually, $|E_x^{(v)}| < |E_z^{(v)}|$ for surface polaritons. Therefore, we will leave only $|\mathcal{E}_z^{(1)}(x,z)|^2$ in expression (3.1) for the situation when $\epsilon_{nl} \simeq 0$. Finally, we have

$$\epsilon_{nl}(x) = \epsilon_1(\omega) + \alpha_e J(x) + O\left(\frac{d}{\lambda}\right) \qquad J(x) = |\mathcal{E}_z^{(1)}(x,0)|^2 \tag{3.13}$$

The expression for the energy flow in a thin film, derived from Eqs. (2.7), (3.3b) and (3.9), is

$$P_x^{(1)}(x) = \frac{1}{2} Re\left[\int_0^d \mathcal{E}_z^{(1)}(x,z) H_z^{(1)*}(x,z) dz\right]$$

$$= \frac{1}{2} Re\left[\frac{i}{\omega\epsilon_0} \frac{d}{\epsilon_{nl}(x)} \frac{\partial \mathcal{H}_y^{(v)}(x,0)}{\partial x} \mathcal{H}_y^{(v)*}(x,0)\right] \tag{3.14}$$

where terms of order $(d^2/\lambda^2)$ have been left out. Thus all the parameters relevant to the nonlinear process are defined but a problem still exists with the unknown x-dependence of $\epsilon_{nl}$.

### 3.2 Fourier-method of definition of the nonlinear eigenmodes

This problem of the dependence of $\epsilon_{nl}(E)$ upon x can be tackled using the Fourier-method that leads to the function $1/\epsilon_{nl}(x)$ being written in the form

$$\frac{1}{\epsilon_{nl}(x)} = C + \left[\frac{1}{\epsilon_{nl}(x)} - C\right] \tag{3.15}$$

where the constant C is

$$C = \lim_{b \to \infty} \frac{1}{b} \left[\int_0^b \frac{1}{\epsilon_{nl}(x)} dx\right] \tag{3.16}$$

Using Eq. (3.3c) the nonlinear contribution to the effective boundary conditions (3.11) can be written as

$$\left[\frac{1}{\epsilon_{nl}(x)} - C\right]\mathcal{E}_z^{(v)}(x,0) = \frac{i}{\omega\epsilon_0}\int_{-\infty}^{+\infty} F(\nu)e^{i\nu x}d\nu \tag{3.17}$$

where the Fourier-harmonic $F(\nu)$ is

$$F(\nu) = -i\frac{1}{2\pi}\int_0^{\infty}\left[\frac{1}{\epsilon_{nl}(x)} - C\right]\frac{\partial\mathcal{H}_y^{(v)}(x,0)}{\partial x}e^{-i\nu x}dx \tag{3.18}$$

From Eqs. (3.15) and (3.18) the effective boundary conditions (3.11a) can be rewritten in the form

$$\mathcal{E}_x^{(2)}(x,0) = \mathcal{E}_x^{(v)}(x,0) + \frac{i}{\omega\epsilon_0}d\left[(1 - C)\frac{\partial^2 H_y^{(v)}(x,0)}{\partial^2 x} + \int_{-\infty}^{+\infty} F(\nu)\nu e^{i\nu x}d\nu\right] \tag{3.19}$$

Equation (3.3c) in the film can be rewritten, using Eqs. (3.15) and (3.17), as

$$\mathcal{E}_z^{(1)}(x,z) = \frac{i}{\omega\epsilon_0}\left[C\frac{\partial\mathcal{H}_y^{(v)}(x,z)}{\partial x} + \int_{-\infty}^{+\infty} F(\nu)e^{i\nu x}d\nu\right] + O\left(\frac{d}{\lambda}\right) \tag{3.20}$$

In the vacuum ($x > d$) and the metal ($x < 0$) the relationships between $\mathcal{E}_z(x,z)$, $\mathcal{E}_x(x,z)$ and $\mathcal{H}_y(x,z)$ are the same as in the linear case. Bearing in mind Eq. (3.13) and the boundary conditions at $x = 0$, we will only be interested in the $\mathcal{E}_z^{(1)}(x,z)$ component in the nonlinear film. Notice that here the boundary conditions at $x = 0$ are the same as in the section preceding Eq. (2.6b).

From Eqs. (3.11b), (3.19) and (3.20) a very important conclusion can be drawn. This is that the total solution of the equations consists of a homogeneous and a nonhomogeneous solution, i.e.

$$\mathcal{H}_y^T(x,z) = [\mathcal{H}_y^h(x,z) + \mathcal{H}_y^{nh}(x,z)]e^{-i\omega t} \qquad \mathcal{E}_{x,z}^t = [\mathcal{E}_{x,z}^h(x,z) + \mathcal{E}_{x,z}^{nh}(x,z)]e^{-i\omega t} \tag{3.21}$$

where $\mathcal{H}_y^h(x,y)$ and $\mathcal{H}_y^{nh}(x,y)$ are, in principle, solutions of one of Eqs. (3.12) but with essentially different effective boundary conditions for the $\{\mathbf{H}^h(x,y), \mathbf{E}^h(x,z)\}$ and $\{\mathbf{H}^{nh}(x,z), \mathbf{E}^{nh}(x,z)\}$ modes. In addition the Maxwell equations in the nonlinear film $0 < z < d$ are

homogeneous modes:

$$\mathcal{E}_z^h(x,z) = \frac{i}{\omega\epsilon_0}C\frac{\partial\mathcal{H}_y^{(v)}(x,z)}{\partial x} + O\left(\frac{d}{\lambda}\right) \tag{3.22a}$$

and

nonhomogeneous modes:

$$\mathcal{E}_z^{nh}(x,z) = \frac{i}{\omega\epsilon_0}\int_{-\infty}^{+\infty} F(\nu)e^{i\nu x}d\nu + O\left(\frac{d}{\lambda}\right) \tag{3.22b}$$

which follows from Eqs. (3.20), (3.21).

### 3.2.1 The homogeneous nonlinear eigenmodes

From Eqs. (3.11b) and (3.19) we derive the solutions

$$\mathcal{E}_x^{h(2)}(x,0) = \mathcal{E}_x^{h(v)}(x,0) + d(1 - C)\frac{\partial \mathcal{E}_z^{h(v)}(x,0)}{\partial x}$$

$$\mathcal{H}_y^{h(2)}(x,0) = \mathcal{H}_y^{h(v)}(x,0) \qquad (3.23)$$

where $\mathcal{H}_y^h(x,z)$ is obtained from the wave equations (3.12) and $\mathcal{E}_x^h(x,z)$ and $\mathcal{E}_z^h(x,z)$ from the relations (3.3b) and (3.3c) in the vacuum and the metal. It is easily shown that these solutions correspond to surface polaritons in the form (2.8) and surface polariton radiation modes in the form (2.16). Repeating the transformations as in section 2 allows us to build the homogeneous nonlinear eigenmode spectrum.

*Nonlinear surface polaritons.* The form of the wave equations (3.12) and the homogeneous effective boundary conditions (3.21) enables us to seek the solution for the nonlinear homogeneous surface polariton, $\mathcal{H}_y^h(x,z)$, $\mathcal{E}_z^h(x,z)$, in the form given by Eqs. (2.12) and (2.8). The nonlinear dispersion law, derived from Eqs. (2.8), (2.12) and (3.23), is

$$\frac{1}{\epsilon_2}k_2^{(j)} + k_1^{(j)} - d(1 - C)(k_s^{(j)})^2 = 0 \qquad (3.24)$$

where the wave number relations are

$$(k_1^{(j)})^2 = (k_s^{(j)})^2 - k_0^2 \qquad (k_2^{(j)})^2 = (k_s^{(j)})^2 - k_0^2\epsilon_2(\omega) \qquad (3.25)$$

In the linear case $C = 1/\epsilon_1(\omega)$ and Eq. (3.24) reduces to a previously published equation.[29] A typical form for the dependence of the frequency $\omega$ of a surface polariton on $k_s$ is presented in Fig. 11. A very interesting result[30] is the existence of a frequency gap $\Delta\omega = \omega_2 - \omega_1$, which contains the frequency $\omega_L$ ($\omega_1 < \omega_L < \omega_2$). There is a situation here for which two surface polaritons exist in this structure for $\omega < \omega_1$. The first surface polariton ($k_s^{(1)}$) is very close to the surface polariton in this structure with the film removed (dashed line in Fig. 11). The second surface polariton has a negative group velocity ($\partial\omega/\partial k_s < 0$) and to satisfy the conditions at $x \to \infty$ it is necessary for the condition $k_s^{(2)} < 0$ to be satisfied.

The transverse envelopes of the nonlinear surface polaritons, derived from Eqs. (2.8) and (3.3), have the form

$$H_y^{(j)}(z) = A_j^h \begin{cases} e^{-k_1^{(j)}z} & z > 0 \\ e^{k_2^{(j)}z} & z < 0 \end{cases} \qquad (3.26a)$$

$$E_z^{(j)}(z) = -\frac{1}{\omega\epsilon_0}k_s^{(j)}A_j^h \begin{cases} e^{-k_1^{(j)}z} & z > d \\ C & 0\ z < 0 \\ \frac{1}{\epsilon_2}e^{k_2^{(j)}z} & z < 0 \end{cases} \qquad (3.26b)$$

where the normalization constant $A_j^h$, determined from the normalization condition (2.26), is

$$A_j^h = \left[\frac{1}{\epsilon_2}\frac{1}{2k_2^{(j)}} + \frac{1}{2k_1^{(j)}} - d(1 - C)(k_s^{(j)})^2\right]^{-1/2} \qquad (3.27)$$

The total dependence of the surface polariton field on the space coordinates is

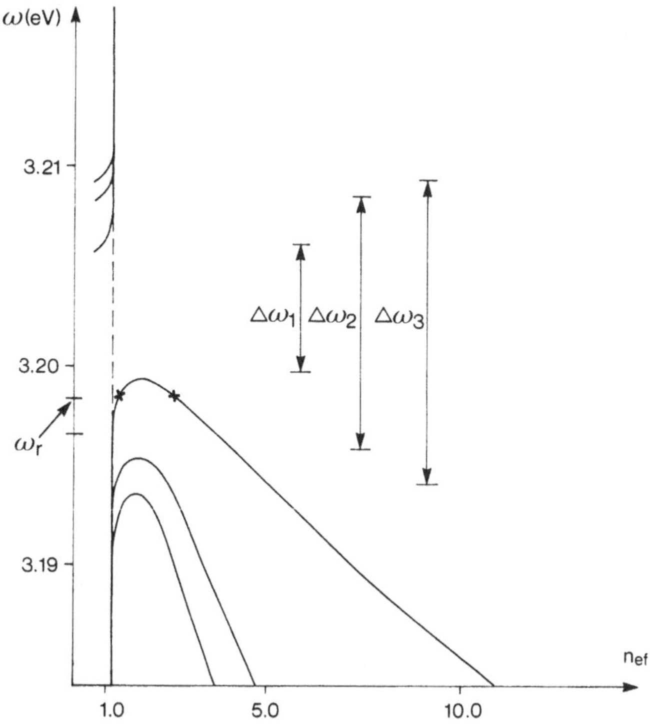


Fig. 11. The dispersion law of a surface polariton in the resonance structure of Fig. 10. The CuCl film is deposited on a surface-active Al crystal with $\epsilon_2(\omega) = 1 - \omega_p^2/\omega^2$, $\omega_p$ = 15.8 eV. For CuCl $\epsilon_\infty = 5.9$, $\omega_L$ = 3.202 eV, $\omega_T$ = 3.122 eV. (a) d = 40 Å, (b) d = 80 Å, (c) d = 100 Å. $\omega_r$ is the frequency of the incident surface polariton.

$$\{\mathscr{H}_y^h(x,z), \mathscr{E}_z^h(x,z)\} = \sum_{j=1}^{N} T_j\{H_y^{(j)}(z), E_z^{(j)}(z)\}e^{ik_s^{(j)}x} \tag{3.28}$$

where N is the number of the surface polaritons in the film and its value is to be determined for each specific case. Let us investigate the process of nonlinear diffraction when $\omega_r < \omega_1$ and there are two surface polaritons in the film (N = 2, Fig. 11). It is easy to see from Eqs. (3.13) and (3.28) that Eq. (3.16) can be transformed to

$$C = \frac{1}{\epsilon_1(\omega) + \alpha_e(\omega)J_\infty} \tag{3.29a}$$

$$J_\infty = |T_1 k_s^{(1)} E_{1z}^{(1)}(0)|^2 + |T_2 k_s^{(2)} E_{2z}^{(1)}(0)|^2 \tag{3.29b}$$

where $J_\infty$ is the intensity of the electric field as $x \to \infty$ (as this limit is reached the intensity of a volume field tends to zero). Note that Eqs. (3.24) and (3.29) show very clearly the nonlinear character of these surface polaritons.

*Nonlinear surface polariton radiation modes.* The solution for the nonlinear surface polariton radiation modes will be sought in the form given by Eqs. (2.16) and (2.21) with homogeneous boundary conditions given by Eq. (3.23). After some simple transformations we can obtain the following z-dependence of the nonlinear surface polariton radiation mode

$$H^{\beta}_{2y}(z) = B^{h}_{\beta} \begin{cases} \cos(\beta z) + [\rho/\epsilon_2(\omega) + d(1 - C)k_x^2]\sin(\beta z)/\beta & z > 0 \\ e^{\rho z} & z < 0 \end{cases} \quad (3.30a)$$

$$E^{\beta}_{2z}(z) = -\frac{1}{\omega\epsilon_0}k_x B^{h}_{\beta} \begin{cases} \cos(\beta z) + [\rho/\epsilon_2(\omega) + d(1 - C)k_x^2]\sin(\beta z)/\beta & z > d \\ C & 0 < z < d \\ [1/\epsilon_2(\omega)]e^{\rho z} & z < 0 \end{cases} \quad (3.30b)$$

where the wave relations are

$$k_x^2 + \beta^2 = k_0^2 \qquad k_x^2 - \rho^2 = k_0^2\epsilon_2(\omega) \quad (3.31)$$

and C is a function of the intensity $J_{\infty}$ [Eq. (3.29)]. It is easily proved that these modes satisfy the orthogonality conditions (2.25) and the normalization-orthogonality conditions (2.27). This enables the normalization constant to be written as

$$B^{h}_{\beta} = \left(\frac{2}{\pi}\frac{\beta^2}{\beta^2 + [\rho/\epsilon_2 + d(1 - C)]^2}\right)^{1/2} \quad (3.32)$$

The orthogonality condition for the surface polariton radiation modes leads to the following expression for the total bulk nonlinear field in the region x > 0 which includes the nonlinear film

$$\{\mathscr{H}^{h}_{2y}(x,z), \mathscr{E}^{h}_{2z}(x,z)\} = \int_0^{m_0} T_{\beta}\{H^{\beta}_{2y}(z), E^{\beta}_{2z}(z)\}e^{ik_x x}d\beta \quad (3.33)$$

where $m_0$ is

$$m_0 = k_0\sqrt{1 + |\epsilon_2(\omega)|} \quad (3.34)$$

### 3.2.2 The nonhomogeneous nonlinear field

In order to satisfy the nonhomogeneous boundary conditions (3.19) the following field in the nonlinear film must be assumed

$$\mathscr{H}^{nh}_{y}(x,z) = \int_{-\infty}^{+\infty} Y_{\nu}(z)e^{i\nu x}d\nu \quad (3.35a)$$

$$\mathscr{E}^{nh}_{x}(x,z) = \int_{-\infty}^{+\infty} W_{\nu}(z)e^{i\nu x}d\nu \quad (3.35b)$$

$$\mathscr{E}^{nh}_{z}(x,z) = \int_{-\infty}^{+\infty} E_{\nu}(z)e^{i\nu x}d\nu \quad (3.35c)$$

Substitution of Eqs. (3.34) into the nonhomogeneous boundary conditions (3.20) gives the following effective boundary conditions for each nonhomogeneous wave

$$W_\nu^{(2)}(0) = W_\nu^{(v)}(0) + \frac{i}{\omega\epsilon_0} d\left[-(1 - C)\nu^2 Y_\nu^{(v)}(0) + \nu d F(\nu)\right] \quad (3.36a)$$

$$Y_\nu^{(2)}(0) = Y_\nu^{(v)}(0) \quad (3.36b)$$

Note that $\mathscr{H}_y^{nh}(x,z)$ must obey the wave equations (3.12) and the homogeneous solution $\mathscr{H}_y^{h}(x,z)$ must satisfy Eq. (3.29). These fields differ only through the effective boundary conditions [i.e. Eq. (3.23) for the homogeneous wave and Eq. (3.36) for the nonhomogeneous wave] and through the relationships between the $\mathscr{E}_z^h$ and $\mathscr{E}_z^{nh}$ components and the magnetic field $\mathscr{H}_y$ in the film [i.e. Eqs. (3.22a) and (3.22b) respectively]. These modes have, in principle, a similar structure in both the vacuum and the metal, because the wave equations in this part of the structure are the same for such fields [Eqs. (3.12)]. There are two types of homogeneous mode [Eqs. (3.21a) and (3.21b)] and two types of solution for $Y_\nu(z)$: nonhomogeneous surface polaritons and nonhomogeneous surface polariton radiation modes.

*Nonhomogeneous surface polaritons.* The first type of solution has the form given by Eq. (2.8) and exists if $\nu \in [-\infty,-k_0]$ & $[k_0,+\infty]$. The relationship between $k_1$ and $k_2$ of Eq. (2.8) and the wave number of the nonhomogeneous surface polariton, $\nu$, is the same as that between $k_1^{(j)}$, $k_2^{(j)}$ and $k_s^{(j)}$ in Eq. (3.25). After some simple transformations, Eqs. (2.8) and (3.36) give the following expression for $A_1$ and $A_2$:

$$A_1 = A_2 = -\frac{\nu F(\nu)}{\frac{1}{\epsilon_2(\omega)}k_2 - k_1 - d(1 - C)\nu^2} \quad (3.37)$$

From Eqs. (3.33a) and (3.35) we have

$$\mathscr{H}_y^{nh}(x,z) = \int_\Omega \left\{ -\frac{\nu F(\nu)}{\frac{1}{\epsilon_2(\omega)}k_2 - k_1 - d(1 - C)\nu^2} \right\} Y_\nu(z) e^{i\nu x} d\nu \quad (3.38)$$

where $\Omega$ is the integration interval $\nu \in [-\infty,-k_0]$ & $[k_0,+\infty]$. The same kind of result can be obtained for the $\mathscr{E}_z^{nh}(x,z)$ component. It can be shown, from Eq. (3.24), that the term

$$\frac{1}{\epsilon_2(\omega)}k_2 - k_1 - d(1 - C)\nu^2 = (\nu - |k_s^{(1)}|)(\nu - |k_s^{(2)}|) \quad (3.39)$$

coincides with the dispersion equation for the homogeneous surface polariton and introduces four poles ($\nu = \pm|k_s^{(1,2)}|$) into the integral (3.38). If the complex variable $V = \nu e^{i\psi}$ is used the integral of Eq. (3.38) becomes the sum of the pole contributions and that of an integral along an upper-half-plane semi-circle $\nu = k_0$, $\psi \in (0,\pi)$. Therefore, the following expressions for $\mathscr{H}_y^{nh}(x,z)$ and $\mathscr{E}_z^{nh}(x,z)$ are obtained from Eq. (3.38) for the pole terms (labelled "a"):

$$\left.\begin{matrix} \mathscr{H}_y^{nh}(x,z)^a \\ \mathscr{E}_z^{nh}(x,z)^a \end{matrix}\right\} = 2\pi i \sum_{j=-2}^{2} \frac{F(k_s^{(j)})}{|k_s^{(n)}| - |k_s^{(j)}|} \left\{\begin{matrix} Y_j(z) \\ E_j(z) \end{matrix}\right\} e^{ik_s^{(j)}x} \quad (3.40)$$

where $n = 1$ if $j = \pm 2$, $n = 2$ if $j = \pm 1$, the functions $Y_j(z)$ and $E_j(z)$ are

$$Y_j(z) = -k_s^{(j)} d \begin{cases} e^{-k_1^{(j)} z} & z > 0 \\ e^{k_2^{(j)} z} & z < 0 \end{cases} \quad (3.41a)$$

$$E_j(z) = \frac{1}{\omega\epsilon_0}\begin{cases} (k_s^{(j)})^2 d e^{-k_1^{(j)}z} & z > d \\ |k_s^{(n)}| - |k_s^{(j)}| & 0 < z < d \\ (k_s^{(j)})^2 \frac{d}{\epsilon_2(\omega)} e^{k_2^{(j)}z} & z < 0 \end{cases} \tag{3.41b}$$

and the wave numbers $k_1^{(j)}$ and $k_2^{(j)}$ are defined in Eq. (3.25). The term labelled "b" obtained from the integration along $\psi$ under $\nu = k_0$ is given by

$$\left.\begin{matrix} \mathscr{H}_y^{nh}(x,z)^b \\ \mathscr{E}_z^{nh}(x,z)^b \end{matrix}\right\} = \int_0^{\pi} \frac{F(k_0 e^{i\psi})}{\frac{1}{\epsilon_2(\omega)}k_2 - k_1 - d(1 - C)k_0^2 e^{2i\psi}} \begin{Bmatrix} Y_\psi(z) \\ E_\psi(z) \end{Bmatrix} e^{ik_0 \exp(i\psi)x} d\psi \tag{3.42}$$

where the functions $Y_\psi(z)$ and $E_\psi(z)$ are

$$Y_\psi(z) = -k_0 e^{i\psi} d \begin{cases} e^{-k_1(\psi)z} & z > 0 \\ e^{k_2(\psi)z} & z < 0 \end{cases} \tag{3.43a}$$

$$E_\psi(z) = \frac{1}{\omega\epsilon_0}\begin{cases} k_0^2 e^{2i\psi} d e^{-k_1(\psi)z} & z > d \\ [1/\epsilon_2(\omega)]k_2 - k_1 - d(1 - C)k_0^2 e^{2i\psi} & 0 < z < d \\ k_0^2 e^{2i\psi}[d/\epsilon_2(\omega)]e^{k_2(\psi)z} & z < 0 \end{cases} \tag{3.43b}$$

and the wave numbers $k_1(\psi)$ and $k_2(\psi)$ are, from Eq. (3.25),

$$k_1(\psi) = k_0\sqrt{1 - e^{2i\psi}} \qquad k_2(\psi) = k_0\sqrt{\epsilon_2(\omega) - e^{2i\psi}} \tag{3.44}$$

*Nonhomogeneous surface polariton radiation modes.* This type of solution follows from the nonhomogeneous boundary conditions (3.36) for the interval $\nu \in [-k_0, k_0]$. The structure of these modes is the same as that of homogeneous surface polariton radiation modes [Eq. (2.16)] but the boundary conditions (3.36) yield other relationships between the constants, i.e.

$$B_1 + B_2 = B \qquad -i\beta(B_1 - B_2) = B\left[\frac{\rho}{\epsilon_2(\omega)} - d(1 - C)\nu^2\right] + \nu d F(\nu) \tag{3.45}$$

Notice that Eqs. (3.45) have two independent parameters B and $F(\nu)$. If $F(\nu) = 0$ we obtain the homogeneous modes of Eqs. (3.30). Conversely, if $B = 0$ we have the nonhomogeneous solution $B_1 = -B_2 = \frac{1}{2}i\nu d F(\nu)/\beta$. Finally, the spectrum of the nonlinear nonhomogeneous surface polariton radiation modes is

$$\{\mathscr{H}_{2y}^{nh}(x,z)^c, \mathscr{E}_{2z}^{nh}(x,z)^c\} = \int_{-k_0}^{k_0} F(\nu)\{Y_\nu(z), E_\nu(z)\}e^{i\nu x} d\nu \tag{3.46}$$

where the functions $Y_\nu(z)$ and $E_\nu(z)$ are

$$Y_\nu(z) = \nu d \begin{cases} \sin(\beta z)/\beta & z > 0 \\ 0 & z < 0 \end{cases} \tag{3.47a}$$

$$E_\nu(z) = -\frac{1}{\omega\epsilon_0}\begin{cases} \nu^2 d\sin(\beta z)/\beta & z > d \\ 1 & 0 < z < 0 \\ 0 & z < 0 \end{cases} \tag{3.47b}$$

where $\beta = k_0\sqrt{1 - \nu^2}$.

### 3.3 The nonlinear diffracted transmitted and reflected fields

It has been assumed that the total solution satisfying the effective boundary conditions (3.11) consists of a homogeneous and a nonhomogeneous part. From Eqs. (3.28) and (3.33) the nonlinear component of the transmitted homogeneous field (x > 0) is

$$\left.\begin{matrix} \mathscr{H}^h_y(x,z) \\ \mathscr{E}^h_z(x,z) \end{matrix}\right\} = \sum_j^N T_j \begin{Bmatrix} H^{(j)}_y(z) \\ E^{(j)}_z(z) \end{Bmatrix} e^{ik^{(j)}_s x} + \int_0^{m_0} T_\beta \begin{Bmatrix} H^\beta_{2y}(z) \\ E^\beta_{2z}(z) \end{Bmatrix} e^{ik_x x} d\beta \tag{3.48}$$

and the nonlinear transmitted nonhomogeneous field is

$$\{\mathscr{H}^{nh}_y(x,z), \mathscr{E}^{nh}_z(x,z)\} = \sum_{\chi=a,b,c} \{\mathscr{H}^{nh}_y(x,z)^\chi, \mathscr{E}^{nh}_z(x,z)^\chi\} \tag{3.49}$$

where the fields $\mathscr{H}^{nh}_y(x,z)^{a,b,c}$ and $\mathscr{E}^{nh}_y(x,z)^{a,b,c}$ are defined in Eqs. (3.40), (3.42) and (3.46).

The form of the reflected field is the same as that in section 2 [Eqs. (2.32)-(2.33)] but with a surface-active crystal with dielectric permittivity $\epsilon_2(\omega)$ and $\epsilon_{xx} = \epsilon_{zz} = 1$. The transverse envelopes of the surface polariton at x < 0 are

$$H^i_y(z) = A^i_0 \begin{cases} e^{-k^i_1 z} & z > 0 \\ e^{k^i_2 z} & z < 0 \end{cases} \tag{3.50a}$$

$$E^i_z(z) = -\frac{1}{\omega\epsilon_0} k^i_s A^i_0 \begin{cases} e^{-k^i_1 z} & z > 0 \\ \frac{1}{\epsilon_2(\omega)} e^{k^i_2 z} & z < 0 \end{cases} \tag{3.50b}$$

The following expression for the normalization constant $A^i_0$ is obtained from Eq. (2.26):

$$A^i_0 = \left(\frac{1}{2k^i_1} + \frac{1}{2k^i_2\epsilon_2(\omega)}\right)^{-1/2} \tag{3.51}$$

where the relationships between $k^i_1$, $k^i_2$ and the wave number $k^i_s$ are the same as those between $k^{(j)}_1$, $k^{(j)}_2$ and $k^{(j)}_s$ in Eq. (3.25). The dispersion equation for surface polaritons at x < 0 follows from Eq. (3.24) in the limit d = 0.

The transverse envelopes of the reflected surface polariton radiation mode, obtained from Eqs. (2.22) and (2.23), are

$$H^\beta_{1y}(z) = B^i_\beta \begin{cases} \cos(\beta z) + \frac{\rho}{\beta}\frac{1}{\epsilon_2(\omega)}\sin(\beta z) & z > 0 \\ e^{\rho z} & z < 0 \end{cases} \tag{3.52a}$$

$$E_{1z}^{\beta}(z) = -\frac{1}{\omega\epsilon_0} k_x B_{\beta}^{i} \begin{cases} \cos(\beta z) + \frac{\rho}{\beta}\frac{1}{\epsilon_2(\omega)}\sin(\beta z) & z > 0 \\ \frac{1}{\epsilon_2(\omega)} e^{\rho z} & z < 0 \end{cases} \tag{3.52b}$$

The normalization condition Eq. (2.27) and Eqs. (3.52) yield

$$B_{\beta} = \left[\frac{2}{\pi}\frac{\beta^2}{\beta^2 + \rho^2/[\epsilon_2(\omega)]^2}\right]^{1/2} \tag{3.53}$$

The eigenmodes of the reflected field are, naturally, linear, because the structure for $x < 0$ is linear. Nevertheless, the total reflected field [Eqs. (2.32)-(2.33)] is nonlinear, since the coefficients R and $R_{\beta}$ arising from the process of nonlinear diffraction depend on the amplitude I of the incident surface polariton.

### 3.4 Nonlinear surface polariton diffraction

The boundary conditions at $x = 0$ that exist during nonlinear diffraction are Eqs. (2.6b), and, as in section 2, we will use Eqs. (2.38) as our starting point for the investigation of the amplitudes of the nonlinear diffracted modes. Using Eqs. (3.21), (2.32)-(2.33) and (3.48)-(3.49), we obtain the following system of equations:

$$(I + R)H_y^{i}(z) + \int_0^{m_0} R_{\beta} H_{1y}^{\beta}(z) d\beta = \sum_1^2 T_j H_y^{j}(z) + \int_0^{m_0} T_{\beta} H_{2y}^{\beta}(z) d\beta + \mathcal{H}_y^{nh}(0,z) \tag{3.54a}$$

$$(I - R)E_z^{i}(z) - \int_0^{m_0} R_{\beta} E_{1z}^{\beta}(z) d\beta = \sum_1^2 T E_z^{\tau}(z) + \int_0^{m_0} T_{\beta} E_{2z}^{\beta}(z) d\beta + \mathcal{E}_z^{nh}(0,z) \tag{3.54b}$$

The method for solving the above equations is the same as that used earlier. Using the orthogonality conditions for the homogeneous modes of the nonlinear film [e.g. Eqs. (2.25)-(2.27)] the following set of equations for the amplitudes of the transmitted homogeneous surface polaritons can be derived from Eq. (3.54b):

$$-\frac{1}{\omega\epsilon_0} k_s^{(j)} T_j + \int_{-\infty}^{+\infty} \mathcal{E}_z^{nh}(0,z) H_y^{(j)*}(z) dz = (I - R)\int_{-\infty}^{+\infty} E_z^{i}(z) H_y^{(j)*}(z) dz$$
$$- \int_0^{m_0} R_{\beta} \left[\int_{-\infty}^{+\infty} E_{1z}^{\beta}(z) H_y^{(j)*}(z) dz\right] d\beta \tag{3.55a}$$

Similarly the amplitudes of the homogeneous surface polariton radiation modes are

$$-\frac{1}{\omega\epsilon_0} k'_x T_{\beta'} + \int_{-\infty}^{+\infty} \mathcal{E}_z^{nh}(0,z) H_{2y}^{\beta'*}(z) dz = (I - R)\int_{-\infty}^{+\infty} E_z^{i}(z) H_{2y}^{\beta'*}(z) dz$$
$$- \int_0^{m_0} R_{\beta} \left[\int_{-\infty}^{+\infty} E_{1z}^{\beta}(z) H_{2y}^{\beta'*}(z) dz\right] d\beta \tag{3.55b}$$

Expressions for the amplitude of the reflected modes follow from the first equation of the set (3.54). For surface polaritons we obtain

$$-\frac{1}{\omega\epsilon_0}k_s^{i*}(I^* + R^*) = \sum_{j=1}^{2} T_j^* \int_{-\infty}^{+\infty} H_y^{(j)*}(z)E_z^i(z)dz$$

$$+ \int_0^{m_0} T_\beta^* \left\{ \int_{-\infty}^{+\infty} H_{2y}^{\beta*}(z)E_z^i(z)dz \right\} d\beta + \int_{-\infty}^{+\infty} \mathscr{H}_y^{nh*}(0,z)E_z^i(z)dz \tag{3.55c}$$

and for the reflected surface polariton radiation modes

$$-\frac{1}{\omega\epsilon_0}k'^*_x R^*_{\beta'} = \sum_{j=1}^{2} T_j^* \int_{-\infty}^{+\infty} H_y^{(j)*}(z)E_{1z}^{\beta'}(z)dz$$

$$+ \int_0^{m_0} T_\beta^* \left\{ \int_{-\infty}^{+\infty} H_{2y}^{\beta*}(z)E_{1z}^{\beta'}(z)dz \right\} d\beta + \int_{-\infty}^{+\infty} \mathscr{H}_y^{nh*}(0,z)E_{1z}^{\beta'}(z)dz \tag{3.55d}$$

Since at the beginning of any calculation the functional dependence of $\epsilon_{nl}(x)$ is not known an iterative method must be used.[31] The first step in the iteration is to set $F(\nu) = 0$ in Eqs. (3.19)-(3.20) and to fix the value of C or $J_\infty$ in Eq. (3.29a). Thus we formally transfer the problem to the linear regime for the first step of the iteration. All the parameters for the structure are known and we can determine the eigenmodes of the nonlinear film. In other words, the nonhomogeneous part of the transmitted field, Eq. (3.21), will be neglected in the first step, i.e. $\mathscr{H}_y^{nh}(x,z) = \mathscr{E}_z^{nh}(x,z) = 0$ initially. Introducing the definitions

$$\int_{-\infty}^{+\infty} E_z^i(z)H_y^{(j)*}(z)dz = -\frac{1}{\omega\epsilon_0}k_s^i(E_z^i, H_y^{(j)})$$

$$\int_{-\infty}^{+\infty} E_z^i(z)H_{2y}^{\beta*}(z)dz = -\frac{1}{\omega\epsilon_0}k_s^i(E_z^i, H_y^{\beta})$$

$$\int_{-\infty}^{+\infty} E_{1z}^{\beta}(z)H_y^{(j)*}(z)dz = -\frac{1}{\omega\epsilon_0}k_x(E_{1z}^{\beta}, H_y^{(j)})$$

$$\int_{-\infty}^{+\infty} E_{1z}^{\beta}(z)H_{2y}^{\beta'*}(z)dz = -\frac{1}{\omega\epsilon_0}k_x(E_{1z}^{\beta}, H_{2y}^{\beta})\delta(\beta - \beta')$$

a system of equations for the nonlinear diffraction process, like Eqs. (2.51)-(2.54) for the linear case, is obtained from Eqs. (3.55)

$$Tk_s^{(j)} = (I - R)k_s^i(E_z^i, H_y^{(j)}) - \int_0^{m_0} R_\beta k_x(E_{1z}^{\beta}, H_y^{(j)})d\beta \tag{3.56a}$$

$$T_\beta k_x = (I - R)k_s^i(E_z^i, H_{2y}^{\beta}) - R_\beta k_x(E_{1z}^{\beta}, H_{2y}^{\beta}) \tag{3.56b}$$

$$I + R = \sum_{j=1}^{2} T_j(H_y^{(j)}, E_z^i) + \int_0^{m_0} T_\beta(H_{2y}^\beta, E_z^i)d\beta \tag{3.56c}$$

$$R_\beta = \sum_{j=1}^{2} T_j(H_y^{(j)}, E_{1z}^\beta) + T_\beta(H_{2y}^\beta, E_{1z}^\beta) \tag{3.56d}$$

where the homogeneous surface polariton mutual overlap integrals are

$$(E_{1z}^\beta, H_{2y}^\beta) = B_\beta^i B_\beta^h \frac{\pi}{2}\left\{1 + \frac{1}{\beta^2}\frac{\rho}{\epsilon_2(\omega)}\left[\frac{\rho}{\epsilon_2(\omega)} + d(1 - C)k_x^2\right]\right\} \tag{3.57}$$

$$(H_y^{(j)}, E_z^i) = A_0^i A_j^h\left\{\frac{1}{k_1^i + k_1^{(j)}} + \frac{1}{\epsilon_2(\omega)}\frac{1}{k_2^i + k_2^{(j)}}\right\} \tag{3.58}$$

and

$$(E_z^i, H_{2y}^\beta) = A_0^i B_\beta^h \frac{d(1 - C)k_x^2}{(k_1^i)^2 + \beta^2} \qquad (E_{1z}^\beta, H_y^{(j)}) = B_\beta^i A_j^h \frac{d(1 - C)(k_s^{(j)})^2}{(k_1^{(j)})^2 + \beta^2} \tag{3.59}$$

The transformations in Eqs. (3.56) are the same as those in Eqs. (2.51)-(2.54). Note that Eqs. (3.56) also have an algebraic character. The results are made more convenient by introducing the two functions

$$\Psi_0(\beta) = \frac{k_s^i(E_z^i, H_{2y}^\beta)}{1 + (E_{1z}^\beta, H_{2y}^\beta)(H_{2y}^\beta, E_{1z}^\beta)} \qquad \Psi_j(\beta) = \frac{(H_{2y}^{(j)}, E_{1z}^\beta)}{1 + (E_{1z}^\beta, H_{2y}^\beta)(H_{2y}^\beta, E_{1z}^\beta)} \tag{3.60}$$

It is also easy to see that the expression $T_j = T_j^0(I - R)$ can be derived from Eqs. (3.56a), (3.56c) and (3.56d) where

$$T_1^0 = \frac{V_{01}(k_s^{(2)} + V_{22}) - V_{02}V_{21}}{(k_s^{(1)} + V_{11})(k_s^{(2)} + V_{22}) - V_{12}V_{21}} \tag{3.61}$$

$$T_2^0 = \frac{V_{02}(k_s^{(1)} + V_{11}) - V_{01}V_{12}}{(k_s^{(1)} + V_{11})(k_s^{(2)} + V_{22}) - V_{12}V_{21}} \tag{3.62}$$

and the constants $V_{0j}$ , $V_{kj}$ are

$$V_{0j} = k_s^i(E_z^i, H_y^{(j)}) - \int_0^{m_0} \Psi_0(\beta)(H_{2y}^\beta, E_{1z}^\beta)(E_{1z}^\beta, H_y^{(j)})d\beta \tag{3.63}$$

$$V_{kj} = \int_0^{m_0} \Psi_k(\beta)(E_{1z}^\beta, H_y^{(j)})k_x d\beta \qquad k,j = 1,2 \tag{3.64}$$

The corresponding expressions for surface polariton radiation modes, $T_\beta = (I - R)T_\beta^0$ and $R_\beta = (I - R)R_\beta^0$, are derived from Eq. (3.56), where

$$T^0_\beta = \frac{1}{k_x}\left[\Psi_0(\beta) - \sum_{j=1}^{2} T^0_j \Psi_j(\beta) k_x (E^\beta_{1z}, H^\beta_{2y})\right] \quad (3.65)$$

$$R^0_\beta = \frac{1}{k_x}\left[\sum_{j=1}^{2} T^0_j \Psi_j(\beta) k_x + \Psi_0(\beta)(H^\beta_{2y}, E^\beta_{1z})\right] \quad (3.66)$$

Finally from Eqs. (3.56c) and (3.66)-(3.67) we have

$$(I + R) = (I - R)N^{(1)}_{ef} \quad (3.67)$$

where $N^{(1)}_{ef}$, the effective reflection factor for the nonlinear film arising from the first iteration step, is

$$N^{(1)}_{ef} = \sum_{j=1}^{2} T^0_j (H^{(j)}_y, E^i_z) + \int_0^m T^0_\beta (H^\beta_{2y}, E^i_z) \quad (3.68)$$

Equation (3.67) is the first one of the two that are necessary for the determination of R and T. The second one, obtained from Eq. (3.29b), is

$$J_\infty = |I - R|^2 J^{(1)}_\infty \qquad J^{(1)}_\infty = C^2\left(|T^0_1 k^{(1)}|^2 + |T^{(2)}_2 k^{(2)}|^2\right) \quad (3.69)$$

where $J^{(1)}_\infty$ is the effective transmission factor of the nonlinear film for a homogeneous field. Notice that the parameters $N^{(1)}_{ef}$ and $J^{(1)}_\infty$ are determined by the value C only. C also determines the other parameters of the diffraction, namely $T^0_j$ [Eqs. (3.61)-(3.62)], $T^0_\beta$ [Eq. (3.65)] and $R^0_\beta$ [Eq. (3.66)]. From Eqs. (3.69)-(3.70), the final expressions for the amplitudes of the reflected and incident surface polaritons obtained in the first iteration step are

$$II^* = \frac{1}{2} J_\infty \frac{N^{(1)}_{ef} + 1}{J^{(1)}_\infty} \qquad RR^* = \frac{1}{2} J_\infty \frac{N^{(1)}_{ef} - 1}{J^{(1)}_\infty} \quad (3.70)$$

This shows the parametric dependence of R upon I. Hence other amplitudes also depend upon I.

The second step in the iteration defines these values more precisely. From Eq. (3.48) we can determine the dependence of $E^h_z$ on x and from Eq. (3.13) find the corresponding dependence for $\epsilon_{nl}(x)$. After transforming Eqs. (3.13) and (2.69) we obtain the following intensity dependence

$$J(x) = \frac{J_\infty}{C^2 \sum_{j=1}^{2} |T_j k^{(j)}_s|^2} \left| \sum_{j=1}^{2} T^0_j k^{(j)}_s e^{ik^{(j)}_s x} + \int_0^{m_0} T^0_\beta k_x e^{ik_x x} d\beta \right|^2 \quad (3.71)$$

The x-dependence of the dielectric permittivity of the nonlinear film $\epsilon_{nl} = \epsilon_{nl}(x)$ can be found from Eq. (3.13). It enables us to include the function $F(\nu)$ [Eq. (3.18)] in the investigation of the diffracted field. Indeed, we know all the necessary characteristics, i.e. the forms of $\mathcal{H}^{(v)}_y(x,z)$ [Eq. (3.48)] and $\epsilon_{nl}(x)$ [Eqs. (3.19) and (3.71)]. From Eq. (3.18) the following expression is obtained:

$$F(\nu) = \sum_{j}^{2} T_j(k_s^{(j)})^2 C(\nu - k_s^{(j)}) + \int_0^{m_0} T_\beta(k_x)^2 C(\nu - k_x) d\beta \tag{3.72}$$

where the Fourier-harmonic $C(\gamma)$ is

$$C(\gamma) = \frac{1}{2\pi} \int_0^{\infty} \left( \frac{1}{\epsilon_{nl}(x)} - C \right) e^{-i\gamma x} dx \tag{3.73}$$

The function $C(\gamma)$ has been singled out because it is defined by the nonhomogeneous part of the dielectric permittivity $\epsilon_{nl}(x)$ along the x-coordinate only [the function $F(\nu)$ contains the magnetic field (3.18)]. In processes such as the propagation of a surface polariton along a rough surface the dielectric permittivity depends upon the coordinates. It can be shown that these processes are also quite well described by the mode-method. Therefore the functions (3.73) will be very important in the study of rough surfaces as well as in the nonlinear case under investigation here.

Including the nonhomogeneous field, Eq. (3.49), in the set of equations (3.56) we can obtain from Eqs. (3.55a) and (3.55c) the following new expressions for I, R and $T_j$

$$T k_s^{(j)} = (I - R) k_s^i (E_z^i, H_y^{(j)}) - \int_0^m R_\beta k_x^{(1)} (E_{1z}^\beta, H_y^{(j)}) d\beta - (I - R)(\mathscr{E}_z^{nh}, H_y^{(j)}) \tag{3.74a}$$

$$I + R = \sum_{j=1}^{2} T^{(j)} (H_y^{(j)}, E_z^i) + \int_0^m T_\beta (H_{2y}^\beta, E_z^i) d\beta + (I - R)(\mathscr{H}_y^{nh}, E_z^i) \tag{3.74b}$$

where

$$(\mathscr{H}_{2y}^{nh}, E_z^i) = A_0^i d \left[ -2\pi i \sum_{j=-2}^{2} \frac{F(k_s^{(j)})}{|k_s^{(i)}| - |k_s^{(j)}|} k_s^{(j)} \left( \frac{1}{k_1^i + k_1^{(j)}} + \frac{1}{\epsilon_2(\omega)} \frac{1}{k_2^i + k_2^{(j)}} \right) \right.$$

$$- \int_0^{\pi} \frac{F(k_0 e^{i\psi}) k_0 e^{i\psi}}{\frac{1}{\epsilon_2(\omega)} k_2(\psi) - k_1(\psi) - d(1 - C) k_0^2 e^{2i\psi}} \left( \frac{1}{k_1(\psi) + k_1^i} + \frac{1}{\epsilon_2(\omega)} \frac{1}{k_2(\psi) + k_2^{(j)}} \right) d\psi$$

$$\left. + \int_{-k_0}^{k_0} \frac{F(\nu)\nu}{(k_1^i)^2 + \beta^2} d\nu \right] \tag{3.75a}$$

and

$$(\mathscr{E}_{2z}^{nh}, H_y^{(j)}) = A_j^h d \left[ 2\pi i \sum_{n=-2}^{2} \frac{F(k_s^{(n)})}{|k_s^{(n)}| - |k_s^{(j)}|} \left( \frac{(k_s^{(n)})^2}{k_1^{(n)} + k_1^{(j)}} + \frac{1}{\epsilon_2(\omega)} \frac{(k_s^{(n)})^2}{k_2^{(n)} + k_2^{(j)}} + |k_s^{(n)}| - |k_s^{(j)}| \right) \right.$$

$$- \int_0^{\pi} \frac{F(k_0 e^{i\psi})}{\frac{1}{\epsilon_2(\omega)} k_2(\psi) - k_1(\psi) - d(1 - C) k_0^2 e^{2i\psi}} \left( \frac{k_0 e^{i\psi}}{k_1(\psi) + k_1^i} + \frac{k_0 e^{i\psi}}{\epsilon_2(\omega)} \frac{1}{k_2(\psi) + k_2^{(j)}} \right) d\psi$$

$$\left. + \int_{-k_0}^{k_0} \frac{F(\nu)\nu^2}{(k_1^{(j)})^2 + \beta^2} d\nu \right] \tag{3.75b}$$

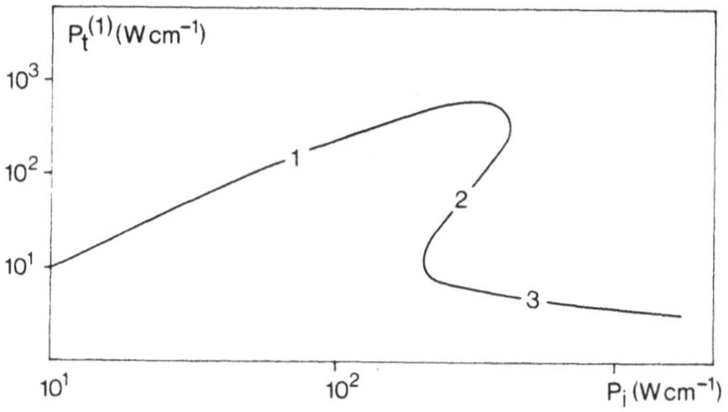


Fig. 12. Optical hysteresis in the dependence of the energy flow of the first surface polariton (Fig. 11) on the energy flow of the incident surface polariton in a structure with d = 80 Å, $\alpha_e$ = 4×10$^{-6}$ cm²/W, $\omega$ = 3.195 eV.

Notice that in this second step of the iteration we are interested in the amplitude of the surface polaritons though it is possible to get an analogous expression for the amplitude of the surface polariton radiation modes. Using the value of the surface polariton radiation mode amplitude from the first step, Eqs. (3.61)-(3.62) and (3.65)-(3.66), the expressions for the parameters of the nonlinear diffraction obtained from Eqs. (3.74) and (3.68) are

$$N_{ef}^{(2)} = N_{ef}^{(1)} - (\mathscr{H}_y^{nh}, E_z^i) \qquad J_\infty^{(2)} = J_\infty^{(1)} + (\mathscr{E}_z^{nh}, H_y^{(j)}) \tag{3.76}$$

The amplitudes of the reflected and incident surface polaritons are

$$II^* = \frac{1}{2} J_\infty \frac{|N_{ef}^{(2)}|^2 + 1}{J_\infty^{(2)}} \qquad RR^* = \frac{1}{2} J_\infty \frac{|N_{ef}^{(2)}|^2 - 1}{J_\infty^{(2)}} \tag{3.77}$$

The expressions for the other amplitudes are

$$\begin{gathered} T_j T_j^* = \frac{1}{2} J_\infty \frac{|T_j^0|^2}{J_\infty^{(2)}} \\ R_\beta R_\beta^* = \frac{1}{2} J_\infty \frac{|R_\beta^0|^2}{J_\infty^{(2)}} \qquad T_\beta T_\beta^* = \frac{1}{2} J_\infty \frac{|T_\beta^0|^2}{J_\infty^{(2)}} \end{gathered} \tag{3.78}$$

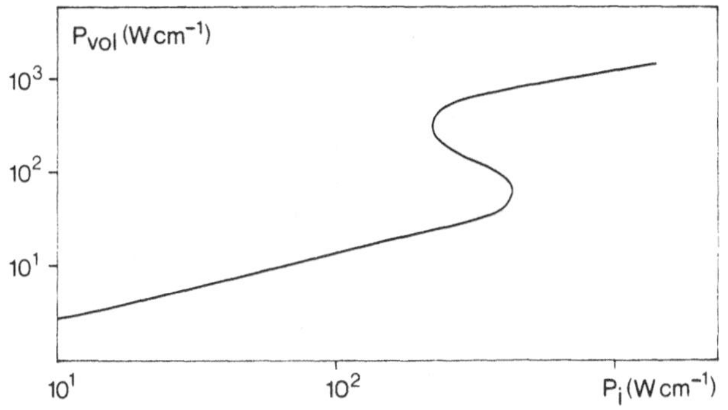


Fig. 13. Same as in Fig. 12 but for the energy flow of the volume diffracted radiation.

Notice that $J_\infty$ is the parameter which is fixed initially in the nonlinear shift $\Delta\epsilon_{nl} = \alpha_e J_\infty$ [$\epsilon_{nl} = \epsilon_1(\omega) + \Delta\epsilon_{nl}$] of the dielectric permittivity $\epsilon_1(\omega)$. The shift $\Delta\epsilon_{nl}$ must be continually updated in order to calculate the dependence of the energy of the diffracted field on the energy of the incident surface polariton.

Numerical solutions for the nonlinear diffraction of surface polaritons are presented in Figs. 12 and 13. The main feature is the optical hysteresis in the dependence of the energy of surface and volume fields upon the energy of the incident surface polariton. The physical explanation of this phenomenon is as follows. The nonlinearity of the resonance film varies as the resonance frequency $\omega_L$ with energy flow, i.e. $\omega_L = \omega_L(J_\infty)$. If we change the energy flow $J_\infty$, Fig. 12 changes as well. If $\alpha_e > 0$ the surface polaritons in the film "rise" to the edge of the gap and the relative value of $T_2$, the amplitude of the second surface polariton, increases. As has been mentioned already the amplitudes of the transmitted surface polaritons $E_z^{(1)}$ and $E_z^{(2)}$ have opposite signs: $Sgn(E_z^{(1)}) = -Sgn(E_z^{(2)})$. Accordingly, the total surface amplitude $E_s = E_z^{(1)} + E_z^{(2)}$ decreases. The process corresponds to the region labelled 2 in Fig. 12. It will continue until the value $E_{2z}$ of the volume field is equal to $E_s$. A further increase in the power of the incident surface polariton corresponds to an increase in the power of the volume field (labelled 3 in Fig. 12).

## 4. CONCLUSION

Although these problems seem to be rather specific, they nevertheless have not only an applied character but also a fundamental one. They are related to the general investigation of the surface wave behaviour in the presence of various types of perturbations on interfaces. These include interface roughness, irregularities, jumps in the medium parameters in the direction of the wave propagation and nonlinear perturbations to the surface. In other words, a whole range of surface diffraction problems is encompassed.

The preceding sections discussed the diffraction of the surface polaritons in linear and nonlinear media at a dielectric barrier. All the investigations here are based upon the mode-method of studying surface diffraction. The general theory of this method was reviewed and it was shown that it could be used for the investigation any diffraction problem in arbitrary multilayered structures. The use of this method enables all the diffracted fields in both linear and nonlinear media to be found and is a useful tool in the investigation of many surface processes.

## REFERENCES


1. D. Marcuse, *Theory of Dielectric Optical Waveguides* (Academic Press, New York, 1974).
2. A. W. Snyder, J. D. Love, *Optical Waveguide Theory* (Chapman, New York, 1983).
3. A. Yariv, P. Yeh, *Optical Waves in Crystals* (Addison-Wesley, New York, 1984).
4. L. M. Brekhovskikh, *Waves in Layered Media* (Academic Press, New York, 1980).
5. *Surface Polaritons*, V. M. Agranovich and D. L. Mills eds. (North Holland, 1982).
6. *Electromagnetic Surface Modes*, A. D. Boardman ed. (John Wiley, 1982).
7. *Integrated Optics*, T. Tamir ed. (Springer-Verlag, Berlin, 1975).
8. A. A. Maradudin, Ztschr. Phys. B. - Condensed Matter **41**, 341, (1981).
9. N. N. Akhmediev, Sov. Phys. JETP **56**, 299 (1982).
10. V. M. Agranovich, V. S. Babichenko, V. Ya. Chernyak, Sov. Phys. JETP Lett. **32**, 512 (1982).
11. K. M. Leung, Phys. Rev. A **31**, 1189 (1975).
12. A. D. Boardman and T. Twardowski, Phys. Rev. A **39**, 2481 (1989).
13. E. Garmire, S. D. Allen, J. Marburger, C. M. Verber, Optics Lett. **3**, 69 (1978).
14. J. L. Jewell, M. C. Rushford, H. M. Gibbs, Appl. Phys. Lett. **48**, 172 (1984).
15. D. L. Mills, A. A. Maradudin, Phys. Rev. Lett. **31**, 372 (1973).
16. A. Otto, Ztschr. Phys. **216**, 398 (1968).
17. E. Kretschman, Ztschr. Phys. **241**, 241 (1971).
18. G. N. Zhizhin, M. A. Moskaleva, E. V. Shomina and V. A. Yakovlev, Sov. Phys. JETP Lett. **29**, 486 (1979).
19. F. DeMartini, Y. R. Shen, Phys. Rev. Lett. **36**, 216 (1976).
20. F. DeMartini, G. Giuliani, P. Mataloni *et al*, Phys. Rev. Lett. **37**, 440 (1976).

21. H. Talaat, W. P. Chen, E. Burstein, J. Schoenwald, 'Scattering of volume and surface electromagnetic waves by surface acoustic waves' in Ultrasonic Symp. Proc., p. 441 (1975).
22. R. Orlowsky, P. Urner, D. Hopnauer, Surface Sci. **82**, 69 (1979).
23. G. I. Stegeman, A. A. Maradudin, T. S. Rahman, Phys. Rev. B **23**, 2576, (1981).
24. J. M. Ziman, *Principles of the Theory of Solids* (Cambridge University Press, 1972).
25. V. V. Shevchenko, *Continuous Transitions in Open Waveguides. Introduction to the Theory* (Golen Press, Boulder, 1971).
26. M. Born, E. Wolf, *Principles of Optics.*
27. A. I. Voronko, L. G. Klimova, G. N. Shkerdin, Solid State Commun. **61**, 361 (1987).
28. A. I. Voronko, L. G. Klimova, G. N. Shkerdin, Sov. Phys. Solid State **28**, 602 (1986).
29. V. M. Agranovich, T. A. Leskova, 'Diffraction methods in the spectroscopy of thin-films in the vicinity of resonances' in Progress in Surface Science **88**, 169-327 (1989).
30. V. M. Agranovich, A. G. Malshukov, Optics Commun. **11**, 169 (1974).
31. V. M. Agranovich, A. I. Voronko, T. A. Leskova, 'Bistability on a Surface in Nonlinear Diffraction', Proc. of the Third USA-USSR Symposium on Laser Optics of Condensed Matter, Leningrad, 1987 (Plenum, New York, 1988), p. 147-55.

# QUANTUM THEORY OF NONLINEAR PLANAR DEVICES

M. Bertolotti and C. Sibilia

Dipartimento di Energetica, Universita di Roma
Via A. Scarpa 16, 00161 Roma, Italy

R. Horak and J. Perina

Joint Laboratory of Optics
Palacky University, Olomouc, Czechoslovakia

J. Janszky

Research Laboratory for Crystal Physics
Budapest, Hungary

## 1. INTRODUCTION

Although problems in optical guided wave systems are usually described simply by using the classical Maxwell equations, there are cases in which a quantum mechanical treatment is necessary. Three purely quantum phenomena that have no classical analogues are known: photon antibunching; sub-Poissonian photon statistics; and squeezing of optical fields. If problems connected with these phenomena or with the evolution of photon statistics are to be dealt with, a quantum mechanical treatment must be used.

A linear system cannot, of course, produce or change these properties, but a nonlinear one can. For this reason, in the following treatment the Hamiltonian for a nonlinear optical waveguide will be derived, and its application to some propagation problems will be considered. Although the Hamiltonian is quite general, for the sake of brevity emphasis will be given to planar structures only.

One of the consequences of having the propagation problem treated in quantum mechanical form is the possibility of studying how purely quantum effects propagate in linear systems. We will show, for example, that a quantum effect such as squeezing is affected by the operation of switching in a linear structure because of the phase changes involved in the operation.

The structure of the remainder of this chapter is as follows. Section 2 covers the basic principles of the quantization of the radiation field and introduces the notation. The coherent states, together with the characteristic functions used to derive the statistics, are introduced in section 3. The quantization of the field in the case of a waveguide, and the derivation of the Hamiltonian describing nonlinear Kerr-like media, is dealt with in section 4. Finally, a few examples of applications are given in the subsequent sections.

## 2. QUANTIZATION OF THE RADIATION FIELD

The starting point of the standard technique for the quantization of electromagnetic radiation *in vacuo* is Maxwell's equations in the absence of charges and currents. In this case, the

*Nonlinear Waves in Solid State Physics*
Edited by A.D. Boardman *et al.*, Plenum Press, New York, 1990

electric field $\mathbf{E}(\mathbf{r},t)$ and the magnetic field $\mathbf{H}(\mathbf{r},t)$ can be derived in terms of the single vector potential $\mathbf{A}(\mathbf{r},t)$, [1] which obeys the wave equation

$$\nabla^2\mathbf{A}(\mathbf{r},t) - \frac{1}{c^2}\frac{\partial^2\mathbf{A}(\mathbf{r},t)}{\partial t^2} = 0 \tag{1}$$

together with the transversality condition

$$\nabla\cdot\mathbf{A}(\mathbf{r},t) = 0 \tag{2}$$

in the form

$$\mathbf{E}(\mathbf{r},t) = -\frac{1}{c}\frac{\partial\mathbf{A}(\mathbf{r},t)}{\partial t} \tag{3a}$$

$$\mathbf{H}(\mathbf{r},t) = \nabla\times\mathbf{A}(\mathbf{r},t) \tag{3b}$$

Thus the electric and magnetic fields are determined by the values of $A_x$, $A_y$, $A_z$ at each point $(x,y,z)$ at time t. It is convenient to describe the fields by a set of discrete variables and, to this end, it is assumed that the radiation field is contained in a cavity with perfectly conducting walls. For simplicity, a cube of total volume $V = L^3$ is used. It turns out that the procedure is independent of the shape of the cavity and, if one is interested in the radiation field of free space, V is simply allowed to go to infinity after the calculations are complete. By imposing boundary conditions on the fields, an infinite discrete set of orthogonal modes $\mathbf{u}_\ell(\mathbf{r})$ is obtained as the solution of the wave equation. The set is complete in the sense that any arbitrary field in the cavity can be expressed as a sum of these modes with suitable amplitudes $q_\ell(t)$.

The general solution of Eqs. (1) and (2) can be written[2]

$$\mathbf{A}(\mathbf{r},t) = \sum_\ell q_\ell(t)\mathbf{u}_\ell(\mathbf{r}) \tag{4}$$

Substitution of Eq. (4) into Eqs. (1) and (2) gives, respectively

$$\nabla^2\mathbf{u}_\ell(\mathbf{r}) + \frac{\omega_\ell^2}{c^2}\mathbf{u}_\ell(\mathbf{r}) = 0 \tag{5}$$

$$\frac{d^2q_\ell(t)}{dt^2} + \omega_\ell^2 q_\ell(t) = 0 \tag{6}$$

where $\omega_\ell^2$ is the separation constant, and

$$\nabla\cdot\mathbf{u}_\ell(\mathbf{r}) = 0 \tag{7}$$

The fact that the $\mathbf{u}_\ell(\mathbf{r})$'s must satisfy the boundary conditions

$$\begin{aligned} \mathbf{u}_\ell\times\mathbf{n} &= 0 \\ (\nabla\times\mathbf{u}_\ell)\cdot\mathbf{n} &= 0 \end{aligned} \tag{8}$$

where $\mathbf{n}$ is the normal to the cavity walls, ensures that Eq. (5) admits a denumerable set of orthogonal eigenfunctions $\mathbf{u}_\ell(\mathbf{r})$ with corresponding eigenvalues $\omega_\ell^2/c^2$.

The total energy of the field

$$H_0 = \frac{1}{8\pi}\int_{\text{cavity}} (\mathbf{E}^2 + \mathbf{H}^2)\, d\mathbf{r} \tag{9}$$

can now easily be evaluated by inserting Eq. (4) into Eqs. (3), giving

$$H_0 = \frac{1}{8\pi c^2}\int \sum_{\ell}\sum_{m} [\dot{q}_\ell(t)\dot{q}_m(t)\mathbf{u}_\ell(\mathbf{r})\mathbf{u}_m(\mathbf{r}) + c^2 q_\ell(t)q_m(t)\nabla\times\mathbf{u}_\ell(\mathbf{r})\cdot\nabla\times\mathbf{u}_m(\mathbf{r})]d\mathbf{r} \tag{10}$$

By using the vector identity

$$(\nabla\times\mathbf{u}_\ell)\cdot(\nabla\times\mathbf{u}_m) = \mathbf{u}_m\cdot(\nabla\times[\nabla\times\mathbf{u}_\ell]) + \nabla\cdot(\mathbf{u}_m\times[\nabla\times\mathbf{u}_\ell]) \tag{11}$$

and the fact that, according to Eqs. (5) and (7),

$$\nabla\times(\nabla\times\mathbf{u}_\ell) = \nabla(\nabla\cdot\mathbf{u}_\ell) - \nabla^2\mathbf{u}_\ell = \frac{\omega_\ell^2}{c^2}\mathbf{u}_\ell \tag{12}$$

Eq. (10) can be written as

$$H_0 = \frac{1}{2}\sum_{\ell}[\dot{q}_\ell^2(t) + \omega_\ell^2 q_\ell^2(t)] + \frac{1}{8\pi}\sum_{\ell}\sum_{m} q_\ell(t)q_m(t)\int_V \nabla\cdot(\mathbf{u}_m\times[\nabla\times\mathbf{u}_\ell])d\mathbf{r} \tag{13}$$

where the orthonormality condition

$$\int_V \mathbf{u}_\ell\cdot\mathbf{u}_m\, d\mathbf{r} = 4\pi c^2\delta_{\ell m} \tag{14}$$

has been introduced. If one observes that the last term on the right-hand side of Eq. (13) vanishes, as can easily be seen by applying Gauss's theorem and remembering Eqs. (8), we finally obtain

$$H_0 = \frac{1}{2}\sum_{\ell}[p_\ell^2(t) + \omega_\ell^2 q_\ell^2(t)] \tag{15}$$

where

$$p_\ell(t) = \frac{dq_\ell(t)}{dt} \tag{16}$$

In this way, the electromagnetic field is described in terms of a set of independent pairs of conjugate variables $q_\ell$ and $p_\ell$ that are related to a set of independent harmonic oscillators.

The quantization of the electromagnetic field is now achieved by regarding the $q_\ell$'s and $p_\ell$'s as Hermitian operators obeying the commutation relations

$$\begin{aligned}[p_\ell, p_m] &= p_\ell p_m - p_m p_\ell = [q_\ell, q_m] = 0 \\ [q_\ell, p_m] &= i\hbar\delta_{\ell m}\end{aligned} \tag{17}$$

according to the basic postulate of quantum mechanics. The $q_\ell$'s and $p_\ell$'s will be either explicitly time independent or not according to whether or not one uses the Schrödinger picture.

The standard procedure for quantization of the harmonic oscillator consists of the introduction of a pair of non-Hermitian operators $\hat{a}_\ell^\dagger$ and $\hat{a}_\ell$ by means of the equations

$$q_\ell = \left(\frac{\hbar}{2\omega_\ell}\right)^{1/2} [\hat{a}_\ell^\dagger + \hat{a}_\ell] \tag{18}$$

$$p_\ell = i\left(\frac{\hbar\omega_\ell}{2}\right)^{1/2} [\hat{a}_\ell^\dagger - \hat{a}_\ell] \tag{19}$$

It is easily seen from Eqs. (18) and (19) that $\hat{a}_\ell$ and $\hat{a}_\ell^\dagger$ are Hermitian conjugate operators, while Eqs. (17) show that they obey the commutation relations

$$\begin{gathered}[\hat{a}_\ell, \hat{a}_m^\dagger] = \delta_{\ell m} \\ [\hat{a}_\ell, \hat{a}_m] = [\hat{a}_\ell^\dagger, \hat{a}_m^\dagger] = 0\end{gathered} \tag{20}$$

The Hamiltonian of the system immediately follows from Eqs. (15) and (18)-(20) as

$$\hat{H}_0 = \sum_\ell \hat{H}_\ell = \sum_\ell \hbar\omega_\ell \left(\hat{a}_\ell^\dagger \hat{a}_\ell + \frac{1}{2}\right) \tag{21}$$

If one now chooses to use the Heisenberg picture, the time evolution of $\hat{a}_\ell(t)$ and $\hat{a}_\ell^\dagger(t)$ is determined by the Heisenberg equations of motion

$$i\hbar \frac{d\hat{a}_\ell(t)}{dt} = [\hat{a}_\ell(t), \hat{H}_0] = \hbar\omega_\ell \hat{a}_\ell(t) \tag{22a}$$

$$i\hbar \frac{d\hat{a}_\ell^\dagger(t)}{dt} = [\hat{a}_\ell^\dagger(t), \hat{H}_0] = -\hbar\omega_\ell \hat{a}_\ell^\dagger(t) \tag{22b}$$

where Eqs. (20) and (21) have been taken into account. One has, according to Eqs. (22)

$$\begin{gathered}\hat{a}_\ell(t) = \hat{a}_\ell e^{-i\omega_\ell t} \\ \hat{a}_\ell^\dagger(t) = \hat{a}_\ell^\dagger e^{i\omega_\ell t}\end{gathered} \tag{23}$$

where $\hat{a}_\ell = \hat{a}_\ell(0)$ and $\hat{a}_\ell^\dagger = \hat{a}_\ell^\dagger(0)$ are the operators in the Schrödinger picture.

As is well known,[2] the eigenvalues $n_\ell$ of the operator associated with the eigenvalue equation

$$\hat{a}_\ell^\dagger \hat{a}_\ell |n_\ell\rangle = n_\ell |n_\ell\rangle \tag{24}$$

can take on all non-negative integer values, so that $n_\ell$ can be interpreted as the number of energy quanta in the mode $\ell$.

This allows us to give $\hat{a}_\ell^\dagger \hat{a}_\ell$ the meaning of a number operator. Furthermore, $\hat{a}_\ell^\dagger$ and $\hat{a}_\ell$ are usually termed *creation* and *annihilation* operators since they can respectively be shown to increase and decrease by one the number of quanta when operating on the eigenstates $|n_\ell\rangle$ of $\hat{a}_\ell^\dagger \hat{a}_\ell$ according to

$$\hat{a}_\ell^\dagger |n_\ell\rangle = \left(n_\ell + 1\right)^{1/2} |n_\ell + 1\rangle \tag{25}$$

$$\hat{a}_\ell |n_\ell\rangle = \left(n_\ell\right)^{1/2} |n_\ell - 1\rangle \tag{26}$$

The preceding discussion summarizes the properties of the vector potential operator, which, according to Eqs. (4) and (18) can be written in the Heisenberg picture as

$$\hat{\mathbf{A}}(\mathbf{r},t) = \sum_\ell \left(\frac{\hbar}{2\omega_\ell}\right)^{1/2} \left(\hat{a}_\ell e^{-i\omega_\ell t} + \hat{a}_\ell^\dagger e^{i\omega_\ell t}\right) \mathbf{u}_\ell(\mathbf{r}) \tag{27}$$

Apart from the possibility of expanding the electromagnetic field in terms of standing waves, it is possible to consider a more general electromagnetic field by introducing a representation of $\mathbf{A}(\mathbf{r},t)$ based on the use of *travelling plane waves.*[2]

More precisely, it can be shown[2] that the general solution of Eq. (1) subject to the transversality condition and to the periodic boundary conditions

$$\mathbf{A}(\mathbf{r}+L\mathbf{i},t) = \mathbf{A}(\mathbf{r}+L\mathbf{j},t) = \mathbf{A}(\mathbf{r}+L\mathbf{h},t) = \mathbf{A}(\mathbf{r},t) \tag{28}$$

can be written in the form of an expansion with complex coefficients

$$\mathbf{A}(\mathbf{r},t) = \sum_\ell \sum_{\sigma=1}^{2} \mathbf{e}_{\ell\sigma}\{c_{\ell\sigma} e^{i(\mathbf{k}_\ell \cdot \mathbf{r} - \omega_\ell t)} + \text{c.c.}\} \tag{29}$$

with

$$\omega_\ell = c k_\ell \tag{30}$$

$\mathbf{k}_\ell$ is the propagation vector and the $\mathbf{e}_{\ell\sigma}$'s are two unit vectors specifying the polarization of each *mode* $(\ell,\sigma)$, which can be chosen to be mutually orthogonal and must satisfy the relation

$$\mathbf{e}_{\ell\sigma} \cdot \mathbf{k}_\ell = 0 \tag{31}$$

The quantization of the vector potential (28) can be performed following a method analogous to the one yielding Eq. (27). One has

$$\hat{\mathbf{A}}(\mathbf{r},t) = \frac{c}{L^{3/2}} \sum_{\ell,\sigma} \left(\frac{\hbar}{\omega_\ell}\right)^{1/2} \mathbf{e}_{\ell\sigma}\{\hat{a}_{\ell\sigma} e^{i(\mathbf{k}_\ell \cdot \mathbf{r} - \omega_\ell t)} + \text{c.c.}\} \tag{32}$$

from which the following relation can be deduced

$$\hat{H}_0 = \sum_{\ell,\sigma} \hbar\omega_\ell \left(\hat{a}_{\ell\sigma}^\dagger \hat{a}_{\ell\sigma} + \frac{1}{2}\right) \tag{33}$$

with

$$\begin{aligned} [\hat{a}_{\ell\sigma}, \hat{a}_{\ell'\sigma'}^\dagger] &= \delta_{\ell\ell'}\delta_{\sigma\sigma'} \\ [\hat{a}_{\ell\sigma}, \hat{a}_{\ell'\sigma'}] &= [\hat{a}_{\ell\sigma}^\dagger, \hat{a}_{\ell'\sigma'}^\dagger] = 0 \end{aligned} \tag{34}$$

where the $\hat{a}_{\ell\sigma}$'s and $\hat{a}_{\ell\sigma}^\dagger$'s can be interpreted as annihilation and creation operators of a photon of energy $\hbar\omega_\ell$, momentum $\mathbf{k}_\ell$, and polarization $\sigma$.

## 3. QUANTUM STATES AND CHARACTERISTIC FUNCTIONS

The electromagnetic field has been quantized by associating a quantum-mechanical harmonic oscillator with each mode $\ell$ of the radiation field. The mode to which a quantum mechanical operator refers is indicated by a subscript: thus $\hat{a}_\ell^\dagger$ and $\hat{a}_\ell$ are the operators that create and destroy a quantum of energy $\hbar\omega_\ell$ in the cavity electromagnetic mode of wavevector $\mathbf{k}_\ell$. These quanta are photons of wavevector $\mathbf{k}_\ell$; the number of photons n excited in the cavity is determined by the eigenvalue $n_\ell$ of the appropriate number operator $\hat{n}_\ell = \hat{a}_\ell^\dagger \hat{a}_\ell$ and has possible values 0,1,2,... . The excitation level of a cavity mode $\ell$ is determined by its eigenstate $|n_\ell\rangle$.

The single-mode number states $|n\rangle$ (Fock-states), where exactly n photons are excited, are energy eigenstates of the harmonic oscillator associated with the electromagnetic field in the mode. They are not of direct significance in the interpretation of experiments since the electromagnetic waves generated by practical light sources do not have definite numbers of photons.

The single-mode states of physical importance correspond not to the individual number states but to linear superpositions of the states $|n\rangle$, named *coherent states* $|\alpha\rangle$. Although the coherent states were first introduced by Schrödinger in 1927,[3] their full utilization in connection with quantum optics is due to Glauber.[4] They have an electric-field variation that in the limit of high excitation approaches that of the classical wave of stable amplitude and fixed phase, and can be defined through the relation

$$|\alpha\rangle = \exp\left(-\frac{1}{2}|\alpha|^2\right) \sum_n \frac{\alpha^n}{(n!)^{1/2}} |n\rangle \tag{35}$$

The $|\alpha\rangle$ form an overcomplete set of states for the harmonic oscillator, and they lack orthogonality. The coherent states are eigenstates of the destruction operator with eigenvalue $\alpha$ which is in general a complex number since $\hat{a}$ is not a Hermitian operator:

$$\hat{a}|\alpha\rangle = \alpha|\alpha\rangle \qquad \langle\alpha|\hat{a}^\dagger = \langle\alpha|\alpha^* \tag{36}$$

Note, however, that the state $|\alpha\rangle$ is not an eigenstate of the creation operator. It can be shown that $|\alpha|^2$ is the mean number of photons in the cavity mode.

The definition shows that the probability of finding n photons in the mode is

$$|\langle n|\alpha\rangle|^2 = e^{-|\alpha|^2} \frac{|\alpha|^{2n}}{n!} \tag{37}$$

i.e. a Poisson probability distribution about a mean value $|\alpha|^2$.

In the following we will adopt the formalism of coherent states. Defining $\Delta\hat{q}$ and $\Delta\hat{p}$ by

$$\Delta\hat{q} = \hat{q} - \langle\hat{q}\rangle \qquad \Delta\hat{p} = \hat{p} - \langle\hat{p}\rangle \tag{38}$$

we have from the commutation rule

$$\frac{\hbar}{2} = \frac{1}{2}|\langle[\hat{q},\hat{p}]\rangle| = \frac{1}{2}|\langle[\Delta\hat{q},\Delta\hat{p}]\rangle| \le |\langle\Delta\hat{q}\Delta\hat{p}\rangle| \le \left[\langle(\Delta\hat{q})^2\rangle\langle(\Delta\hat{p})^2\rangle\right]^{1/2} \tag{39}$$

Making use of Eq. (36) we obtain for the expectation values of Eqs. (18) and (19) in terms of the coherent states

$$\langle\hat{q}\rangle = \left(\frac{\hbar}{2\omega}\right)^{1/2}(\alpha + \alpha^*) \qquad \langle\hat{p}\rangle = -i\left(\frac{\hbar\omega}{2}\right)^{1/2}(\alpha - \alpha^*) \tag{40a}$$

$$\langle(\Delta\hat{q})^2\rangle = \frac{\hbar}{2\omega} \qquad \langle(\Delta\hat{p})^2\rangle = \frac{\hbar\omega}{2} \tag{40b}$$

Consequently, for the coherent states the inequality (39) reduces to the equality

$$\langle(\Delta\hat{q})^2\rangle\langle(\Delta\hat{p})^2\rangle = \frac{\hbar^2}{4} \tag{41}$$

Thus, the coherent states are the closest to the classical states that quantum theory allows.

We can also write[5]

$$\frac{1}{2}[\langle(\Delta\hat{p})^2\rangle + \omega^2\langle(\Delta\hat{q})^2\rangle] \geq \omega\left[\langle(\Delta\hat{q})^2\rangle\langle(\Delta\hat{p})^2\rangle\right]^{1/2} = \frac{\hbar\omega}{2} \tag{42}$$

which reduces to the equality for the coherent states. The quantity

$$\frac{1}{2}[\langle(\Delta\hat{p})^2\rangle + \omega^2\langle(\Delta\hat{q})^2\rangle] = \frac{1}{2}[\langle\hat{p}^2\rangle + \omega^2\langle\hat{q}^2\rangle] - \frac{1}{2}[\langle\hat{p}\rangle^2 + \omega^2\langle\hat{q}\rangle^2] \tag{43}$$

represents the difference between the total energy of the field and its coherent energy, i.e. it represents the incoherent energy of the field. Therefore, it may be said that in the coherent state the incoherent energy takes on its minimum value of $\hbar\omega/2$ which represents only the energy of zero point fluctuations.

Squeezed states are defined by the property that $\langle(\Delta\hat{q})^2\rangle$ or $\langle(\Delta\hat{p})^2\rangle$ is less than $\hbar^2/4$. For the sake of brevity, we will usually define

$$\hat{Q} = \hat{a} + \hat{a}^\dagger \qquad \text{and} \qquad \hat{P} = -i(\hat{a} - \hat{a}^\dagger) \tag{44}$$

normalizing with respect to $\hbar$ so that squeezing requires $\langle(\Delta\hat{Q})^2\rangle$ or $\langle(\Delta\hat{P})^2\rangle$ less than unity.

To derive the statistics, we will introduce the normal characteristic function[6]

$$C_N(\beta, t) = Tr\{\hat{\rho}\exp(\beta\hat{a}^\dagger)\exp(-\beta^*\hat{a})\}$$

$$= \int \Phi_N(\alpha)\exp(\alpha^*\beta - \alpha\beta^*)\, d^2\alpha \tag{45}$$

where the Glauber-Sudarshan representation of the density matrix has been used[4,7]

$$\hat{\rho} = \int \Phi_N(\alpha)|\alpha\rangle\langle\alpha|\, d^2\alpha \tag{46}$$

This allows us to reduce the quantum expectation values of normally ordered operators (q-numbers) to expectation values of eigenvalues of coherent states (c-numbers).

From the knowledge of the normal characteristic function we can derive all the moments of $\Phi_N$ and $\Phi_A$ (which are normally and antinormally ordered quasi-distribution functions respectively) so that we can define the statistics of the radiation during the interaction

$$\frac{\partial^k}{\partial\beta^k}\frac{\partial^\ell}{\partial(-\beta^*)^\ell}C_N(\beta)|_{\beta=\beta^*=0} = Tr\{\hat{\rho}\hat{a}^{\dagger k}\hat{a}^\ell\} = \int \Phi_N(\alpha)\alpha^{*k}\alpha^\ell d^2\alpha = \langle\alpha^{*k}\alpha^\ell\rangle_N \tag{47}$$

## 4. QUANTUM MECHANICAL DESCRIPTION OF PROPAGATION IN PLANAR WAVEGUIDES

A planar dielectric waveguide is a medium whose dielectric permittivity depends on one direction, which we shall assume to be parallel to the x-axis (Fig. 1). If this medium does not contain absorbing centres, if there is no amplification of radiation, and if the permittivity is weakly dependent on the field frequency $\omega$, the electromagnetic field can be analyzed within the second quantization representation. Therefore, it can be expanded in terms of normal modes in the following form,[8,9]

$$\mathbf{A}_\ell(\mathbf{r}) = f_\ell(x)\, e^{i\mathbf{k}_\ell \cdot \mathbf{r}} \tag{48}$$

where $\mathbf{k}_\ell$ is the wave-vector with components $k_{\ell y}$ and $k_{\ell z}$ ($\mathbf{k}_\ell \cdot \mathbf{r} = k_{\ell y} y + k_{\ell z} z$) and $f_\ell(x)$ is a function dependent only on x and determined over all space.

The vector-potential operator of the field is then

$$\hat{\mathbf{A}}(\mathbf{r},t) = \sum_\ell \{\mathbf{A}_\ell(\mathbf{r})\hat{a}_\ell(t) + \mathbf{A}_\ell^*(\mathbf{r})\hat{a}_\ell^\dagger(t)\}$$

$$= \sum_\ell \left(\frac{\hbar}{2\omega_\ell}\right)^{1/2} [\mathbf{u}_\ell(\mathbf{r})\hat{a}_\ell(t) + \mathbf{u}_\ell^*(\mathbf{r})\hat{a}_\ell^\dagger(t)] \tag{49}$$

where $\mathbf{u}_\ell(\mathbf{r})$ are modal orthogonal functions which characterize the particular waveguide geometry. They are a complete set determined by the boundary conditions and geometry; $\ell$ is an abbreviation for the set of numbers which characterize the mode. The $\mathbf{u}_\ell$'s are completely specified known functions. The amplitude of each normal mode needed to specify a particular field configuration is $\hat{a}_\ell$. In a waveguide the mode expansion of the field also includes the modes characterized by the same frequency but with different values of the $\mathbf{k}_\ell$ vectors.

In a nonlinear optical waveguide, the permittivity is field dependent. In this case, assuming the same second quantization representation of the field as given in Eq. (49), we can write the following radiation Hamiltonian

$$H_{rad} = \frac{1}{8\pi}\int_V [\mathbf{E}^2(\mathbf{r},t) + \mathbf{H}^2(\mathbf{r},t)]d\mathbf{r} + \frac{1}{8\pi}\int_V \epsilon'(\mathbf{r},t)\mathbf{E}^2(\mathbf{r},t)d\mathbf{r} \tag{50}$$

where $\epsilon'(\mathbf{r},t)$ is the nonlinear part of the dielectric permittivity. Assuming a Kerr-type nonlinearity we have

$$\epsilon'(\mathbf{r},t) = \epsilon_1(x)\mathbf{E}^2(\mathbf{r},t) \tag{51}$$

$\epsilon_1(x)$ being a space dependent function.

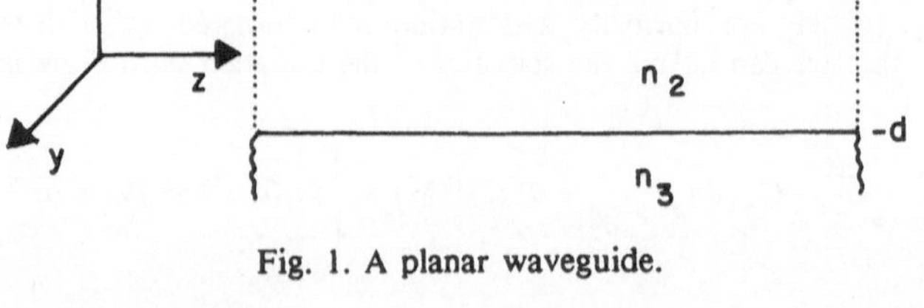

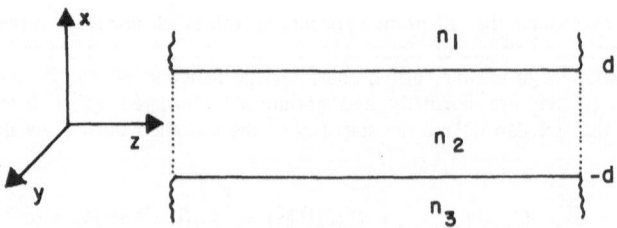

Fig. 1. A planar waveguide.

Equation (51) expresses the fact that the nonlinear medium modulates the beam by the induced changes of the index of refraction, which leads to self-modulation of the beam. Substituting Eqs. (51) and (3) into (50), we obtain[10]

$$\hat{H}_{rad} = \sum_{\ell} \hbar\omega_{\ell}\left(\hat{a}_{\ell}^{\dagger}\hat{a}_{\ell} + \frac{1}{2}\right) + \sum_{k,\ell,m,n} \hbar U_{k\ell mn}\hat{a}_{k}^{\dagger}\hat{a}_{\ell}^{\dagger}\hat{a}_{m}\hat{a}_{n} \tag{52}$$

where

$$U_{k\ell mn} \sim \int_{V} \epsilon_1(x)\mathbf{u}_k^*(\mathbf{r})\mathbf{u}_{\ell}^*(\mathbf{r})\mathbf{u}_m(\mathbf{r})\mathbf{u}_n(\mathbf{r})d\mathbf{r} \tag{53}$$

In Eq. (52) quickly varying terms, such as $\hat{a}_k^{\dagger}\hat{a}_{\ell}\hat{a}_m\hat{a}_n$, $\hat{a}_k\hat{a}_{\ell}\hat{a}_m\hat{a}_n$, etc., have been neglected. Terms containing $\hat{a}_k^{\dagger}\hat{a}_k$, arising from the fourth order terms by the commutation rule, are neglected in comparison with $\omega_k\hat{a}_k^{\dagger}\hat{a}_k$; the off-diagonal terms $V_{kl}\hat{a}_k^{\dagger}\hat{a}_{\ell}$, with $k \neq \ell$, provide only a linear exchange of energy among modes and lead to no change of photon number statistics (but they change the degree of coherence[8]). Here we include these terms in the loss mechanism.

The Hamiltonian of Eq. (52) is our basic result. It is the sum of the usual oscillator part of the free radiation plus an interaction term which contains the product of four annihilation or creation operators. It can be used to solve many different problems. In the following we will consider some examples:

(i) single-mode propagation in a nonlinear guide;
(ii) two-mode propagation in a single waveguide;
(iii) propagation in a directional coupler in both the linear and nonlinear cases.

## 5. SINGLE-MODE PROPAGATION IN A PLANAR WAVEGUIDE

The radiation Hamiltonian suitable for this problem can be written from Eq. (52) as

$$\hat{H} = \hbar\omega\left(\hat{a}^{\dagger}\hat{a} + \frac{1}{2}\right) + \hbar\chi\hat{a}^{\dagger 2}\hat{a}^{2} \tag{54}$$

where $\chi$ is a real coupling constant of the self-interaction, and the mode labelling has been dropped since there is only one mode.

We wish to consider the one-mode radiation field interacting with an infinite reservoir boson system which may represent losses in the guided mode (radiation modes of the waveguide). This can be done working in the Heisenberg picture of the annihilation and creation operators $\hat{a}$ and $\hat{a}^{\dagger}$ and with an infinite boson system with annihilation and creation operators $\hat{b}_j$ and $\hat{b}_j^{\dagger}$.[5] Starting from the total Hamiltonian of the process

$$\hat{H} = \hat{H}_0 + \hat{H}_{int} \tag{55}$$

where

$$\hat{H}_0 = \hbar\omega\left(\hat{a}^{\dagger}\hat{a} + \frac{1}{2}\right) + \sum_{j} \hbar\psi_j\left(\hat{b}_j^{\dagger}\hat{b}_j + \frac{1}{2}\right) \tag{56a}$$

$$\hat{H}_{int} = \sum_{j} (\hbar\kappa_j\hat{b}_j\hat{a}^{\dagger} + \hbar\kappa_j^*\hat{b}_j^{\dagger}\hat{a}) \tag{56b}$$

with $\psi_j$ the frequency of the boson system and $\kappa$ the coupling constant of the interaction, we can write the Heisenberg equations for the interaction as

$$\frac{d\hat{a}}{dt} = -\frac{\gamma}{2}\hat{a} - 2i\chi\hat{a}^\dagger\hat{a}^2 + \hat{L}$$
$$\frac{d\hat{a}^\dagger}{dt} = -\frac{\gamma}{2}\hat{a}^\dagger + 2i\chi\hat{a}^{\dagger 2}\hat{a} + \hat{L} \qquad (57)$$

where losses are introduced[5] through the damping constant $\gamma$ and the Langevin force $\hat{L}$, defined by

$$\gamma = 2\pi|\kappa(\omega)|^2\rho(\omega) \qquad (58)$$

$\rho(\omega)$ being the density function of the reservoir oscillators and

$$\hat{L}(t) = -i\sum_{\ell}\kappa_\ell\hat{b}_\ell(0)\,e^{-i\psi_\ell t} \qquad (59)$$

In Eqs. (57), it must be noted that in the propagation problems the time t is given by $t = z/v$, where z is the distance travelled and v is the wave velocity.

The solutions of Eqs. (57) are mostly impossible to find in closed form. Neglecting losses, approximate solutions will now be discussed. Solutions with losses must be obtained by computer calculations.

Neglecting losses (i.e. $\gamma = 0$ and $\hat{L} = 0$), it is easy to see that the solution of Eqs. (57) is given by

$$\hat{a}(t) = e^{-2i\chi t\hat{a}^\dagger\hat{a}}\hat{a} \qquad (60)$$

where $\hat{a} = \hat{a}(0)$.

A first explicit form of the solution given by Eq. (60) can be obtained by developing it up to terms in $(\chi t)^2$, using the commutation rules for the field operators. We obtain

$$\hat{a}(t) = [1-2i\chi t\hat{a}^\dagger\hat{a} - 2\chi^2t^2(\hat{a}^{\dagger 2}\hat{a}^2 + \hat{a}^\dagger\hat{a})]\hat{a} \qquad (61)$$

Starting initially from the coherent state $|\xi(0)\rangle$, we may use this solution to arrive at the approximate statistics described by the normal characteristic function (45)

$$C_N(\beta,t) = \langle\exp[\beta\hat{a}^\dagger(t)]\exp[-\beta^*\hat{a}(t)]\rangle \simeq \exp\left[-B(t)|\beta|^2 + \frac{1}{2}C(t)\beta^{*2} + \frac{1}{2}C^*(t)\beta^2 + \beta\xi^*(t) - \beta^*\xi(t)\right] \qquad (62)$$

where $\beta$ is a parameter,

$$\xi(t) = \langle\hat{a}(t)\rangle \simeq \xi\left[1-2i\chi t|\xi|^2 - 2\chi^2t^2|\xi|^2\left(|\xi|^2 + 1\right)\right] \qquad (63)$$

is the time dependent complex amplitude of the field, and $\xi = \xi(0)$

$$B(t) = \langle\Delta\hat{a}^\dagger(t)\Delta\hat{a}(t)\rangle \simeq 4\chi^2t^2|\xi|^4$$
$$C(t) = \langle(\Delta\hat{a})^2\rangle \simeq -2i\chi t\xi^2 - 2\chi^2t^2\xi^2\left(1 + 6|\xi|^2\right) \qquad (64)$$

From the previous equations we get

$$|\xi(t)|^2 \simeq |\xi|^2 \left(1 - 4\chi^2 t^2 |\xi|^2\right) \tag{65}$$

which represents the field intensity. Hence

$$B(t) + |\xi(t)|^2 = |\xi|^2 = \text{const} \tag{66}$$

is a conservation law.

The characteristic function, Eq. (62), corresponds to the one obtained for the propagation of coherent radiation through a random medium; this means that in our case this characteristic function represents a generalized superposition of coherent field $\xi(t)$ and noise, i.e. quantum noise characterized by B(t) and C(t).[5]

For studying complete solutions we may start from the equation of motion for the density matrix within the framework of the Schrödinger picture

$$i\hbar \frac{\partial \rho}{\partial t} = [\hat{H}_0 + \hat{H}_{int}, \rho] \tag{67}$$

The Gaussian property of the reservoir, leading to the Markoffian $\hat{L}(t)$, makes it possible to perform two iterations when solving Eq. (67) for the density matrix traced over reservoir operators, and arrive at the master equation for the reduced density matrix[5]

$$\frac{\partial \rho}{\partial t} = -i\omega\{[\hat{a}, \hat{\rho}\hat{a}^\dagger] + [\hat{a}^\dagger, \hat{a}\hat{\rho}]\} + \frac{\gamma}{2}(\langle n_\ell\rangle + 1)\{[\hat{a}\hat{\rho}, \hat{a}^\dagger] + [\hat{a}, \hat{\rho}\hat{a}^\dagger]\} + \frac{\gamma}{2}\langle n_\ell\rangle\{[\hat{a}^\dagger\hat{\rho}, \hat{a}] + [\hat{a}^\dagger, \hat{\rho}\hat{a}]\}$$

$$= \left(\frac{\gamma}{2} - i\omega\right)[\hat{a}, \hat{\rho}\hat{a}^\dagger] - \left(\frac{\gamma}{2} + i\omega\right)[\hat{a}^\dagger, \hat{a}\hat{\rho}] - \frac{\gamma}{2}\langle n_\ell\rangle([\hat{a}^\dagger, [\hat{a}, \hat{\rho}]] + [\hat{a}, [\hat{a}^\dagger, \hat{\rho}]]) \tag{68}$$

Performing antinormal ordering in this equation, we obtain after some algebra the generalized Fokker-Planck equation for the Glauber-Sudarshan quasidistribution function

$$\frac{\partial \Phi_N}{\partial t} = -i\chi\left\{4\alpha \frac{\partial \Phi_N}{\partial \alpha} - 4\alpha^* \frac{\partial \Phi_N}{\partial \alpha^*} - 2|\alpha|^2 \alpha \frac{\partial \Phi_N}{\partial \alpha} + 2|\alpha|^2 \alpha^* \frac{\partial \Phi_N}{\partial \alpha^*} + \alpha^2 \frac{\partial^2 \Phi_N}{\partial \alpha^2} - \alpha^{*2} \frac{\partial^2 \Phi_N}{\partial \alpha^{*2}}\right\}$$

$$+ \frac{\gamma}{2}\left[\frac{\partial}{\partial \alpha}(\alpha \Phi_N) + \frac{\partial}{\partial \alpha^*}(\alpha^* \Phi_N)\right] + \gamma \bar{n} \frac{\partial^2 \Phi_N}{\partial \alpha \partial \alpha^*} \tag{69}$$

where $\bar{n}$ is the number of reservoir oscillators, i.e. the mean number of damping oscillators coupled with the field mode, related to the Langevin force through the following identity[5]

$$\langle \hat{L}^\dagger(t)\hat{a}(t) + \hat{a}^\dagger(t)\hat{L}(t)\rangle = \gamma \bar{n} \tag{70}$$

The corresponding moment equations read

$$\frac{d}{dt}\langle \alpha^{*k} \alpha^\ell\rangle = i\chi[(k-\ell)(k+\ell-1)\langle \alpha^{*k}\alpha^\ell\rangle + 2(k-\ell)\langle \alpha^{*k+1}\alpha^{\ell+1}\rangle] - \frac{\gamma}{2}(\ell+k)\langle \alpha^{*k}\alpha^\ell\rangle + \gamma\bar{n}k\ell\langle \alpha^{*k-1}\alpha^{\ell-1}\rangle \tag{71}$$

where k and $\ell$ are integers. It is easy to see that the following conservation law holds if losses are neglected ($\gamma = \bar{n} = 0$),

$$\frac{d}{dt}\langle |\alpha|^{2k}\rangle = 0 \tag{72}$$

Now, assuming that the statistical properties of light are described (at least as an approximation at short times or distances) by the characteristic function given by Eq. (62), we can calculate the following quantities

$$\langle\alpha\rangle = \frac{\partial}{\partial(-\beta^*)} C_N|_{\beta=0} = \xi(t)$$

$$\langle\alpha^2\rangle = \frac{\partial^2}{\partial(-\beta^*)^2} C_N|_{\beta=0} = C(t) + \xi^2(t)$$

$$\langle|\alpha|^2\rangle = \frac{\partial^2}{\partial\beta\partial(-\beta^*)} C_N|_{\beta=0} = B(t) + |\xi(t)|^2 \tag{73}$$

$$\langle\alpha^*\alpha^2\rangle = \frac{\partial^3}{\partial\beta\partial(-\beta^*)^2} C_N|_{\beta=0} = 2B(t)\xi(t) + [C(t) + \xi^2(t)]\xi^*(t)$$

$$\langle\alpha^*\alpha^3\rangle = \frac{\partial^4}{\partial\beta\partial(-\beta^*)^3} C_N|_{\beta=0} = |\xi(t)|^2\xi^2(t) + 3B(t)\xi^2(t) + 3C(t)|\xi(t)|^2 + 3B(t)C(t)$$

Equations (73), when substituted into the first of Eqs. (71), are just sufficient to determine all the unknown functions $\xi(t)$, $\xi^*(t)$, $B(t)$, $C(t)$, $C^*(t)$ and this procedure provides a basis for a physical truncation of the system of equations (71).

Substituting Eqs. (73) into Eqs. (71), we arrive at the final system of motion equations

$$\frac{d\xi}{dt} = -\left\{\frac{\gamma}{2} + 2i\chi\left(2|\xi_0|^2 e^{-\gamma t} - |\xi|^2\right) + 4i\chi\bar{n}[1 - e^{-\gamma t}]\xi\right\} - 2i\chi C\xi^*$$

$$\frac{dC}{dt} = C\left\{-\gamma + 2i\chi\left(2|\xi|^2 - 6|\xi_0|^2 e^{-\gamma t} - 1\right) - 12i\chi\bar{n}[1 - e^{-\gamma t}]\right\}$$
$$+ \xi^2\left\{2i\chi\left(2|\xi|^2 - 2|\xi_0|^2 e^{-\gamma t} - 1\right) - 4i\chi\bar{n}(1 - e^{-\gamma t})\right\} \tag{74}$$

$$B + |\xi|^2 = |\xi_0|^2 e^{-\gamma t} + \bar{n}[1 - e^{-\gamma t}]$$

where now $\xi_0 = \xi(0)$.

If we neglect losses ($\gamma = \bar{n} = 0$) in Eq. (74) by substituting $C(t) = 0$ as first order in the iteration, we obtain

$$\xi(t) \simeq \xi_0 \exp\left\{-2i\chi t\left(2|\xi_0|^2 - |\xi_0|^2\right)\right\} = \xi_0 \exp\left(-2i\chi t|\xi_0|^2\right) \tag{75}$$

which corresponds to Eq. (60). Substituting the first order solution in the second of Eqs. (74), we simply obtain $C(t)$ as given in Eq. (64), and so on.

The photon-number distribution $p(n,t)$ and its factorial momenta $\langle W^k\rangle$ are expressed by

$$p(n,t) = (EF)^{-1/2}\left(1 - \frac{1}{F}\right)^n \exp\left(-\frac{A_1}{E} - \frac{A_2}{F}\right)\sum_{k=0}^{n}\frac{1}{\Gamma(k+1/2)\Gamma(n-k+1/2)}\left(\frac{1 - 1/E}{1 - 1/F}\right)^k$$
$$\times L_k^{-1/2}\left(-\frac{A_1}{E(E-1)}\right)L_{n-k}^{-1/2}\left(-\frac{A_2}{F(F-1)}\right) \tag{76}$$

and

$$\langle W^k \rangle = k!(F-1)^k \sum_{\ell=0}^{k} \frac{1}{\Gamma(\ell + 1/2)\Gamma(k - \ell + 1/2)} \left(\frac{E-1}{F-1}\right)^{\ell} L_{\ell}^{-1/2}\left(-\frac{A_1}{E-1}\right) L_{k-\ell}^{-1/2}\left(-\frac{A_2}{F-1}\right) \tag{77}$$

where

$$A_{1,2} = \frac{1}{2}\left\{ |\xi(t)|^2 \mp \frac{1}{2|C(t)|}[C^*(t)\xi^2(t) + c.c.] \right\}$$

$$E = B(t) - C(t) + 1$$

$$F = B(t) + C(t) + 1$$

$\Gamma$ is the gamma function and $L_k^{-1/2}$ the Laguerre polynomial.

The squeezing properties are described by

$$\begin{aligned} \langle(\Delta\hat{Q})^2\rangle &= 1 + 2[B(t) + \mathit{Re}C(t)] \\ \langle(\Delta\hat{P})^2\rangle &= 1 + 2[B(t) - \mathit{Re}C(t)] \end{aligned} \tag{78}$$

However neglecting losses we can easily show that this interaction leads to the conservation of photon statistics; due to the fact that only the phase in Eq. (60) is nonlinearly modified, only fluctuations in $\hat{Q}$ and $\hat{P}$ are created by the dynamics.

Moreover, it can be easily shown that the generating function $C(\lambda, t)$ reduces to

$$C(\lambda, t) = \frac{1}{\pi\lambda}\int \exp\left(-\frac{|\beta|^2}{\lambda}\right) C_N(\beta, t)d^2\beta = \exp\left(-\lambda|\xi_0|^2\right) \tag{79}$$

which is independent of time t. This is the property of the Poissonian statistics; therefore, if we neglect losses, this interaction does not modify the Poissonian statistics of the field.

If we include losses, then the superposition of chaotic noise will be appropriate with the mean number of coherent photons given by

$$\langle n_c \rangle = |\xi_0|^2 e^{-\gamma t} \tag{80}$$

and the mean number of chaotic photons given by

$$\langle n_{ch} \rangle = \bar{n}(1-e^{-\gamma b}) \tag{81}$$

Unfortunately, an analytical solution valid at all times is not easy to find, and computer calculations are necessary in this case. To show the effect of noise, the nonlinear motion equations (74) are solved subject to the initial conditions $\xi(0) = \xi_0$ and $B(0) = C(0) = 0$, and $p(n,t)$ and $\langle(\Delta W)^2\rangle/\langle W^2\rangle$ are calculated using Eqs. (76) and (77). Fluctuations in $\hat{Q}$ and $\hat{P}$ are obtained using Eqs. (78).

Figures 2 and 3 refer to squeezing. When losses are zero the field is squeezed (Fig. 2) but the damping reduces this behaviour. In particular, for high values of the damping constant, the oscillator behaviour, as a function of space, is reduced. This is shown in Fig. 3 for $\langle(\Delta\hat{Q})^2\rangle$; similar results can be obtained for $\langle(\Delta\hat{P})^2\rangle$. Thus, the loss mechanism smears out the nonclassical behaviour of light. In general, by increasing the mean number of reservoir oscillators, the damping of the oscillation of the squeezing operator increases.

Figure 4 represents the photon number distribution when losses are included: the Poissonian distribution of the zero loss case is modified.

Finally, in Fig. 5, the behaviour of the normalized factorial moment $\langle(\Delta W)^2\rangle/\langle W^2\rangle$ demonstrates that, when losses are considered, the initial coherent light fast approaches a chaotic state, i.e. the initial Poisson distribution is transformed into a Bose-Einstein distribution.

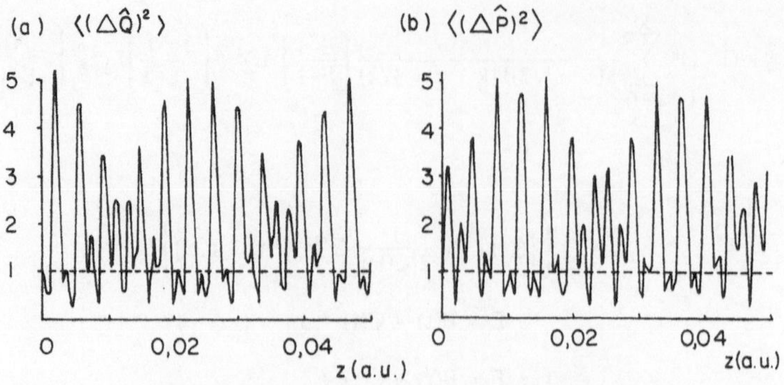


Fig. 2. Space dependence of (a) $\langle(\Delta\hat{Q})^2\rangle$ and (b) $\langle(\Delta\hat{P})^2\rangle$ for $\chi = 100$, $\gamma = \bar{n} = 0$, $|\xi_0|^2 = 4$, $\phi = 0$.

## 6. TWO-MODE PROPAGATION IN A PLANAR WAVEGUIDE

We wish now to study the interaction of two modes of different frequencies by writing the Hamiltonian (52) in the form[10]

$$\hat{H}_{rad} = \hbar\omega_1\left(\hat{a}_1^\dagger\hat{a}_1 + \frac{1}{2}\right) + \hbar\omega_2\left(\hat{a}_2^\dagger\hat{a}_2 + \frac{1}{2}\right) + \hbar g_1\hat{a}_1^{\dagger 2}\hat{a}_1^2 + \hbar g_2\hat{a}_2^{\dagger 2}\hat{a}_2^2 + \hbar\chi\hat{a}_1^\dagger\hat{a}_2^\dagger\hat{a}_1\hat{a}_2 \qquad (82)$$

where the coupling constants are $g_j \sim U_{jjjj}$, $j = 1,2$ and $\chi \sim U_{1212}$.

The Hamiltonian expressed by Eq. (82), although very general because it also includes the degenerate case, could describe, for example, the excitation of a TE mode at frequency $\omega_1$ and a TM mode at frequency $\omega_2$.

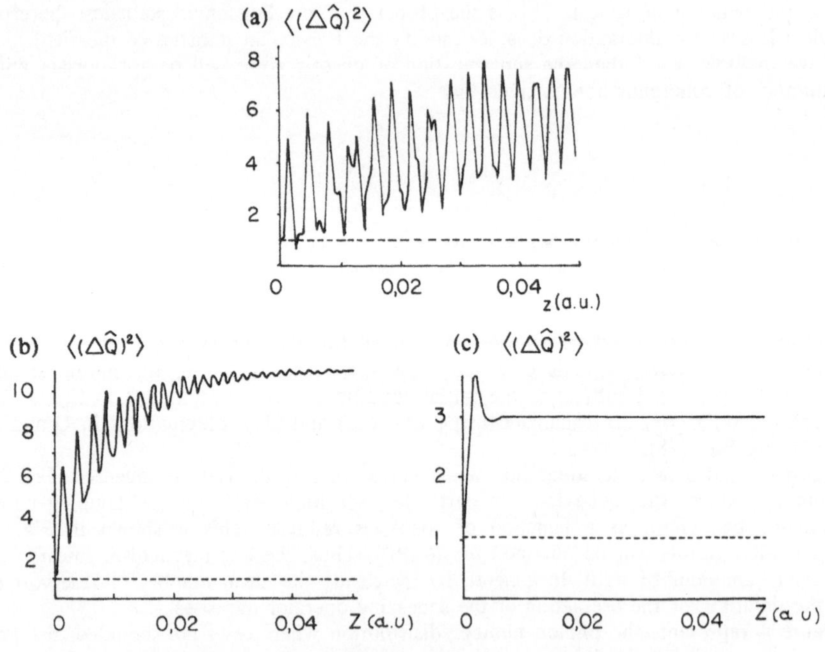


Fig. 3. $\langle(\Delta\hat{Q})^2\rangle$ for $\chi = 100$, $|\xi_0|^2 = 4$, $\phi = -\pi/4$ and (a) $\gamma = 10$, $\bar{n} = 5$, (b) $\gamma = 100$, $\bar{n} = 5$ and (c) $\gamma = 1000$, $\bar{n} = 1$.

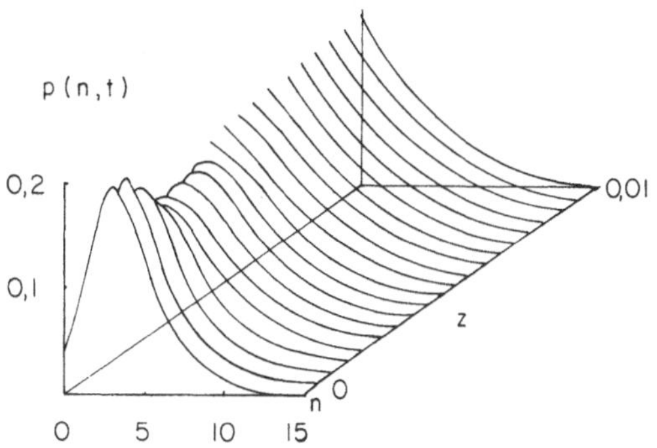


Fig. 4. Space dependence of the photon number distribution p(n, t) for the same data as in Fig. 3.

If the nondegenerate case (i.e. different frequency modes) is considered, the last interaction term in Eq. (82) is dominant compared to the terms $\hat{a}_1^{\dagger 2}\hat{a}_2^2$ and $\hat{a}_1^2\hat{a}_2^{\dagger 2}$ which are quickly varying and can therefore be neglected.

By including the loss mechanism in a similar manner to that used for the single-mode case, we can write down the Heisenberg-Langevin equations[5]

$$\frac{d\hat{a}_1}{dt} = -\left(i\omega_1 + \frac{\gamma_1}{2}\right)\hat{a}_1 - 2ig_1\hat{a}_1^\dagger\hat{a}_1^2 - i\chi\hat{a}_1\hat{a}_2^\dagger\hat{a}_2 - i\sum_\ell \chi_{1\ell}\hat{b}_{1\ell}(0)\exp[-i\psi_{1\ell}(t)]$$

$$\frac{d\hat{a}_2}{dt} = -\left(i\omega_2 + \frac{\gamma_2}{2}\right)\hat{a}_2 - 2ig_2\hat{a}_2^\dagger\hat{a}_2^2 - i\chi\hat{a}_1^\dagger\hat{a}_1\hat{a}_2 - i\sum_\ell \chi_{2\ell}\hat{b}_{2\ell}(0)\exp[-i\psi_{2\ell}(t)] \qquad (83)$$

where $\gamma_j$ are the damping constants, $\chi_{j\ell} \sim V_{j\ell}$ are the reservoir coupling constants, $\hat{b}_{j\ell}(0)$ are the initial reservoir operators, and $\psi_{j\ell}$ are the frequencies of the reservoir oscillators. Some of the radiative modes can be interpreted as reservoir modes with respect to the guided modes. Their influence can be included through the damping constants $\gamma_j$ and the mean numbers of the reservoir oscillators

$$\bar{n}_j = \langle\hat{b}_j^\dagger(0)\hat{b}_j(0)\rangle$$

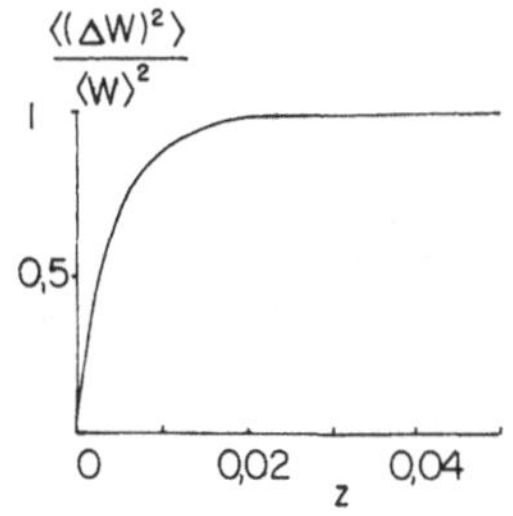


Fig. 5. Space dependence of $\langle(\Delta\hat{W})^2\rangle/\langle W\rangle^2$ for the same data as in Fig. 3.

It is convenient to introduce the interaction picture

$$\hat{a}_j(t) = \hat{A}_j(t)e^{-i\omega_j t}$$

where the upper case operators obey the equations

$$\frac{d\hat{A}_1}{dt} = \left(-\frac{\gamma_1}{2}\right)\hat{A}_1 - 2ig_1(\hat{A}_1^\dagger\hat{A}_1)\hat{A}_1 - i\chi\hat{A}_1(\hat{A}_2^\dagger\hat{A}_2) - i\sum_\ell \chi_{1\ell}\hat{b}_{1\ell}(0)e^{-i(\psi_{1\ell} - \omega_1)t}$$

$$\frac{d\hat{A}_2}{dt} = \left(-\frac{\gamma_2}{2}\right)\hat{A}_2 - 2ig_2(\hat{A}_2^\dagger\hat{A}_2)\hat{A}_2 - i\chi\hat{A}_2(\hat{A}_1^\dagger\hat{A}_1) - i\sum_\ell \chi_{2\ell}\hat{b}_{2\ell}(0)e^{-i(\psi_{2\ell} - \omega_2)t} \quad (84)$$

Photon statistics and squeezing can be studied via a procedure similar to the one used for the single-mode case. Neglecting losses, after long calculations the generating function can be written as

$$C(\lambda, t) = \exp\left[-\lambda\left(\left|\xi_{10}\right|^2 + \left|\xi_{20}\right|^2\right)\right] \quad (85)$$

Thus, the initial coherent light remains coherent for all times and the photon statistics are Poissonian with the mean number of coherent photons given by

$$\langle n_c\rangle = \left|\xi_{10}\right|^2 + \left|\xi_{20}\right|^2 \quad (86)$$

If losses are included, then the photon statistics of the superposition of coherent and chaotic fields are appropriate, with the mean number of coherent photons given by

$$\langle n_{cj}\rangle = \left|\xi_{j0}\right|^2 e^{-\gamma_j t} \quad (87)$$

and the mean number of chaotic photons given by

$$\langle n_{chj}\rangle = \bar{n}_j[1 - e^{-\gamma_j t}] \qquad j = 1,2 \quad (88)$$

When losses are included, the photon statistics again evolve from Poisson to Bose-Einstein distributions. Squeezing can be more easily studied by defining

$$\hat{Q}_j = \hat{A}_j + \hat{A}_j^\dagger \qquad \hat{P}_j = -i(\hat{A}_j - \hat{A}_j^\dagger)$$

$$\hat{Q}_{12} = \hat{Q}_1 + \hat{Q}_2 \qquad \hat{P}_{12} = \hat{P}_1 + \hat{P}_2 \quad (89)$$

In this case, squeezing exists for single modes if $\langle(\Delta\hat{Q}_j)^2\rangle$ or $\langle(\Delta\hat{P}_j)^2\rangle$ are less than 1 and for coupled modes if $\langle(\Delta\hat{Q}_{12})^2\rangle$ or $\langle(\Delta\hat{P}_{12})^2\rangle$ are less than 2.

Also in this case it can be shown that the squeezing effect for each mode exhibits the same characteristics as already discussed. The squeezing effect for both modes is shown in Figs. 6, which show the space dependence of the variances $\langle(\Delta\hat{Q}_{12})^2\rangle$ and $\langle(\Delta\hat{P}_{12})^2\rangle$, including effects caused by the coupling of modes (the nonlinear coupling between the modes influences the phase of each mode). Note the oscillatory behaviour which was not present in the single-mode case. The squeezing effect is shown in greater detail in Figs. 7 where it is seen that $\langle(\Delta\hat{Q}_{12})^2\rangle < 2$ for short distances, whereas no squeezing occurs in $\langle(\Delta\hat{P}_{12})^2\rangle$, as seen in Fig. 6b.

The effect of losses can be seen in Figs. 8, where losses are chosen large enough to destroy squeezing completely. Similarly to the single-mode case, oscillations are damped proportionally to the damping constants $\gamma_j$ and the fluctuations are enhanced proportionally to the mean number of reservoir oscillators $\bar{n}_j$.

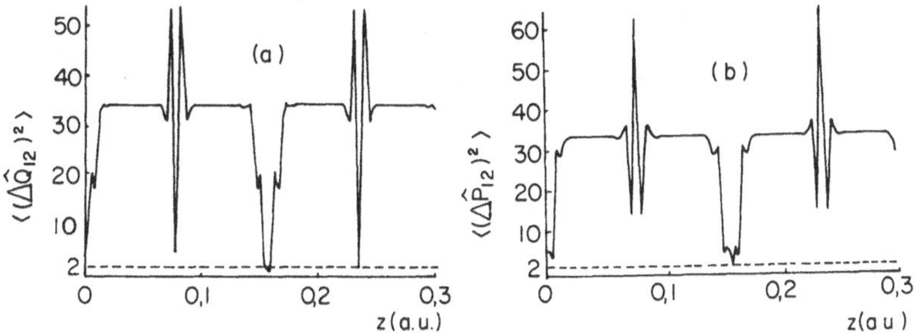


Fig. 6. Space dependence of the variances for $\chi = 40$, $g_1 = g_2 = 20$, $|\xi_{10}|^2 = |\xi_{20}|^2 = 4$, $\phi_1 = \phi_2 = 0$.

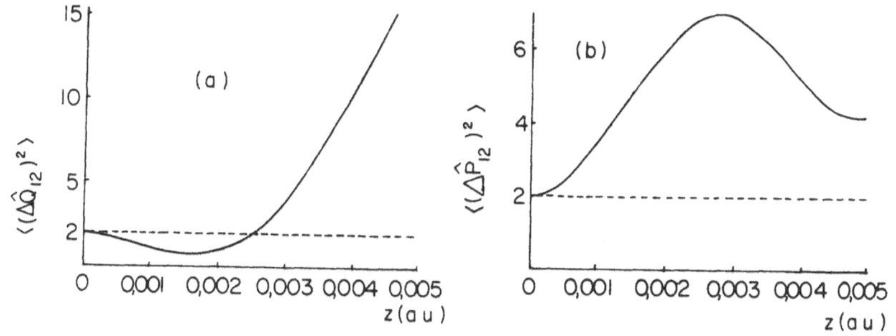


Fig. 7. As Fig. 6 but on a much enlarged scale.

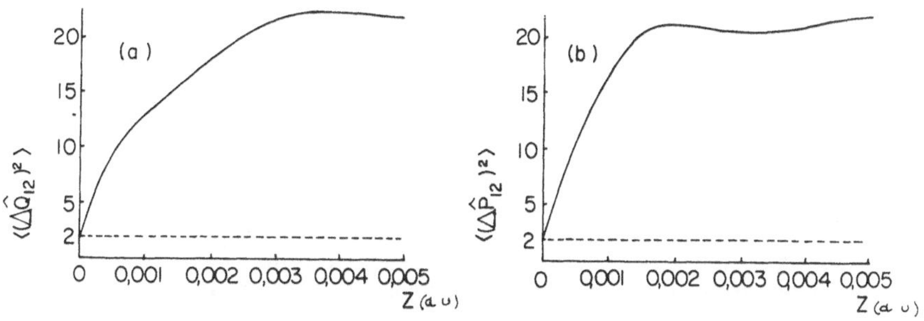


Fig. 8. As Fig. 7 with $\gamma_1 = \gamma_3 = 10^3$, $\bar{n}_1 = \bar{n}_2 = 5$.

## 7. BEHAVIOUR OF A LINEAR DIRECTIONAL COUPLER WHEN NONCLASSICAL STATES ARE INVOLVED (SQUEEZING)

The linear directional coupler consists of two adjacent and parallel waveguides.[12] When radiation goes through the structure, exchange of power between the channels is possible because of the evanescent field which is present in the region between them.

Within the framework of classical theory the coupler is studied by using the *coupled mode theory*,[12,13] in which a perturbation polarization responsible for the coupling contains the refractive index of the guides. From the differences between the refractive indices we get the coupling constant K of the structure. Complete power transfer occurs in a distance

$$L = \left(\frac{\pi}{2}\right)K$$

if the detuning parameter $\delta$ is zero, i.e. in the case of complete phase matching,[13] where

$$\delta = \frac{1}{2}(\beta_a - \beta_b)$$

and $\beta_a$, $\beta_b$ are the wavevectors of two modes propagating inside the structure. If $\delta$ is not zero the maximum fraction of power that can be transferred is proportional to

$$\frac{K^2}{K^2 + \delta^2}$$

To study the propagation of nonclassical fields through this structure we will use the following characteristic function

$$C_N(\beta) = Tr\{\rho\exp[\beta\hat{a}^\dagger]\exp[-\beta^*\hat{a}]\}$$

$$= \exp\left[-M|\beta|^2 + \frac{1}{2}(S^*\beta^2 + S\beta^{*2}) + \beta W^* - \beta^* W\right] \tag{90}$$

which is able to describe a field which is not a pure coherent or squeezed state, but which simultaneously has squeezed, coherent and chaotic features.[14] In Eq. (90), $W = We^{i\psi}$ is the coherent signal, and M and S are related to the noncoherent part of the field. For the vacuum state we have $M = S = W = 0$; in the pure coherent state $M = S = 0$; and for the chaotic field $W = S = 0$.

The state is a pure squeezed state if

$$M = Q = \frac{1}{2}\left((4|S|^2 + 1)^{1/2} - 1\right)$$

$$S = e^{i\Phi}\cosh(r)\sinh(r) \tag{91}$$

where r is the squeezing parameter.[14]

A mixed state is obtained from a superposition of a pure squeezed state with coherent signal W and a chaotic field described by the normally ordered characteristic function given by Eq. (90) if

$$M = Q + N \tag{92}$$

where N is the noise photon number.

From the classical, coupled-mode theory, equations for the complex amplitude of the directional coupler, we get with obvious generalization the following Heisenberg equations[15]

$$\frac{d\hat{a}}{dz} = -iK\hat{b}e^{2i\delta z}$$

$$\frac{d\hat{b}}{dz} = -iK\hat{a}e^{-2i\delta z} \tag{93}$$

where $\hat{a}$ and $\hat{b}$ are the field annihilation operators in channels $a$ and $b$ respectively, K is the real coupling constant and $\delta$ is the detuning parameter. For the sake of simplicity we neglect damping terms because we are interested in the coupling effect only; this is a good approximation at low temperatures and optical frequencies.

In this way we get the following solutions of Eqs. (93)

$$\hat{a} = C_a \hat{a}_0 + G_a \hat{b}_0$$
$$\hat{b} = C_b \hat{b}_0 + G_b \hat{a}_0 \tag{94}$$

where $\hat{a}_0$ and $\hat{b}_0$ are the input annihilation operators and

$$C_a = e^{i\delta z}\left[\cos(\gamma z) - i\frac{\delta}{\gamma}\sin(\gamma z)\right]$$

$$C_b = e^{-i\delta z}\left[\cos(\gamma z) + i\frac{\delta}{\gamma}\sin(\gamma z)\right]$$

$$G_a = -i\frac{K}{\gamma}\sin(\gamma z)e^{i\delta z} \tag{95}$$

$$G_b = -i\frac{K}{\gamma}\sin(\gamma z)e^{-i\delta z}$$

where $\gamma^2 = K^2 + \delta^2$.

We shall suppose that the input statistics of light in both modes can be described by the normally ordered characteristic function given by Eq. (90). Putting solutions (94) into Eq. (90) we can see that the truncated normally ordered output characteristic functions will have the same functional form as the input ones with new terms

$$M_a = M_a^{(0)}\left|C_a\right|^2 + M_b^{(0)}\left|G_a\right|^2$$

$$S_a = S_a^{(0)}C_a^2 + S_b^{(0)}G_a^2 \tag{96}$$

$$W_a = W_a^{(0)}C_a + W_b^{(0)}G_a$$

where the superscript (0) labels the input quantities. Similar expressions can be found for the b-mode, by interchanging the subscripts a and b.

We are interested in finding expressions for the variances $\langle(\Delta\hat{Q})^2\rangle$ and $\langle(\Delta\hat{P})^2\rangle$ defined through Eq. (44) and for the noise level. It can be shown that[14]

$$\langle(\Delta\hat{Q})^2\rangle = 1 + 2M + S + S^*$$
$$\langle(\Delta\hat{P})^2\rangle = 1 + 2M - S - S^* \tag{97}$$

Several interesting cases can be considered which depend on the way the coupler is fed. Let us suppose first that a pure squeezed state enters channel *b* and a coherent (or vacuum) state channel *a*. It can be shown in this case that

$$\left\langle \begin{matrix}(\Delta\hat{Q})^2\\(\Delta\hat{P})^2\end{matrix}\right\rangle_a = \left(\frac{\delta}{\gamma}\right)^2 + \left[1 - \left(\frac{\delta}{\gamma}\right)^2\right]\left[\cos^2(\gamma L) + \left\langle \begin{matrix}(\Delta\hat{P})^2\\(\Delta\hat{Q})^2\end{matrix}\right\rangle_{b0}\sin(\gamma L)\right]$$

$$\left\langle \begin{matrix}(\Delta\hat{Q})^2\\(\Delta\hat{P})^2\end{matrix}\right\rangle_b = \left[1 - \left(\frac{\delta}{\gamma}\right)^2\right]\sin^2(\gamma L) + \left\langle \begin{matrix}(\Delta\hat{Q})^2\\(\Delta\hat{P})^2\end{matrix}\right\rangle_{b0}\left[\cos^2(\gamma L) + \left(\frac{\delta}{\gamma}\right)^2\sin^2(\gamma L)\right] \tag{98}$$

where L is the length of each channel.

For $L = \pi/2k$ and $\delta = 0$ we have

$$\langle(\Delta\hat{Q})^2\rangle_a = \langle(\Delta\hat{P})^2\rangle_{b0} \qquad \langle(\Delta\hat{P})^2\rangle_a = \langle(\Delta\hat{Q})^2\rangle_{b0} \qquad \langle(\Delta\hat{Q})^2\rangle_b = \langle(\Delta\hat{P})^2\rangle_b = 1 \tag{99}$$

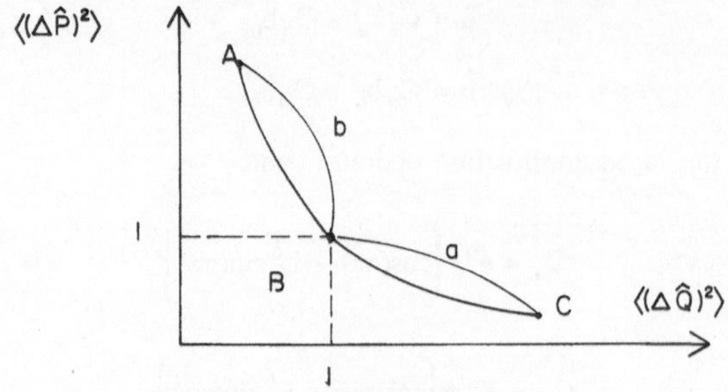


Fig. 9. Behaviour of the linear directional coupler for $L = \pi/2k$, $\delta = 0$.

This means that at the output of channel $a$ we have a squeezing opposite to that at the input of channel $b$, while the output in channel $b$ shows no squeezing.

A related situation is obtained when two oppositely squeezed fields enter the two channels under the same conditions as in the previous case. In this case squeezing is preserved in both channels because the field entering channel $a$ comes out of channel $b$ with opposite squeezing and *vice versa* for the field entering channel $b$. At intermediate lengths of the coupler the squeezing is not completely preserved.

To finish this point let us discuss the role of the detuning parameter $\delta$. We can see from Eqs. (98) that in general if $\delta \neq 0$ some noise is added to both channels and squeezing is reduced. It is worth noting however that for some special values of $\delta$ noise is absent.

Let us consider for example the case in which $\delta = \sqrt{3}k$. In this case $\gamma = 2k$ and for the same coupler length $L = \pi/2k$, if a squeezed field enters channel $b$ and a coherent or vacuum field channel $a$ then we have at the output

$$\langle(\Delta\hat{Q})^2\rangle_a = \langle(\Delta\hat{P})^2\rangle_a = 1 \qquad \langle(\Delta\hat{Q})^2\rangle_b = \langle(\Delta\hat{Q})^2\rangle_{b0} \qquad \langle(\Delta\hat{P})^2\rangle_b = \langle(\Delta\hat{P})^2\rangle_{b0} \tag{100}$$

We can see that changing the detuning parameter from zero to $\sqrt{3}k$ switches from one channel to another. This result is rather interesting for the purpose of measurement. The squeezed state is detected by interfering it with a coherent reference light and looking at fluctuations. The switching behaviour just described allows both the squeezed state and its reference beam to be preserved.

In Figs. 9 and 10 the behaviour of the switching is shown in terms of $\langle(\Delta\hat{Q})^2\rangle$ and $\langle(\Delta\hat{P})^2\rangle$. In Fig. 9 the initial squeezed state A in channel $b$ is converted to coherent state B, while the initial coherent state B turns into a squeezed state C in channel $a$.

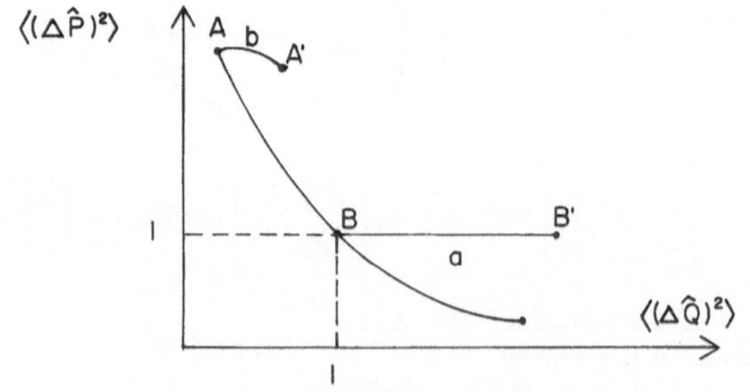


Fig. 10. As Fig. 9 but with $\delta = \sqrt{3}k$.

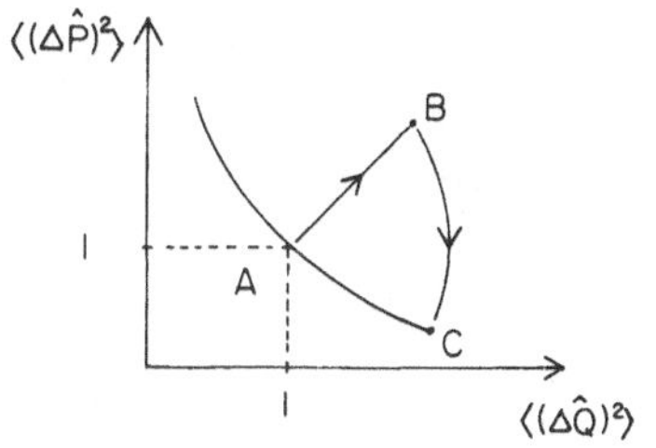


Fig. 11. Fate of a pure coherent signal A on passing through a four wave mixer (B) and linear coupler (C).

The case of detuning with $\delta = \sqrt{3}k$ is shown in Fig. 10. Here at half the distance along the channels ($z = \pi/4k$) the initial squeezed state A in channel $b$ is converted to a noisy state A′, while in channel $a$ the state B becomes state B′. These noisy states are not independent of each other; they have complementary noises disappearing during the remaining path when A′ goes back to A and B′ to B.

The complementary nature of the noise created in the channels has suggested an interesting application with correlated input noises.[16] Let us consider a degenerate backward four-wave mixing process. A coherent signal enters the mixer and goes through it while a conjugate field emerges from the interaction, coming out of the mixer in the opposite direction. During the interaction some noise is added to the signal beam.[17] In the conjugate beam there is also a noise which is exactly complementary to the noise in the signal beam. If these two beams from the output of the four-wave mixer enter the two different channels of a linear coupler with length $L = \pi/4k$ and zero detuning, at the output we get two pure states with opposite squeezing.

We can see this effect illustrated in Fig. 11. A coherent signal A emerges from the four-wave mixer as a mixed state B and becomes a pure squeezed state C while travelling through the linear coupler together with its conjugate beam.

## 8. THE NONLINEAR COHERENT COUPLER

If a nonlinear medium is placed between the two waveguides of a directional coupler, the nonlinear interaction modifies the exchange of power and leads to strongly nonlinear transmission characteristics. Nonlinear couplers have been studied within the framework of classical theory by several authors.[18] Here we are interested in a quantum-mechanical approach. We suppose that the fields of the individual waveguides can be described as two different modes of a common field. The fields of these modes exchange energy due to a linear interaction and in addition they are modulated by the Kerr-type nonlinear interaction.

The Hamiltonian suitable for this problem is[19]

$$H = \hbar\omega(\hat{a}_1^\dagger\hat{a}_1 + \hat{a}_2^\dagger\hat{a}_2) - \hbar\kappa(\hat{a}_1^\dagger\hat{a}_2 + \hat{a}_2^\dagger\hat{a}_1) - \frac{1}{2}\hbar g(\hat{a}_1^{\dagger 2}\hat{a}_1^2 + \hat{a}_2^{\dagger 2}\hat{a}_2^2 + 2\epsilon\hat{a}_1^\dagger\hat{a}_2^\dagger\hat{a}_1\hat{a}_2) + \hat{V}_R \tag{101}$$

where the constant $\kappa$ describes a linear coupling between waveguides, the terms involving g describe the nonlinear self-interaction of the field with itself in the corresponding channel, and the term containing $g\epsilon$ reflects the mutual nonlinear interaction between the channels. Finally the term $\hat{V}_R$ represents the usual reservoir, introduced to include losses and various kinds of noise.

The Heisenberg-Langevin equations (in the interaction picture) are

$$\frac{d\hat{a}_1}{dt} = -\gamma\hat{a}_1 + ig(\hat{a}_1^\dagger\hat{a}_1 + \epsilon\hat{a}_2^\dagger\hat{a}_2)\hat{a}_1 + i\kappa\hat{a}_2 + \hat{L}_1 \tag{102a}$$

$$\frac{d\hat{a}_2}{dt} = -\gamma\hat{a}_2 + ig(\hat{a}_2^\dagger\hat{a}_2 + \epsilon\hat{a}_1^\dagger\hat{a}_1)\hat{a}_2 + i\kappa\hat{a}_1 + \hat{L}_2 \tag{102b}$$

The Langevin forces $\hat{L}_j$ have the following properties

$$\langle L_j \rangle = 0 \qquad \langle \hat{L}_j\hat{L}_k \rangle = 0 \qquad j,k = 1,2$$

$$\langle \hat{L}_j^\dagger(t)\hat{L}_k(t') \rangle = 2\gamma\bar{n}\delta_{ik}\delta(t-t') \tag{103}$$

$$\langle \hat{L}_j(t)\hat{L}_k^\dagger(t') \rangle = 2\gamma(\bar{n} + 1)\delta_{ik}\delta(t-t')$$

where $\bar{n}$ is the mean number of noise photons.

Unfortunately, exact general solutions of Eqs. (102) are not known. In order to find an approximate solution in the case of strong linear coupling of the waveguides, we introduce new operators $\hat{c}_j$ by means of the following transformations

$$\hat{a}_1 = \frac{1}{\sqrt{2}}[e^{i\kappa t}\hat{c}_1 - e^{-i\kappa t}\hat{c}_2]$$

$$\hat{a}_2 = \frac{1}{\sqrt{2}}[e^{i\kappa t}\hat{c}_1 + e^{-i\kappa t}\hat{c}_2] \tag{104}$$

The operators $\hat{a}$ and $\hat{c}$ obey the same commutation relations

$$[\hat{a}_j, \hat{a}_k^\dagger] = [\hat{c}_j, \hat{c}_k^\dagger] = \delta_{ik} \qquad j,k = 1,2 \tag{105}$$

Using relations (104) and supposing that $1/\kappa << 1/g$ we have

$$\frac{d\hat{c}_1}{dt} = -\gamma\hat{c}_1 + iG(\hat{c}_1^\dagger\hat{c}_1 + B\hat{c}_2^\dagger\hat{c}_2)\hat{c}_1 + \hat{K}_1$$

$$\frac{d\hat{c}_2}{dt} = -\gamma\hat{c}_2 + iG(\hat{c}_2^\dagger\hat{c}_2 + B\hat{c}_1^\dagger\hat{c}_1)\hat{c}_2 + \hat{K}_2 \tag{106}$$

where

$$G = \frac{1}{2}g(1 + \epsilon) \qquad B = \frac{2}{1 + \epsilon} \qquad \hat{K}_{1,2} = \frac{1}{\sqrt{2}}(\hat{L}_1 \pm \hat{L}_2)e^{\mp i\kappa t} \tag{107}$$

Solutions of Eqs. (106) have been derived in the form

$$\hat{c}_j(t) = e^{-\gamma t - i\Phi_j(t)}[\hat{c}_{0j} + \hat{I}_j(t)] \qquad j = 1,2 \tag{108}$$

where

$$\hat{I}_j(t) = \int_0^t e^{\gamma\tau + i\Phi_j(\tau)}\hat{K}_j(\tau)d\tau$$

$$\Phi_j(t) = D(t)[\hat{c}_{0j}^\dagger\hat{c}_{0j} + B\hat{c}_{0,3-j}^\dagger\hat{c}_{0,3-j}] + X(t)[\hat{b}_j^\dagger\hat{b}_j + B\hat{b}_{3-j}^\dagger\hat{b}_{3-j}]$$

$$D(t) = \frac{G}{2\gamma}(1 - e^{-2\gamma t})$$

$$X(t) = Gt - D(t)$$

and $\hat{c}_{0j} \equiv \hat{c}_j(0)$. The operators $\hat{b}_j$ and $\hat{b}_j^\dagger$ are the annihilation and creation operators respectively of a reservoir with the properties

$$\langle b_j^\dagger b_j \rangle = \bar{n} \qquad\qquad j = 1,2$$

If we suppose that the initial field is in the coherent state $|\alpha_{01},\alpha_{02}\rangle$ with $\alpha_0 = 0$ and $\alpha_{02} = \alpha_{02}^*$ (i.e. input power is launched into the second waveguide only), using Eqs. (108) and (104), after some algebra, we obtain

$$\langle \hat{a}_1^\dagger \hat{a}_1 \rangle = \frac{1}{2} e^{-2\gamma t} \alpha_{02}^2 [1 - U^2(D(1-B)) V^2(X(1-B)) \cos(2\kappa t)] + \bar{n}[1 - e^{-2\gamma t}] \qquad (109a)$$

$$\langle \hat{a}_2^\dagger \hat{a}_2 \rangle = \frac{1}{2} e^{-2\gamma t} \alpha_{02}^2 [1 + U^2(D(1-B)) V^2(X(1-B)) \cos(2\kappa t)] + \bar{n}[1 - e^{-2\gamma t}] \qquad (109b)$$

$$\left\langle \begin{matrix} (\Delta\hat{Q}_1)^2 \\ (\Delta\hat{P}_1)^2 \end{matrix} \right\rangle = 1 + \bar{n}[1 - e^{-2\gamma t}] + e^{-2\gamma t}\alpha_{02}^2 \Big\{ 1 \pm U(2D)U(2BD)V(2X)V(2BX)\cos(2\kappa t)\cos(\psi_1)$$
$$- U^2(D)U^2(BD)V^2(X)V^2(BX)[1 \pm \cos(2\kappa t)\cos(\psi_2) - \cos(2\kappa t) \mp \cos(\psi_2)]$$
$$- U^2(D(1-B))V^2(X(1-B))\cos(2\kappa t) \mp U^2(D(1+B))V^2(X(1+B))\cos(\psi_3) \Big\} \qquad (109c)$$

$$\left\langle \begin{matrix} (\Delta\hat{Q}_2)^2 \\ (\Delta\hat{P}_2)^2 \end{matrix} \right\rangle = 1 + \bar{n}[1 - e^{-2\gamma t}] + e^{-2\gamma t}\alpha_{02}^2 \Big\{ 1 \pm U(2D)U(2BD)V(2X)V(2BX)\cos(2\kappa t)\cos(\psi_1)$$
$$- U^2(D)U^2(BD)V^2(X)V^2(BX)[1 \pm \cos(2\kappa t)\cos(\psi_2) + \cos(2\kappa t) \pm \cos(\psi_2)]$$
$$+ U^2(D(1-B))V^2(X(1-B))\cos(2\kappa t) \pm U^2(D(1+B))V^2(X(1+B))\cos(\psi_3) \Big\} \qquad (109d)$$

where

$$\psi_1 = -D - \frac{1}{2}\alpha_{02}^2[\sin(2D) + \sin(2BD)] + Ac(2X) + Ac(2BX)$$

$$\psi_2 = - \frac{1}{2}\alpha_{02}^2[\sin(D) + \sin(BD)] + 2Ac(X) + 2Ac(BX)$$

$$\psi_3 = -BD - \alpha_{02}^2 \sin(D(1 + B)) + 2Ac(X(1 + B))$$

$$Ac(y) = \arccos\{V(y)[1 + \bar{n}(1 - \cos(y))]\}$$

$$U(y) = \exp\left\{ -\frac{1}{2}\left|\alpha_{02}\right|^2 [1 - \cos(y)] \right\}$$

$$V(y) = \Big(1 + 2\bar{n}(\bar{n} + 1)[1 - \cos(y)]\Big)^{-1/2}$$

We may now discuss the expressions for the mean number of photons. Supposing there are no losses in the coupler, we obtain from Eqs. (109a) and (109b)

$$\begin{matrix} \langle \hat{a}_1^\dagger \hat{a}_1 \rangle \\ \langle \hat{a}_2^\dagger \hat{a}_2 \rangle \end{matrix} = \frac{1}{2}\alpha_{02}^2\{1 \mp U^2[Gt(1-B)]\}\cos(2\kappa t) \qquad (110)$$

The influence of quantum effects in the nonlinear interaction is included in the function U. If G = 0, then we obtain the correct relations for a linear coupler.

The waveguides of a linear coupler exchange photons (power) harmonically with period

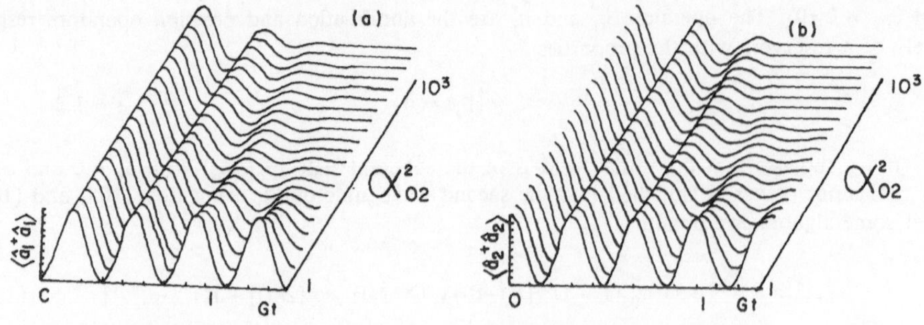


Fig. 12. Behaviour of (a) the first and (b) the second waveguide of a nonlinear coherent coupler for B = 2, $\kappa = 1000$, G = 20, $\gamma = 0.1$ and $\bar{n} = 1$.

$T = \pi/\kappa$. The behaviour of a nonlinear coupler is the same if the input power is well below a critical value.[18] If the input power is greater than a critical value, then the power remains in each waveguide, being slightly modulated only.

This conclusion is not correct from the quantum point of view. It follows from Eqs. (110) that the harmonic exchange of photons is modulated by the function U. The amplitude of the modulation of the mean photon numbers is attenuated and again amplified as a consequence of the presence of the function U[Gt(1-B)]. This quantum nonlinear effect is shown in Fig 12. One can see that the attenuation of the oscillation of the mean photon numbers is faster if the initial field increases.

From the point of view of the classical theory, the coupler behaves as though it has passed from the below-critical power region to the above-critical region. For high input power, the coupler remains mainly in the above-critical region.

If losses are nonzero ($\gamma \neq 0$), then the total number of photons in the coupler decreases. Noise in the coupler ($\bar{n} \neq 0$) has the tendency to smooth out the oscillating exchange of photons between the waveguides.

Squeezing properties are shown with the help of Figs. 13 in which the two quantities $\langle(\Delta\hat{P}_1)^2\rangle$ and $\langle(\Delta\hat{Q}_2)^2\rangle$ are drawn only in the regions where they are less than one. Note that squeezing occurs for the field in both waveguides but for quadrature components [i.e. $\langle(\Delta P_1)^2\rangle$ in the first waveguide, and $\langle(\Delta Q_2)^2\rangle$ for the second waveguide].

Finally an interesting comparison is shown with experimental results in Fig. 14. The behaviour of mean photon numbers (powers) is shown. Crosses are experimental results obtained in Ref. 20. Full lines have been calculated from Eqs. (110). The deviation of the character of the transfer of photons from the sinusoidal form discussed in Ref. 20 appears to be caused by purely quantum effects rather than imperfections in the waveguides. This is one of the few cases in which a purely quantum effect needs a high photon number to be seen.

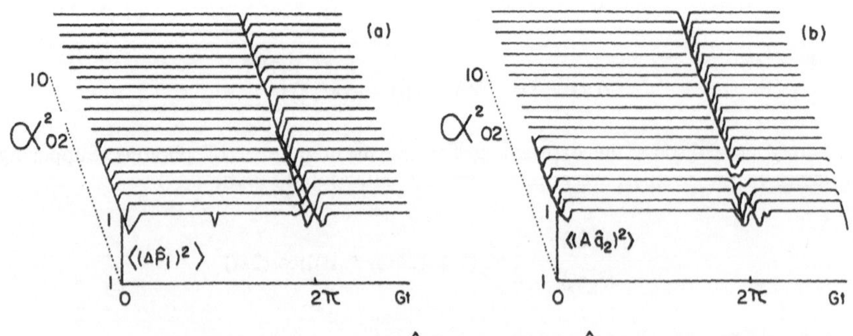


Fig. 13. Detailed picture of (a) $\langle(\Delta\hat{P}_1)^2\rangle$ and (b) $\langle(\Delta\hat{Q}_2)^2\rangle$ for the same data as in Fig. 12. The straight lines correspond to the value 1.

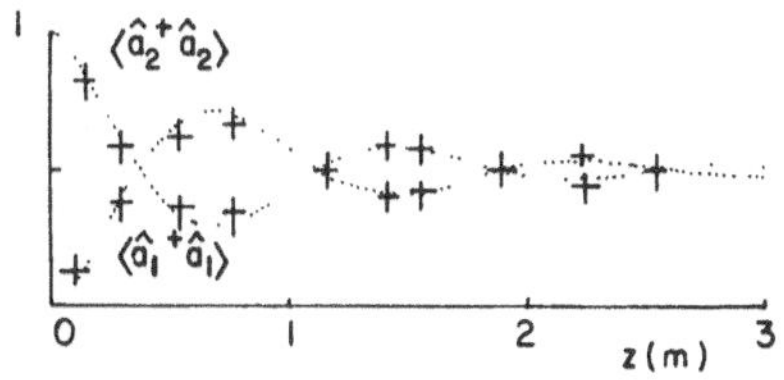


Fig. 14. The behaviour of the photon field when the damping term is zero but the Langevin force is non-zero.

With regard to the results shown in Fig. 14, the following points should be made:

(i) the c-number equations describing the evolution of the fields inside the structure, derived from the Heisenberg-Langevin equation (101), are obtained using the commutation rules. Without commutation rules for the operators, the U function in Eq. (110) gives unit value, i.e. it does not fit the experimental values; and

(ii) the Heisenberg-Langevin equations used to describe the propagation of photons inside the structure contain the Langevin forces and damping terms. The Langevin force plays the role of a noise generator which generates fluctuations in the system.

In conclusion, there exists a recent classical analysis of the limitations imposed by linear and nonlinear absorption and saturation on all optical switching, which shows that the main limitation on power switching is imposed by linear absorption.[21] Nevertheless, it is interesting to see how well the quantum analysis presented in this chapter explains the experimental results.

## REFERENCES


1. M. Born and E. Wolf, *Principles of Optics* (Pergamon, Oxford, 1970).
2. W. H. Louisell, *Radiation and Noise in Quantum Electronics* (McGraw-Hill, New York, 1964), ch. IV.
3. E. Schrödinger, Naturwiss. **14**, 644 (1927).
4. R. Glauber, Phys. Rev. **130**, 2529 (1963); ibid **131**, 2766 (1963).
5. J. Perina, *Quantum Statistics of Linear and Nonlinear Optical Phenomena* (D. Reidel, Boston, 1984).
6. R. Glauber, *Physics of Quantum Electronics*, P. L. Kelly, B. Lax and P. E. Tannenwald, eds. (McGraw-Hill, New York, 1966), p. 788.
7. E. C. G. Sudarshan, Phys. Rev. Lett. **10**, 277 (1963).
8. I. A. Deryugin, S. S. Abdullaev, A. T. Mirzaev, Sov. J. Quantum Electr. **7**, 1243 (1977).
9. V. S. Pekar, Sov. Phys. JETP **40**, 233 (1975).
10. J. Perina, R. Horak, Z. Hradil, C. Sibilia and M. Bertolotti, J. Mod. Opt. **36**, 571 (1989).
11. V. Perinova and A. Luks, J. Mod. Opt. **35**, 1513 (1988).
12. D. Marcuse, *Theory of Optical Dielectric Waveguides* (Academic Press, New York, 1974).
13. A. Yariv and P. Yeh, *Optical Waves in Crystals* (Wiley Interscience, New York, 1984).
14. J. Janszky and Y. Yushin, Phys. Rev. A **36**, 1288 (1987).
15. J. Janszky, C. Sibilia, M. Bertolotti and Y. Yushin, J. Mod. Opt. **35**, 1757 (1988).
16. M. Bertolotti, J. Janszky, C. Sibilia, Y. Yushin, *Proc. OPTIKA '88* (Budapest, 1988), vol. II, p. 500.
17. J. Janszky and Y. Yushin, Opt. Commun. **49**, 290 (1984).
18. S. M. Jensen, IEEE J. Quantum Electron. **QE-18**, 1580 (1982);
A. A. Maier, Sov. J. Quantum Electron. **12**, 1490 (1982); ibid **16**, 892 (1986);
S. Trillo and S. Wabnitz, Appl. Phys. Lett. **49**, 752 (1986);
E. Caglioti, S. Trillo, S. Wabnitz, B. Daino, G. I. Stegeman, J. Opt. Soc. Am. B **5**, 472 (1988).
19. R. Horak, M. Bertolotti, C. Sibilia and J. Perina, J. Opt. Soc. Am. B **6**, 199 (1989).
20. D. D. Gusovskii, E. M. Dianov, A. A. Maier, V. B. Neustruev, E. I. Shklovskii, I. A. Shcherbakov, Sov. J. Quantum Electron. **15**, 1523 (1985).
21. E. Caglioti, S. Trillo, S. Wabnitz, G. I. Stegeman, J. Opt. Soc. Am. B **5**, 472 (1988).

# ELECTROMAGNETIC RESONANCE INDUCED NONLINEAR OPTICAL PHENOMENA

R. Reinisch and G. Vitrant

LEMO URA CNRS n° 833
ENSERG, 23 Avenue des Martyrs, BP 257
38016 Grenoble Cédex, France

M. Nevière

Laboratoire d'Optique Electromagnétique - UA CNRS n° 843
Faculté des Sciences et Techniques - Centre de Saint Jérôme
13397 Marseille Cédex 13, France

## 1. INTRODUCTION

It is well known that efficient nonlinear optical interactions require strong nonlinear polarizations.[1,2] This is achieved by increasing the nonlinear susceptibility and/or the electromagnetic field inside the nonlinear medium. The former requires a material study whereas the latter, in which we are interested, requires an electromagnetic study.

In this chapter we will show that electromagnetic resonances associated with the resonant excitation of guided waves or surface plasmons lead to an enhancement of the pump field(s) and, as a result, of the nonlinear polarization. Prism or grating couplers allow the excitation of these electromagnetic resonances which can be either linear or nonlinear, depending on the type of optical nonlinear effect. Two nonlinear optical phenomena will be considered:

(i) Second harmonic generation, whose enhancement involves linear surface plasmon or guided wave resonances through bare or coated grating couplers. Thus we are faced with the problem of diffraction in nonlinear optics. Numerous papers have been published dealing with second harmonic generation at grating surfaces,[3-21] from both an experimental and a theoretical point of view. In the theory presented in Refs. 5-8 and 12-17, the groove depth of the grating is rigorously accounted for. This allows the study of strongly corrugated surfaces and leads to the prediction of a new effect at the second harmonic frequency, namely, the existence of an optimum groove depth, linked with the resonant excitation at the pump frequency of surface plasmons or guided waves, for which the second harmonic efficiency is greatest. This effect has been observed experimentally.[3-8,18-21]

(ii) The optical Kerr effect, which involves nonlinear electromagnetic resonances. We are interested in the electromagnetic resonance induced response of nonlinear optical resonators such as:
  - nonlinear Fabry Pérot cavities where nonlinear Airy resonances[22] are involved, and
  - nonlinear prism couplers which allow the excitation of nonlinear guided modes.[23]

## 2. GENERAL FORMALISM OF DIFFRACTION FOR SECOND HARMONIC GENERATION

Our aim is to derive propagation equations valid whatever the nature of the nonlinear medium may be: e.g. non-centrosymmetric dielectrics or metals.

Thus, consider a nonlinear medium with an entrance face, S, having a nonlinear polarization at the second harmonic frequency, $2\omega$, of the general form

$$\mathbf{P}^{NL}(2\omega) = \mathbf{P}_V^{NL}(2\omega) + \mathbf{P}_S^{NL}(2\omega)\delta_S \tag{1}$$

Equation (1) shows that $\mathbf{P}^{NL}$ is the sum of two terms:

(i) $\mathbf{P}_V^{NL}$: a bulk term which accounts for the nonlinear polarization in the nonlinear medium.

(ii) $\mathbf{P}_S^{NL}\delta_S$: a surface term located on the surface S of the nonlinear medium, where $\delta_S$ is the distribution[24-26] located on S, defined by

$$\langle \delta_S, \phi \rangle = \iint_S \phi dS$$

with $\phi$ being an infinitely differentiable function with compact support.

Due to the existence of $\mathbf{P}_S^{NL}\delta_S$, the usual boundary conditions at $2\omega$ (continuity of the tangential components of the electric and magnetic fields at $2\omega$) are no longer valid. New boundary conditions have to be established at frequency $2\omega$. This calculation is developed in an orthogonal curvilinear coordinate system[24] $(u, v, z)$. The equation of S is given by $v(x, y, z) = 0$.

Assuming an $e^{-2j\omega t}$ time dependence of the signal and the magnetic permittivity of the nonlinear medium to be equal to $\mu_0$, Maxwell's equations read as follows:

$$\nabla\times\mathbf{E}(2\omega) = j2\omega\mu_0\mathbf{H}(2\omega) \tag{2a}$$

$$\nabla\times\mathbf{H}(2\omega) = -j2\omega\epsilon_0\epsilon(2\omega, v)\mathbf{E}(2\omega) - j2\omega\mathbf{P}^{NL}(2\omega) \tag{2b}$$

where $\mathbf{P}^{NL}(2\omega)$ is given by Eq. (1) and

$$\epsilon(2\omega, v) = \begin{cases} 1 & \text{if } v > 0 \\ \epsilon_2(2\omega) & \text{if } v < 0 \end{cases} \tag{3}$$

Assuming Maxwell's equations to be valid in the sense of distributions,[25,26] Eq. (2) can be rewritten in the sense of distributions as

$$\{\nabla\times\mathbf{H}(2\omega)\} + (\hat{n}\times J[\mathbf{H}(2\omega)])\delta_S = -j2\omega\epsilon_0\epsilon(2\omega, v)\mathbf{E}(2\omega) - j2\omega\mathbf{P}_V^{NL}(2\omega) - j2\omega\mathbf{P}_S^{NL}(2\omega)\delta_S \tag{4}$$

where $\hat{n}$ is the unit vector of the normal to S oriented from metal to vacuum, J( ) is the jump of ( ) across S in the direction of $\hat{n}$ and { } is the operator written in the usual sense of functions[25,26] (see Appendix A).

Before going further, let us consider Eq. (4). It involves classical terms and distributive terms (including $\delta_S$). It is well known[25,26] that Eq. (4) implies the equality of the functions present on the left and right hand sides and that the same thing holds for the distributive parts.

At this stage, the following important remark should be made: the distribution $(\hat{n}\times J[\mathbf{H}(2\omega)])\delta_S$ has only a tangential component, whereas $\mathbf{P}_S^{NL}(2\omega)\delta_S$ has not only a tangential component but also a component perpendicular to S. Thus, on the right-hand side of Eq. (4), $\mathbf{E}(2\omega)$ cannot be a normal function. It must also have a distributive part, which has, at least, a component perpendicular to S. Therefore, the electric field $\mathbf{E}(2\omega)$ in Eqs. (2) and (4), which represents the total electric field, is the sum of two terms, a classical (volume) part $\mathbf{E}_V(2\omega)$ and a distribution $\mathbf{E}_S(2\omega)\delta_S$

$$\mathbf{E}(2\omega) = \mathbf{E}_V(2\omega) + \mathbf{E}_S(2\omega)\delta_S \tag{5}$$

where $\mathbf{E}_S(2\omega)\delta_S$ has a component normal to S.

It follows that Eqs. (2a) and (4) written in the sense of distributions become

$$\{\nabla\times\mathbf{E}_V(2\omega)\} + \nabla\times[\mathbf{E}_S(2\omega)\delta_S] + (\hat{n}\times J[\mathbf{E}_V(2\omega)])\delta_S = j2\omega\mu_0\mathbf{H}(2\omega) \tag{6}$$

$$\{\nabla\times\mathbf{H}(2\omega)\} + (\hat{n}\times J[\mathbf{H}(2\omega)])\delta_S =$$
$$-j2\omega\epsilon_0\epsilon(2\omega,v)\mathbf{E}_V(2\omega) - j2\omega\epsilon_0\epsilon(2\omega,v)\mathbf{E}_S(2\omega)\delta_S - j2\omega\mathbf{P}_V^{NL}(2\omega) - j2\omega\mathbf{P}_S^{NL}(2\omega)\delta_S \tag{7}$$

From Eqs. (6) and (7) the following set of equations is obtained:

$$\{\nabla\times\mathbf{E}_V(2\omega)\} = j2\omega\mu_0\mathbf{H}(2\omega) \tag{8}$$

$$\{\nabla\times\mathbf{H}(2\omega)\} = -j2\omega\epsilon_0\epsilon(2\omega,v)\mathbf{E}_V(2\omega) - j2\omega\mathbf{P}_V^{NL}(2\omega) \tag{9}$$

$$\nabla\times[\mathbf{E}_S(2\omega)\delta_S] + (\hat{n}\times J[\mathbf{E}_V(2\omega)])\delta_S = 0 \tag{10}$$

$$\hat{n}\times J[\mathbf{H}(2\omega)]\delta_S = -j2\omega\epsilon_0\epsilon(2\omega,v)\mathbf{E}_S(2\omega)\delta_S - j2\omega\mathbf{P}_S^{NL}(2\omega)\delta_S \tag{11}$$

It is shown in Appendix B that the solution of Eq. (10) is characterized by the two conditions

$$\mathbf{E}_S(2\omega) = \hat{n}\,E_S(2\omega) \tag{12}$$

i.e., $\mathbf{E}_S(2\omega)$ is perpendicular to surface S, and

$$J[\mathbf{E}_{V,t}(2\omega)] = \nabla_t E_S(2\omega) \tag{13}$$

In Eq. (13), $\mathbf{E}_{V,t}(2\omega)$ is the tangential component of $\mathbf{E}_V(2\omega)$ on the surface S. The $\nabla_t$ operator is the two-dimensional gradient taken in the plane tangent to S. Since $E_S(2\omega)$ is defined on S, $\nabla_t E_S(2\omega)$ is well defined.

Projecting Eq. (11) onto a plane perpendicular to $\hat{n}$ gives

$$\hat{n}\times J[\mathbf{H}(2\omega)] = -j2\omega\hat{n}\times[\mathbf{P}_S^{NL}(2\omega)\times\hat{n}] \tag{14}$$

Projecting Eq. (11) onto $\hat{n}$ leads to

$$\epsilon_0 E_S(2\omega)\delta_S = -\frac{1}{\epsilon(2\omega,v)}[\hat{n}\cdot\mathbf{P}_S^{NL}(2\omega)]\delta_S \tag{15}$$

Note that Eqs. (8) and (9) are Maxwell's equations written in the usual sense of functions.

Equation (14) is the usual boundary condition that relates the discontinuity of the tangential component of the magnetic field $\mathbf{H}(2\omega)$ to the component of the surface nonlinear polarization parallel to S.

From Eq. (15), we see that the normal component of the surface nonlinear polarization gives rise to a surface electric field at the second harmonic frequency. Unfortunately Eq. (15) is ambiguous since it contains the product of a step function $\epsilon(v)$ with $\delta_S$ [this is also true for Eq. (11)]. This difficulty is solved below by considering a specific nonlinear medium.

Throughout this chapter, it is assumed that $\partial/\partial z = 0$. Therefore the signal at $2\omega$ is either TM or TE. Use of the nonambiguous equations (9), (8), (2), (10), (12) and (13) allows the propagation equations at the second harmonic frequency to be derived.

### 2.1 TM polarization at $2\omega$ ($2\omega_{TM}$)

For the TM polarization

$$\mathbf{H}(2\omega) = H\mathbf{e}_z \tag{16}$$

where $\mathbf{e}_z$ is a unit vector along the z-axis. Let

$$\epsilon(v) = \epsilon(2\omega,v) \tag{17}$$

According to Eq. (16), Eq. (9) becomes

$$\mathbf{E}_V = \frac{1}{j2\omega\epsilon_0\epsilon(v)}\mathbf{e}_z \times \{\nabla H\} - \frac{1}{\epsilon_0\epsilon(v)}\mathbf{P}_V^{NL} \tag{18}$$

Using Eqs. (2a), (10) and (18) leads to

$$\nabla\cdot\left[\frac{1}{k^2}\{\nabla H\}\right] + H = \frac{j}{2\omega\mu_0}\mathbf{e}_z\cdot[\hat{n}\times J(\mathbf{E}_V)\delta_S] + j2\omega\mathbf{e}_z\cdot\nabla\times\left[\frac{1}{k^2}\mathbf{P}_V^{NL}\right] \tag{19}$$

$$k^2 = k^2(v) = 4\omega^2\epsilon_0\mu_0\epsilon(v)$$

The following remarks should be made. In Eq. (19)

(i) the operators $\nabla\cdot$ and $\nabla\times$ are written in the sense of distributions,
(ii) $\{\nabla\}$ is the $\nabla$ operator written in the sense of functions (see Appendix A), thus $\{\nabla H\} = \nabla H - \mathbf{n}J(H)\delta_S$ where $J(H)$ is given by Eq. (14), and
(iii) $J(\mathbf{E}_V)$ is given by Eqs. (13) and (15).

The right hand sides of Eqs. (14) and (15) depend on the nature of the nonlinear medium.

### 2.2 TE polarization at $2\omega$ ($2\omega_{TE}$)

For the TE polarization we set

$$\mathbf{E}(2\omega) = E\mathbf{e}_z \tag{20}$$

According to Eq. (12), there is no surface second harmonic electric field when dealing with the $2\omega_{TE}$ case. Thus

$$\mathbf{E} = E_V\mathbf{e}_z \tag{21}$$

Use of Eqs. (8), (13), (20), (21) and (2b) yields

$$\Delta E_V + k^2 E_V = -4\omega^2\mu_0 P_z^{NL} \tag{22}$$

In Eq. (22), $\Delta$ is written in the sense of distributions and $P_z^{NL}$ is the z component of $\mathbf{P}^{NL}$ given by Eq. (1).

Equations (19) and (22) are the fundamental propagation equations for the H and E fields at the second harmonic frequency. The source term at $2\omega$ is expressed by the right hand side of these equations. Since Eqs. (19) and (22) are valid in the sense of distributions, they include the boundary conditions across S.

### 2.3 Nonlinear materials: non-centrosymmetric dielectrics

For non-centrosymmetric dielectrics, $\mathbf{P}^{NL}(2\omega)$ is given by[1,2]

$$P_j^{NL}(2\omega) = \epsilon_0\chi_{jih}(2\omega)E_i(\omega)E_h(\omega)$$

where $\mathbf{E}(\omega)$ is the electric field at $\omega$ in the nonlinear dielectric.

Thus $\mathbf{P}^{NL}(2\omega)$, Eq. (1), has only a bulk contribution, i.e. $\mathbf{P}_S^{NL} = \mathbf{0}$ and Eq. (15) shows that there is no surface electric field at $2\omega$. According to Eqs. (13) and (14)

$$J(\mathbf{E}_V) = 0 \qquad J(\mathbf{H}) = 0$$

As a result, Eqs. (19) and (22) can be rewritten as

$$\nabla\cdot\left[\frac{1}{k^2}\nabla H\right] + H = j2\omega\mathbf{e}_z\cdot\nabla\times\left[\frac{1}{k^2}\mathbf{P}^{NL}\right] \tag{23}$$

for the TM polarization and

$$\Delta E + k^2E = -4\omega^2\mu_0 P_z^{NL} \quad (24)$$

for the TE polarization. From Eqs. (23) and (24), it is seen that the polarization of the signal is TE or TM depending on whether $\mathbf{P}^{NL}$ is perpendicular to the plane of incidence or within this plane. The orientation of $\mathbf{P}^{NL}$ is obviously related to the crystalline class of the dielectric and the polarization of the pump beam.

## 2.4 Nonlinear materials: metals

In the case of metals, we are faced with the problem of second harmonic generation in centrosymmetric media. As is well known, second harmonic generation results from the breaking of inversion symmetry at the metal surface.[8,15,17,27-30] Following the work of Bloembergen *et al*,[27] the pump beam at frequency $\omega$ gives rise to a nonlinear polarization in the metal, given by

$$\mathbf{P}^{NL}(2\omega) = \gamma\nabla[\mathbf{E}(\omega)\cdot\mathbf{E}(\omega)] + \beta\mathbf{E}(\omega)\nabla\cdot\mathbf{E}(\omega) \quad (25)$$

where

$$\beta = \frac{e\epsilon_0}{2m\omega^2}$$

$$\gamma = \frac{\beta}{4}[1-\epsilon_2(\omega)]$$

$$\epsilon_2(\omega) = 1 - \frac{\omega_p^2}{\omega^2}$$

and e is the charge of the electron, m is the conduction band electron mass, $\epsilon_2(\omega)$ is the permittivity of the metal at the pump circular frequency $\omega$, $\omega_p$ is the plasma frequency, and $\mathbf{E}(\omega)$ is the electric field at the pump circular frequency $\omega$ in the metal.

Inside the bulk material, $\nabla\cdot\mathbf{E}(\omega) = 0$. Thus the second term in Eq. (25) can only give a contribution at the metal surface. $\mathbf{P}^{NL}(2\omega)$ includes not only a bulk term but also a surface term, coming from both terms of Eq. (25), which behaves like a Dirac $\delta$-function. To this end, the metal surface is considered as the limit, when $\eta \to 0$, of a continuous transition region of thickness $\eta$ between metal and vacuum. We calculate $\mathbf{P}^{NL}$ [Eq. (25)] in this transition region and let $\eta \to 0$. Since Maxwell's equations are valid in the sense of distributions,[25,26] the limit, when $\eta \to 0$, of the terms in Eq. (25) is calculated in the sense of distributions.

When the limit of $\mathbf{P}^{NL}$ as $\eta \to 0$ is calculated, bounded and unbounded terms are obtained. The bounded quantities are of no interest since they do not contribute to the second harmonic field when $\eta \to 0$. The interesting quantities are the unbounded terms. In the limit $\eta \to 0$, they behave like Dirac $\delta$-functions and thus give rise to the surface nonlinear polarization.

The transition region, of thickness $\eta$, extends between surface S and surface S′, placed in a vacuum, whose equation is $v(x,y,z) = \eta$. We assume that $\epsilon$ depends only on v, and make the substitution $\epsilon_2(\omega) \to \epsilon(\omega, v)$ in Eq. (25).

It is convenient to calculate first the limit L, when $\eta \to 0$, of the quantity

$$\epsilon^p(v)\,\frac{\partial\epsilon^q(v)}{\partial v} \quad (26)$$

It is shown in Appendix C that

$$L = \begin{cases} \dfrac{q}{p+q}e_2(1-\epsilon_2^{p+q})\delta_S & \text{if } p+q \neq 0 \\ -qe_2(ln\epsilon_2)\delta_S & \text{if } p+q = 0 \end{cases} \quad (27)$$

where $e_2$ is the local unit length along v.

The derivation of the limit of $\mathbf{P}^{NL}(2\omega)$ involves the quantities

$$L_1 = \lim_{\eta \to 0} \gamma \nabla[\mathbf{E}(\omega)\cdot\mathbf{E}(\omega)]$$

$$L_2 = \lim_{\eta \to 0} \beta \mathbf{E}(\omega)\nabla\cdot\mathbf{E}(\omega) \tag{28}$$

In calculating $L_1$ and $L_2$ we note that (a) only the unbounded terms are of interest and (b) the tangential component of $\mathbf{E}(\omega)$ and the perpendicular component of $\mathbf{D}(\omega)$, $D_V(\omega) = \epsilon(\omega, v)E_V(\omega)$, are continuous across S.

Using the reminders (a) and (b) together with Eq. (27) yields

$$L_1 = \hat{n}\, D_V^2 \frac{\beta}{4\epsilon_2^2}[\epsilon_2(\omega) - 1]^2 \delta_S$$

$$L_2 = E_t D_V \left\{1 - \frac{1}{\epsilon_2(\omega)}\right\}\delta_S + \hat{n}\, D_V^2 \frac{1}{2}\left\{1 - \frac{1}{\epsilon_2^2(\omega)}\right\}\delta_S \tag{29}$$

where $E_t(\omega)$ are the tangential components of $\mathbf{E}(\omega)$. From a nonrigorous but more intuitive point of view, the two-dimensional distribution $\delta_S$ is merely the classical Dirac function $\delta(n)$, where n represents the abscissa along the normal to S.

Equations (29) show that there indeed exists a surface nonlinear polarization, given by

$$\mathbf{P}_S^{NL}(2\omega)\delta_S = E_t \beta E_V^-[\epsilon_2(\omega)-1]\delta_S + \hat{n}\, \frac{3}{4}\beta E_V^{-2}[\epsilon_2(\omega)-1]\left\{\frac{1}{3}\epsilon_2(\omega) + 1\right\}\delta_S \tag{30}$$

where $E_V^-$ is the normal component of the pump electric field just below S.

Finally, Eq. (25) may be rewritten as

$$\mathbf{P}^{NL}(2\omega) = \mathbf{P}_V^{NL}(2\omega) + \mathbf{P}_S^{NL}(2\omega)\delta_S \tag{31}$$

where $\mathbf{P}_V^{NL}$ is the nonlinear polarization in the metal. Then

$$\mathbf{P}_V^{NL}(2\omega) = \frac{\beta}{4}[1 - \epsilon_2(\omega)]\{\nabla[\mathbf{E}(\omega)\cdot\mathbf{E}(\omega)]\} \tag{32}$$

Note that the surface nonlinear polarization, Eq. (30), is located on S and not above it.[28,29] Moreover, the expression for $\mathbf{P}_S^{NL}\delta_S$ is independent of the profile $\epsilon(v)$ in the transition region.

Thus Eq. (1) appears as the expression for the nonlinear polarization of nonlinear metallic media. The existence of a surface nonlinear polarization leads to the existence of a surface electric field at $2\omega$ and to the boundary condition equations (13) and (14). It is shown in Appendix D that

$$E_S(2\omega)\delta_S = -\frac{\beta}{3\epsilon_0}E_V^{-2}(\omega)\left\{\frac{4\epsilon_2^2(\omega)}{3} ln \frac{\epsilon_2(\omega)}{\epsilon_2(2\omega)} + [\epsilon_2(\omega)-1][3-\epsilon_2(\omega)]\right\}\delta_S \tag{33}$$

and

$$J[E_{V,t}(2\omega)] = -\frac{\beta}{3\epsilon_0}\left\{\frac{4\epsilon_2^2(\omega)}{3} ln \frac{\epsilon_2(\omega)}{\epsilon_2(2\omega)} + [\epsilon_2(\omega)-1][3-\epsilon_2(\omega)]\right\}\nabla_t E_V^{-2}(\omega) \tag{34}$$

where $\epsilon_2(2\omega)$ is the permittivity of the metal at frequency $2\omega$, $E_{V,t}(2\omega)$ is the tangential component of the classical (volume) part of $E(2\omega)$ and $\nabla_t$ is the two-dimensional gradient taken

Table 1. States of polarization and corresponding origin of the $2\omega$ signal

| Polarisation of the pump beam | $2\omega_{TE}$ | $2\omega_{TM}$ |
|---|---|---|
| $\psi = \frac{\pi}{2} : \omega_{TE}$ | No | Yes, origin $\mathbf{P}_V^{NL}$ |
| $\psi = 0 : \omega_{TM}$ | No | Yes, origin $\mathbf{P}_V^{NL}$, $\mathbf{P}_S^{NL}$ |
| $\psi \neq 0, \frac{\pi}{2}$ (neither TE nor TM) | Yes, origin $\mathbf{P}_{S,z}^{NL}$ | Yes, origin $\mathbf{P}_V^{NL}$, $\mathbf{P}_S^{NL}$ |

in the plane tangent to S. In Eq. (34), the term $\nabla_t E_V^{-2}(\omega)$ comes from the existence of a normal component in the surface nonlinear polarization $\mathbf{P}_S^{NL}$.

The jump of H is obtained from Eqs. (14) and (30). Equation (14) is a known result of electromagnetism, but Eq. (34) is not. Equation (34) shows that the tangential component of the electric field at $2\omega$ is no longer continuous when there exists a surface nonlinear polarization having a component perpendicular to S.

We consider now the polarization properties of the signal field at $2\omega$. Let $\psi$ be the angle between $\mathbf{E}(\omega)$ and the plane of incidence. We consider the three situations $\psi = \pi/2$ (TE), $\psi = 0$ (TM), and $\psi \neq 0, \pi/2$ (neither TE nor TM).

For each case the existence of the second harmonic signal, its origin (surface source term and/or volume source term), and its state of polarization, are sought. The results are shown in Table 1. The columns $2\omega_{TM}$ and $2\omega_{TE}$ are filled using Eqs. (19) and (22) together with Eqs. (30)-(32) and (34).

It should be kept in mind that the free-electron model may be questioned from a physical point of view.[28,29,31] Nevertheless the method developed here to rigorously derive the new boundary conditions is not linked to a particular model and can be reutilized as soon as a more realistic expression for $\mathbf{P}^{NL}$ (taking into account non-local effects for example) is derived.

## 3. FORMALISM OF DIFFRACTION FOR ENHANCED SECOND HARMONIC GENERATION

### 3.1 General considerations

Figure 1 describes the principal grating configurations (periodicity d, groove depth $\delta$) that may be used to increase second harmonic generation through electromagnetic resonances. In Figs. 1a and 1b the pump field (frequency $\omega$, angle of incidence $\theta$) will have to be TM polarized in order to be able to excite surface plasmons or surface polaritons. In Fig. 1c the incident field will be TM or TE/TM polarized depending on whether surface plasmons or guided waves are excited.

Thus we have to consider the interaction $(\omega,\omega) \rightarrow 2\omega = \omega + \omega$ in modulated structures (gratings). This interaction corresponds to a multistep process: a pump beam of frequency $\omega$,

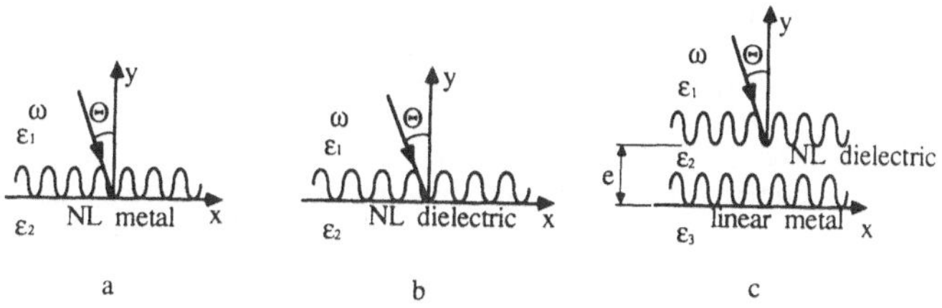


Fig. 1 Main grating configurations.

incident on a bare or coated grating, is diffracted by the grating. The electromagnetic resonance involves one of the evanescent diffracted orders. The interaction among all the diffracted orders at frequency $\omega$ gives rise to a nonlinear polarization at frequency $2\omega$. This polarization radiates an electromagnetic field at $2\omega$, which, in turn, is diffracted by the grating. Since the frequency of the diffracted signal ($2\omega$) differs from that of the incident field ($\omega$), this kind of enhanced second harmonic generation appears as a special case of diffraction in nonlinear optics. We compute the enhancement of the signal with respect to the associated smooth structure, i.e. deduce it from the modulated structure by letting the groove depth go to zero.

Our aims are as follows:

(a) to determine the directions of diffraction of the second harmonic field at circular frequency $2\omega$, i.e. to derive the grating equation in nonlinear optics, equivalent to the well known Fraunhofer equation in linear optics;
(b) to compute the diffracted efficiencies in the different spectral orders at the second harmonic frequency in the absence of any electromagnetic resonance;
(c) to do the same as in (b) but in the presence of different kinds of electromagnetic resonances (surface plasmons, guided waves, etc.) and compute the induced electromagnetic resonance enhancement in order to determine the most efficient resonance; and
(d) to study the influence of the grating parameters (groove spacing, groove depth, groove shape, grating material, etc.) on the second harmonic intensity.

The usual undepleted pump approximation is made for the pump beam. Thus, Maxwell's equations at frequency $\omega$ (with an $e^{-j\omega t}$ time dependence) read

$$\nabla\times\mathbf{E}(\omega) = j\omega\mu_0\mathbf{H}(\omega) \tag{35a}$$

$$\nabla\times\mathbf{H}(\omega) = -j\omega\epsilon_0\epsilon(\omega)\mathbf{E}(\omega) \tag{35b}$$

These two equations show that the pump beam is linearly diffracted by the grating and that this diffraction occurs independently from that of the signal at $2\omega$. Consequently, the diffraction of the pump beam can be studied using standard methods: differential, integral, Yasuura and modal methods.[26,32-34]

Now let us focus on the diffraction process at $2\omega$. The starting point is:

(i) Eqs. (23) and (24) when dealing with Figs. 1b, 1c;
(ii) Eqs. (19) and (22) when dealing with Fig. 1a.

### 3.2 Diffraction at $2\omega$: nonlinear dielectrics (Figs. 1b, 1c)

Included in this section are dielectric gratings (Fig. 1b) and linear metallic gratings overcoated by a nonlinear dielectric (Fig. 1c).

Equations (23) and (24) describe the diffraction process at $2\omega$. Taking into account the periodicity of $\epsilon(2\omega, x, y)$ and the pseudo-periodicity of E and H with respect to x, which allows these functions to be represented by their Fourier series, Eqs. (23) and (24) are transformed into an infinite set of coupled differential equations. This set, after truncation, is numerically integrated with respect to y in the modulated region, and the numerical solution is matched with the Rayleigh expansions of the fields outside the modulated region. The fields and the efficiencies in the different spectral orders at the second harmonic frequency are then obtained.[13] This method of solving the problem is a generalization of the differential formalism[34] for diffraction gratings, previously developed in linear optics.

### 3.3 Diffraction at $2\omega$: nonlinear metals (Fig. 1a)

A differential method similar to that of section 3.2 can be used, except that it requires the calculation of the Fourier coefficients of distribution.[15] The method has been developed and implemented numerically. However, it suffers from numerical instabilities when the pump wavelength belongs to the visible or infrared region and can only be used for very shallow gratings in these spectral domains (this limitation does not hold for UV and X-ray regions).

Thus we seek an alternative method. Contrary to what happens with nonlinear dielectrics, Eqs. (23) and (24), it follows from Eqs. (19), (22) and (32), together with $\partial/\partial z = 0$, that the only source terms are located on the grating surface, i.e. there are no bulk source terms. This may be surprising since $\mathbf{P}^{NL}$ includes a volume term $\mathbf{P}_V^{NL}$. But, as can be deduced from Maxwell's equations, $\mathbf{P}_V^{NL}$ only contributes through its jump across the grating surface. This is consistent with the well known fact that a centrosymmetric medium does not exhibit second harmonic generation.

Accordingly, Eqs. (19) and (22) can be cast into the form

$$\Delta F_q + k_q^2(2\omega)F_q = 0 \qquad (q = 1,2 \text{ depending on the medium})$$

$$F = \begin{cases} H(2\omega) & \text{for } 2\omega_{TM} \\ E(2\omega) & \text{for } 2\omega_{TE} \end{cases} \tag{36a}$$

$$J(F) = \begin{cases} \text{right-hand side of Eq. (14) for } 2\omega_{TM} \\ 0 \text{ for } 2\omega_{TE} \end{cases} \tag{36b}$$

$$J\left[c\frac{\partial F}{\partial n}\right] = \begin{cases} -j2\omega\epsilon_0\hat{t}\cdot J[\mathbf{E}_V(2\omega)] - j2\omega\hat{t}\cdot J\left[\dfrac{\mathbf{P}_V^{NL}(2\omega)}{\epsilon(2\omega)}\right] & c=\dfrac{1}{\epsilon(2\omega)} \text{ for } 2\omega_{TM} \\ -4\omega^2\mu_0 P_{S,z}^{NL}(2\omega) & c=1 \text{ for } 2\omega_{TE} \end{cases} \tag{36c}$$

where $\hat{t}$ is the unit vector in the plane of incidence, tangent to the grating surface.

Note that this boundary value problem is the same as in linear optics[32,33] except for the values of J(F) and J[c(∂F/∂n)]. Thus the integral formalism of linear optics[32,33] can be used for metallic gratings in nonlinear optics. In nonlinear optics, the same computer code has to be called twice, the first time to calculate the diffracted pump field, the second time for the signal. In the meantime, the jumps of F and c(∂F/∂n) have to be calculated from Eqs. (36b) and (36c).

### 3.4 Summary of the theory and general remarks

The formalism developed in the previous sections is summarized in Fig. 2 which shows that this formalism corresponds to a three-step theory.

The following remarks should be made: some diffracted orders at ω are radiated; the rest remain bounded to the bare or coated grating. The electromagnetic resonance, which involves normal modes (surface plasmon or guided wave) of the associated smooth structure, is associated with the resonant excitation of one of the bounded diffracted orders[13] at ω. This leads to the enhancement of the intensity of the resonantly excited diffracted order. For surface plasmon resonance, the pump field must be TM polarized, whereas there is no requirement concerning the polarization of the pump beam for guided wave resonance. Since there is a coupling between the diffracted orders at ω, the resonant excitation of a given bounded dif-

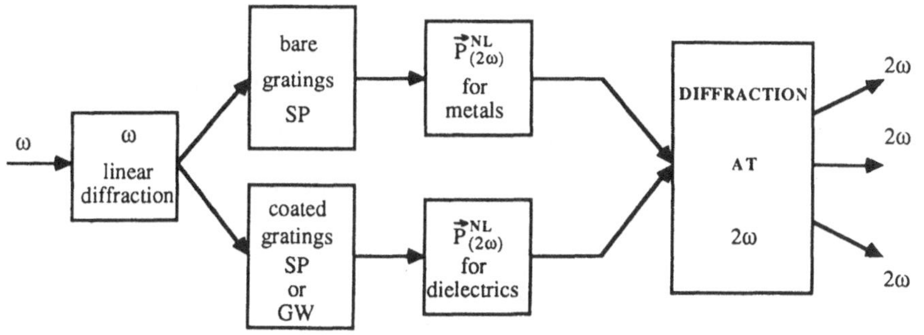


Fig. 2 Successive steps in the theory of diffraction at the second harmonic frequency.

fracted order[32-34] increases the intensity of the neighbouring diffracted ones. The increase of the intensity of the resonantly excited diffracted order and its neighbours induces an enhancement of the magnitude of the source terms at $2\omega$, leading to an increase in the intensity at $2\omega$ compared to the flat case ($\delta = 0$), where no electromagnetic resonance takes place.

The integral and differential methods presented here are rigorous in the sense that the groove depth $\delta$ of the grating is not considered as a perturbative parameter, neither at $\omega$ nor at $2\omega$. These methods allow the derivation of the directions of propagation of the diffracted orders at $2\omega$ which are not bounded to the grating[13] (whatever the geometry of diffraction may be: Figs. 1). In the case of second harmonic generation, one obtains

$$\frac{d}{\lambda/2}[n \sin\psi_p - n_1(\lambda) \sin\theta] = p \tag{37}$$

where $\omega = 2\pi c/\lambda$, $n_1 = \sqrt{\epsilon_1}$ and $\psi_p$ is the diffraction angle of the diffracted order p at $2\omega$ in the outside medium of index n at $2\omega$.

Equation (37) is very important since it represents the nonlinear grating equation. This equation allows us to determine which diffracted orders p at $2\omega$ are radiated in one, or both, of the outside media. The directions of propagation of these orders are given by $\psi_p$.

### 3.5 Numerical results for surface plasmon enhanced second harmonic generation

The enhancement of the second harmonic intensity is computed with respect to the associated smooth structure, which, in this case, is a flat vacuum/metal interface. The reflectivity at $2\omega$ of a flat metallic medium is calculated first and then results for surface plasmon enhanced second harmonic generation are presented.

Calculation of the reflectivity at $2\omega$ of a flat metallic medium is achieved using Eq. (19) and the results of section 2.4. The geometry is that of Fig. 1a with $\delta = 0$.

The nonlinear reflectivity of the metallic mirror is given by the ratio

$$R(2\omega) = \frac{D(2\omega)}{D^2(\omega)}$$

where $D(\omega)$ and $D(2\omega)$ are the power densities at frequencies $\omega$ and $2\omega$, respectively.

From the boundary conditions at the metal-vacuum interface, the following expression for the reflectivity at $2\omega$ of a unit power TM polarized pump beam is obtained

$$R(2\omega) = \frac{1}{2}\left(\frac{\mu_0}{\epsilon_0}\right)^{3/2} \left| \frac{-2\omega\epsilon_0\epsilon_2(2\omega)J(E_x) + \alpha_{2,2}J(H_z) + 2\omega P^{NL}_{V,x}}{\alpha_{2,2} + \alpha_{1,2}\epsilon_2(2\omega)} \right|^2 \tag{38}$$

where $\alpha_{1,2}$, $\alpha_{2,2}$ are perpendicular components of the wave vector at the second harmonic frequency in vacuum and in the metal, respectively, and $P^{NL}_{V,x}$ is the x-component of the volume part of $\mathbf{P}^{NL}$ calculated at $y = 0^-$.

The reflectivity of a silver mirror at the second harmonic frequency has been measured[7,17,35] as a function of the angle of incidence $\theta$ using a pump wavelength of 1.064 $\mu$m. It has been found that $R(2\omega)$ calculated from Eq. (38) is 20 times higher than the measured reflectivity at $2\omega$.

This leads us to make a slight modification to Bloembergen's expression for $\mathbf{P}^{NL}(2\omega)$ given in Eq. (25), by introducing[7,8,17] two phenomenological coefficients A and B

$$\mathbf{P}^{NL}(2\omega) = A\gamma\nabla[\mathbf{E}(\omega)\cdot\mathbf{E}(\omega)] + B\beta\mathbf{E}(\omega)\nabla\cdot\mathbf{E}(\omega)$$

$R(2\omega)$ is still given by Eq. (38), but now it depends on A and B through the quantities $J(E_x)$, $J(H_z)$ and $P^{NL}_{V,x}$. Use of the method of section 2 yields

$$\hat{n} \times J[\mathbf{H}(2\omega)] = -Bj2\omega\hat{n} \times [\mathbf{P}^{NL}_S(2\omega)\times\hat{n}]$$

instead of Eq. (14) and

Table 2. Computed (theoretical) and measured (experimental, when available) optimum groove depth and optimum surface plasmon induced enhancement of the intensity of a given diffracted order p at $2\omega$ for different grating periodicities and profiles. $\lambda = 1.064\ \mu m$.

| Profile | Optimum groove depth Å | | Optimum enhancement | |
|---|---|---|---|---|
| | Theoretical | Experimental | Theoretical | Experimental |
| d (Å) | p | p | p | p |
| Sinusoidal | 250 | 300 | 47 | 36 |
| 5556 | -1 | -1 | -1 | -1 |
| Sinusoidal | 900 | 1000 | 2370 | 2500 |
| 15300 | +1 | +1 | +1 | +1 |
| Trapezoidal | 670 | | $\simeq 10^5$ | |
| 15300 | 0 | | 0 | |
| Triangular | 780 | | 7784 | |
| 15300 | +1 | | +1 | |

$$J[\mathbf{E}_{V,t}(2\omega)] = \frac{1}{3}(2A+B)J_0[\mathbf{E}_{V,t}(2\omega)] - \frac{\beta}{3\epsilon_0}(B-A)[\epsilon_2^2(\omega)-1]\nabla_t[\bar{E}_V^2(\omega)]$$

where $J_0[\mathbf{E}_{V,t}(2\omega)]$ is given by Eq. (34). Also, $\mathbf{P}_V^{NL}(2\omega) = A\mathbf{P}_{0,V}^{NL}(2\omega)$, where $\mathbf{P}_{0,V}^{NL}(2\omega)$ is given by Eq. (32).

The two unknown parameters A and B are determined by fitting (using a least squares routine) the theoretical and experimental second harmonic reflectivities $R(2\omega)$. The best agreement[7,17,35] is obtained for A = 2.2 and B = 0.57, with the corresponding values of $\epsilon_2(\omega)$ and $\epsilon_2(2\omega)$ for silver, measured by ellipsometry[7,17,35]

$$\epsilon_2(\omega) = -36 + 2j \qquad \epsilon_2(2\omega) = -7.9 + 0.1j$$

Note that the computation of the enhancement of the second harmonic intensity as compared to the flat case is performed using these two values of A and B. That is, in computing the enhancement at $2\omega$ there are *no unknown* parameters.

The computation of the second harmonic enhancement has been done for a TM polarized pump field exciting a surface plasmon resonance on a sinusoidal, trapezoidal or triangular silver grating.[17] The results are summarized in Table 2.

As predicted by theory, the data shows the existence of an optimum value $\delta_{opt}$ linked with the surface plasmon resonance at $\omega$ for which the enhancement E is the greatest: $E(\delta=\delta_{opt}) = E_{opt}$. This value of $\delta_{opt}$ depends on the number p of the diffracted order at $2\omega$, and also on $\omega$. Indeed, in the case of a sinusoidal silver grating,

(i) for d = 0.5556 $\mu m$, $\delta_{opt}(\omega)$ = 110 Å, whereas $\delta_{opt,p=0}(2\omega)$ = 600 Å and $\delta_{opt,p=-1}(2\omega)$ = 250 Å; and

(ii)[36] for d = 1.53 $\mu m$, $\delta_{opt}(\omega)$ = 300 Å, whereas $\delta_{opt,p=0}(2\omega)$ = 1700 Å and $\delta_{opt,p=+1}(2\omega)$ = 900 Å.

The value of $E_{opt}$ depends strongly on d, as shown in Table 2. An increase in d from 0.5556 $\mu m$ to 1.53 $\mu m$ leads to a strong increase in $E_{opt}$, from 36 to 2500. The reason can be understood as follows: when d increases, the ratio $\lambda/d$ decreases. Thus, the diffracted orders at $\omega$ become closer and closer, i.e., more coupled to each other and especially to the resonantly excited order. Of course, the number of propagating diffracted orders, and therefore the radiation losses, increase with d, inducing a less efficient electromagnetic resonance at $\omega$.

Table 3. Computed values of $\delta_{opt}$ and $E_{opt}$ in the case of guided wave enhanced second harmonic generation

| Diffracted order at $2\omega$ | $\delta_{opt}$ | $E_{opt}$ |
|---|---|---|
| 0 | 335 Å | $7.2 \times 10^3$ |
| +1 | 333 Å | $1.7 \times 10^5$ |

### 3.6 Numerical results for guided wave enhanced second harmonic generation

The numerical results[16] refer to a ZnS coated grating with a sinusoidal profile, where the substrate is silver and the upper medium is vacuum (Fig. 1c). Since ZnS belongs to the $\bar{4}3m$ symmetry class, the pump beam must be TM polarized to generate a nonlinear polarization at $2\omega$. This nonlinear polarization has a single component along the z axis,

$$P_z^{NL}(2\omega) = 2\epsilon_0\chi_{xyz}(2\omega)E_x(\omega)E_y(\omega)$$

which gives rise [Eqs. (23) and (24)] to a TE polarized diffracted signal at $2\omega$. The enhancement at $2\omega$, due to the guided wave resonance at $\omega$, is computed with respect to the associated smooth structure (deduced from the modulated structure by letting $\delta \rightarrow 0$).

The guided wave resonance occurs at $\omega$ for the -1 diffracted order, the corresponding incidence angle being $\theta = \theta_{res} = 28.92°$. The optimum value of the groove depth at $\omega$ is $\delta_{opt}(\omega) = 321$ Å.

As for surface plasmon enhanced second harmonic generation, there exists an optimum value $\delta_{opt}$ linked with the guided wave resonance at $\omega$ for which the peak enhancement takes its greatest value $E_{opt}$. Table 3 summarizes the main results.

Given a periodicity of 0.400 $\mu$m, the value $E_{opt} = 1.7\times10^5$ obtained for guided wave enhanced second harmonic generation must be compared to the value 47 obtained from surface plasmon resonance (see Table 2).

As predicted in Ref. 12, when enhanced second harmonic generation is desired, it is worth using the guided wave rather than the surface plasmon resonance. The fact that the guided wave resonance leads to a much greater increase in the second harmonic efficiency than does the surface plasmon resonance can be easily understood. Indeed, with guided waves, the greatest part of the pump field lies in the nonlinear guiding layer, whereas, with surface plasmons, the electromagnetic field extends much farther into the vacuum than into the metal (which is the nonlinear medium).

## 4. ELECTROMAGNETIC RESONANCE INDUCED RESPONSE OF NONLINEAR OPTICAL RESONATORS: LOCAL KERR EFFECT

Let us now consider a nonlinear thin film which exhibits the optical Kerr effect. We are interested in the geometry of Fig. 3 which allows the study of the nonlinear prism coupler and the nonlinear Fabry Pérot cavity (the latter is obtained by letting $\epsilon_2 = \epsilon_1$). The aim here is to study the response of these nonlinear optical resonators, taking into account the transverse effects due to the finite width of the incident pump beam (frequency $\omega$) under normal and non-normal incidence $\theta$. This problem of transverse effects has already received much attention in the literature.[37-55]

A TE incident pump beam and isotropic nonlinear medium are assumed. Thus there is no TE-TM coupling and the solution is also TE polarized.

Let $\mathbf{E}_q = E_q\mathbf{e}_z$, where $\mathbf{e}_z$ is the unit vector along the z-axis (q = 1,..., 4). In the nonlinear medium, Maxwell's equations read

$$\nabla\times\mathbf{E}_3(x,y) = j\omega\mu_0\mathbf{H}_3(x,y) \tag{39a}$$

$$\nabla\times\mathbf{H}_3(x,y) = -j\omega\epsilon_0\epsilon_3\mathbf{E}_3(x,y) - j\omega\mathbf{P}^{NL}(x,y) \tag{39b}$$

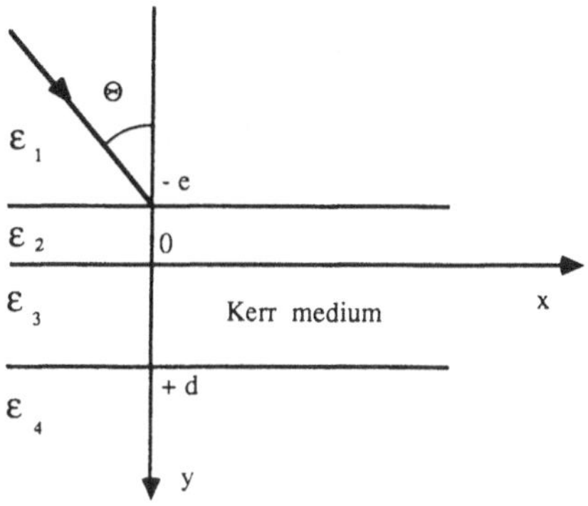


Fig. 3 A nonlinear optical resonator.

If medium 3 is assumed to be isotropic, then[56]

$$\mathbf{P}^{NL} = \mathbf{e}_z P^{NL}$$

where

$$P^{NL} = \epsilon_0 \chi^{(3)} |E_3(x,y)|^2 E_3(x,y)$$
$$\chi^{(3)} = \chi_{zzzz} \tag{40}$$

From Eqs. (39), we derive the equation obeyed by the electric field in the nonlinear medium

$$\Delta E_3 + k_3^2 E_3 = -\frac{\omega^2}{c^2}\chi^{(3)}|E_3|^2 E_3 \tag{41}$$

In media 1, 2 and 4 the electric field $E_q$ ($q = 1,2,4$) obeys the following linear equation

$$\Delta E_q + k_q^2 E_q = 0 \tag{42}$$

where in Eqs. (41) and (42)

$$k_q^2 = \frac{\omega^2}{c^2}\epsilon_q \qquad (q = 1,\ldots,4)$$

It is convenient to express $E_3(x,y)$ as a sum of a downward and an upward wave

$$E_3(x,y) = [E_D(x,y)\, e^{-jk_{3y}y} + E_U(x,y)\, e^{jk_{3y}y}]\, e^{jk_x x} \tag{43}$$

with

$$k_{3y}^2 + k_x^2 = k_q^2 \qquad k_x = k_1 \sin\theta$$

On substituting Eq. (43) into Eq. (41) and making the usual slowly varying envelope approximation, we obtain

$$\frac{\partial F_3}{\partial Y} = j\eta\left(|F_3|^2 + 2|B_3|^2\right)F_3 - r\frac{\partial F_3}{\partial X} + j\frac{\partial^2 F_3}{\partial X^2} \tag{44a}$$

$$\frac{\partial B_3}{\partial Y} = -j\eta\left(|B_3|^2 + 2|F_3|^2\right)B_3 - r\frac{\partial B_3}{\partial X} + j\frac{\partial^2 B_3}{\partial X^2} \tag{44b}$$

where the dimensionless quantities $F_3$, $B_3$, r, X and Y are defined according to

$$F_3 = \frac{\omega}{c}\left|\chi^{(3)}\right|^{1/2} \frac{E_D}{2k_{3y}} \tag{45a}$$

$$B_3 = \frac{\omega}{c}\left|\chi^{(3)}\right|^{1/2} \frac{E_U}{2k_{3y}} \tag{45b}$$

$$r = \frac{k_x}{k_{3y}} \tag{45c}$$

$$X = 2k_{3y}\,x \tag{45d}$$

$$Y = 2k_{3y}\,y \tag{45e}$$

$$\eta = \mathrm{sgn}(\chi^{(3)}) \qquad \eta = \begin{cases} +1 & \text{for focusing media} \\ -1 & \text{for defocusing media} \end{cases} \tag{45f}$$

In Eqs. (44), the term $r(\partial/\partial X)$ is equal to zero under normal incidence. When $\theta \neq 0$, $r \neq 0$ and this quantity describes the lateral beam displacement due to the zig-zag path inside the cavity. The term $\partial^2/\partial X^2$ accounts for the diffraction process.

Equations (44) have to be solved with the boundary conditions at y = 0, d and X = $(R_\ell, R_r)$. These boundary conditions, which result from the continuity of E and $\partial E/\partial y$, have a very complicated form since one does not know the form of the electromagnetic field outside (and inside) the nonlinear medium. For the sake of simplicity, boundary conditions at y = 0, d corresponding to the linear plane wave situation will be used.[48] Thus, for an angle of incidence corresponding to the nonlinear prism coupler regime, the nonlinear Goos-Hänchen shift is not taken into account.[57]

The boundary conditions at y = 0, d are given by

$$F_3(X, Y{=}0) = T_{13}F_i(X) + R_{31}B_3(X, Y{=}0) \tag{46a}$$

$$B_3(X, Y{=}2\Phi_0) = r_{34}e^{j2\Phi_0}F_3(X, Y{=}2\Phi_0) \tag{46b}$$

where

$$\Phi_0 = k_{3y}\,d \tag{47}$$

$\Phi_0$ represents the linear phase shift inside the cavity (Fabry Pérot or prism coupler type).

The boundary conditions at X = $R_\ell$, $R_r$ are

$$F_3(R_{\ell,r}, Y) = 0 \tag{48a}$$

$$B_3(R_{\ell,r}, Y) = 0 \tag{48b}$$

The incident pump beam is assumed to be Gaussian and, in dimensionless form, is written as

$$F_i(X) = F_1 \exp\left[-\left(\frac{X\cos\theta}{X_0}\right)^2\right] \tag{49a}$$

where

$$X_0 = 2k_{3y}\,w_0 \tag{49b}$$

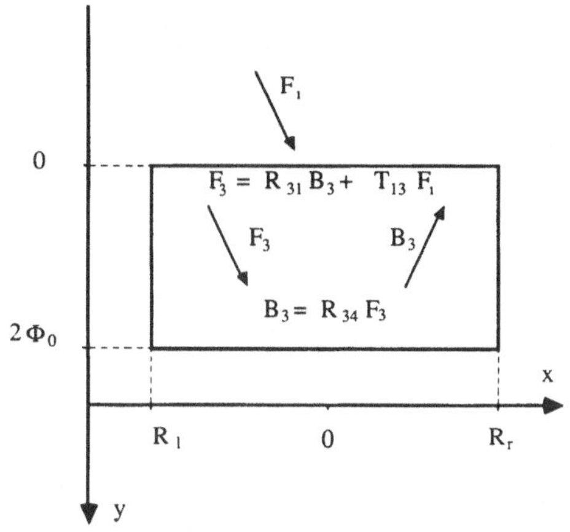


Fig. 4 Equivalent representation of Fig. 3. The numerical integration is performed only in the nonlinear layer. Valid for the nonlinear Fabry Pérot cavity and the nonlinear prism coupler, $R_{34} = r_{34}\exp(j2\Phi_0)$.

and $w_0$ is the waist of the incident beam. It is assumed that the plane Y = 0 is placed at the beam waist. The situation described by Eqs. (44), (46), (48) and (49) is depicted in Fig. 4.

The normalisation equations (45a), (45b), (45d) and (45e) are the ones used by Maneuf *et al.*[58] They are different from those of Refs. 47 and 48, but seem more appropriate since they do not lead to quantities that depend on the width of the input beam.

In the following, it is convenient to introduce the Fresnel number, $\mathcal{F}$ defined as

$$\mathcal{F} = \frac{d}{k_{3y} w_0^2} \tag{50}$$

Use of Eq. (50) leads to

$$X_0 = 2\left(\frac{\Phi_0}{\mathcal{F}}\right)^{1/2} \tag{51}$$

Equation (50) shows that the influence of the diffraction process not only depends on the waist $w_0$ but also on the thickness of the nonlinear film.

Let us first consider the plane wave case. Although it does not correspond to any realistic situation, plane wave excitation is pedagogically interesting, if only because it is always very tempting to identify a $w_0 \rightarrow \infty$ Gaussian beam with a plane wave. As we shall see, it is worth refraining from doing this.

### 4.1 Plane wave excitation

The corresponding solution can be obtained in two ways:

(i) either by a numerical integration setting $F_i(X) = F_1$ in Eq. (49a), or
(ii) by noting that $\partial/\partial X = \partial^2/\partial X^2 = 0$ in Eqs. (44).

Approach (ii) is interesting because it leads to an analytical solution of Eqs. (44) and (46).

$$F_3 = \frac{T_{13}F_1}{1-R_{31}R_{34}e^{j2\Phi^{NL}}} \tag{52a}$$

with

$$\Phi^{NL} = 3\Phi_0\eta\left(1+|R_{34}|^2\right)|F_3|^2$$

$$R_{34} = r_{34}e^{j2\Phi_0} \quad (52b)$$

Equations (52) apply whatever type of device Fig. 3 may be: i.e. a nonlinear Fabry Pérot cavity or nonlinear prism coupler. Equation (52b) gives the nonlinear phase shift arising from the self index variation associated with the optical Kerr effect. According to Eqs. (52), the usual graphical construction used to get the plane wave response of the nonlinear Fabry Pérot cavity[59] also applies to the nonlinear prism coupler. As is well known,[60] the plane wave response depends on $\Phi_0$, i.e. on the detuning of the resonator. Since $\Phi^{NL}$ depends on $F_1$, an increase of the incident intensity leads to a tuning of the resonator and, as a result, to the resonant excitation of normal modes of this nonlinear optical resonator. It is this resonant excitation which gives rise to optical bistability, optical transistor or optical limiter behaviour.

The electromagnetic resonance involved when considering the nonlinear Fabry Pérot cavity is the nonlinear Airy resonance,[22] and when considering the nonlinear prism coupler it is the nonlinear guided mode[23] since it is associated with the resonant excitation of a guided mode of the nonlinear waveguide ($\epsilon_{2,3,4}$). In that case $\epsilon_1 > \epsilon_3 > \epsilon_{2,4}$ and $\theta$ has to be larger than the critical angle of total reflection at the interface $\epsilon_3$, $\epsilon_4$.

It can be seen that the nonlinear Fabry Pérot cavity and the nonlinear prism coupler are similar. However, it is worth noting that the nonlinear prism coupler only works under oblique incidence. As we shall see, this feature strongly modifies the finite width response of the nonlinear prism coupler as compared to the plane wave one.

## 4.2 Gaussian beam excitation

This situation corresponds to the problem of diffraction in Kerr type optical resonators. To get the response of the device shown in Fig. 3, the solution of Eqs. (44) with Eqs. (49a), (46) and (48) is required.

Basically, two numerical methods have been developed:

(i) A transverse iterative approach introduced by Firth *et al.*[42] Weaire *et al*[47,48] replace the quasi fast Hankel transform[42] by an elementary finite difference method. Olin[50] improves Weaire *et al*'s method by computing the input power as a function of the output one. This allows getting the negative slope part of the optical bistability cycle. Using Hartmuth's method,[61] it can be shown that these transverse numerical approaches are stable only within a small range of parameters. This has also been noticed by Weaire *et al.*[48] In particular, transverse numerical integration fails for high finesse cavities.

(ii) A continuation method introduced by Kubicek[62] and developed, along with a longitudinal integration, for the first time to our knowledge, by one of us (GV) in the study of transverse effects in optical Kerr resonators.[63] The advantages of such an approach are two-fold:
   (a) the continuation method allows choosing the most suited variable $\tau$ so that the solution is always a monovalued function of $\tau$; and
   (b) the numerical longitudinal integration can always be made stable in the nonlinear prism coupler configuration. Under normal, or quasi normal, incidence, this method is unconditionally unstable. Nevertheless since this approach is not an iterative one, the numerical integration can be performed (i.e. there is no divergence) by dividing the interval $(R_\ell, R_r)$ into N sections of length L

$$L = \frac{R_r - R_\ell}{N}$$

The minimum value of N, $N_{min}$, comes from the fact that one requires $L < L_{max}$ in order to have a non divergent numerical scheme. Of course, the greater the N, the longer the computation time.

## 4.3 Numerical results

The influence of the sign of the nonlinearity under normal incidence will be considered first. Since only positive Kerr media exhibit self-focusing, it can be expected that the

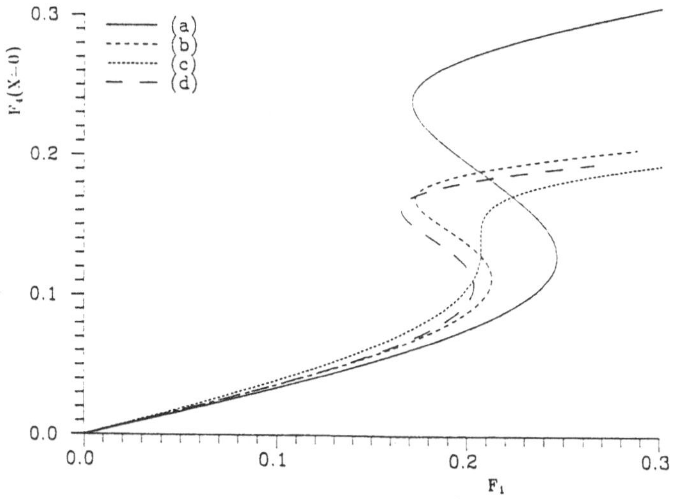


Fig. 5 Transmitted amplitude $F_4(X=0)$ as a function of $F_1$. (a) $\eta = +1$, $\mathcal{F} = 10^{-2}$; (b) $\eta = +1$, plane wave response; (c) $\eta = -1$, $\mathcal{F} = 10^{-2}$; (d) $\eta = -1$, plane wave response. Other data: $R_{34} = R_{31} = 0.95$, $T_{34} = 1.95$.

response of the nonlinear device (Fig. 3) to a Gaussian beam depends on whether one deals with a positive or negative Kerr nonlinearity.

The results are reported in Fig. 5 which is a plot of the normalized transmitted amplitude in medium 4 at $X = 0$, $F_4$, as a function of $F_1$. The absolute value of the detuning $\Delta\phi_0$, is the same for positive and negative Kerr nonlinearity

$$|\Delta\phi_0| = \pi - \phi_0 = 8°$$

$$\Rightarrow \phi_0 = \begin{cases} 172° & \eta = +1 \\ 188° & \eta = -1 \end{cases}$$

and the nonlinear Airy resonance occurs for $\phi^{NL} = \pi$. The values of $R_{34}$ and $R_{31}$ correspond to a high finesse cavity. The plane wave response is also plotted for the sake of comparison.

It is seen that the nonlinear Airy resonance induced response of the Fabry Pérot cavity is very different depending on whether $\eta = \mp 1$

(i) when $\eta = -1$, there is no optical bistability cycle, only a curve reminiscent of that of an optical transistor,

(ii) the case $\eta = +1$ is very interesting. Indeed, it appears that the bistable loop widens as compared to the plane wave one:[40,49] the switch-up values, $F_1 \uparrow$, and the upper level are larger than for a plane wave excitation.

These results may be understood by considering Figs. 6a and 6b which represent $F_3(X, Y = 2\Phi_0)$ as a function of X for $\eta = \pm 1$. These figures show that the effect (ii) is related to the existence of a self-focused peak which, in a certain way, compensates for the linear diffraction. On the other hand, when $\eta = -1$ the nonlinearity enhances the diffraction phenomenon.

The effect of the Fresnel number $\mathcal{F}$ on the response of the nonlinear Fabry Pérot cavity for the case $\eta = +1$ is shown in Fig. 7a. As the width of the input beam increases, $F_1 \uparrow$ decreases towards a limit which is close to the plane wave value. In the range of $\mathcal{F}$ considered here, the switch-down values of $F_1$, $F_1 \downarrow$, and the upper level are rather insensitive to $\mathcal{F}$.

The curves plotted in Fig. 7a show the existence of a limiting bistable loop as $\mathcal{F}$ decreases. It is worth noting that this loop does not correspond to the plane wave solution. Figure 7b shows that this comes from the fact that self-focusing occurs even for a large value of the incident beam width: the size of the self-focused peak is a constant independent of $\mathcal{F}$.

Results demonstrating the effect of the angle of incidence are given in Figs. 8a and 8b for two Fresnel numbers: 0.01 and 0.001. Since $\theta \neq 0$, it is the peak value $F_{3M}$ of $F_3$ (which does not correspond to a given X) that is reported on the vertical axis.

Several features appear:

(i) as $\theta$ increases, the size of the "main" part of the bistable loop decreases. Simultaneously, additional small loops appear for $\mathscr{F} = 0.01$ and $\theta = 5°, 7°$;

(ii) there exists a limiting value of $\theta$, $\theta_{\lim}$, beyond which there is no optical bistability. For $\theta > \theta_{\lim}$, the response curve is rather smooth; and

(iii) $\theta_{\lim}$ is smaller for wide beams than for narrow ones. This result is presumably due to the fact that as $\mathscr{F}$ decreases, the Fourier transform of the input beam in k-space becomes narrower.

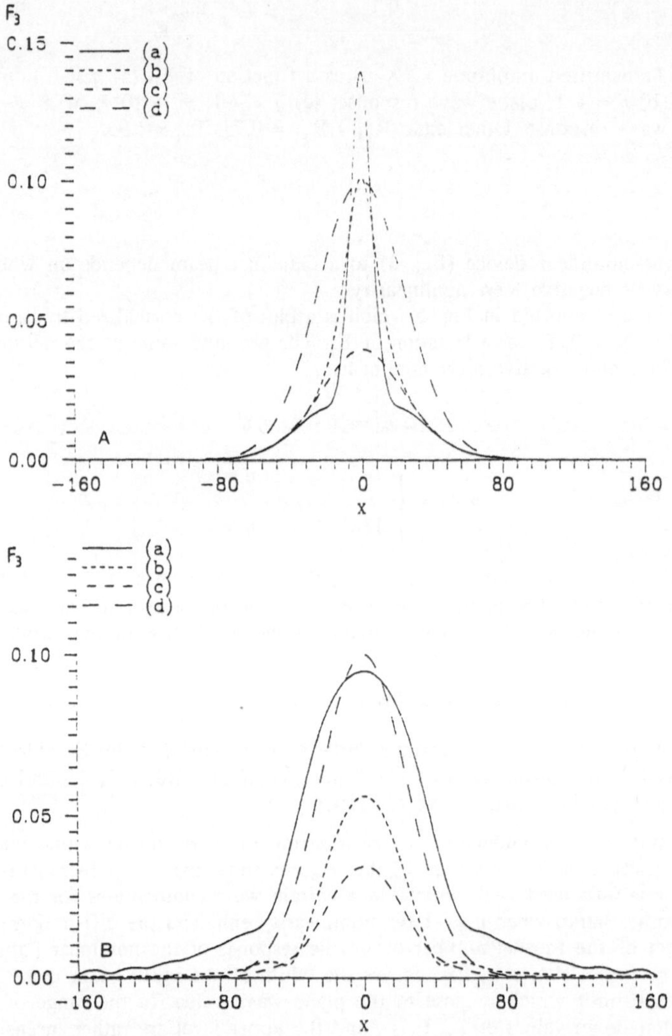


Fig. 6 (A) Field map $F_3(X)$ for $F_1 = 0.2$ in Fig. 5 and $\eta = +1$: (a) upper level, (b) negative slope part, (c) lower level, (d) Gaussian input (divided by two); (B) same as (A) but with $\eta = -1$: (a) $F_1 = 0.25$, (b) $F_1 = 0.20$, (c) $F_1 = 0.15$, (d) Gaussian input (divided by two). $R_{34} = R_{31} = 0.95$, $T_{34} = 1.95$.

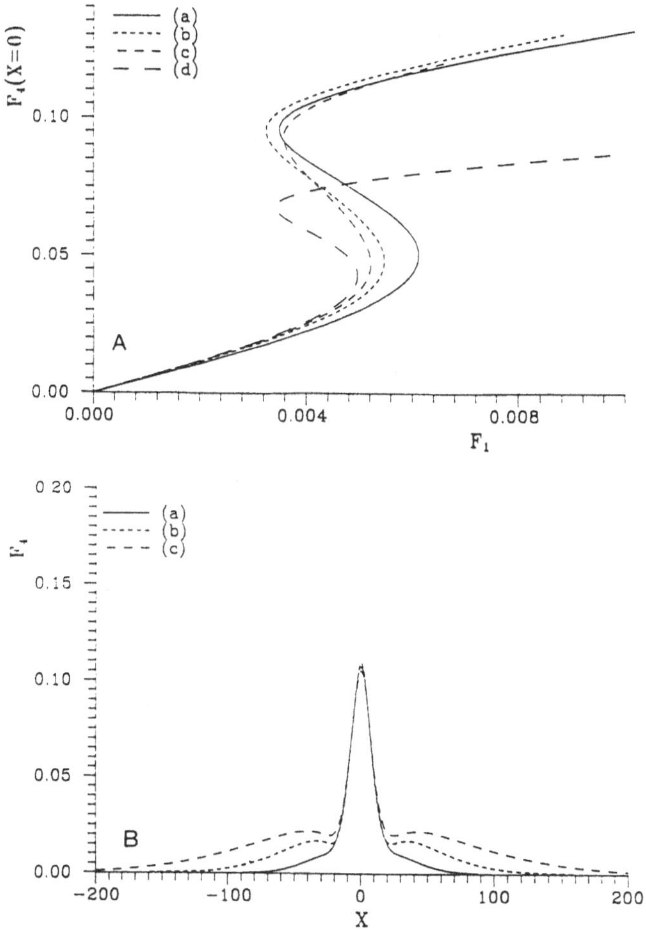


Fig. 7 (A) Transmitted amplitude $F_4(X=0)$ as a function of $F_1$ for different Fresnel numbers in a positive Kerr medium: (a) $\mathcal{F} = 10^{-2}$, (b) $\mathcal{F} = 3 \times 10^{-3}$, (c) $\mathcal{F} = 10^{-3}$, (d) plane wave; (B) field map $F_4(X)$ for $F_1 = 0.0045$ upper level: (a) $\mathcal{F} = 10^{-2}$, (b) $\mathcal{F} = 3 \times 10^{-3}$, (c) $\mathcal{F} = 10^{-3}$. $R_{34} = T_{34} = T_{13} = 1$, $R_{31} = 0.95$.

The direct consequence of these results is that the nonlinear prism coupler exhibits no optical bistability when the local Kerr effect is involved, a conclusion already published by Carter and Chen[51] within the framework of a coupled mode theory.[64,65]

## 5. NONLOCAL THEORY OF THE NONLINEAR PRISM COUPLER

Since optical bistability has been observed in nonlinear prism coupler experiments,[66-70] there seems to be a discrepancy between theory and experiments. However, it should be remembered that the results of section 4 apply to local Kerr media whereas the experiments, Refs. 66-70, involve nonlocal nonlinearity (i.e. heat transfer, electron diffusion, etc.). We now consider the nonlinear prism coupler with a nonlocal nonlinearity.

The theory[54] is performed using a coupled mode approach[64,65] together with a diffusion equation for the nonlinear process.

The growth of the guided wave amplitude, $A_{gw}(x)$, along the longitudinal coordinate direction is given by[51]

$$\frac{d}{dx}A_{gw}(x) = j[-\Delta\beta + j\beta_i + U(x)]A_{gw}(x) + tA_iG(x) \tag{53}$$

where the nonlinear contribution to the guided wave wavevector is given by U(x). The second term describes the coupling (transfer coefficient t) between the Gaussian incident field

$$E_i(x) = A_i\ G(x)\ \exp(jk_x x) \tag{54}$$

where $G(x) = \exp(-x^2/w_0^2)$, and the guided wave

$$E_{gw}(x) = A_{gw}(x)\psi(y)\exp[j(\beta_r + j\beta_i)x] \tag{55}$$

where the guided wave wavevector $\beta_r + j\beta_i$ is complex because the guided wave is leaky due to the presence of the prism. $\psi(y)$ is the normalized transverse field distribution.[71] The wavevector detuning between the two waves is $\Delta\beta = k_x - \beta_r$.

The spatial variation of the nonlinear term U(x) depends on the model assumed for the nonlinearity. For a local Kerr-law nonlinearity, first order perturbation theory gives the local nonlinear change in the refractive index n:

$$\Delta n(x,y) = n^{NL}(y)|\psi(y)|^2\ |A_{gw}(x)|^2$$

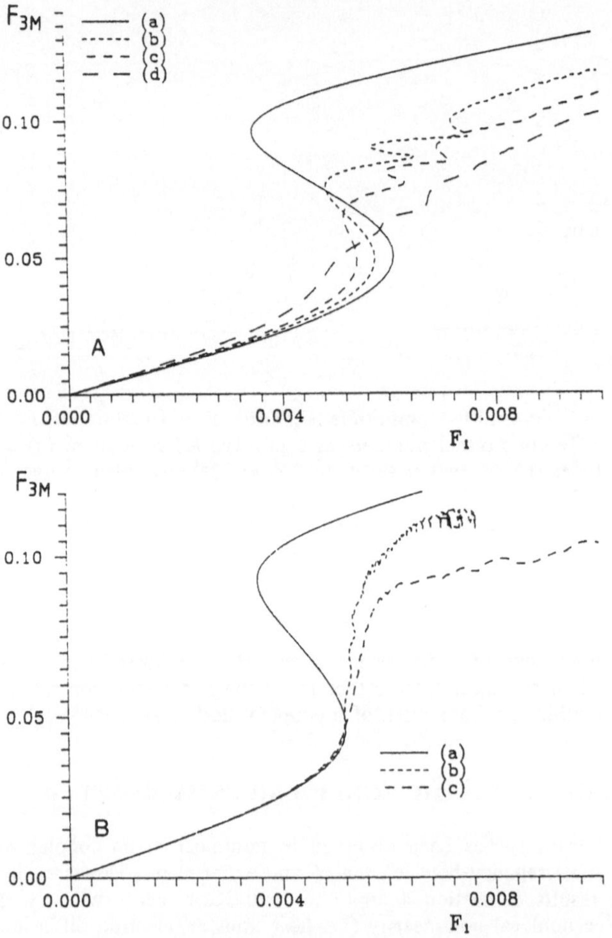


Fig. 8 (A) Peak value of $F_3$, $F_{3M}$, as a function of $F_1$ for $\mathcal{F} = 0.01$: (a) $\theta = 0°$, (b) $\theta = 5°$, (c) $\theta = 7°$, (d) $\theta = 10°$; (B) same as in (A) but with $\mathcal{F} = 0.001$: (a) $\theta = 0°$, (b) $\theta = 5°$, (c) $\theta = 7°$. $R_{34} = T_{34} = T_{13} = 1$, $R_{31} = 0.95$.

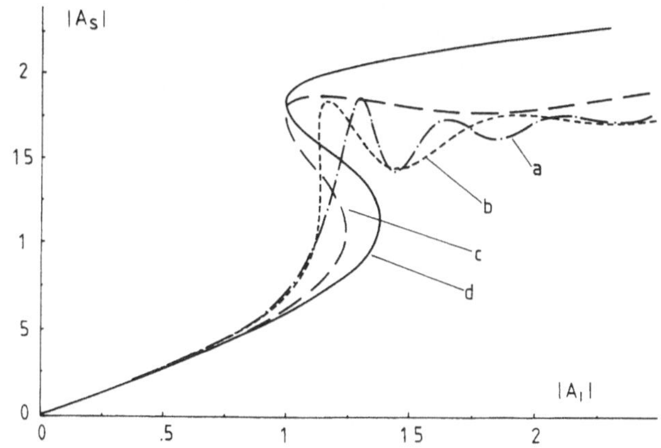


Fig. 9 The influence of the diffusion length D on the response of a nonlinear prism coupler. (a) D = 0 (local case), (b) D = 1.5, (c) D = 10 and (d) D = 100. The last case corresponds to a complete delocalisation of the nonlinear effect. Other data: $\Delta\beta$ = 1.8, $\beta_i$ = 0.5, $w_0$ = 10, $\ell_1$ = -25, $\ell_2$ = 0, $n^{NL}$ = 1 and t = 1.

If furthermore the field profile is independent of the guided wave power:[53]

$$U(x) = \frac{4\pi^2}{\beta_r \lambda^2} \int dy\ n_3(y) n^{NL}(y) |\psi(y)|^4 . |A_{gw}(x)|^2$$

$$= \Delta\beta_0 |A_{gw}(x)|^2 \tag{56}$$

where $n_3(y)$ and $n^{NL}(y)$ describe the transverse variation in the linear refractive index and nonlinear coefficient respectively. If the nonlinearity is non-local, for example thermal, the nonlinear index change is effectively spread over the sample volume. Since it is longitudinal feedback which is required for bistability, we consider only a one dimensional diffusion equation along the propagation axis, and Eq. (56) is replaced by

$$D^2 \frac{d^2U(x)}{dx^2} = U(x) - \Delta\beta_0 |A_{gw}(x)|^2 \tag{57}$$

which reduces to Eq. (56) in the limit $D \to 0$, where D is the effective diffusion length.

We assume that the prism extends from $x = -\infty$ to $\ell_2$ so that the boundary conditions for the guided wave are $A_{gw}(-\infty) = 0$ and $A_{gw}(x) = A_{gw}(\ell_2) = A_S$ for $x > \ell_2$. This means that $U(\infty)$ and $U(-\infty)$ are both finite.

The numerical integrations were performed between $x = -\ell_1$ and $x = \ell_2$ where $\ell_1$ is chosen large enough so that $A_{gw}(-\ell_1)$ is negligible. The boundary conditions on U(x) at $-\ell_1$ and $\ell_2$ can be obtained from Eq. (57) as

$$U(-\ell_1) - D\frac{dU(x)}{dx} = 0 \qquad \text{at } x = -\ell_1 \tag{58a}$$

$$U(\ell_2) + D\frac{dU(x)}{dx} = \Delta\beta_0 |A_{gw}(\ell_2)|^2 \qquad \text{at } x = \ell_2 \tag{58b}$$

The input and output parameters of interest are $A_i$ and $A_S$ respectively. The problem is to find $U_1 \simeq U(-\ell_1)$ so that after integrating self-consistently Eqs. (53) and (57) with the boundary conditions $A_{gw}(-\ell_1) = 0$ and Eq. (58a), the final boundary condition equation (58b) is satisfied. Equations (53) and (57), together with the boundary condition equations (58), are solved using a continuation method.

The influence of the diffusion length D on the response of the nonlinear prism coupler is shown in Fig. 9 for a given width, $w_0$, of the input beam. The delocalisation of the nonlinear optical properties of the guiding layer induces a longitudinal feedback which gives rise to optical bistability provided D is greater than a threshold value $D_{th}$. Other results[72] not reported here show that $D_{th}$ is of the order of magnitude of the coupling length of the coupler, i.e. $1/\beta_i$. This means that the spatial delocalisation of the nonlinear effect must be of the order of the beam width $w_0$ in order to get optical bistability, since in an optimized coupler $w_0$ is close to $1/\beta_i$.

## 6. CONCLUSION

In order to investigate the role played by electromagnetic resonances in nonlinear optics, two phenomena have been considered: second harmonic generation and the optical Kerr effect. It has been found that electromagnetic resonance induced nonlinear optical phenomena lead to a rich variety of effects.

In the case of second harmonic generation, the undepleted pump approximation applies and, as a result, linear surface plasmon or guided wave resonances are involved. The enhancement of the second harmonic intensity not only depends on the groove depth, but also on the profile of the grating, its periodicity and the nature of the electromagnetic resonance. This electromagnetic resonance induced enhancement can be optimized: values as high as $10^5$ can be reached which are associated with very shallow modulation ($\delta_{opt}/d \simeq 10^{-1}$). Hence the great sensitivity of the second harmonic efficiency on the groove depth $\delta$ when electromagnetic resonances are excited. The results reported here show that gratings act as powerful electromagnetic field amplifiers for second harmonic generation which corresponds to a non spectroscopic use of gratings.

The optical Kerr effect leads to the possibility of "controlling light with light".[55] A correct investigation of this possibility requires a study of the nonlinear electromagnetic resonance induced response of nonlinear optical resonators, such as the nonlinear Fabry Pérot cavity or nonlinear prism coupler, taking fully into account the transverse effects due to the finite width of the incident beam. Indeed, the plane wave response may lead to erroneous results in the sense that optical bistability predicted by a plane wave study may not exist: this is, for example, the case for the nonlinear prism coupler when the local Kerr effect is involved. The lack of longitudinal feedback explains such a result. The key point is that in any case, the "finite width" response of nonlinear optical resonators is very different from the plane wave one: for defocusing media the bistable loop disappears whereas for self-focusing media the size of this loop increases when decreasing the width of the pump beam. These results are very important with regard to the optimization of nonlinear optical microresonators that can be run simultaneously for the purpose of parallel processing.

## APPENDIX A: EXPRESSION OF DIVERGENCE, GRADIENT AND CURL IN DISTRIBUTION THEORY

Let us recall the expressions for the differential operators used in the theory of distributions. According to Refs. 25 and 26, we get

$$\nabla\cdot\mathbf{A} = \{\nabla\cdot\mathbf{A}\} + \hat{n}\cdot J(\mathbf{A})\delta_S \tag{A1}$$

$$\nabla b = \{\nabla b\} + \hat{n}\, J(b)\delta_S \tag{A2}$$

$$\nabla\times\mathbf{A} = \{\nabla\times\mathbf{A}\} + \hat{n}\times J(\mathbf{A})\delta_S \tag{A3}$$

**A** and b are functions that are discontinuous on the surface S. The operators $\nabla\cdot$, $\nabla$ and $\nabla\times$, on the left-hand side of Eqs. (A1)-(A3), are taken in the sense of distributions; the operators $\{\nabla\cdot\}$, $\{\nabla\}$ and $\{\nabla\times\}$ on the right-hand side are taken in the usual sense of functions; $\hat{n}$ is the unit vector normal to S oriented from metal to vacuum and J( ) is the jump of ( ) across S in the direction of $\hat{n}$.

## APPENDIX B: DERIVATION OF EQUATIONS (12) AND (13)

We recall[25] that a quantity $V\delta_S$ is the distribution of $D'^3$, a set of linear and continuous functionals on $D^3$, where $D^3$ is a space of functions $\phi(u,v,z)$ having the following properties: (i) $\phi(u,v,z)$ is indefinitely differentiable, and (ii) $\phi(u,v,z)$ has a bounded support. The distribution $V\delta_S$ is defined by

$$\langle \mathbf{V}\delta_S, \phi \rangle = \iint_S \mathbf{V}\phi dS \tag{B1}$$

Note that the right-hand side integral may be performed even though V is not defined outside S. An important consequence is that any derivative of $V\delta_S$ exists.[25] We recall, for instance, that

$$\langle \frac{\partial \mathbf{V}\delta_S}{\partial v}, \phi \rangle = -\langle \mathbf{V}\delta_S, \frac{\partial \phi}{\partial v} \rangle = -\iint_S \mathbf{V}\frac{\partial \phi}{\partial v} dS \tag{B2}$$

According to Eq. (B1), it is easy to show that

$$\langle \nabla \times (\mathbf{E}_S \delta_S), \phi \rangle = \iint_S (\mathbf{E}_S \times \nabla \phi) dS \tag{B3}$$

Use of Eqs. (B1) and (B3) leads to the following expression of Eq. (10)

$$\iint_S (\mathbf{E}_S \times \nabla \phi) dS + \iint_S [\hat{n} \times J(\mathbf{E}_V)] \phi dS = 0 \tag{B4}$$

The dot product of Eq. (B4) by vector $\hat{n}$ shows that

$$\iint_S (\nabla \phi) \cdot (\hat{n} \times \mathbf{E}_S) dS = 0 \tag{B5}$$

whatever the function $\phi \in D^3$. This leads to

$$\mathbf{E}_S = \hat{n} \cdot E_S \tag{B6}$$

Equation (B6) is Eq. (12): it shows that $E_S$ is perpendicular to S.

Use of Eq. (B6) shows that Eq. (B4) can be rewritten as

$$\iint_S \{(\hat{n} \times E_S \nabla_t \phi) + [\hat{n} \times J(\mathbf{E}_V)]\} \phi dS = 0 \tag{B7}$$

According to property (ii), we get for Eq. (B7)

$$\iint_S \hat{n} \times [-\nabla_t E_S + J(\mathbf{E}_V)] \phi dS = 0 \tag{B8}$$

Since Eq. (B8) must be satisfied for any function $\phi$,

$$J(E_{V,t}) = \nabla_t E_S \tag{B9}$$

Equation (B9) is Eq. (13).

## APPENDIX C: EXPRESSIONS FOR L

We wish to determine the limit L

$$L = \lim_{\eta \to 0} \epsilon^p(v) \frac{\partial \epsilon^q(v)}{\partial v} \tag{C1}$$

From Eq. (C1), we get

$$L = \begin{cases} \dfrac{q}{p+q} \lim\limits_{\eta \to 0} \dfrac{\partial \epsilon^{p+q}(v)}{\partial v} & \text{if } p+q \neq 0 \\[2ex] q \lim\limits_{\eta \to 0} \dfrac{\partial ln[\epsilon(v)]}{\partial v} & \text{if } p+q = 0 \end{cases} \tag{C2}$$

Let us pursue the case $p+q \neq 0$. It can be shown[25,26] that, in the sense of distributions,

$$\lim_{\eta \to 0} \frac{\partial \epsilon^{p+q}(v)}{\partial v} = \frac{\partial}{\partial v} \lim_{\eta \to 0} \epsilon^{p+q}(v) \tag{C3}$$

Thus,

$$L = \frac{q}{p+q} \frac{\partial}{\partial v} \lim_{\eta \to 0} \epsilon^{p+q}(v) \tag{C4}$$

In the limit $\eta \to 0$, $\epsilon(v)$ behaves like a step function

$$\epsilon(v) = \begin{cases} 1 & \text{if } v > 0 \\ \epsilon_2 & \text{if } v < 0 \end{cases} \tag{C5}$$

Thus, we get[25,26] the following expression for L

$$L = \frac{q}{p+q} J(\epsilon^{p+q}) \delta(v) \tag{C6}$$

where $J(\epsilon^{p+q})$ is the jump of $\epsilon^{p+q}$ across S. According to Eq. (C5),

$$J(\epsilon^{p+q}) = 1 - \epsilon_2^{p+q} \tag{C7}$$

It is easily shown that

$$\delta(v) = e_2 \delta_S \tag{C8}$$

where $e_2$ is the local unit length along v, for instance by intuitively bearing in mind that $\delta_S = \delta(n)$ with n being the curvilinear abscissa measured along the normal $\hat{n}$ to S and $dn = e_2 dv$. Thus, we obtain

$$L = \begin{cases} \dfrac{q}{p+q} e_2 (1 - \epsilon_2^{p+q}) \delta_S & \text{if } p + q \neq 0 \\ -q e_2 (ln \epsilon_2) \delta_S & \text{if } p + q = 0 \end{cases} \tag{C9}$$

## APPENDIX D: DERIVATION OF EQUATIONS (33) AND (34)

As already pointed out, Eqs. (11) and (15) are ambiguous. This difficulty is solved by considering that S is the limit, when $\eta \to 0$, of a transition region of thickness $\eta$. Thus, we have to rewrite the right-hand side of Eq. (15) in the transition region and look to its limit $L_3$ when $\eta \to 0$

$$L_3 = -\lim \frac{1}{\epsilon(2\omega, v)}(\hat{n} \cdot \mathbf{P}_{ub}^{NL}) \tag{D1}$$

In Eq. (D1), $\hat{n} \cdot \mathbf{P}_{ub}^{NL}$ is the normal component of the unbounded part of the nonlinear polarization in the transition region. We get

$$\hat{n} \cdot \mathbf{P}_{ub}^{NL} = \frac{\beta}{2}\frac{1}{e_2}D_V^2\left(\frac{3}{2}\frac{\partial[1/\epsilon^2(\omega, v)]}{\partial v} - \frac{\partial[1/\epsilon(\omega, v)]}{\partial v}\right) \tag{D2}$$

since[29]

$$\epsilon(2\omega, v) = \frac{1}{4}[3 + \epsilon(\omega, v)] \tag{D3}$$

Equation (D1) can be rewritten as

$$L_3 = -\frac{2}{e_2}\beta D_V^2 \lim_{\eta \to 0}\left[\frac{1}{3+\epsilon(\omega, v)}\left(\frac{3}{2}\frac{\partial[1/\epsilon^2(\omega, v)]}{\partial v} - \frac{\partial[1/\epsilon(\omega, v)]}{\partial v}\right)\right] \tag{D4}$$

Using the method of Appendix A together with Eqs. (29), we get, after a straightforward but tedious calculation

$$L_3 = -\frac{\beta}{3}E_V^{-2}\left\{\frac{4\epsilon_2^2(\omega)}{3}ln\left(\frac{\epsilon_2(\omega)}{\epsilon_2(2\omega)}\right) + [\epsilon_2(\omega)-1][3-\epsilon_2(\omega)]\right\}\delta_S \tag{D5}$$

Use of Eqs. (D5) and (15) leads to

$$E_S(2\omega)\delta_S = -\frac{\beta}{3\epsilon_0}E_V^{-2}(\omega)\left\{\frac{4\epsilon_2^2(\omega)}{3}ln\left(\frac{\epsilon_2(\omega)}{\epsilon_2(2\omega)}\right) + [\epsilon_2(\omega)-1][3-\epsilon_2(\omega)]\right\}\delta_S \tag{D6}$$

Equation (D6) is the expression for the surface electric field at $2\omega$ that is located at the entrance face S of the metal.

The ambiguity appearing in Eq. (15) has been removed. The price to be paid is that $E_S(2\omega)\delta_S$ can no longer be expressed in terms of the normal component of the surface nonlinear polarization [second term of Eq. (30)]. Nevertheless, according to Eq. (D1) it is this normal component that gives rise to $E_S(2\omega)\delta_S$.

Equation (D6) together with Eq. (13) gives the value of the discontinuity of the tangential component of the volume electric field at the second harmonic frequency

$$J(E_{V,t}(2\omega)) = -\frac{\beta}{3\epsilon_0}\left\{\frac{4\epsilon_2^2(\omega)}{3}ln\left(\frac{\epsilon_2(\omega)}{\epsilon_2(2\omega)}\right) + [\epsilon_2(\omega)-1][3-\epsilon_2(\omega)]\right\}\nabla_t E_V^{-2}(\omega) \tag{D7}$$

## REFERENCES


1. N. Bloembergen, *Nonlinear optics* (Benjamin, New York, 1965)
2. R. Shen, *The principles of nonlinear optics* (Wiley Interscience, New York, 1984)
3. J. L. Coutaz, M. Nevière, E. Pic and R. Reinisch, Phys. Rev. B **32**, 2227 (1985)
4. J. L. Coutaz, J. Opt. Soc. Am. B **4**, 105 (1987)
5. R. Reinisch and M. Nevière, "Nonlinear surface polariton interactions: surface enhanced nonlinear optical effects" in *Electromagnetic Surface Excitations*, R. F. Wallis and G. I. Stegeman eds. (Springer Verlag, New York, 1986), pp. 232-260
6. J. L. Coutaz, D. Maystre, M. Nevière and R. Reinisch, Proc. 14th Congress of the International Commission for Optics (Québec), pp. 149-150, ICO (1987)
7. M. Nevière, J. L. Coutaz, D. Maystre, E. Pic and R. Reinisch, CLEO'86-IQEC'86 (San Francisco, Sept. 1986), p. 68
8. D. Maystre, M. Nevière R. Reinisch and J. L. Coutaz, J. Opt. Soc. Am. B **5**, 338 (1988)
9. G. S. Agarwal and S. S. Jha, Phys. Rev. B **26**, 482 (1982)
10. K. Arya, Phys. Rev. B **29**, 4451 (1984)
11. G. A. Farias and A. A. Maradudin, Phys. Rev. B **30**, 3002 (1984)
12. R. Reinisch and M. Nevière, Phys. Rev. B **26**, 5987 (1982)
13. R. Reinisch and M. Nevière, Phys. Rev. B **28**, 1870 (1983)
14. M. Nevière, R. Reinisch and D. Maystre, Phys. Rev. B **32**, 3634 (1985)
15. M. Nevière, P. Vincent, D. Maystre, R. Reinisch and J. L. Coutaz, J. Opt. Soc. Am. B **5**, 330 (1988)
16. H. Akhouayri, M. Nevière, P. Vincent and R. Reinisch, Proc. 14th Congress of the International Commission for Optics (Québec), pp. 239-240 (1987)
17. R. Reinisch, M. Nevière, H. Akhouayri, J. L. Coutaz, D. Maystre and E. Pic, Opt. Engineering **27**, 961 (1988)
18. J. C. Quail and H. J. Simon, J. Opt. Soc. Am. B **5**, 325 (1988)
19. H. J. Simon, C. Huang, J. C. Quail and Z. Chen, Phys. Rev. B **38**, 7408 (1988)
20. Z. Chen and H. J. Simon, Opt. Lett. **13**, 1008 (1988)
21. H. J. Simon and Z. Chen, Phys. Rev. B **39**, (1989)
22. R. Reinisch and G. Vitrant, Phys. Rev. B **39**, 5775 (1989)
23. G. I. Stegeman, IEEE J. Quantum Electron. **QE-18**, 1610 (1982);
    G. I. Stegeman, C. T. Seaton, W. M. Hetherington III, A. D. Boardman and P. Egan, in *Electromagnetic Surface Excitations*, R. F. Wallis and G. I. Stegeman eds. (Springer-Verlag, New York, 1986), p. 261;
    A. D. Boardman and P. Egan, ibid. p. 301;
    R. Reinisch, P. Arlot, G. Vitrant and E. Pic, Appl. Phys. Lett. **47**, 1248 (1985)
24. G. Arfken, *Mathematical Methods for Physicists* (Academic Press, New York, 1970)
25. L. S. Schwartz, *Mathematical Methods for Physical Sciences* (Addison-Wesley, Reading, Mass., 1966)
26. R. Petit, "A tutorial introduction" in *Electromagnetic Theory of Gratings*, R. Petit ed. (Springer-Verlag, New York, 1980), pp. 1-50
27. N. Bloembergen, R. K. Chang, S. S. Jha and C. H. Lee, Phys. Rev. **174**, 813 (1968)
28. J. E. Sipe, V. C. Y. So, M. Fukui and G. I. Stegeman, Phys. Rev. B **21**, 4389 (1980)
29. J. E. Sipe and G. I. Stegeman, "Nonlinear optical response of metal surfaces" in *Surface polaritons, electromagnetic waves at surfaces and interfaces*, V. M. Agranovich and D. L. Mills eds. (North Holland, Amsterdam, 1982), pp. 661-701
30. D. Maystre, M. Nevière and R. Reinisch, Appl. Phys. A **39**, 115 (1986)
31. H. R. Jensen, K. Pedersen and D. Keller, Proceedings of the Int. Conf. on nonlinear optics NLO'88, Ireland 1988
32. D. Maystre, "Integral Methods" in *Electromagnetic theory of gratings*, R. Petit ed. (Springer-Verlag, New York, 1980), pp. 63-100
33. D. Maystre, "General study of grating anomalies from electromagnetic surface modes" in *Electromagnetic surface modes*, A. D. Boardman ed. (Wiley, New York, 1982), pp. 661-724
34. M. Nevière, P. Vincent and R. Petit, Rev. Optique **5**, 65 (1974);
    P. Vincent, "Differential methods" in *Electromagnetic theory of gratings*, R. Petit ed. (Springer-Verlag, New York, 1980), pp. 101-121
35. J. L. Coutaz, D. Maystre, M. Nevière and R. Reinisch, J. Appl. Phys. **62**, 1529 (1987)
36. The characteristics of silver corresponding to all the experiments performed with d = 1.53

$\mu$m periodicity gratings lead to: $\epsilon_2(\omega)$ = -43.6 + 2j; $\epsilon_2(2\omega)$ = -8.4 + 0.1j. The best fit is obtained for A = 2.9 and B = 0.4245.
37. J. V. Moloney and H. M. Gibbs, Phys. Rev. Lett. **48**, 1607 (1982)
38. J. V. Moloney, M. R. Belic and H. M. Gibbs, Opt. Commun. **41**, 379 (1982)
39. J. V. Moloney, M. Sargent III and H. M. Gibbs, Opt. Commun. **44**, 289 (1983)
40. J. V. Moloney, Opt. Acta **29**, 1503 (1982)
41. E. M. Wright, W. J. Firth and I. Galbraith, J. Opt. Soc. Am. B **2**, 383 (1985)
42. W. J. Firth, I. Galbraith and E. M. Wright, J. Opt. Soc. Am. B **2**, 1005 (1985)
43. N. N. Rozanov, Sov. Phys. JETP **53**, 47 (1981)
44. N. N. Rozanov and V. E. Semenov, Opt. Spectrosc. **48**, 59 (1980)
45. N. N. Rozanov and V. E. Semenov, Opt. Commun. **38**, 435 (1981)
46. N. N. Rozanov, V. E. Semenov and G. V. Khodova, Sov. J. Quantum Electron. **12**, 193 and 198 (1982)
47. D. Weaire, J. P. Kermode and V. M. Dwyer, Opt. Commun. **55**, 3 (1985)
48. D. Weaire and J. P. Kermode, J. Opt. Soc. Am. B **3**, 1706 (1986)
49. U. Olin and O. Sahlén, J. Opt. Soc Am. B **4**, 319 (1987)
50. U. Olin, J. Opt. Soc. Am. B **5**, 20 (1988)
51. G. M. Carter and Y. J. Chen, Appl. Phys. Lett. **42**, 643, (1983)
52. G. Vitrant, P. Arlot and R. Reinisch, SPIE **800**, 169 (1987)
53. G. I. Stegeman, G. Assanto, R. Zanoni, C. T. Seaton, E. Garmire, A. A. Maradudin, R. Reinisch and G. Vitrant, Appl. Phys. Lett. **52**, 869 (1988)
54. G. Vitrant, R. Reinisch, J. Cl. Paumier, G. Assanto and G. Stegeman, Nonlinear guided wave phenomena: physics and applications, Topical Meeting, p. 167, (Houston, USA) (1989)
55. H. M. Gibbs, *Controlling light with light* (Academic Press, New York, 1985)
56. J. W. Nibler and G. V. Knighten, in *Raman Spectroscopy of Gases and Liquids*, A. Weber ed. (Springer-Verlag, New York, 1979), p. 243
57. G. Stegeman, private communication
58. S. Maneuf, A. Barthelemy and Cl. Froehly, J. Optics (Paris) **17**, 139 (1986)
59. J. Marburger and F. Felber, Phys. Rev. A **17**, 335 (1978)
60. F. A. P. Tooley, S. D. Smith and C. T. Seaton, Appl. Phys. Lett. **43**, 807 (1983)
61. H. F. Harmuth, Journal of Mathematics and Physics **36**, 269 (1957)
62. M. Kubicek, Algorithm 502, ACM TOMS **2**, 98 (1976)
63. G. Vitrant, Thèse d'Etat, INP Grenoble (France) (1989)
64. A. Yariv, IEEE J. Quantum Electron. **QE-9**, 919 (1973)
65. H. Kogelnik, in *Integrated Optics*, T. Tamir ed. (Springer-Verlag, New York, 1979), Chap. 2
66. G. Vitrant and P. Arlot, J. Appl. Phys. **61**, 4744 (1987)
67. H. Chelli, A. Koster, N. Paraire, F. Pardo, H. Sauer, M. Carton and S. Laval, Rev. Phys. Appl. **22**, 1273 (1987)
68. W. Lukosz, P. Pirani and V. Briguet, in *Optical Bistability III*, H. M. Gibbs, P. Mandel, N. Peyghambarian and S. D. Smith eds. (Springer-Verlag, Berlin, 1986), p. 109
69. G. Assanto, B. Svensson, D. Kuchibhatla, U. J. Gibson, C. T. Seaton and G. I. Stegeman, Opt. Lett. **11**, 644 (1986)
70. P. Martinot, A. Koster and S. Laval, IEEE J. Quantum Electron. **QE-21**, 1140 (1985)
71. C. Liao and G. I. Stegeman, Appl. Phys. Lett. **44**, 164, (1984);
C. Liao, G. I. Stegeman, C. T. Seaton, R. L. Shoemaker, J. D. Valera and H. G. Winful, J. Opt. Soc. Am. A **2**, 590 (1985)
72. G. Vitrant, R. Reinisch, J. Cl. Paumier, G. Assanto and G. Stegeman, Opt. Lett. **14**, 898 (1989)

# OVERVIEW OF NONLINEAR INTEGRATED OPTICS

George I. Stegeman

Optical Sciences Center
University of Arizona
Tucson, AZ 85721, USA

## 1. INTRODUCTION

Nonlinear integrated optics is defined as the implementation of nonlinear optical interactions in planar and channel waveguides. This field has grown dramatically in the past few years for two reasons. First, for optical data storage and xerography, there is a pressing need for efficient doubling of semiconductor lasers. This has led to increased interest in second harmonic generation in waveguides, the goal being to produce a few milliwatts of deep blue light with 40 mW input. Second, a concept of ultrafast processing of optically coded information has emerged. It involves using an intensity-dependent refractive index to produce all-optical control of switching functions. In both cases, the goal is to be able to use semiconductor lasers.

There are essentially two properties of waveguides which make them attractive for low power nonlinear integrated optics. For planar and channel waveguides, the beam cross-sectional dimensions are limited to typically the wavelength of light in one and two dimensions respectively. In fact, the minimum possible beam cross-section is obtained in a waveguide. Since it is optimizing the intensity (power/area) which is important for efficient nonlinear optical interactions, waveguide geometries minimize the power needed. (Note, however, that in the plane of the surface, light in a planar waveguide diffracts in the usual way.) The second key property is that this high beam confinement is maintained for distances limited by scattering and/or absorption losses, typically millimeters to centimeters. This allows long interaction lengths, again improving the efficiency.

These attractive characteristics were recognized in the late 1960s and led to pioneering experiments on second harmonic generation in GaAs at $CO_2$ laser wavelengths.[1] Progress in waveguide second harmonic generation, and in second order parametric processes in general has grown steadily since then.[2-4] Interest in third order interactions has been more recent.[5] Optical bistability[6] led to an overall awareness in all-optical signal processing and indirectly to concepts of ultrafast all-optical switching in waveguides.[3,5,7] Third order nonlinear integrated optics started in the early 1980's and it is only in the last few years that all-optical devices have been reported.[7]

In this chapter we will present an overview of the field of nonlinear integrated optics. A similar overview can be found in Ref. 8 and here we will concentrate primarily (but not wholly) on the most recent developments. Discussed first will be brief summaries of linear integrated optics and nonlinear optics, the two disciplines which are combined in nonlinear integrated optics. Then in succession, second and third order nonlinear guided wave interactions and devices will be developed.

## 2. SUMMARY OF INTEGRATED AND NONLINEAR OPTICS

We start our discussion of nonlinear guided wave interactions with a brief review of the properties of guided waves.[9,10] In this section we assume that the fields are normal modes and

*Nonlinear Waves in Solid State Physics*
Edited by A.D. Boardman *et al.*, Plenum Press, New York, 1990

that their properties do not depend on the power of the guided wave. This is adequate for most current applications of nonlinear guided waves. The exceptions in which more general solutions to the nonlinear wave equation are required are discussed in another chapter of this book.

The guided wave fields are written as

$$\mathbf{E}^{(m)}(\mathbf{r},t) = \frac{1}{2}\mathbf{E}^{(m)}(x,y)a^{(m)}(z)e^{i(\omega t-\beta^{(m)}z)} + \text{c.c.} \tag{1}$$

where "m" defines the mode number and $\beta$ is the guided wave wavevector. a(z) is the amplitude coefficient with the detailed cross-sectional dependence of the guided wave given by $\mathbf{E}(x,y)$, normalized so that $|a(z)|^2$ is the guided wave power in watts. This requires that

$$\frac{\beta^{(m)}}{2k_0c\mu_0}\int_{-\infty}^{\infty}dx\int_{-\infty}^{\infty}dy\ \mathbf{E}^{(m)}(x,y)\cdot\mathbf{E}^{*(r)}(x,y) = \delta_{m,r}\ . \tag{2}$$

Note that it is primarily this normalization which makes the subsequent formulae appear different from the well-known plane wave cases. For a planar waveguide, it is useful to assume that the guided wave beam is very wide ($D >> \lambda$) and uniform along the y-axis so that the field distribution is independent of y, and the integral over y just produces D, the beam width. The term $\beta^{(m)}/k_0$ plays the role of a refractive index for propagation along the x-axis and is called "the effective index", $n_{eff}$. Its value is obtained from an eigenvalue equation or dispersion relation by satisfying the boundary conditions across each interface.

Essentially two types of waveguides will be of interest here, namely planar and channel waveguides depicted in Fig 1. The detailed form of the fields and the dispersion relations can become very complex when optically anisotropic media are involved. For our purposes, we will normally assume that the linear optical properties are isotropic. In this limit, the modes for a planar waveguide can be separated into a pure TE (E-field polarized along the y-axis) and a pure TM (H-field polarized along the y-axis, and E-field components along the x- and z-axes). A finite number of discrete modes occur for a given film thickness, and the field distributions associated with the first few $TE_m$ and $TM_m$ waves, shown in Fig. 1, exhibit oscillatory behaviour in the film, and decay exponentially with distance into the cladding and

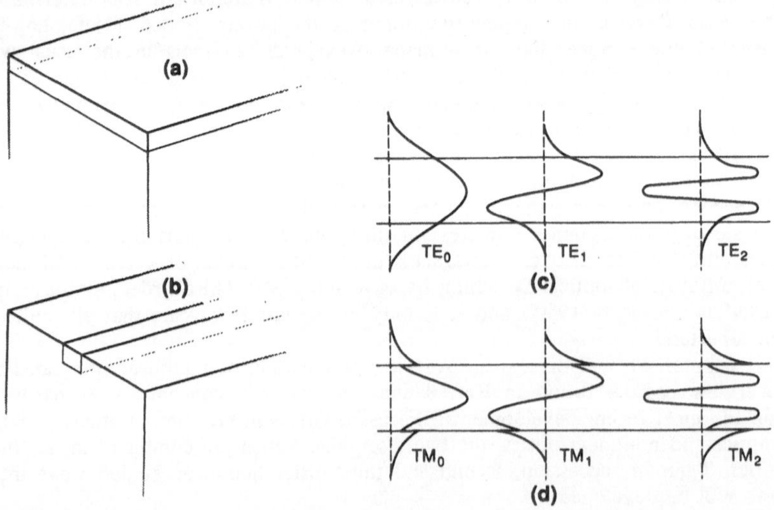


Fig. 1 Examples of guided wave geometries and field distributions: (a) planar or slab waveguide; (b) channel waveguides; (c) $TE_m$ field distributions and; (d) $TM_m$ fields.

substrate. Therefore nonlinear interactions can take place in any one of the three media. One of the unique features of thin film guided waves (relative to plane waves) is that the $\beta^{(m)}$ for each mode depend on film thickness. See for example Fig. 2. The TM waves have similar characteristics, but the dispersion relations are different leading to different variation with film thickness.

The situation is more complicated for channel waveguides. There are still two sets of orthogonal normal modes in the sense of Eq. (2). However, both modes contain all three electric field components [$E_x$, $E_y$ and $E_z$ (usually small)]. The mode with $E_y$ as the dominant field component is designated as $TE_{m,n}$ and the mode for which $E_x$ is dominant is called $TM_{m,n}$. Note that the modes are now designated by two integers because the field is confined in 2 dimensions, in contrast to the planar 1-dimensionally confined modes described by a single integer. The dispersion relations are very complicated and cannot be expressed in analytical form, requiring numerical techniques for evaluation.

Nonlinear optics consists of the mixing of one (with itself in some cases) or more fields to produce a nonlinear polarization source term, which in turn can radiate a new electromagnetic wave.[11,12] Historically the nonlinear polarization fields are usually expanded in products of the mixing fields. A nonlinear polarization field of frequency $\omega_s$ can be generated in each of the media in which the guided wave fields exist. It is usually written as

$$\mathbf{P}^{NL}(\mathbf{r},t) = \frac{1}{2}\mathbf{P}^{NL}(\mathbf{r},\omega_s)e^{i(\omega_s t-\boldsymbol{\beta}_p\cdot\mathbf{r})} + c.c. \tag{3a}$$

If we now assume up to three input guided wave fields of frequency $\omega_a$, $\omega_b$ and $\omega_c$ (of which two or three may be degenerate)

$$\mathbf{P}^{NL}(\mathbf{r},\omega_s) = \epsilon_0\chi^{(2)}(-\omega_s;\omega_a,\pm\omega_b):\mathbf{E}^{(m,a)}(x,y)\mathbf{E}^{(m',b)}(x,y)a^{(m,a)}(z)a^{(m',b)}(z)$$
$$+ \epsilon_0\chi^{(3)}(-\omega_s;\omega_a,\pm\omega_b,\pm\omega_c)\vdots\ \mathbf{E}^{(m,a)}(x,y)\mathbf{E}^{(m',b)}(x,y)\mathbf{E}^{(m'',c)}(x,y)a^{(m,a)}(z)a^{(m',b)}(z)a^{(m'',c)}(z) \tag{3b}$$

where $\chi^{(2)}$ and $\chi^{(3)}$ are the second and third order susceptibilities (material parameters), and a minus sign for a frequency corresponds to taking the complex conjugate of the appropriate field. Note that $\beta_p = \beta^{(m,a)} \pm \beta^{(m',b)} \pm \beta^{(m'',c)}$ is the wavevector associated with the nonlinear polarization source field, and it is *not necessarily* equal to $\beta^{(n,s)}$ which is the value appropriate to a propagating field of that frequency ($\omega_s$). The case $\beta_p = \beta^{(n,s)}$ corresponds to phase-matching, as will be discussed later. For the second order processes there are two possible input waves with frequencies $\omega_a$ and $\omega_b$ which produce polarization and signal fields at $\omega_s = \omega_a \pm \omega_b$. For third order processes, the nonlinear polarizations can occur at the frequencies $\omega_s = \omega_a \pm \omega_b \pm \omega_c$. Many of these interactions have now been demonstrated in nonlinear guided wave experiments.

It is noteworthy that this power law expansion is an assumption whose validity can be improved for any particular case by including progressively more terms. In some cases, it is more physically meaningful to calculate directly nonlinear index changes, for example in semiconductors, without using the standard nonlinear optics expansion.

It is important to note that the existence of the nonlinear polarization field does not ensure the generation of significant signal fields. With the exception of phenomena based on an intensity-dependent refractive index, the generation of the nonlinearly produced signal waves at frequency $\omega_s$ can be treated in the slowly varying amplitude approximation using well-known guided wave coupled mode theory.[9,10] As already explicitly assumed in Eq. (1), the amplitudes of the waves are allowed to vary slowly with propagation distance z, that is

$$\frac{d^2}{dz^2}a^{(n',s)}(z) << \beta^{(n',s)}\frac{d}{dz}a^{(n',s)}(z)\ ,$$

which leads to

$$\frac{d}{dz}a^{(n,s)}(z) = i\frac{\omega_s}{4}\int_{-\infty}^{\infty}dx\int_{-\infty}^{\infty}dy\ \mathbf{P}^{NL}(\mathbf{r},\omega_s)\cdot\mathbf{E}^{*(n,s)}(x,y)e^{-i(\beta_p-\beta^{(n,s)})z}\ , \tag{4}$$

where we have generalized the result to also include TM guided wave field interactions. Once the nonlinear polarization field has been found for a particular process, Eq. (4) allows the amplitude of the generated guided mode to be evaluated.

## 3. SECOND HARMONIC GENERATION

The simplest case is a single fundamental beam producing a collinear second harmonic generation wave via the material nonlinearity. The previous formalism simplifies to two waves present, the fundamental ($\omega_a = \omega = k_0 c$) and the second harmonic ($\omega_s = 2\omega$). In terms of the $d_{ijk}$ tensor (which is usually used instead of the $\chi_{ijk}$ tensor), the nonlinear polarization fields contain terms due to the mixing of the fundamental wave with itself, and with the second harmonic which is generated. The pertinent terms are:[11]

$$P_i^{NL}(\mathbf{r},2\omega) = \epsilon_0 d_{ijk}(2\omega;\omega,\omega)E_j^{(m,\omega)}(x,y)E_k^{(m,\omega)}(x,y)[a^{(m,\omega)}(z)]^2 \,, \tag{5a}$$

$$P_i^{NL}(\mathbf{r},\omega) = 2\epsilon_0 d_{ijk}(\omega;2\omega,-\omega)E_j^{(n,2\omega)}(x,y)E_k^{*(m,\omega)}(x,y)a^{(n,2\omega)}(z)a^{*(m,\omega)}(z) \,. \tag{5b}$$

Substituting into Eq. (4) gives

$$\frac{d}{dz}a^{(n,2\omega)}(z) = i\frac{\omega\epsilon_0}{2}\int_{-\infty}^{\infty}dx\int_{-\infty}^{\infty}dy\; d_{ijk}(2\omega;\omega,\omega)E_i^{*(n,2\omega)}(x,y)E_j^{(m,\omega)}(x,y)E_k^{(m,\omega)}(x,y) \times [a^{(m,\omega)}(z)]^2\, e^{-2i\Delta\beta z} \,, \tag{6a}$$

$$\frac{d}{dz}a^{(m,\omega)}(z) = i\frac{\omega\epsilon_0}{2}\int_{-\infty}^{\infty}dx\int_{-\infty}^{\infty}dy\; d_{ijk}(\omega;2\omega,-\omega)E_i^{*(m,\omega)}(x,y)E_j^{(n,2\omega)}(x,y)E_k^{*(m,\omega)}(x,y) \times a^{(n,2\omega)}(z)a^{*(m,\omega)}(z)\, e^{2i\Delta\beta z} \,, \tag{6b}$$

which are the two equations which must be solved with $\Delta\beta = \frac{1}{2}(2\beta^{(m,\omega)} - \beta^{(n,2\omega)})$. Based on the plane wave formula, one does not expect the $\epsilon_0$ to appear in the numerator - it is a consequence of the normalization used. At frequencies far from any material resonances, the usual assumption that the three indices ijk can be interchanged is used and $d_{ijk}(2\omega;\omega,\omega) = d_{ijk}(\omega;2\omega,-\omega)$. In this limit the integrals in Eqs. (6), called the *overlap integrals*, yield the same value, to within a phase factor.

For most practical cases the fundamental beam is essentially undepleted and the conversion to the harmonic power after an interaction distance L is small. Then

$$\left|a^{(n,2\omega)}(L)\right|^2 = (k_0 L)^2\left(\frac{c\epsilon_0}{2}\right)^2 |K|^2 \frac{\sin^2(\Delta\beta L)}{(\Delta\beta L)^2}\left|a^{(m,\omega)}(0)\right|^2 , \tag{7a}$$

$$K = \int_{\infty}^{\infty}dx\int_{-\infty}^{\infty}dy\; d_{ijk}E_j^{(m,\omega)}(x,y)E_k^{(m,\omega)}(x,y)E_i^{*(n,2\omega)}(x,y) \,. \tag{7b}$$

Note that we have also assumed that there is no loss due to guided wave attenuation. The $\sin^2\phi/\phi^2$ term describes the effect of phase-mismatch, that is $\Delta\beta L \neq 0$. Efficient conversion can only be obtained when the phase-mismatch $\Delta\beta L < \pi/4$. The optimum occurs for $\Delta\beta = 0$, the phase-matched case.

From Eqs. (7), it is clear that there are a number of key factors which govern efficient conversion. They are (1) phase-matching, (2) optimization of the overlap integral, (3) large nonlinear coefficients consistent with phase-matching and (4) low waveguide losses (large L). These factors will be discussed in succeeding paragraphs.

Phase-matching is primarily a question of finding a geometry for which $\beta^{(m,2\omega)} = \beta^{(n,\omega)}$. The difficulty with phase-matching is shown schematically for a slab waveguide in Fig. 2.

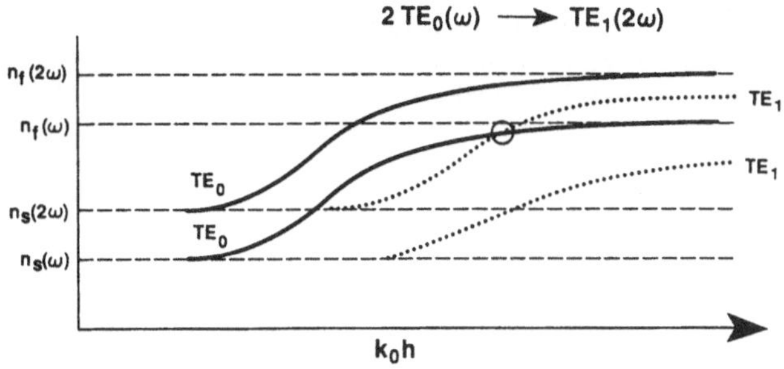


Fig. 2 The guided mode dispersion curves used to determine phase-matching possibilities (intersection of the fundamental and harmonic dispersion curves).

Plotted there is the effective index $N = \beta/k_0$ for both fundamental and harmonic waves of both polarizations. The problem is to find an intersection between the dispersion curves for a fundamental and harmonic guided wave field. In the absence of material birefringence, such crossings effectively only occur for $\beta^{(m,\omega)} = \beta^{(n,2\omega)}$ when $n > m$. That is, the mode numbers are different for the fundamental and harmonic waves, indicating that the respective field distributions have a different number of zeros inside the film. However, given thick enough films (and enough modes), phase-matching conditions can always be found.

The tolerances on waveguide dimensions can also be understood from Fig. 2. The propagation wavevector varies with the waveguide dimensions through the dispersion relations. The ideal situation in terms of thickness tolerances and fluctuations is for the relative angle between the crossing dispersion relations to be small and for the curves to be parallel to the thickness axis. In terms of index uniformity, again a small crossing angle is advantageous with the curves approximately parallel to the index axis at the phase-matching condition. The key point is the small crossing angle, which implies small dispersion in index with wavelength as being desireable. Typically tolerances of the order of 5 nm on waveguide dimensions are sufficient.[13] Uniformity of refractive index and waveguide dimensions are both problems in technology and are not easily solved.

The overlap integral (K) can reduce dramatically the efficiency of a doubling phase matching configuration if the mode numbers of the interacting fields are different. This is clear from the sketch in Fig. 3. K is proportional to the integral over the three field distributions involved in the mixing interaction. For example, as shown in Fig. 3, interference effects occur when $m \neq n$, reducing K. This negates an apparent advantage in using guided waves, namely the availability of a range of $\beta$'s at a given frequency for phase-matching. A solution to this problem is also shown in Fig. 3. By making the guiding film a layered combination of linear and nonlinear films, the interference effects can be eliminated with a reduction (in Fig. 3) in K of only a factor of 2.[14]

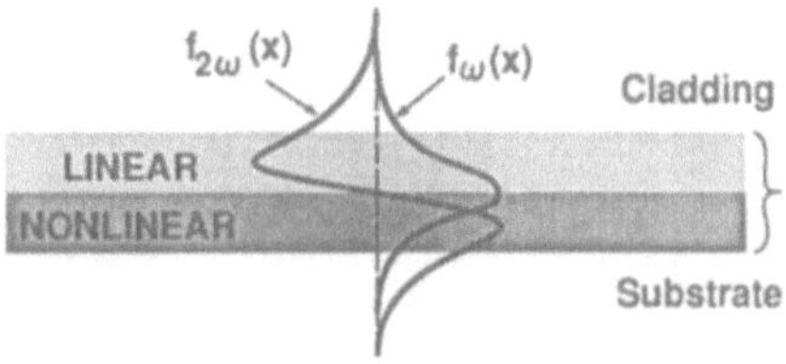


Fig. 3 Guided wave field overlap for a two film waveguide for harmonic generation for $TE_0 + TE_0 \rightarrow TM_1$. If the film consists totally of the nonlinear film, interference effects occur in the integration over the depth coordinate and K is small. Note that there are no interference effects in the second harmonic generation signal if only half of the effective guiding layer is nonlinear.

Material birefringence can also be used to obtain phase-matching with orthogonally polarized modal fields of the same mode number.[2] An example is shown in Fig. 4 where phase-matching between the $TE_0$ and $TM_0$ modes is possible. (There material dispersion in the film material only was assumed for simplicity - inclusion of dispersion in the substrate index does not change the general conclusion.) Small birefringence, of the order of the material dispersion with wavelength, is desirable. The problem with this aproach is that it requires off-diagonal tensor elements which unfortunately tend to be smaller than the diagonal elements.

Although a large variety of nonlinear materials have been used in prototype harmonic doublers,[3] only waveguides in $LiNbO_3$ have been of potentially usable quality and we will limit our discussions to this material system.[3,15-22] A list of previous waveguide doublers can be found in Ref. 3 and the improvements since then have been primarily limited to $LiNbO_3$ devices.

Table 1 is a compilation of the best results obtained to date in $LiNbO_3$ channel waveguides. (Also included is a recent very promising report of doubling in KTP.[23]) There are two figures of merit given,

$$\eta' = 100 \frac{|a^{(2\omega)}(L)|^2}{|a^{(\omega)}(L)|^2} \,\% \, W^{-1} \,, \tag{8a}$$

$$\eta = 100 \frac{|a^{(2\omega)}(L)|^2}{|a^{(\omega)}(L)|^2 L^2} \,\% \, W^{-1}cm^{-2} \,. \tag{8b}$$

The first gives the conversion efficiency for a given device, while the second quantifies the intrinsic conversion efficiency independent of the device length. For the Taniuchi result, the second harmonic appears as Čerenkov radiation and hence $\eta$ scales linearly with the length, not quadratically, i.e. the units are % $W^{-1}cm^{-1}$.[19] The best results correspond to $\eta \simeq 40\%$ $W^{-1}cm^{-2}$. This implies that for typical 40 mW semiconductor laser powers, 1.6 mW of blue can be generated.

The results of Uesugi and Sohler were obtained in channel waveguides using the $d_{13}$ coefficient, and material birefringence for phase-matching between the $TE_{00}$ and $TM_{00}$ modes.[15-17] For this case, $\eta \simeq 4\%$ $W^{-1}cm^{-2}$ is the best value obtained to date. Sohler *et al* used coated ends on their sample to form a resonator for the fundamental beam, raising $\eta$ to $\simeq 12\%$ $W^{-1}cm^{-2}$ by increasing the fields in the waveguide at the fundamental frequency.[16] By also making a resonator at the harmonic frequency, an $\eta$ of 45% $W^{-1}cm^{-2}$ was achieved.[17] This result was less than the theoretically predicted value and further improvements appear possible.

The most promising recent developments involve using the $d_{33} = 7d_{13}$ coefficient in $LiNbO_3$ which is not phase-matchable in the usual ways: the geometries are shown in Fig. 5. Taniuchi and co-workers allowed the second harmonic to leave the waveguide region in the

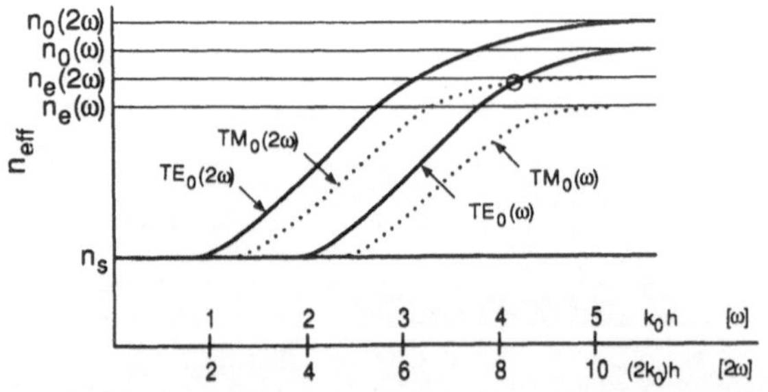


Fig. 4 The guided mode dispersion curves for a birefringent film and an optically isotropic substrate. Both the fundamental and harmonic curves are shown. The TE mode utilizes the ordinary refractive index and TM primarily the extraordinary index. Note the change in horizontal axis needed to plot both the fundamental and harmonic dispersion curves. Phase-matching of the $TE_0(\omega)$ to the $TM_0(2\omega)$ is obtained at the intersection of the appropriate fundamental and harmonic curves.

Table 1. Summary of second harmonic generation results for lithium niobate waveguides. $\ell_{eff}$ is the phase-matching distance, $P(\omega)$ the fundamental power, and $\eta$ and $\eta'$ are the figures of merit defined in the text. Also included is a recent value for KTP

| Reference | Phase-Matching Conditions | $\ell_{eff}$ cm | $P(\omega)$ mW | $\eta'$ $\%W^{-1}$ | $\eta$ $\%W^{-1}cm^{-2}$ |
|---|---|---|---|---|---|
| Uesugi (1978) | Birefringence | 2 | 60 | 12 | 4 |
| Sohler (1978) | Birefringence | 4.7 | 0.5 | 28 | 1.3 |
| Sohler (1983) | Birefringence† | 2.4 | 1.5 | 100 | 12 |
| Sohler (1988) | Birefringence† | 4.7 | 0.5 | 40 | 1.8 |
| Sohler (1988) | Birefringence‡ | 4.7 | 0.1 | 1000 | 45 |
| Taniuchi (1986) | Čerenkov | 0.6 | 40 | 12 | 42 |
| Fejer (1989) | Quasi-phase-matching | 0.1 | 14.7 | 0.37 | 37 |
| Burlein (1988) | KTP | | | | 4 |

form of Čerenkov radiation.[19] When channel waveguides are used this leads to the harmonic light appearing in an arc, not all of which can be focused down to the diffraction limit. The approach in the Lieu *et al* and Webjorn *et al* cases was different and relied on quasi-phase-matching.[20,21] The poling direction in the surface region of a z-cut lithium niobate surface can be changed by Ti in-diffusion or $LiO_2$ out-diffusion. By carrying out these processes through a mask, a periodic structure in $d_{33}$ is formed with the coefficient reversing direction every half period. For masks of the right periodicity, this allows phase matching via $\beta(2\omega) = 2\beta(\omega) + \kappa$ where $\kappa$ is the grating wavevector. The theoretical value for $\eta$ is > 300% $W^{-1}cm^{-2}$, making this approach very promising. Both the Čerenkov and quasi-phase-matching techniques allow any wavelength to be doubled, as long as both the fundamental and harmonic fall within the material's transparency band.

It is useful to recognize the reasons for the successes with $LiNbO_3$ because many factors have played key roles. First of all, the second harmonic generation developments were able to utilize waveguide technologies developed primarily for electro-optical applications. The material is birefringent, and the birefringence is of the same order of magnitude as the dispersion with wavelength. Furthermore, the material is originally single crystal and the waveguide formation does not change the waveguide crystallinity or lead to high waveguiding losses. This is not to say that other materials should have the same properties to be useful, just that a material with a good nonlinearity is not enough.

There are a number of potentially useful materials systems for doubling in waveguide formats. They are listed in Table 2 for comparison with $LiNbO_3$ based on its $d_{13}$ coefficient. The figure of merit is the one usually used for bulk materials and is adequate for this purpose. It is clear what a large improvement both the Čerenkov and quasi-phase-matching schemes allow. Both MNA and NPP have been crystallized in bulk form, but not yet in acceptable waveguide form.[24]

The last category corresponds to poled molecules in polymer films, a very recent development.[25-27] This particular approach is being driven by the electro-optic properties of such films, and provides a good opportunity for rapid technological evolution. The molecules required for preparing such films have both a large dipole moment and a large second order hyperpolarizability. A polymer host, charged with the molecules, is heated above its glass

† Resonator at fundamental wavelength
‡ Resonator at both fundamental and harmonic wavelengths

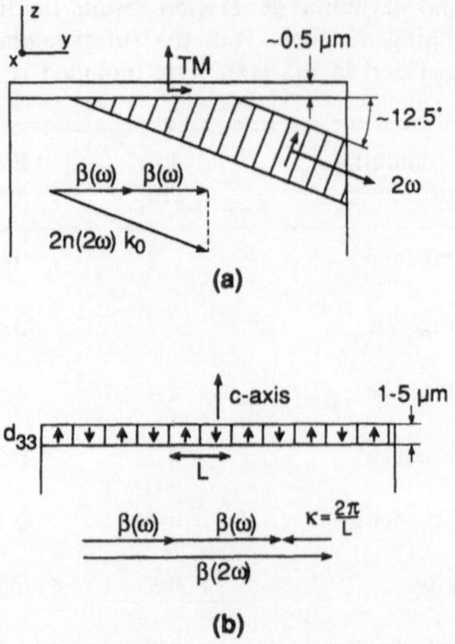


Fig. 5 Two recent geometries for producing efficient second harmonic generation using the $d_{33}$ nonlinear coefficient in lithium niobate waveguides. (a) TM modes generate harmonic Čerenkov radiation at an angle of 12° from the surface. (b) Periodic reversal of the sign of the $d_{33}$ nonlinearity leads to phase-matched second harmonic generation of the TM mode.

Table 2. Second harmonic generation figures of merit relative to $LiNbO_3$ ($d_{13}$ coefficient) for materials with potential for waveguide applications

| Material | $d_{eff}$ (esu) | $d^2/n^3$ |
|---|---|---|
| $LiNbO_3$ | $1.2\times10^{-8}$ | 1 |
| $LiNbO_3$ ($d_{33}$) | $8.5\times10^{-8}$ | 50 |
| KTP | $\simeq 10^{-8}$ | 1.5 |
| MNA | $7\times10^{-8}$ | 75 |
| NPP | $2\times10^{-7}$ | 600 |
| (PS)O-NPP | $2\times10^{-8}$ | |
| DCV/PMMA | $7\times10^{-8}$ | 85 |
| HCC#1232 | $2.5\times10^{-7}$ | |

MNA - metanitroaniline
NPP - N-(4-nitrophenyl)-L-prolinol
(PS)O-NPP - chromophore functionalized polymer
DCV/PMMA - dicyanovinyl azo dye in PMMA

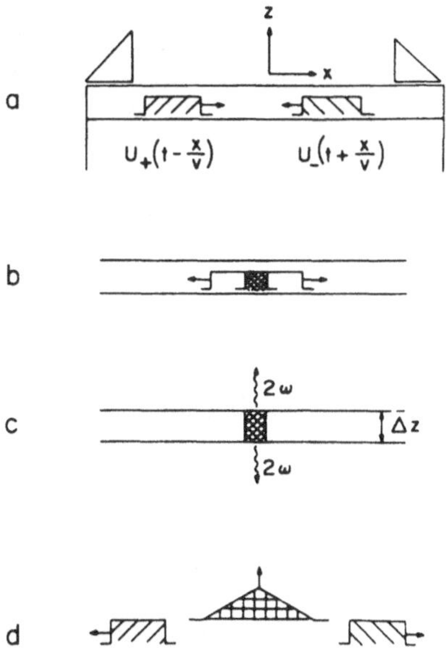


Fig. 6 The nonlinear mixing of two counter-propagating guided modes. (a) The pulse geometry; (b) partial overlap of the pulses; (c) radiation at the harmonic frequency; (d) shape of the second harmonic signal radiated normal to the surface.

transition temperature and a field is applied to orientate the molecules via their dipole moment. The field is maintained through the cooling phase, locking in the molecular orientation. Initially it was found that a fraction of the molecular orientation decayed with time, but recently chemical bonding has been used to "fix" the molecular orientation. Promising results have already been obtained, including quasi-phase-matching via periodic poling, and there are high expectations for this approach.

Second harmonic generation can also occur via the mixing of two counterpropagating guided waves in a slab waveguide.[28-30] The geometry is shown in Fig. 6. For input $TE_0$ waves, the pertinent induced nonlinear polarization field has the form

$$P_y^{NL}(\mathbf{r},t) = \frac{1}{2}2\epsilon_0 d_{222}(2\omega;\omega,\omega)\, E_y^{2(0)}(x)a^{2(0)}(z)e^{2i\omega t} + \text{c.c.} \tag{9}$$

This polarization field has no wavevector and can only radiate normal to the waveguide surface at the harmonic frequency. Because the interaction length L is effectively the waveguide thickness, the conversion efficiency into the second harmonic is too small to be useful as a doubler. The radiated field has been observed for pulses of 1.06 $\mu$m radiation prism coupled into opposite ends of a Ti-indiffused $LiNbO_3$ waveguide,[28,29] and more recently for mode-locked pulses in a GaAs waveguide.[30] The experimental and theoretical results were in good agreement.

## 4. PARAMETRIC PROCESSES

Other second order nonlinear optical processes have also been implemented in waveguides, namely sum and difference frequency generation, parametric amplification and parametric oscillation.[3] To date, the only waveguides of sufficient quality for producing parametric amplification and oscillation have been Ti:in-diffused $LiNbO_3$ waveguides.

### 4.1 Sum frequency generation

In this case there are two separate incident guided waves at the frequencies $\omega_a$ and $\omega_b$ ($\omega_a \geq \omega_b$) which mix via the second order susceptibility to produce nonlinear polarization fields at the frequency $\omega_s = \omega_a + \omega_b$. The sum frequency polarization field is

$$P_i^{NL}(\mathbf{r},\omega_a+\omega_b) = 2\epsilon_0 d_{ijk}(\omega_a+\omega_b;\omega_a,\omega_b)E_j^{(m,a)}(x,y)E_k^{(m',b)}(x,y)a^{(m,a)}(z)a^{(m',b)}(z) , \quad (10)$$

Since the reported conversion efficiencies in waveguides have been small, only the weak generation case is of current interest. For sum frequency generation with phase-matching over a distance L

$$\left|a^{(n,s)}(L)\right|^2 = \frac{[(\omega_a+\omega_b)L\epsilon_0]^2}{4}|K|^2|a^{(m,a)}(0)|^2|a^{(m',b)}(0)|^2 , \quad (11a)$$

$$K = \int_{-\infty}^{\infty} dx \int_{-\infty}^{\infty} dy \; d_{ijk}(\omega_s;\omega_a,\omega_b)E_j^{(m,a)}(x,y)E_k^{(m',b)}(x,y)E_i^{*(n,s)}(x,y) . \quad (11b)$$

The same comments about phase-matching, the overlap integral etc. made for the second harmonic generation case are also valid here.

Two experiments have been reported to date, both in $LiNbO_3$ waveguides. Sum frequency generation has been reported by Uesugi *et al*[31] in in-diffused $LiNbO_3$ channel waveguides and by Reutov and Tarashchenko[32] in thin $LiNbO_3$ platelets. Uesugi *et al* used either a HeNe (1.19 $\mu$m) or a Nd:YAG laser in conjunction with an optical parametric oscillator operating in the near infrared to produce tunable sum frequency radiation from 0.532 $\mu$m to 0.545 $\mu$m.

### 4.2 Difference frequency generation, parametric amplification and oscillation

Difference frequency generation, usually in the form of parametric amplification or oscillation, is a very useful technique for producing tunable radiation in the infrared. In this case, a different nomenclature has evolved. The signal, "pump" and "idler" waves have the frequencies $\omega_s$ ($\omega_c$), $\omega_p$ ($\omega_a$) and $\omega_i$ ($\omega_b$) respectively. The pertinent nonlinear polarizations are

$$P_i^{NL}(\mathbf{r},\omega_c) = 2\epsilon_0 d_{ijk}(\omega_c;\omega_a,-\omega_b)E_j^{(m,a)}(x,y)E_k^{*(m',b)}(x,y)a^{(m,a)}(z)a^{*(m',b)}(z) , \quad (12a)$$

$$P_i^{NL}(\mathbf{r},\omega_a) = 2\epsilon_0 d_{ijk}(\omega_a;\omega_c,\omega_b)E_j^{(n,c)}(x,y)E_k^{(m',b)}(x,y)a^{(n,c)}(z)a^{(m',b)}(z) , \quad (12b)$$

$$P_i^{NL}(\mathbf{r},\omega_b) = 2\epsilon_0 d_{ijk}(\omega_b;\omega_a,-\omega_c)E_j^{(m,a)}(x,y)E_k^{*(n,c)}(x,y)a^{*(n,c)}(z)a^{(m,a)}(z) . \quad (12c)$$

In parametric amplification, a weak input signal at $\omega_s$ is amplified by the phase-matched conversion of the pump wave into both a signal and an idler wave, that is $\omega_p = \omega_s + \omega_i$ and $\beta^{(m,p)} = \beta^{(n,s)} + \beta^{(m',i)}$. For parametric oscillation, only the pump beam is incident and the signal and idler build up from noise in the resonator cavity. In this case, the frequencies which are generated are fixed by the wavevector matching and frequency sum conditions.

The appropriate nonlinear polarization fields are given by Eqs. (12). Substituting into Eq. (4), assuming phase-matching and no significant depletion of the pump beam, the signal beam is given by

$$a^{(n,c)}(z) = a^{(n,c)}(0)\cosh\left(\frac{z}{\ell_{opa}}\right) , \quad (13a)$$

$$\frac{1}{\ell_{opa}} = \sqrt{(\omega_b\omega_c)}\,\frac{\epsilon_0}{2}|K|\,|a^{(m,a)}(0)| . \quad (13b)$$

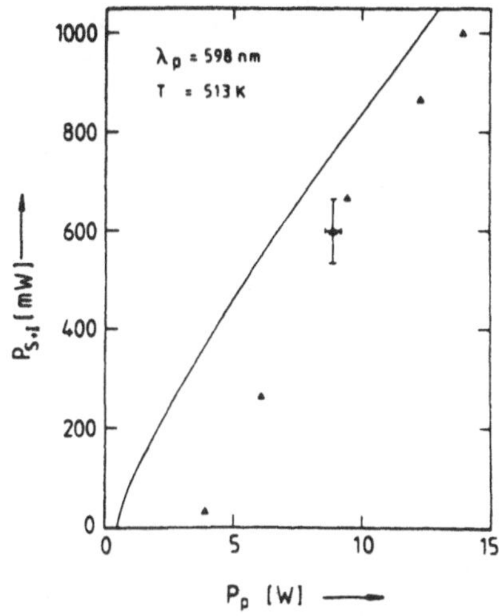


Fig. 7 Sum of the signal and idler powers versus pump wavelength for optical parametric oscillation in a Ti:in-diffused $LiNbO_3$ channel waveguide. The solid line is the theoretical curve.[20]

$$K = \int_{-\infty}^{\infty} dx \int_{-\infty}^{\infty} dy\ d_{ijk} E_i^{*(n,c)}(x,y) E_j^{(m,a)}(x,y) E_k^{*(m',b)}(x,y)\ . \qquad (13c)$$

In the limit $z >> \ell_{opa}$, the hyperbolic function degenerates into an exponential with gain coefficient $1/\ell_{opa}$. It is this gain coefficient which determines the quality of the amplification.

Optical parametric amplification and the corresponding difference frequency generation have been studied experimentally in $LiNbO_3$ waveguides by two groups.[33,34] Uesugi[33] reported the generation of a difference frequency signal when an idler wave from a cw Nd:YAG ($\lambda \simeq 1.318$ $\mu$m) laser and a dye laser ($\lambda \simeq 0.58$ $\mu$m), which served as the pump, were mixed in a Ti:in-diffused $LiNbO_3$ channel waveguide. Using the mode combination $TE_0 + TM_0 \rightarrow TM_0$ with pump and idler powers of 70 mW and 1.7 mW respectively, a generation efficiency of 0.014% was obtained at $\lambda \simeq 1.035$ $\mu$m. By tuning the temperature and adjusting the dye laser wavelength appropriately, the phase matching condition was tuned and the resulting difference frequency signal was varied from 1.035 $\mu$m to 1.19 $\mu$m.

A much broader tuning range, 2.5 $\rightarrow$ 3.0 $\mu$m was obtained in a recent experiment by Herrmann and Sohler.[34] They used channel waveguides, 20 and 30 $\mu$m wide and 3.9 cm long to mix a 3.39 $\mu$m He-Ne laser with the output from a tunable KCl color center laser. For 50 mW and 100 $\mu$W respectively input powers, an output of 10 nW was obtained.

In other experiments, Sohler and Suche managed to reduce their waveguide losses sufficiently so that they were able to measure parametric signal gain. In the initial experiments,[16] the signal source was a cw He-Ne laser ($\lambda = 1.15$ $\mu$m) and the pump was derived from a tunable pulsed dye-laser operating in the visible ($\lambda \simeq 0.65$ $\mu$m). The best single-pass parametric gain observed was 1.75 obtained with a peak pump power of about 200 watts and a $TE_{10}$ (pump) $\rightarrow$ $TM_{00}$ (signal, idler) coupling. Much improved performance was obtained in a later experiment[16] with a 20 $\mu$m wide channel, 48 mm long. With 150 watts of peak power, a gain of 16 dB, which corresponds to a signal amplification of 43 was achieved.

The quality of the 10 $\mu$m wide channel waveguides made by Sohler and Suche was so good that parametric oscillation was obtained when the ends of the channels were coated to a reflectivity of 0.96 at the signal wavelength.[16,35] At high enough pump powers, the signal and idler gain surpasses the resonator losses and simultaneous oscillation at the signal and idler frequencies will start from noise. This will of course only happen under phase-matching conditions. The results for the signal and idler power generated are shown in Fig. 7 as a function of pump power (at $\lambda = 0.598$ $\mu$m). For a pump power of 14 watts, 1 watt is converted into the sum of signal and idler power. Varying the input pump wavelength results in tuning of the signal idler wavelengths as phase-matching is maintained. This is illustrated in Fig. 8. By also varying temperature, the signal and idler wavelengths can be tuned from 0.587 $\mu$m to 0.616 $\mu$m.

## 5. DEGENERATE FOUR WAVE MIXING

We now turn our attention to nonlinear phenomena which depend on the third order susceptibility [$\chi^{(3)}$ in Eq. (3)]. Here there are many possibilities for the mixing of three input beams to produce one output beam. In fact, however, only a few of all of the possible combinations have been of interest in guided wave geometries (or plane wave interactions as well). These are degenerate four-wave mixing, Coherent Raman Scattering and an intensity-dependent refractive index.

The classic degenerate four wave mixing interaction involves three input beams, and one output beam, all at the same frequency.[36] In bulk media, as long as there are two pump beams (1,2) aligned exactly for contra-directional propagation, the process is automatically phase-matched for any direction of incidence of the third input ("signal") beam (3), and the conjugate output beam (4) is exactly contra-directional with the signal beam. This same condition is required for the guided wave version. This wavevector conservation condition for guided waves is given by $\beta^{(m,1)} + \beta^{(m',2)} + \beta^{(m'',3)} + \beta^{(n,4)} = 0$ and requires that $m = m'$ [$\beta^{(m,1)} = -\beta^{(m,2)}$] and $m'' = n$ [$\beta^{(n,4)} = -\beta^{(n,3)}$]. Note that although this process requires that beams 1 and 2 have the same mode order m, and that beam 4 is radiated into the same mode n as beam 3, the mode numbers m and n are not necessarily equal.[37] (An efficiency reason will be given later for having m = n.)

The nonlinear polarization field corresponding to degenerate four wave mixing is given by Eqs. (3). For an isotropic material with equal mode indices for all of the planar guided wave input beams, and for frequencies far from any material resonances so that $\chi^{(3)}_{1111}(\omega;\omega,-\omega,\omega) = 3\chi^{(3)}_{1122}(\omega;\omega,-\omega,\omega) = 3\chi^{(3)}_{1221}(\omega;\omega,-\omega,\omega) = 3\chi^{(3)}_{eff}$,[37]

$$P^{NL}_{\gamma i}(\mathbf{r},\omega) = 2c\epsilon_0^2 n_\gamma^2 n_{2\gamma}\left[\frac{2}{3}E^{(n,3)}_{\gamma i}(x)E^{(n,1)}_{\gamma j}(x)E^{*(n,2)}_{\gamma j}(x) + \frac{1}{3}E^{(n,3)}_{\gamma j}(x)E^{(n,2)}_{\gamma j}(x)E^{*(n,1)}_{\gamma i}(x)\right], \quad (14)$$

in each medium (labelled $\gamma$) where $n_{2\gamma} \neq 0$. Here we have defined $3\chi^{(3)}_{eff} = n_{2\gamma} n_\gamma^2 \epsilon_0 c^2$, the justification for which will be discussed later. Assuming that TE beams 1 and 2 propagate along z′ which is oriented at an angle $\theta$ to the z-axis along which TE beams 3 and 4 are travelling,

$$\mathbf{E}^{(n,1 \text{ or } 2)}(x) = E^{(n,1)}(x)[\sin\theta, \cos\theta, 0] \qquad \mathbf{E}^{(n,3 \text{ or } 4)}(x) = E^{(n,3)}(x)[0,1,0] . \quad (15)$$

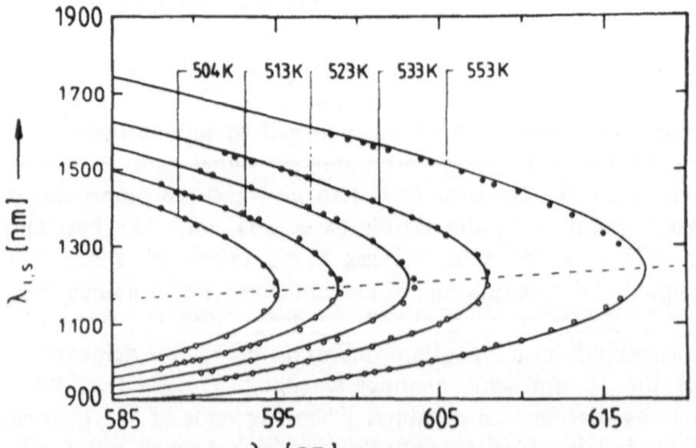


Fig. 8 The signal and idler wavelengths versus pump wavelength for optical parametric oscillation in a Ti:in-diffused $LiNbO_3$ channel waveguide at a series of temperatures.[20]

In the limit of no attenuation or depletion for the input beams, and assuming that the beams 1, 2 and 3 are introduced into the interaction region at $z' = 0$, $z' = L'$ and $z = L$ respectively, substituting into Eq. (7) and integrating the signal beam from $0 \rightarrow L$ gives

$$a^{(n,4)}(L) = (k_0L)Ka^{(n,1)}(0)a^{(n,2)}(L')a^{(n,3)}(0) , \tag{16}$$

$$K = i\left(\frac{c^2\epsilon_0^2}{6}\right)(2+\cos^2\theta)\int dy\int_{-\infty}^{\infty} dx\; n_\gamma^2(x)n_{2\gamma}(x)\left|E_\gamma^{(n,1)}(x)\right|^4 . \tag{17}$$

The integral over the y-coordinate has been left unspecified because it determines the overlap between the beams, which in turn depends on $\theta$. If in actual fact $m \neq n$, then the field product in the overlap integral, Eq. (17), would have been of the form $|E_\gamma^{(m,1)}(x)|^2|E_\gamma^{(n,3)}(x)|^2$. In that case, the overlap integral K would have been reduced from its value when $m = n$ because the two different modes would have had their maxima and minima at different points across the film. This is essentially the reason why $m = n$ is preferable. In the most general case, it is necessary to take incident beam depletion into account which leads to a series of coupled mode equations between the amplitudes of the various beams.

A number of experiments on degenerate four-wave mixing in planar waveguides have been reported,[38-40] none yet in channels. The initial work utilized liquid $CS_2$ as the nonlinear cladding medium.[38] More recent work has used small inclusions of a nonlinear material in a glassy host, with the nonlinearity in effectively the guiding (high index) region of the waveguides. Such experiments are difficult because at least two, and sometimes three coupling prisms are required for the four beams needed in the interaction, leading to difficult alignment problems. For $\theta = 90°$ which minimizes the stray scattering, maximum reflectivities of 1% and 10% were obtained in semiconductor doped glass waveguides with 10 ns and 80 ps pulses respectively, see Fig. 9.[39] The most recent work has utilized gelatins with colloidal suspensions of gold and silver particles, and again reflectivities in the few percent were obtained.[40]

## 6. COHERENT ANTI-STOKES RAMAN SCATTERING (CARS)

This interaction involves the mixing of two guided wave beams of different frequencies ($\omega_a \geq \omega_b$ and guided wave wavevectors $\beta^{(m,a)}$ and $\beta^{(m',b)}$) in an almost codirectional geometry.[11,41] The signal emerges at the frequency $2\omega_a - \omega_b$ (Coherent Anti-Stokes Raman Scattering) or $2\omega_b - \omega_a$ (Coherent Stokes Raman Scattering, CSRS). The two laser beams illuminate a sample containing molecules characterized by a vibrational frequency $\Omega_r$. When the difference between the two laser frequencies is tuned to approximately the resonance frequency, that is $\omega_a - \omega_b \simeq \Omega_r$, the molecules are excited to vibrate at this difference frequency. This results in a coherent (in space and time) excitation of the vibrational normal mode. Light incident onto the excited molecules is now scattered via the usual Raman process. In nonlinear optics language, this corresponds to the creation of nonlinear polarization fields oscillating at the frequencies $\omega_s = 2\omega_a - \omega_b$ and $\omega_s = 2\omega_b - \omega_a$. For coherent anti-Stokes Raman scattering the spatial periodicity is given by $\beta_p = 2\beta^{(m,a)} - \beta^{(m',b)}$. The angle between the two intersecting beams is usually small, typically a few degrees, and can be considered zero for the purpose of calculating the signal output. The corresponding coherent anti-Stokes Raman scattering nonlinear polarization field is

$$P_i^{NL}(\mathbf{r},\omega_s) = \epsilon_0\left[\chi^{(3)}_{b\;ijk\ell}(\omega_s;\omega_a,-\omega_b,\omega_a) + \sum_r \chi^{(3)}_{r\;ijk\ell}(\omega_s;\omega_a,-\omega_b,\omega_a)\left(\frac{\Gamma_r/\pi}{\omega_a-\omega_b-\Omega_r-i\Gamma_r}\right)\right]$$
$$\times E^{2(m,a)}(x)E^{*(m',b)}(x)a^{2(m,a)}(z)a^{(m',b)}(z) , \tag{18}$$

The $\chi_b^{(3)}$ term corresponds to the background third order susceptibility term due to the electronic degrees of freedom etc. It can be considered to be independent of frequency over the resonance regions. The magnitude of the resonant contributions, $\chi_r$, is determined by factors such as density of the Raman active species, oscillator strength etc., and the larger the ratio $\chi_r/\chi_b$ the better.

The amplitude of the guided waves generated at the frequencies $2\omega_a - \omega_b$ and $2\omega_b - \omega_a$ is calculated from coupled mode theory via Eq. (4). Assuming an interaction distance L, a coherent anti-Stokes Raman scattering guided wave signal field (mode number n) which starts with zero amplitude at z = 0, and assuming no depletion of the incident beams

$$|a^{(n,s)}(L)|^2 = (k_s L)^2 W |K_{2a-b}|^2 \frac{\sin^2\phi}{\phi^2} |a^{2(m,a)}(0)|^2 |a^{(m',b)}(0)|^2 , \tag{19}$$

$$K_{2a-b} = \int_{-\infty}^{\infty} dx\, c\epsilon_0 \left[ \chi^{(3)}_{b\ ijk\ell} + \sum_r \chi^{(3)}_{r\ ijk\ell} \left( \frac{\Gamma_r/\pi}{\omega_a - \omega_b - \Omega_r - i\Gamma_r} \right) \right] E^{2(m,a)}(x) E^{*(m',b)}(x) , \tag{20}$$

where W is the width of the interacting beams. Here $\phi = \frac{1}{2}(\beta^{(n,s)} - \boldsymbol{\beta}_p \cdot \hat{\mathbf{z}})L$ and it has been assumed that the coherent anti-Stokes Raman scattering signal is phase-matched along the z-axis, with $\hat{\mathbf{z}}$ being the unit vector along the z-axis. Phase-matching occurs for $\phi = 0$. Because of both waveguide and material dispersion, it can be angularly tuned for only the coherent anti-Stokes Raman scattering or coherent Stokes Raman scattering case, but not for both simultaneously. Typically the angles between the two incident beams are only a few degrees.

How this phenomenon can be used for Raman spectroscopy is clear from the resonance denominators associated with the $\chi_r$ terms. If the background signal is negligible, or can be eliminated, then the coherent anti-Stokes Raman scattering power is

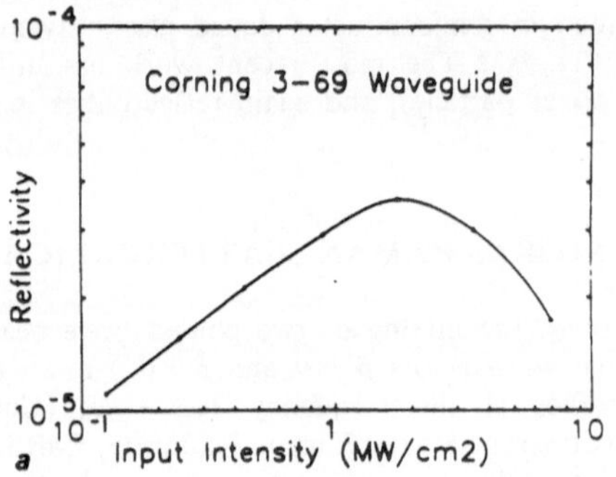


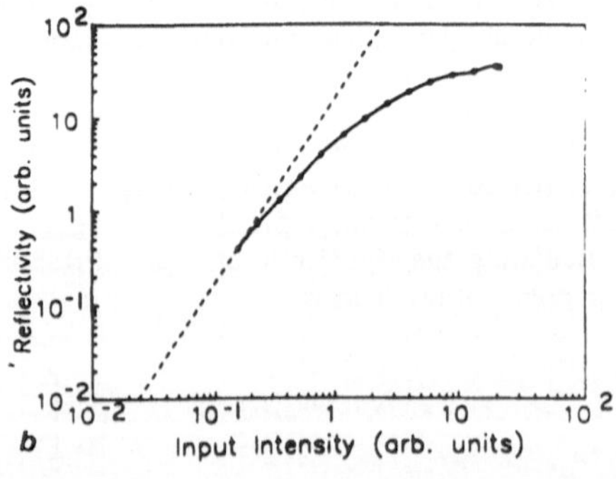


Fig. 9 Degenerate four wave mixing in planar semiconductor-doped glass waveguides with signal and pump beams at right angles to each other. (a) The reflectivity of the conjugate signal related to the powers incident onto the coupling prisms using 10 nanosecond pulses at $\lambda = 0.53\ \mu m$. Saturation in the reflectivity corresponds to 1% in the waveguide and occurs at 10 watt peak power. (b) The reflectivity for 80 picosecond pulses, peaking at 10% conversion.

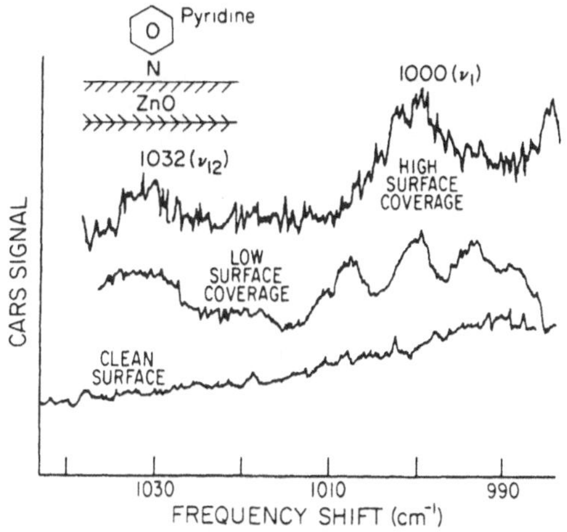


Fig. 10 Coherent anti-Stokes Raman scattering signal obtained from pyridine molecules on a ZnO waveguide. The uppermost curve corresponds to many monolayers on the waveguide surface. The middle curve is obtained after pumping on the system and corresponds to less than a monolayer coverage, and the lowest curve is obtained after prolonged pumping which almost totally removes the pyridine molecules.

$$P_s \propto \sum_r \left| \chi_r^{(3)} \right|^2 \frac{\Gamma_r/\pi}{[(\omega_a-\omega_b-\Omega_r)^2 + \Gamma_r^2]} \, . \tag{21}$$

The signal power is a maximum whenever $\omega_a - \omega_b = \Omega_r$. Therefore, by tuning the frequency difference $\omega_a - \omega_b$ it is possible to map out all of the vibrational transitions. Even when the background term cannot be eliminated, the signal power undergoes dispersion of some kind with frequency difference and the characteristic $\Omega_r$ can be identified.

Initial experiments on 2.0 $\mu$m thick polystyrene thin film waveguides indicated large signal levels, in good agreement with theoretical calculations based on the above formulae.[42] Using two tunable dye lasers with 100 ps long pulses and peak power densities in the film of 30 and 60 MW/cm$^2$, the conversion efficiency was 0.2%, in excellent agreement with the theoretical calculations.

The remarkable efficiency of this process suggested applications to the investigation of monolayers deposited on film surfaces. The problem is that the background term for the film can lead to a large signal which can mask the desired monolayer signal, that is under normal conditions the guiding film can be thousands of angstroms thick whereas a monolayer has a thickness of say five angstroms. The goal, therefore, is to eliminate the background term in $K_{2a-b}$ as completely as possible. This is the opposite to the usual goal of making this term as large as possible which is achieved by using modes with the same mode number. This suggests that minimizing $K_{2a-b}$ can be achieved with a judicious choice of interacting modes with different mode numbers.[41] If at least one of the input fields corresponds to a higher order mode, that is $TE_m$ with $m > 0$, interference effects occur and hence the film contribution can be minimized leaving the monolayer term as the dominant contribution. This approach was first used under vacuum conditions to observe the bonding of Ti atoms to oxygen bonds on the surface of a $Nb_2O_5$ film.[3] Since then a number of monolayer problems have been addressed in this way. An example is shown in Fig. 10 of pyridine molecules adsorbed onto the surface of a high quality ZnO epitaxial thin film waveguide.[43] The $\nu_1$ (ring breathing mode, symmetric stretch) mode now splits into a number of peaks corresponding to different bonding sites on the ZnO surface, a beautiful example of the power of this spectroscopic technique.

## 7. THEORY OF ALL-OPTICAL INTERACTIONS AND DEVICES

The fastest growing areas in nonlinear integrated optics deal with phenomena based on an intensity-dependent refractive index.[5,7] Such an optically induced refractive index change can occur due to many different possible mechanisms, for example distortion of the electron cloud distribution within molecules and atoms, electron excitation via absorption from the valence to the conduction band in semiconductors etc. The simplest form of the nonlinearity is the pure cubic nonlinearity associated with an ideal Kerr-law medium which leads to a nonlinear polarization field for plane waves of the form

$$P_i^{NL}(\mathbf{r},\omega_1) = \epsilon_0 \chi_{ijji}^{(3)}(\omega_1;\omega_2,-\omega_2,\omega_1)E_j E_j^* E_i \ . \tag{22}$$

Including the usual linear polarization term, the total polarization can be written as

$$P_i(\mathbf{r},\omega_1) = \epsilon_0[\chi_{ii}^{(1)}(\omega_1;\omega_1) + \chi_{ijji}^{(3)}(\omega_1;\omega_2,-\omega_2,\omega_1)|E_j|^2]E_i \tag{23}$$

where the quantity in the square brackets can be interpreted as an intensity-dependent dielectric constant, i.e. $n^2$. For example, assuming an isotropic medium with two codirectional beams of frequency $\omega_1$ and $\omega_2$ respectively, the refractive index "seen" by beam $\omega_1$ is given by

$$n(\omega_1) = n_0 + n_2(\omega_1;\omega_1)I_1 + n_2(\omega_1;\omega_2)I_2 \ , \tag{24a}$$

$$n_2(\omega_1,\omega_1) = 3\,\frac{\chi_{1111}^{(3)}(\omega_1;\omega_1,-\omega_1,\omega_1)}{n_0^2 c\epsilon_0} \ , \tag{24b}$$

$$n_2(\omega_1,\omega_2) = 6\,\frac{\chi_{1111}^{(3)}(\omega_1;\omega_2,-\omega_2,\omega_1)}{n_0^2 c\epsilon_0} \qquad \text{[co-polarized]}, \tag{24c}$$

$$n_2(\omega_1,\omega_2) = 6\,\frac{\chi_{1221}^{(3)}(\omega_1;\omega_2,-\omega_2,\omega_1)}{n_0^2 c\epsilon_0} \qquad \text{[orthogonally polarized]}. \tag{24d}$$

If $\omega_1 = \omega_2$ and if beam 2 is orthogonally polarized to beam 1, Eq. (24d) is valid. For isotropic media, $\chi_{1111}^{(3)} = 3\chi_{1221}^{(3)}$ far from any resonance. The essential meaning of these equations is that the refractive index of a given beam can be modified by its own intensity, a beam orthogonally polarized to it at the same frequency, and by intense beams at other frequencies. Thus there are many ways of all-optically inducing a change in refractive index.

There are two limits in which such an optically induced refractive index change can be utilized in waveguide geometries.[5,7] The waveguide is defined in the first place by refractive index differences between the central guiding region and the bounding media. Defining $\Delta n_0$ as a typical refractive index difference between the guiding and neighbouring linear media at low powers, and $\Delta n$ as the optically induced refractive index change, then different behaviour is obtained for $\Delta n/\Delta n_0 \geq 1$ and for $\Delta n/\Delta n_0 \ll 1$. In the second case, the guided wave field distribution is to a good approximation independent of power and corresponds to that of the normal mode, and only a small change in the propagation wavevector is induced. On the other hand, when $\Delta n \geq \Delta n_0$, the field distributions depend on power and it is necessary to solve the nonlinear wave equation along with the boundary conditions, this time with a field dependent refractive index.

Most of the experiments reported to date have involved small changes in refractive index. Over the last 20 years a number of integrated optics devices have been developed for branching, switching etc.[44] These include both passive and active devices, the active ones usually based on the electro-optic effect and tuned by changing an applied electric field. The operating characteristics of all of these devices can in principle be tuned by changing the optical power in the incidence channel, if the waveguide contains materials with intensity-dependent refractive indices. The power-dependent refractive index change is obtained by a suitable average of the nonlinearity over the intensity distribution associated with the guided wave. In general, the propagation wavevector can be written as[7]

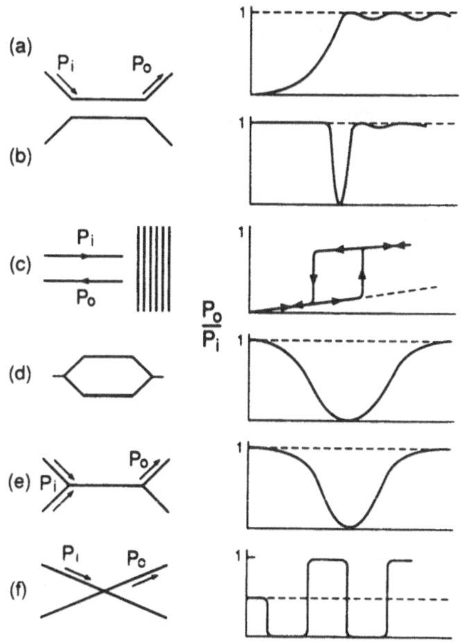


Fig. 11 A number of all-optical guided wave devices and their responses to increasing power. (a) Half beat length directional coupler. (b) One beat length directional coupler. (c) Distributed feedback grating reflector. (d) Nonlinear Mach-Zehnder interferometer. (e) Nonlinear mode mixer. (f) Nonlinear X-switch.

$$\beta = \beta_0 + \Delta\beta_{11}P_1 + \Delta\beta_{12}P_2 , \tag{25}$$

where $P_2$ ($\equiv |a_2^{(m,2)}|^2$) refers to a guided wave of the orthogonal polarization to $P_1$ ($\equiv |a_1^{(n,1)}|^2$), or to a guided wave of different frequency of any polarization. To a good approximation,

$$\Delta\beta_{ij} = \omega_1 D_{ij}\epsilon_0 \int_{-\infty}^{\infty} dx \int_{-\infty}^{\infty} dy \ \chi_{ijji}^{(3)} \left| E_j^{(m,2)}(x,y) \right|^2 \left| E_i^{(n,1)}(x,y) \right|^2 , \tag{26}$$

where $D_{ij}$ is either 3 or 6, depending on the details of the interaction as discussed above.[11] An equivalent and simpler form is

$$\Delta\beta_{ij} = \frac{\omega_1}{c} \frac{n_2(\omega_1, \omega_2)}{A_{eff}} , \tag{27}$$

in which case the effect of the integral in Eq. (26) is replaced by an effective cross-sectional area over which the nonlinear interaction occurs.

A number of all-optical integrated optics devices are summarized in Fig. 11. Clearly the output can be tuned by changing the input power. Although all of these devices require a power-dependent $\beta$, they can further be subdivided into two interaction geometries: (1) in which two guided wave modes interact with propagation distance and the nonlinearity affects this interaction; and (2) in which a guided mode undergoes a nonlinear phase shift which changes its interference condition with another optical field. The first category includes the nonlinear directional coupler[45,46] and its variants,[47-49] nonlinear distributed feedback gratings,[50] and the nonlinear X- and Y-junctions.[51,52] Belonging to category two are the nonlinear Mach-Zehnder interferometer and the nonlinear distributed (prism or grating) coupler. It is noteworthy that it is the nonlinear coupled mode devices which exhibit the sharpest switching characteristics.

## 7.1 Simple theory of nonlinear coupled mode devices

These interactions typically involve two coupled co-directional or contra-directional guided wave modes which exchange power. They are governed by the nonlinear coupled mode equations (m=n)[5,7]

$$-i\frac{d}{dz}a_1(z) = \Gamma a_2(z)e^{i\Delta\beta z} + [\Delta\beta_{11}|a_1(z)|^2 + G\Delta\beta_{12}|a_2(z)|^2]a_1(z) + G'\Delta\beta_{12}a_1^*(z)a_2^2(z)e^{-2i\Delta\beta z}\ , \tag{28a}$$

$$\pm i\frac{d}{dz}a_2(z) = \Gamma a_1(z)e^{-i\Delta\beta z} + [\Delta\beta_{22}|a_2(z)|^2 + G\Delta\beta_{21}|a_1(z)|^2]a_2(z) + G'\Delta\beta_{21}a_2^*(z)a_1^2(z)e^{2i\Delta\beta z}\ , \tag{28b}$$

where the ± signs refer to contradirectional and codirectional guided wave interactions respectively. The parameter $\Gamma$ quantifies the strength of the field overlap and hence the linear coupling between the modes, and $\Delta\beta$ is the linear wavevector mismatch between the modes. The terms in the square brackets are the nonlinear contributions to the propagation wavevectors and those proportional to G′ describe the scattering between two orthogonal modes of the structure, when appropriate. Finally, G and G′ are constants which depend on the nature of the interaction and nonlinearity, and the polarization of the guided modes. For example, for two TE polarized waves and a Kerr-law nonlinearity, G = 2.

The nonlinear directional coupler is potentially a useful device because it has four ports, two input and two output. Optimally the two channels are identical ($\beta_1 = \beta_2 \rightarrow \Delta\beta = 0$) and the coupling occurs via field overlap quantified by

$$\Gamma = \omega\epsilon_0 \int_{-\infty}^{\infty} dx \int_{-\infty}^{\infty} dy \mathbf{E}_1^*(\mathbf{r})\cdot\mathbf{E}_2(\mathbf{r})\ . \tag{29}$$

As a result, when only one of the channels is excited with low powers at the input, the power oscillates between the two channels with a beat length $L_b$, as shown in Fig. 12. As the input power is increased, a mismatch is induced in the wavevectors ($\beta_1 + \Delta\beta_{11}P_1(0)$, $\beta_2 \neq \beta_1$) which decreases the rate (with z) of the power transfer. This leads to an increase in the effective beat length. There is a critical power associated with the solutions,[5,6]

$$P_c = \frac{8\Gamma}{\Delta\beta_{11} + \Delta\beta_{22} - 2G\Delta\beta_{12}}\ , \tag{30}$$

for which an infinitely long, lossless nonlinear directional coupler acts as a 50:50 splitter, i.e. $L_b \rightarrow \infty$. For higher input powers, the initial wavevector mismatch is too large to overcome and the power effectively stays in the input channel, Fig. 12. Therefore, if the device length is terminated at $L_b/2$, then at low powers the signal comes out of channel 2 and at high powers out of channel 1. This is essentially an all-optical switch. The net result is the response sketched in Fig. 11a. The nonlinear directional coupler response is essentially the same for all nonlinear coupled mode interactions which are co-directional.[7] This includes using gratings for coupling modes with m ≠ n,[49] etc.

In order to couple contra-propagating modes, it is necessary to essentially turn the propagation wavevector of the incident mode right around. This can be achieved with a grating with wavevector $\kappa \simeq 2\beta_1$. The coupling coefficient is written as

$$\Gamma = \omega\epsilon_0 \int_{-\infty}^{\infty} dx \int_{-\infty}^{\infty} dy \Delta\epsilon \mathbf{E}_1^*(\mathbf{r})\cdot\mathbf{E}_2(\mathbf{r})\ . \tag{31}$$

where $\Delta\epsilon$ is the modulation produced by the grating. Here $\Delta\beta = \beta_1 + \beta_2 - \kappa$. Again there is a "critical" power given by

$$P_c = \frac{8\Gamma}{3(\Delta\beta_{11} + \Delta\beta_{22})k_0 L}\ , \tag{32}$$

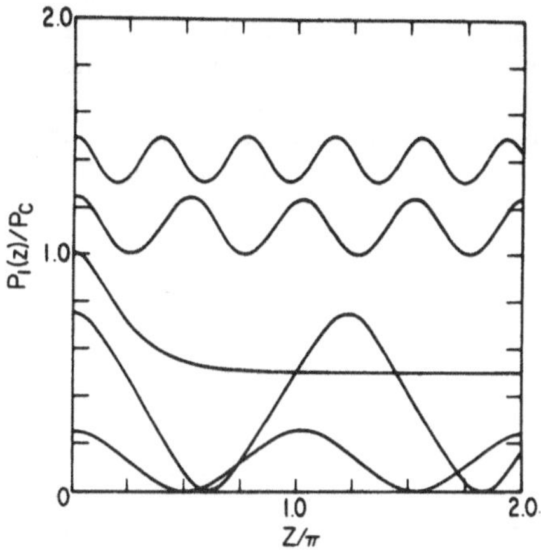


Fig. 12 Response of a half beat length nonlinear directional coupler. Evolution of the power in the incidence channel with propagation distance for various input powers.

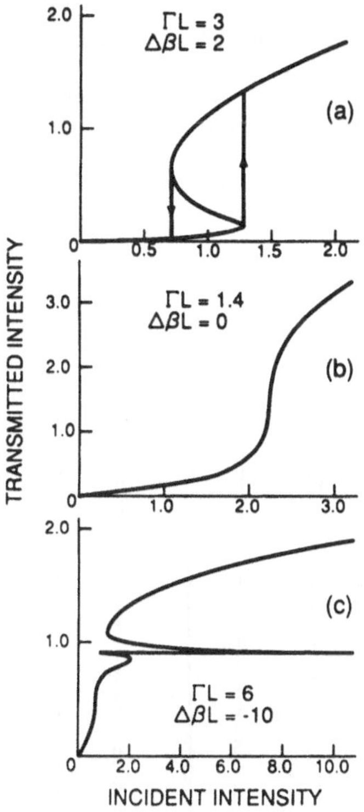


Fig. 13 All-optical response of a nonlinear distributed feedback grating under different conditions of grating strength (ΓL) and initial detuning (ΔβL). (a) Bistability, (b) differential gain and (c) optical limiting.

at which the response of the grating reflector becomes power-dependent. The response also varies with the initial detuning $\Delta\beta L$, and a selection of responses under different operating conditions is shown in Fig. 13. Note the sharp bistable switching which occurs under the appropriate conditions.

The last case to be discussed in this category is the nonlinear X-switch shown in Fig. 11f.[52] The modes here are the supermodes of the structure, that is for any given waveguide separation there is a symmetric mode in which the fields are positive in both branches and one which is antisymmetric with the field in both branches $\pi$ out of phase with one another. The total field is a linear combination of the two supermodes. This problem cannot be treated analytically since the varying waveguide separation means that the structure of the supermodes varies with propagation distance. Using numerical field propagation techniques the response of the X-switch can be calculated. The result shown in Fig. 14 looks promising for digital switching.

## 7.2 Simple theory of nonlinear interferometric devices

This category includes devices in which the nonlinear phase shift of a guided wave mode accumulates with propagation distance, changing a subsequent interference condition with another field. There are three devices of this type, a nonlinear Mach-Zehnder interferometer,[53] nonlinear grating and prism couplers, and a nonlinear mode mixer.[54]

The simplest device is a nonlinear Mach-Zehnder interferometer as depicted in Fig. 11d. Consider the two channels to be of equal length, with a nonlinearity only in channel 2. They are sufficiently well separated so that there is essentially zero field overlap. Thus the differential phase shift between the two channels is $\Delta\phi^{NL} = \Delta\beta_{22}LP_{in}/2$ where L is the length of the nonlinear region and $P_{in}$ is the input power (half of which propagates in each channel). Whenever $\Delta\phi^{NL}$ changes by $\pi$, the output changes from a maximum to a minimum, producing the response sketched in Fig. 11d. In practice, it may not be possible to make only one channel nonlinear so that there is an accumulated nonlinear phase shift in both channels. In that case, the cross-sectional area of the two channels can differ by as much as a factor of two and still maintain single mode structures. This would give $\Delta\beta_{22} \simeq 2\Delta\beta_{11}$ so that $\Delta\phi^{NL} = (\Delta\beta_{22} - \Delta\beta_{11})LP_{in}$. Again a minimum nonlinear phase difference of $\pi$ between the channels is required for all-optical switching.

Another interferometric device is the nonlinear mode-mixer.[54] It relies on small power-dependent changes in the beat length between two modes in a two-mode waveguide, leading

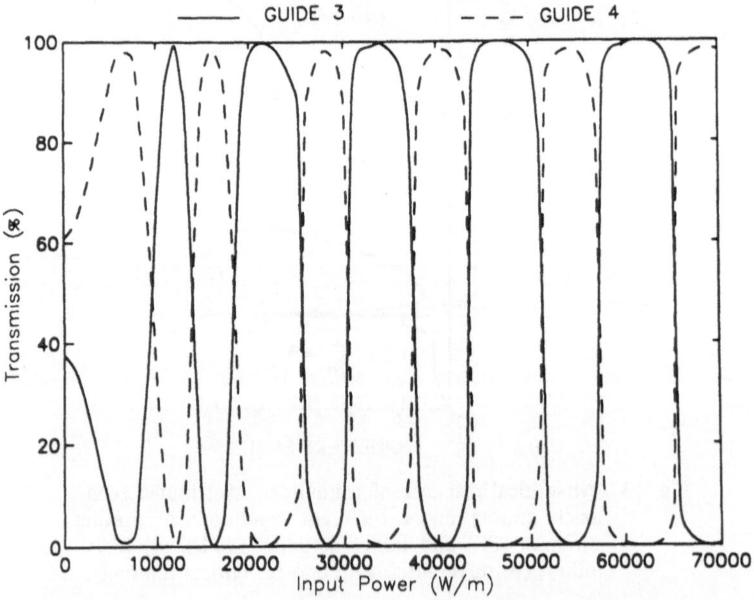


Fig. 14 Output response versus input power for a nonlinear X-switch.

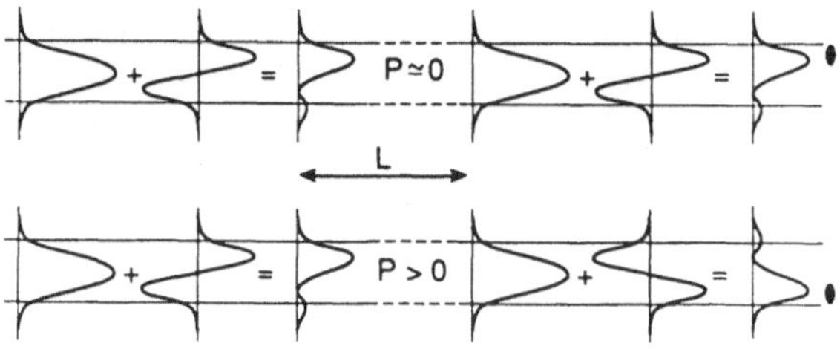


Fig. 15 Power induced changes in the interference pattern between two equally excited lowest order modes in a single nonlinear waveguide. The change in the propagation constants with power is different for the two modes leading to a change in the coherent superposition of the two modes.

to changes in the spatial interference between the two modal fields with increasing power. If the $TE_0$ and $TE_1$ modes of a waveguide are excited with approximately equal powers ($P_0 = P_1$), the maximum in the total field oscillates in position with propagation distance, as shown in Fig. 15. Noting that $\Delta\beta_{01} = \Delta\beta_{10}$, the differential phase change between the modes is given by

$$\delta\phi = (\beta_0 - \beta_1)L + P(\Delta\beta_{00} - \Delta\beta_{11})L \ . \tag{33}$$

When $P(\Delta\beta_{00} - \Delta\beta_{11})L = \pi$, the interference between the two modes changes the position of the field maximum as indicated in Fig. 15. This change can be turned into a switch by using a Y-branch to provide two alternative output paths for the field. The response is similar to that of the Mach-Zehnder interferometer.

The third case is a nonlinear prism (or grating) coupler.[55] This is the subject of another chapter in this book and will only be briefly discussed here for completeness. Radiation is incident from inside onto the base of a prism which is separated from the waveguide surface by sub-micron distances. The growth of the guided mode under the prism is given by[56]

$$\frac{d}{dz}a_1(z) = i[-\Delta\beta + i\beta_i + \Delta\beta_{11}(z)]a_1(z) + tA_iG(z) \ , \tag{34}$$

where the second term describes the coupling (transfer coefficient t) between the gaussian incident field $[G(z) = \exp(-z^2/\omega_0^2)]$,

$$E_i(z) = A_iG(z)e^{i\beta_0 z} \ , \tag{35}$$

and the guided wave. The guided wave wavevector $\beta_1 + i\beta_i$ is complex because the guided wave is leaky due to the presence of the prism, and the low-power wavevector detuning between the two waves is $\Delta\beta = \beta_0 - \beta_1$. For a Kerr-medium

$$\Delta\beta_{11}(z) = \Delta\beta_{11}|a_1(z)|^2 \ . \tag{36}$$

However, in most of the experiments reported to date the nonlinearity has been nonlocal, that is diffusive in space. In that case the change in propagation wavevector $\Delta\beta(z)$ is the solution to the equation

$$D^2\frac{d^2}{dz^2}\Delta\beta(z) = \Delta\beta(z) - \Delta\beta_{11}|a_1(z)|^2 \ , \tag{37}$$

where D is the diffusion constant. Details of the solutions are given in the chapter by Reinisch *et al.* The key results are (1) the reduction in the coupling efficiency with increasing power, and (2) switching and bistability under appropriate conditions of angular detuning from the optimum low power coupling condition. It is now clear that for local nonlinearities (D = 0) only switching can occur, but bistability can occur if D > 0, see Fig. 16.[56]

A case commonly encountered experimentally with coupling pulsed lasers into waveguides is that of a thermal nonlinearity whose relaxation time is much longer than the pulse duration.[57] In this case, the thermo-optic change in the refractive index does not have time to diffuse and the index change at each point in the coupling region can accumulate over the pulse duration. Typical waveguide thermal relaxation times are $\tau_t \simeq 1\ \mu s$ and the thermal $n_2$ is given by[58]

$$n_2 = \frac{dn}{dT}\frac{\alpha\tau_t}{\rho C_p}, \tag{38}$$

where $\alpha$ is the absorption coefficient, $\rho$ is the density, $C_p$ is the specific heat at constant pressure and dn/dT is the thermo-optic coefficient. For pulses with widths $\Delta t << \tau_t$, $n_{2eff} \simeq n_2\Delta t/\tau_t$, where clearly the effective nonlinearity for the thermal nonlinearities is reduced by the ratio of the pulse width to the relaxation time. The principal characteristics in this domain are (1) asymmetric pulse distortion, (2) decrease in coupling efficiency with increasing pulse energy and (3) a shift in the optimum coupling away from its low power value.

## 8. EXPERIMENTS WITH ALL-OPTICAL DEVICES

A large number of experiments have been reported on all-optical devices in the past few years as interest in this field has grown. The principal activity has concentrated on nonlinear distributed and coherent couplers.

### 8.1 Nonlinear grating and prism couplers

The nonlinear properties of prism and grating couplers were initially used to measure the nonlinearities in waveguides via the power-dependent shift in the optimum coupling angle.[59-61] The earliest work actually utilized surface plasmon modes[59] and only later were integrated optics waveguides used.[60,61] Subsequently the optical limiting properties were studied for both cw and pulsed excitation, as well as pulse distortion.[62-64] However, the major issue for many years revolved around the question of whether bistability was or was not possible

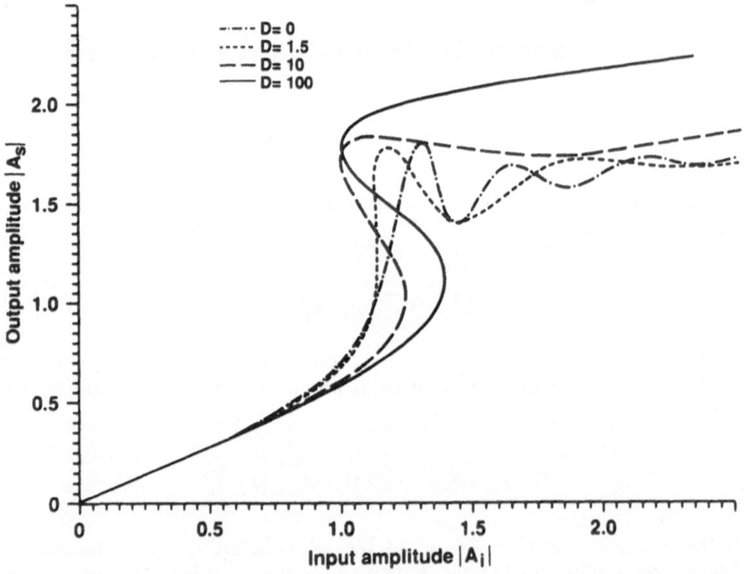


Fig. 16 The influence of the diffusion length D on the response of a nonlinear prism coupler. The numerical parameters are: $\Delta\beta = 1.8$, $\beta_i = 0.5$, $\omega_0 = 10.0$, $\delta = 1.0$ and t = 1.

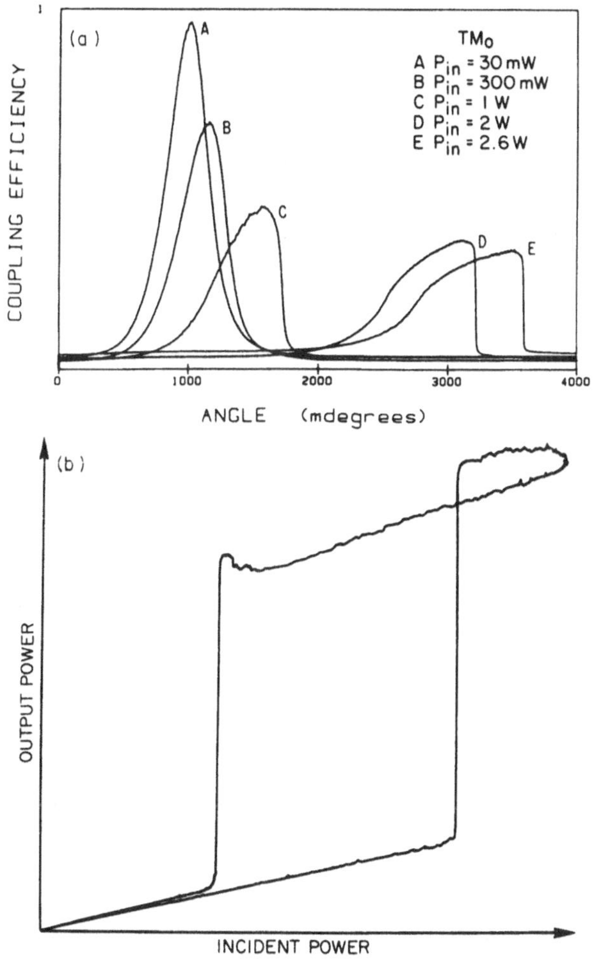


Fig. 17 (a) The angular dependence of the coupling efficiency with increasing incident cw power of wavelength 0.515 $\mu$m for incoupling into a ZnS waveguide. (b) Bistablity obtained by increasing the incident power at a fixed positive angular detuning.

with a local nonlinearity.[65-67] As stated previously, on theoretical grounds it has been shown that a nonlocal nonlinearity is needed for bistability.[56]

Examples of experimental results are shown in Figs. 17 and 18 for cw and pulsed nonlinear prism and grating excitation, both for thermal nonlinearities. There is (Fig. 17a) both a shift in the peak coupling angle and an increasing line distortion with increasing power.[68] In fact the distortion eventually develops a switching edge. Under such angular switching conditions, bistability in the output versus input response is obtained, Fig. 17b. All of these results are in excellent agreement with theory.

Excellent agreement with theory was also obtained with pulsed laser sources. The nonlinearities used have been thermal,[57,68] a combination of thermal and electronic,[69] and purely electronic.[70] The results in Fig. 18 are for a strictly thermal response.[57] Note the asymmetric pulse distortion for this case. By using different pulse widths, the relative contributions of the thermal and electronic terms can be varied, allowing the two nonlinearities to be isolated.[69] Alternatively, a pump-probe technique can be used to isolate the fast component and hence evaluate the electronic component.[70]

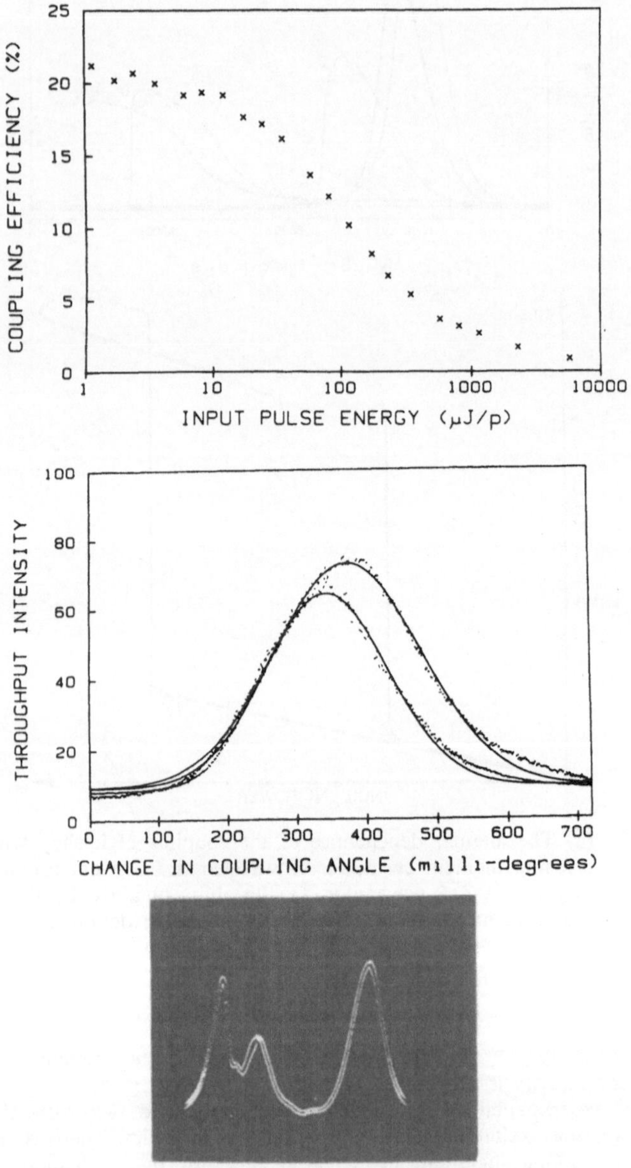


Fig. 18 The response of a nonlinear prism coupler for coupling into a ZnO waveguide with 10 ns pulses at 0.53 μm. (a) The decrease in coupling efficiency with increasing pulse energy. (b) The variation in the coupling efficiency with incidence angle at two different power levels. (c) The temporal profile of the incoupled pulse at low and high pulse energies.

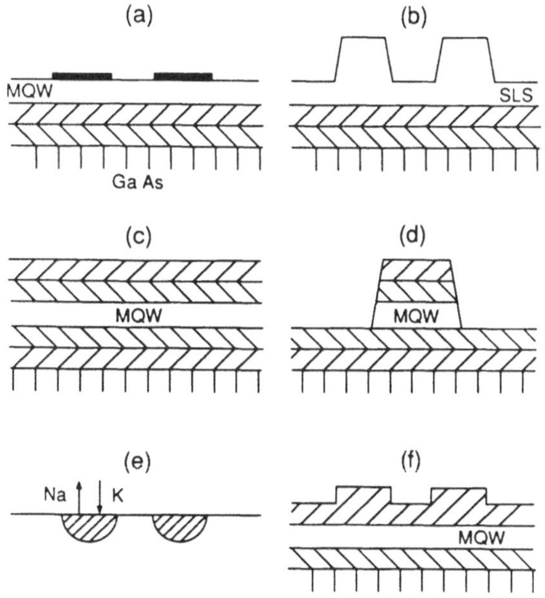


Fig. 19 Schematics of the half beat length nonlinear directional coupler geometries investigated to date in semiconductor systems. (a) Strain induced waveguides (along each edge of the gold strips) in multiple quantum well waveguides; (b) strained layer superlattice strip-loaded waveguides; (c) planar waveguides separated by a nonlinear multiple quantum well layer; (d) vertically confined channel waveguides separated by a nonlinear multiple quantum well layer; (e) channel waveguides in CdSSe doped glass made by Na-K ion exchange; (f) strip-loaded multiple quantum well waveguide. The vertical lines indicate GaAs substrates. The slanted lines indicate GaAlAs layers with different concentrations, except in (b) where these layers are InGaAs.

## 8.2 Nonlinear directional couplers

Nonlinear directional couplers have been implemented in integrated optics structures exclusively using semiconductor-based nonlinearities.[71-76] (In fact, the cleanest and fastest nonlinear directional coupler responses have been obtained in dual core fibers,[77,78] and their equivalents.[47,48]) The geometries are summarized in Fig. 19, most using some form of GaAlAs nonlinearity. Of these, channel waveguides were used except for 19c where the nonlinear coupling between two planar waveguides was used.[73] Cases 19c and 19d are distinguished by the fact that the nonlinearity identified by the authors was excitonic, located in the medium separating the waveguides.[72,74] Their cw switching powers were impressively low, typically a few milliwatts. Similarly low cw switching powers were also reported for geometry 19a for which case band-filling was assumed to be the appropriate nonlinearity.[71] The most recent of the GaAlAs waveguide experiments involved using picosecond pulses which were switched with a device recovery time (carrier recombination time) in the 10's of nanoseconds.[76] In all cases, absorption played a dominant role and the throughputs were low. It is not clear whether any of these cases actually corresponded to a Kerr-nonlinearity (local response and index decay much faster than the pulse width), although most used some form of electronic (versus thermal) nonlinearity. The best switching characteristics in terms of complete switching appear to be those shown in Fig. 20.[74]

## 8.3 Other nonlinear devices

Only two of the other all-optical switching devices depicted in Fig. 11 have been implemented, namely the nonlinear mode mixer[79] and the nonlinear distributed feedback grating.[80,81] The mode mixer was implemented in a two mode GaAlAs waveguide at the wave-

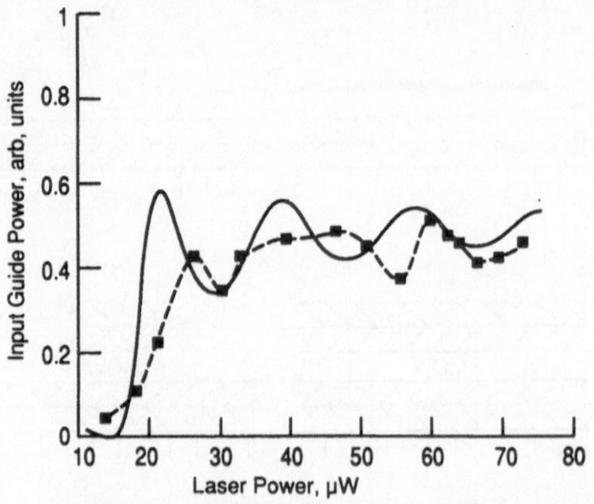


Fig. 20 The switching characteristics for vertically confined channel waveguides separated by a nonlinear multiple quantum well layer.

length 865 nm near the material bandgap. Shown in Fig. 21 are the outputs obtained at two different power levels differing by ≃ 5 mW. The output clearly switches from one waveguide region to the other, as expected. The problem in this device, just as with the other nonlinear GaAlAs all-optical devices, is the low throughput.

Nonlinear grating devices have also been demonstrated, probably based on thermal nonlinearities. Figure 22 shows an example of all-optical tuning of a distributed feedback grating. Note the optically tunable change in the reflectivity in an InSb waveguide.[81] A form of bistability was also observed in a waveguide grating overcoated with a polydiacetylene material.[80]

## 8.4 Figures of merit and materials choices

The key to the development of all-optical switching devices is the identification of appropriate materials. To understand the requirements, we examine the case of the nonlinear directional coupler implemented in semiconductor-doped glass waveguides.[75] There, it was clear that the fraction of power switched reached a saturation value, that is increasing the power past a certain point did not lead to an increase in the fraction of the light switched

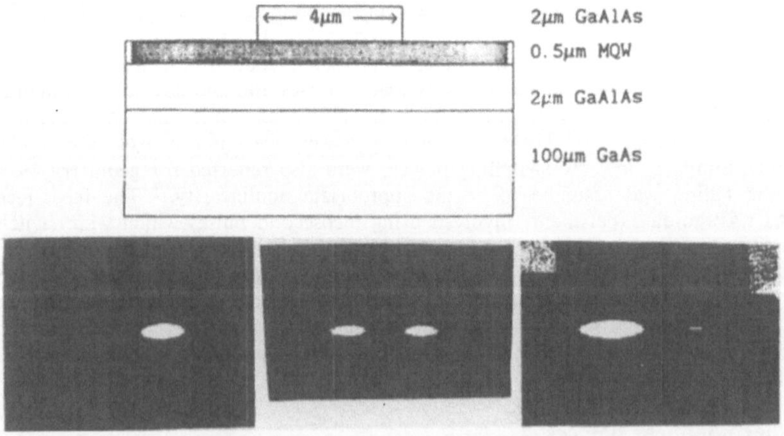


Fig. 21 Observed power-dependent change in the position of the field maximum in a two mode GaAlAs waveguide. The top view shows the waveguide geometry. The bottom pictures show mode patterns (rotated 90° with respect to the top geometry) for low powers (left), medium power (middle) and high powers (right).

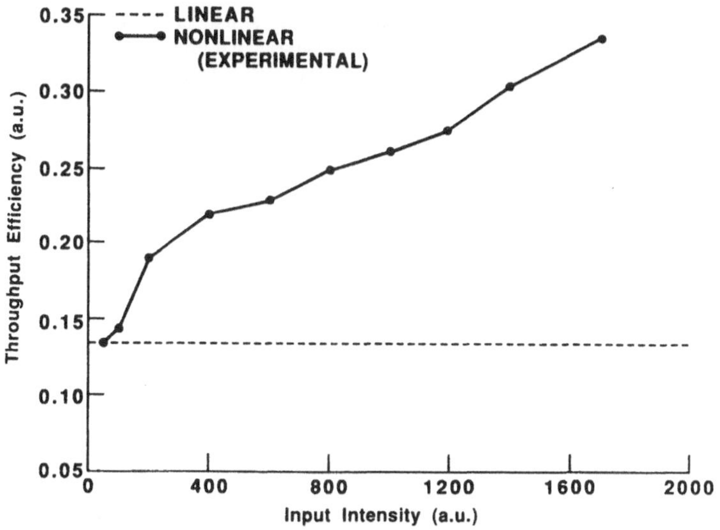


Fig. 22 The optically induced tuning of the throughput of a distributed feedback grating excited by a guided wave at $CO_2$ laser wavelengths in an InSb waveguide.

from one output channel to the other. To expedite the modelling, one can use a saturable two level nonlinearity at some detuning from resonance of the form[82]

$$\Delta n = -\frac{\Delta n_{sat}}{1 + P/P_{sat}}, \tag{39a}$$

$$\alpha = \frac{\alpha_0}{1 + P/P_{sat}}, \tag{39b}$$

where $\alpha_0$ and $\Delta n_{sat}$ are fixed by the detuning from the line center. The variation in the fraction of light switched in a nonlinear directional coupler as a function of the figure of merit

$$w = \Delta n_{sat} \frac{L}{\lambda} \simeq \Delta\beta_{sat} \frac{k_0 L}{2\pi} \tag{40}$$

is shown in Fig. 23. The key point is that a change in w from 1 to 2 corresponds to essentially a change from zero to 100% switching.[82] In fact, $\Delta\phi^{NL} = 2\pi w$ with $w = 2$ is the maximum phase shift which one of the channels should be *capable of* for switching to be possible. Note that this is not the phase shift experienced by each channel (which is $2\pi$). The nonlinear phase shifts required for a range of devices is reproduced in Table 3.[52,53,82,83] What is perhaps surprising are the large values of phase shift required.

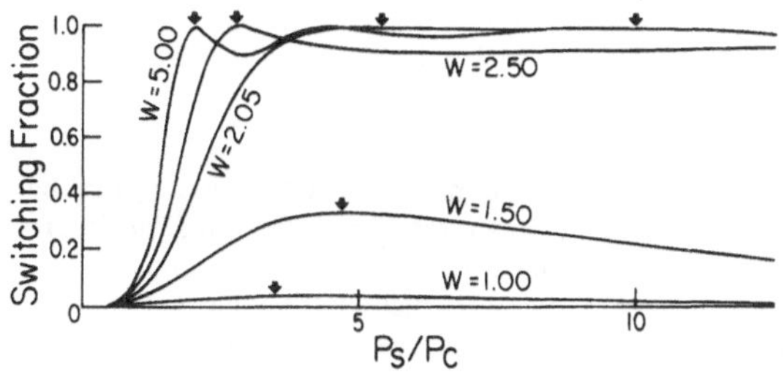


Fig. 23 Fraction of light emerging from the input ("bar") channel in a half-beat length nonlinear directional coupler versus w. $P_c$ is the critical power and $P_s$ is the input power for this case. A saturable two-level model was used for the nonlinearity. The switching power is labelled by the small arrows.

Table 3. The minimum nonlinear phase shift, $\Delta\phi^{NL}$, and the minimum dimensionless material parameter W (for > 80% transmission) required for various nonlinear guided wave devices

| Nonlinear Device | $\Delta\phi^{NL}$ | W |
|---|---|---|
| X-Switch | $7.4\pi$ | 18.5 |
| Directional Coupler 1/2 beat length | $4\pi$ | 10 |
| Directional Coupler 1 beat length | $\simeq 3.3\pi$ | 8 |
| Mach-Zehnder Interferometer | $2\pi$ | 5 |
| Distributed Feedback Grating | $\pi$ | 2.5 |

However, based on Eq. (40), it is apparently possible to increase w simply by increasing the device length L. This is in fact true, if the waveguide attenuation coefficient satisfies $L << \alpha^{-1}$. Otherwise, the throughput is decreased since $P(L)/P(0) \propto \exp[-\alpha L]$. Taking into account increasing loss with increasing L, w is divided by $\alpha L$ to produce a materials figure of merit

$$W = \frac{w}{\alpha L} = \frac{\Delta n_{sat}}{\alpha\lambda} . \quad (41)$$

The values required of W for switching with 80% throughput are also listed for a spectrum of devices in Table 3. We note that the devices with the sharpest switching characteristics require the largest values of W.

Table 4 lists a number of materials systems, and the coresponding value of W. Because of their very low losses, glasses in fiber form exhibit the largest values of W. However, long propagation distances are needed to utilize glasses, much too long for applications to integrated optics devices. The second most promising system appears to be nonlinear organic materials. However, little is really known about such materials and it is too early to tell if they will indeed fulfill their initial promise. This leaves semiconductor materials. The near-resonance response can lead to large enough values of W, as shown by the theoretical results in Fig. 24.[84] However the response time appears to be too slow to be useful at this time. As far as the AC Kerr effect is concerned, this coefficient appears to be of the order of the nonlinearity in $CS_2$, maybe too small.[85] In addition, there are other problems such as two photon absorption still to be addressed.[86]

## 9. SUMMARY

An overview of recent progress in nonlinear integrated optics has been presented. The implementation of nonlinear optics phenomena in integrated optics formats has led to both efficient versions of well-known plane wave nonlinear interactions, as well as new devices unique to guided wave structures. In particular, second harmonic generation has been developed to the point that efficient doubling of semiconductor lasers is now possible. Most of the progress has occurred in lithium niobate waveguides, especially with the discoveries that non-phase-matched generation of Čerenkov radiation and quasi-phase-matched conversion via periodic domain inversion could utilize the large $d_{33}$ coefficient in this material. Other recent breakthroughs involve the partial orientation of highly nonlinear molecules in polymeric glass

Table 4. Figure of merit, $\Delta n_{sat}/\alpha\lambda$, of different nonlinear materials for application of third order nonlinearities to guided wave devices. The label "r" refers to resonant and "nr" to detuned from resonance

| Material System | $n_2$ $m^2/W$ | $\tau$ sec | $\alpha$ $cm^{-1}$ | $\Delta n_{sat}$ | W |
|---|---|---|---|---|---|
| *Semiconductors* | | | | | |
| GaAlAs (r) | $-10^{-8}$ | $10^{-8}$ | $10^4$ | 0.1 | |
| (nr, theory) | $-10^{-13}$ | $10^{-8}$ | 10 | 0.01 | 10 |
| *Organics* | | | | | |
| PTS (r) | $2\times10^{-15}$ | $2\times10^{-12}$ | $10^5$ | $\simeq 0.1$ | |
| (nr) | $10^{-16}$ | | 0.1 | $> 10^{-3}$ | $> 100$ |
| *Glasses* | | | | | |
| $SiO_2$ (nr) | $10^{-20}$ | $10^{-14}$ | $10^{-5}$ | $> 10^{-6}$ | $> 10^{-3}$ |

hosts via electric field poling. Potentially in the future, single crystal waveguide films of highly nonlinear organic materials may be possible, although there are many technological barriers to be surmounted.

The development of all-optical switching devices utilizing integrated optics channel waveguides has created a great deal of excitement recently and progress has been rapid. In the last few years a number of devices, most specifically nonlinear directional couplers have been reported and some of the limitations of such switching concepts identified. An important step forward has been the analysis of such devices implemented with non-ideal Kerr materials, and the identification of both materials and device figures of merit. This has allowed the suitability of different materials systems to be critically assessed with the not surprising conclusion that much more materials development is needed.

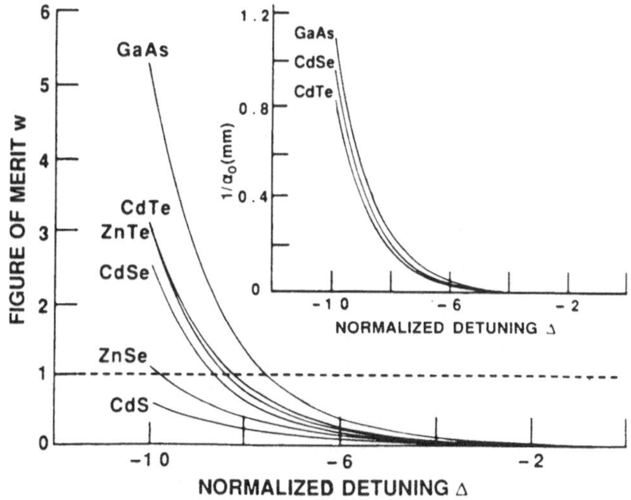


Fig. 24 Figure of merit w as a function of the normalized detuning Δ for a variety of semiconductors. The inset shows the coupling length $\ell_c$ for a nonlinear directional coupler which is equal to the inverse linear absorption coefficient.

Not discussed here is the branch of all-optical phenomena which relies on large index changes, namely $\Delta n/\Delta n_0 > 1$.[5,7] This topic is discussed in the chapter by Boardman *et al.*


## ACKNOWLEDGEMENT

This research has been supported by the Joint Services Optics Program of ARO and AFOSR, and the Optical Circuitry Cooperative.


## REFERENCES


1. D. B. Anderson and J. T. Boyd, Appl. Phys. Lett. **19**, 266, (1971).
2. Review of guided wave second harmonic generation: W. Sohler in *New Directions in Guided Wave and Coherent Optics*, D. B. Ostrowsky and E. Spitz eds. (Nijhoff, The Hague, 1984), NATO ASI Series Vol 79, Part II, pp 449-479.
3. G. I. Stegeman and C. T. Seaton, Applied Physics Reviews (J. Appl. Physics) **58**, R57 (1985).
4. Material requirements for nonlinear waveguide optics: G. I. Stegeman and R. H. Stolen, J. Opt. Soc. Am. B **6**, 652 (1989).
5. Review of third order nonlinear integrated optics: G. I. Stegeman, R. Zanoni, N. Finlayson, E. M. Wright and C. T. Seaton, J. Lightwave Tech. **6**, 953 (1988).
6. H. M. Gibbs, *Optical Bistability: Controlling Light with Light* (Academic Press, Orlando, 1985)
7. G. I. Stegeman and E. M. Wright, "All-Optical Waveguide Switching", review paper for J. Optical and Quantum Electron., in press
8. G. I. Stegeman, C. T. Seaton, A. Boardman and P. Egan, "Nonlinear guided waves" in *Proceedings of Summer School on Surface Electromagnetic Excitations*, R. F. Wallis and G. I. Stegeman eds. (Springer-Verlag, Berlin, 1986), p261-300
9. D. Marcuse, *Theory of Dielectric Optical Waveguides* (Academic Press, New York, 1974)
10. Numerous articles in *Integrated Optics, Vol. 7 of Topics in Applied Physics*, T. Tamir ed. (Springer-Verlag, Berlin, 1975).
11. F. Hopf and G. I. Stegeman, *Advanced Classical Electrodynamics Vol. II: Nonlinear Optics* (John Wiley and Sons, 1986)
12. Y. R. Shen, *The Principles of Nonlinear Optics* (J. Wiley, New York, 1984)
13. Y. Suematsu, Y. Sasaki, K. Furuya, K. Shibata and S. Ibukuro, IEEE J. Quantum Electron. **10**, 222 (1974).
14. H. Ito and H. Inaba, Opt. Lett. **2**, 139 (1978).
15. N. Uesugi, K. Daikoku and M. Fukuma, J. Appl. Phys. **49**, 4945 (1978)
16. W. Sohler and H. Suche in *Integrated Optics III*, L. D. Hutcheson and D. G. Hall eds., Proc. SPIE **408**, 163 (1983).
17. R. Regener and W. Sohler, J. Opt. Soc. Am. B **5**, 267 (1988).
18. F. Laurell and G. Arvidsson, J. Opt. Soc. Am. B **5**, 292 (1988).
19. T. Taniuchi and K. Yamamoto, Digest of CLEO'86 (Optical Society of America, Washington, 1986), paper WR3.
20. E. J. Lim, M. M. Fejer and R. L. Byer, Electron. Lett. **25**, 174 (1989);
    E. J. Lim, M. M. Fejer, R. L. Byer and W. J. Kozlovsky, Electron. Lett., in press (1989).
21. J. Webjorn, F. Laurell and G. Arvidsson, IEEE J. Lightwave Tech., in press (1989);
    J. Webjorn, F. Laurell and G. Arvidsson, CLEO'89, postdeadline paper PD10.
22. M. M. Fejer, M. J. F. Digonnet and R. L. Byer, Opt. Lett. **11**, 230 (1986).
23. J. D. Bierlein and H. Vanherzeele, J. Opt. Soc. Am. B **6**, 622 (1989).
24. I. Ledoux, D. Josse, P. Vidakovic and J. Zyss, Opt. Eng. **25**, 202, (1986).
25. C. Ye, N. Minami, T. J. Marks, J. Yang and G. K. Wong, Proceedings of NATO Advanced Research Workshop on *Nonlinear Optical Effects in Organic Polymers*, Series E: Applied Sciences, Vol 162 (Kluwer Acad. Pub., London, 1989), pp 173-83
26. J. E. Sohn, K. D. Singer, M. G. Kuzyk, W. R. Holland, H. E. Katz, C. W. Dirk, M. L. Schilling and R. B. Comizzoli, ibid, pp 291-7
27. G. Khanarian, D. Haas, R. Keosian, D. Karim and P. Landi, Technical Digest for CLEO'89 (Opt. Soc. Am., Washington, 1989), paper ThB1, pp 254-5
28. R. Normandin and G. I. Stegeman, Opt. Lett. **4**, 58 (1979)
29. R. Normandin and G. I. Stegeman, Appl. Phys. Lett. **40**, 759 (1982)

30. D. Vakhshoori and S. Wang, Appl. Phys. Lett. **53**, 347 (1988).
31. N. Uesugi, K. Daikoku and M. Fukuma, Appl. Phys. Lett. **49**, 4945 (1978).
32. A. T. Reutov and P. P. Tarashchenko, Opt. Spectrosc. **37**, 447 (1974)
33. N. Uesugi, Appl. Phys. Lett. **30**, 178 (1980).
34. H. Herrmann and W. Sohler, J. Opt. Soc. Am. B **5**, 267 (1988).
35. B. Hampel, R. Regener, R. Ricken, H. Suche and R. Volk, IEEE J. Lightwave Tech. **LT-4**, 772 (1986).
36. Reviewed in the book *Optical Phase Conjugation*, R. A. Fisher ed. (Academic Press, N. Y., 1982)
37. C. Karaguleff and G. I. Stegeman, IEEE J. Quantum Electron. **QE-20**, 716 (1984).
38. C. Karaguleff, G. I. Stegeman, R. Zanoni and C. T. Seaton, Appl. Physics Lett. **7**, 621 (1985)
39. A. Gabel, K. W. Delong, C. T. Seaton and G. I. Stegeman, Appl. Phys. Lett. **51**, 1682 (1987).
40. M. J. Bloemer and P. R. Ashley, Digest of Topical Meeting on Nonlinear Guided Wave Phenomena: Physics and Applications (Opt. Soc. Am., Washington, 1989), pp 98-101.
41. G. I. Stegeman, R. Fortenberry, C. Karaguleff, R. Moshrefzadeh, W. M. Hetherington III, and N. E. Van Wyck, Opt. Lett. **8**, 295 (1983).
42. W. M. Hetherington III, N. E. Van Wyck, E. W. Koening, G. I. Stegeman and R. M. Fortenberry, Opt. Lett. **9**, 88 (1984).
43. W. M. Hetherington III, Z. Z. Ho, E. W. Koenig, G. I. Stegeman and R. M. Fortenberry, Chem. Phys. Lett. **128**, 150 (1986);
W. M. Hetherington III, Z. Z. Ho, E. W. Koenig, G. I. Stegeman and R. M. Fortenberry, Proc. SPIE **620**, 102 (1986).
44. Reviews of electro-optic switching devices: L. Thylen, J. Lightwave Tech. **6**, 847 (1988); A. Selvarajan and J. E. Midwinter, Opt. and Quantum Electron. **21**, 1 (1989).
45. S. M. Jensen, IEEE J. Quantum Electron. **QE-18**, 1580 (1982).
46. A. Maier, Sov. J. Quantum Electron. **12**, 1490 (1982).
47. S. Trillo, S. Wabnitz, R. H. Stolen, G. Assanto, C. T. Seaton and G. I. Stegeman, Appl. Phys. Lett. **49**, 1224 (1986).
48. S. Trillo, S. Wabnitz, W. C. Banyai, N. Finlayson, C. T. Seaton, G. I. Stegeman and R. H. Stolen, IEEE J. Quantum Electron. **QE-25**, 104 (1989).
49. S. Trillo, S. Wabnitz and G. I. Stegeman, J. Lightwave Tech. **6**, 971 (1988).
50. H. G. Winful, J. H. Marburger and E. Garmire, Appl. Phys. Lett. **35**, 379 (1979).
51. Y. Silberberg and B. G. Sfez, Opt. Lett. **13**, 1132 (1988).
52. J. P. Sabini, N. Finlayson, C. T. Seaton, and G. I. Stegeman, Appl. Phys. Lett., in press
53. H. Kawaguchi, Opt. Lett. **10**, 411 (1985);
L. Thylen, N. Finlayson, C. T. Seaton and G. I. Stegeman, Appl. Phys. Lett. **51**, 1304 (1987).
54. Y. Silberberg and G. I. Stegeman, Appl. Phys. Lett. **50**, 801 (1987).
55. Y. J. Chen and G. M. Carter, Solid State Commun. **45**, 277 (1983);
G. M. Carter and Y. J. Chen, Appl. Phys. Lett. **42**, 643 (1983)
56. G. Vitrant, R. Reinisch, J. Cl. Paumier, G. Assanto, and G. I. Stegeman, Opt. Lett. **14**, 898 (1989).
57. R. M. Fortenberry, G. Assanto, R. Moshrefzadeh, C. T. Seaton and G. I. Stegeman, J. Opt. Soc. Am. B **5**, 425 (1988).
58. G. Assanto, R. M. Fortenberry, C. T. Seaton and G. I. Stegeman, J. Opt. Soc. Am. B **5**, 432 (1988).
59. G. M. Carter and Y. J. Chen, Appl. Phys. Lett. **42**, 643 (1983)
60. G. M. Carter, Y. J. Chen, S. K. Tripathy, Appl. Phys. Lett. **43**, 891 (1983).
61. Y. J. Chen, G. M. Carter, G. J. Sonek and J. M. Ballantyne, Appl. Phys. Lett. **48**, 272 (1986)
62. J. D. Valera, C. T. Seaton, G. I. Stegeman, R. L. Shoemaker, Xu Mai and C. Liao, Appl. Phys. Lett. **45**, 1013 (1984).
63. R. M. Fortenberry, R. Moshrefzadeh, G. Assanto, Xu Mai, E. M. Wright, C. T. Seaton and G. I. Stegeman, Appl. Phys. Lett. **49**, 6987 (1986)
64. F. Pardo, H. Chelli, A. Koster, N. Paraire, and S. Laval, IEEE J. Quantum Electron. **QE-23**, 545 (1987).
65. P. Vincent, N. Paraire, M. Nevière, A. Koster and R. Reinisch, J. Opt. Soc. Am. B **2**, 1106 (1985).
66. V. J. Montemayor and R. T. Deck, J. Opt. Soc. Am. B **2**, 1010 (1985); ibid **3**, 1211 (1986).

67. C. Liao and G. I. Stegeman, Appl. Phys. Lett. **44**, 164 (1984);
C. Liao, G. I. Stegeman, C. T. Seaton, R. L. Shoemaker, J. D. Valera and H. G. Winful, J. Opt. Soc. Am. A 2, 590 (1985).
68. G. Assanto, B. Svensson, D. Kuchibhatla, U. J. Gibson, C. T. Seaton and G. I. Stegeman, Opt. Lett. **11**, 644 (1986).
69. R. Burzynski, P. N. Prasad, R. Zanoni, G. I. Stegeman, Appl. Phys. Lett. **53**, 2011 (1988).
70. M. Sinclair, D. McBranch, D. Moses and A. J. Heeger, Appl. Phys. Lett. **53**, 2374 (1988).
71. P. Li Kam Wa, J. E. Stich, N. J. Mason, J. S. Roberts, and P. N. Robson, Electron. Lett. **21**, 26 (1985);
P. Li Kam Wa, J. H. Marsh, P. N. Robson, J. S. Roberts, N. J. Mason, Proc. SPIE **578**, 110 (1985).
72. U. Das, Yi Chen, and P. Bhattacharya, Appl. Phys. Lett. **51**, 1679 (1987).
73. M. Cada, B. P. Keyworth, J. M. Glinski, A. J. Springthorpe, and P. Mandeville, J. Opt. Soc. Am. B **5**, 462 (1988).
74. P. R. Berger, Yi Chem, P. Bhattacharya, J. Pamulapati, and G. C. Vezzoli, Appl. Phys. Lett. **52**, 1125 (1988).
75. N. Finlayson, W. C. Banyai, E. M. Wright, C. T. Seaton, G. I. Stegeman, T. J. Cullen and C. N. Ironside, Appl. Phys. Lett. **53**, 1144 (1988).
76. R. Jin, C. L. Chuang, H. M. Gibbs, S. W. Koch, J. N. Polky, and G. A. Pubanz, Appl. Phys. Lett. **53**, 1791 (1988).
77. D. D. Gusovskii, E. M. Dianov, A. A. Maier, V. B. Neustreuev, E. I. Shklovsii and I. A. Shcherbakov, Sov. J. Quantum Electron. **15**, 1523 (1985).
78. S. R. Friberg, A. M. Weiner, Y. Silberberg, B. G. Sfez and P. S. Smith, Opt. Lett. **13**, 904 (1988).
79. P. Li Kam Wa, P. N. Robson, J. S. Roberts, M. A. Pate and J. P. R. David, Appl. Phys. Lett. **52**, 2013 (1988);
P. Li Kam Wa, S. V. Burke, J. S. Roberts, M. A. Pate and P. N. Robson, Digest of Topical Meeting on Nonlinear Guided Wave Phenomena: Physics and Applications (Opt. Soc. Am., Washington, 1989), pp 214-7.
80. K. Sasaki, K. Fujii, T. Tomioka and T. Kinoshita, J. Opt. Soc. Am. **5**, 457 (1988).
81. G. I. Stegeman, R. Zanoni, E. M. Wright, N. Finlayson, K. Rochford, J. Ehrlich and C. T. Seaton in *Organic Materials for Nonlinear Optics*, R. A. Hann and D. Bloor eds. (Royal Soc. Chem., London, 1989), pp369-81.
82. S. Trillo, S. Wabnitz, E. Caglioti and G. I. Stegeman, Optics Commun. **63**, 281 (1987).
83. A. Mecozzi, S. Trillo, S. Wabnitz and G. I. Stegeman, Proceedings of ECOC'87, 391-4 (1987).
84. E. M. Wright, S. W. Koch, J. E. Ehrlich, C. T. Seaton and G. I. Stegeman, Appl. Phys. Lett. **52**, 2127 (1988).
85. J. J. Wynne, Phys. Rev. **178**, 1295 (1969).
86. V. Mizrahi, K. W. DeLong, G. I. Stegeman, M. A. Saifi and M. J. Andrejco, Opt. Lett., in press

# INDEX